VALIDATION COMPLIANCE BIANNUAL

1996–1997

VALIDATION COMPLIANCE BIANNUAL

1996–1997

INTERNATIONAL
VALIDATION
FORUM, INC.

Key West, Florida

Marcel Dekker, Inc. New York • Basel • Hong Kong

ISBN: 0-8247- 9746-9

The publisher offers discounts on this book when ordered in bulk quantities. For more information, write to Special Sales/Professional Marketing at the address below.

This book is printed on acid-free paper.

MARCEL DEKKER, INC.
270 Madison Avenue, New York, New York 10016

Current printing (last digit):
10 9 8 7 6 5 4 3 2 1

PRINTED IN THE UNITED STATES OF AMERICA

PREFACE TO THE 1996–1997 EDITION

Validation issues are not the most important problems facing the pharmaceutical, medical device, biologics, food, and cosmetic industries. Consumers are more concerned with the accuracy of blood tests, the cleanliness of food products, the purity and effectiveness of drugs, the functionality of medical devices, and the stability of biologics than with the processes and systems that underlie the manufacturing, testing, and analyzing of these products.

Scientists, however, are well aware of the critical importance of computers conducting those blood tests, controlling that food inventory, researching those pharmaceutical properties, manufacturing those medical devices, tracking those biologics, and analyzing data in environmental laboratories. The industry awareness of the importance and vulnerability of those systems has been created largely by the effective emphasis that agencies like the Food and Drug Administration and the Environmental Protection Agency are placing on the validation of those systems.

Why should validation be on the regulatory agenda? Two important facts can re-emphasize what those in the regulated industries already know. Fact one: We have yet to see any system that has met all appropriate standards for control, documentation, and performance prior to the initial audit of those systems. This suggests that validation does not merely provide assurances; it concretely and demonstrably improves the quality of data and of the processes which systems control.

Fact two: The dependence of the industries upon computers is overwhelming. We all find computers dominating the laboratory testing facility, the Quality Assurance laboratory, the manufacturing floor, the inventory control process, the clinical design and study process, and the statistical analysis of data. In fact, every aspect of researching, testing, manufacturing,and distribution is computer-dominated. The managers who are responsible for the safe and effective control of those processes are more and more computer-dependent.

There is a side issue. In the last three years, a new issue has captured our national health agenda -- cost effectiveness. Every manufacturer is forced now to focus on lowering the costs of discovering, perfecting, and delivering products. Their other goals have not changed, however. Manufacturers must still focus on the quality and safety of those products.

Are these goals conflicting? Does quality decline when cost effectiveness is maximized. Probably not, as long as controls are maintained. With proper quality controls built in, cost effectiveness can be maximized through high quality and safe products.

In the modern company, that control is a function of systems -- watching inventory, improving testing, maximizing manufacturing and processing efficiency, and tracking distribution. Systems that are proven to be effective assure safety and quality in a cost-effective world.

This brings us to the reason for this volume. The *Validation Compliance Annual* is intended to give those individuals involved in system validation a compendium of the regulations, guidance documents, issues, compliance tools, terminology, and literature involved in computer system validation. This is the second update of this compendium, and we intend to continue to update bi-annually. The field of system validation is still relatively young and rapidly growing. Regulations change worldwide, new procedures are devised, and new issues emerge regularly, and we want to keep up with this changing world.

The preparation of this book required the assistance of many individuals. In particular, Dr. Sandy Weinberg, Janie Spelton, and John English of Weinberg, Spelton & Sax, Inc., provided invaluable support in collecting information. Additional assistance and contributions were received from Debbie Cenci.

Finally, our grateful appreciation to Stein Research Associates of Westminster, Maryland, whose skills in coordinating, compiling, writing, re-writing and editing brought it to a second successful fruition.

CONTENTS

Preface iii

Section I: Overview

Chapter 1 -- Behind Validation 1
Chapter 2 -- The Regulatory World 22

Section II: Government Regulation

Chapter 3 FDA Regulations 28
Chapter 4 EPA Regulations 159
Chapter 5 European Union Regulations 173

Section III: Situational Review

Chapter 6 Situational Review by Industry 207
Chapter 7 Situational Review in the U.S.A. 231
Chapter 8 International Situational Review 241

Section IV: Tools for Validation

Chapter 9 Drug Inspections 347
Chapter 10 Biologics Inspections 458
Chapter 11 Medical Device Inspections 563

Section V: Field Findings

Chapter 12 FDA Adverse Findings 721

Section VI: Future Developments

Chapter 13 Future Developments 737

Appendices

Appendix A: Glossary 747
Appendix B: Validation Bibliography 807

Index 842

Chapter 1

BEHIND VALIDATION

REGULATORY CONCERNS

Systems affect almost every aspect of our daily lives. It is difficult to imagine any modern industry that is not system-oriented or system-controlled to some extent. This holds true for today's pharmaceutical, medical device, and biologics industries and environmental laboratories. The proliferation of computerized systems used in these industries -- to manufacture drugs and medical devices (and as components of medical devices), or to process blood for transfusion, or to analyze laboratory data -- has led to greater regulatory scrutiny of these systems. Investigators from the U.S. Food and Drug Administration (FDA) and the U.S. Environmental Protection Agency (EPA) are finding more and more computer-attributable errors resulting from erroneous programming, hardware flaws, calibration errors, or electronically corrupted data. For example, errors in logic have become increasingly significant in findings by FDA investigators in the medical device industry. The result of this is FDA's expanded emphasis on the design phase of system development; the agency has used these findings and that emphasis as part of its rationale for the new Good Manufacturing Practice regulations for medical devices.

The findings of these investigators have led to a great deal of skepticism surrounding the accuracy of computer operations, and those agencies that regulate some of these industries -- the FDA for medical devices, pharmaceuticals, and biologics and the EPA for environmental laboratories -- continually demand more and more proof that a wide range of controls are present in any computerized operation.

This skepticism which underlies the demand for proof of systems control is not alien to either the scientist or the regulatory professional, yet it often emerges as a personal affront when representatives from the two camps interact. Perhaps this resentment springs from the history of interaction between the two groups. Pharmaceutical and medical device manufacturers and laboratory scientists have seen regulatory demands grow beyond levels that seem reasonable. Regulators, on the other hand, have seen too many dubious data submissions claiming to be solid evidence.

In the computer automation field, that skepticism may graduate into full-scale cynicism. The technical complexities involved in computerization may exceed the expertise of both scientists and regulators, who have grown increasing uncomfortable with the jargon-filled, non-explanations of computer professionals.

In the earliest days of computer systems, highly inflated estimates of the power, potential, and accuracy of systems created a strong faith in their reliability. "The computer says so" became the rallying cry and defense of billing agents, government clerks, and bureaucrats the world over. But as stories of enormous computer errors -- both humorous and tragic -- flooded the media in later years, a "computer error" became as common a punchline as "the check's in the mail." Computer professionals fell from god-like status to a reputation probably far below the reasonable norm of accurate and reliable system function.

Those computer professionals contribute to the atmosphere of cynicism, too, with their resentments. Their world has never previously had to surrender the shroud of authority for the ego-reducing discipline of double-check and confirmation.

Finally, experience has created the need for evidence supporting computer accuracy. Too many systems have failed in the past, despite all the best promises of controls and safeguards.

The result of this combination of history, reality, and attitude is a general regulatory dismissal of a presumption of system control. The "default situation," the unproven norm expectation, is that a system is not adequately controlled. Until firm evidence of control is actually provided, an automated manufacturing process or laboratory system is considered to be without appropriate controls, and both industry management and the data that is presented to the regulatory agencies are suspect.

Validation defines the proofs of systems control that are necessary to establish regulatory compliance.

IMPORTANCE OF VALIDATION

A validation study translates to a directed total quality effort. Validation guides a manufacturer towards the total quality goal of moving away from defect detection and towards DEFECT PREVENTION. Validation can be defined as:

> Establishing documented evidence which provides a high degree of assurance that a specific process will consistently produce a product meeting its pre-determined specifications and quality attributes.

This definition is taken right from the definition of "validation" in federal regulations.[1] It is the baseline from which all aspects of validation have arisen.

---

[1]Food and Drug Administration, "Guideline on General Principles of Process Validation" (May, 1987; reprinted May 1990).

We interpret this definition as the accumulation of evidence -- what we see as a series of proofs -- in support of expert testimony demonstrating present and continuing control of a system utilized in manufacturing, research, quality assurance, or in the function of a medical device. These proofs can be described as:

Evidence of Current Accuracy and Reliability

The validation study must present clear and compelling evidence that the system as configured can be considered to provide consistent and faithful processing of entered data and activated programs.

A system user has a responsibility to design or adopt an appropriate function test, to conduct that test periodically, and to use normal scientific procedures of establishing acceptance criteria and performance expectations in advance of the testing. That advance definition of acceptability is generally established as the "needs statement" or "system goal," and is the first documentary step in the effective utilization of a system life cycle approach for system management, the industry standard for control of system development and implementation.

Evidence of Continued Accuracy and Reliability

The validation study must present clear and compelling evidence that the system as configured will continue to provide consistent and faithful processing of entered data and activated programs until such time as an internal auditing or monitoring system provides warning of aberration.

System users or reviewers have a responsibility to look into the software and confirm the appropriateness of the formulae and algorithms used. Not to do so is to defer decision responsibility to an unknown -- and unaccountable -- programmer. The results may be serious: There are LIMS products with inaccurate statistical formulae, laboratory equipment with erroneous calibration algorithms, data acquisition systems that misassign data fields, and decision systems that simply make erroneous decisions. Unless the source code is examined, these problems are undetectable, and unless that code review is included, the validation study has simply failed to assume proper responsibility and control of the system.

Evidence of Management Awareness and Control

The validation study must present clear and compelling evidence that the organization and/or system management maintains appropriate control mechanisms to assure proper operations, use, maintenance, and procedural supervision of the system.

This is such a common industry requirement as to be considered self-obvious, but there are environments in which, despite the most rigorous attention to testing and reporting Standard Operating Procedure (SOP) documentation, instructions for using, controlling, and recovering the computer system is casual and unwritten. SOPs for system control are as important as SOPs in any other area, and they are a necessary demonstration of the awareness and control of effective management.

Evidence of Auditability

The validation study must present clear and compelling evidence that the system contains within its infrastructure the constructs, tools, and controls necessary to allow full and complete audit of functions and activities.

A manual system of notebooks and laboratory records provides an audit trail of changes and deletions. Unless a system is carefully constructed, that audit trail may not be maintained in an electronic system. Yet a record of all changes, and the rationale for those changes, is an important control issue. Audit trail evidence assures that control and assures that a data change would not be inadvertently or fraudulently unrecorded.

Evidence of Data Integrity

The validation study must present clear and compelling evidence that the system receives, stores, transmits, operates upon, processes, recalls, sorts, and otherwise manipulates user-provided or -generated data without system imposed inappropriate modification, distortion, addition, or deletion.

Confidence that a system is not inappropriately or spontaneously changing data points is critical to reliance upon that system. Data integrity evidence provides that confidence, demonstrating that a system has not truncated, misassigned, or otherwise modified a key entry.

Evidence of Reviewer Credentials

The validation study must present clear and compelling evidence that the expert judgements upon which its conclusions are based will pass Expert Witness credibility review for technical and theoretical expertise, appropriate experience, and complete independence.

This is a critical issue, as will be discussed in further detail below. It is beyond the scope of regulatory agencies to have investigators examine source code, test systems, or analyze complex protocols. A user must assume that responsibility. The investigator's role is to determine that an effective, independent study has been conducted. Not surprisingly, the majority of all validation 483 observations issued by the FDA in the last two years were issued for failure to provide a validation study or summary certificate. To be useful, that study must be examined as expert testimony, with a review of the testifying person or agency to assure independence and qualification.

Well-trained, internal Quality Assurance staff, Certified Validation Professionals (CVPs), outside auditors, or a combination of persons may be involved, but their joint credentials demonstrating independence of reporting responsibility, system hardware and software expertise, and knowledge of appropriate applications and regulations, are a critical part of the investigator's review.

MANAGERIAL COMMITMENT

Regulatory pressures to validate computer systems are now clear and widely accepted. The U.S. Food and Drug Administration (FDA), the Bundesgesundheitsamtes (BGA) in Germany, the UK Department of Health, the Canadian Health Protection Bureau, the U.S. Environmental Protection Agency (EPA), and other international bodies have issued unambiguous statements establishing that the development of evidence in support of system accuracy and reliability -- *VALIDATION* -- is a top priority.

But how does one most effectively manage the process of validating systems? Once a commitment has been made to the testing and review of computer control and quality, what is the most effective method of implementing system validation?

A successful, six-step approach to the development of a validation management process has evolved:

1. securing timely ***information*** concerning regulatory intent;
2. developing internal ***guidelines*** or standards;
3. establishing an appropriate validation team ***organizational structure***;
4. determining the appropriate ***priority*** for validation studies;
5. completing the ***follow-through*** efforts validation reviews may determine to be appropriate; and
6. ***auditing and certifying*** the final results.

Information

Validation is an extremely diverse and complex area of regulatory concern, impacting all areas of pharmaceutical, medical device, and biologic research, manufacturing, and clinical testing. In the United States, validation clarification and evolutionary conceptualization is continuously released by more than a dozen different FDA spokespersons and offices and interpreted in twenty-two field offices. Simultaneously, nine different EPA offices are addressing laboratory validation issues. In Europe, two European Union (EU) offices and national agencies in the United Kingdom, Germany, France, the Netherlands, Switzerland, Sweden, Belgium, Spain, Italy, and Ireland have all issued statements on validation policy and requirements. In other parts of the world, regulatory agencies in Canada, Japan, Israel, Australia, Brazil, and India have addressed validation concerns.

This constant flow of important information is critical to the validation process, since it serves to dynamically define the two limiting dimensions of validation activity: How much evidence is required? What elements affect the credibility of that evidence? Without answers to these two questions, validation can be an all but endless process, cost inefficient and hopelessly convoluted. Sorting and digesting the data that provides these limited dimensions is the most significant first step in managing the validation process.

Newsletters can serve an important function in summarizing that informational morass. There are specific newsletters covering the pharmaceutical, medical device, biologics, and environmental laboratory fields. These provide monthly access to the key documents and policy statements addressing validation and other issues. A new publication by the International Validation Forum, the *IVF Newsletter*, specifically addresses all areas of validation. A European publication, *The Regulatory Affairs Journal*, provides similar international information. There are also newsletters published by industry organizations, such as the Drug Information Association (DIA), the Parenteral Drug Association (PDA), and the Regulatory Affairs Professional Society (RAPS). Similarly, a number of books have recently been released to summarize publication date state-of-the-art policies.

Guidelines

Using the information identified in step 1, and updating that information regularly in response to an evolving environment, an organization addressing validation issues must develop a set of clear guidelines for evaluation and control. One source of these guidelines, of course, is the formal regulations themselves, found in the CFR, the Code of Federal Regulations (see chart, below). But the regulations of the FDA and other regulatory groups only indirectly address validation concerns. The situation is not unlike traffic control: the law requires "safe conduct." A transportation agency is empowered to establish clear definitions of safe practices and posts speed limits, warning signs, and other devices. Ultimately, though, it is the field police officers, in deciding the speeds to enforce, the deviation from norms to accept, and the regulations to cite who define through their activities the actual traffic rules.

Rather than relying strictly upon the vague and general principles contained in the CFR, guidelines should be formulated upon the self-regulating question: "What kinds of evidence are appropriate in this environment of application, risk, software, hardware, and product to provide assurances of the continuing control of this system?" Generally, the answer will include a combination of system testing (in both normal and stress modes), source code review or analysis, documentation review (including the development of appropriate user Standard Operating Procedures), and an audit of the data or results produced by the system.

For the greatest efficiency in developing appropriate guidelines, many organizations begin with an established set of standards, such as the System Validation Standards (Weinberg, 1990); the EPA's Good Automated Laboratory Practices Implementation Guide, the most recent edition of which was issued by the EPA in August 1995; or System Validation Automated Checklist (Weinberg, 1993). By adapting one or more of these documents to the operating conditions and risk levels of a specific environment, a clear set of widely-accepted standards can be very rapidly and cost effectively developed.

Those standards, developed "from scratch" or through adoption and adaptation, serve as the basis for validation studies throughout the organization.

Organization

The organization of validation teams is complicated by a single factor: A level of independence, similar to the Quality Assurance role in manufacturing, is fundamental to the credibility of a validation study. Vendors should not validate their own products; users should not be providing assurances that they are developing and following appropriate Standard Operating Procedures. The unconfirmed statement by a programmer that proper design and testing procedures have been followed does not meet validation criteria. The audit of such claims is critical to the credibility of the validation process.

FDA AND EPA REGULATIONS

Requirement	Register	GMP	Validation
Human Medical Devices	21 CFR 807	21 CFR 820	21 CFR 820 812
Human Pharmaceuticals	21 CFR 207/314	21 CFR 210/211	21 CFR 211/312
Animal Pharmaceuticals	21 CFR 207/514	21 CFR 210/211	21 CFR 211/511
Human Biologics	21 CFR 601	21 CFR 606/640	21 CFR 606/312
Animal Biologics	21 CFR 102	21 CFR 108	21 CFR 108/103
Human In-Vitro Diagnostics	21 CFR 807	21 CFR 820	21 CFR 820/812
Animal In-Vitro Diagnostics	9 CFR 102	9 CFR 108	9 CFR 103/108
Pesticides (EPA)	40 CFR 167	NONE	40 CFR 160
Toxic Substances Control (EPA)	40 CFR 720	NONE	40 CFR 792

Most organizations have evolved a team model for validation. Representatives from the user department, along with representatives from Information Systems, provide the important background, application, and technical expertise for a validation study. But the team is not complete, and its conclusions not credible, until an independent element is added.

In some organizations, the team is completed or enhanced with the addition of an outside consultant who can bring experience, a methodology, regulatory information, valuable review tools, and highly credible credentials (including a Certified Validation Professional or equivalent credential review). In other organizations, a representative of the Quality Assurance department brings this level of independence, experience, and expertise to the validation team. In either variation, the joint team model seems an effective approach to managing the validation process.

Priority

A team has been formed; it has been equipped with a continuing source of up-to-date information of regulatory guidelines and requirements; it carries with it a set of validation standards or a Standard Operating Procedure on Validation that coincides with those regulatory positions. One step remains before the actual testing, review, and analysis that constitutes the validation process.

A major pharmaceutical company may have one hundred or more systems that require validation: Laboratory Management Information Systems, computerized manufacturing systems, computerized laboratory and manufacturing equipment, labeling systems, inventory and MRP systems, and so forth. To effectively manage validation it is necessary to establish some sort of logical system for prioritizing these projects.

Having a logical priority list not only tells the team where to begin and how to proceed. It also shows regulators that, while not all systems may yet have been completely validated, the organization is in clear control of the process. With a good prioritization model, it is possible to tell a regulator exactly when (to within a month) every system in house will be validated; it will, as well, demonstrate the progress already made on the most critical or vulnerable systems.

Follow Through

While actually conducting the validation study is, of course, the most important part of the process, only a few managerial suggestions will be offered here.

First, focus is a key issue. Organizing and conducting a validation study requires a concentrated, uninterrupted period of time. Tests can not be halted in mid data set. If weeks pass between one test and the next, a true picture of the system at a single point in time is not provided. A review of the source code should be concentrated to avoid system down time. If a clear priority focus is not a practicality for the validation team, consideration should be given to the cost effective use of outside validation specialists.

Second, do not waste time in "reinventing the wheel." Have the team use pre-developed validation tools, guidelines, and procedures wherever possible. These templates will need modification and customization, of course, but they can provide an excellent and efficient starting point.

Finally, maintain careful records. Documentation is the key to any regulatory process, including validation. Signed and dated notes can document code walk-through analyses. Test result printouts will prove useful evidence. SOPs should be accompanied by inspection reports documenting confirmatory observations. As far as possible, document everything.

Auditing and Certifying

Certification, or confirmatory auditing, is not a required step in system validation under the regulatory requirements of most agencies. It may, however, be a cost effective method of quality control and "insurance" in high risk situations.

In the certifying process, a recognized and credible outside party reviews the validation study and audits the conclusions. In confirming the findings, a certificate under the name and responsibility of the auditing group is issued. This process serves to review the efforts of the validation team, reviews the quality of the team, provides a recognized (and demonstrable) assurance of validation, and provides a level of assurance of regulatory acceptance of the result.

Under the guidelines of the ISO 9000 program, for example, one group prepares for the audit (in parallel to the validation team role), while an outside group (BSI, Underwriters Laboratory, etc.) certifies the result. This outside group can serve as the validation team itself (or as a member of the team), or it can serve as the certifying group, adding the outside independence or credibility of compliance with FDA or EPA requirements; equivalent requirements from other regulatory agencies; or with international guidelines of groups like ISO, World Health Organization, and CITECH.

A standardized certification process, such as that established by the International Validation Forum for its Certified Validation Professionals (CVPs) is an additional source of confidence in determining whether the professional performing the validation is actually a qualified expert in this complex field.

When is it cost effective to have a system certified? Most companies conclude that the risk of regulatory rejection or questioning of a validation study is too great with a system critical to a New Drug Application, or fundamental to the manufacture or control of a high risk product (controlled substances, blood products, etc.). Other organizations have used the certification process as an important managerial step in controlling the costs, time, and quality of validation teams, using outside auditors in much the same role as a CPA is used to audit the effectiveness and efficiency of an in-house accounting department. Still other companies have added the certification process in order to have an advertisable standard of quality: Hewlett-Packard, Gettinge, Pharmacia, and other companies, for example, advertise the certification of their laboratory instruments.

Summary

These six steps of validation management, identified above as information, guidelines, organization, prioritization, follow through, and certifying, provide a sound structure for controlling the process of gathering evidence in support of a system. Together, this framework can serve as the fundamental principles for developing and implementing a validation plan.

The key to developing such a plan, and the key to the implementation of that plan, is commitment. Validation is an efficient process, but it still represents an expenditure. That investment produces significant savings in lost data, erroneous conclusions, regulatory delays, and potential fines. It produces a positive cost/benefit ratio, but it still represents an out-of-pocket corporate expenditure.

In an era of cost consciousness, such expenditures are still necessary but must be carefully analyzed. The managerial controls suggested above can provide the necessary efficiencies to minimize expenditures, eliminate waste, and improve the effective use of automated systems in regulated industries.

BENEFITS OF VALIDATION

A pharmaceutical, medical device, or biologics product manufacturer is in business to obtain satisfactory market acceptance of existing and new products in a cost-effective and profitable manner. Computer-system validation can meet this goal.

The FDA has stated its intention to consider all data derived directly or indirectly from systems not validated through a policy meeting the above guidelines to be "suspect data," presumably delaying all applications and approvals and triggering additional inspection visits. Furthermore, the FDA has stated its intention in the future to reject automatically as "un-calibrated" all data derived directly or indirectly from systems not validated through a conforming policy.

The cost of **NOT** validating can be immense. For example, if an FDA inspector wants to see evidence of validation of a computer system utilized in a pharmaceutical manufacturing facility, and that evidence is not available, that manufacturer will receive a warning letter from the FDA. In the United States, a company's stock price drops three to six points when a warning letter becomes public information. But what happens if the inspector chooses to halt operations until the manufacturer takes corrective action? A company can lose thousands of dollars for every day of delay during the submission process.

Compliance with regulatory guidelines is a major justification for performing computer validation, but there are other reasons why validation is simply good business. More and more people -- in industry and in the regulatory agencies as well -- are recognizing that validation is good for business in an economic sense and in a quality sense. In fact, he believes that if the regulatory agencies entirely did away with their regulations requiring validation, 99% of the manufacturing firms would still validate.

From a business point of view, regulatory and industry spokespersons have recently expressed their conviction that computer validation will also save a company money. Validation makes a positive contribution in several areas, such as:

- Higher reliability of systems, resulting in higher output quality:
  -- A decrease in the number of errors, which are sometimes expensive to correct
  -- Better acceptance among customers and users
- Improved control over process efficiency and timing, leading to faster output
- Defense against potential adverse events and timely and comprehensive recovery when an adverse event occurs.

For example, have you ever come in contact with a new employee at your company trying to use a computer system to perform a task, only to be frustrated because of inadequate documentation on how to use the system? It is very common for systems to lack appropriate documentation, or the software version is wrong, or it is hard to understand, or controlling Standard Operating Procedures (SOPs) are out of date. What does it cost a company in lost productivity and decreased learning curve because of inadequate documentation and no planned training? Validation addresses these issues.

Have you seen someone who is your company's computer expert and has all of the SOPs and testing documentation ***MENTALLY*** recorded? What happens when this person wins the lottery, gets a major inheritance or gets run over by a bus -- in all cases leaving you with no documentation. How much will it cost you to recreate this documentation and how many best practices will you lose in the process? This is not a fanciful situation; a recent (10/28/93) FDA Warning Letter to a medical device manufacturer cited the following violation: "The validation protocol is inadequate.... The data is maintained in an employee's laboratory notebook and only he can explain it."

QUALIFICATIONS OF VALIDATORS

The validation of computer systems, like the validation of any product or procedure, is essentially a process of collecting credible quality-related evidence of the control and functionality of that system. That evidence may include test results, analyses of Standard Operating Procedures, inspections, and a variety of other inputs that provide a comprehensive picture of the system in question and the way in which it is being utilized.

The credibility of that evidentiary picture resides in four key factors:

1. the methodology of standards employed;
2. the breadth and depth of the test and analytical data generated under that methodology;
3. the degree of documentation available for review of that methodology and data; and,
4. the least tangible factor in the quadrad: the credibility of the validator or validating group itself.

It is this fourth critical, but ephemeral, foundation of system validation that forms a major focus in validation.

The credibility of a validation study should not be viewed from a law enforcement perspective: the antithesis of a credible study is not fraud but rather sloppy or incomplete work. It is possible, no doubt, to fabricate data, to forge review documents, or to otherwise distort analyses to provide convincing but spurious evidence. To some degree, the credible reputation that the validators provide proves convincing in the rebuttal of fraud. But the appropriate and realized regulatory position is not a prior assumption of dishonesty, but rather skepticism that a validation study meets two critical tests for credibility: first, that the certifying or signing validator has the knowledge and skills to conduct an appropriate quality assurance review of the system; second, that the validator had the appropriate level of independence to assure the non-contamination of that review. These two criteria of expertise and independence together establish the basic credibility of the validation study. A deficiency in either area may throw the value of that study into question. If the validators lacked the knowledge to determine and conduct the appropriate reviewing, testing and analyses; or if the validators failed the test of independence designed to assure that their conclusions are unbiased, the value of the study is reduced below acceptable levels.

Expertise Required

The expertise required to give a validation study the weight of credibility may reside in either the validation team (or person) itself, or in an audit group (or person) responsible for the review and certification or approval of the validation study. If an audit group is the primary expert source, its review should focus upon the methodologies utilized as well as the results of the validation study. Examining only the results may lead to erroneous acceptance of conclusions based upon inadequate or inappropriate sampling, limiting testing approaches, or other methodological inadequacies.

Expertise can be further divided into six competency areas: system hardware, system software, applicational environment, validation methods, statistical techniques, and experience. While these six competencies need not be represented by any one member of the validation (or audit) team, all six should be incorporated in the union of skills of that team. Strength in one or more areas is not sufficient to compensate for absence or significant weakness of other skill areas.

Although there have been attempts made in recent years to identify academic or training programs that represent proficiency in these six areas, to date there is no single certification or training program that will assure, upon completion, expertise in all six skill sets. A combination of training courses, however, coupled with academic or experiential education, may together provide appropriate confidence in the mastery of validation competencies. The CVP certification program of the International Validation Forum, providing as it does a dual certification (either the basic Fundamental Certified Validation Professional or the Advanced Certified Validation Professional) provides the combined training and skills competency necessary for validation experts.

Hardware

The design of appropriate testing, and the identification of key elements to be tested, often requires an engineering-level understanding of the computer equipment (hardware) that runs a particular system. This requirement is maximized in the cases of robotics, medical device, or manufacturing control systems and is probably minimized in centralized data oriented analytical and reporting systems. In distributed data systems, engineering concerns generally focus upon the networking and communication elements in the systems.

To the degree that system validation may overlap process validation (for diagnostic or production equipment, for example), the hardware expertise may be reflected most strongly in the design and implementation of Installation Qualification, Operations Qualification, and Performance Qualification testing.

Hardware expertise may be demonstrated experientially, or it may be "certified" through a variety of channels. Most vendors offer training programs for customers that may include hardware issues in their curricula. Academic programs, generally in engineering, provide strong credibility. Engineering credentials, through licensure and/or membership in IEEE, AIIE, and equivalent professional organizations, provide even stronger evidence of hardware competence. Particular care must be taken in the review of qualifications of American academically prepared persons in "Computer Science,"

"Information Systems," or "Communication Science," whose skills may be limited to software or mathematical application areas without a grounding in system hardware.

Software

Software skills parallel hardware skills in specificality, demonstration, and concerns. Unfortunately, the software arena lacks even the limited hardware "certification" represented by an engineering society's membership; ACM, SIM, DPMA, and the other software associations are generally "open membership," demonstrating little more than an interest and a willingness to pay an annual fee. Many software professionals are self-trained, either in reality or claim, and may have skills tightly focused upon a single programming language, technique, or sub-method.

Generally, software experience should include three key elements. First, the validator must have the ability to read the software language(s) utilized. This literacy must be of sufficient depth to permit a trace or walkthrough of the code, including mastery of all symbols, positioning codes, and syntax features. Second, the software expert must be qualified to diagnose, develop, and evaluate algorithms. This skill includes the ability to applicationally understand the various software approaches to a given problem, and to selectively evaluate those approaches. Finally, the software expertise required for validation includes an understanding of the innate characteristics of various computer languages, incorporating syntax checking, interpretation and compilation characteristics, and internal diagnostic capabilities and limitations.

U.S. academic programs in computer science, information systems, communication science, data management, and programming generally provide strong software skills, although care must be taken to evaluate the match between mastered languages and code languages in use for a specific validation project. Again, care must be taken in assuming that a computer engineering curriculum includes applicational software topics. Many do not.

Applicational Environment

Expertise in just computer hardware and software is not sufficient; a validator must also understand the applicational environment in which the system operates. Consider a blood testing system, for example: without at least a rudimentary grasp of the blood factors under analysis it is not possible to assure that the system in question is properly interpreting the results received. Identification of the critical inputs that may appropriately require double entry, or the life-threatening impact of power fluctuation in a pacemaker, or of any other situationally-specific conditions that may identify significant control issues, is fundamental to the validation process.

Generally, system validation environments can be placed in one of three categories: data, control, and device. The data systems, generally seen in GLP environments, receive, analyze, interpret, report, and potentially evaluate based upon information analysis. Control systems (usually in GMP environments), often incorporating A- to D-converters, add a feedback mechanism to the operation of fermentation tanks, mixers, stamping machines, and other manufacturing equipment. Computerized devices, such as laboratory diagnostic equipment, pacemakers, infusions pumps, and similar types of equipment may incorporate both kinds of systems, sensing and analyzing data while controlling or operating processes. All three environments, and the myriad subcategories possible, have unique issues that require at least a fundamental understanding of functionality.

Expertise in the applicational area is generally experiential, supplemented where appropriate by academic training (bioengineering, chemistry, medicine/nursing, biology, etc.).

Validation Methods

Validation of a system provides confidence to Quality Assurance, management, and the regulatory community. It is not intended to simply satisfy a regulator, but rather as an integral part of the entire quality picture of an operation. As a relatively new element in that picture, however, validation is still open to interpretation. While the essence has been distilled, the ramifications of that essence are still evolving. An important part of the validation process, then, centers around "why" questions. Why is this testing sufficient? Why have you chosen this approach? Why have you decided to validate this system while another is assigned a lower priority? In effect, these questions ask for a validation methodology, and the value placed upon the responses is largely a matter of the credibility of the validator providing answers. If those answers are "method free," that credibility is immediately suspect; in effect, the lack of a rationale and an a priori definition of the validation methods is as glaring a weakness as the absence of a system design document. Both problems suggest that no one thought through the issues and goals before starting.

Expertise in validation methods is generally demonstrated through the completion of a validation training program built around a specific methodology. Currently, such system validation courses are available from a number of organizations including the Center for Professional Advancement (New Jersey and Amsterdam), and the International Validation Forum (Key West, Florida).

Statistics and Testing

Advanced academic education in most fields provides a methodological emphasis: doctoral degrees in the sciences, social sciences, and some humanities disciplines include testing and statistical analysis in their disciplines. Other sources of testing and analysis credentials include undergraduate academic programs in statistics, research methods courses and practicums, and special training programs in statistics methods (often sponsored by vendors of statistical packages such as SAS, SPSS, and BMD). A Quality Assurance or Quality Control background is likely to incorporate and demonstrate experiential training in testing, particularly if coupled with training in sampling, test design, or statistics.

All of these credential sources are designed to provide a clear demonstration of expertise in the development and implementation of testing methodologies and in the analysis and interpretation of the results of those methodologies. Most validation testing utilizes unique stratified sampling scenarios. Statistical interpretation is generally limited to descriptive tools such as standard deviations, ranges, and chi-square analyses (except, of course, when the statistical programs form the subject rather than the method of the study, as in validating statistical analysis computer systems).

Test design is perhaps the most complex of the methodology expertises. Determining the appropriate test strategy, and the logical testing construct, is based upon a complicated ability to adapt theory to practice. Many organizations have attempted to standardize testing for all systems (often with a resulting "overkill" approach) to minimize the testing expertise necessary. This approach is possible, although it may lead to billing abuses with consulting firms operating on an hourly billing basis.

Experience

The "experience" criterion is the clearest to demonstrate, the easiest to define, and the hardest to theoretically justify. Every non-parent psychologist is a master of child-rearing theory, but he or she may not be so successful when faced with a real tantrum-throwing two-year-old. If a validator can demonstrate all five other areas of expertise, experience may not be apparently critical. But the gap between the hypothetical and the real is huge, and a validator with experience in that real environment carries more credible weight than one who has studied but never applied.

The degree of experience necessary is not a constant. Since there are three distinct kinds of systems (data, control, and device), a validator should have a minimum of three experiences in order to provide credible testimony in all three areas. Some jobs, however, may limit responsibility to a single environment. In the same manner, a validator providing guidance for a GLP environment LIMS may extrapolate from a database system of another type, but will be more credible with more direct applicability. In short, more is better, and some experience is critical. The validator who has never applied his or her skills before is to be avoided (or, better yet, paired with an experienced validator); the value of the testimony increases with the similarity and extent of the experience, until some reasonable asymptote is reached.

Required Independence

The basic quality assurance precept of "independence of review" is appropriately carried to the qualification of a validation expert. If a coder checks his or her own work, that check is suspect for three fundamental reasons: a) there is a natural bias toward NOT finding fault internally; b) there is the human tendency, perhaps most evident in proofreading, not to see errors that one is "too close to;" and c) since the role of a validation (or any other quality review) is in part to see if rules and procedures have been followed, self-examination falls to a logical fallacy. That is, if validation is necessary (there is a possibility that procedures are not followed), self-validation is impossible (the validation procedures are likely not to be followed either); if self-validation is possible, it is not necessary. For all three reasons, validation can only carry the "second check" conformational credibility of its intention if the validators meet an independence criterion.

Validation studies may be conducted by either internal or external sources, each with different independence criteria. In both cases, however, independence of source provides an aura for a complete team. It is perfectly possible, and often appropriate, to include in a validation team representatives of the users (who may have a dependency upon the design and functionality of the system) and of the developers, whose independence may be suspect because of financial gain or work limitations flowing from acceptance. Their expertise is important, even if their independence is restricted. The validation team, however, may meet the criteria of independence through the inclusion within it (in a veto-potential role) of an independent member who may review and evaluate the contributions of the rest of the team. This individual must have sufficient expertise to understand, if of lesser depth than that defined above.

In-House Independence

The general rule of thumb for in-house independence is chain-of-command branching. As with any other quality assurance officer, the career path of the validator should not be influenced by the results of a validation study. That is, the validator should not directly report to the "owner" or developer of the system, nor to the manager to whom that owner or developer reports. In small shops this independence may by necessity be compromised, but overlapping reporting relations may affect the credibility of the validation report.

Incidentally, commercial vendors of systems used in high risk or high hazard environments may be subject to even more stringent independence requirements. Because of sales pressures, all members of the vendor organization may be suspect, effectively requiring outside validation.

The most recent (August 1995) edition of the U.S. EPA's Good Automated Laboratory Practices (GALPs) recommendations provide a very clear guideline for internal independence. The GALPs introduce the concept of "Responsible Person" (RP), an individual of any title or position who ultimately is charged with the effective use of the system. The GALPs recommend a clear distinction in person and reporting structure between the RP and the quality assurance or validation role, defining the validation as the recheck or review of the RP system control role.

External Independence

When external validators are used, independence can be assumed as long as three criteria are met:

a) *Conflict of Interest.* The validators should not be drawn from a division or company that also markets services of software development, system analysis, or other conflicting activities. A software company offering validation services does not pass the presumption of independence. Similarly, the company that consulted in the design or development of the system should not also be charged with the validation of that design.
b) *Conflict of Incentive.* The audit and testing of a system may be suspect if the outside validators operate on a basis where it is perceived that validators are trying to find or avoid problems based upon financial considerations. This may subjectively compromise independence.

c) *Conflict of Objectivity*. A bias toward a problem solution, computer brand, or software language may bring into question the independence of a validator. The independent validator is consistently committed to a methodology, not to an environment, computer, program, or device. In effect, this criterion of independence is best demonstrated through extensive experience.

Of course, it may be practical for an outside validator to demonstrate independence despite a failure in one or more of these three areas, but the presumption is carried only with full conformity. Of particular suspicion are outside validators who combine their services with a wide range of non-validation offerings and the validation groups too small in numbers to offer depth of expertise or experience.

Summary

The value of validation reports rests largely with the credibility of the validator; the credentials of that validator (or validation team) should form an integral part of that report. In fact, many field investigators have confidentially reported that the credentials of the validators are the single most important part of that report; two separately confessed that the credentials were the only part of the report they reviewed.

That credibility rests upon two foundation stones: expertise and independence. The most credible validation, providing the largest degree of managerial, quality assurance, and regulatory confidence and acceptability, will be signed and certified by a person or group with demonstrable expertise in the specific hardware and software involved, in an appropriate validation methodology, in statistical analysis and testing, and in the specific applicational environment. That expertise will include extensive prior experience, as closely related as possible, and completion of a recognized validation professional certification program. And, finally, that experience, expertise, and certification will be supported by internal or external independence, assuring the credibility of the conclusions and appropriately maximizing that credibility.

VALIDATION RESPONSIBILITY

The current regulatory requirement is that all computer systems utilized in the support of data submission, in the manufacture of regulated products, or in the monitoring of regulated processes be regularly validated and tested to assure accuracy, reliability, and appropriateness of control. The question of who should assume responsibility for that validation process, however, has been left largely unaddressed.

Should a pharmaceutical, blood processing, or medical device company follow the current Quality Assurance Model, establishing an oversight group to perform retrospective and periodic validations? Can the internal data processing or management information system staff perform the validation function as a part of the development and installation process? Will FDA or EPA inspectors in effect validate a suspect system through their review processes? Or is another model appropriate to insure the objectivity and accuracy of system validations?

There are five possible models corresponding to these options: the Quality Assurance approach, the MIS model, the FDA Review alternative, a Consultant model, and a proposed "SEC" model. Each approach must be evaluated on criteria of appropriateness, performance capability, and -- perhaps most importantly -- objectivity.

In-House Alternatives

Most internal checks on data submitted to the regulatory agencies are provided by an in-house department or staff, generally called Quality Assurance, Quality Control, or Compliance. This internal group acts as a second check on experimental design, data collection, statistical analysis, and related functions, working closely with appropriate departments while simultaneously remaining aloof from the operations of those departments. Quality Assurance staffs are often (inappropriately) perceived as "big brother" inspectors. If appropriately utilized, however, they confirm the quality control measures employed on all levels and serve as an internal resource in the design of those measures.

The addition of system validation responsibility places a bi-directional strain upon Quality Assurance operations, however. First, appropriate validation procedures require a high level of technical expertise, including the ability to read and evaluate source code, experience in the evaluation of security control and disaster recovery procedures, and a hardware background that includes reliability testing and performance standards. Few, if any, existing Quality Assurance departments possess such expertise, which is acquired only through lengthy professional training. It is, of course, possible to add experienced system validation personnel to the Quality Assurance operation, although such individuals are likely to prove scarce and expensive.

One of the key organizational constructs of Quality Assurance is an independent reporting relationship. Quality Assurance personnel should (by regulation) report through a separate channel of responsibility, independent of the chain of supervision controlling the departments they audit. This independence, assuring a freedom from undue influence, is critical to their objectivity and, in turn, their credibility.

The Quality Assurance Model, then, meets the test of objectivity despite the lack of pure independence inherent in an in-house operation. Unfortunately, however, it is unlikely that the Quality Assurance staff will cost-effectively possess the depth of expertise and experience necessary for effective system testing and validation.

That expertise is likely to be present, however, on at least the top levels of the management information systems staff. The system management personnel are presumably fluent in the coding languages in use; they know at least the rudimentary details of the hardware configuration and functionality; and they have been exposed to or can master the techniques of effective system testing and review. In terms of the ability to perform validations, they stand temptingly as the ideal group for the assignment.

But the MIS model fails to meet the critical test of objectivity. Every effective programmer, supervisor of programmers, and user accepting a new program performs routine tests of the functionality of the code. In fact, the most commonly taught and used programming procedure relies upon a process of coding, testing, "debugging," and retesting. However, to rely upon the assurances of that programmer, both psychologically and politically biased in the evaluation of that program, is a chancy proposition at best. While the reliance upon a supervisor to double check work represents a slight improvement, it too fails to meet the independence criterion. Within a single department, reporting to a single line of responsibility, no true objectivity of validation is assumable.

With the dilemma of expertise generally concentrated in the MIS department, while the objectivity criterion is met only through the traditional Quality Assurance formulation, these two tempting in-house models are jointly doomed. It has been argued that Quality Assurance people do not require great depth of expertise in their other areas of responsibility (statistics, for example), but both the sad history of computer validation (or its lack) and the depth of knowledge inherent in current regulatory guidelines suggests that the analogy does not hold. Both expertise and objectivity are required, but the combination is sometimes difficult to find in-house.

Outside Alternatives

In the past, the validation function has tacitly been performed by the regulatory agencies, and at least in the short term this model can be continued. For example, when FDA reviewers find data discrepancies, they track down the cause, placing new drug applications on hold in the interim. Should the cause prove computer related, an increased schedule of inspections will be instituted, presumably including the assignment of an FDA computer specialist (the agency has some individuals who have been trained as computer specialists, but not enough to be included in every inspection team). While this model may cause increased discomfort for the organization -- and result in costly delays as problems are discovered -- it can not be disqualified as a current validation alternative. The FDA does have inspectors with credentials indicating sufficient expertise to perform at least a preliminary validation, and this clearly meets the criterion of objectivity. The model, however, has a very short-term future.

There are an estimated 1,200 pharmaceutical companies with computer systems and an additional 240 computerized blood processing organizations in the United States, with approximately another one-third that number under FDA regulation but operating in Europe and Japan. Each of those 1,880 systems requires ongoing validation, testing, or review, and the number of systems is growing at a rate of more than 12% per year. It is clear that the number of FDA computer-qualified inspectors will prove vastly inadequate to the task, forcing the FDA to require adoption of another model. One FDA computer inspector, pointing out that he now has a workload allowing less than one day per system review (a process generally requiring forty-five to fifty person-days), suggested that the process of removing the FDA from direct validation has already begun and that within two years the FDA will require that all data submitted be generated by pre-validated systems. Clearly, the FDA model has little or no life expectancy, and does not represent a viable option.

The remaining model available today is the Consultant alternative. In effect, outside consultants specializing in system validation and testing, using a methodology pre-approved by the regulatory agencies, act as adjuncts to the Quality Assurance operation. A consulting team or individual conducts a validation, providing an in-house report to Quality Assurance, and presumably increasing the expertise in the Quality Assurance department until it is capable of assuming increasing shares of the validation responsibility. While the consultants may be expensive, their services are generally cost effective if the consultants bring both a high level of experience in validation and a highly focused commitment to validation procedures in the regulatory environment. The Consultant Model clearly meets the joint effectiveness and objectivity criteria, although a further improvement may be possible.

The "SEC" Model

The one potential weakness in the Consultant Model lies in the area of responsibility. Unless the consulting group brings high levels of credibility and a previous reputation with the regulatory agencies, the regulated industry is left in the situation of having its Quality Assurance department assume responsibility for an area in which they admittedly lack expertise, relying wholly on the advice of outsiders.

The Security and Exchange Commission has evolved a model designed to cope with the same issues of expertise, objectivity, and responsibility in its own area of regulatory responsibility. This SEC paradigm provides a very powerful model that the FDA and EPA should consider carefully. By all inside accounts, in fact, that consideration is already underway.

A public company uses its in-house controller to accumulate regulatory data and relies on its corporate treasurer to act as a quality control check on that process. But more than the in-house reports are submitted to the SEC. The company is required to employ an outside, independent group -- in this case, a public accounting firm -- to audit and verify the submitted data. The accounting firm applies clear standards, draws a conclusion, and assumes responsibility and liability for its conclusions (with reasonable constraints).

While the public accounting firms are not appropriate to the pharmaceutical, medical device, biologics, and environmental laboratory situations, both due to a lack of system validation expertise and a conflict of interest as financial auditors, the model has high applicability. If the regulatory agencies can establish guidelines for validation consulting firms, including credentialing of validators, review of methodology, evaluation of previous experience, and similar criteria, the system validation equivalent of CPAs can be identified. These validators would then presumably be invited into a client company to conduct a highly objective, independent evaluation of a system and to certify, with assumption of (perhaps limited) legal responsibility, the conformity of that system to the established criteria. This is the level of education and certification being provided by the International Validation Forum's Certified Validation Professional program.

In the interim, as the regulators and industry are jointly designing the institutional mechanisms to establish an "SEC" Model for System Validation and Testing, the executive responsible for system control can best serve the organization by following the Consultant Model strategy, taking care to select an outside consulting firm with the credentials necessary to move into the certification role. Briefly, those credentials include a strongly defined validation focus, experience with regulatory guidelines of validation, and utilization of a regulatory-reviewed methodology.

As the "SEC" Model evolves for the industry, true validation control will be achieved. Executives and managers will have the assurance that systems have been tested and approved according to established standards, in a process conducted by experts meeting the joint criteria of objectivity and knowledge. That assurance will be reinforced by the assignment of responsibility (and, presumably, liability) to the independent group providing the review and certification. Regulators, in turn, will be able to fulfill their appropriate mission within their resource constraints, in effect transferring their focus from the system to the certifying system inspectors, much as the SEC concentrates upon the auditors rather than attempting to randomly re-audit the vast number of organizations under their regulatory umbrellas.

Chapter 2

OVERVIEW: THE REGULATORY WORLD

PHILOSOPHY: AUDITED SELF-REGULATION

If you drive along a major highway in any part of the world, you are likely to encounter a road sign telling you the maximum legal speed limit: 100 kilometers per hour, or 55 miles per hour, or whatever the local limit may be. Those signs serve a dual purpose: They give the driver a general idea of the safe speed for that road, and they serve as an enforcement warning for police action in restricting that limit. Of course, the posted limit is only a rough approximation. Bad weather, a poorly tuned automobile, or an inexperienced driver may all argue for a lower safe speed. And in many areas, the upper limit is really much in excess of the posted maximum. Traffic may routinely move much faster than the posted speed, and traffic officers may well enforce a limit well above the maximum allowed.

Despite these variations of condition and of reasonableness, the posted limit does provide the driver with some general guidelines, and with a defensible position in the event of capricious or inappropriate enforcement. While the prudent driver will strive to understand the "real" enforced top speed, the posted limit will help in arriving at a reasoned judgement of the proper rate for unique combination of enforcement, weather, and road conditions.

Effective regulation works much as a speed limit sign. It provides a general guideline, around which experts ought to be able to make reasoned decisions. It provides a framework for those decisions. And it serves as a definition for the limitations of regulatory authority. But effective regulation can not and should not replace that reasoned and professional judgement. High allowable speeds may not be safe in some conditions of snow or ice, and unusual circumstances may justify the risk of a still safe limit in some reasoned excess of posted warnings.

The effective road sign, and effective regulation, serve the primary purposes of establishing an agenda for decision; evaluative criteria for that decision; a financial justification for assessing the risks and benefits of that decision; and a framework for the reasoned defense of that decision.

As increased industry reliance upon computerized systems necessitates regulation of data and the software systems that generate that data, so effective road signs are appropriate to define the guidelines for system validation. Today, those signs are increasingly being posted, and the fuzzy limits of interpretation and application of those guidelines are increasingly being brought into sharp focus.

The United States Food and Drug Administration (FDA) has taken the leading role in setting the international agenda for system validation, and industry awareness of the importance and vulnerability of computer systems has been created largely by the effective agenda-setting function of the FDA. Working closely with colleagues throughout Europe, North America, and Japan, the FDA has used its public speeches and forums to establish an agenda for proving control of systems used throughout the industry. Private companies, as well as associations like the Pharmaceutical Research and Manufacturers of America (PhRMA), the Health Industry Manufacturers Association (HIMA), the Organization for Economic Cooperation and Development (OECD), the Drug Information Association (DIA) and others have responded. And international regulatory agencies worldwide have provided the reasoned argument, the analytical data, and the legal incentive to make system validation a high priority.

We now have road signs in many areas, defining appropriate standards of testing, control, documentation, and planning for laboratory systems, medical devices, manufacturing systems, and controllers. As field visits examine the validation studies that confirm those controls, the fuzzy limits and interpretations of validation are coming into focus. Here are a few examples of the clear consensus issues that have emerged:

Application of Validation Guidelines

System Validation guidelines clearly apply to all computerized systems in a regulated environment, with the isolated exceptions of financial and personnel systems. There is not any persuasive rationale for exclusion of any system operating in a regulated environment. The depth of controlling evidence, however, can be reasonably limited by criteria of system hazard, increasing test samples when human life is at risk; by the universality of the system, limiting review efforts in the case of operating systems or standardized packages; and by the credibility of the system developer, established through independent validation, ISO certification, and audit of the development and support organization. These three criteria may call for pre-established but varying levels of evidence of control, all encompassing the totality of regulated systems in an environment. All systems equally need to be validated, but all validation studies need not be equal in extent or rigor.

Credibility

All validations must be equal, however, in the credibility of the effort. That credibility is supported by two factors: the expertise of the validation team, and the independence of that team. Defining the fuzzy limits for a particular environment, and reasonably justifying any variation from the posted guideline, requires an expertise intersecting computer system knowledge, experience in the application area, and an understanding of the detail and philosophy of regulatory-legal requirements. Appropriately, the credentials of the validation team or the Certified Validation Professional should be scrutinized for evidence of these three areas of expertise and certification before accepting their collective professional opinion of the validation of a system.

That opinion, however, is suspect if it is biased. A regulator, or an industry executive, has a right to know that the opinion on the validation of a system is not tainted by vested interests. Vendors who self-validate will always face difficulty in justifying their independence. The use of outside, third-party experts to validate systems has emerged as a common sense necessity, and commonplace reality. Within regulated organizations, a validation team should be assembled with all the care inherent in any quality assurance role. We would not ask a bench scientist to validate the accuracy of his own research, nor can we ask programmers or information systems professionals to validate the computer system that their own department has constructed, selected, or manages. Similarly, the users of a system can not self-assess their own use and operating procedures without compromising their credibility and their capability. Independence, either through the use of an outside auditing group or an expert in-house validation department, is a practical, theoretical, and regulatory necessity.

Software Code Review

Yet another consensus issue surrounds the software code itself. Functional testing can determine current performance of a system. But there exists to our knowledge no way of assessing the future performance, the supportability, reliability, and formulae level of accuracy, of any system without reviewing the source code of that system. A review need not include a character-by-character assessment, but should examine the code structure, the internal organization and documentation, the appropriateness of the algorithms employed, and the general professionalism of the construction. Inability to obtain access to the code of a system immediately raises suspicions. Failure to review that code once access is obtained appropriately leads to serious questions about the expertise and professionalism of the validation team.

Classification of Software Code

The last point of consensus is a logical extension of the previous discussion. Clearly, computer software is utilized in the laboratory for the interpretation of results, in the manufacturing site for the production of biologics and pharmaceuticals, and in the clinical setting for the analysis of findings. The classification of such software as "medical devices" is an extension of reason without any unnecessary stretching. Even HIMA has chosen not to disagree with such a designation. This was emphasized most forcefully by the FDA's Center for Biologics Evaluation and Research. On March 31, 1994, CBER's director, Dr. Kathryn Zoon, wrote a letter to 23 computer software manufacturers in the United States and Canada, stating "that the Food and Drug Administration (FDA) considers software products intended for use in the manufacture of blood and blood components or for the maintenance of data that personnel use in making decisions regarding the suitability of donors and the release of blood or blood components for transfusion or further manufacture to be devices under section 201(h) of the Federal Food, Drug, and Cosmetic Act"

Clearly, consensus has been reached, and federal agencies will vigorously enforce their regulations. In more and more applications, the software utilized in testing and production falls under the regulatory guidelines applying to medical devices. While an educational period will be necessary, and time will be required to organize the inspection and enforcement mechanisms, there is no doubt that within the next five years the international Good Manufacturing Practices will be extended to the developers and vendors of application software. Already a number of responsible vendors, including Hewlett-Packard, Johnson-Yokagawa, Pharmacia LKB, and others are arranging the preparation of independent validation certification for their software products.

BENEFITS OF REGULATION

In addition to setting an agenda, there are four good reasons why regulation really works. There are probably some other good reasons why they do not work, and there are probably some bad reasons why they work, too.

Defining Responsibility

The first good reason for regulation in the medical device, pharmaceutical, biologics and blood processing industries and in environmental laboratories is that regulation -- that clear guidance -- has the effect of defining and limiting responsibility. In effect, regulation limits the liability of any manufacturer or laboratory manager by defining good common practice.

Without that definition, a group of non-peers, often a lay jury, has to make determinations of whether proper protections have been made, whether proper analytical procedures have been followed, and whether proper kinds of standards have been enforced. With some kind of regulation, the fact that a medical device manufacturer, for example, is compliant provides a clear definition, usually decided by the right level of professionals, defining what responsibility and liability is. A manufacturer of a LIMS system, for example, has almost infinite liability if that particular system malfunctions and gives a false reading that ultimately hurts the public. However, with regulation restricting liability, that particular device no longer has unlimited liability. For example, if a product is registered as a medical device under the 510(k) submission process, the federal courts have ruled that such registration limits a manufacturer's liability to pure responsibility and liability.

This is not only a legal definition, but probably a moral one as well. For a manufacturer or laboratory manager to know, in their altruistic goals, that they have done everything practical and possible to assure quality is a very positive gain. With regulation, we can define what kind of quality is practical and possible.

Balancing Competing Requirements

A second advantage of the regulatory process is that it balances competing public requirements. At the same time that we have a public requirement for certain kinds of safety guidelines or health standards, there are competing public requirements for price controls and for reasonable cost effectiveness.

As nations throughout the world debate health-care reform, those particular interests are being thrown out of balance. Suddenly, the importance of cost effectiveness is overshadowing the importance of ascertaining that health care is actually being delivered. Pharmaceutical companies, medical device manufacturers, and environmental and biological laboratories are aware of those competing areas, and they are being pressured to find ways of providing less expensive products, or conducting fewer and less expensive tests, even if it means that the products or tests are not quite as effective. But at what point do we say that we will take a greater risk to save greater dollars?

The need to balance these competing priorities can be defined best by clear regulations. With them, we can know when we need to be cost effective, and when we need to put safety forward as our predominant consideration.

Leveling the Playing Field

A third reason why regulation works is that regulation has the effect of leveling the playing field. If we are placed in a position in which some companies put a great deal of money and effort into quality control and safety maintenance, while other companies make a decision to perhaps cut corners and skimp in that part of their operation, we end up with an unfair competitive advantage for low-quality production.

By having the quality of production clearly defined, we then have a level playing field throughout the United States and throughout the world.

Rationale for Quality

The fourth reason why regulation works is that compliance really has the effect of making quality cost effective, and that provides a clear rationale for quality. By having requirements for validating computer systems, it becomes cost-effective to go ahead and do that kind validation. It becomes cost effective very specifically because doing that kind of validation reduces the economic risks of potential fines, potential seizures of property, potential recalls, or potential legal action that will add greater cost to companies that are not compliant.

SUMMARY

The problem, of course, is that this discussion is based on an idealized kind of regulation. Reality, perhaps, is failing in all four areas. The reality is that we do not have clear definitions, or clear regulations on liability. We do not have a very tight balance between competing public concerns, because we often have different agencies, and sometimes different portions of agencies, emphasizing different kinds of regulatory concerns. Because we have those differences, our playing field is far from level; it is filled with valleys and peaks so that regulations often affect one company, while another is left to go ahead and use a lower level of standards. Or we are left in a situation where we do not have good, clear cost-effectiveness guidelines, because we have not had a clear definition of what is and what is not really required. Often, some companies decide to be pro-active, and they go ahead and invest more money and more effort in quality issues, while companies without those financial resources are forced to wait and see whether they will be cited by the regulatory agencies. And there is no certainty that that citation will come.

Finally, despite the record of the FDA and other agencies in setting the agenda for system validation, we are often left without a very precise agenda and, therefore, without a very real incentive to go ahead and take some kind of action. It is like driving down the turnpike where we do not know what the real speed limit is. We only know what the posted limit is, and that is clearly not what is being enforced. We do not know how often or how diligently the police might enforce whatever speed limit there is. If we are stopped, we do not know whether we will get a fine. If we do get a fine, we are almost certain that that fine is not proportionate to our particular income or ability to pay; while a $20.00 fine may not be too significant for some, it may break the bank of some other driver.

In effect, that lack of standardization, that lack of certainty, that lack of clear definition, leaves us with some very serious problems in this whole regulatory area.

In response to this kind of need, as well as in response to detailed problems in the regulated industries, there are some regulatory incentives and initiatives being undertaken that are intended to try and solve those very problems. The current inability to readily evaluate the qualifications of a validator has lead the International Validation Forum to develop a standardized certification process -- the Certified Validatin Professional -- to assist in demonstrating that validators do, in fact, meet standards of expertise in this increasingly complex field.

The recognition that we need clearer standards comes not just from industry, however, but from the regulators themselves. If they want to emphasize cost effectiveness, and be cost effective themselves, they realize that they, too, need to establish some standard that gives that kind of uniformity. To achieve this, we are seeing two actions -- two kinds of standardization -- coming out, one from the EPA and one from the FDA.

The EPA has established the Good Automated Laboratory Practices (GALPs), a combination of a variety of standards that are intended to provide one, concrete definition for an automated laboratory doing EPA work. The FDA, on its own initiative, is rewriting and reinterpreting its Good Manufacturing Practice (GMP) and Good Laboratory Practice (GLP) regulations to try and get some broader picture in terms of what is happening in terms of automation in a broad spectrum of its regulated industries.

This trend will no doubt call for the placement of more road signs. Similarly, the speeds posted will be subject to adjustment as experience provides greater warning of dangerous curves and a comparative relaxation in less threatening areas. No doubt, too, the pharmaceutical, biologics, and medical devices industries' sophistication in calculating the risks of exceeding posted limits will grow, and the range considered tolerable will be more clearly defined.

There is no doubt, however, that regulatory system validation "speed limits" will continue to be set; that those limits will provide expert guidelines with tight and parsimonious standards of exactness; and that we will all continue to test, expand, and learn from the practical and reasonable limits under which the proper control of regulated computer systems can be demonstrated.

Chapter 3

FDA REGULATIONS

THE FDA APPROACH

The U.S. Food and Drug Administration, an agency within the Department of Health and Human Services, Public Health Service, regulates the manufacture and distribution of foods, human and veterinary drugs, cosmetics, medical devices, and biological products by authority of the Food, Drug and Cosmetic Act (21 U.S.C.) and the Public Health Service Act (42 U.S.C.), as amended.

FDA requirements for validation generally fall under the following categories of FDA regulations, all of which can be found in Title 21 of the *Code of Federal Regulations* (CFR):

- *Good Manufacturing Practices* (GMPs) for drugs, biologics, medical devices, and veterinary products.
- *Good Laboratory Practices* (GLPs) for nonclinical laboratory studies.
- *Good Clinical Practices* (GCPs) for clinical investigations supporting research applications or marketing permits.

The FDA's Center for Drug Evaluation and Research (CDER) supervises regulatory compliance for drug (both prescription and non-prescription) manufacturers. The Center for Devices and Radiological Health (CDRH) supervises regulatory compliance for manufacturers of medical devices used in the diagnosis, treatment, prevention, and cure of disease, as well as *in vitro* diagnostic products used to test tissue specimens. The Center for Biologics Evaluation and Research (CBER) supervises regulatory compliance for manufacturers of biological products, including serums, vaccines, blood, and blood components. The Center for Veterinary Medicine supervises regulatory compliance for manufacturers of animal drugs and feeds.

The primary authority for the FDA mission of overseeing the production of safe and effective drugs and medical devices is the Food, Drug and Cosmetic Act. The text of pertinent parts of that act follows:

FEDERAL FOOD, DRUG AND COSMETIC ACT

[Bracketed citations are to Title 21, United States Code]

CHAPTER II--DEFINITIONS

SEC. 201 [21 U.S.C. 321] Definitions

For the purposes of this Act--

(a)(1) The term "State", except as used in the last sentence of section 702(a), means any State or Territory of the United States, the District of Columbia, and the Commonwealth of Puerto Rico.

(2) The term "Territory" means any Territory or possession of the United States, including the District of Columbia, and excluding the Commonwealth of Puerto Rico and the Canal Zone.

(b) The term "interstate commerce" means (1) commerce between any State or Territory and any place outside thereof, and (2) commerce within the District of Columbia or within any other Territory not organized with a legislative body.

(c) The term "Department" means the U.S. Department of Health and Human Services.

(d) The term "Secretary" means the Secretary of Health and Human Services.

(e) The term "person" includes individual, partnership, corporation, and association.

(f) The term "food" means (1) articles used for food or drink for man or other animals, (2) chewing gum, and (3) articles used for components of any such article.

(g)(1) The term "drug" means (A) articles recognized in the official United States Pharmacopeia, official Homeopathic Pharmacopeia of the United States, or official National Formulary, or any supplement to any of them; and (B) articles intended for use in the diagnosis, cure, mitigation, treatment, or prevention of disease in man or other animals; and (C) articles (other than food) intended to affect the structure or any function of the body of man or other animals; and (D) articles intended for use as a component of any articles specified in clause (A), (B), or (C). A food for which a claim, subject to sections 403(r)(1)(B) and 403(r)(3) or sections 403(r)(1)(B) and 403(r)(5)(D), is made in accordance with the requirements of section 403(r) is not a drug under clause (B) solely because the label or labeling contains such a claim.

(2) The term "counterfeit drug" means a drug which, or the container or labeling of which, without authorization, bears the trademark, trade name, or other identifying mark, imprint, or device, or any likeness thereof, of a drug manufacturer, processor, packer, or distributor other than the person or persons who in fact manufactured, processed, packed, or distributed such drug and which thereby falsely purports or is represented to be the product of, or to have been packed or distributed by, such other drug manufacturer, processor, packer, or distributor.

(h) The term "device" (except when used in paragraph (n) of this section and in sections 301(i), 403(f), 502(c), and 602(c)) means an instrument, apparatus, implement, machine, contrivance, implant, in vitro reagent, or other similar or related article, including any component, part, or accessory, which is--

(1) recognized in the official National Formulary, or the United States Pharmacopeia, or any supplement to them,

(2) intended for use in the diagnosis of disease or other conditions, or in the cure, mitigation, treatment, or prevention of disease, in man or other animals, or

(3) intended to affect the structure or any function of the body of man or other animals, and which does not achieve its primary intended purposes through chemical action within or on the body of man or other animals and

which is not dependent upon being metabolized for the achievement of its primary intended purposes.

(i) The term "cosmetic" means (1) articles intended to be rubbed, poured, sprinkled, or sprayed on, introduced into, or otherwise applied to the human body or any part thereof for cleansing, beautifying, promoting attractiveness, or altering the appearance, and (2) articles intended for use as a component of any such articles, except that such term shall not include soap.

(j) The term "official compendium" means the official United States Pharmacopeia, official Homeopathic Pharmacopeia of the United States, official National Formulary, or any supplement to any of them.

(k) The term "label" means a display of written, printed, or graphic matter upon the immediate container of any article; and a requirement made by or under authority of this Act that any word, statement, or other information appear on the label shall not be considered to be complied with unless such word, statement, or other information also appears on the outside container or wrapper, if any there be, of the retail package of such article, or is easily legible through the outside container or wrapper.

(l) The term "immediate container" does not include package liners.

(m) The term "labeling" means all labels and other written, printed, or graphic matter (1) upon any article or any of its containers or wrappers, or (2) accompanying such article.

(n) If an article is alleged to be misbranded because the labeling or advertising is misleading, then in determining whether the labeling or advertising is misleading there shall be taken into account (among other things) not only representations made or suggested by statement, word, design, device, or any combination thereof, but also the extent to which the labeling or advertising fails to reveal facts material in the light of such representations or material with respect to consequences which may result from the use of the article to which the labeling or advertising relates under the conditions of use prescribed in the labeling or advertising thereof or under such conditions of use as are customary or usual.

(o) The representation of a drug, in its labeling, as an antiseptic shall be considered to be a representation that it is a germicide, except in the case of a drug purporting to be, or represented as, an antiseptic for inhibitory use as a wet dressing, ointment, dusting powder, or such other use as involves prolonged contact with the body.

(p) The term "new drug" means--

(1) Any drug (except a new animal drug or an animal feed bearing or containing a new animal drug) the composition of which is such that such drug is not generally recognized, among experts qualified by scientific training and experience to evaluate the safety and effectiveness of drugs, as safe and effective for use under the conditions prescribed, recommended, or suggested in the labeling thereof, except that such a drug not so recognized shall not be deemed to be a "new drug" if at any time prior to the enactment of this Act it was subject to the Food and Drugs Act of June 30, 1906, as amended, and if at such time its labeling contained the same representations concerning the conditions of its use; or

(2) Any drug (except a new animal drug or an animal feed bearing or containing a new animal drug) the composition of which is such that such drug, as a result of investigations to determine its safety and effectiveness for use under such conditions, has become so recognized but which has not, otherwise than in such investigations, been used to a material extent or for a material time under such conditions.

(q) The term "pesticide chemical" means any substance which, alone in chemical combination or in formulation with one or more other substances, is a pesticide within the meaning of the Federal Insecticide, Fungicide, and Rodenticide Act (7 U.S.C., secs. 135-135k) as now in force or as hereafter amended, and which is used in the production, storage, or transportation of raw agricultural commodities.

(r) The term "raw agricultural commodity" means any food in its raw or natural state, including all fruits that are washed, colored, or otherwise treated in their unpeeled natural form prior to marketing.

(s) The term "food additive" means any

substance the intended use of which results or may reasonably be expected to result, directly or indirectly, in its becoming a component or otherwise affecting the characteristics of any food (including any substance intended for use in producing, manufacturing, packing, processing, preparing, treating, packaging, transporting, or holding food; and including any source of radiation intended for any such use), if such substance is not generally recognized, among experts qualified by scientific training and experience to evaluate its safety, as having been adequately shown through scientific procedures (or, in the case of a substance used in food prior to January 1, 1958, through either scientific procedures or experience based on common use in food) to be safe under the conditions of its intended use; except that such term does not include--

(1) a pesticide chemical in or on a raw agricultural commodity; or

(2) a pesticide chemical to the extent that it is intended for use or is used in the production, storage, or transportation of any raw agricultural commodity; or

(3) a color additive; or

(4) any substance used in accordance with a sanction or approval granted prior to the enactment of this paragraph pursuant to this Act, the Poultry Products Inspection Act (21 U.S.C. 451 0, and the following or the Meat Inspection Act of March 4, 1907 (34 Stat. 1260), as amended and extended (21 U.S.C. 71 and the following); or

(5) a new animal drug.

(t)(1) The term "color additive" means a material which--

(A) is a dye, pigment, or other substance made by a process of synthesis or similar artifice, or extracted, isolated, or otherwise derived, with or without intermediate or final change of identity, from a vegetable, animal, mineral, or other source, and

(B) when added or applied to a food, drug, or cosmetic, or to the human body or any part thereof, is capable (alone or through reaction with other substance) of imparting color thereto; except that such term does not include any material which the Secretary, by regulation, determines is used (or intended to be used) solely for a purpose or purposes other than coloring.

(2) The term "color" includes black, white, and intermediate grays.

(3) Nothing in subparagraph (1) of this paragraph shall be construed to apply to any pesticide chemical, soil or plant nutrient, or other agricultural chemical solely because of its effect in aiding, retarding, or otherwise affecting, directly or indirectly, the growth or other natural physiological processes of produce of the soil and thereby affecting its color, whether before or after harvest. (u) The term "safe," as used in paragraph (s) of this section and in sections 409, 512, and 721, has reference to the health of man or animal.

(v) Repealed by sec. 701 of P.L. 91-513.

(w) The term "new animal drug" means any drug intended for use for animals other than man, including any drug intended for use in animal feed but not including such animal feed--

(1) the composition of which is such that such drug is not generally recognized, among experts qualified by scientific training and experience to evaluate the safety and effectiveness of animal drugs, as safe and effective for use under the conditions prescribed, recommended, or suggested in the labeling thereof; except that such a drug not so recognized shall not be deemed to be a "new animal drug" if at any time prior to June 25, 1938, it was subject to the Food and Drug Act of June 30, 1906, as amended, and if at such time its labeling contained the same representations concerning the conditions of its use; or

(2) the composition of which is such that such drug, as a result of investigations to determine its safety and effectiveness for use under such conditions, has become so recognized but which has not, otherwise than in such investigations, been used to a material extent or for a material time under such conditions.

(x) The term "animal feed", as used in paragraph (w) of this section, in section 512, and in provisions of this Act referring to such paragraph or section, means an article which is intended for use for food for animals other than man and which is intended for use as a substantial source of nutrients in the diet of the animal, and is not limited to a mixture intended to be the sole ration of the animal.

(y) The term ''informal hearing'' means a hearing which is not subject to section 554, 556, or 557 of title 5 of the United States Code and which provides for the following:

(1) The presiding officer in the hearing shall be designated by the Secretary from officers and employees of the Department who have not participated in any action of the Secretary which is the subject of the hearing and who are not directly responsible to an officer or employee of the Department who has participated in any such action.

(2) Each party to the hearing shall have the right at all times to be advised and accompanied by an attorney.

(3) Before the hearing, each party to the hearing shall be given reasonable notice of the matters to be considered at the hearing, including a comprehensive statement of the basis for the action taken or proposed by the Secretary which is the subject of the hearing and a general summary of the information which will be presented by the Secretary at the hearing in support of such action.

(4) At the hearing the parties to the hearing shall have the right to hear a full and complete statement of the action of the Secretary which is the subject of the hearing together with the information and reasons supporting such action, to conduct reasonable questioning, and to present any oral or written information relevant to such action.

(5) The presiding officer in such hearing shall prepare a written report of the hearing to which shall be attached all written material presented at the hearing. The participants in the hearing shall be given the opportunity to review and correct or supplement the presiding officer's report of the hearing.

(6) The Secretary may require the hearing to be transcribed. A party to the hearing shall have the right to have the hearing transcribed at his expense. Any transcription of a hearing shall be included in the presiding officer's report of the hearing.

(z) The term ''saccharin'' includes calcium saccharin, sodium saccharin, and ammonium saccharin.

(aa) The term ''infant formula'' means a food which purports to be or is represented for special dietary use solely as a food for infants by reason of its simulation of human milk or its suitability as a complete or partial substitute for human milk.

(bb) The term ''abbreviated drug application'' means an application submitted under section 505(j) or 507 for the approval of a drug that relies on the approved application of another drug with the same active ingredient to establish safety and efficacy, and--

(1) in the case of section 306, includes a supplement to such an application for a different or additional use of the drug but does not include a supplement to such an application for other than a different or additional use of the drug, and

(2) in the case of sections 307 and 308, includes any supplement to such an application.

(cc) The term ''knowingly'' or ''knew'' means that a person, with respect to information--

(1) has actual knowledge of the information, or

(2) acts in deliberate ignorance or reckless disregard of the truth or falsity of the information.

(dd) For purposes of section 306, the term ''high managerial agent''--

(1) means--

(A) an officer or director of a corporation or an association,

(B) a partner of a partnership, or

(C) any employee or other agent of a corporation, association, or partnership, having duties such that the conduct of such officer, director, partner, employee, or agent may fairly be assumed to represent the policy of the corporation, association, or partnership, and

(2) includes persons having management responsibility for--

(A) submissions to the Food and Drug Administration regarding the development or approval of any drug product,

(B) production, quality assurance, or quality control of any drug product, or

(C) research and development of any drug product.

(ee) For purposes of sections 306 and 307, the term ''drug product'' means a drug subject to regulation under section 505, 507, 512, or 802 of this Act or under section 351 of the Public Health Service Act.

(ff) The term "Commissioner" means the Commissioner of Food and Drugs.

CHAPTER III -- PROHIBITED ACTS AND PENALTIES

SEC. 301 [21 U.S.C. 331] Prohibited acts

The following acts and the causing thereof are hereby prohibited:

(a) The introduction or delivery for introduction into interstate commerce of any food, drug, device, or cosmetic that is adulterated or misbranded.

(b) The adulteration or misbranding of any food, drug, device, or cosmetic in interstate commerce.

(c) The receipt in interstate commerce of any food, drug, device, or cosmetic that is adulterated or misbranded, and the delivery or proffered delivery thereof for pay or otherwise.

(d) The introduction or delivery for introduction into interstate commerce of any article in violation of section 404 or 505.

(e) The refusal to permit access to or copying of any record as required by section 412 or 703; or the failure to establish or maintain any record, or make any report, required under section 412, 505 (i) or (k), 507(d) or (g), 512 (j), (l) or (m), 515(f), or 519 or the refusal to permit access to or verification or copying of any such required record.

(f) The refusal to permit entry or inspection as authorized by section 704.

(g) The manufacture within any Territory of any food, drug, device, or cosmetic that is adulterated or misbranded.

(h) The giving of a guaranty or undertaking referred to in section 333(c)(2) of this title which guaranty or undertaking is false, except by a person who relied upon a guaranty or undertaking to the same effect signed by, and containing the name and address of, the person residing in the United States from whom he received in good faith the food, drug, device, or cosmetic; or the giving of a guaranty or undertaking referred to in section 333 (c)(3) of this title which guaranty or undertaking is false.

(i)(1) Forging, counterfeiting, simulating, or falsely representing, or without proper authority using any mark, stamp, tag, label, or other identification device authorized or required by regulations promulgated under the provisions of section 404, 506, 507, or 721.

(2) Making, selling, disposing of, or keeping in possession, control, or custody, or concealing any punch, die, plate, stone, or other thing designed to print, imprint, or reproduce the trademark, trade name, or other identifying mark, imprint, or device of another or any likeness of any of the foregoing upon any drug or container or labeling thereof so as to render such drugs a counterfeit drug.

(3) The doing of any act which causes a drug to be a counterfeit drug, or the sale or dispensing, or the holding for sale or dispensing, of a counterfeit drug.

(j) The using by any person to his own advantage, or revealing, other than to the Secretary or officers or employees of the Department, or to the courts when relevant in any judicial proceeding under this Act, any information acquired under authority of section 404, 409, 412, 505, 506, 507, 510, 512, 513, 514, 515, 516, 518, 519, 520, 704, 721, or 708 concerning any method or process which as a trade secret is entitled to protection. This paragraph does not authorize the withholding of information from either House of Congress or from, to the extent of matter within its jurisdiction, any committee or subcommittee of such committee or any joint committee of Congress or any subcommittee of such joint committee.

(k) The alteration, mutilation, destruction, obliteration, or removal of the whole or any part of the labeling of, or the doing of any other act with respect to, a food, drug, device, or cosmetic, if such act is done while such article is held for sale (whether or not the first sale) after shipment in interstate commerce and results in such article being adulterated or misbranded.

(l) The using, on the labeling of any drug or device or in any advertising relating to such drug or device, of any representation or suggestion that approval of an application with respect to such drug or device is in effect under section 505, 515, or 520(g), as the case may be, or that such drug or device complies with the provisions of such section.

(m) The sale or offering for sale of colored oleomargarine or colored margarine, or the possession or serving of colored oleomargarine or colored margarine in violation of section 407(b) or 407(c).

(n) The using, in labeling, advertising or other sales promotion of any reference to any report or analysis furnished in compliance with section 704.

(o) In the case of a prescription drug distributed or offered for sale in interstate commerce, the failure of the manufacturer, packer, or distributor thereof to maintain for transmittal, or to transmit, to any practitioner licensed by applicable State law to administer such drug who makes written request for information as to such drug, true and correct copies of all printed matter which is required to be included in any package in which that drug is distributed or sold, or such other printed matter as is approved by the Secretary. Nothing in this paragraph shall be construed to exempt any person from any labeling requirement imposed by or under other provisions of this Act.

(p) The failure to register in accordance with section 510, the failure to provide any information required by section 510(j) or 510(k); or the failure to provide a notice required by section 510(j)(2).

(q)(1) The failure or refusal to (A) comply with any requirement prescribed under section 518 or 520(g), (B) furnish any notification or other material or information required by or under section 519 or 520(g), or (C) comply with a requirement under section 522.

(2) With respect to any device, the submission of any report that is required by or under this Act that is false or misleading in any material respect.

(r) The movement of a device in violation of an order under section 304(g) or the removal or alteration of any mark or label required by the order to identify the device as detained.

(s) The failure to provide the notice required by section 412(c) or 412(d) , the failure to make the reports required by section 412(f)(1)(B), the failure to retain the records required by section 412(b)(4), or the failure to meet the requirements prescribed under section 412(f)(3).

(t) The importation of a drug in violation of section 801(d)(1), the sale, purchase, or trade of a drug or drug sample or the offer to sell, purchase, or trade a drug or drug sample in violation of section 503(c), the sale, purchase, or trade of a coupon, the offer to sell, purchase, or trade such a coupon, or the counterfeiting of such a coupon in violation of section 503(c)(2), the distribution of a drug sample in violation of section 503(d) or the failure to otherwise comply with the requirements of section 503(d), or the distribution of drugs in violation of section 503(e) or the failure to otherwise comply with the requirements of section 503(e).

SEC. 304 [21 U.S.C. 334] Seizure

(a)(1) Any article of food, drug, or cosmetic that is adulterated or misbranded when introduced into or while in interstate commerce or while held for sale (whether or not the first sale) after shipment in interstate commerce, or which may not, under the provisions of section 404 or 505, be introduced into interstate commerce, shall be liable to be proceeded against while in interstate commerce, or at any time thereafter, on libel of information and condemned in any district court of the United States or United States court of a Territory within the jurisdiction of which the article is found: *Provided*, however, That no libel for condemnation shall be instituted under this Act, for any alleged misbranding if there is pending in any court a libel for condemnation proceeding under this Act based upon the same alleged misbranding, and not more than one such proceeding shall be instituted if no such proceeding is so pending, except that such limitations shall not apply (A) when such misbranding has been the basis of a prior judgment in favor of the United States, in a criminal, injunction, or libel for condemnation proceeding under this Act, or (B) when the Secretary has probable cause to believe from facts found, without hearing, by him or any officer or employee of the Department that the misbranded article is dangerous to health, or that the labeling of the misbranded article is fraudulent, or would be in a material respect misleading to the injury or damage of the purchaser or consumer. In any case where the number of libel for condemnation proceedings is limited as above provided the proceeding pending or instituted shall, on application of the claimant, seasonably made, be removed for trial to any district agreed upon by stipulation between the parties, or, in case of failure to so stipulate within a reasonable time, the claimant may apply to the court of the district in which the seizure has been made and such court (after giving the United States attorney for such district reasonable notice and opportunity to be heard) shall by order, unless good cause to the contrary is shown, specify a district of reasonable proximity to the claimant's principal place of business to which the case shall be removed for trial.

(2) The following shall be liable to be proceeded against at any time on libel of information and condemned in any district court of the United States or United States court of a Territory within the jurisdiction of which they are found: (A) Any drug that is a counterfeit drug, (B) Any container of a counterfeit drug, (C) Any punch, die, plate, stone, labeling, container, or other thing used or designed for use in making a counterfeit drug or drugs, and (D) Any adulterated or misbranded device.

(3)(A) Except as provided in subparagraph (B), no libel for condemnation may be instituted under paragraph (1) or (2) against any food which--

(i) is misbranded under section 403(a)(2) because of its advertising, and

(ii) is being held for sale to the ultimate consumer in an establishment other than an establishment owned or operated by a manufacturer, packer, or distributor of the food.

(B) A libel for condemnation may be instituted under paragraph (1) or (2) against a food described in subparagraph (A) if--

(i)(I) the food's advertising which resulted in the food being misbranded under section 403(a)(2) was disseminated in the establishment in which the food is being held for sale to the ultimate consumer,

(II) such advertising was disseminated by, or under the direction of the owner or operator of such establishment, or

(III) all or part of the cost of such advertising was paid by such owner or operator; and

(ii) the owner or operator of such establishment used such advertising in the establishment to promote the sale of the food.

(b) The article, equipment, or other thing proceeded against shall be liable to seizure by process pursuant to the libel, and the procedure in cases under this section shall conform, as nearly as may be, to the procedure in admiralty; except that on demand of either party any issue of fact joined in any such case shall be tried by jury. When libel for condemnation proceedings under this section, involving the same claimant and the same issues of adulteration or misbranding, are pending in two or more jurisdictions, such pending proceedings, upon application of the claimant seasonably made to the court of one such jurisdiction, shall be consolidated for trial by order of such court, and tried in (1) any district selected by the claimant where one of such proceedings is pending; or (2) a district agreed upon by stipulation between the parties. If no order for consolidation is so made within a reasonable time, the claimant may apply to the court of one such jurisdiction, and such court (after giving the United States attorney for such district reasonable notice and opportunity to be heard) shall by order, unless good cause to the contrary is shown, specify a district of reasonable proximity to the claimant's principal place of business, in which all such pending proceedings shall be consolidated for trial and tried. Such order of consolidation shall not apply so as to require the removal of any case the date for trial of which has been fixed. The court granting such order shall give prompt notification thereof to the other courts having jurisdiction of the cases covered thereby.

(c) The court at any time after seizure up to a reasonable time before trial shall by order allow any party to a condemnation proceeding, his attorney or agent, to obtain a representative sample of the article seized and a true copy of the analysis, if any, on which the proceeding is based and the identifying marks or numbers, if any, of the packages from which the samples analyzed were obtained.

(d)(1) Any food, drug, device, or cosmetic condemned under this section shall, after entry of the decree, be disposed of by destruction or sale as the court may, in accordance with the provisions of this section, direct and the proceeds thereof, if sold, less the legal costs and charges, shall be paid into the Treasury of the United States; but such article shall not be sold under such decree contrary to the provisions of this Act or the laws of the jurisdiction in which sold: *Provided*, That after entry of the decree and upon the payment of the costs of such proceedings and the execution of a good and sufficient bond conditioned that such article shall not be sold or disposed of contrary to the provisions of this Act or the laws of any State or Territory in which sold, the court may by order direct that such article be delivered to the owner thereof to be destroyed or brought into compliance with the provisions of this Act under the supervision of an officer or employee duly designated by the Secretary, and the expenses of such supervision shall be paid by the person obtaining release of the article under bond. If the article was imported into the United States and the person seeking its release establishes (A) that the adulteration, misbranding, or violation did not occur after the article was imported, and (B) that he had no cause for believing that it was adulterated, misbranded, or in violation before it was released from customs custody, the court may permit the article to be delivered to the owner for exportation in lieu of destruction upon a showing by the owner that all of the conditions of section 801(e) can and will be met: *Provided*, however, That the provisions of this sentence shall not apply where condemnation is based upon violation of section 402(a) (1), (2), or (6), section 501(a)(3), section 502(j), or section 601(a) or (d): And provided further, That where such exportation is made to the original foreign supplier, then paragraphs (1) and (2) of section 801(e) and the foregoing proviso shall not be applicable; and in all cases of exportation the bond shall be conditioned that the article shall not be sold or disposed of until the applicable conditions of section 801(e) have been met. Any article condemned by reason of its being an article which may not, under section 404 or 505, be introduced into interstate commerce, shall be disposed of by destruction.

(2) The provisions of paragraph (1) of this subsection shall, to the extent deemed appropriate by the court, apply to any equipment or other thing which is not otherwise within the scope of such paragraph and which is referred to in paragraph (2) of subsection (a).

(3) Whenever in any proceeding under this section, involving paragraph (2) of subsection (a), the condemnation of any equipment or thing (other than a drug) is decreed, the court shall allow the claim of any claimant, to the extent of such claimant's interest, for remission or mitigation of such forfeiture if such claimant proves to the satisfaction of the court (i) that he has not committed or caused to be committed any prohibited act referred to in such paragraph (2) and has no interest in any drug referred to therein, (ii) that he has an interest in such equipment or other thing as owner or lienor or otherwise, acquired by him in good faith, and (iii) that he at no time had any knowledge or reason to believe that such equipment or other thing was being or would be used in, or to facilitate, the violation of laws of the United States relating to counterfeit drugs.

(e) When a decree of condemnation is entered against the article, court costs and fees, and storage and other proper expenses, shall be awarded against the person, if any, intervening as claimant of the article.

(f) In the case of removal for trial of any case as provided by subsection (a) or (b)--

(1) The clerk of the court from which removal is made shall promptly transmit to the court in which the case is to be tried all records in the case necessary in order that such court may exercise jurisdiction

(2) The court to which such case was removed shall have the powers and be subject to the duties for purposes of such case, which the court from which removal was made would have had, or to which such court would have been subject, if such case had not been removed.

(g)(1) If during an inspection conducted under section 704 of a facility or a vehicle, a device which the officer or employee making the inspection has reason to believe is adulterated or misbranded is found in such facility or vehicle, such officer or employee may order the device detained (in accordance with regulations prescribed by the Secretary) for a reasonable period which may not exceed twenty days unless the Secretary determines that a period of detention greater than twenty days is required to institute an action under subsection (a) or section 302, in which case he may authorize a detention period of not to exceed thirty days. Regulations of the Secretary prescribed under this paragraph shall require that before a device may be ordered detained under this paragraph the Secretary or an officer or employee designated by the Secretary approve such order. A detention order under this paragraph may require the labeling or marking of a device during the period of its detention for the purpose of identifying the device as detained. Any person who would be entitled to claim a device if it were seized under subsection (a) may appeal to the Secretary a detention of such device under this paragraph. Within five days of the date an appeal of a detention is filed with the Secretary, the Secretary shall after affording opportunity for an informal hearing by order confirm the detention or revoke it.

(2)(A) Except as authorized by subparagraph (B), a device subject to a detention order issued under paragraph (1) shall not be moved by any person from the place at which it is ordered detained until--

(i) released by the Secretary, or

(ii) the expiration of the detention period applicable to such order,

whichever occurs first.

(B) A device subject to a detention order under paragraph (1) may be moved--

(i) in accordance with regulations prescribed by the Secretary, and

(ii) if not in final form for shipment, at the discretion of the manufacturer of the device for the purpose of completing the work required to put it in such form.

SEC. 308 [21 U.S.C. 335c] Authority to withdraw approval of abbreviated drug applications

(a) In General. The Secretary--

(1) shall withdraw approval of an abbreviated drug application if the Secretary finds that the approval was obtained, expedited, or otherwise facilitated through bribery, payment of an illegal gratuity, or fraud or material false statement, and

(2) may withdraw approval of an abbreviated drug application if the Secretary finds that the applicant has repeatedly demonstrated a lack of ability to produce the drug for which the application was submitted in accordance with the formulations or manufacturing practice set forth in the abbreviated drug application and has introduced, or attempted to introduce, such adulterated or misbranded drug into commerce.

(b) Procedure. The Secretary may not take any action under subsection (a) with respect to any person unless the Secretary has issued an order for such action made on the record after opportunity for an agency hearing on disputed issues of material fact. In the course of any investigation or hearing under this subsection, the Secretary may administer oaths and affirmations, examine witnesses, receive evidence, and issue subpoenas requiring the attendance and testimony of witnesses and the production of evidence that relates to the matter under investigation.

(c) Applicability. Subsection (a) shall apply with respect to offenses or acts regardless of when such offenses or acts occurred.

(d) Judicial Review. Any person that is the subject of an adverse decision under subsection (a) may obtain a review of such decision by the United States Court of Appeals for the District of Columbia or for the circuit in which the person resides, by filing in such court (within 60 days following the date the person is notified of the Secretary's decision) a petition requesting that the decision be modified or set aside.

CHAPTER V -- DRUGS AND DEVICES

SUBCHAPTER A -- Drugs and Devices

SEC. 501 [21 U.S.C. 351] Adulterated drugs and devices

A drug or device shall be deemed to be

adulterated--

(a)(1) If it consists in whole or in part of any filthy, putrid, or decomposed substance; or (2)(A) if it has been prepared, packed, or held under insanitary conditions whereby it may have been contaminated with filth, or whereby it may have been rendered injurious to health; or (B) if it is a drug and the methods used in, or the facilities or controls used for, its manufacture, processing, packing, or holding do not conform to or are not operated or administered in conformity with current good manufacturing practice to assure that such drug meets the requirements of this Act as to safety and has the identity and strength, and meets the quality and purity characteristics, which it purports or is represented to possess; or (3) if its container is composed, in whole or in part, of any poisonous or deleterious substance which may render the contents injurious to health; or (4) if (A) it bears or contains, for purposes of coloring only, a color additive which is unsafe within the meaning of section 721(a), or (B) it is a color additive the intended use of which in or on drugs or devices is for purposes of coloring only and is unsafe within the meaning of section 721(a); or (5) if it is a new animal drug which is unsafe within the meaning of section 512; or (6) if it is an animal feed bearing or containing a new animal drug, and such animal feed is unsafe within the meaning of section 512.

(b) If it purports to be or is represented as a drug the name of which is recognized in an official compendium, and its strength differs from, or its quality or purity falls below, the standards set forth in such compendium. Such determination as to strength, quality, or purity shall be made in accordance with the tests or methods of assay set forth in such compendium, except that whenever tests or methods of assay have not been prescribed in such compendium, or such tests or methods of assay as are prescribed are, in the judgment of the Secretary, insufficient for the making of such determination, the Secretary shall bring such fact to the attention of the appropriate body charged with the revision of such compendium, and if such body fails within a reasonable time to prescribe tests or methods of assay which, in the judgment of the Secretary, are sufficient for purposes of this paragraph, then the Secretary shall promulgate regulations prescribing appropriate tests or methods of assay in accordance with which such determination as to strength, quality, or purity shall be made. No drug defined in an official compendium shall be deemed to be adulterated under this paragraph because it differs from the standard of strength, quality, or purity therefor set forth in such compendium if its difference in strength, quality, or purity from such standards is plainly stated on its label. Whenever a drug is recognized in both the United States Pharmacopeia and the Homeopathic Pharmacopeia of the United States it shall be subject to the requirements of the United States Pharmacopeia unless it is labeled and offered for sale as a homeopathic drug, in which case it shall be subject to the provisions of the Homeopathic Pharmacopeia of the United States and not to those of the United States Pharmacopeia.

(c) If it is not subject to the provisions of paragraph (b) of this section and its strength differs from, or its purity or quality falls below, that which it purports or is represented to possess.

(d) If it is a drug and any substa nce has been (1) mixed or packed therewith so as to reduce its quality or strength or (2) substituted wholly or in part therefor.

(e) If it is, or purports to be or is represented as, a device which is subject to a performance standard established under section 514, unless such device is in all respects in conformity with such standard.

(f)(1) If it is a class III device--

(A)(i) which is required by a regulation promulgated under subsection (b) of section 515 to have an approval under such section of an application for premarket approval and which is not exempt from section 515 under section 520(g), and

(ii)(I) for which an application for premarket approval or a notice of completion of a product development protocol was not filed with the Secretary within the ninety-day period beginning on the date of the promulgation of such regulation, or

(II) for which such an application was filed

and approval of the application has been denied, suspended, or withdrawn, or such a notice was filed and has been declared not completed or the approval of the device under the protocol has been withdrawn;

(B)(i) which was classified under section 513(f) into class III, which under section 515(a) is required to have in effect an approved application for premarket approval, and which is not exempt from section 515 under section 520(g), and

(ii) which has an application which has been suspended or is otherwise not in effect; or

(C) which was classified under section 520(l) into class III, which under such section is required to have in effect an approved application under section 515, and which has an application which has been suspended or is otherwise not in effect.

(2)(A) In the case of a device classified under section 513(f) into class III and intended solely for investigational use, paragraph (1)

(B) shall not apply with respect to such device during the period ending on the ninetieth day after the date of the promulgation of the regulations prescribing the procedures and conditions required by section 520(g)(2).

(B) In the case of a device subject to a regulation promulgated under subsection (b) of section 515, paragraph (1) shall not apply with respect to such device during the period ending--

(i) on the last day of the thirtieth calendar month beginning after the month in which the classification of the device in class III became effective under section 513, or

(ii) on the ninetieth day after the date of the promulgation of such regulation, whichever occurs later.

(g) If it is a banned device.

(h) If it is a device and the methods used in, or the facilities or controls used for, its manufacture, packing, storage, or installation are not in conformity with applicable requirements under section 520(f)(1) or an applicable condition prescribed by an order under section 520(f)(2).

(i) If it is a device for which an exemption has been granted under section 520(g) for investigational use and the person who was granted such exemption or any investigator who uses such device under such exemption fails to comply with a requirement prescribed by or under such section.

SEC. 502 [21 U.S.C. 352] Misbranded drugs and devices

A drug or device shall be deemed to be misbranded--

(a) If its labeling is false or misleading in any particular.

(b) If in a package form unless it bears a label containing (1) the name and place of business of the manufacturer, packer, or distributor; and (2) an accurate statement of the quantity of the contents in terms of weight, measure, or numerical count: *Provided*, That under clause (2) of this paragraph reasonable variations shall be permitted, and exemptions as to small packages shall be established, by regulations prescribed by the Secretary.

(c) If any word, statement, or other information required by or under authority of this Act to appear on the label or labeling is not prominently placed thereon with such conspicuousness (as compared with other words, statements, designs, or devices, in the labeling) and in such terms as to render it likely to be read and understood by the ordinary individual under customary conditions of purchase and use.

(d) If it is for use by man and contains any quantity of the narcotic or hypnotic substance alpha-eucaine, barbituric acid, beta-eucaine, bromal, cannabis, carbromal, chloral, coca, cocaine, codeine, heroin, marihuana, morphine, opium, paraldehyde, peyote, or sulfonmethane; or any chemical derivative of such substance, which derivative has been by the Secretary, after investigation, found to be, and by regulations designated as, habit forming; unless its label bears the name, and quantity or proportion of such substance or derivative and in juxtaposition there with the statement "Warning -- May be habit forming."

(e)(1) If it is a drug, unless (A) its label bears, to the exclusion of any other nonproprietary name (except the applicable systematic chemical name or the chemical formula), (i) the established name (as defined in subparagraph (3)) of the drug, if such there be, and (ii) in case it is fabricated from two or more ingredients, the established name and quantity of each active ingredient, including the quantity, kind, and proportion of any alcohol, and also including whether active or not, the established name and quantity or proportion of any bromides, ether, chloroform, acetanilide, acetophenetidin, amidopyrine, antipyrine, atropine, hyoscine, hyoscyamine, arsenic, digitalis, digitalis glucosides, mercury, ouabain, strophanthin, strychnine, thyroid, or any derivative or preparation of any such substances, contained therein. *Provided*, That the requirement for stating the quantity of the active ingredients, other than the quantity of those specifically named in this paragraph, shall apply only to prescription drugs; and (B) for any prescription drug the established name of such drug or ingredient, as the case may be, on such label (and on any labeling on which a name for such drug or ingredient is used) is printed prominently and in type at least half as large as that used thereon for any proprietary name or designation for such drug or ingredient: and *Provided*, That to the extent that compliance with the requirements of clause (A)(ii) or clause (B) of this subparagraph is impracticable exemptions shall be established by regulations promulgated by the Secretary.

(2) If it is a device and it has an established name, unless its label bears, to the exclusion of any other nonproprietary name, its established name (as defined in subparagraph (4)) prominently printed in type at least half as large as that used thereon for any proprietary name or designation for such device, except that to the extent compliance with the requirements of this subparagraph is impracticable, exemptions shall be established by regulations promulgated by the Secretary.

(3) As used in subparagraph (1), the term "established name", with respect to a drug or ingredient thereof, means (A) the applicable official name designated pursuant to section 508, or (B) if there is no such name and such drug, or such ingredient, is an article recognized in an official compendium, then the official title thereof in such compendium, or (C) if neither clause (A) nor clause (B) of this subparagraph applies, then the common or usual name, if any, of such drug or of such ingredient: *Provided further*, That where clause (B) of this subparagraph applies to an article recognized in the United States Pharmacopeia and in the Homeopathic Pharmacopeia under different official titles, the official title used in the United States Pharmacopeia shall apply unless it is labeled and offered for sale as a homeopathic drug, in which case the official title used in the Homeopathic Pharmacopeia shall apply.

(4) As used in subparagraph (2), the term "established name" with respect to a device means (A) the applicable official name of the device designated pursuant to section 508, (B) if there is no such name and such device is an article recognized in an official compendium, then the official title thereof in such compendium, or (C) if neither clause (A) nor clause (B) of this subparagraph applies, then any common or usual name of such device.

(f) Unless its labeling bears (1) adequate directions for use; and (2) such adequate warnings against use in those pathological conditions or by children where its use may be dangerous to health, or against unsafe dosage or methods or duration of administration or application, in such manner and form, as are necessary for the protection of users: *Provided*, That where any requirement of clause (1) of this paragraph, as applied to any drug or device, is not necessary for the protection of the public health, the Secretary shall promulgate regulations exempting such drug or device from such requirement.

(g) If it purports to be a drug the name of which is recognized in an official compendium, unless it is packaged and labeled as prescribed therein: *Provided*, That the method of packing may be modified with the consent of the Secretary. Whenever a drug is recognized in both the United States Pharmacopeia and the Homeopathic Pharmacopeia of the United States, it shall be subject to the requirements of the United States Pharmacopeia with respect to packaging, and labeling unless it is labeled and offered for sale as a homeopathic drug, in which case it shall be subject to the provisions of the Homeopathic Pharmacopeia of the United States, and not to those of the United States Pharmacopeia: *Provided further*, That, in the event of inconsistency between the requirements of this paragraph and those of paragraph (e) as to the name by which the drug or its ingredients shall be designated, the requirements of paragraph (e) shall prevail.

(h) If it has been found by the Secretary to be a drug liable to deterioration, unless it is packaged in such form and manner, and its label bears a statement of such precautions, as the Secretary shall by regulations require as necessary for the protection of the public health. No such regulation shall be established for any drug recognized in an official compendium until the Secretary shall have informed the appropriate body charged with the revision of such compendium of the need for such packaging or labeling requirements and such body shall have failed within a reasonable time to prescribe such requirements.

(i)(1) If it is a drug and its container is so made, formed, or filled as to be misleading; or (2) if it is an imitation of another drug; or (3) if it is offered for sale under the name of another drug.

(j) If it is dangerous to health when used in the dosage or manner; or with the frequency or duration prescribed, recommended, or suggested in the labeling thereof.

(k) If it is, or purports to be, or is represented as a drug composed wholly or partly of insulin, unless (1) it is from a batch with respect to which a certificate or release has been issued pursuant to section 506, and (2) such certificate or release is in effect with respect to such drug.

(l) If it is, or purports to be, or is represented as a drug (except a drug for use in animals other than man) composed wholly or partly of any kind of penicillin, streptomycin, chlortetracycline, chloramphenicol, bacitracin, or any other antibiotic drug, or any derivative thereof, unless (1) it is from a batch with respect to which a certificate or release has been issued pursuant to section 507, and (2) such certificate or release is in effect with respect to such drug: *Provided*, That this paragraph shall not apply to any drug or class of drugs exempted by regulations promulgated under section 507 (c) or (d).

(m) If it is a color additive the intended use of which is for the purpose of coloring only, unless its packaging and labeling are in conformity with such packaging and labeling requirements applicable to such color additive, as may be contained in regulations issued under section 721.

(n) In the case of any prescription drug distributed or offered for sale in any State, unless the manufacturer, packer, or distributor thereof includes in all advertisements and other descriptive printed matter issued or caused to be issued by the manufacturer, packer, or distributor with respect to that drug a true statement of (1) the established name as defined in section 502(e), printed prominently and in type at least half as large as that used for any trade or brand name thereof, (2) the formula showing quantitatively each ingredient of such drug to the extent required for labels under section 502(e), and (3) such other information in brief summary relating to side effects, contraindications, and effectiveness as shall be required in regulations which shall be issued by the Secretary in accordance with the procedure specified in section 701(e) of this Act: *Provided*, That (A) except in extraordinary circumstances, no regulation issued under this paragraph shall require prior approval by the Secretary of the content of any advertisement, and (B) no advertisement of a prescription drug, published after the effective date of regulations issued under this paragraph applicable to advertisements of prescription drugs, shall, with respect to the matters specified in this paragraph or covered by such regulations, be subject to the provisions of sections 12 through 17 of the Federal Trade Commission Act, as amended (15 U.S.C. 52-57). This paragraph (n) shall not be applicable to any printed matter which the Secretary determines to be labeling as defined in section 201(m) of this Act. Nothing in the Convention on Psychotropic Substances, signed at Vienna, Austria, on February 21, 1971, shall be construed to prevent drug price communications to consumers.

(o) If it was manufactured, prepared, propagated, compounded, or processed in an establishment in any State not duly registered under section 510, if it was not included in a list required by section 510(j), if a notice or other information respecting it was not provided as required by such section or section 510(k), or if it does not bear such symbols from the uniform system for identification of devices prescribed under section 510(e) as the Secretary by regulation requires.

(p) If it is a drug and its packaging or labeling is in violation of an applicable regulation issued pursuant to section 3 or 4 of the Poison Prevention Packaging Act of 1970.

(q) In the case of any restricted device distributed or offered for sale in any State, if (1) its advertising is false or misleading in any particular, or (2) it is sold, distributed, or used in violation of regulations prescribed under section 520(e).

(r) In the case of any restricted device distributed or offered for sale in any State, unless the manufacturer, packer, or distributor thereof includes in all advertisements and other descriptive printed matter issued or caused to be issued by the manufacturer, packer, or distributor with respect to that device (1) a true statement of the device's established name as defined in section 502(e), printed prominently and in type at least half as large as that used for any trade or brand name thereof, and (2) a brief statement of the intended uses of the device and relevant warnings, precautions, side effects, and contraindications and, in the case of specific devices made subject to a finding by the Secretary after notice and opportunity for comment that such action is necessary to protect the public health, a full description of the components of such device or the formula showing quantitatively each ingredient of such device to the extent required in regulations which shall be issued by the Secretary after an opportunity for a hearing. Except in extraordinary circumstances, no regulation issued under this paragraph shall require prior approval by the Secretary of the content of any advertisement and no advertisement of a restricted device, published after the effective date of this paragraph shall, with respect to the matters specified in this paragraph or covered by regulations issued hereunder, be subject to the provisions of sections 12 through 15 of the Federal Trade Commission Act (15 U.S.C. 52-55). This paragraph shall not be applicable to any printed matter which the Secretary determines to be labeling as defined in section 201(m).

(s) If it is a device subject to a performance standard established under section 514, unless it bears such labeling as may be prescribed in such performance standard.

(t) If it is a device and there was a failure or refusal (1) to comply with any requirement prescribed under section 518 respecting the device, (2) to furnish any material or information required by or under section 519 respecting the device, or (3) to comply with a requirement under section 522.

SEC. 503 [21 U.S.C. 353] Exemptions and consideration for certain drugs, devices, and biological products

(a) The Secretary is hereby directed to promulgate regulations exempting from any labeling or packaging requirement of this Act drugs and devices which are, in accordance with the practice of the trade, to be processed, labeled, or repacked in substantial quantities at establishments other than those where originally processed or packed, on condition that such drugs and devices are not adulterated or misbranded, under the provisions of this Act upon removal from such processing, labeling, or repacking establishment.

(b)(1) A drug intended for use by man which--

(A) is a habit-forming drug to which section 502(d) applies; or

(B) because of its toxicity or other potentiality for harmful effect, or the method of its use, or the collateral measures necessary to its use, is not safe for use except under the supervision of a practitioner licensed by law to administer such drug; or

(C) is limited by an approved application under section 505 to use under the professional supervision of a practitioner licensed by law to administer such drug; shall be dispensed only (i) upon a written prescription of a practitioner licensed by law to administer such drug, or (ii) upon an oral prescription of such practitioner which is reduced promptly to writing and filed by the pharmacist, or (iii) by refilling any such written or oral prescription if such refilling is authorized by the prescriber either in the original prescription or by oral order which is reduced promptly to writing and filed by the pharmacist. The act of dispensing a drug contrary to the provisions of this paragraph shall be deemed to be an act which results in the drug being misbranded while held for sale.

(2) Any drug dispensed by filling or refilling a written or oral prescription of a practitioner licensed by law to administer such drug shall be exempt from the requirements of section 502, except paragraphs (a), (i) (2) and (3), (k), and (l), and the packaging requirements of paragraphs (g), (h), and (p), if the drug bears a label containing the name and address of the dispenser, the serial number and date of the prescription or of its filling, the name of the prescriber, and, if stated in the prescription, the name of the patient, and the directions for use and cautionary statements, if any contained in such prescription. This exemption shall not apply to any drug dispensed in the course of the conduct of a business of dispensing drugs pursuant to diagnosis by mail, or to a drug dispensed in violation of paragraph (1) of this subsection.

(3) The Secretary may by regulation remove drugs subject to section 502(d) and section 505 from the requirements of paragraph (1) of this subsection when such requirements are not necessary for the protection of the public health.

(4) A drug which is subject to paragraph (1) of this subsection shall be deemed to be misbranded if at any time prior to dispensing its label fails to bear the statement "Caution: Federal law prohibits dispensing without prescription." A drug to which paragraph (1) of this subsection does not apply shall be deemed to be misbranded if at any time prior to dispensing its label bears the caution statement quoted in the preceding sentence.

(5) Nothing in this subsection shall be construed to relieve any person from any requirement prescribed by or under authority of law with respect to drugs now included or which may hereafter be included within the classifications stated in section 3220 of the Internal Revenue Code (26 U.S.C. 3220), or to marihuana as defined in section 3238(b) of the Internal Revenue Code (26 U.S.C. 3238(b)).

(c)(1) No person may sell, purchase, or trade or offer to sell, purchase, or trade any drug sample. For purposes of this paragraph and subsection (d), the term "drug sample" means a unit of a drug, subject to subsection (b), which is not intended to be sold and is intended to promote the sale of the drug. Nothing in this paragraph shall subject an officer or executive of a drug manufacturer or distributor to criminal liability solely because of a sale, purchase, trade, or offer to sell, purchase, or trade in violation of this paragraph by other employees of the manufacturer or distributor.

(2) No person may sell, purchase, or trade, offer to sell, purchase, or trade, or counterfeit any coupon. For purposes of this paragraph, the term "coupon" means a form which may be redeemed, at no cost or at a reduced cost, for a drug which is prescribed in accordance with subsection (b).

(3)(A) No person may sell, purchase, or trade, or offer to sell, purchase, or trade, any drug--

(i) which is subject to subsection (b), and

(ii)(I) which was purchased by a public or private hospital or other health care entity, or

(II) which was donated or supplied at a reduced price to a charitable organization described in section 501(c)(3) of the Internal Revenue Code of 1954.

(B) Subparagraph (A) does not apply to--

(i) the purchase or other acquisition by a hospital or other health care entity which is a member of a group purchasing organization of a drug for its own use from the group purchasing organization or from other hospitals or health care entities which are members of such organization,

(ii) the sale, purchase, or trade of a drug or an offer to sell, purchase, or trade a drug by an organization described in subparagraph (A)(ii)(II) to a nonprofit affiliate of the organization to the extent otherwise permitted by law,

(iii) a sale, purchase, or trade of a drug or an offer to sell, purchase, or trade a drug among hospitals or other health care entities which are under common control,

(iv) a sale, purchase, or trade of a drug or

an offer to sell, purchase, or trade a drug for emergency medical reasons, or

(v) a sale, purchase, or trade of a drug, an offer to sell, purchase, or trade a drug, or the dispensing of a drug pursuant to a prescription executed in accordance with subsection (b). For purposes of this paragraph, the term "entity" does not include a wholesale distributor of drugs or a retail pharmacy licensed under State law and the term "emergency medical reasons" includes transfers of a drug between health care entities or from a health care entity to a retail pharmacy undertaken to alleviate temporary shortages of the drug arising from delays in or interruptions of regular distribution schedules.

(d)(1) Except as provided in paragraphs (2) and (3), no person may distribute any drug sample. For purposes of this subsection, the term "distribute" does not include the providing of a drug sample to a patient by a--

(A) practitioner licensed to prescribe such drug,

(B) health care professional acting at the direction and under the supervision of such a practitioner, or

(C) pharmacy of a hospital or of another health care entity that is acting at the direction of such a practitioner and that received such sample pursuant to paragraph (2) or (3).

(2)(A) The manufacturer or authorized distributor of record of a drug subject to subsection (b) may, in accordance with this paragraph, distribute drug samples by mail or common carrier to practitioners licensed to prescribe such drugs or, at the request of a licensed practitioner, to pharmacies of hospitals or other health care entities. Such a distribution of drug samples may only be made--

(i) in response to a written request for drug samples made on a form which meets the requirements of subparagraph (B), and

(ii) under a system which requires the recipient of the drug sample to execute a written receipt for the drug sample upon its delivery and the return of the receipt to the manufacturer or authorized distributor of record.

(B) A written request for a drug sample required by subparagraph (A)(i) shall contain--

(i) the name, address, professional designation, and signature of the practitioner making the request,

(ii) the identity of the drug sample requested and the quantity requested,

(iii) the name of the manufacturer of the drug sample requested, and

(iv) the date of the request.

(C) Each drug manufacturer or authorized distributor of record which makes distributions by mail or common carrier under this paragraph shall maintain, for a period of 3 years, the request forms submitted for such distributions and the receipts submitted for such distributions and shall maintain a record of distributions of drug samples which identifies the drugs distributed and the recipients of the distributions. Forms, receipts, and records required to be maintained under this subparagraph shall be made available by the drug manufacturer or authorized distributor of record to Federal and State officials engaged in the regulation of drugs and in the enforcement of laws applicable to drugs.

(3) The manufacturer or authorized distributor of record of a drug subject to subsection (b) may, by means other than mail or common carrier, distribute drug samples only if the manufacturer or authorized distributor of record makes the distributions in accordance with subparagraph (A) and carries out the activities described in subparagraphs (B) through (F) as follows:

(A) Drug samples may only be distributed--

(i) to practitioners licensed to prescribe such drugs if they make a written request for the drug samples, or

(ii) at the written request of such a licensed practitioner, to pharmacies of hospitals or other health care entities.

A written request for drug samples shall be made on a form which contains the practitioner's name, address, and professional designation, the identity of the drug sample requested, the quantity of drug samples requested, the name of the manufacturer or authorized distributor of record of the drug sample, the date of the request and signature of the practitioner making the request.

(B) Drug manufacturers or authorized distributors of record shall store drug samples under conditions that will maintain their stability, integrity, and effectiveness and will assure that the drug samples will be free of contamination, deterioration, and adulteration.

(C) Drug manufacturers or authorized distributors of record shall conduct, at least annually, a complete and accurate inventory of all drug samples in the possession of representatives of the manufacturer or authorized distributor of record. Drug manufacturers or authorized distributors of record shall maintain lists of the names and address of each of their representatives who distribute drug samples and of the sites where drug samples are stored. Drug manufacturers or authorized distributors of record shall maintain records for at least 3 years of all drug samples distributed, destroyed, or returned to the manufacturer or authorized distributor of record, of all inventories maintained under this subparagraph, of all thefts or significant losses of drug samples, and of all requests made under subparagraph (A) for drug samples. Records and lists maintained under this subparagraph shall be made available by the drug manufacturer or authorized distributor of record to the Secretary upon request.

(D) Drug manufacturers or authorized distributors of record shall notify the Secretary of any significant loss of drug samples and any known theft of drug samples.

(E) Drug manufacturers or authorized distributors of record shall report to the Secretary any conviction of their representatives for violations of subsection (c)(1) or a State law because of the sale, purchase, or trade of a drug sample or the offer to sell, purchase, or trade a drug sample.

(F) Drug manufacturers or authorized distributors of record shall provide to the Secretary the name and telephone number of the individual responsible for responding to a request for information respecting drug samples.

(e)(1)(A) Each person who is engaged in the wholesale distribution of a drug subject to subsection (b) and who is not the manufacturer or an authorized distributor of record of such drug shall, before each wholesale distribution of such drug (including each distribution to an authorized distributor of record or to a retail pharmacy), provide to the person who receives the drug a statement (in such form and containing such information as the Secretary may require) identifying each prior sale, purchase, or trade of such drug (including the date of the transaction and the names and addresses of all parties to the transaction).

(B) Each manufacturer of a drug subject to subsection (b) shall maintain at its corporate offices a current list of the authorized distributors of record of such drug.

(2)(A) No person may engage in the wholesale distribution in interstate commerce of drugs subject to subsection (b) in a State unless such person is licensed by the State in accordance with the guidelines issued under subparagraph (B) or has registered with the Secretary in accordance with paragraph (3).

(B) The Secretary shall by regulation issue guidelines establishing minimum standards, terms, and conditions for the licensing of persons to make wholesale distributions in interstate commerce of drugs subject to subsection (b). Such guidelines shall prescribe requirements for the storage and handling of such drugs and for the establishment and maintenance of records of the distributions of such drugs.

(3) Any person who engages in the wholesale distribution in interstate commerce of drugs that are subject to subsection (b) in a State that does not have a program that meets the guidelines established under paragraph (2)(B) shall register with the Secretary the following:

(A) The person's name and place of business.

(B) The name of each establishment the person owns or operates that is engaged in the wholesale distribution of drugs in a State that does not have a program to license persons engaged in such distribution.

(4) For the purposes of this subsection and subsection (d)--

(A) the term "authorized distributors of record" means those distributors with whom a manufacturer has established an ongoing relationship to distribute such manufacturer's products, and

(B) the term "wholesale distribution" means distribution of drugs subject to subsection (b) to other than the consumer or patient but does not include intracompany sales and does not include distributions of drugs described in subsection (c)(3)(B).

(f)(1)(A) A drug intended for use by animals other than man which--

(i) because of its toxicity or other potentiality for harmful effect or the method of its use, or the collateral measures necessary for its use, is not safe for animal use except under the professional supervision of a licensed veterinarian, or

(ii) is limited by an approved application under subsection (b) of section 512 to use under the professional supervision of a licensed veterinarian, shall be dispensed only by or upon the lawful written or oral order of a licensed veterinarian in the course of the veterinarian's professional practice.

(B) For purposes of subparagraph (A), an order is lawful if the order--

(i) is a prescription or other order authorized by law,

(ii) is, if an oral order, promptly reduced to writing by the person lawfully filling the order, and filed by that person, and

(iii) is refilled only if authorized in the original order or in a subsequent oral order promptly reduced to writing by the person lawfully filling the order, and filed by that person.

(C) The act of dispensing a drug contrary to the provisions of this paragraph shall be deemed to be an act which results in the drug being misbranded while held for sale.

(2) Any drug when dispensed in accordance with paragraph (1) of this subsection--

(A) Shall be exempt from the requirements of section 502, except subsections (a), (g), (h), (i)(2), (i)(3), and (p) of such section, and

(B) shall be exempt from the packaging requirements of subsections (g), (h), and (p) of such section, if--

(i) when dispensed by a licensed veterinarian, the drug bears a label containing the name and address of the practitioner and any directions for use and cautionary statements specified by the practitioner, or

(ii) when dispensed by filling the lawful order of a licensed veterinarian, the drug bears a label containing the name and address of the dispenser, the serial number and date of the order or of its filing, the name of the licensed veterinarian, and the directions for use and cautionary statements, if any, contained in such order.

The preceding sentence shall not apply to any drug dispensed in the course of the conduct of a business of dispensing drugs pursuant to diagnosis by mail.

(3) The Secretary may by regulation exempt drugs for animals other than man subject to section 512 from the requirements of paragraph (1) when such requirements are not necessary for the protection of the public health.

(4) A drug which is subject to paragraph (1) shall be deemed to be misbranded if at any time prior to dispensing its label fails to bear the statement "Caution: Federal law restricts this drug to use by or on the order of a licensed veterinarian.". A drug to which paragraph (1) does not apply shall be deemed to be misbranded if at any time prior to dispensing its label bears the statement specified in the preceding sentence.

(g)(1) The Secretary shall designate a component of the Food and Drug Administration to regulate products that constitute a combination of a drug, device, or biological product. The Secretary shall determine the primary mode of action of the combination product. If the Secretary determines that the primary mode of action is that of--

(A) a drug (other than a biological product), the persons charged with premarket review of drugs shall have primary jurisdiction,

(B) a device, the persons charged with premarket review of devices shall have primary jurisdiction, or

(C) a biological product, the persons charged with premarket review of biological products shall have primary jurisdiction.

(2) Nothing in this subsection shall prevent the Secretary from using any agency resources of the Food and Drug Administration necessary to ensure adequate review of the safety, effectiveness, or substantial equivalence of an article.

(3) The Secretary shall promulgate regulations to implement market clearance procedures in accordance with paragraphs (1) and (2) not later than 1 year after the date of enactment of this subsection.

(4) As used in this subsection:

(A) The term ''biological product'' has the meaning given the term in section 351(a) of the Public Health Service Act (42 U.S.C. 262(a)).

(B) The term ''market clearance'' includes--

(i) approval of an application under section 505, 507, 515, or 520(g),

(ii) a finding of substantial equivalence under this subchapter, and

(iii) approval of a product or establishment license under subsection (a) or (d) of section 351 of the Public Health Service Act (42 U.S.C. 262).

SEC. 505 [21 U.S.C. 355] New drugs

(a) No person shall introduce or deliver for introduction into interstate commerce any new drug, unless an approval of an application filed pursuant to subsection (b) or (j) is effective with respect to such drug.

(b)(1) Any person may file with the Secretary an application with respect to any drug subject to the provisions of subsection (a). Such persons shall submit to the Secretary as a part of the application (A) full reports of investigations which have been made to show whether or not such drug is safe for use and whether such drug is effective in use; (B) a full list of the articles used as components of such drug; (C) a full statement of the composition of such drug; (D) a full description of the methods used in, and the facilities and controls used for, the manufacture, processing, and packing of such drug; (E) such samples of such drug and of the articles used as components thereof as the Secretary may require; and (F) specimens of the labeling proposed to be used for such drug. The applicant shall file with the application the patent number and the expiration date of any patent which claims the drug for which the applicant submitted the application or which claims a method of using such drug and with respect to which a claim of patent infringement could reasonably be asserted if a person not licensed by the owner engaged in the manufacture use, or sale of the drug. If a application is filed under this subsection for a drug and a patent which claims such drug or a method of using such drug is issued after the filing date but before approval of the application, the applicant shall amend the application to include the information required by the preceding sentence. Upon approval of the application, the Secretary shall publish information submitted under the two preceding sentences.

(2) An application submitted under paragraph (1) for a drug for which the investigations described in clause (A) of such paragraph and relied upon by the applicant for approval of the application were not conducted by or for the applicant and for which the applicant has not obtained a right of reference or use from the person by or for whom the investigations were conducted shall also include--

(A) a certification, in the opinion of the applicant and to the best of his knowledge, with respect to each patent which claims the drug for which such investigations were conducted or which claims a use for such drug for which the applicant is seeking approval under this subsection and for which information is required to be filed under paragraph (1) or subsection (c)--

(i) that such patent information has not been filed,

(ii) that such patent has expired,

(iii) of the date on which such patent will expire, or

(iv) that such patent is invalid or will not be infringed by the manufacture, use, or sale of the new drug for which the application is submitted; and

(B) if with respect to the drug for which investigations described in paragraph (1)(A) were conducted information was filed under paragraph (1) or subsection (c) for a method of use patent which does not claim a use for which the applicant is seeking approval under this subsection, a statement that the method of use patent does not claim such a use.

(3)(A) An applicant who makes a certification described in paragraph (2)(A)(iv) shall include in the application a statement that the applicant will give the notice required by subparagraph (B) to--

(i) each owner of the patent which is the subject of the certification or the representative of such owner designated to receive such notice, and

(ii) the holder of the approved application under subsection (b) for the drug which is claimed by the patent or a use of which is claimed by the patent or the representative of such holder designated to receive such notice.

(B) The notice referred to in subparagraph (A) shall state that an application has been submitted under this subsection for the drug with respect to which the certification is made to obtain approval to engage in the commercial manufacture, use, or sale of the drug before the expiration of the patent referred to in the certification. Such notice shall include a detailed statement of the factual and legal basis of the applicant's opinion that the patent is not valid or will not be infringed.

(C) If an application is amended to include a certification described in paragraph (2)(A)(iv), the notice required by subparagraph (B) shall be given when the amended application is submitted.

(c)(1) Within one hundred and eighty days after the filing of an application under subsection (b), or such additional period as may be agreed upon by the Secretary and the applicant, the Secretary shall either--

(A) approve the application if he then finds that none of the grounds for denying approval specified in subsection (d) applies, or

(B) give the applicant notice of an opportunity for a hearing before the Secretary under subsection (d) on the question whether such application is approvable. If the applicant elects to accept the opportunity for hearing by written request within thirty days after such notice, such hearing shall commence not more than ninety days after the expiration of such thirty days unless the Secretary and the applicant otherwise agree. Any such hearing shall thereafter be conducted on an expedited basis and the Secretary's order thereon shall be issued within ninety days after the date fixed by the Secretary for filing final briefs.

(2) If the patent information described in subsection (b) could not be filed with the submission of an application under subsection (b) because the application was filed before the patent information was required under subsection (b) or a patent was issued after the application was approved under such subsection, the holder of an approved application shall file with the Secretary, the patent number and the expiration date of any patent which claims the drug for which the application was submitted or which claims a method of using such drug and with respect to which a claim of patent infringement could reasonably be asserted if a person not licensed by the owner engaged in the manufacture, use, or sale of the drug. If the holder of an approved application could not file patent information under subsection (b) because it was not required at the time the application was approved, the holder shall file such information under this subsection not later than thirty days after the date of the enactment of this sentence, and if the holder of an approved application could not file patent information under subsection (b) because no patent had been issued when an application was filed or approved, the holder shall file such information under this subsection not later than thirty days after after the date the patent involved is issued. Upon the submission of patent information under this subsection, the Secretary shall publish it.

(3) The approval of an application filed under subsection (b) which contains a certification required by paragraph (2) of such subsection shall be made effective on the last applicable date determined under the following:

(A) If the applicant only made a certification described in clause (i) or (ii) of subsection (b)(2)(A) or in both such clauses, the approval may be made effective immediately.

(B) If the applicant made a certification described in clause (iii) of subsection (b)(2)(A), the approval may be made effective on the date certified under clause (iii).

(C) If the applicant made a certification described in clause (iv) of subsection (b)(2)(A), the approval shall be made effective immediately unless an action is brought for infringement of a patent which is the subject of the certification before the expiration of forty-five days from the date the notice provided under paragraph (3)(B) is received. If such an action is brought before the expiration of such days, the approval may be made effective upon the expiration of the thirty-month period beginning on the date of the receipt of the notice provided under paragraph (3)(B) or such shorter or longer period as the court may order because either party to the action failed to reasonably cooperate in expediting the action, except that--

(i) if before the expiration of such period the court decides that such patent is invalid or not infringed, the approval may be made effective on the date of the court decision,

(ii) if before the expiration of such period, the court decides that such patent has been infringed, the approval may be made effective on such date as the court orders under section 271(e)(4)(A) of title 35, United States Code, or

(iii) if before the expiration of such period the court grants a preliminary injunction prohibiting the applicant from engaging in the commercial manufacture or sale of the drug until the court decides the issues of patent validity and infringement and if the court decides that such patent is invalid or not infringed, the approval shall be made effective on the date of such court decision. In such an action, each of the parties shall reasonably cooperate in expediting the action. Until the expiration of forty-five days from the date the notice made under paragraph (3)(B) is received, no action may be brought under section 2201 of title 28, United States Code, for a declaratory judgment with respect to the patent. Any action brought under such section 2201 shall be brought in the judicial district where the defendant has its principal place of business or a regular and established place of business.

(D)(i) If an application (other than an abbreviated new drug application) submitted under subsection (b) for a drug, no active ingredient (including any ester or salt of the active ingredient) of which has been approved in any other application under subsection (b) was approved during the period beginning January 1, 1982, and ending on the date of the enactment of this subsection, the Secretary may not make the approval of another application for a drug for which the investigations described in clause (A) of subsection (b)(1) and relied upon by the applicant for approval of the application were not conducted by or for the applicant and for which the applicant has not obtained a right of reference or use from the person by or for whom the investigations were conducted effective before the expiration of ten years from the date of the approval of the application previously approved under subsection (b).

(ii) If an application submitted under subsection (b) for a drug, no active ingredient (including any ester or salt of the active ingredient) of which has been approved in any other application under subsection (b), is approved after the date of the enactment of this clause, no application which refers to the drug for which the subsection (b) application was submitted and for which the investigations described in clause (A) of subsection (b)(1) and relied upon by the applicant for approval of the application were not conducted by or for the applicant and for which the applicant has not obtained a right of reference or use from the person by or for whom the investigations were conducted may be submitted under subsection (b) before the expiration of five years from the date of the approval of the application under subsection (b), except that such an application may be submitted under subsection (b) after the expiration of four years from the date of the approval of the subsection (b) application if it contains a certification of patent invalidity or noninfringement described in clause (iv) of subsection (b)(2)(A). The approval of such an application shall be made effective in accordance with this paragraph except that, if an action for patent infringement is commenced during the one-year period beginning forty-eight months after the date of the approval of the subsection (b) application, the thirty-month period referred to in subparagraph (C) shall be extended by such amount of time (if any) which is required for seven and one-half years to have elapsed from the date of approval of the subsection (b) application.

(iii) If an application submitted under subsection (b) for a drug, which includes an active ingredient (including any ester or salt of the active ingredient) that has been approved in another application approved under subsection (b), is approved after the date of the enactment of this clause and if such application contains reports of new clinical investigations (other than bioavailability studies) essential to the approval of the application and conducted or sponsored by the applicant, the Secretary may not make the approval of an application submitted under subsection (b) for the conditions of approval of such drug in the approved subsection (b) application effective before the expiration of three years from the date of the approval of the application under subsection (b) if the investigations described in clause (A) of subsection (b)(1) and relied upon by the applicant for approval of the application were not conducted by or for the applicant and if the applicant has not obtained a right of reference or use from the person by or for whom the investigations were conducted.

(iv) If a supplement to an application approved under subsection (b) is approved after the date of enactment of this clause and the supplement contains reports of new clinical investigations (other than bioavailability studies) essential to the approval of the supplement and conducted or sponsored by the person submitting the supplement, the Secretary may not make the approval of an application submitted under subsection (b) for a change approved in the supplement effective before the expiration of three years from the date of the approval of the supplement under subsection (b) if the investigations described in clause (A) of subsection (b)(1) and relied upon by the applicant for approval of the application were not conducted by or for the applicant and if the applicant has not obtained a right of reference or use from the person by or for whom the investigations were conducted.

(v) If an application (or supplement to an application) submitted under subsection (b) for a drug, which includes an active ingredient (including any ester or salt of the active ingredient) that has been approved in another application under subsection (b), was approved during the period beginning January 1, 1982, and ending on the date of the enactment of this clause, the Secretary may not make the approval of an application submitted under this subsection and for which the investigations described in clause (A) of subsection (b)(1) and relied upon by the applicant for approval of the application were not conducted by or for the applicant and for which the applicant has not obtained a right of reference or use from the person by or for whom the investigations were conducted and which refers to the drug for which the subsection (b) application was submitted effective before the expiration of two years from the date of enactment of this clause.

(d) If the Secretary finds, after due notice to the applicant in accordance with subsection (c) and giving him an opportunity for a hearing, in accordance with said subsection, that (1) the investigations, reports of which are required to be submitted to the Secretary pursuant to subsection (b), do not include adequate tests by all methods reasonably applicable to show whether or not such drug is safe for use under the conditions prescribed, recommended, or suggested in the proposed labeling thereof; (2) the results of such tests show that such drug is unsafe for use under such conditions or do not show that such drug is safe for use under such conditions; (3) the methods used in, and the facilities and controls used for, the manufacture, processing, and packing of such drug are inadequate to preserve its identity, strength, quality, and purity; (4) upon the basis of the information submitted to him as part of the application, or upon the basis of any other information before him with respect to such drug, he has insufficient information to determine whether such drug is safe for use under such conditions; or (5) evaluated on the basis of the information submitted to him as part of the application and any other information before him with respect to such drug, there is a lack of substantial evidence that the drug will have the effect it purports or is represented to have under the conditions of use prescribed, recommended, or suggested in the proposed labeling thereof; or (6) the application failed to contain the patent information prescribed by subsection (b); or (7) based on a fair evaluation of all material facts, such labeling is false or misleading in any particular; he shall issue an order refusing to approve the application. If, after such notice and opportunity for hearing, the Secretary finds that clauses (1) through (6) do not apply, he shall issue an order approving the application. As used in this subsection and subsection (e), the term "substantial evidence" means evidence consisting of adequate and well-controlled investigations, including clinical investigations, by experts qualified by scientific training and experience to evaluate the effectiveness of the drug involved, on the basis of which it could fairly and responsibly be concluded by such experts that the drug will have the effect it purports or is represented to have under the conditions of use prescribed, recommended, or suggested in the labeling or proposed labeling thereof.

(e) The Secretary shall, after due notice and opportunity for hearing to the applicant, withdraw approval of an application with respect to any drug under this section if the Secretary finds (1) that clinical or other experience, tests, or other scientific data show that such drug is unsafe for use under the conditions of use upon the basis of which the application was approved; (2) that new evidence of clinical experience, not contained in such application or not available to the Secretary until after such application was approved, or tests by new methods, or tests by methods not deemed reasonably applicable when such application was approved, evaluated together with the evidence available to the Secretary when the application was approved, shows that such drug is not shown to be safe for use under the conditions of use upon the basis of which the application was approved; or (3) on the basis of new information before him with respect to such drug, evaluated together with the evidence available to him when the application was approved, that there is a lack of substantial evidence that the drug will have the effect it purports or is represented to have under the conditions of use prescribed, recommended, or suggested in the labeling thereof; or (4) the patent information prescribed by subsection (c) was not filed within thirty days after the receipt of written notice from the Secretary specifying the failure to file such information; or (5) that the application contains any untrue statement of a material fact *Provided,* That if the Secretary (or in his absence the officer acting as Secretary) finds that there is an imminent hazard to the public health, he may suspend the approval of such application immediately, and give the applicant prompt notice of his action and afford the applicant the opportunity for an expedited hearing under this subsection; but the authority conferred by this proviso to suspend the approval of an application shall not be delegated. The Secretary may also, after due notice and opportunity for hearing to the applicant, withdraw the approval of an application submitted under subsection (b) or (j) with respect to any drug under this section if the Secretary finds (1) that the applicant has failed to establish a system for maintaining required records, or has repeatedly or deliberately failed to maintain such records or to make required reports, in accordance with a regulation or order under subsection (k) or to comply with the notice requirements of section 510(k)(2), or the applicant has refused to permit access to, or copying or verification of, such records as required by paragraph (2) of such subsection; or (2) that on the basis of new information before him, evaluated together with the evidence before him when the application was approved, the methods used in, or the facilities and controls used for, the manufacture, processing, and packing of such drug are inadequate to assure and preserve its identity, strength, and were not made adequate within a reasonable time after receipt of written notice from the Secretary specifying the matter complained of, or (3) that on the basis of new information before him, evaluated together with the evidence before him when the application was approved, the labeling of such drug, based on a fair evaluation of all material facts, is false or misleading in any particular and was not corrected within a reasonable time after receipt of written notice from the Secretary specifying the matter complained of. Any order under this subsection shall state the findings upon which it is based.

(f) Whenever the Secretary finds that the facts so require, he shall revoke any previous order under subsection (d) or (e) refusing, withdrawing, or suspending approval of an application and shall approve such application or reinstate such approval, as may be appropriate.

(g) Orders of the Secretary issued under this

section shall be served (1) in person by any officer or employee of the Department designated by the Secretary or (2) by mailing the order by registered mail or by certified mail addressed to the applicant or respondent at his last-known address in the records of the Secretary.

(h) An appeal may be taken by the applicant from an order of the Secretary refusing or withdrawing approval of an application under this section. Such appeal shall be taken by filing in the United States court of appeals for the circuit wherein such applicant resides or has his principal place of business, or in the United States Court of Appeals for the District of Columbia Circuit, within sixty days after the entry of such order, a written petition praying that the order of the Secretary be set aside. A copy of such petition shall be forthwith transmitted by the clerk of the court to the Secretary, or any officer designated by him for that purpose, and thereupon the Secretary shall certify and file in the court the record upon which the order complained of was entered, as provided in section 2112 of title 28, United States Code. Upon the filing of such petition such court shall have exclusive jurisdiction to affirm or set aside such order, except that until the filing of the record the Secretary may modify or set aside his order. No objection to the order of the Secretary shall be considered by the court unless such objection shall have been urged before the Secretary or unless there were reasonable grounds for failure so to do. The finding of the Secretary as to the facts, if supported by substantial evidence, shall be conclusive. If any person shall apply to the court for leave to adduce additional evidence, and shall show to the satisfaction of the court that such additional evidence is material and that there were reasonable grounds for failure to adduce such evidence in the proceeding before the Secretary, the court may order such additional evidence to be taken before the Secretary and to be adduced upon the hearing in such manner and upon such terms and conditions as to the court may seem proper. The Secretary may modify his findings as to the facts by reason of the additional evidence so taken, and he shall file with the court such modified findings which, if supported by substantial evidence, shall be conclusive, and his recommendation, if any, for the setting aside of the original order. The judgment of the court affirming or setting aside any such order of the Secretary shall be final, subject to review by the Supreme Court of the United States upon certiorari or certification as provided in section 1254 of title 28 of the United States Code. The commencement of proceedings under this subsection shall not, unless specifically ordered by the court to the contrary, operate as a stay of the Secretary's order.

This amendment shall not apply to any appeal taken prior to the date of enactment of the Drug Amendments of 1962, enacted Oct. 10, 1962.

(i) The Secretary shall promulgate regulations for exempting from the operation of the foregoing subsections of this section drugs intended solely for investigational use by experts qualified by scientific training and experience to investigate the safety and effectiveness of drugs. Such regulations may, within the discretion of the Secretary, among other conditions relating to the protection of the public health, provide for conditioning such exemption upon--

(1) the submission to the Secretary, before any clinical testing of a new drug is undertaken, of reports, by the manufacturer or the sponsor of the investigation of such drug, or preclinical tests (including tests on animals) of such drug adequate to justify the proposed clinical testing;

(2) the manufacturer or the sponsor of the investigation of a new drug proposed to be distributed to investigators for clinical testing obtaining a signed agreement from each of such investigators that patients to whom the drug is administered will be under his personal supervision, or under the supervision of investigators responsible to him, and that he will not supply such drug to any other investigator, or to clinics, for administration to human beings; and

(3) the establishment and maintenance of such records, and the making of such reports to the Secretary, by the manufacturer or the sponsor of the investigation of such drug, of data (including but not limited to analytical

reports by investigators) obtained as the result of such investigational use of such drug, as the Secretary finds will enable him to evaluate the safety and effectiveness of such drug in the event of the filing of an application pursuant to subsection (b).

Such regulations shall provide that such exemption shall be conditioned upon the manufacturer, or the sponsor of the investigation, requiring that experts using such drugs for investigational purposes certify to such manufacturer or sponsor that they will inform any human beings to whom such drugs, or any controls used in connection therewith, are being administered, or their representatives, that such drugs are being used for investigational purposes and will obtain the consent of such human beings or their representatives, except where they deem it not feasible or, in their professional judgment, contrary to the best interests of such human beings. Nothing in this subsection shall be construed to require any clinical investigator to submit directly to the Secretary reports on the investigational use of drugs.

(j)(1) Any person may file with the Secretary an abbreviated application for the approval of a new drug.

(2)(A) An abbreviated application for a new drug shall contain--

(i) information to show that the conditions of use prescribed, recommended, or suggested in the labeling proposed for the new drug have been previously approved for a drug listed under paragraph (6) (hereinafter in this subsection referred to as a "listed drug");

(ii)(I) if the listed drug referred to in clause (i) has only one active ingredient, information to show that the active ingredient of the new drug is the same as that of the listed drug;

(II) if the listed drug referred to in clause (i) has more than one active ingredient, information to show that the active ingredients of the new drug are the same as those of the listed drug, or

(III) if the listed drug referred to in clause (i) has more than one active ingredient and if one of the active ingredients of the new drug is different and the application is filed pursuant to the approval of a petition filed under subparagraph (C), information to show that the other active ingredients of the new drug are the same as the active ingredients of the listed drug, information to show that the different active ingredient is an active ingredient of a listed drug or of a drug which does not meet the requirements of section 201(p), and such other information respecting the different active ingredient with respect to which the petition was filed as the Secretary may require;

(iii) information to show that the route of administration, the dosage form, and the strength of the new drug are the same as those of the listed drug referred to in clause (i), or, if the route of administration, the dosage form, or the strength of the new drug is different and the application is filed pursuant to the approval of a petition filed under subparagraph (C), such information respecting the route of administration, dosage form, or strength with respect to which the petition was filed as the Secretary may require;

(iv) information to show that the new drug is bioequivalent to the listed drug referred to in clause (i), except that if the application is filed pursuant to the approval of a petition filed under subparagraph (C), information to show that the active ingredients of the new drug are of the same pharmacological or therapeutic class as those of the listed drug referred to in clause (i) and the new drug can be expected to have the same therapeutic effect as the listed drug when administered to patients for a condition of use referred to in clause (i);

(v) information to show that the labeling proposed for the new drug is the same as the labeling approved for the listed drug referred to in clause (i) except for changes required because of differences approved under a petition filed under subparagraph (C) or because the new drug and the listed drug are produced or distributed by different manufacturers;

(vi) the items specified in clauses (B) through (F) of subsection (b)(1);

(vii) a certification, the opinion of the applicant and to the best of his knowledge, with respect to each patent which claims the listed drug referred to in clause (i) or which claims a use for such listed drug for which the applicant is seeking approval under this subsection and for which information is required to be filed under subsection (b) or (c)--

(I) that such patent information has not been filed,

(II) that such patent has expired,

(III) of the date on which such patent will expire, or

(IV) that such patent is invalid or will not be infringed by the manufacture, use, or sale of the new drug for which the application is submitted; and

(viii) if with respect to the listed drug referred to in clause (i) information was filed under subsection (b) or (c) for a method of use patent which does not claim a use for which the applicant is seeking approval under this subsection, a statement that the method of use patent does not claim such a use.

The Secretary may not require that an abbreviated application contain information in addition to that required by clauses (i) through (viii).

(B)(i) An applicant who makes a certification described in subparagraph (A)(vii)(IV) shall include in the application a statement that the applicant will give the notice required by clause (ii) to--

(I) each owner of the patent which is the subject of the certification or the representative of such owner designated to receive such notice, and

(II) the holder of the approved application under subsection (b) for the drug which is claimed by the patent or a use of which is claimed by the patent or the representative of such holder designated to receive such notice.

(ii) The notice referred to in clause (i) shall state that an application, which contains data from bioavailability or bioequivalence studies, has been submitted under this subsection for the drug with respect to which the certification is made to obtain approval to engage in the commercial manufacture, use, or sale of such drug before the expiration of the patent referred to in the certification. Such notice shall include a detailed statement of the factual and legal basis of the applicant's opinion that the patent is not valid or will not be infringed.

(iii) If an application is amended to include a certification described in subparagraph (A)(vii)(IV), the notice required by clause (ii) shall be given when the amended application is submitted.

(C) If a person wants to submit an abbreviated application for a new drug which has a different active ingredient or whose route of administration, dosage form, or strength differ from that of a listed drug, such person shall submit a petition to the Secretary seeking permission to file such an application. The Secretary shall approve or disapprove a petition submitted under this subparagraph within ninety days of the date the petition is submitted. The Secretary shall approve such a petition unless the Secretary finds--

(i) that investigations must be conducted to show the safety and effectiveness of the drug or of any of its active ingredients, the route of administration, the dosage form, or strength which differ from the listed drug; or

(ii) that any drug with a different active ingredient may not be adequately evaluated for approval as safe and effective on the basis of the information required to be submitted in an abbreviated application.

(3) Subject to paragraph (4), the Secretary shall approve an application for a drug unless the Secretary finds--

(A) the methods used in, or the facilities and controls used for, the manufacture, processing, and packing of the drug are inadequate to assure and preserve its identity, strength, quality, and purity;

(B) information submitted with the application is insufficient to show that each of the proposed conditions of use have been previously approved for the listed drug referred to in the application;

(C)(i) if the listed drug has only one active ingredient, information submitted with the application is insufficient to show that the active ingredient is the same as that of the listed drug;

(ii) if the listed drug has more than one active ingredient, information submitted with the application is insufficient to show that the active ingredients are the same as the active ingredients of the listed drug, or

(iii) if the listed drug has more than one active ingredient and if the application is for a drug which has an active ingredient different from the listed drug, information submitted with the application is insufficient to show--

(I) that the other active ingredients are the same as the active ingredients of the listed drug, or (II) that the different active ingredient is an active ingredient of a listed drug or a drug which does not meet the requirements of section 201(p).

or no petition to file an application for the drug with the different ingredient was approved under paragraph (2)(C);

(D)(i) if the application is for a drug whose route of administration, dosage form, or strength of the drug is the same as the route of administration, dosage form, or strength of the listed drug referred to in the application, information submitted in the application is insufficient to show that the route of administration, dosage form, or strength is the same as that of the listed drug, or

(ii) if the application is for a drug whose route of administration, dosage form, or strength of the drug is different from that of the listed drug referred to in the application, no petition to file an application for the drug with the different route of administration, dosage form, or strength was approved under paragraph (2)(C);

(E) if the application was filed pursuant to the approval of a petition under paragraph (2)(C), the application did not contain the information required by the Secretary respecting the active ingredient, route of administration, dosage form, or strength which is not the same;

(F) information submitted in the application is insufficient to show that the drug is bioequivalent to the listed drug referred to in the application or, if the application was filed pursuant to a petition approved under paragraph (2)(C) information submitted in the application is insufficient to show that the active ingredients of the new drug are of the same pharmacological or therapeutic class as those of the listed drug referred to in paragraph (2)(A)(i) and that the new drug can be expected to have the same therapeutic effect as the listed drug when administered to patients for a condition of use referred to in such paragraph;

(G) information submitted in the application is insufficient to show that the labeling proposed for the drug is the same as the labeling approved for the listed drug referred to in the application except for changes required because of differences approved under a petition filed under paragraph (2)(C) or because the drug and the listed drug are produced or distributed by different manufacturers;

(H) information submitted in the application or any other information available to the Secretary shows that (i) the inactive ingredients of the drug are unsafe for use under the conditions prescribed, recommended, or suggested in the labeling proposed for the drug, or (ii) the composition of the drug is unsafe under such conditions because of the type or quantity of inactive ingredients included or the manner in which the inactive ingredients are included; (I) the approval under subsection (c) of the listed drug referred to in the application under this subsection has been withdrawn or suspended for grounds described in the first sentence of subsection (e), the Secretary has published a notice of opportunity for hearing to withdraw approval of the listed drug under subsection (c) for grounds described in the first sentence of subsection (e), the approval under this subsection of the listed drug referred to in the application under this subsection has been withdrawn or suspended under paragraph (5), or the Secretary has determined that the listed drug has been withdrawn from sale for safety or effectiveness reasons;

(J) the application does not meet any other requirement of paragraph (2)(A); or

(K) the application contains an untrue statement of material fact.

(4)(A) Within one hundred and eighty days of the initial receipt of an application under paragraph (2) or within such additional period as may be agreed upon by the Secretary and the applicant, the Secretary shall approve or disapprove the application.

(B) The approval of an application submitted under paragraph (2) shall be made effective on the last applicable date determined under the following:

(i) If the applicant only made a certification described in subclause (I) or (II) of paragraph (2)(A)(vii) or in both such subclauses, the approval may be made effective immediately.

(ii) If the applicant made a certification described in subclause (III) of paragraph (2)(A)(vii), the approval may be made effective on the date certified under subclause (III).

(iii) If the applicant made a certification described in subclause (IV) of paragraph (2)(A)(vii), the approval shall be made effective immediately unless an action is brought for infringement of a patent which is the subject of the certification before the expiration of forty-five days from the date the notice provided under paragraph (2)(B)(i) is received. If such an action is brought before the expiration of such days, the approval shall be made effective upon the expiration of the thirty-month period beginning on the date of the receipt of the notice provided under paragraph (2)(B)(i) or such shorter or longer period as the court may order because either party to the action failed to reasonably cooperate in expediting the action, except that--

(I) if before the expiration of such period the court decides that such patent is invalid or not infringed, the approval shall be made effective on the date of the court decision.

(II) if before the expiration of such period the court decides that such patent has been infringed, the approval shall be made effective on such date as the court orders under section 271(e)(4)(A) of title 35, United States Code, or

(III) if before the expiration of such period the court grants a preliminary injunction prohibiting the applicant from engaging in the commercial manufacture or sale of the drug until the court decides the issues of patent validity and infringement and if the court decides that such patent is invalid or not infringed, the approval shall be made effective on the date of such court decision. In such an action, each of the parties shall reasonably cooperate in expediting the action. Until the expiration of forty-five days from the date the notice made under paragraph (2)(B)(i) is received, no action, may be brought under section 2201 of title 28, United States Code, for a declaratory judgment with respect to the patent. Any action brought under section 2201 shall be brought in the judicial district where the defendant has its principal place of business or a regular and established place of business.

(iv) If the application contains a certification described in subclause (IV) of paragraph (2)(A)(vii) and is for a drug for which a previous application has been submitted under this subsection continuing such a certification, the application shall be made effective not earlier than one hundred and eighty days after--

(I) the date the Secretary receives notice from the applicant under the previous application of the first commercial marketing of the drug under the previous application, or

(II) the date of a decision of a court in action described in clause (ii) holding the patent which is the subject of the certification to be invalid or not infringed, whichever is earlier.

(C) If the Secretary decides to disapprove an application, the Secretary shall give the applicant notice of an opportunity for a hearing before the Secretary on the question of whether such application is approvable. If the applicant elects to accept the opportunity for hearing by written request within thirty days after such notice, such hearing shall commence not more than ninety days after the expiration of such thirty days unless the Secretary and the applicant otherwise agree. Any such hearing shall thereafter be conducted on an expedited basis and the Secretary's order thereon shall be issued within ninety days after the date fixed by the Secretary for filing final briefs.

(D)(i) If an application (other than an abbreviated new drug application) submitted under subsection (b) for a drug, no active ingredient (including any ester or salt of the active ingredient) of which has been approved in any other application under subsection (b) was approved during the period beginning January 1, 1982, and ending on the date of the enactment of this subsection, the Secretary may not make the approval of an application submitted under this subsection which refers to the drug for which the subsection (b) application was submitted effective before the expiration of ten years from the date of the approval of the application under subsection (b).

(ii) If an application submitted under subsection (b) for a drug, no active ingredient (including any ester or salt of the active ingredient) of which has been approved in any other application under subsection (b), is approved after the date of the enactment of this subsection, no application may be submitted under this subsection which refers to the drug for which the subsection (b) application was submitted before the expiration of five years from the date of the approval of the application under subsection (b), except that such an application may be submitted under this subsection after the expiration of four years from the date of the approval of the subsection (b) application if it contains a certification of patent invalidity or noninfringement described in subclause (IV) of paragraph (2)(A)(vii). The approval of such an application shall be made effective in accordance with subparagraph (B) except that, if an action for patent infringement is commenced during the one-year period beginning forty-eight months after the date of the approval of the subsection (b) application, the thirty-month period referred to in subparagraph (B)(iii) shall be extended by such amount of time (if any) which is required for seven and one-half years to have elapsed from the date of approval of the subsection (b) application.

(iii) If an application submitted under subsection (b) for a drug, which includes an active ingredient (including any ester or salt of the active ingredient) that has been approved in another application approved under subsection (b), is approved after the date of enactment of this subsection and if such application contains reports of new clinical investigations (other than bioavailability studies) essential to the approval of the application and conducted or sponsored by the applicant, the Secretary may not make the approval of an application submitted under this subsection for the conditions of approval of such drug in the subsection (b) application effective before the expiration of three years from the date of the approval of the application under subsection (b) for such drug.

(iv) If a supplement to an application approved under subsection (b) is approved after the date of enactment of this subsection and the supplement contains reports of new clinical investigations (other than bioavailability studies) essential to the approval of the supplement and conducted or sponsored by the person submitting the supplement, the Secretary may not make the approval of an application submitted under this subsection for a change approved in the supplement effective before the expiration of three years from the date of the approval of the supplement under subsection (b).

(v) If an application (or supplement to an application) submitted under subsection (b) for a drug, which includes an active ingredient (including any ester or salt of the active ingredient) that has been approved in another application under subsection (b), was approved during the period beginning January 1, 1982, and ending on the date of the enactment of this subsection, the Secretary may not make the approval of an application submitted under this subsection which refers to the drug for which the subsection (b) application was submitted or which refers to a change approved in a supplement to the subsection (b) application effective before the expiration of two years from the date of enactment of this subsection.

(5) If a drug approved under this subsection refers in its approved application to a drug the approval of which was withdrawn or suspended for grounds described in the first sentence of subsection (e) or was withdrawn or suspended under this paragraph or which, as determined by the Secretary, has been withdrawn from sale for safety or effectiveness reasons, the approval of the drug under this subsection shall be withdrawn or suspended--

(A) for the same period as the withdrawal or suspension under subsection (e) or this paragraph, or

(B) if the listed drug has been withdrawn from sale, for the period of withdrawal from sale or, if earlier, the period ending on the date the Secretary determines that the withdrawal from sale is not for safety or effectiveness reasons.

(6)(A)(i) Within sixty days of the date of the enactment of this subsection, the Secretary shall publish and make available to the public--

(I) a list in alphabetical order of the official and proprietary name of each drug which has been approved for safety and effectiveness under subsection (c) before the date of the enactment of this subsection;

(II) the date of approval if the drug is approved after 1981 and the number of the application which was approved; and

(III) whether in vitro or in vivo bioequivalence studies, or both such studies, are required for applications filed under this subsection which will refer to the drug published.

(ii) Every thirty days after the publication of the first list under clause (i) the Secretary shall revise the list to include each drug which has been approved for safety and effectiveness under subsection (c) or approved under this subsection during the thirty-day period.

(iii) When patent information submitted under subsection (b) or (c) respecting a drug included on the list is to be published by the Secretary, the Secretary shall, in revisions made under clause (ii), include such information for such drug.

(B) A drug approved for safety and effectiveness under subsection (c) or approved under this subsection shall, for purposes of this subsection, be considered to have been published under subparagraph (A) on the date of its approval or the date of enactment, whichever is later.

(C) If the approval of a drug was withdrawn or suspended for grounds described in the first sentence of subsection (e) or was withdrawn or suspended under paragraph (5) or if the Secretary determines that a drug has been withdrawn from sale for safety or effectiveness reasons it may not be published in the list under subparagraph (A) or, if the withdrawal or suspension occurred after its publication in such list, it shall be immediately removed from such list--

(i) for the same period as the withdrawal or suspension under subsection (e) or paragraph (5), or

(ii) if the listed drug has been withdrawn from sale, for the period of withdrawal from sale or, if earlier, the period ending on the date the Secretary determines that the withdrawal from sale is not for safety or effectiveness reasons.

A notice of the removal shall be published in the Federal Register.

(7) For purposes of this subsection:

(A) The term "bioavailability" means the rate and extent to which the active ingredient or therapeutic ingredient is absorbed from a drug and becomes available at the site of drug action.

(B) A drug shall be considered to be bioequivalent to a listed drug if--

(i) the rate and extent of absorption of the drug do not show a significant difference from the rate and extent of absorption of the listed drug when administered at the same molar dose of the therapeutic ingredient under similar experimental conditions in either a single dose or multiple doses; or

(ii) the extent of absorption of the drug does not show a significant difference from the extent of absorption of the listed drug when administered at the same molar dose of the therapeutic ingredient under similar experimental conditions in either a single dose or multiple doses and the difference from the listed drug in the rate of absorption of the drug is intentional, is reflected in its proposed labeling, is not essential to the attainment of effective body drug concentrations on chronic use, and is considered medically insignificant for the drug.

(8) The Secretary shall, with respect to each application submitted under this subsection, maintain a record of--

(A) the name of the applicant,

(B) the name of the drug covered by the application,

(C) the name of each person to whom the review of the chemistry of the application was assigned and the date of such assignment, and

(D) the name of each person to whom the bioequivalence review for such application was assigned and the date of such assignment.

The information the Secretary is required to maintain under this paragraph with respect to an application submitted under this subsection shall be made available to the public after the approval of such application.

(k)(1) In the case of any drug for which an approval of an application filed under subsection (b) or (j) is in effect, the applicant shall establish and maintain such records, and make such reports to the Secretary, of data relating to clinical experience and other data or information, received or otherwise obtained by such applicant with respect to such drug, as the Secretary may by general regulation, or by order with respect to such application, prescribe on the basis of a finding that such records and reports are necessary in order to enable the Secretary to determine, or facilitate a determination, whether there is or may be ground for invoking subsection (e) of this section: *Provided*, however, That regulations and orders issued under this subsection and under subsection (i) shall have due regard for the professional ethics of the medical profession and the interests of patients and shall provide, where the Secretary deems it to be appropriate, for the examination, upon request, by the persons to whom such regulations or orders are applicable, of similar information received or otherwise obtained by the Secretary.

(2) Every person required under this section to maintain records, and every person in charge or custody thereof, shall, upon request of an officer or employee designated by the Secretary, permit such officer or employee at all reasonable times to have access to and copy and verify such records.

(l) Safety and effectiveness data and information which has been submitted in an application under subsection (b) for a drug and which has not previously been disclosed to the public shall be made available to the public, upon request, unless extraordinary circumstances are shown--

(1) if no work is being or will be undertaken to have the application approved,

(2) if the Secretary has determined that the application is not approvable and all legal appeals have been exhausted,

(3) if approval of the application under subsection (c) is withdrawn and all legal appeals have been exhausted,

(4) if the Secretary has determined that such drug is not a new drug, or

(5) upon the effective date of the approval of the first application under subsection (j) which refers to such drug or upon the date upon which the approval of an application under subsection (j) which refers to such drug could be made effective if such an application had been submitted.

(m) For purposes of this section, the term "patent" means a patent issued by the Patent and Trademark Office of the Department of Commerce.

SEC. 506 [21 U.S.C. 356] Certification of drugs containing insulin

(a) The Secretary, pursuant to regulations promulgated by him, shall provide for the certification of batches of drugs composed wholly or partly of insulin. A batch of any such drug shall be certified if such drug has such characteristics of identity and such batch has such characteristics of strength, quality, and purity, as the Secretary prescribes in such regulations as necessary to adequately insure safety and efficacy of use, but shall not otherwise be certified. Prior to the effective date of such regulations the Secretary, in lieu of certification, shall issue a release for any batch which, in his judgment, may be released without risk as to the safety and efficacy of its use. Such release shall prescribe the date of its expiration and other conditions under which it shall cease to be effective as to such batch and as to portions thereof.

(b) Regulations providing for such certification shall contain such provisions as are necessary to carry out the purposes of this section, including provisions prescribing (1) standards of identity and of strength, quality, and purity; (2) tests and methods of assay to determine compliance with such standards; (3) effective periods for certificates, and other conditions under which they shall cease to be effective as to certified batches and as to portions thereof; (4) administration and procedure; and (5) such fees, specified in such regulations, as are necessary to provide, equip, and maintain an adequate certification service. Such regulations shall prescribe no standard of identity or of strength, quality, or purity for any drug different from the standard of identity, strength, quality, or purity set forth for such drug in an official compendium.

(c) Such regulations, insofar as they prescribe tests or methods of assay to determine strength, quality, or purity of any drug, different from the tests or methods of assay set forth for such drug in an official compendium, shall be prescribed, after notice and opportunity for revision of such compendium, in the manner provided in the second sentence of section 501(b). The provisions of subsections (e), (f), and (g) of section 701 shall be applicable to such portion of any regulation as prescribes any such different test or method, but shall not be applicable to any other portion of any such regulation.

SEC. 507 [21 U.S.C. 357] Certification of antibiotics

(a) The Secretary, pursuant to regulations promulgated by him, shall provide for the certification of batches of drugs (except drugs for use in animals other than man) composed wholly or partly of any kind of penicillin, streptomycin,chlortetracycline,chloramphenicol, bacitracin, or any other antibiotic drug, or any derivative thereof. A batch of any such drug shall be certified if such drug has such characteristics of identity and such batch has such characteristics of strength, quality, and purity, as the Secretary prescribes in such regulations as necessary to adequately insure safety and efficacy of use, but shall not otherwise be certified. Prior to the effective date of such regulations the Secretary, in lieu of certification, shall issue a release for any batch which, in his judgment, may be released without risk as to the safety and efficacy of its use. Such release shall prescribe the date of its expiration and other conditions under which it shall cease to be effective as to such batch and as to portions thereof. For purposes of this section and of section 502(l), the term "antibiotic drug" means any drug intended for use by man containing any quantity of any chemical substance which is produced by a micro-organism and which has the capacity to inhibit or destroy micro-organisms in dilute solution (including the chemically synthesized equivalent of any such substance).

(b) Regulations providing for such certifications shall contain such provisions as are necessary to carry out the purposes of this section, including provisions prescribing (1) standards of identity and of strength, quality, and purity; (2) tests and methods of assay to determine compliance with such standards; (3) effective periods for certificates, and other conditions under which they shall cease to be effective as to certified batches and as to portions thereof; (4) administration and procedure; and (5) such fees, specified in such regulations, as are necessary to provide, equip, and maintain an adequate certification service. Such regulations shall prescribe only such tests and methods of assay as will provide for certification or rejection within the shortest time consistent with the purposes of this section.

(c) Whenever in the judgment of the Secretary, the requirements of this section and of section 502(l) with respect to any drug or class of drugs are not necessary to insure safety and efficacy of use, the Secretary shall promulgate regulations exempting such drug or class of drugs from such requirements. In deciding whether an antibiotic drug, or class of antibiotic drugs, is to be exempted from the requirement of certification the Secretary shall give consideration, among other relevant factors, to--

(1) whether such drug or class of drugs is manufactured by a person who has, or hereafter shall have, produced fifty consecutive batches of such drug or class of drugs in compliance with the regulations for the certification thereof within a period of not more than eighteen calendar months, upon the application by such person to the Secretary, or

(2) whether such drug or class of drugs is manufactured by any person who has otherwise demonstrated such consistency in the production of such drug or class of drugs, in compliance with the regulations for the certification thereof, as in the judgment of the Secretary is adequate to insure the safety and efficacy of use thereof.

When an antibiotic drug or a drug manufacturer has been exempted from the requirement of certification, the manufacturer may still obtain certification of a batch or batches of that drug if he applies for and meets the requirements for certification. Nothing in this Act shall be deemed to prevent a manufacturer or distributor of an antibiotic drug from making a truthful statement in labeling or advertising of the product as to whether it has been certified or exempted from the requirement of certification.

(d) The Secretary shall promulgate regulations exempting from any requirement of this section and of section 502(l), (1) drugs which are to be stored, processed, labeled, or repacked at establishments other than those where manufactured, on condition that such drugs comply with all such requirements upon removal from such establishments; (2) drugs which conform to applicable standards of identity, strength, quality, and purity prescribed by these regulations and are intended for use in manufacturing other drugs; and (3) drugs which are intended solely for investigational use by experts qualified by scientific training and experience to investigate the safety and efficacy of drugs. Such regulations may, within the discretion of the Secretary, among other conditions relating to the protection of the public health, provide for conditioning the exemption under clause (3) upon--

(1) the submission to the Secretary, before any clinical testing of a new drug is undertaken, of reports, by the manufacturer or the sponsor of the investigation of such drug, of preclinical tests (including tests on animals) of such drug adequate to justify the proposed clinical testing;

(2) the manufacturer or the sponsor of the investigation of a new drug proposed to be distributed to investigators for clinical testing obtaining a signed agreement from each of such investigators that patients to whom the drug is administered will be under his personal supervision, or under the supervision of investigators responsible to him, and that he will not supply such drug to any other investigator, or to clinics, for administration to human beings; and (3) the establishment and maintenance of such records, and the making of such reports to the Secretary, by the manufacturer or the sponsor of the investigation of such drug, of data (including but not limited to analytical reports by investigators) obtained as the result of such investigational use of such drug, as the Secretary finds will enable him to evaluate the safety and effectiveness of such drug in the event of the filing of an application for certification or release pursuant to subsection (a).

Such regulations shall provide that such exemption shall be conditioned upon the manufacturer, or the sponsor of the investigation, requiring that experts using such drugs for investigational purposes certify to such manufacturer or sponsor that they will inform any human beings to whom such drugs, or any controls used in connection therewith, are being administered, or their representatives, that such drugs are being used for investigational purposes and will obtain the consent of such human beings or their representatives, except where they deem it not feasible or, in their professional judgment, contrary to the best interests of such human beings. Nothing in this subsection shall be construed to require any clinical investigator to submit directly to the Secretary reports on the investigational use of drugs.

(e) No drug which is subject to section 507 shall be deemed to be subject to any provision of section 505 except a new drug exempted from the requirements of this section and of section 502(l) pursuant to regulations promulgated by the Secretary: *Provided*, That, for purposes of section 505, the initial request for certification, as thereafter duly amended, pursuant to section 507, of a new drug so exempted shall be considered a part of the application filed pursuant to section 505(b) with respect to the person filing such request and to such drug as of the date of the exemption. Compliance of any drug subject to section 502(l) or 507 with section 501(b) and 502(g) shall be determined by the application of the standards of strength, quality, and purity, the tests and methods of assay, and the requirements of packaging, and labeling, respectively, prescribed by regulations promulgated under section 507.

(f) Any interested person may file with the Secretary a petition proposing the issuance, amendment, or repeal of any regulation contemplated by this section. The petition shall set forth the proposal in general terms and shall state reasonable grounds therefor. The Secretary shall give public notice of the proposal and an opportunity for all interested persons to present their views thereon orally or in writing, and as soon as practicable thereafter shall make public his action upon such proposal. At any time prior to the thirtieth day after such action is made public any interested person may file objections to such action, specifying with particularity the changes desired, stating reasonable grounds therefor, and requesting a public hearing upon such objections. The Secretary shall thereupon, after due notice, hold such public hearing. As soon as practicable after completion of the hearing, the Secretary shall by order make public his action on such objections. The Secretary shall base his order only on substantial evidence of record at the hearing and shall set forth as part of the order detailed findings of fact on which the order is based. The order shall be subject to the provision of section 701 (f) and (g).

(g)(1) Every person engaged in manufacturing, compounding, or processing any drug within the purview of this section with respect to which a certificate or release has been issued pursuant to this section shall establish and maintain such records, and make such reports to the Secretary, of data relating to clinical experience and other data or information, received or otherwise obtained by such person with respect to such drug, as the Secretary may by general regulation, or by order with respect to such certification or release prescribe on the basis of a finding that such records and reports are necessary in order to enable the Secretary to make, or to facilitate, a determination as to whether such certification or release should be rescinded or whether any regulation issued under this section should be amended or repealed: *Provided*, however, That regulations and orders issued under this subsection and under clause (3) of subsection (d) shall have due regard for the professional ethics of the medical profession and the interests of patients and shall provide, where the Secretary deems it to be appropriate, for the examination, upon request, by the persons to whom such regulations or orders are applicable, of similar information received or otherwise obtained by the Secretary.

(2) Every person required under this section to maintain records, and every person having charge or custody thereof, shall, upon request of an officer or employee designated by the Secretary, permit such officer or employee at all reasonable times to have access to and copy and verify such records.

(h) In the case of a drug for which, on the day immediately preceding the effective date of this subsection, a prior approval of an application under section 505 had not been withdrawn under section 505(e), the initial issuance of regulations providing for certification or exemption of such drug under this section 507 shall, with respect to the conditions of use prescribed, recommended, or suggested in the labeling covered by such application, not be conditioned upon an affirmative finding of the efficacy of such drug. Any subsequent amendment or repeal of such regulations so as no longer to provide for such certification or exemption on the ground of a lack of efficacy of such drug for use under such conditions of use may be effected only on or after that effective date of clause (3) of the first sentence of section 505(e) which would be applicable to such drug under such conditions of use if such drug were subject to section 505(e), and then only if (1) such amendment or repeal is made in accordance with the procedure specified in subsection (f) of this section (except that such amendment or repeal may be initiated either by a proposal of the Secretary or by a petition of any interested person) and (2) the Secretary finds, on the basis of new information with respect to such drug evaluated together with the information before him when the application under section 505 became effective or was approved, that there is a lack of substantial evidence (as defined in section 505(d)) that the drug has the effect it purports or is represented to have under such conditions of use.

SEC. 508 [21 U.S.C. 358] Authority to designate official names

(a) The Secretary may designate an official name for any drug or device if he determines that such action is necessary or desirable in the interest of usefulness and simplicity. Any official name designated under this section for any drug or device shall be the only official name of that drug or device used in any official compendium published after such name has been prescribed or for any other purpose of this Act. In no event, however, shall the Secretary establish an official name so as to infringe a valid trademark.

(b) Within a reasonable time after the effective date of this section, and at such other times as he may deem necessary, the Secretary shall cause a review to be made of the official names by which drugs are identified in the official United States Pharmacopeia the official Homeopathic Pharmacopeia of the United States, and the official National Formulary, and all supplements thereto, and at such times as he may deem necessary shall cause a review to be made of the official names by which devices are identified in any official compendium (and all supplements thereto) to determine whether revision of any of those names is necessary or desirable in the interest of usefulness and simplicity.

(c) Whenever he determines after any such review that (1) any such official name is unduly complex or is not useful for any other reason, (2) two or more official names have been applied to a single drug or device, or to two or more drugs which are identical in chemical structure and pharmacological action and which are substantially identical in strength, quality, and purity, or to two or more devices which are substantially equivalent in design and purpose or (3) no official name has been applied to a medically useful drug or device, he shall transmit in writing to the compiler of each official compendium in which that drug or drugs or device are identified and recognized his request for the recommendation of a single official name for such drug or drugs or device which will have usefulness and simplicity. Whenever such a single official name has not been recommended within one hundred and eighty days after such request, or the Secretary determines that any name so recommended is not useful for any reason, he shall designate a single official name for such drug or drugs or device. Whenever he determines that the name so recommended is useful, he shall designate that name as the official name of such drug or drugs or device. Such designation shall be made as a regulation upon public notice and in accordance with the procedure set forth in section 4 of the Administrative Procedure Act (5 U.S.C. 1003).

(d) After each such review, and at such other times as the Secretary may determine to be necessary or desirable, the Secretary shall cause to be compiled, published, and publicly distributed a list which shall list all revised official names of drugs or devices designated under this section and shall contain such descriptive and explanatory matter as the Secretary may determine to be required for the effective use of those names.

(e) Upon a request in writing by any compiler of any official compendium that the Secretary exercise the authority granted to him under section 508(a), he shall upon public notice and in accordance with the procedure set forth in section 4 of the Administrative Procedure Act (5 U.S.C. 1003) designate the official name of the drug or device for which the request is made.

SEC. 510 [21 U.S.C. 360] Registration of producers of drugs and devices

(a) As used in this section--

(1) the term ''manufacture, preparation, propagation, compounding, or processing'' shall include repackaging or otherwise changing the container, wrapper, or labeling of any drug package or device package in furtherance of the distribution of the drug or device from the original place of manufacture to the person who makes final delivery or sale to the ultimate consumer or user; and

(2) the term ''name'' shall include in the case of a partnership the name of each partner and, in the case of a corporation, the name of each corporate officer and director, and the State of incorporation.

(b) On or before December 31 of each year every person who owns or operates any establishment in any State engaged in the manufacture, preparation, propagation, compounding, or processing of a drug or drugs or a device or devices shall register with the Secretary his name, places of business, and all such establishments.

(c) Every person upon first engaging in the manufacture, preparation propagation, compounding, or processing of a drug or drugs or a device or devices in any establishment which he owns or operates in any State shall immediately register with the Secretary his name, place of business, and such establishment.

(d) Every person duly registered in accordance with the foregoing subsections of this section shall immediately register with the Secretary any additional establishment which he owns or operates in any State and in which he begins the manufacture, preparation, propagation, compounding, or processing of a drug or drugs or a device or devices.

(e) The Secretary may assign a registration number to any person or any establishment registered in accordance with this section. The Secretary may also assign a listing number to each drug or class of drugs listed under subsection (j). Any number assigned pursuant to the preceding sentence shall be the same as that assigned pursuant to the National Drug Code. The Secretary may by regulation prescribe a uniform system for the identification of devices intended for human use and may require that persons who are required to list such devices pursuant to subsection (j) shall list such devices in accordance with such system.

(f) The Secretary shall make available for inspection, to any person so requesting, any registration filed pursuant to this section; except that any list submitted pursuant to paragraph (3) of subsection (j) and the information accompanying any list or notice filed under paragraph (1) or (2) of that subsection shall be exempt from such inspection unless the Secretary finds that such an exemption would be inconsistent with protection of the public health.

(g) The foregoing subsections of this section shall not apply to--

(1) pharmacies which maintain establishments in conformance with any applicable local laws regulating the practice of pharmacy and medicine and which are regularly engaged in dispensing prescription drugs or devices, upon prescriptions of practitioners licensed to administer such drugs or devices to patients under the care of such practitioners in the course of their professional practice, and which do not manufacture, prepare, propagate, compound, or process drugs or devices for sale other than in the regular course of their business of dispensing or selling drugs or devices at retail;

(2) practitioners licensed by law to prescribe or administer drugs or devices and who manufacture, prepare, propagate, compound, or process drugs or devices solely for use in the course of their professional practice;

(3) persons who manufacture, prepare, propagate, compound, or process drugs or devices solely for use in research, teaching, or chemical analysis and not for sale;

(4) such other classes of persons as the Secretary may by regulation exempt from the application of this section upon a finding that registration by such classes of persons in accordance with this section is not necessary for the protection of the public health.

(h) Every establishment in any State registered with the Secretary pursuant to this section shall be subject to inspection pursuant to section 704 and every such establishment engaged in the manufacture, propagation, compounding, or processing of a drug or drugs or of a device or devices classified in class II or III shall be so inspected by one or more officers or employees duly designated by the Secretary at least once in the 2-year period beginning with the date of registration of such establishment pursuant to this section and at least once in every successive 2-year period thereafter.

(i) Any establishment within any foreign country engaged in the manufacture, preparation, propagation, compounding, or processing of a drug or drugs or a device or devices, shall be permitted to register under this section pursuant to regulations promulgated by the Secretary. Such regulations shall require such establishment to provide the information required by subsection (j) and shall require such establishment to provide information required by subsection (j) in the case of a device or devices and shall include provisions for registration of any such establishment upon condition that adequate and effective means are available, by arrangement with the government of such foreign country or otherwise, to enable the Secretary to determine from time to time whether drugs or devices manufactured, prepared, propagated, compounded, or processed in such establishment, if imported or offered for import into the

United States, shall be refused admission on any of the grounds set forth in section 801(a) of this Act.

(j)(1) Every person who registers with the Secretary under subsection (b), (c), or (d) shall, at the time of registration under any such subsection, file with the Secretary a list of all drugs and a list of all devices and a brief statement of the basis for believing that each device included in the list is a device rather than a drug (with each drug and device in each list listed by its established name (as defined in section 502(e)) and by any proprietary name) which are being manufactured, prepared, propagated, compounded, or processed by him for commercial distribution and which he has not included in any list of drugs or devices filed by him with the Secretary under this paragraph or paragraph (2) before such time of registration. Such list shall be prepared in such form and manner as the Secretary may prescribe and shall be accompanied by--

(A) in the case of a drug contained in the applicable list and subject to section 505, 506, 507, or 512, or a device intended for human use contained in the applicable list with respect to which a performance standard has been established under section 514 or which is subject to section 515, a reference to the authority for the marketing of such drug or device and a copy of all labeling for such drug or device;

(B) in the case of any other drug or device contained in an applicable list--

(i) which drug is subject to section 503(b)(1), or which device is a restricted device, a copy of all labeling for such drug or device, a representative sampling of advertisements for such drug or device, and, upon request made by the Secretary for good cause, a copy of all advertisements for a particular drug product or device, or

(ii) which drug is not subject to section 503(b)(1) or which device is not a restricted device, the label and package insert for such drug or device and a representative sampling of any other labeling for such drug or device;

(C) in the case of any drug contained in an applicable list which is described in subparagraph (B), a quantitative listing of its active ingredient or ingredients, except that with respect to a particular drug product the Secretary may require the submission of a quantitative listing of all ingredients if he finds that such submission is necessary to carry out the purposes of this Act; and

(D) if the registrant filing a list has determined that a particular drug product or device contained in such list is not subject to section 505, 506, 507, or 512, or the particular device contained in such list is not subject to a performance standard established under section 514 or to section 515 or is not a restricted device, a brief statement of the basis upon which the registrant made such determination if the Secretary requests such a statement with respect to that particular drug product or device.

(2) Each person who registers with the Secretary under this section shall report to the Secretary once during the month of June of each year and once during the month of December of each year the following information:

(A) A list of each drug or device introduced by the registrant for commercial distribution which has not been included in any list previously filed by him with the Secretary under this subparagraph or paragraph (1) of this subsection. A list under this subparagraph shall list a drug or device by its established name (as defined in section 502(e)) and by any proprietary name it may have and shall be accompanied by the other information required by paragraph (1).

(B) If since the date the registrant last made a report under this paragraph (or if he has not made a report under this paragraph, since the effective date of this subsection) he has discontinued the manufacture, preparation, propagation, compounding, or processing for commercial distribution of a drug or device included in a list filed by him under subparagraph (A) or paragraph (1); notice of such discontinuance, the date of such discontinuance, and the identity (by established name (as defined in section 502(e)) and by any proprietary name) of such drug or device.

(C) If since the date the registrant reported pursuant to subparagraph (B) a notice of discontinuance he has resumed the manufacture, preparation, propagation, compounding, or processing for commercial distribution of the drug or device with respect to which such notice of discontinuance was reported; notice of such resumption, the date of such resumption, the identity of such drug or device (by established name (as defined in section 502(e)) and by any proprietary name), and the other information required by paragraph (1), unless the registrant has previously reported such resumption to the Secretary pursuant to this subparagraph.

(D) Any material change in any information previously submitted pursuant to this paragraph or paragraph (1).

(3) The Secretary may also require each registrant under this section to submit a list of each drug product which (A) the registrant is manufacturing, preparing, propagating, compounding, or processing for commercial distribution, and (B) contains a particular ingredient. The Secretary may not require the submission of such a list unless he has made a finding that the submission of such a list is necessary to carry out the purposes of this Act.

(k) Each person who is required to register under this section and who proposes to begin the introduction or delivery for introduction into interstate commerce for commercial distribution of a device intended for human use shall, at least ninety days before making such introduction or delivery, report to the Secretary (in such form and manner as the Secretary shall by regulation prescribe)--

(1) the class in which the device is classified under section 513 or if such person determines that the device is not classified under such section, a statement of that determination and the basis for such person's determination that the device is or is not so classified, and

(2) action taken by such person to comply with requirements under section 514 or 515 which are applicable to the device.

SEC. 513 [21 U.S.C. 360c] Classification of devices intended for human use

(a)(1) There are established the following classes of devices intended for human use:

(A) Class I, General Controls.

(i) A device for which the controls authorized by or under section 501, 502, 510, 516, 518, 519, or 520 or any combination of such sections are sufficient to provide reasonable assurance of the safety and effectiveness of the device.

(ii) A device for which insufficient information exists to determine that the controls referred to in clause (i) are sufficient to provide reasonable assurance of the safety and effectiveness of the device or to establish special controls to provide such assurance, but because it--

(I) is not purported or represented to be for a use in supporting or sustaining human life or for a use which is of substantial importance in preventing impairment of human health, and

(II) does not present a potential unreasonable risk of illness or injury, is to be regulated by the controls referred to in clause (i).

(B) Class II, Special Controls. A device which cannot be classified as a class I device because the general controls by themselves are insufficient to provide reasonable assurance of the safety and effectiveness of the device, and for which there is sufficient information to establish special controls to provide such assurance, including the promulgation of performance standards, postmarket surveillance, patient registries, development and dissemination of guidelines (including guidelines for the submission of clinical data in premarket notification submissions in accordance with section 510(k)), recommendations, and other appropriate actions as the Secretary deems necessary to provide such assurance. For a device that is purported or represented to be for a use in supporting or sustaining human life, the Secretary shall examine and identify the special controls, if any, that are necessary to provide adequate assurance of safety and effectiveness and describe how such controls provide such assurance.

(C) Class III, Premarket Approval. A device which because--

(i) it (I) cannot be classified as a class I device because insufficient information exists to determine that the application of general controls are sufficient to provide reasonable assurance of the safety and effectiveness of the device, and (II) cannot be classified as a class II device because insufficient information exists to determine that the special controls described in subparagraph (B) would provide reasonable assurance of its safety and effectiveness, and

(ii)(I) is purported or represented to be for a use in supporting or sustaining human life or for a use which is of substantial importance in preventing impairment of human health, or

(II) presents a potential unreasonable risk of illness or injury,

is to be subject, in accordance with section 515, to premarket approval to provide reasonable assurance of its safety and effectiveness. If there is not sufficient information to establish a performance standard for a device to provide reasonable assurance of its safety and effectiveness, the Secretary may conduct such activities as may be necessary to develop or obtain such information.

(2) For purposes of this section and sections 514 and 515, the safety and effectiveness of a device are to be determined--

(A) with respect to the persons for whose use the device is represented or intended,

(B) with respect to the conditions of use prescribed, recommended, or suggested in the labeling of the device, and

(C) weighing any probable benefit to health from the use of the device against any probable risk of injury or illness from such use.

(3)(A) Except as authorized by subparagraph (B), the effectiveness of a device is, for purposes of this section and sections 514 and 515 to be determined, in accordance with regulations promulgated by the Secretary, on the basis of well-controlled investigations, including clinical investigations where appropriate, by experts qualified by training and experience to evaluate the effectiveness of the device, from which investigations it can fairly and responsibly be concluded by qualified experts that the device will have the effect it purports or is represented to have under the conditions of use prescribed, recommended, or suggested in the labeling of the device.

(B) If the Secretary determines that there exists valid scientific evidence (other than evidence derived from investigations described in subparagraph (A))--

(i) which is sufficient to determine the effectiveness of a device, and

(ii) from which it can fairly and responsibly be concluded by qualified experts that the device will have the effect it purports or is represented to have under the conditions of use prescribed, recommended, or suggested in the labeling of the device,

then, for purposes of this section and sections 514 and 515, the Secretary may authorize the effectiveness of the device to be determined on the basis of such evidence.

(b)(1) For purposes of--

(A) determining which devices intended for human use should be subject to the requirements of general controls, performance standards, or premarket approval, and

(B) providing notice to the manufacturers and importers of such devices to enable them to prepare for the application of such requirements to devices manufactured or imported by them, the Secretary shall classify all such devices (other than devices classified by subsection (f)) into the classes established by subsection (a). For the purpose of securing recommendations with respect to the classification of devices, the Secretary shall establish panels of experts or use panels of experts established before the date of the enactment of this section, or both. Section 14 of the Federal Advisory Committee Act shall not apply to the duration of a panel established under this paragraph.

(2) The Secretary shall appoint to each panel established under paragraph (1) persons who are qualified by training and experience to evaluate the safety and effectiveness of the devices to be referred to the panel and who, to the extent feasible, possess skill in the use of, or experience in the development, manufacture, or utilization of, such devices. The Secretary shall make appointments to each panel so that each panel shall consist of members with adequately diversified expertise in such fields as clinical and administrative medicine, engineering, biological and physical sciences, and other related professions. In addition, each panel shall include as nonvoting members a representative of consumer interests and a representative of interests of the device manufacturing industry. Scientific, trade, and consumer organizations shall be

afforded an opportunity to nominate individuals for appointment to the panels. No individual who is in the regular full-time employ of the United States and engaged in the administration of this Act may be a member of any panel. The Secretary shall designate one of the members of each panel to serve as chairman thereof.

(3) Panel members (other than officers or employees of the United States), while attending meetings or conferences of a panel or otherwise engaged in its business, shall be entitled to receive compensation at rates to be fixed by the Secretary, but not at rates exceeding the daily equivalent of the rate in effect for grade GS-18 of the General Schedule, for each day so engaged, including traveltime; and while so serving away from their homes or regular places of business each member may be allowed travel expenses (including per diem in lieu of subsistence) as authorized by section 5703(b) of title 5, United States Code, for persons in the Government service employed intermittently.

(4) The Secretary shall furnish each panel with adequate clerical and other necessary assistance.

(c)(1) The Secretary shall organize the panels according to the various fields of clinical medicine and fundamental sciences in which devices intended for human use are used. The Secretary shall refer a device to be classified under this section to an appropriate panel established or authorized to be used under subsection (b) for its review and for its recommendation respecting the classification of the device. The Secretary shall by regulation prescribe the procedure to be followed by the panels in making their reviews and recommendations. In making their reviews of devices, the panels, to the maximum extent practicable, shall provide an opportunity for interested persons to submit data and views on the classification of the devices.

(2)(A) Upon completion of a panel's review of a device referred to it under paragraph (1), the panel shall, subject to subparagraphs (B) and (C), submit to the Secretary its recommendation for the classification of the device. Any such recommendation shall (i) contain (I) a summary of the reasons for the recommendation, (II) a summary of the data upon which the recommendation is based, and (III) an identification of the risks to health (if any) presented by the device with respect to which the recommendation is made, and (ii) to the extent practicable, include a recommendation for the assignment of a priority for the application of the requirements of section 514 or 515 to a device recommended to be classified in class II or class III.

(B) A recommendation of a panel for the classification of a device in class I shall include a recommendation as to whether the device should be exempted from the requirements of section 510, 519, or 520(f).

(C) In the case of a device which has been referred under paragraph (1) to a panel, and which--

(i) is intended to be implanted in the human body or is purported or represented to be for a use in supporting or sustaining human life, and

(ii)(I) has been introduced or delivered for introduction into interstate commerce for commercial distribution before the date of enactment of this section, or

(II) is within a type of device which was so introduced or delivered before such date and is substantially equivalent to another device within that type,

such panel shall recommend to the Secretary that the device be classified in class III unless the panel determines that classification of the device in such class is not necessary to provide reasonable assurance of its safety and effectiveness. If a panel does not recommend that such a device be classified in class III, it shall in its recommendation to the Secretary for the classification of the device set forth the reasons for not recommending classification of the device in such class.

(3) The panels shall submit to the Secretary within one year of the date funds are first appropriated for the implementation of this section their recommendations respecting all devices of a type introduced or delivered for introduction into interstate commerce for commercial distribution before the date of the enactment of this section.

(d)(1) Upon receipt of a recommendation from a panel respecting a device, the Secretary shall publish in the Federal Register the panel's recommendation and a proposed regulation classifying such device and shall provide interested persons an opportunity to submit comments on such recommendation and the proposed regulation. After reviewing such comments, the Secretary shall, subject to paragraph (2), by regulation classify such device.

(2)(A) A regulation under paragraph (1) classifying a device in class I shall prescribe which, if any, of the requirements of section 510, 519 or 520(f) shall not apply to the device. A regulation which makes a requirement of section 510, 519, or 520(f) inapplicable to a device shall be accompanied by a statement of the reasons of the Secretary for making such requirement inapplicable.

(B) A device described in subsection (c)(2)(C) shall be classified in class III unless the Secretary determines that classification of the device in such class is not necessary to provide reasonable assurance of its safety and effectiveness. A proposed regulation under paragraph (1) classifying such a device in a class other than class III shall be accompanied by a full statement of the reasons of the Secretary (and supporting documentation and data) for not classifying such device in such class and an identification of the risks to health (if any) presented by such device.

(3) In the case of devices classified in class II and devices classified under this subsection in class III and described in section 515(b)(1) the Secretary may establish priorities which, in his discretion, shall be used in applying sections 514 and 515, as appropriate, to such devices.

(e)(1) Based on new information respecting a device, the Secretary may, upon his own initiative or upon petition of an interested person by regulation (A) change such device's classification, and (B) revoke, because of the change in classification, any regulation or requirement in effect under section 514 or 515 with respect to such device. In the promulgation of such a regulation respecting a device's classification, the Secretary may secure from the panel to which the device was last referred pursuant to subsection (c) a recommendation respecting the proposed change in the device's classification and shall publish in the Federal Register any recommendation submitted to the Secretary by the panel respecting such change. A regulation under this subsection changing the classification of a device from class III to class II may provide that such classification shall not take effect until the effective date of a performance standard established under section 514 for such device.

(2) By regulation promulgated under paragraph (1), the Secretary may change the classification of a device from class III--

(A) to class II if the Secretary determines that special controls would provide reasonable assurance of the safety and effectiveness of the device and that general controls would not provide reasonable assurance of the safety and effectiveness of the device, or

(B) to class I if the Secretary determines that general controls would provide reasonable assurance of the safety and effectiveness of the device.

(f)(1) Any device intended for human use which was not introduced or delivered for introduction into interstate commerce for commercial distribution before the date of the enactment of this section is classified in class III unless--

(A) the device--

(i) is within a type of device (I) which was introduced or delivered for introduction into interstate commerce for commercial distribution before such date and which is to be classified pursuant to subsection (b), or (II) which was not so introduced or delivered before such date and has been classified in class I or II, and

(ii) is substantially equivalent to another device within such type, or

(B) the Secretary in response to a petition submitted under paragraph (2) has classified such device in class I or II.

A device classified in class III under this paragraph shall be classified in that class until the effective date of an order of the Secretary under paragraph (2) classifying the device in class I or II.

(2)(A) The Secretary may initiate the reclassification of a device classified into class III under paragraph (1) of this subsection or the manufacturer or importer of a device classified under paragraph (1) may petition the Secretary (in such form and manner as he shall prescribe) for the issuance of an order classifying the device in class I or class II. Within thirty days of the filing of such a petition, the Secretary shall notify the petitioner of any deficiencies in the petition which prevent the Secretary from making a decision on the petition.

(B)(i) Upon determining that a petition does not contain any deficiency which prevents the Secretary from making a decision on the petition, the Secretary may for good cause shown refer the petition to an appropriate panel established or authorized to be used under subsection (b). A panel to which such a petition has been referred shall not later than ninety days after the referral of the petition make a recommendation to the Secretary respecting approval or denial of the petition. Any such recommendation shall contain (I) a summary of the reasons for the recommendation, (II) a summary of the data upon which the recommendation is based, and (III) an identification of the risks to health (if any) presented by the device with respect to which the petition was filed. In the case of a petition for a device which is intended to be implanted in the human body or which is purported or represented to be for a use in supporting or sustaining human life, the panel shall recommend that the petition be denied unless the panel determines that the classification in class III of the device is not necessary to provide reasonable assurance of its safety and effectiveness. If the panel recommends that such petition be approved, it shall in its recommendation to the Secretary set forth its reasons for such recommendation.

(ii) The requirements of paragraphs (1) and (2) of subsection (c) (relating to opportunities for submission of data and views and recommendations respecting priorities and exemptions from sections 510, 519, and 520(f)) shall apply with respect to consideration by panels of petitions submitted under subparagraph (A).

(C)(i) Within ninety days from the date the Secretary receives the recommendation of a panel respecting a petition (but not later than 210 days after the filing of such petition) the Secretary shall by order deny or approve the petition. If the Secretary approves the petition, the Secretary shall order the classification of the device into class I or class II in accordance with the criteria prescribed by subsection (a)(1)(A) or (a)(1)(B). In the case of a petition for a device which is intended to be implanted in the human body or which is purported or represented to be for a use in supporting or sustaining human life, the Secretary shall deny the petition unless the Secretary determines that the classification in class III of the device is not necessary to provide reasonable assurance of its safety and effectiveness. An order approving such petition shall be accompanied by a full statement of the reasons of the Secretary (and supporting documentation and data) for approving the petition and an identification of the risks to health (if any) presented by the device to which such order applies.

(ii) The requirements of paragraphs (1) and (2)(A) of subsection (d) (relating to publication of recommendations, opportunity for submission of comments, and exemption from sections 510, 519, and 520(f)) shall apply with respect to action by the Secretary on petitions submitted under subparagraph (A).

(3) If a manufacturer reports to the Secretary under section 510(k) that a device is substantially equivalent to another device--

(A) which the Secretary has classified as a class III device under subsection (b),

(B) which was introduced or delivered for introduction into interstate commerce for commercial distribution before December 1, 1990, and

(C) for which no final regulation requiring premarket approval has been promulgated under section 515(b),

the manufacturer shall certify to the Secretary that the manufacturer has conducted a reasonable search of all information known or otherwise available to the manufacturer respecting such other device and has included in the report under section 510(k) a summary of and a citation to all adverse safety and effectiveness data respecting such other device and respecting the device for which the section 510(k) report is being made and which has not been submitted to the Secretary under section 519. The Secretary may require the manufacturer to submit the adverse safety and effectiveness data described in the report.

(g) Within sixty days of the receipt of a written request of any person for information respecting the class in which a device has been classified or the requirements applicable to a device under this Act, the Secretary shall provide such person a written statement of the classification (if any) of such device and the requirements of this Act applicable to the device.

(h) For purposes of this section and sections 501, 510, 514, 515, 516, 519, and 520--

(1) a reference to "general controls" is a reference to the controls authorized by or under sections 501, 502, 510, 516, 518, 519, and 520,

(2) a reference to "class I," "class II," or "class III" is a reference to a class of medical devices described in subparagraph (A), (B), or (C) of subsection (a)(1), and

(3) a reference to a "panel under section 513" is a reference to a panel established or authorized to be used under this section.

(i)(1)(A) For purposes of determinations of substantial equivalence under subsection (f) and section 520(l), the term "substantially equivalent" or "substantial equivalence" means, with respect to a device being compared to a predicate device, that the device has the same intended use as the predicate device and that the Secretary by order has found that the device--

(i) has the same technological characteristics as the predicate device, or

(ii)(I) has different technological characteristics and the information submitted that the device is substantially equivalent to the predicate device contains information, including clinical data if deemed necessary by the Secretary, that demonstrates that the device is as safe and effective as a legally marketed device, and (II) does not raise different questions of safety and efficacy than the predicate device.

(B) For purposes of subparagraph (A), the term "different technological characteristics" means, with respect to a device being compared to a predicate device, that there is a significant change in the materials, design, energy source, or other features of the device from those of the predicate device.

(2) A device may not be found to be substantially equivalent to a predicate device that has been removed from the market at the initiative of the Secretary or that has been determined to be misbranded or adulterated by a judicial order.

(3)(A) As part of a submission under section 510(k) respecting a device, the person required to file a premarket notification under such section shall provide an adequate summary of any information respecting safety and effectiveness or state that such information will be made available upon request by any person.

(B) Any summary under subparagraph (A) respecting a device shall contain detailed information regarding data concerning adverse health effects and shall be made available to the public by the Secretary within 30 days of the issuance of a determination that such device is substantially equivalent to another device.

SEC. 514 [21 U.S.C. 360d] Performance standards

(a)(1) The special controls required by section 513(a)(1)(B) shall include performance standards for a class II device if the Secretary determines that a performance standard is necessary to provide reasonable assurance of the safety and effectiveness of the device. A class III device may also be considered a class II device for purposes of establishing a standard for the device under this section if the device has been reclassified as a class II device under a regulation under section 513(e) but such regulation provides that the reclassification is not to take effect until the effective date of such a standard for the device.

(2) A performance standard established under this section for a device--

(A) shall include provisions to provide reasonable assurance of its safe and effective performance;

(B) shall, where necessary to provide reasonable assurance of its safe and effective performance, include--

(i) provisions respecting the construction, components, ingredients, and properties of the device and its compatibility with power systems and connections to such systems,

(ii) provisions for the testing (on a sample basis or, if necessary, on an individual basis) of the device or, if it is determined that no other more practicable means are available to the Secretary to assure the conformity of the device to the standard, provisions for the testing (on a sample basis or, if necessary, on an individual basis) by the Secretary or by another person at the direction of the Secretary,

(iii) provisions for the measurement of the performance characteristics of the device,

(iv) provisions requiring that the results of each or of certain of the tests of the device required to be made under clause (ii) show that the device is in conformity with the portions of the standard for which the test or tests were required, and

(v) a provision requiring that the sale and distribution of the device be restricted but only to the extent that the sale and distribution of a device may be restricted under a regulation under section 520(e); and

(C) shall, where appropriate, require the use and prescribe the form and content of labeling for the proper installation, maintenance, operation, and use of the device.

(3) The Secretary shall provide for periodic evaluation of performance standards established under this section to determine if such standards should be changed to reflect new medical, scientific, or other technological data.

(4) In carrying out his duties under this section, the Secretary shall, to the maximum extent practicable--

(A) use personnel, facilities, and other technical support available in other Federal agencies,

(B) consult with other Federal agencies concerned with standard-setting and other nationally or internationally recognized standard-setting entities, and

(C) invite appropriate participation, through joint or other conferences, workshops, or other means, by informed persons representative of scientific, professional, industry, or consumer organizations who in his judgment can make a significant contribution.

(b)(1)(A) The Secretary shall publish in the Federal Register a notice of proposed rulemaking for the establishment, amendment, or revocation of any performance standard for a device.

(B) A notice of proposed rulemaking for the establishment or amendment of a performance standard for a device shall--

(i) set forth a finding with supporting justification that the performance standard is appropriate and necessary to provide reasonable assurance of the safety and effectiveness of the device,

(ii) set forth proposed findings with respect to the risk of illness or injury that the performance standard is intended to reduce or eliminate,

(iii) invite interested persons to submit to the Secretary, within 30 days of the publication of the notice, requests for changes in the classification of the device pursuant to section 513(e) based on new information relevant to the classification, and

(iv) invite interested persons to submit an existing performance standard for the device, including a draft or proposed performance standard, for consideration by the Secretary.

(C) A notice of proposed rulemaking for the revocation of a performance standard shall set forth a finding with supporting justification that the performance standard is no longer necessary to provide reasonable assurance of the safety and effectiveness of a device.

(D) The Secretary shall provide for a comment period of not less than 60 days.

(2) If, after publication of a notice in accordance with paragraph (1), the Secretary receives a request for a change in the classification of the device, the Secretary shall, within 60 days of the publication of the notice, after consultation with the appropriate panel under section 513, either deny the request or give notice of an intent to initiate such change under section 513(e).

(3)(A) After the expiration of the period for comment on a notice of proposed rulemaking published under paragraph (1) respecting a performance standard and after consideration of such comments and any report from an advisory committee under paragraph (5), the Secretary shall (i) promulgate a regulation establishing a performance standard and publish in the Federal Register findings on the matters referred to in paragraph (1), or (ii) publish a notice terminating the proceeding for the development of the standard together with the reasons for such termination. If a notice of termination is published the Secretary shall (unless such notice is issued because the device is a banned device under section 516) initiate a proceeding under section 513(e) to reclassify the device subject to the proceeding terminated by such notice.

(B) A regulation establishing a performance standard shall set forth the date or dates upon which the standard shall take effect, but no such regulation may take effect before one year after the date of its publication unless (i) the Secretary determines that an earlier effective date is necessary for the protection of the public health and safety, or (ii) such standard has been established for a device which, effective upon the effective date of the standard, has been reclassified from class III to class II. Such date or dates shall be established so as to minimize, consistent with the public health and safety, economic loss to, and disruption or dislocation of, domestic and international trade.

(4)(A) The Secretary, upon his own initiative or upon petition of an interested person may by regulation, promulgated in accordance with the requirements of paragraphs (1), (2), and (3)(B) of this subsection, amend or revoke a performance standard.

(B) The Secretary may declare a proposed amendment of a performance standard to be effective on and after its publication in the Federal Register and until the effective date of any final action taken on such amendment if he determines that making it so effective is in the public interest. A proposed amendment of a performance standard made so effective under the preceding sentence may not prohibit, during the period in which it is so effective, the introduction or delivery for introduction into interstate commerce of a device which conforms to such standard without the change or changes provided by such proposed amendment.

(5)(A) The Secretary--

(i) may on his own initiative refer a proposed regulation for the establishment, amendment, or revocation of a performance standard, or

(ii) shall, upon the request of an interested person which demonstrates good cause for referral and which is made before the expiration of the period for submission of comments on such proposed regulation

refer such proposed regulation, to an advisory committee of experts, established pursuant to subparagraph (B) for a report and recommendation with respect to any matter involved in the proposed regulation which requires the exercise of scientific judgment. If a proposed regulation is referred under this subparagraph to an advisory committee, the Secretary shall provide the advisory committee with the data and information on which such proposed regulation is based. The advisory committee shall, within sixty days of the referral of a proposed regulation and after independent study of the data and information furnished to it by the Secretary and other data and information before it, submit to the Secretary a report and recommendation respecting such regulation, together with all underlying data and information and a statement of the reason or basis for the recommendation. A copy of such report and recommendation shall be made public by the Secretary.

(B) The Secretary shall establish advisory committees (which may not be panels under section 513) to receive referrals under subparagraph (A). The Secretary shall appoint as members of any such advisory committee persons qualified in the subject matter to be referred to the committee and of appropriately diversified professional background, except that the Secretary may not appoint to such a committee any individual who is in the regular full-time employ of the United States and engaged in the administration of this Act. Each such committee shall include as nonvoting members a representative of consumer interests and a representative of interests of the device manufacturing industry. Members of an advisory committee who are not officers or employees of the United States, while attending conferences or meetings of their committee or otherwise serving at the request of the Secretary, shall be entitled to receive compensation at rates to be fixed by the Secretary, which rates may not exceed the daily equivalent of the rate in effect for grade GS-18 of the General Schedule, for each day (including traveltime) they are so engaged; and while so serving away from their homes or regular places of business each member may be allowed travel expenses including per diem in lieu of subsistence, as authorized by section 5703 of title 5 of the United States Code for persons in the Government service employed intermittently. The Secretary shall designate one of the members of each advisory committee to serve as chairman thereof. The Secretary shall furnish each advisory committee with clerical and other assistance, and shall by regulation prescribe the procedures to be followed by each such committee in acting on referrals made under subparagraph (A).

SEC. 515 [21 U.S.C. 360e] Premarket approval

(a) A class III device--

(1) which is subject to a regulation promulgated under subsection (b); or

(2) which is a class III device because of section 513(f),

is required to have, unless exempt under section 520(g), an approval under this section of an application for premarket approval.

(b)(1) In the case of a class III device which--

(A) was introduced or delivered for introduction into interstate commerce for commercial distribution before the date of enactment of this section; or

(B) is (i) of a type so introduced or delivered, and (ii) is substantially equivalent to another device within that type;

the Secretary shall by regulation, promulgated in accordance with this subsection, require that such device have an approval under this section of an application for premarket approval.

(2)(A) A proceeding for the promulgation of a regulation under paragraph (1) respecting a device shall be initiated by the publication in the Federal Register of a notice of proposed rulemaking. Such notice shall contain--

(i) the proposed regulation;

(ii) proposed findings with respect to the degree of risk of illness or injury designed to be eliminated or reduced by requiring the device to have an approved application for premarket approval and the benefit to the public from use of the device;

(iii) opportunity for the submission of comments on the proposed regulation and the proposed findings; and

(iv) opportunity to request a change in the classification of the device based on new information relevant to the classification of the device.

(B) If, within fifteen days after publication of a notice under subparagraph (A), the Secretary receives a request for a change in the classification of a device, he shall, within sixty days of the publication of such notice and after consultation with the appropriate panel under section 513, by order published in the Federal Register, either deny the request for change in classification or give notice of his intent to initiate such a change under section 513(e).

(3) After the expiration of the period for comment on a proposed regulation and proposed findings published under paragraph (2) and after consideration of comments submitted on such proposed regulation and findings, the Secretary shall (A) promulgate such regulation and publish in the Federal Register findings on the matters referred to in paragraph (2)(A)(ii), or (B) publish a notice terminating the proceeding for the promulgation of the regulation together with the reasons for such termination. If a notice of termination is published the Secretary shall (unless such notice is issued because the device is a banned device under section 516) initiate a proceeding under section 513(e) to reclassify the device subject to the proceeding terminated by such notice.

(4) The Secretary, upon his own initiative or upon petition of an interested person, may by regulation amend or revoke any regulation promulgated under this subsection. A regulation to amend or revoke a regulation under this subsection shall be promulgated in accordance with the requirements prescribed by this subsection for the promulgation of the regulation to be amended or revoked.

(c)(1) Any person may file with the Secretary an application for premarket approval for a class III device. Such an application for a device shall contain--

(A) full reports of all information, published or known to or which should reasonably be known to the applicant, concerning investigations which have been made to show whether or not such device is safe and effective;

(B) a full statement of the components, ingredients, and properties and of the principle or principles of operation, of such device;

(C) a full description of the methods used in, and the facilities and controls used for, the manufacture, processing, and, when relevant, packing and installation of, such device;

(D) an identifying reference to any performance standard under section 514 which would be applicable to any aspect of such device if it were a class II device, and either adequate information to show that such aspect of such device fully meets such performance standard or adequate information to justify any deviation from such standard;

(E) such samples of such device and of components thereof as the Secretary may reasonably require, except that where the submission of such samples is impracticable or unduly burdensome, the requirement of this subparagraph may be met by the submission of complete information concerning the location of one or more such devices readily available for examination and testing;

(F) specimens of the labeling proposed to be used for such device; and

(G) such other information relevant to the subject matter of the application as the Secretary, with the concurrence of the appropriate panel under section 513, may require.

(2) Upon receipt of an application meeting the requirements set forth in paragraph (1), the Secretary--

(A) may on the Secretary's own initiative refer such application, or

(B) shall, upon the request of an applicant unless the Secretary finds that the information in the application which would be reviewed by a panel substantially duplicates information which has previously been reviewed by a panel appointed under section 513,

refer such application to the appropriate panel under section 513 for study and for submission (within such period as he may establish) of a report and recommendation respecting approval of the application together with all underlying data and the reasons or basis for the recommendation.

(d)(1)(A) As promptly as possible, but in no event later than one hundred and eighty days after the receipt of an application under subsection (c) (except as provided in section 520(l)(3)(D)(ii) or unless, in accordance with subparagraph (B)(i), an additional period as agreed upon by the Secretary and the applicant), the Secretary, after considering the report and recommendation submitted under paragraph (2) of such subsection, shall--

(i) issue an order approving the application if he finds that none of the grounds for denying approval specified in paragraph (2) of this subsection applies; or

(ii) deny approval of the application if he finds (and sets forth the basis for such finding as part of or accompanying such denial) that one or more grounds for denial specified in paragraph (2) of this subsection apply.

(B)(i) The Secretary may not enter into an agreement to extend the period in which to take action with respect to an application submitted for a device subject to a regulation promulgated under subsection (b) unless he finds that the continued availability of the device is necessary for the public health.

(ii) An order approving an application for a device may require as a condition to such approval that the sale and distribution of the device be restricted but only to the extent that the sale and distribution of a device may be restricted under a regulation under section 520(e).

(2) The Secretary shall deny approval of an application for a device if, upon the basis of the information submitted to the Secretary as part of the application and any other information before him with respect to such device, the Secretary finds that--

(A) there is a lack of a showing of reasonable assurance that such device is safe under the conditions of use prescribed, recommended, or suggested in the proposed labeling thereof;

(B) there is a lack of a showing of reasonable assurance that the device is effective under the conditions of use prescribed, recommended, or suggested in the proposed labeling thereof;

(C) the methods used in, or the facilities or controls used for, the manufacture, processing, packing, or installation of such device do not conform to the requirements of section 520(f);

(D) based on a fair evaluation of all material facts, the proposed labeling is false or misleading in any particular; or

(E) such device is not shown to conform in all respects to a performance standard in effect under section 514 compliance with which is a condition to approval of the application and there is a lack of adequate information to justify the deviation from such standard.

Any denial of an application shall, insofar as the Secretary determines to be practicable, be accompanied by a statement informing the applicant of the measures required to place such application in approvable form (which measures may include further research by the applicant in accordance with one or more protocols prescribed by the Secretary).

(3) An applicant whose application has been denied approval may, by petition filed on or before the thirtieth day after the date upon which he receives notice of such denial, obtain review thereof in accordance with either paragraph (1) or (2) of subsection (g), and any interested person may obtain review, in accordance with paragraph (1) or (2) of subsection (g), of an order of the Secretary approving an application.

(e)(1) The Secretary shall, upon obtaining, where appropriate, advice on scientific matters from a panel or panels under section 513 and after due notice and opportunity for informal hearing to the holder of an approved application for a device, issue an order withdrawing approval of the application if the Secretary finds--

(A) that such device is unsafe or ineffective under the conditions of use prescribed, recommended, or suggested in the labeling thereof;

(B) on the basis of new information before him with respect to such device, evaluated together with the evidence available to him when the application was approved, that there is a lack of a showing of reasonable assurance that the device is safe or effective under the conditions of use prescribed, recommended, or suggested in the labeling thereof;

(C) that the application contained or was accompanied by an untrue statement of a material fact;

(D) that the applicant (i) has failed to establish a system for maintaining records, or has repeatedly or deliberately failed to maintain records or to make reports, required by an applicable regulation under section 519(a), (ii) has refused to permit access to or copying or verification of, such records as required by section 704, or (iii) has not complied with the requirements of section 510;

(E) on the basis of new information before him with respect to such device, evaluated together with the evidence before him when the application was approved, that the methods used in, or the facilities and controls used for, the manufacture, processing, packing, or installation of such device do not conform with the requirements of section 520(f) and were not brought into conformity with such requirements within a reasonable time after receipt of written notice from the Secretary of nonconformity;

(F) on the basis of new information before him, evaluated together with the evidence before him when the application was approved, that the labeling of such device, based on a fair evaluation of all material facts, is false or misleading in any particular and was not corrected within a reasonable time after receipt of written notice from the Secretary of such fact; or

(G) on the basis of new information before him, evaluated together with the evidence before him when the application was approved, that such device is not shown to conform in all respects to a performance standard which is in effect under section 514 compliance with which was a condition to approval of the application and that there is a lack of adequate information to justify the deviation from such standard.

(2) The holder of an application subject to an order issued under paragraph (1) withdrawing approval of the application may, by peti-

tion filed on or before the thirtieth day after the date upon which he receives notice of such withdrawal, obtain review thereof in accordance with either paragraph (1) or (2) of subsection (g).

(3) If, after providing an opportunity for an informal hearing, the Secretary determines there is reasonable probability that the continuation of distribution of a device under an approved application would cause serious, adverse health consequences or death the Secretary shall by order temporarily suspend the approval of the application approved under this section. If the Secretary issues such an order, the Secretary shall proceed expeditiously under paragraph (1) to withdraw such application.

(f)(1) In the case of a class III device which is required to have an approval of an application submitted under subsection (c), such device shall be considered as having such an approval if a notice of completion of testing conducted in accordance with a product development protocol approved under paragraph (4) has been declared completed under paragraph (6).

(2) Any person may submit to the Secretary a proposed product development protocol with respect to a device. Such a protocol shall be accompanied by data supporting it. If, within thirty days of the receipt of such a protocol, the Secretary determines that it appears to be appropriate to apply the requirements of this subsection to the device with respect to which the protocol is submitted, he shall refer the proposed protocol to the appropriate panel under section 513 for its recommendation respecting approval of the protocol.

(3) A proposed product development protocol for a device may be approved only if--

(A) the Secretary determines that it is appropriate to apply the requirements of this subsection to the device in lieu of the requirement of approval of an application submitted under subsection (c); and

(B) the Secretary determines that the proposed protocol provides--

(i) a description of the device and the changes which may be made in the device,

(ii) a description of the preclinical trials (if any) of the device and a specification of (I) the results from such trials to be required before the commencement of clinical trials of the device, and (II) any permissible variations in preclinical trials and the results therefrom,

(iii) a description of the clinical trials (if any) of the device and a specification of (I) the results from such trials to be required before the filing of a notice of completion of the requirements of the protocol, and (II) any permissible variations in such trials and the results therefrom,

(iv) a description of the methods to be used in, and the facilities and controls to be used for, the manufacture, processing, and when relevant, packing and installation of the device,

(v) an identifying reference to any performance standard under section 514 to be applicable to any aspect of such device,

(vi) if appropriate, specimens of the labeling proposed to be used for such device,

(vii) such other information relevant to the subject matter of the protocol as the Secretary, with the concurrence of the appropriate panel or panels under section 513, may require, and

(viii) a requirement for submission of progress reports and, when completed, records of the trials conducted under the protocol which records are adequate to show compliance with the protocol.

(4) The Secretary shall approve or disapprove a proposed product development protocol submitted under paragraph (2) within one hundred and twenty days of its receipt unless an additional period is agreed upon by the Secretary and the person who submitted the protocol. Approval of a protocol or denial of approval of a protocol is final agency action subject to judicial review under chapter 7 of title 5, United States Code.

(5) At any time after a product development protocol for a device has been approved pursuant to paragraph (4), the person for whom the protocol was approved may submit a notice of completion--

(A) stating (i) his determination that the requirements of the protocol have been fulfilled and that, to the best of his knowledge, there is no reason bearing on safety or effectiveness why the notice of completion should not become effective, and (ii) the data and other information upon which such determination was made, and

(B) setting forth the results of the trials required by the protocol and all the information required by subsection (c)(1).

(6)(A) The Secretary may, after providing the person who has an approved protocol an opportunity for an informal hearing and at any time prior to receipt of notice of completion of such protocol, issue a final order to revoke such protocol if he finds that--

(i) such person has failed substantially to comply with the requirements of the protocol,

(ii) the results of the trials obtained under the protocol differ so substantially from the results required by the protocol that further trials cannot be justified, or

(iii) the results of the trials conducted under the protocol or available new information do not demonstrate that the device tested under the protocol does not present an unreasonable risk to health and safety.

(B) After the receipt of a notice of completion of an approved protocol the Secretary shall, within the ninety-day period beginning on the date such notice is received, by order either declare the protocol completed or declare it not completed. An order declaring a protocol not completed may take effect only after the Secretary has provided the person who has the protocol opportunity for an informal hearing on the order. Such an order may be issued only if the Secretary finds--

(i) such person has failed substantially to comply with the requirements of the protocol,

(ii) the results of the trials obtained under the protocol differ substantially from the results required by the protocol, or

(iii) there is a lack of a showing of reasonable assurance of the safety and effectiveness of the device under the conditions of use prescribed, recommended, or suggested in the proposed labeling thereof.

(C) A final order issued under subparagraph (A) or (B) shall be in writing and shall contain the reasons to support the conclusions thereof.

(7) At any time after a notice of completion has become effective, the Secretary may issue an order (after due notice and opportunity for an informal hearing to the person for whom the notice is effective) revoking the approval of a device provided by a notice of completion which has become effective as provided in subparagraph (B) if he finds that any of the grounds listed in subparagraphs (A) through (G) of subsection (e)(1) of this section apply. Each reference in such subparagraphs to an application shall be considered for purposes of this paragraph as a reference to a protocol and the notice of completion of such protocol, and each reference to the time when an application was approved shall be considered for purposes of this paragraph as a reference to the time when a notice of completion took effect.

(8) A person who has an approved protocol subject to an order issued under paragraph (6)(A) revoking such protocol, a person who has an approved protocol with respect to which an order under paragraph (6)(B) was issued declaring that the protocol had not been completed, or a person subject to an order issued under paragraph (7) revoking the approval of a device may, by petition filed on or before the thirtieth day after the date upon which he receives notice of such order, obtain review thereof in accordance with either paragraph (1) or (2) of subsection (g).

(g)(1) Upon petition for review of--

(A) an order under subsection (d) approving or denying approval of an application or an order under subsection (e) withdrawing approval of an application, or

(B) an order under subsection (f)(6)(A) revoking an approved protocol, under subsection (f)(6)(B) declaring that an approved protocol has not been completed, or under subsection (f)(7) revoking the approval of a device,

the Secretary shall, unless he finds the petition to be without good cause or unless a petition for review of such order has been submitted under paragraph (2), hold a hearing, in accordance with section 554 of title 5 of the United States Code, on the order. The panel or panels which considered the application, protocol, or device subject to such order shall designate a member to appear and testify at any such hearing upon request of the Secretary, the petitioner, or the officer conducting the hearing, but this requirement does not preclude any other member of the panel or panels from appearing and testifying at any such hearing.

Upon completion of such hearing and after considering the record established in such hearing, the Secretary shall issue an order either affirming the order subject to the hearing or reversing such order and, as appropriate, approving or denying approval of the application, reinstating the application's approval, approving the protocol, or placing in effect a notice of completion.

(2)(A) Upon petition for review of--

(i) an order under subsection (d) approving or denying approval of an application or an order under subsection (e) withdrawing approval of an application, or

(ii) an order under subsection (f)(6)(A) revoking an approved protocol, under subsection (f)(6)(B) declaring that an approved protocol has not been completed, or under subsection (f)(7) revoking the approval of a device,

the Secretary shall refer the application or protocol subject to the order and the basis for the order to an advisory committee of experts established pursuant to subparagraph (B) for a report and recommendation with respect to the order. The advisory committee shall, after independent study of the data and information furnished to it by the Secretary and other data and information before it, submit to the Secretary a report and recommendation, together with all underlying data and information and a statement of the reasons or basis for the recommendation. A copy of such report shall be promptly supplied by the Secretary to any person who petitioned for such referral to the advisory committee.

(B) The Secretary shall establish advisory committees (which may not be panels under section 513) to receive referrals under subparagraph (A). The Secretary shall appoint as members of any such advisory committee persons qualified in the subject matter to be referred to the committee and of appropriately diversified professional backgrounds, except that the Secretary may not appoint to such a committee any individual who is in the regular full-time employ of the United States and engaged in the administration of this Act. Members of an advisory committee (other than officers or employees of the United States), while attending conferences or meetings of their committee or otherwise serving at the request of the Secretary, shall be entitled to receive compensation at rates to be fixed by the Secretary which rates may not exceed the daily equivalent for grade GS-18 of the General Schedule for each day (including traveltime) they are so engaged; and while so serving away from their homes or regular places of business each member may be allowed travel expenses including per diem in lieu of subsistence, as authorized by section 5703 of title 5 of the United States Code for persons in the Government service employed intermittently. The Secretary shall designate the chairman of an advisory committee from its members. The Secretary shall furnish each advisory committee with clerical and other assistance, and shall by regulation prescribe the procedures to be followed by each such committee in acting on referrals made under subparagraph (A).

(C) The Secretary shall make public the report and recommendation made by an advisory committee with respect to an application and shall by order, stating the reasons therefor, either affirm the order referred to the advisory committee or reverse such order and, if appropriate, approve or deny approval of the application, reinstate the application's approval, approve the protocol, or place in effect a notice of completion.

(h) Orders of the Secretary under this section shall be served (1) in person by any officer or employee of the department designated by the Secretary, or (2) by mailing the order by registered mail or certified mail addressed to the applicant at his last known address in the records of the Secretary.

(i)(1) Before December 1, 1995, the Secretary shall by order require manufacturers of devices, which were introduced or delivered for introduction into interstate commerce for commercial distribution before May 28, 1976, and which are subject to revision of classification under paragraph (2), to submit to the Secretary a summary of and citation to any information known or otherwise available to the manufacturer respecting such devices, including adverse safety or effectiveness information which has not been submitted under section 519. The Secretary may require the manufacturer to submit the adverse safety or effectiveness data for which a summary and citation were submitted, if such data are available to the manufacturer.

(2) After the issuance of an order under paragraph (1) but before December 1, 1995, the Secretary shall publish a regulation in the Federal Register for each device--

(A) which the Secretary has classified as a class III device, and

(B) for which no final regulation has been promulgated under section 515(b),

revising the classification of the device so that the device is classified into class I or class II, unless the regulation requires the device to remain in class III. In determining whether to revise the classification of a device or to require a device to remain in class III, the Secretary shall apply the criteria set forth in section 513(a). Before the publication of a regulation requiring a device to remain in class III or revising its classification, the Secretary shall publish a proposed regulation respecting the classification of a device under this paragraph and provide reasonable opportunity for the submission of comments on any such regulation. No regulation requiring a device to remain in class III or revising its classification may take effect before the expiration of 90 days from the date of its publication in the Federal Register as a proposed regulation.

(3) The Secretary shall, as promptly as is reasonably achievable, but not later than 12 months after the effective date of the regulation requiring a device to remain in class III, establish a schedule for the promulgation of a section 515(b) regulation for each device which is subject to the regulation requiring the device to remain in class III.

SEC. 516 [21 U.S.C. 360f] Banned devices

(a) Whenever the Secretary finds, on the basis of all available data and information, that--

(1) a device intended for human use presents substantial deception or an unreasonable and substantial risk of illness or injury; and

(2) in the case of substantial deception or an unreasonable and substantial risk of illness or injury which the Secretary determined could be corrected or eliminated by labeling or change in labeling and with respect to which the Secretary provided written notice to the manufacturer specifying the deception or risk of illness or injury, the labeling or change in labeling to correct the deception or eliminate or reduce such risk, and the period within which such labeling or change in labeling was to be done, such labeling or change in labeling was not done within such period;

he may initiate a proceeding to promulgate a regulation to make such device a banned device.

(b) The Secretary may declare a proposed regulation under subsection (a) to be effective upon its publication in the Federal Register and until the effective date of any final action taken respecting such regulation if (1) he determines, on the basis of all available data and information, that the deception or risk of illness or injury associated with the use of the device which is subject to the regulation presents an unreasonable, direct, and substantial danger to the health of individuals, and (2) before the date of the publication of such regulation, the Secretary notifies the manufacturer of such device that such regulation is to be made so effective. If the Secretary makes a proposed regulation so effective, he shall, as expeditiously as possible, give interested persons prompt notice of his action under this subsection, provide reasonable opportunity for an informal hearing on the proposed regulation, and either affirm, modify, or revoke such proposed regulation.

SEC. 517 [21 U.S.C. 360g] Judicial review

(a) Not later than thirty days after--

(1) the promulgation of a regulation under section 513 classifying a device in class I or changing the classification of a device to class I or an order under subsection (f)(2) of such section reclassifying a device or denying a petition for reclassification of a device,

(2) the promulgation of a regulation under section 514 establishing, amending, or revoking a performance standard for a device,

(3) the issuance of an order under section 514(b)(2) or 515(b)(2)(B) denying a request

for reclassification of a device,

(4) the promulgation of a regulation under paragraph (3) of section 515(b) requiring a device to have an approval of a premarket application, a regulation under paragraph (4) of that section amending or revoking a regulation under paragraph (3), or an order pursuant to section 515(g)(1) or 515(g)(2)(C),

(5) the promulgation of a regulation under section 516 (other than a proposed regulation made effective under subsection (b) of such section upon the regulation's publication) making a device a banned device,

(6) the issuance of an order under section 520(f)(2),

(7) an order under section 520(g)(4) disapproving an application for an exemption of a device for investigational use or an order under section 520(g)(5) withdrawing such an exemption for a device,

(8) an order pursuant to section 513(i),

(9) a regulation under section 515(i)(2) or 520(l)(5)(B), or

(10) an order under section 520(h)(4)(B),

any person adversely affected by such regulation or order may file a petition with the United States Court of Appeals for the District of Columbia or for the circuit wherein such person resides or has his principal place of business for judicial review of such regulation or order. A copy of the petition shall be transmitted by the clerk of the court to the Secretary or other officer designated by him for that purpose. The Secretary shall file in the court the record of the proceedings on which the Secretary based his regulation or order as provided in section 2112 of title 28, United States Code. For purposes of this section, the term "record" means all notices and other matter published in the Federal Register with respect to the regulation or order reviewed, all information submitted to the Secretary with respect to such regulation or order, proceedings of any panel or advisory committee with respect to such regulation or order, any hearing held with respect to such regulation or order, and any other information identified by the Secretary, in the administrative proceeding held with respect to such regulation or order, as being relevant to such regulation or order.

(b) If the petitioner applies to the court for leave to adduce additional data, views, or arguments respecting the regulation or order being reviewed and shows to the satisfaction of the court that such additional data, views, or arguments are material and that there were reasonable grounds for the petitioner's failure to adduce such data, views, or arguments in the proceedings before the Secretary, the court may order the Secretary to provide additional opportunity for the oral presentation of data, views, or arguments and for written submissions. The Secretary may modify his findings, or make new findings by reason of the additional data, views, or arguments so taken and shall file with the court such modified or new findings, and his recommendation, if any, for the modification or setting aside of the regulation or order being reviewed, with the return of such additional data, views, or arguments.

(c) Upon the filing of the petition under subsection (a) of this section for judicial review of a regulation or order, the court shall have jurisdiction to review the regulation or order in accordance with chapter 7 of title 5, United States Code, and to grant appropriate relief, including interim relief, as provided in such chapter. A regulation described in paragraph (2) or (5) of subsection (a) and an order issued after the review provided by section 515(g) shall not be affirmed if it is found to be unsupported by substantial evidence on the record taken as a whole.

(d) The judgment of the court affirming or setting aside, in whole or in part, any regulation or order shall be final, subject to review by the Supreme Court of the United States upon certiorari or certification, as provided in section 1254 of title 28 of the United States Code.

(e) The remedies provided for in this section shall be in addition to and not in lieu of any other remedies provided by law.

(f) To facilitate judicial review under this section or under any other provision of law of a regulation or order issued under section 513, 514, 515, 516, 518, 519, 520, or 521 each such regulation or order shall contain a statement of the reasons for its issuance and the basis, in the record of the proceedings held in connection with its issuance, for its issuance.

SEC. 518 [21 U.S.C. 360h] Notification and other remedies

(a) If the Secretary determines that--

(1) a device intended for human use which is introduced or delivered for introduction into interstate commerce for commercial distribution presents an unreasonable risk of substantial harm to the public health, and

(2) notification under this subsection is necessary to eliminate the unreasonable risk of such harm and no more practicable means is available under the provisions of this Act (other than this section) to eliminate such risk,

the Secretary may issue such order as may be necessary to assure that adequate notification is provided in an appropriate form, by the persons and means best suited under the circumstances involved, to all health professionals who prescribe or use the device and to any other person (including manufacturers, importers, distributors, retailers, and device users) who should properly receive such notification in order to eliminate such risk. An order under this subsection shall require that the individuals subject to the risk with respect to which the order is to be issued be included in the persons to be notified of the risk unless the Secretary determines that notice to such individuals would present a greater danger to the health of such individuals than no such notification. If the Secretary makes such a determination with respect to such individuals the order shall require that the health professionals who prescribe or use the device provide for the notification of the individuals whom the health professionals treated with the device of the risk presented by the device and of any action which may be taken by or on behalf of such individuals to eliminate or reduce such risk. Before issuing an order under this subsection, the Secretary shall consult with the persons who are to give notice under the order.

(b)(1)(A) If, after affording opportunity for an informal hearing, the Secretary determines that--

(i) a device intended for human use which is introduced or delivered for introduction into interstate commerce for commercial distribution presents an unreasonable risk of substantial harm to the public health.

(ii) there are reasonable grounds to believe that the device was not properly designed or manufactured with reference to the state of the art as it existed at the time of its design or manufacture,

(iii) there are reasonable grounds to believe that the unreasonable risk was not caused by failure of a person other than a manufacturer, importer, distributor, or retailer of the device to exercise due care in the installation, maintenance, repair, or use of the device, and

(iv) the notification authorized by subsection (a) would not by itself be sufficient to eliminate the unreasonable risk and action described in paragraph (2) of this subsection is necessary to eliminate such risk,

the Secretary may order the manufacturer, importer, or any distributor of such device, or any combination of such persons, to submit to him within a reasonable time a plan for taking one or more of the actions described in paragraph (2). An order issued under the preceding sentence which is directed to more than one person shall specify which person may decide which action shall be taken under such plan and the person specified shall be the person who the Secretary determines bears the principal, ultimate financial responsibility for action taken under the plan unless the Secretary cannot determine who bears such responsibility or the Secretary determines that the protection of the public health requires that such decision be made by a person (including a device user or health professional) other than the person he determines bears such responsibility.

(B) The Secretary shall approve a plan submitted pursuant to an order issued under subparagraph (A) unless he determines (after affording opportunity for an informal hearing) that the action or actions to be taken under the plan or the manner in which such action or actions are to be taken under the plan will not assure that the unreasonable risk with respect to which such order was issued will be eliminated. If the Secretary disapproves a plan, he shall order a revised plan to be submitted to him within a reasonable time. If the Secretary determines (after affording opportunity for an informal hearing) that the revised plan is unsatisfactory or if no revised plan or no initial plan has been submitted to the Secretary within the prescribed time, the Secretary shall (i) prescribe a plan to be carried out by the person or persons to whom the order issued under subparagraph (A) was directed, or (ii) after affording an opportunity for an informal hearing, by order prescribe a plan to be carried out by a person who is a manufacturer, importer, distributor, or retailer of the device with respect to which the order was issued but to whom the order under subparagraph (A) was not directed.

(2) The actions which may be taken under a plan submitted under an order issued under paragraph (1) are as follows:

(A) To repair the device so that it does not present the unreasonable risk of substantial harm with respect to which the order under paragraph (1) was issued.

(B) To replace the device with a like or equivalent device which is in conformity with all applicable requirements of this Act.

(C) To refund the purchase price of the device (less a reasonable allowance for use if such device has been in the possession of the device user for one year or more--

(i) at the time of notification ordered under subsection (a), or

(ii) at the time the device user receives actual notice of the unreasonable risk with respect to which the order was issued under paragraph (1), whichever first occurs).

(3) No charge shall be made to any person (other than a manufacturer importer, distributor or retailer) for availing himself of any remedy, described in paragraph (2) and provided under an order issued under paragraph (1), and the person subject to the order shall reimburse each person (other than a manufacturer, importer, distributor, or retailer) who is entitled to such a remedy for any reasonable and foreseeable expenses actually incurred by such person in availing himself of such remedy.

(c) An order issued under subsection (b) with respect to a device may require any person who is a manufacturer, importer, distributor, or retailer of the device to reimburse any other person who is a manufacturer, importer, distributor, or retailer of such device for such other person's expenses actually incurred in connection with carrying out the order if the Secretary determines such reimbursement is required for the protection of the public health. Any such requirement shall not affect any rights or obligations under any contract to which the person receiving reimbursement or the person making such reimbursement is a party.

(d) Compliance with an order issued under this section shall not relieve any person from liability under Federal or State law. In awarding damages for economic loss in an action brought for the enforcement of any such liability, the value to the plaintiff in such action of any remedy provided him under such order shall be taken into account.

(e)(1) If the Secretary finds that there is a reasonable probability that a device intended for human use would cause serious, adverse health consequences or death, the Secretary shall issue an order requiring the appropriate person (including the manufacturers, importers, distributors, or retailers of the device)--

(A) to immediately cease distribution of such device, and

(B) to immediately notify health professionals and device user facilities of the order and to instruct such professionals and facilities to cease use of such device.

The order shall provide the person subject to the order with an opportunity for an informal hearing, to be held not later than 10 days after the date of the issuance of the order, on the actions required by the order and on whether the order should be amended to require a recall of such device. If, after providing an opportunity for such a hearing, the Secretary determines that inadequate grounds exist to support the actions required by the order, the Secretary shall vacate the order.

(2)(A) If, after providing an opportunity for an informal hearing under paragraph (1), the Secretary determines that the order should be amended to include a recall of the device with respect to which the order was issued, the Secretary shall, except as provided in subparagraphs (B) and (C), amend the order to require a recall. The Secretary shall specify a timetable in which the device recall will occur and shall require periodic reports to the Secretary describing the progress of the recall.

(B) An amended order under subparagraph (A)--

(i) shall--

(I) not include recall of a device from individuals, and

(II) not include recall of a device from device user facilities if the Secretary determines that the risk of recalling such device from the facilities presents a greater health risk than the health risk of not recalling the device from use, and

(ii) shall provide for notice to individuals subject to the risks associated with the use of such device.

In providing the notice required by clause (ii), the Secretary may use the assistance of health professionals who prescribed or used such a device for individuals. If a significant number of such individuals cannot be identified, the Secretary shall notify such individuals pursuant to section 705(b).

(3) The remedy provided by this subsection shall be in addition to remedies provided by subsections (a), (b), and (c).

SEC. 519 [21 U.S.C. 360i] Records and reports on devices

(a) Every person who is a manufacturer, importer, or distributor of a device intended for human use shall establish and maintain such records, make such reports, and provide such information, as the Secretary may by regulation reasonably require to assure that such device is not adulterated or misbranded and to otherwise assure its safety and effectiveness. Regulations prescribed under the preceding sentence--

(1) shall require a device manufacturer or importer to report to the Secretary whenever the manufacturer or importer receives or otherwise becomes aware of information that reasonably suggests that one of its marketed devices--

(A) may have caused or contributed to a death or serious injury, or

(B) has malfunctioned and that such device or a similar device marketed by the manufacturer or importer would be likely to cause or contribute to a death or serious injury if the malfunction were to recur;

(2) shall define the term "serious injury" to mean an injury that--

(A) is life threatening,

(B) results in permanent impairment of a body function or permanent damage to a body structure, or

(C) necessitates medical or surgical intervention to preclude permanent impairment of a body function or permanent damage to a body structure;

(3) shall require reporting of other significant adverse device experiences as determined by the Secretary to be necessary to be reported;

(4) shall not impose requirements unduly burdensome to a device manufacturer, importer, or distributor taking into account his cost of complying with such requirements and the need for the protection of the public health and the implementation of this Act;

(5) which prescribe the procedure for making requests for reports or information shall require that each request made under such regulations for submission of a report or information to the Secretary state the reason or purpose for such request and identify to the fullest extent practicable such report or information;

(6) which require submission of a report or information to the Secretary shall state the reason or purpose for the submission of such report or information and identify to the fullest extent practicable such report or information;

(7) may not require that the identity of any patient be disclosed in records, reports, or information required under this subsection unless required for the medical welfare of an individual, to determine the safety or effectiveness of a device, or to verify a record, report, or information submitted under this Act;

(8) may not require a manufacturer, importer, or distributor of a class I device to--

(A) maintain for such a device records respecting information not in the possession of the manufacturer, importer, or distributor, or

(B) to submit for such a device to the Secretary any report or information--

(i) not in the possession of the manufacturer,

importer, or distributor, or

(ii) on a periodic basis,

unless such report or information is necessary to determine if the device should be reclassified or if the device is adulterated or misbranded; and

(9) shall require distributors who submit such reports to submit copies of the reports to the manufacturer of the device for which the report was made.

In prescribing such regulations, the Secretary shall have due regard for the professional ethics of the medical profession and the interests of patients. The prohibitions of paragraph (4) of this subsection continue to apply to records, reports, and information concerning any individual who has been a patient, irrespective of whether or when he ceases to be a patient.

(b)(1)(A) Whenever a device user facility receives or otherwise becomes aware of information that reasonably suggests that a device has or may have caused or contributed to the death of a patient of the facility, the facility shall, as soon as practicable but not later than 10 working days after becoming aware of the information, report the information to the Secretary and, if the identity of the manufacturer is known, to the manufacturer of the device. In the case of deaths, the Secretary may by regulation prescribe a shorter period for the reporting of such information.

(B) Whenever a device user facility receives or otherwise becomes aware of--

(i) information that reasonably suggests that a device has or may have caused or contributed to the serious illness of, or serious injury to, a patient of the facility, or

(ii) other significant adverse device experiences as determined by the Secretary by regulation to be necessary to be reported,

the facility shall, as soon as practicable but not later than 10 working days after becoming aware of the information, report the information to the manufacturer of the device or to the Secretary if the identity of the manufacturer is not known.

(C) Each device user facility shall submit to the Secretary on a semi-annual basis a summary of the reports made under subparagraphs (A) and (B). Such summary shall be submitted on January 1 and July 1 of each year. The summary shall be in such form and contain such information from such reports as the Secretary may require and shall include--

(i) sufficient information to identify the facility which made the reports for which the summary is submitted,

(ii) in the case of any product which was the subject of a report, the product name, serial number, and model number,

(iii) the name and the address of the manufacturer of such device, and

(iv) a brief description of the event reported to the manufacturer.

The Secretary may by regulation alter the frequency and timing of reports required by this subparagraph.

(D) For purposes of subparagraphs (A), (B), and (C), a device user facility shall be treated as having received or otherwise become aware of information with respect to a device of that facility when medical personnel who are employed by or otherwise formally affiliated with the facility receive or otherwise become aware of information with respect to that device in the course of their duties

(2) The Secretary may not disclose the identity of a device user facility which makes a report under paragraph (1) except in connection with--

(A) an action brought to enforce section 301(q),

(B) a communication to a manufacturer of a device which is the subject of a report under paragraph (1), or

(C) a disclosure required under subsection (a).

This paragraph does not prohibit the Secretary from disclosing the identity of a device user facility making a report under paragraph (1) or any information in such a report to employees of the Department of Health and Human Services, to the Department of Justice, or to the duly authorized committees and subcommittees of the Congress.

(3) No report made under paragraph (1) by--

(A) a device user facility,

(B) an individual who is employed by or otherwise formally affiliated with such a facility, or

(C) a physician who is not required to make

such a report,
shall be admissible into evidence or otherwise used in any civil action involving private parties unless the facility, individual, or physician who made the report had knowledge of the falsity of the information contained in the report.

(4) A report made under paragraph (1) does not affect any obligation of a manufacturer who receives the report to file a report as required under subsection (a).

(5) For purposes of this subsection:

(A) The term "device user facility" means a hospital, ambulatory surgical facility, nursing home, or outpatient treatment facility which is not a physician's office. The Secretary may by regulation include an outpatient diagnostic facility which is not a physician's office in such term.

(B) The terms "serious illness" and "serious injury" mean illness or injury, respectively, that--

(i) is life threatening,

(ii) results in permanent impairment of a body function or permanent damage to a body structure, or

(iii) necessitates medical or surgical intervention to preclude permanent impairment of a body function or permanent damage to a body structure. (c) Subsection (a) shall not apply to--

(1) any practitioner who is licensed by law to prescribe or administer devices intended for use in humans and who manufactures or imports devices solely for use in the course of his professional practice;

(2) any person who manufactures or imports devices intended for use in humans solely for such person's use in research or teaching and not for sale (including any person who uses a device under an exemption granted under section 520(g)); and

(3) any other class of persons as the Secretary may by regulation exempt from subsection (a) upon a finding that compliance with the requirements of such subsection by such class with respect to a device is not necessary to (A) assure that a device is not adulterated or misbranded or (B) otherwise to assure its safety and effectiveness.

(d) Each manufacturer, importer, and distributor required to make reports under subsection (a) shall submit to the Secretary annually a statement certifying that--

(1) the manufacturer, importer, or distributor did file a certain number of such reports, or

(2) the manufacturer, importer, or distributor did not file any report under subsection (a).

(e) Every person who registers under section 510 and is engaged in the manufacture of--

(1) a device the failure of which would be reasonably likely to have serious adverse health consequences and which is (A) a permanently implantable device, or (B) a life sustaining or life supporting device used outside a device user facility, or

(2) any other device which the Secretary may designate,
shall adopt a method of device tracking.

(f)(1) Except as provided in paragraph (2), the Secretary shall by regulation require a manufacturer, importer, or distributor of a device to report promptly to the Secretary any correction or removal of a device undertaken by such manufacturer, importer, or distributor if the removal or correction was undertaken--

(A) to reduce a risk to health posed by the device, or

(B) to remedy a violation of this Act caused by the device which may present a risk to health. A manufacturer, importer, or distributor of a device who undertakes a correction or removal of a device which is not required to be reported under this paragraph shall keep a record of such correction or removal.

(2) No report of the corrective action or removal of a device may be required under paragraph (1) if a report of the corrective action or removal is required and has been submitted under subsection (a).

(3) For purposes of paragraphs (1) and (2), the terms "correction" and "removal" do not include routine servicing.

SEC. 520 [21 U.S.C. 360j] General provisions respecting control of devices intended for human use

(a) Any requirement authorized by or under section 501, 502, 510, or 519 applicable to a device intended for human use shall apply to such device until the applicability of the requirement to the device has been changed by action taken under section 513, 514, or 515 or under subsection (g) of this section, and any requirement established by or under section 501, 502, 510, or 519 which is inconsistent with a requirement imposed on such device under section 514 or 515 or under subsection (g) of this section shall not apply to such device.

(b) Sections 514 and 515 do not apply to any device which, in order to comply with the order of an individual physician or dentist (or any other specially qualified person designated under regulations promulgated by the Secretary after an opportunity for an oral hearing necessarily deviates from an otherwise applicable performance standard or requirement prescribed by or under section 515 if (1) the device is not generally available in finished form for purchase or for dispensing upon prescription and is not offered through labeling or advertising by the manufacturer, importer, or distributor thereof for commercial distribution, and (2) such device--

(A)(i) is intended for use by an individual patient named in such order of such physician or dentist (or other specially qualified person so designated) and is to be made in a specific form for such patient, or

(ii) is intended to meet the special needs of such physician or dentist (or other specially qualified person so designated) in the course of the professional practice of such physician or dentist (or other specially qualified person so designated), and

(B) is not generally available to or generally used by other physicians or dentists (or other specially qualified persons so designated).

(c) Any information reported to or otherwise obtained by the Secretary or his representative under section 513, 514, 515, 516, 518, 519, or 704 or under subsection (f) or (g) of this section which is exempt from disclosure pursuant to subsection (a) of section 552 of title 5, United States Code, by reason of subsection (b)(4) of such section shall be considered confidential and shall not be disclosed and may not be used by the Secretary as the basis for the reclassification of a device from class III to class II or class I or as the basis for the establishment or amendment of a performance standard under section 514 for a device reclassified from class III to class II, except (1) in accordance with subsection (h), and (2) that such information may be disclosed to other officers or employees concerned with carrying out this Act or when relevant in any proceeding under this Act (other than section 513 or 514 thereof).

(d) Each notice of proposed rulemaking under section 513, 514, 515, 516, 518, or 519, or under this section, any other notice which is published in the Federal Register with respect to any other action taken under any such section and which states the reasons for such action, and each publication of findings required to be made in connection with rulemaking under any such section shall set forth--

(1) the manner in which interested persons may examine data and other information on which the notice or findings is based, and

(2) the period within which interested persons may present their comments on the notice or findings (including the need therefor) orally or in writing, which period shall be at least sixty days but may not exceed ninety days unless the time is extended by the Secretary by a notice published in the Federal Register stating good cause therefor.

(e)(1) The Secretary may by regulation require that a device be restricted to sale, distribution, or use--

(A) only upon the written or oral authorization of a practitioner licensed by law to administer or use such device, or

(B) upon such other conditions as the Secretary may prescribe in such regulation,

if, because of its potentiality for harmful effect or the collateral measures necessary to its use, the Secretary determines that there cannot otherwise be reasonable assurance of its safety and effectiveness. No condition prescribed under subparagraph (B) may restrict the use of a device to persons with specific training or experience in its use or to persons for use in certain facilities unless the Secretary determines that such a restriction is required for the safe and effective use of the device. No such condition may exclude a person from using a device solely because the person does not have the training or experience to make him eligible for certification by a certifying board recognized by the American Board of Medical Specialties or has not been certified by such a Board. A device subject to a regulation under this subsection is a restricted device.

(2) The label of a restricted device shall bear such appropriate statements of the restrictions required by a regulation under paragraph (1) as the Secretary may in such regulation prescribe.

(f)(1)(A) The Secretary may, in accordance with subparagraph (B), prescribe regulations requiring that the methods used in, and the facilities and controls used for, the manufacture, pre-production design validation (including a process to assess the performance of a device but not including an evaluation of the safety or effectiveness of a device), packing, storage, and installation of a device conform to current good manufacturing practice, as prescribed in such regulations, to assure that the device will be safe and effective and otherwise in compliance with this Act.

(B) Before the Secretary may promulgate any regulation under subparagraph (A) he shall--

(i) afford the advisory committee established under paragraph (3) an opportunity to submit recommendations to him with respect to the regulation proposed to be promulgated, and

(ii) afford opportunity for an oral hearing.

The Secretary shall provide the advisory committee a reasonable time to make its recommendation with respect to proposed regulations under subparagraph (A).

(2)(A) Any person subject to any requirement prescribed by regulations under paragraph (1) may petition the Secretary for an exemption or variance from such requirement. Such a petition shall be submitted to the Secretary in such form and manner as he shall prescribe and shall--

(i) in the case of a petition for an exemption from a requirement, set forth the basis for the petitioner's determination that compliance with the requirement is not required to assure that the device will be safe and effective and otherwise in compliance with this Act,

(ii) in the case of a petition for a variance from a requirement, set forth the methods proposed to be used in, and the facilities and controls proposed to be used for, the manufacture, packing, storage, and installation of the device in lieu of the methods, facilities, and controls prescribed by the requirement, and

(iii) contain such other information as the Secretary shall prescribe. (B) The Secretary may refer to the advisory committee established under paragraph (3) any petition submitted under subparagraph (A). The advisory committee shall report its recommendations to the Secretary with respect to a petition referred to it within sixty days of the date of the petition's referral. Within sixty days after--

(i) the date the petition was submitted to the Secretary under subparagraph (A), or

(ii) if the petition was referred to an advisory committee, the expiration of the sixty-day period beginning on the date the petition was referred to the advisory committee, whichever occurs later, the Secretary shall by order either deny the petition or approve it.

(C) The Secretary may approve--

(i) a petition for an exemption for a device from a requirement if he determines that compliance with such requirement is not required to assure that the device will be safe and effective and otherwise in compliance with this Act, and

(ii) a petition for a variance for a device from a requirement if he determines that the methods to be used in, and the facilities and controls to be used for, the manufacture, packing, storage, and installation of the device in lieu of the methods, controls, and facilities prescribed by the requirement are sufficient to assure that the device will be safe and effective and otherwise in compliance with this Act.

An order of the Secretary approving a petition for a variance shall prescribe such conditions respecting the methods used in, and the facilities and controls used for, the manufacture, packing, storage, and installation of the device to be granted the variance under the petition as may be necessary to assure that the device will be safe and effective and otherwise in compliance with this Act.

(D) After the issuance of an order under

subparagraph (B) respecting a petition, the petitioner shall have an opportunity for an informal hearing on such order.

(3) The Secretary shall establish an advisory committee for the purpose of advising and making recommendations to him with respect to regulations proposed to be promulgated under paragraph (1)(A) and the approval or disapproval of petitions submitted under paragraph (2). The advisory committee shall be composed of nine members as follows:

(A) Three of the members shall be appointed from persons who are officers or employees of any State or local government or of the Federal Government.

(B) Two of the members shall be appointed from persons who are representative of interests of the device manufacturing industry; two of the members shall be appointed from persons who are representative of the interests of physicians and other health professionals; and two of the members shall be representative of the interests of the general public.

Members of the advisory committee who are not officers or employees of the United States, while attending conferences or meetings of the committee or otherwise engaged in its business, shall be entitled to receive compensation at rates to be fixed by the Secretary, which rates may not exceed the daily equivalent of the rate in effect for grade GS-18 of the General Schedule, for each day (including traveltime) they are so engaged; and while so serving away from their homes or regular places of business each member may be allowed travel expenses, including per diem in lieu of subsistence, as authorized by section 5703 of title 5 of the United States Code for persons in the Government service employed intermittently. The Secretary shall designate one of the members of the advisory committee to serve as its chairman. The Secretary shall furnish the advisory committee with clerical and other assistance. Section 14 of the Federal Advisory Committee Act shall not apply with respect to the duration of the advisory committee established under this paragraph.

(g)(1) It is the purpose of this subsection to encourage to the extent consistent with the protection of the public health and safety and with ethical standards, the discovery and development of useful devices intended for human use and to that end to maintain optimum freedom for scientific investigators in their pursuit of that purpose

(2)(A) The Secretary shall, within the one hundred and twenty-day period beginning on the date of the enactment of this section, by regulation prescribe procedures and conditions under which devices intended for human use may upon application be granted an exemption from the requirements of section 502, 510, 514, 515, 516, 519, or 721 or subsection (e) or (f) of this section or from any combination of such requirements to permit the investigational use of such devices by experts qualified by scientific training and experience to investigate the safety and effectiveness of such devices.

(B) The conditions prescribed pursuant to subparagraph (A) shall include the following:

(i) A requirement that an application be submitted to the Secretary before an exemption may be granted and that the application be submitted in such form and manner as the Secretary shall specify.

(ii) A requirement that the person applying for an exemption for a device assure the establishment and maintenance of such records, and the making of such reports to the Secretary of data obtained as a result of the investigational use of the device during the exemption, as the Secretary determines will enable him to assure compliance with such conditions, review the progress of the investigation, and evaluate the safety and effectiveness of the device.

(iii) Such other requirements as the Secretary may determine to be necessary for the protection of the public health and safety.

(C) Procedures and conditions prescribed pursuant to subparagraph (A) for an exemption may appropriately vary depending on (i) the scope and duration of clinical testing to be conducted under such exemption (ii) the number of human subjects that are to be involved in such testing, (iii) the need to permit changes to be made in the device subject to the exemption during testing conducted in accordance with a clinical testing plan required under paragraph (3)(A), and (iv) whether the clinical testing of such device is for the purpose of developing data to obtain approval for the commercial distribution of such device.

(3) Procedures and conditions prescribed pursuant to paragraph (2)(A) shall require, as a condition to the exemption of any device to be the subject of testing involving human subjects, that the person applying for the exemption--

(A) submit a plan for any proposed clinical testing of the device and a report of prior investigations of the device (including, where appropriate, tests on animals) adequate to justify the proposed clinical testing--

(i) to the local institutional review committee which has been established in accordance with regulations of the Secretary to supervise clinical testing of devices in the facilities where the proposed clinical testing is to be conducted, or

(ii) to the Secretary, if--

(I) no such committee exists, or

(II) the Secretary finds that the process of review by such committee is inadequate (whether or not the plan for such testing has been approved by such committee),

for review for adequacy to justify the commencement of such testing; and, unless the plan and report are submitted to the Secretary, submit to the Secretary a summary of the plan and a report of prior investigations of the device (including, where appropriate, tests on animals);

(B) promptly notify the Secretary (under such circumstances and in such manner as the Secretary prescribes) of approval by a local institutional review committee of any clinical testing plan submitted to it in accordance with subparagraph (A);

(C) in the case of a device to be distributed to investigators for testing, obtain signed agreements from each of such investigators that any testing of the device involving human subjects will be under such investigator's supervision and in accordance with subparagraph (D) and submit such agreements to the Secretary; and

(D) assure that informed consent will be obtained from each human subject (or his representative) of proposed clinical testing involving such device, except where subject to such conditions as the Secretary may prescribe, the investigator conducting or supervising the proposed clinical testing of the device determines in writing that there exists a life threatening situation involving the human subject of such testing which necessitates the use of such device and it is not feasible to obtain informed consent from the subject and there is not sufficient time to obtain such consent from his representative.

The determination required by subparagraph (D) shall be concurred in by a licensed physician who is not involved in the testing of the human subject with respect to which such determination is made unless immediate use of the device is required to save the life of the human subject of such testing and there is not sufficient time to obtain such concurrence.

(4)(A) An application, submitted in accordance with the procedures prescribed by regulations under paragraph (2), for an exemption for a device (other than an exemption from section 516) shall be deemed approved on the thirtieth day after the submission of the application to the Secretary unless on or before such day the Secretary by order disapproves the application and notifies the applicant of the disapproval of the application.

(B) The Secretary may disapprove an application only if he finds that the investigation with respect to which the application is submitted does not conform to procedures and conditions prescribed under regulations under paragraph (2). Such a notification shall contain the order of disapproval and a complete statement of the reasons for the Secretary's disapproval of the application and afford the applicant opportunity for an informal hearing on the disapproval order.

(5) The Secretary may by order withdraw an exemption granted under this subsection for a device if the Secretary determines that the conditions applicable to the device under this subsection for such exemption are not met. Such an order may be issued only after opportunity for an informal hearing, except that such an order may be issued before the provision of an opportunity for an informal hearing if the Secretary determines that the continuation of testing under the exemption with respect to which the order is to be issued will

result in an unreasonable risk to the public health.

(h)(1) The Secretary shall promulgate regulations under which a detailed summary of information respecting the safety and effectiveness of a device which information was submitted to the Secretary and which was the basis for--

(A) an order under section 515(d)(1)(A) approving an application for premarket approval for the device or denying approval of such an application or an order under section 515(e) withdrawing approval of such an application for the device,

(B) an order under section 515(f)(6)(A) revoking an approved protocol for the device, an order under section 515(f)(6)(B) declaring a protocol for the device completed or not completed, or an order under section 515(f)(7) revoking the approval of the device, or

(C) an order approving an application under subsection (g) for an exemption for the device from section 516 or an order disapproving, or withdrawing approval of, an application for an exemption under such subsection for the device,

shall be made available to the public upon issuance of the order. Summaries of information made available pursuant to this paragraph respecting a device shall include information respecting any adverse effects on health of the device.

(2) The Secretary shall promulgate regulations under which each advisory committee established under section 515(g)(2)(B) shall make available to the public a detailed summary of information respecting the safety and effectiveness of a device which information was submitted to the advisory committee and which was the basis for its recommendation to the Secretary made pursuant to section 515(g)(2)(A). A summary of information upon which such a recommendation is based shall be made available pursuant to this paragraph only after the issuance of the order with respect to which the recommendation was made and each summary shall include information respecting any adverse effect on health of the device subject to such order.

(3) Except as provided in paragraph (4), any information respecting a device which is made available pursuant to paragraph (1) or (2) of this subsection (A) may not be used to establish the safety or effectiveness of another device for purposes of this Act by any person other than the person who submitted the information so made available, and (B) shall be made available subject to subsection (c) of this section.

(4)(A) Any information contained in an application for premarket approval filed with the Secretary pursuant to section 515(c), including clinical and preclinical tests or studies, but excluding descriptions of methods of manufacture and product composition, that demonstrates the safety and effectiveness of a device shall be available 1 year after the original application for the fourth device of a kind has been approved by the Secretary, for use by the Secretary in approving devices, or determining whether a product development protocol has been completed, under section 515, establishing a performance standard under section 514, and reclassifying devices under subsections (e) and (f) of section 513, and subsection (l)(2). The Secretary shall deem devices that incorporate the same technologies, have the same principles of operation, and are intended for the same use or uses to be within a kind of device.

(B) The Secretary, contemporaneously with the approval of the fourth device of a kind, shall publish an order in the Federal Register identifying the four devices of a kind that have been approved under section 515 and the date on which the data contained in premarket approval applications for the devices will be available to the Secretary for use, as described in subparagraph (A).

(C) The publicly available detailed summaries of information respecting the safety and effectiveness of devices required by paragraph (1)(A) shall be available for use by the Secretary as the evidentiary basis for the regulatory action described in subparagraph (A).

(D)(i) This paragraph shall become effective--

(I) on November 15, 1990, for devices for which four devices of a kind were approved on or before December 31, 1987, and

(II) on November 15, 1991, for devices not described in subclause (I).

(ii) For each device described in clause (i)(I), the Secretary shall publish a notice in the Federal Register setting forth the date, which shall be not earlier than 1 year after the date of the notice, that data identified in subparagraph (A) shall be available for the use of the Secretary.

(E)(i) Except as provided in clause (ii), the approval date of a device, for purposes of this paragraph, shall be the date of the letter of the Secretary to the applicant approving a device under section 515 and permitting the applicant to commercially distribute the device.

(ii) For each device described in subparagraph (D)(i)(II) for which the original application for a fourth device of a kind is approved by the Secretary before November 1, 1991, the approval date of the fourth device of a kind shall be deemed to be November 15, 1991.

(F) Any challenge to an order under subparagraph (B) shall be made not later than 30 days after the date of the Federal Register notice referred to in such subparagraph.

(i) Each panel under section 513 and each advisory committee established under section 514(b)(5)(B) or 515(g) or under subsection (f) of this section shall make and maintain a transcript of any proceeding of the panel or committee. Each such panel and committee shall delete from any transcript made pursuant to this subsection information which under subsection (c) of this section is to be considered confidential.

(j) Except as provided in section 519(e), no regulation under this Act may impose on a type or class of device requirements for the traceability of such type or class of device unless such requirements are necessary to assure the protection of the public health.

(k) The Secretary may enter into contracts for research, testing, and demonstrations respecting devices and may obtain devices for research, testing, and demonstration purposes without regard to sections 3648 and 3709 of the Revised Statutes (31 U.S.C. 529, 41 U.S.C. 5).

(l)(1) Any device intended for human use--

(A) for which on the date of enactment of the Medical Device Amendments of 1976 (hereinafter in this subsection referred to as the "enactment date") an approval of an application submitted under section 505(b) was in effect;

(B) for which such an application was filed on or before the enactment date and with respect to which application no order of approval or refusing to approve had been issued on such date under subsection (c) or (d) of such section;

(C) for which on the enactment date an exemption under subsection (i) of such section was in effect;

(D) which is within a type of device described in subparagraph (A), (B), or (C) and is substantially equivalent to another device within that type;

(E) which the Secretary in a notice published in the Federal Register before the enactment date has declared to be a new drug subject to section 505; or

(F) with respect to which on the enactment date an action is pending in a United States court under section 302, 303, or 304 for an alleged violation of a provision of section 301 which enforces a requirement of section 505 or for an alleged violation of section 505(a), is classified in class III unless the Secretary in response to a petition submitted under paragraph (2) has classified such device in class I or II.

(2) The Secretary may initiate the reclassification of a device classified into class III under paragraph (1) of this subsection or the manufacturer or importer of a device classified under paragraph (1) may petition the Secretary (in such form and manner as he shall prescribe) for the issuance of an order classifying the device in class I or class II. Within thirty days of the filing of such a petition, the Secretary shall notify the petitioner of any deficiencies in the petition which prevent the Secretary from making a decision on the petition. Except as provided in paragraph (3)(D)(ii), within one hundred and eighty days after the filing of a petition under this paragraph, the Secretary shall, after consultation with the appropriate panel under section 513, by order either deny the petition or order the classification, in accordance with the criteria prescribed by section 513(a)(1)(A) or 513(a)(1)(B), of the device in class I or

class II.

(3)(A) In the case of a device which is described in paragraph (1)(A) and which is in class III--

(i) such device shall on the enactment date be considered a device with an approved application under section 515, and

(ii) the requirements applicable to such device before the enactment date under section 505 shall continue to apply to such device until changed by the Secretary as authorized by this Act.

(B) In the case of a device which is described in paragraph (1)(B) and which is in class III, an application for such device shall be considered as having been filed under section 515 on the enactment date. The period in which the Secretary shall act on such application in accordance with section 515(d)(1) shall be one hundred and eighty days from the enactment date (or such greater period as the Secretary and the applicant may agree upon after the Secretary has made the finding required by section 515(d)(1)(B)(i)) less the number of days in the period beginning on the date an application for such device was filed under section 505 and ending on the enactment date. After the expiration of such period such device is required, unless exempt under subsection (g), to have in effect an approved application under section 515.

(C) A device which is described in paragraph (1)(C) and which is in class III shall be considered a new drug until the expiration of the ninety-day period beginning on the date of the promulgation of regulations under subsection (g) of this section. After the expiration of such period such device is required, unless exempt under subsection (g), to have in effect an approved application under section 515.

(D)(i) Except as provided in clauses (ii) and (iii), a device which is described in subparagraph (D), (E), or (F) of paragraph (1) and which is in class III is required, unless exempt under subsection (g) of this section, to have on and after sixty days after the enactment date in effect an approved application under section 515.

(ii) If--

(I) a petition is filed under paragraph (2) for a device described in subparagraph (D), (E), or (F) of paragraph (1), or

(II) an application for premarket approval is filed under section 515 for such a device, within the sixty-day period beginning on the enactment date (or within such greater period as the Secretary, after making the finding required under section 515(d)(1)(B), and the petitioner or applicant may agree upon), the Secretary shall act on such petition or application in accordance with paragraph (2) or section 515 except that the period within which the Secretary must act on the petition or application shall be within the one hundred and twenty-day period beginning on the date the petition or application is filed. If such a petition or application is filed within such sixty-day (or greater) period, clause (i) of this subparagraph shall not apply to such device before the expiration of such one hundred and twenty-day period, or if such petition is denied or such application is denied approval, before the date of such denial, whichever occurs first.

(iii) In the case of a device which is described in subparagraph (E) of paragraph (1), which the Secretary in a notice published in the Federal Register after March 31, 1976, declared to be a new drug subject to section 505, and which is in class III--

(I) the device shall, after eighteen months after the enactment date have in effect an approved application under section 515 unless exempt under subsection (g) of this section, and

(II) the Secretary may, during the period beginning one hundred and eighty days after the enactment date and ending eighteen months after such date, restrict the use of the device to investigational use by experts qualified by scientific training and experience to investigate the safety and effectiveness of such device, and to investigational use in accordance with the requirements applicable under regulations under subsection (g) of this section to investigational use of devices granted an exemption under such subsection. If the requirements under subsection (g) of this section are made applicable to the investigational use of such a device, they shall be made applicable in such a manner that the device shall be made reasonably available to physicians meeting appropriate qualifications prescribed by the Secretary.

(4) Any device intended for human use which on the enactment date was subject to the requirements of section 507 shall be subject to such requirements as follows:

(A) In the case of such a device which is classified into class I, such requirements shall apply to such device until the effective date of the regulation classifying the device into such class.

(B) In the case of such a device which is classified into class II, such requirements shall apply to such device until the effective date of a performance standard applicable to the device under section 514.

(C) In the case of such a device which is classified into class III, such requirements shall apply to such device until the date on which the device is required to have in effect an approved application under section 515.

(5)(A) Before December 1, 1991, the Secretary shall by order require manufacturers of devices described in paragraph (1), which are subject to revision of classification under subparagraph (B), to submit to the Secretary a summary of and citation to any information known or otherwise available to the manufacturers respecting the devices, including adverse safety or effectiveness information which has not been submitted under section 519. The Secretary may require a manufacturer to submit the adverse safety or effectiveness data for which a summary and citation were submitted, if such data are available to the manufacturer.

(B) Except as provided in subparagraph (C), after the issuance of an order under subparagraph (A) but before December 1, 1992, the Secretary shall publish a regulation in the Federal Register for each device which is classified in class III under paragraph (1) revising the classification of the device so that the device is classified into class I or class II, unless the regulation requires the device to remain in class III. In determining whether to revise the classification of a device or to require a device to remain in class III, the Secretary shall apply the criteria set forth in section 513(a). Before the publication of a regulation requiring a device to remain in class III or revising its classification, the Secretary shall publish a proposed regulation respecting the classification of a device under this subparagraph and provide an opportunity for the submission of comments on any such regulation. No regulation under this subparagraph requiring a device to remain in class III or revising its classification may take effect before the expiration of 90 days from the date of the publication in the Federal Register of the proposed regulation.

(C) The Secretary may by notice published in the Federal Register extend the period prescribed by subparagraph (B) for a device for an additional period not to exceed 1 year.

(m)(1) To the extent consistent with the protection of the public health and safety and with ethical standards, it is the purpose of this subsection to encourage the discovery and use of devices intended to benefit patients in the treatment and diagnosis of diseases or conditions that affect fewer than 4,000 individuals in the United States.

(2) The Secretary may grant a request for an exemption from the effectiveness requirements of sections 514 and 515 for a device for which the Secretary finds that--

(A) the device is designed to treat or diagnose a disease or condition that affects fewer than 4,000 individuals in the United States,

(B) the device would not be available to a person with a disease or condition referred to in subparagraph (A) unless the Secretary grants such an exemption and there is no comparable device, other than under this exemption, available to treat or diagnose such disease or condition, and

(C) the device will not expose patients to an unreasonable or significant risk of illness or injury and the probable benefit to health from the use of the device outweighs the risk of injury or illness from its use, taking into account the probable risks and benefits of currently available device or alternative forms of treatment.

(3) No person granted an exemption under paragraph (2) with respect to a device may sell the device for an amount that exceeds the costs of research and development, fabrication, and distribution of the device.

(4) Devices granted an exemption under

paragraph (2) may only be used--

(A) in facilities that have established, in accordance with regulations of the Secretary, a local institutional review committee to supervise clinical testing of devices in the facilities, and

(B) if, before the use of a device, an institutional review committee approves the use in the treatment or diagnosis of a disease or condition referred to in paragraph (2)(A).

(5) An exemption under paragraph (2) shall be for a term of 18 months and may only be initially granted in the 5-year period beginning on the date regulations under paragraph (6) take effect. The Secretary may extend such an exemption for a period of 18 months if the Secretary is able to make the findings set forth in paragraph (2) and if the applicant supplies information demonstrating compliance with paragraph (3). An exemption may be extended more than once and may be extended after the expiration of such 5-year period.

(6) Within one year of the date of the enactment of this subsection, the Secretary shall issue regulations to implement this subsection.

SEC. 521 [21 U.S.C. 360k] State and local requirements respecting devices

(a) Except as provided in subsection (b), no State or political subdivision of a State may establish or continue in effect with respect to a device intended for human use any requirement--

(1) which is different from, or in addition to, any requirement applicable under this Act to the device, and

(2) which relates to the safety or effectiveness of the device or to any other matter included in a requirement applicable to the device under this Act.

(b) Upon application of a State or a political subdivision thereof, the Secretary may, by regulation promulgated after notice and opportunity for an oral hearing, exempt from subsection (a), under such conditions as may be prescribed in such regulation, a requirement of such State or political subdivision applicable to a device intended for human use if--

(1) the requirement is more stringent than a requirement under this Act which would be applicable to the device if an exemption were not in effect under this subsection; or

(2) the requirement--

(A) is required by compelling local conditions, and

(B) compliance with the requirement would not cause the device to be in violation of any applicable requirement under this Act.

SEC. 522 [21 U.S.C. 360l Postmarket surveillance

(a) In General

(1) Required surveillance. The Secretary shall require a manufacturer to conduct postmarket surveillance for any device of the manufacturer first introduced or delivered for introduction into interstate commerce after January 1, 1991, that--

(A) is a permanent implant the failure of which may cause serious, adverse health consequences or death,

(B) is intended for a use in supporting or sustaining human life, or

(C) potentially presents a serious risk to human health.

(2) Discretionary surveillance. The Secretary may require a manufacturer to conduct postmarket surveillance for a device of the manufacturer if the Secretary determines that postmarket surveillance of the device is necessary to protect the public health or to provide safety or effectiveness data for the device.

(b) Surveillance Approval. Each manufacturer required to conduct a surveillance of a device under subsection (a)(1) shall, within 30 days of the first introduction or delivery for introduction of such device into interstate commerce, submit, for the approval of the Secretary, a protocol for the required surveillance. Each manufacturer required to conduct a surveillance of a device under subsection (a)(2) shall, within 30 days after receiving notice that the manufacturer is required to conduct such surveillance, submit, for the approval of the Secretary, a protocol for the required surveillance. The Secretary, within 60 days of the receipt of such protocol, shall determine if the principal investigator proposed to be used in the surveillance has sufficient qualifications and experience to conduct such surveillance and if such protocol will result in collection of useful data or other information necessary to protect the public health and to provide safety and effectiveness information for the device. The Secretary may not approve such a protocol until it has been reviewed by an appropriately qualified scientific and technical review committee established by the Secretary.

SUBCHAPTER B -- Drugs for Rare Diseases or Conditions

SEC. 525 [21 U.S.C. 360 aa] Recommendations for investigations of drugs for rare diseases or conditions

(a) The sponsor of a drug for a disease or condition which is rare in the States may request the Secretary to provide written recommendations for the nonclinical and clinical investigations which must be conducted with the drug before--

(1) it may be approved for such disease or condition under section 505,

(2) if the drug is an antibiotic, it may be certified for such disease or condition under section 507, or

(3) if the drug is a biological product, it may be licensed for such disease or condition under section 351 of the Public Health Service Act.

If the Secretary has reason to believe that a drug for which a request is made under this section is a drug for a disease or condition which is rare in the States, the Secretary shall provide the person making the request written recommendations for the nonclinical and clinical investigations which the Secretary believes, on the basis of information available to the Secretary at the time of the request under this section, would be necessary for approval of such drug for such disease or condition under section 505, certification of such drug for such disease or condition under section 507, or licensing of such drug for such disease or condition under section 351 of the Public Health Service Act.

(b) The Secretary shall by regulation promulgate procedures for the implementation of subsection (a).

SEC. 526 [21 U.S.C. 360bb] Designation of drugs for rare diseases or conditions

(a)(1) The manufacturer or the sponsor of a drug may request the Secretary to designate the drug as a drug for a rare disease or condition. A request for designation of a drug shall be made before the submission of an application under section 505(b) for the drug, the submission of an application for certification of the drug under section 507, or the submission of an application for licensing of the drug under section 351 of the Public Health Service Act. If the Secretary finds that a drug for which a request is submitted under this subsection is being or will be investigated for a rare disease or condition and--

(A) if an application for such drug is approved under section 505,

(B) if a certification for such drug is issued under section 507, or

(C) if a license for such drug is issued under section 351 of the Public Health Service Act,

the approval, certification, or license would be for use for such disease or condition, the Secretary shall designate the drug as a drug for such disease or condition. A request for a designation of a drug under this subsection shall contain the consent of the applicant to notice being given by the Secretary under subsection (b) respecting the designation of the drug.

(2) For purposes of paragraph (1), the term "rare disease or condition" means any disease or condition which (A) affects less than 200,000 persons in the United States, or (B) affects more than 200,000 in the United States and for which there is no reasonable expectation that the cost of developing and making available in the United States a drug for such disease or condition will be recovered from sales in the United States of such drug. Determinations under the preceding sentence with respect to any drug shall be made on the basis of the facts and circumstances as of the date the request for designation of the drug under this subsection is made.

(b) A designation of a drug under subsection (a) shall be subject to the condition that--

(1) if an application was approved for the drug under section 505(b) a certificate was

issued for the drug under section 507, or a license was issued for the drug under section 351 of the Public Health Service Act, the manufacturer of the drug will notify the Secretary of any discontinuance of the production of the drug at least one year before discontinuance, and

(2) if an application has not been approved for the drug under section 505(b), a certificate has not been issued for the drug under section 507, or a license has not been issued for the drug under section 351 of the Public Health Service Act and if preclinical investigations or investigations under section 505(i) are being conducted with the drug, the manufacturer or sponsor of the drug will notify the Secretary of any decision to discontinue active pursuit of approval of an application under section 505(b), approval of an application for certification under section 507, or approval of a license under section 351 of the Public Health Service Act.

(c) Notice respecting the designation of a drug under subsection (a) shall be made available to the public.

(d) The Secretary shall by regulation promulgate procedures for the implementation of subsection (a).

SEC. 527 [21 U.S.C. 360cc] Protection for drugs for rare diseases or conditions

(a) Except as provided in subsection (b), if the Secretary--

(1) approves an application filed pursuant to section 505,

(2) issues a certification under section 507, or

(3) issues a license under section 351 of the Public Health Service Act for a drug designated under section 526 for a rare disease or condition, the Secretary may not approve another application under section 505, issue another certification under section 507, or issue another license under section 351 of the Public Health Service Act for such drug for such disease or condition for a person who is not the holder of such approved application, of such certification, or of such license until the expiration of seven years from the date of the approval of the approved application, the issuance of the certification, or the issuance of the license. Section 505(c)(2) does not apply to the refusal to approve an application under the preceding sentence.

(b) If an application filed pursuant to section 505 is approved for a drug designated under section 526 for a rare disease or condition, if a certification is issued under section 507 for such a drug, or if a license is issued under section 351 of the Public Health Service Act for such a drug, the Secretary may, during the seven-year period beginning on the date of the application approval, of the issuance of the certification under section 507, or of the issuance of the license, approve another application under section 505, issue another certification under section 507,, or issue a license under section 351 of the Public Health Service Act, for such drug for such disease or condition for a person who is not the holder of such approved application, of such certification, or of such license if--

(1) The Secretary finds, after providing the holder notice and opportunity for the submission of views, that in such period the holder of the approved application, of the certification, or of the license cannot assure the availability of sufficient quantities of the drug to meet the needs of persons with the disease or condition for which the drug was designated; or

(2) such holder provides the Secretary in writing the consent of such holder for the approval of other applications, issuance of other certifications, or the issuance of other licenses before the expiration of such seven-year period.

SEC. 528 [21 U.S.C. 360dd] Open protocols for investigations of drugs for rare diseases or conditions

If a drug is designated under section 526 as a drug for a rare disease or condition and if notice of a claimed exemption under section 505(i) or regulations issued thereunder is filed for such drug, the Secretary shall encourage the sponsor of such drug to design protocols for clinical investigations of the drug which may be conducted under the exemption to permit the addition to the investigations of persons with the disease or condition who need the drug to treat the disease or condition and who cannot be satisfactorily treated by available alternative drugs.

SUBCHAPTER C -- Electronic Product Radiation Control

SEC. 531 [21 U.S.C. 360hh] Definitions

As used in this subchapter--

(A) any ionizing or non-ionizing electromagnetic or particulate radiation, or

(B) any sonic, infrasonic, or ultrasonic wave, which is emitted from an electronic product as the result of the operation of an electronic circuit in such product;

(2) the term ''electronic product'' means (A) any manufactured or assembled product which, when in operation, (i) contains or acts as part of an electronic circuit and (ii) emits (or in the absence of effective shielding or other controls would emit) electronic product radiation, or (B) any manufactured or assembled article which is intended for use as a component, part, or accessory of a product described in clause (A) and which when in operation emits (or in the absence of effective shielding or other controls would emit) such radiation;

(3) the term ''manufacturer'' means any person engaged in the business of manufacturing, assembling, or importing of electronic products;

(4) the term ''commerce'' means (A) commerce between any place in any State and any place outside thereof; and (B) commerce wholly within the District of Columbia; and

(5) the term ''State'' includes the District of Columbia, the Commonwealth of Puerto Rico, the Northern Mariana Islands, the Virgin Islands, Guam, and American Samoa.

SEC. 532 [21 U.S.C. 360ii] Electronic product radiation control program

(a) The Secretary shall establish and carry out an electronic product radiation control program designed to protect the public health and safety from electronic product radiation. As a part of such program, he shall--

(1) pursuant to section 534, develop and administer performance standards for electronic products;

(2) plan, conduct, coordinate, and support research, development, training, and operational activities to minimize the emissions of and the exposure of people to, unnecessary electronic product radiation;

(3) maintain liaison with and receive information from other Federal and State departments and agencies with related interests, professional organizations, industry, industry and labor associations and other organizations on present and future potential electronic product radiation;

(4) study and evaluate emissions of, and conditions of exposure to, electronic product radiation and intense magnetic fields;

(5) develop, test, and evaluate the effectiveness of procedures and techniques for minimizing exposure to electronic product radiation; and

(6) consult and maintain liaison with the Secretary of Commerce, the Secretary of Defense, the Secretary of Labor, the Atomic Energy Commission, and other appropriate Federal departments and agencies on (A) techniques, equipment, and programs for testing and evaluating electronic product radiation, and (B) the development of performance standards pursuant to section 534 to control such radiation emissions.

(b) In carrying out the purposes of subsection (a), the Secretary is authorized to--

(1)(A) collect and make available, through publications and other appropriate means, the results of, and other information concerning, research and studies relating to the nature and extent of the hazards and control of electronic product radiation; and (B) make such recommendations relating to such hazards and control as he considers appropriate;

(2) make grants to public and private agencies, organizations, and institutions, and to individuals for the purposes stated in paragraphs (2), (4), and (5) of subsection (a) of this section;

(3) contract with public or private agencies, institutions, and organizations, and with individuals, without regard to section 3324 of title 31, United States Code, and section 3709

of the Revised Statutes of the United States (41 U.S.C. 5); and

(4) procure (by negotiation or otherwise) electronic products for research and testing purposes, and sell or otherwise dispose of such products.

(c)(1) Each recipient of assistance under this subchapter pursuant to grants or contracts entered into under other than competitive bidding procedures shall keep such records as the Secretary shall prescribe, including records which fully disclose the amount and disposition by such recipient of the proceeds of such assistance, the total cost of the project or undertaking in connection with which such assistance is given or used, and the amount of that portion of the cost of the project or undertaking supplied by other sources, and such other records as will facilitate an effective audit.

(2) The Secretary and the Comptroller General of the United States, or any of their duly authorized representatives, shall have access for the purpose of audit and examination to any books, documents, papers, and records of the recipient that are pertinent to the grants or contracts entered into under this subchapter under other than competitive bidding procedures.

SEC. 533 [21 U.S.C. 360jj] Studies by the Secretary

(a) The Secretary shall conduct the following studies, and shall make a report or reports of the results of such studies to the Congress on or before January 1, 1970, and from time to time thereafter as he may find necessary, together with such recommendations for legislation as he may deem appropriate:

(1) A study of present State and Federal control of health hazards from electronic product radiation and other types of ionizing radiation, which study shall include, but not be limited to--

(A) control of health hazards from radioactive materials other than materials regulated under the Atomic Energy Act of 1954;

(B) any gaps and inconsistencies in present controls;

(C) the need for controlling the sale of certain used electronic products, particularly antiquated X-ray equipment, without upgrading such products to meet the standards for new products or separate standards for used products;

(D) measures to assure consistent and effective control of the aforementioned health hazards;

(E) measures to strengthen radiological health programs of State governments; and

(F) the feasibility of authorizing the Secretary to enter into arrangements with individual States or groups of States to define their respective functions and responsibilities for the control of electronic product radiation and other ionizing radiation;

(2) A study to determine the necessity for the development of standards for the use of nonmedical electronic products for commercial and industrial purposes; and

(3) A study of the development of practicable procedures for the detection and measurement of electronic product radiation which may be emitted from electronic products manufactured or imported prior to the effective date of any applicable standard established pursuant to this subchapter.

(b) In carrying out these studies, the Secretary shall invite the participation of other Federal departments and agencies having related responsibilities and interests, particularly those of States which regulate radioactive materials under section 274 of the Atomic Energy Act of 1954, as amended, and interested professional, labor, and industrial organizations. Upon request from congressional committees interested in these studies, the Secretary shall keep these committees currently informed as to the progress of the studies and shall permit the committees to send observers to meetings of the study groups.

(c) The Secretary or his designee shall organize the studies and the participation of the invited participants as he deems best. Any dissent from the findings and recommendations of the Secretary shall be included in the report if so requested by the dissenter.

SEC. 534 [21 U.S.C. 360kk] Performance standards for electronic products (a)(1) The Secretary shall by regulation prescribe performance standards for electronic products to control the emission of electronic product radiation from such products if he determines that such standards are necessary for the protection of the public health and safety. Such standards may include provisions for the testing of such products and the measurement of their electronic product radiation emissions, may require the attachment of warning signs and labels, and may require the provision of instructions for the installation, operation, and use of such products. Such standards may be prescribed from time to time whenever such determinations are made, but the first of such standards shall be prescribed prior to January 1, 1970. In the development of such standards, the Secretary shall consult with Federal and State departments and agencies having related responsibilities or interests and with appropriate professional organizations and interested persons, including representatives of industries and labor organizations which would be affected by such standards, and shall give consideration to--

(A) the latest available scientific and medical data in the field of electronic product radiation;

(B) the standards currently recommended by (i) other Federal agencies having responsibilities relating to the control and measurement of electronic product radiation, and (ii) public or private groups having an expertise in the field of electronic product radiation;

(C) the reasonableness and technical feasibility of such standards as applied to a particular electronic product;

(D) the adaptability of such standards to the need for uniformity and reliability of testing and measuring procedures and equipment; and

(E) in the case of a component, or accessory described in paragraph (2)(B) of section 531, the performance of such article in the manufactured or assembled product for which it is designed.

(2) The Secretary may prescribe different and individual performance standards, to the extent appropriate and feasible, for different electronic products so as to recognize their different operating characteristics and uses.

(3) The performance standards prescribed under this section shall not apply to any electronic product which is intended solely for export if (A) such product and the outside of any shipping container used in the export of such product are labeled or tagged to show that such product is intended for export, and (B) such product meets all the applicable requirements of the country to which such product is intended for export.

(4) The Secretary may by regulation amend or revoke any performance standard prescribed under this section.

(5) The Secretary may exempt from the provisions of this section any electronic product intended for use by departments or agencies of the United States provided such department or agency has prescribed procurement specifications governing emissions of electronic product radiation and provided further that such product is of a type used solely or predominantly by departments or agencies of the United States.

(b) The provisions of subchapter II of chapter 5 of title 5 of the United States Code (relating to the administrative procedure for rulemaking), and of chapter 7 of such title (relating to judicial review), shall apply with respect to any regulation prescribing, amending, or revoking any standard prescribed under this section.

(c) Each regulation prescribing, amending, or revoking a standard shall specify the date on which it shall take effect which, in the case of any regulation prescribing, or amending any standard, may not be sooner than one year or not later than two years after the date on which such regulation is issued, unless the Secretary finds, for good cause shown, that an earlier or later effective date is in the public interest and publishes in the Federal Register his reason for such finding, in which case such earlier or later date shall apply.

(d)(1) In a case of actual controversy as to the validity of any regulation issued under this section prescribing, amending, or revoking a performance standard, any person who will be adversely affected by such regulation when it is effective may at any time prior to the sixtieth day after such regulation is issued file a petition with the United States court of appeals for the circuit wherein such person resides or has his principal place of business, for a judicial review of such regulation. A copy of the petition shall be forthwith transmitted by the clerk of the court to the Secretary or other officer designated by him for that purpose. The Secretary thereupon shall file in the court the record of the proceedings on which the Secretary based the regulation, as provided in section 2112 of title 28 of the United States Code.

(2) If the petitioner applies to the court for leave to adduce additional evidence, and shows to the satisfaction of the court that such additional evidence is material and that there were reasonable grounds for the failure to adduce such evidence in the proceeding before the Secretary, the court may order such additional evidence (and evidence in rebuttal thereof) to be taken before the Secretary, and to be adduced upon the hearing, in such manner and upon such terms and conditions as to the court may seem proper. The Secretary may modify his findings, or make new findings, by reason of the additional evidence so taken, and he shall file such modified or new findings, and his recommendations, if any, for the modification or setting aside of his original regulation, with the return of such additional evidence.

(3) Upon the filing of the petition referred to in paragraph (1) of this subsection, the court shall have jurisdiction to review the regulation in accordance with chapter 7 of title 5 of the United States Code and to grant appropriate relief as provided in such chapter.

(4) The judgment of the court affirming or setting aside, in whole or in part, any such regulation of the Secretary shall be final, subject to review by the Supreme Court of the United States upon certiorari or certification as provided in section 1254 of title 28 of the United States Code.

(5) Any action instituted under this subsection shall survive, notwithstanding any change in the person occupying the office of Secretary or any vacancy in such office.

(6) The remedies provided for in this subsection shall be in addition to and not in substitution for any other remedies provided by law.

(e) A certified copy of the transcript of the record and administrative proceedings under this section shall be furnished by the Secretary to any interested party at his request, and payment of the costs thereof, and shall be admissible in any criminal, exclusion of imports, or other proceeding arising under or in respect of this subchapter, irrespective of whether proceedings with respect to the regulation have previously been initiated or become final under this section.

(f)(1)(A) The Secretary shall establish a Technical Electronic Product Radiation Safety Standards Committee (hereafter in this subchapter referred to as the "Committee") which he shall consult before prescribing any standard under this section. The Committee shall be appointed by the Secretary, after consultation with public and private agencies concerned with the technical aspect of electronic product radiation safety, and shall be composed of fifteen members each of whom shall be technically qualified by training and experience in one or more fields of science or engineering applicable to electronic product radiation safety, as follows:

(i) Five members shall be selected from governmental agencies, including State and Federal Governments;

(ii) Five members shall be selected from the affected industries after consultation with industry representatives; and

(iii) Five members shall be selected from the general public, of which at least one shall be a representative of organized labor.

(B) The Committee may propose electronic product radiation safety standards to the Secretary for his consideration. All proceedings of the Committee shall be recorded and the record of each such proceeding shall be available for public inspection.

(2) Payments to members of the Committee who are not officers or employees of the United States pursuant to subsection (c) of section 208 of this the Public Health Service Act shall not render members of the Committee officers or employees of the United States for any purpose.

(g) The Secretary shall review and evaluate on a continuing basis testing programs carried out by industry to assure the adequacy of safeguards against hazardous electronic product radiation and to assure that electronic products comply with standards prescribed under this section.

(h) Every manufacturer of an electronic product to which is applicable a standard in effect under this section shall furnish to the distributor or dealer at the time of delivery of such product, in the form of a label or tag permanently affixed to such product or in such manner as approved by the Secretary, the certification that such product conforms to all applicable standards under this section. Such certification shall be based upon a test, in accordance with such standard, of the individual article to which it is attached or upon a testing program which is in accord with good manufacturing practice and which has not been disapproved by the Secretary (in such manner as he shall prescribe by regulation) on the grounds that it does not assure the adequacy of safeguards against hazardous electronic product radiation or that it does not assure that electronic products comply with the standards prescribed under this section.

SEC. 535 [21 U.S.C. 360ll] Notification of defects in, and repair or replacement of, electronic products

(a)(1) Every manufacturer of electronic products, who discovers that an electronic product produced, assembled, or imported by him has a defect which relates to the safety of use of such product by reason of the emission of electronic product radiation, or that an electronic product produced, assembled, or imported by him on or after the effective date of an applicable standard prescribed pursuant to section 534 fails to comply with such standard, shall immediately notify the Secretary of such defect or failure to comply if such product has left the place of manufacture and shall (except as authorized by paragraph (2)) with reasonable promptness furnish notification of such defect or failure to the persons (where known to the manufacturer) specified in subsection (b) of this section.

(2) If, in the opinion of such manufacturer, the defect or failure to comply is not such as to create a significant risk of injury, including genetic injury, to any person, he may, at the time of giving notice to the Secretary of such defect or failure to comply, apply to the Secretary for an exemption from the requirement of notice to the persons specified in subsection (b). If such application states reasonable grounds for such exemption, the Secretary shall afford such manufacturer an opportunity to present his views and evidence in support of the application, the burden of proof being on the manufacturer. If, after such presentation, the Secretary is satisfied that such defect or failure to comply is not such as to create a significant risk of injury, including genetic injury, to any person, he shall exempt such manufacturer from the requirement of notice to the persons specified in subsection (b) of this section and from the requirements of repair or replacement imposed by subsection (f) of this section.

(b) The notification (other than to the Secretary) required by paragraph (1) of subsection (a) of this section shall be accomplished--

(1) by certified mail to the first purchaser of such product for purposes other than resale, and to any subsequent transferee of such product; and

(2) by certified mail or other more expeditious means to the dealers or distributors of such manufacturer to whom such product was delivered.

(c) The notifications required by paragraph (1) of subsection (a) of this section shall contain a clear description of such defect or failure to comply with an applicable standard, an evaluation of the hazard reasonably related to such defect or failure to comply, and a statement of the measures to be taken to repair such defect. In the case of a notification to a person referred to in subsection (b) of this section, the notification shall also advise the person of his rights under subsection (f) of this section.

(d) Every manufacturer of electronic products shall furnish to the Secretary a true or representative copy of all notices, bulletins, and other communications to the dealers or distributors of such manufacturer or to purchasers (or subsequent transferees) of electronic products of such manufacturer regarding any such defect in such product or any such failure to comply with a standard applicable to such product. The Secretary shall disclose to the public so much of the information contained in such notice or other information obtained under section 537 as he deems will assist in carrying out the purposes of this subchapter, but he shall not disclose any information which contains or relates to a trade secret or other matter referred to in section 1905 of title 18 of the United States Code unless he determines that it is necessary to carry out the purposes of this subchapter.

(e) If through testing, inspection, investigation, or research carried out pursuant to this subchapter, or examination of reports submitted pursuant to section 537, or otherwise, the Secretary determines that any electronic product--

(1) does not comply with an applicable standard prescribed pursuant to section 534; or

(2) contains a defect which relates to the safety of use of such product by reason of the emission of electronic product radiation;

he shall immediately notify the manufacturer of such product of such defect or failure to comply. The notice shall contain the findings of the Secretary and shall include all information upon which the findings are based. The Secretary shall afford such manufacturer an opportunity to present his views and evidence in support thereof, to establish that there is no failure of compliance or that the alleged defect does not exist or does not relate to safety of use of the product by reason of the emission of such radiation hazard. If after such presentation by the manufacturer the Secretary determines that such product does not comply with an applicable standard prescribed pursuant to section 534, or that it contains a defect which relates to the safety of use of such product by reason of the emission of electronic product radiation, the Secretary shall direct the manufacturer to furnish the notification specified in subsection (c) of this section to the persons specified in paragraphs (1) and (2) of subsection (b) of this section (where known to the manufacturer), unless the manufacturer has applied for an exemption from the requirement of such notification on the ground specified in paragraph (2) of subsection (a) and the Secretary is satisfied that such noncompliance or defect is not such as to create a significant risk of injury, including genetic injury, to any person.

(f) If any electronic product is found under subsection (a) or (e) to fail to comply with an applicable standard prescribed under this subchapter or to have a defect which relates to the safety of use of such product, and the notification specified in subsection (c) is required to be furnished on account of such failure or defect, the manufacturer of such product shall (1) without charge, bring such product into conformity with such standard or remedy such defect and provide reimbursement for any expenses for transportation of such product incurred in connection with having such product brought into conformity or having such defect remedied, (2) replace such product with a like or equivalent product which complies with each applicable standard prescribed under this subchapter and which has no defect relating to the safety of its use, or (3) make a refund of the cost of such product. The manufacturer shall take the action required by this subsection in such manner, and with respect to such persons, as the Secretary by regulations shall prescribe.

(g) This section shall not apply to any electronic product that was manufactured before the date of the enactment of this subchapter.

SEC. 536 [21 U.S.C. 360mm] Imports

(a) Any electronic product offered for importation into the United States which fails to comply with an applicable standard prescribed under this subchapter, or to which is not affixed a certification in the form of a label or tag in conformity with section 534(h) shall be refused admission into the United States. The Secretary of the Treasury shall deliver to the Secretary of Health and Human Services, upon the latter's request, samples of electronic products which are being imported or offered for import into the United States, giving notice thereof to the owner or consignee, who may have a hearing before the Secretary of Health and Human Services. If it appears from an examination of such samples or otherwise that any electronic product fails to comply with applicable standards prescribed pursuant to section 534, then, unless subsection (b) of this section applies and is complied with, (1) such electronic product shall be refused admission, and (2) the Secretary of the Treasury shall cause the destruction of such electronic product unless such article is exported, under regulations prescribed by the Secretary of the Treasury, within 90 days after the date of notice of refusal of admission or within such additional time as may be permitted by such regulations.

(b) If it appears to the Secretary of Health and Human Services that any electronic product refused admission pursuant to subsection (a) of this section can be brought into compliance with applicable standards prescribed pursuant to section 534, final determination as to admission of such electronic product may be deferred upon filing of timely written application by the owner or consignee and the execution by him of a good and sufficient bond providing for the payment of such liquidated damages in the event of default as the Secretary of Health and Human Services may by regulation prescribe. If such application is filed and such bond is executed the Secretary of Health and Human Services may, in accordance with rules prescribed by him, permit the applicant to perform such operations with respect to such electronic product as may be specified in the notice of permission.

(c) All expenses (including travel, per diem or subsistence, and salaries of officers or employees of the United States) in connection with the destruction provided for in subsection (a) of this section and the supervision of operations provided for in subsection (b) of this section, and all expenses in connection with the storage, cartage, or labor with respect to any electronic product refused admission pursuant to subsection (a) of this section, shall be paid by the owner or consignee, and, in event of default, shall constitute a lien against any future importations made by such owner or consignee.

(d) It shall be the duty of every manufacturer offering an electronic product for importation into the United States to designate in writing an agent upon whom service of all administrative and judicial processes, notices, orders, decisions, and requirements may be made for and on behalf of said manufacturer, and to file such designation with the Secretary, which designation may from time to time be changed by like writing, similarly filed. Service of all administrative and judicial processes, notices, orders, decisions, and requirements may be made upon said manufacturer by service upon such designated agent at his office or usual place of residence with like effect as if made personally upon said manufacturer, and in default of such designation of such agent, service of process, notice order, requirement, or decision in any proceeding before the Secretary or in any judicial proceeding for enforcement of this subchapter or any standards prescribed pursuant to this subchapter may be made by posting such process, notice, order, requirement, or decision in the Office of the Secretary or in a place designated by him by regulation.

SEC. 537 [21 U.S.C. 360nn] Inspection and reports

(a) If the Secretary finds for good cause that the methods, tests, or programs related to electronic product radiation safety in a particular factory, warehouse, or establishment in which electronic products are manufactured or held, may not be adequate or reliable, officers or employees duly designated by the Secretary, upon presenting appropriate credentials and a written notice to the owner, operator, or agent in charge, are thereafter authorized (1) to enter, at reasonable times, any area in such factory, warehouse, or establishment in which the manufacturer's tests (or testing programs) required by section 534(h) are carried out, and (2) to inspect, at reasonable times and within reasonable limits and in a reasonable manner, the facilities and procedures within such area which are related to electronic product radiation safety. Each such inspection shall be commenced and completed with reasonable promptness. In addition to other grounds upon which good cause may be found for purposes of this subsection, good cause will be considered to exist in any case where the manufacturer has introduced into commerce any electronic product which does not comply with an applicable standard prescribed under this subchapter and with respect to which no exemption from the notification requirements has been granted by the Secretary under section 535(a)(2) or 535(e).

(b) Every manufacturer of electronic products shall establish and maintain such records (including testing records), make such reports, and provide such information, as the Secretary may reasonably require to enable him to determine whether such manufacturer has acted or is acting in compliance with this subchapter and standards prescribed pursuant to this subchapter and shall, upon request of an officer or employee duly designated by the Secretary, permit such officer or employee to inspect appropriate books, papers, records, and documents relevant to determining whether such manufacturer has acted or is acting in compliance with standards prescribed pursuant to this subchapter.

(c) Every manufacturer of electronic products shall provide to the Secretary such performance data and other technical data related to safety as may be required to carry out the purposes of this subchapter. The Secretary is authorized to require the manufacturer to give such notification of such performance and technical data at the time of original purchase to the ultimate purchaser of the electronic product, as he determines necessary to carry out the purposes of this subchapter after consulting with the affected industry.

(d) Accident and investigation reports made under this subchapter by any officer, employee, or agent of the Secretary shall be available for use in any civil, criminal, or other judicial proceeding arising out of such accident. Any such officer, employee, or agent may be required to testify in such proceedings as to the fact developed in such investigations. Any such report shall be made available to the public in a manner which need not identify individuals. All reports on research projects, demonstration projects, and other related activities shall be public information.

(e) The Secretary or his representative shall not disclose any information reported to or otherwise obtained by him, pursuant to subsection (a) or (b) of this section, which concerns any information which contains or relates to a trade secret or other matter referred to in section 1905 of title 18 of the United States Code, except that such information may be disclosed to other officers or employees of the Department and of other agencies concerned with carrying out this subchapter or when relevant in any proceeding under this subchapter. Nothing in this section shall authorize the withholding of information by the Secretary, or by any officers or employees under his control, from the duly authorized committees of the Congress.

(f) The Secretary may by regulation (1) require dealers and distributors of electronic products, to which there are applicable standards prescribed under this subchapter and the retail prices of which is not less than $50, to furnish manufacturers of such products such information as may be necessary to identify and locate, for purposes of section 535, the first purchasers of such products for purposes other than resale, and (2) require manufacturers to preserve such information. Any regulation establishing a requirement pursuant to clause (1) of the preceding sentence shall (A) authorize such dealers and distributors to elect, in lieu of immediately furnishing such information to the manufacturer, to hold and preserve such information until advised by the manufacturer or Secretary that such information is needed by the manufacturer for purposes of section 535 and (B) provide that the dealer or distributor shall, upon making such election, give prompt notice of such election (together with information identifying the notifier and the product) to the manufacturer and shall, when advised by the manufacturer or Secretary of the need therefor for the purposes of section 535, immediately furnish the manufacturer with the required information. If a dealer or distributor discontinues the dealing in or distribution of electronic products, he shall turn the information over to the manufacturer. Any manufacturer receiving information pursuant to this subsection concerning first purchasers of products for purposes other than resale shall treat it as confidential and may use it only if necessary for the purpose of notifying persons pursuant to section 535(a).

SEC. 538 [21 U.S.C. 360oo] Prohibited acts

(a) It shall be unlawful--

(1) for any manufacturer to introduce, or to deliver for introduction, into commerce, or to import into the United States, any electronic product which does not comply with an applicable standard prescribed pursuant to section 534;

(2) for any person to fail to furnish any notification or other material or information required by section 535 or 537; or to fail to comply with the requirements of section 535(f);

(3) for any person to fail or to refuse to establish or maintain records required by this subchapter or to permit access by the Secretary or any of his duly authorized representatives to, or the copying of, such records, or to permit entry or inspection, as required by or pursuant to section 537;

(4) for any person to fail or to refuse to make any report required pursuant to section 537(b) or to furnish or preserve any information required pursuant to section 537(f); or

(5) for any person (A) to fail to issue a certification as required by section 534(h), or (B) to issue such a certification when such certification is not based upon a test or testing program meeting the requirements of section 534(h) or when the issuer, in the exercise of due care, would have reason to know that such certification is false or misleading in a material respect.

(b) The Secretary may exempt any electronic product, or class thereof, from all or part of subsection (a), upon such conditions as he may find necessary to protect the public health or welfare, for the purpose of research, investigations, studies, demonstrations, or training, or for reasons of national security.

SEC. 539 [21 U.S.C. 360pp] Enforcement

(a) The district courts of the United States shall have jurisdiction, for cause shown, to restrain violations of section 538 and to restrain dealers and distributors of electronic products from selling or otherwise disposing of electronic products which do not conform to an applicable standard prescribed pursuant to section 534 except when such products are disposed of by returning them to the distributor or manufacturer from whom they were obtained. The district courts of the United States shall also have jurisdiction in accordance with section 1355 of title 28 of the United States Code to enforce the provisions of subsection (b) of this section.

(b)(1) Any person who violates section 538 shall be subject to a civil penalty of not more than $1,000. For purposes of this subsection, any such violation shall with respect to each electronic product involved, or with respect to each act or omission made unlawful by section 538, constitute a separate violation, except that the maximum civil penalty imposed on any person under this subsection for any related series of violations shall not exceed $300,000.

(2) Any such civil penalty may on application be remitted or mitigated by the Secretary. In determining the amount of such penalty or whether it should be remitted or mitigated and in what amount, the appropriateness of such penalty to the size of the business of the person charged and the gravity of the violation shall be considered. The amount of such penalty, when finally determined, may be deducted from any sums owing by the United States to the person charged.

(c) Actions under subsections (a) and (b) of this section may be brought in the district court of the United States for the district wherein any act or omission or transaction constituting the violation occurred, or in such court for the district where the defendant is found or transacts business, and process in such cases may be served in any other district of which the defendant is an inhabitant or wherever the defendant may be found.

(d) Nothing in this subchapter shall be construed as requiring the Secretary to report for the institution of proceedings minor violations of this subchapter whenever he believes that the public interest will be adequately served by a suitable written notice or warning.

(e) Except as provided in the first sentence of section 542, compliance with this subchapter or any regulations issued thereunder shall not relieve any person from liability at common law or under statutory law.

(f) The remedies provided for in this subchapter shall be in addition to and not in substitution for any other remedies provided by law.

SEC. 540 [21 U.S.C. 360qq] Annual report

(a) The Secretary shall prepare and submit to the President for transmittal to the Congress on or before April 1 of each year a comprehensive report on the administration of this subchapter for the preceding calendar year. Such report shall include--

(1) a thorough appraisal (including statistical analyses, estimates, and long-term projections) of the incidence of biological injury and effects, including genetic effects, to the population resulting from exposure to electronic product radiation, with a breakdown, insofar as practicable, among the various sources of such radiation;

(2) a list of Federal electronic product radiation control standards prescribed or in effect in such year, with identification of standards newly prescribed during such year;

(3) an evaluation of the degree of observance of applicable standards, including a list of enforcement actions, court decisions, and compromises of alleged violations by location and company name;

(4) a summary of outstanding problems confronting the administration of this subchapter in order of priority;

(5) an analysis and evaluation of research activities completed as a result of Government and private sponsorship, and technological progress for safety achieved during such year;

(6) a list, with a brief statement of the issues, of completed or pending judicial actions under this subchapter;

(7) the extent to which technical information was disseminated to the scientific, commercial, and labor community and consumer-oriented information was made available to the public; and

(8) the extent of cooperation between Government officials and representatives of industry and other interested parties in the implementation of this subchapter including a log or summary of meetings held between Government officials and representatives of industry and other interested parties.

(b) The report required by subsection (a) shall contain such recommendations for additional legislation as the Secretary deems necessary to promote cooperation among the several States in the improvement of electronic product radiation control and to strengthen the national electronic product radiation control program.

SEC. 541 [21 U.S.C. 360rr] Federal-State cooperation

The Secretary is authorized (1) to accept from State and local authorities engaged in activities related to health or safety or consumer protection, on a reimbursable basis or otherwise, any assistance in the administration and enforcement of this subchapter which he may request and which they may be able and willing to provide and, if so agreed, may pay in advance or otherwise for the reasonable cost of such assistance, and (2) he may, for the purpose of conducting examinations, investigations, and inspections, commission any officer or employee of any such authority as an officer of the Department.

SEC. 542 [21 U.S.C. 360ss] Effect on State standards

Whenever any standard prescribed pursuant to section 534 with respect to an aspect of performance of an electronic product is in effect, no State or political subdivision of a State shall have any authority either to establish, or to continue in effect, any standard which is applicable to the same aspect of performance of such product and which is not identical to the Federal standard. Nothing in this subchapter shall be construed to prevent the Federal Government or the government of any State or political subdivision thereof from establishing a requirement with respect to emission of radiation from electronic products procured for its own use if such requirement imposes a more restrictive standard than that required to comply with the otherwise applicable Federal standard.

SEC. 708 [21 U.S.C. 379] Confidential information

The Secretary may provide any information which is exempt from disclosure pursuant to subsection (a) of section 552 of title 5, United States Code, by reason of subsection (b)(4) of such section to a person other than an officer or employee of the Department if the Secretary determines such other person requires the information in connection with an activity which is undertaken under contract with the Secretary, which relates to the administration of this Act, and with respect to which the Secretary (or an officer or employee of the Department) is not prohibited from using such information. The Secretary shall require as a condition to the provision of information under this section that the person receiving it take such security precautions respecting the information as the Secretary may by regulation prescribe.

SEC. 709 [21 U.S.C. 379a] Presumption

In any action to enforce the requirements of this Act respecting a device the connection with interstate commerce required for jurisdiction in such action shall be presumed to exist.

SEC. 710 [21 U.S.C. 379b] Consolidated administrative and laboratory facility

(a) Authority. The Secretary, in consultation with the Administrator of the General Services Administration, shall enter into contracts for the design, construction, and operation of a consolidated Food and Drug Administration administrative and laboratory facility.

(b) Awarding of Contract. The Secretary shall solicit contract proposals under subsection (a) from interested parties. In awarding contracts under such subsection, the Secretary shall review such proposals and give priority to those alternatives that are the most cost effective for the Federal Government and that allow for the use of donated land, federally owned property, or lease-purchase arrangements. A contract under this subsection shall not be entered into unless such contract results in a net cost savings to the Federal Government over the duration of the contract, as compared to the Government purchase price including borrowing by the Secretary of the Treasury.

(c) Donations. In carrying out this section, the Secretary shall have the power, in connection with real property, buildings, and facilities, to accept on behalf of the Food and Drug Administration gifts or donations of services or property, real or personal, as the Secretary determines to be necessary.

(d) Authorization of Appropriations. There are authorized to be appropriated to carry out this section $100,000,000 for fiscal year 1991, and such sums as may be necessary for each of the subsequent fiscal years, to remain available until expended.

SEC. 711 [21 U.S.C. 379d] Automation of Food and Drug Administration

(a) In General. The Secretary, acting through the Commissioner of Food and Drugs, shall automate appropriate activities of the Food and Drug Administration to ensure timely review of activities regulated under this Act.

(b) Authorization of Appropriations. There are authorized to be appropriated each fiscal year such sums as are necessary to carry out this section.

SUBPART C--Fees

PART 2--Fees relating to drugs

SEC. 735 [21 U.S.C. 379g] Definitions

For purposes of this subchapter:

(1) The term "human drug application" means an application for--

(A) approval of a new drug submitted under section 505(b)(1),

(B) approval of a new drug submitted under section 505(b)(2) after September 30, 1992, which requests approval of--

(i) a molecular entity which is an active ingredient (including any salt or ester of an active ingredient), or

(ii) an indication for a use,

that had not been approved under an application submitted under section 505(b),

(C) initial certification or initial approval of an antibiotic drug under section 507, or

(D) licensure of a biological product under section 351 of the Public Health Service Act.

Such term does not include a supplement to such an application, does not include an application with respect to whole blood or a blood component for transfusion, does not include an application with respect to a bovine blood product for topical application licensed before September 1, 1992, an allergenic extract product, or an in vitro diagnostic biologic product licensed under section 351 of the Public Health Service Act, and does not include an application with respect to a large volume parenteral drug product approved before September 1, 1992.

(2) The term "supplement" means a request to the Secretary to approve a change in a human drug application which has been approved.

(3) The term "prescription drug product" means a specific strength or potency of a drug in final dosage form--

(A) for which a human drug application has been approved, and

(B) which may be dispensed only under prescription pursuant to section 503(b).

Such term does not include whole blood or a blood component for transfusion, does not include a bovine blood product for topical application licensed before September 1, 1992, an allergenic extract product, or an in vitro diagnostic biologic product licensed under section 351 of the Public Health Service Act, and does not include a large volume parenteral drug product approved before September 1, 1992.

(4) The term "final dosage form" means, with respect to a prescription drug product, a finished dosage form which is approved for administration to a patient without further manufacturing.

(5) The term "prescription drug establishment" means a foreign or domestic place of business which is--

(A) at one general physical location consisting of one or more buildings all of which are within 5 miles of each other, at which one or more prescription drug products are manufactured in final dosage form, and

(B) under the management of a person that is listed as the applicant in a human drug application for a prescription drug product with respect to at least one such product.

For purposes of this paragraph, the term "manufactured" does not include packaging.

(6) The term "process for the review of human drug applications" means the following activities of the Secretary with respect to the review of human drug applications and supplements:

(A) The activities necessary for the review of human drug applications and supplements.

(B) The issuance of action letters which approve human drug applications or which set forth in detail the specific deficiencies in such applications and, where appropriate, the actions necessary to place such applications in condition for approval.

(C) The inspection of prescription drug establishments and other facilities undertaken as part of the Secretary's review of pending human drug applications and supplements.

(D) Activities necessary for the review of applications for licensure of establishments subject to section 351 of the Public Health Service Act and for the release of lots of biologics under such section.

(E) Monitoring of research conducted in connection with the review of human drug applications.

(7) The term "costs of resources allocated for the process for the review of human drug applications" means the expenses incurred in connection with the process for the review of human drug applications for--

(A) officers and employees of the Food and Drug Administration, employees under contract with the Food and Drug Administration who work in facilities owned or leased for the Food and Drug Administration, advisory committees, and costs related to such officers, employees, and committees,

(B) management of information, and the acquisition, maintenance, and repair of computer resources,

(C) leasing, maintenance, renovation, and repair of facilities and acquisition, maintenance, and repair of fixtures, furniture, scientific equipment, and other necessary materials and supplies, and

(D) collecting fees under section 736 and accounting for resources allocated for the review of human drug applications and supplements.

(8) The term ''adjustment factor'' applicable to a fiscal year is the lower of--

(A) the Consumer Price Index for all urban consumers (all items; United States city average) for August of the preceding fiscal year divided by such Index for August 1992, or

(B) the total of discretionary budget authority provided for programs in the domestic category for the immediately preceding fiscal year (as reported in the Office of Management and Budget sequestration preview report, if available, required under section 254(d) of the Balanced Budget and Emergency Deficit Control Act of 1985) divided by such budget authority for fiscal year 1992 (as reported in the Office of Management and Budget final sequestration report submitted after the end of the 102d Congress, 2d Session).

The terms ''budget authority'' and ''category'' in subparagraph (B) are as defined in the Balanced Budget and Emergency Deficit Control Act of 1985, as in effect as of September 1, 1992.

SEC. 736 [21 U.S.C. 379h] Authority to assess and use drug fees

(a) Types of Fees. Beginning in fiscal year 1993, the Secretary shall assess and collect fees in accordance with this section as follows:

(1) Human drug application and supplement fee.

(A) In general. Each person that submits, on or after September 1, 1992, a human drug application or a supplement shall be subject to a fee as follows:

(i) A fee established in subsection (b) for a human drug application for which clinical data (other than bioavailability or bioequivalence studies) with respect to safety or effectiveness are required for approval.

(ii) A fee established in subsection (b) for a human drug application for which clinical data with respect to safety or effectiveness are not required or a supplement for which clinical data (other than bioavailability or bioequivalence studies) with respect to safety or effectiveness are required.

(B) Payment schedule.

(i) First payment. 50 percent of the fee required by subparagraph (A) shall be due upon submission of the application or supplement.

(ii) Final payment. The remaining 50 percent of the fee required by subparagraph (A) shall be due upon--

(I) the expiration of 30 days from the date the Secretary sends to the applicant a letter designated by the Secretary as an action letter described in section 735(6)(B), or

(II) the withdrawal of the application or supplement after it is filed unless the Secretary waives the fee or a portion of the fee because no substantial work was performed on such application or supplement after it was filed.

The designation under subclause (I) or the waiver under subclause (II) shall be solely in the discretion of the Secretary and shall not be reviewable.

(C) Exception for previously filed application or supplement. If human drug application or supplement was submitted by a person that paid the fee for such application or supplement, was accepted for filing, and was not approved or was withdrawn (without a waiver), the submission of a human drug application or a supplement for the same product by the same person (or the person's licensee, assignee, or successor) shall not be subject to a fee under subparagraph (A).

(D) Refund of fee if application not accepted for filing. The Secretary shall refund 50 percent of the fee paid under subparagraph (B)(i) for any application or supplement which is not accepted for filing.

(2) Prescription drug establishment fee. Each person that--

(A) owns a prescription drug establishment, at which is manufactured at least 1 prescription drug product which is not the, or not the same as a, product approved under an application filed under section 505(b)(2) or 505(j), and

(B) after September 1, 1992, had pending

before the Secretary a human drug application or supplement,

shall be subject to the annual fee established in subsection (b) for each such establishment, payable on or before January 31 of each year.

(3) Prescription drug product fee.

(A) In general. Except as provided in subparagraph (B), each person--

(i) who is named as the applicant in a human drug application for a prescription drug product which is listed under section 510, and

(ii) who, after September 1, 1992, had pending before the Secretary a human drug application or supplement,

shall pay for each such prescription drug product the annual fee established in subsection (b). Such fee shall be payable at the time of the first such listing of such product in each calendar year. Such fee shall be paid only once each year for each listed prescription drug product irrespective of the number of times such product is listed under section 510.

(B) Exception. The listing of a prescription drug product under section 510 shall not require the person who listed such product to pay the fee prescribed by subparagraph (A) if such product is the same product as a product approved under an application filed under section 505(b)(2) or 505(j).

(b) Fee Amounts.--

(1) Schedule. Except as provided in paragraph (2) and subsections (c), (d), (f), and (g), the fees required under subsection (a) shall be paid in accordance with the following schedule:

	Fiscal Year 1993	Fiscal Year 1994	Fiscal Year 1995	Fiscal Year 1996	Fiscal Year 1997
Drug application fee:					
Subsection (a)(1)(A)(i) fee . . .	$100,000	$150,000	$208,000	$217,000	$233,000
Subsection (a)(1)(A)(ii) fee . . .	$50,000	$75,000	$104,000	$108,000	$116,000
Fee revenue	$12,000,000	$18,000,000	$25,000,000	$26,000,000	$28,000,000
Annual establishment fee:					
Fee per establishment	$60,000	$88,000	$126,000	$131,000	$138,000
Fee revenue	$12,000,000	$18,000,000	$25,000,000	$26,000,000	$28,000,000
Annual product fee:					
Fee per product . . .	$6,000	$9,000	$12,500	$13,000	$14,000
Fee revenue	$12,000,000	$18,000,000	$25,000,000	$26,000,000	$28,000,000
Total fee revenues..	$16,000,000	$54,000,000	$75,000,000	$78,000,000	$84,000,000

(2) Small business exception. Any business which has fewer than 500 employees, including employees of affiliates, and which does not have a prescription drug product introduced or delivered for introduction into interstate commerce shall pay one-half the amount of the fee for human drug applications it submits and shall pay the entire amount of the fee for supplements it submits. Such a business shall not be required to pay any portion of any fee required under subsection (a)(1)(A) until 1 year after the date of the submission of the application involved. For purposes of this paragraph, one business is an affiliate of another business when, directly or indirectly, one business controls, or has the power to control, the other business or a third party controls, or has the power to control, both businesses.

(c) Increases and Adjustments.

(1) Revenue increase. The total fee revenues established by the schedule in subsection

(b)(1) shall be increased by the Secretary by notice, published in the Federal Register, for a fiscal year to reflect the greater of--

(A) the total percentage increase that occurred during the preceding fiscal year in the Consumer Price Index for all urban consumers (all items; U.S. city average), or

(B) the total percentage increase for such fiscal year in basic pay under the General Schedule in accordance with section 5332 of title 5 United States Code, as adjusted by any locality-based comparability payment pursuant to section 5304 of such title for Federal employees stationed in the District of Columbia.

(2) Annual fee adjustment. Subject to the amount appropriated for a fiscal year under subsection (g), the Secretary shall, within 60 days after the end of each fiscal year beginning after October 1, 1992, adjust the fees established by the schedule in subsection (b)(1) for the following fiscal year to achieve the total fee revenues, as may be increased under paragraph (1). Such fees shall be adjusted under this paragraph to maintain the proportions established in such schedule.

(3) Limit. The total amount of fees charged, as adjusted under paragraph (2), for a fiscal year may not exceed the total costs for such fiscal year for the resources allocated for the process for the review of human drug applications.

(d) Fee Waiver or Reduction. The Secretary shall grant a waiver from or a reduction of 1 or more fees under subsection (a) where the Secretary finds that--

(1) such waiver or reduction is necessary to protect the public health,

(2) the assessment of the fee would present a significant barrier to innovation because of limited resources available to such person or other circumstances,

(3) the fees to be paid by such person will exceed the anticipated present and future costs incurred by the Secretary in conducting the process for the review of human drug applications for such person, or

(4) assessment of the fee for an application or a supplement filed under section 505(b)(1) pertaining to a drug containing an active ingredient would be inequitable because an application for a product containing the same active ingredient filed by another person under section 505(b)(2) could not be assessed fees under subsection (a)(1).

In making the finding in paragraph (3), the Secretary may use standard costs.

(e) Effect of Failure To Pay Fees. A human drug application or supplement submitted by a person subject to fees under subsection (a) shall be considered incomplete and shall not be accepted for filing by the Secretary until all fees owed by such person have been paid.

(f) Assessment of Fees.

(1) Limitation. Fees may not be assessed under subsection (a) for a fiscal year beginning after fiscal year 1993 unless appropriations for salaries and expenses of the Food and Drug Administration for such fiscal year (excluding the amount of fees appropriated for such fiscal year) are equal to or greater than the amount of appropriations for the salaries and expenses of the Food and Drug Administration for the fiscal year 1992 multiplied by the adjustment factor applicable to the fiscal year involved.

(2) Authority. If the Secretary does not assess fees under subsection (a) during any portion of a fiscal year because of paragraph (1) and if at a later date in such fiscal year the Secretary may assess such fees, the Secretary may assess and collect such fees, without any modification in the rate, for human drug applications and supplements, prescription drug establishments, and prescription drug products at any time in such fiscal year notwithstanding the provisions of subsection (a) relating to the date fees are to be paid.

(g) Crediting and Availability of Fees.

(1) In general. Fees collected for a fiscal year pursuant to subsection (a) shall be credited to the appropriation account for salaries and expenses of the Food and Drug Administration and shall be available in accordance with appropriation Acts until expended without fiscal year limitation.

(2) Collections and appropriation acts. The fees authorized by this section--

(A) shall be collected in each fiscal year in an amount equal to the amount specified in appropriation acts for such fiscal year, and

(B) shall only be collected and available to defray increases in the costs of the resources allocated for the process for the review of human drug applications (including increases in such costs for an additional number of full-time equivalent positions in the Department of Health and Human Services to be engaged in such process) over such costs for fiscal year 1992 multiplied by the adjustment factor.

(3) Authorization of appropriations. There are authorized to be appropriated for fees under this section--

(A) $36,000,000 for fiscal year 1993,

(B) $54,000,000 for fiscal year 1994,

(C) $75,000,000 for fiscal year 1995,

(D) $78,000,000 for fiscal year 1996, and

(E) $84,000,000 for fiscal year 1997,

as adjusted to reflect increases in the total fee revenues made under subsection (c)(1).

(h) Collection of Unpaid Fees. In any case where the Secretary does not receive payment of a fee assessed under subsection (a) within 30 days after it is due, such fee shall be treated as a claim of the United States Government subject to subchapter II of chapter 37 of title 31, United States Code.

(i) Construction. This section may not be construed to require that the number of full--time equivalent positions in the Department of Health and Human Services, for officers, employers, and advisory committees not engaged in the process of the review of human drug applications, be reduced to offset the number of officers, employees, and advisory committees so engaged.

CHAPTER VIII--Imports and Exports

SEC. 801 [21 U.S.C. 381] Imports and exports

(a) The Secretary of the Treasury shall deliver to the Secretary of Health and Human Services, upon his request, samples of food, drugs, devices, and cosmetics which are being imported or offered for import into the United States, giving notice thereof to the owner or consignee, who may appear before the Secretary of Health and Human Services and have the right to introduce testimony. The Secretary of Health and Human Services shall furnish to the Secretary of the Treasury a list of establishments registered pursuant to subsection (i) of section 510 and shall request that if any drugs or devices manufactured, prepared, propagated, compounded, or processed in an establishment not so registered are imported or offered for import into the United States, samples of such drugs or devices be delivered to the Secretary of Health and Human Services, with notice of such delivery to the owner or consignee, who may appear before the Secretary of Health and Human Services and have the right to introduce testimony. If it appears from the examination of such samples or otherwise that (1) such article has been manufactured, processed, or packed under insanitary conditions or, in the case of a device, the methods used in, or the facilities or controls used for, the manufacture, packing, storage, or installation of the device do not conform to the requirements of section 520(f), or (2) such article is forbidden or restricted in sale in the country in which it was produced or from which it was exported, or (3) such article is adulterated, misbranded, or in violation of section 505, then such article shall be refused admission, except as provided in subsection (b) of this section. The Secretary of the Treasury shall cause the destruction of any such article refused admission unless such article is exported, under regulations prescribed by the Secretary of the Treasury, within ninety days of the date of notice of such refusal or within such additional time as may be permitted pursuant to such regulations. Clause (2) of the third sentence of this paragraph shall not be construed to prohibit the admission of narcotic drugs the importation of which is permitted under the Controlled Substances Import and Export Act.

(b) Pending decision as to the admission of an article being imported or offered for import, the Secretary of the Treasury may authorize delivery of such article to the owner or consignee upon the execution by him of a good and sufficient bond providing for the payment of such liquidated damages in the event of default as may be required pursuant to regulations of the Secretary of the Treasury. If it appears to the Secretary of Health and Human Services that an article included within the provisions of clause (3) of subsection (a) of this section can, by relabeling or other action, be brought into compliance with the Act or rendered other than a food, drug, device, or cosmetic, final determination as to admission of such article may be deferred and, upon filing of timely written application by the owner or consignee and the execution by him of a bond as provided in the preceding provisions of this subsection, the Secretary may, in accordance with regulations, authorize the applicant to perform such relabeling or other action specified in such authorization (including destruction or export of rejected articles or portions thereof, as may be specified in the Secretary's authorization). All such relabeling or other action pursuant to such authorization shall in accordance with regulations be under the supervision of an officer or employee of the Department of Health and Human Services designated by the Secretary, or an officer or employee of the Department of the Treasury designated by the Secretary of the Treasury.

(c) All expenses (including travel, per diem or subsistence, and salaries of officers or employees of the United States) in connection with the destruction provided for in subsection (a) of this section and the supervision of the relabeling or other action authorized under the provisions of subsection (b) of this section, the amount of such expenses to be determined in accordance with regulations, and all expenses in connection with the storage, cartage, or labor with respect to any article refused admission under subsection (a) of this section, shall be paid by the owner or consignee and, in default of such payment, shall constitute a lien against any future importations made by such owner or consignee.

(d)(1) Except as provided in paragraph (2), no drug subject to section 503(b) which is manufactured in a State and exported may be imported into the United States unless the drug is imported by the manufacturer of the drug.

(2) The Secretary may authorize the importation of a drug the importation of which is prohibited by paragraph (1) if the drug is required for emergency medical care.

(e)(1) A food, drug, device, or cosmetic intended for export shall not be deemed to be adulterated or misbranded under this Act if it--

(A) accords to the specifications of the foreign purchaser,

(B) is not in conflict with the laws of the country to which it is intended for export,

(C) is labeled on the outside of the shipping package that it is intended for export, and

(D) is not sold or offered for sale in domestic commerce.

This paragraph does not authorize the exportation of any new animal drug, or an animal feed bearing or containing a new animal drug, which is unsafe within the meaning of section 512.

(2) Paragraph (1) does not apply to any device--

(A) which does not comply with an applicable requirement of section 514 or 515,

(B) which under section 520(g) is exempt from either such section, or

(C) which is a banned device under section 516,

unless, in addition to the requirements of paragraph (1), the Secretary has determined that the exportation of the device is not contrary to public health and safety and has the approval of the country to which it is intended for export.

SEC. 802 [21 U.S.C. 382] Exports of certain unapproved products

(a) A drug (including a biological product) intended for human or animal use--

(1) which--

(A) requires approval by the Secretary under section 505 or section 512, or

(B) requires licensing by the Secretary under section 351 of the Public Health Service Act or by the Secretary of Agriculture under the Act of March 4, 1913 (known as the Virus Serum Toxin Act),

before it may be introduced or delivered for introduction into interstate commerce to a country, and

(2) which does not have such approval or license, which is not exempt from such sections or Act, and which is introduced or delivered for introduction into interstate commerce to a country, is adulterated, misbrand-

ed, and in violation of such sections or Act unless the export of the drug is authorized under subsection (b).

(b)(1) A drug (including a biological product) may, upon approval of an application submitted under paragraph (3), be exported if--

(A) the drug contains the same active ingredient as a--

(i) new drug--

(I) which has an exemption under section 505(i), and

(II) for which approval is actively being pursued by the person who has the exemption,

(ii) biological product for human use--

(I) which has an exemption under section 505(i), and

(II) for which licensing of the biological product under section 351 of the Public Health Service Act is actively being pursued by the person who has the exemption,

(iii) biological product for animal use--

(I) for which authority has been granted under the Virus-Serum Toxin Act for the preparation of an experimental drug product, and

(II) for which the licensing of the biological product under such Act is actively being pursued by the person who has the authority, or

(iv) new animal drug--

(I) which has an exemption under section 512(j), and

(II) for which approval is actively being pursued by the person who has the exemption,

(B) except as provided in paragraph (2), the drug is exported to a country which is listed under paragraph (4) and in which the drug is approved and has not been withdrawn from sale,

(C) an application for the drug under section 505 or 512, section 351 of the Public Health Service Act, or the Virus- Serum Toxin Act has not been disapproved by an order of the Secretary under section 505(d) or 512(d) or 351 of the Public Health Service Act or by the Secretary of Agriculture in the case of an application under the Virus-Serum Toxin Act,

(D) the drug is manufactured, processed, packaged, and held in conformity with current good manufacturing practice and is not adulterated under paragraph (a)(1), (a)(2)(A), (a)(3), (c), or (d) of section 501,

(E) the outside of the shipping package is labeled with the following statement: "This drug may be sold or offered for sale only in the following countries: ", the blank space being filled with a list of the countries to which export of the drug is authorized under this subsection,

(F) the drug is not the subject of a notice by the Secretary or the Secretary of Agriculture of a determination that the manufacture of the drug in the United States for export to a country is contrary to the public health and safety of the United States,

(G) the requirements of subparagraphs (A) through (D) of section 801(d)(1) have been met. The Secretary shall determine that an applicant is actively pursuing the approval or licensing of a drug if the applicant has demonstrated that degree of attention and continuous directed effort as may reasonably be expected from, and are ordinarily exercised by, a person before approval or licensing of a drug, such as the preparation for and the conduct of preclinical or clinical investigations, the analysis of the results of such investigations, conferences on such investigations with government officials, and the preparation of an application of approval or licensing for the drug.

(2) The Secretary may permit the export of a drug under paragraph (1) to a country which is listed under paragraph (4) and in which the drug is not approved if the drug is exported to such country solely for the purpose of further export to a country listed in paragraph (4) in which the drug is approved.

(3)(A) Any person may apply to have a drug exported under paragraph (1). Such an application shall be filed at least 90 days before the date the applicant proposes to export the drug for which the application is submitted. Before the expiration of 10 days from the date of the submission of such an application, the Secretary shall publish a notice in the Federal Register which identifies the applicant under such application, the drug proposed to be exported under such application, and the country to which the drug is proposed to be exported.

(B) An application for the export of a drug

shall--

(i) identify the drug to be exported,

(ii) list each country to which the drug is to be exported and list the persons in each such country to which the drug is to be exported,

(iii) contain a certification by the applicant that--

(I) the applicant will export the drug only to a country which is listed in paragraph (4) and in which the drug is approved unless the drug is authorized to be exported under paragraph (2) and will export only those quantities of the drug which may reasonably be sold in each country to which it is to be exported,

(II) the drug is approved by each country to which it is to be exported unless the drug is authorized to be exported under paragraph (2) and the drug has not been withdrawn from sale in such country,

(III) the drug meets the requirements of paragraph (1)(D),

(IV) the drug will be labeled in accordance with paragraph (1)(E), and

(V) the drug meets the requirements of paragraphs (1)(C) and (1)(G),

(iv) contain a certification by the holder of the exemption or authority for such drug described in paragraph (1)(A) that the holder will actively pursue the approval or licensing of the drug,

(v) identify the exemption or authority for an experimental drug product in effect for such drug under the laws referred to in paragraph (1)(A),

(vi) identify the establishments in which the drug is manufactured, and

(vii) include a written agreement from each importer to whom the drug is to be exported from the United States that such importer will not export the drug to a country which is not listed under paragraph (4) and will provide notice to the applicant of any knowledge of an export of the drug to such a country by any person and will maintain records of the drug wholesale distributors to which the drug is sold.

(C)(i) Before the expiration of 30 days after the date an application is submitted to the Secretary, the Secretary shall review the application to determine if the application meets the requirements of clauses (i), (ii), (iv), (v), (vi), and (vii) of subparagraph (B) and contains the certifications described in clauses (iii) (other than the certification required by clause (iii)(II)) and (iv). If the Secretary determines that the application meets such requirements and contains such certifications the Secretary shall conditionally approve the application. An application which is so conditionally approved shall be finally approved within 5 days of the submission of the certification required by clause (iii)(II).

(ii) If the Secretary proposes to disapprove an application, the Secretary shall provide the applicant with a written statement specifying--

(I) the deficiencies which the applicant must correct in order to enable the Secretary to approve the application, and

(II) that the applicant has 60 days after receiving the statement to correct such deficiencies.

(D) If the holder of an application approved under subparagraph (C) for the export of a drug intends to export such drug to a country listed in paragraph (4) which is not listed in such application, such holder shall submit an amendment to such application to the Secretary not later than 30 days before the date of the proposed export to such country identifying the country to which the holder intends to export such drug and containing information sufficient to show that the drug is approved by such country and has not been withdrawn from sale in such country. The Secretary shall approve or disapprove the export of such drug to such country within 15 days of the receipt of the notice required by this subparagraph.

(4)(A) The countries to which a drug may be exported under paragraph (1) are--

(i) Australia,

(ii) Austria,

(iii) Belgium,

(iv) Canada,

(v) Denmark,

(vi) Federal Republic of Germany,

(vii) Finland,

(viii) France,

(ix) Iceland,

(x) Ireland,

(xi) Italy,

(xii) Japan,

(xiii) Luxembourg,

(xiv) The Netherlands,
(xv) New Zealand,
(xvi) Norway,
(xvii) Portugal,
(xviii) Spain,
(xix) Sweden,
(xx) Switzerland, and
(xxi) The United Kingdom.

(B) Changes in the list contained in subparagraph (A) shall be based on the following criteria:

(i) Statutory or regulatory requirements which require the review of drugs for safety and effectiveness by an entity of the government of such country and which authorize the approval of only those drugs which have been determined to be safe and effective by experts employed by or acting on behalf of such entity and qualified by scientific training and experience to evaluate the safety and effectiveness of drugs on the basis of adequate and well--controlled investigations, including clinical investigations, conducted by experts qualified by scientific training and experience to evaluate the safety and effectiveness of drugs.

(ii) Statutory or regulatory requirements that the methods used in, and the facilities and controls used for, the manufacture, processing and packing of drugs in the country are adequate to preserve their identity, quality, purity, and strength.

(iii) Statutory or regulatory requirements for the reporting of adverse reactions to drugs and procedures to withdraw approval and remove drugs found not to be safe or effective.

(iv) Statutory or regulatory requirements that the labeling and promotion of drugs must be in accordance with the approval of the drug.

(c)(1) The holder of an approved application under subsection (b) authorizing the export of a drug shall report to the Secretary--

(A) any withdrawal of an approval of the drug by any country to which it has been exported,

(B) any withdrawal of the drug from sale in any such country,

(C) the withdrawal of an application by the holder under section 505 or 512, section 351 of the Public Health Service Act, or the Virus-Serum Toxin Act, and

(D) the receipt of any credible information indicating that the drug is being or may have been exported from a country listed under subsection (b)(4) to a country which is not listed under such subsection.

The reporting of an event described in subparagraph (A), (B), or (C) shall be made within 15 days of the occurrence of the event and the reporting of the receipt of information under subparagraph (D) shall be made within 15 days of the receipt of such information.

(2) The holder of an approved application under subsection (b) authorizing the export of a drug shall report annually to the Secretary after the date of the approval of the application of the actions taken by the holder in pursuit of the approval of such drug during the year reported on. Not later than 90 days from the date of the receipt of a report under this paragraph the Secretary shall determine if the holder is actively pursuing the approval of such drug.

(d) A drug authorized to be exported to a country under an application approved under subsection (b) may not be exported to such country if--

(1) an approval of such drug is withdrawn by such country,

(2) the drug is withdrawn from sale in such country,

(3) the Secretary issues an order refusing to approve an application of the holder of such application under section 505 or 512, section 351 of the Public Health Service Act, or the Virus-Serum Toxin Act, or

(4) an application for such drug under such section or Act is withdrawn or if an exemption for such drug under section 505(i) or 512(j) or the authority granted for such drug to prepare an experimental drug product under the Virus-Serum Toxin Act is withdrawn and no application for approval of such drug has been submitted under section 505 or 512, section 351 of the Public Health Service Act, or the Virus-Serum Toxin Act.

(e)(1) If the Secretary determines that--

(A) a drug for which an application was approved under subsection (b) no longer complies with subparagraphs (A), (D), (E), and (G) of paragraph (1) of such subsection or with paragraph (2) of such subsection or the holder of such application has not made the reports required by subsection (c), or

(B) the manufacture of a drug in the United States for export is contrary to the public health and safety of the United States and an application for the export of such drug has been approved under subsection (b),

then before taking action against the holder of an application for which a determination was made under subparagraph (A) or (B), the Secretary shall notify the holder in writing of the determination and provide the holder 30 days to take such action as may be required so that the Secretary would be unable to make such determination. When the Secretary takes action against such holder because of such a determination, the Secretary shall provide the holder a written statement specifying the reasons for such determination and provide the person, on request, an opportunity for an informal hearing with respect to such determination.

(2) If the Secretary determines that the approval of a drug is not being actively pursued as required by subsection (b)(1) (A), the Secretary shall give the holder of the application authorizing the export of such drug 60 days to assure that actions are taken to actively pursue such approval. During the 60-day period the Secretary shall give the holder an opportunity for an informal hearing on the determination of the Secretary. If upon the expiration of such 60-day period the Secretary determines that approval of such drug is not being actively pursued, the Secretary shall prohibit the export of such drug.

(3)(A) If at any time the Secretary, or in the absence of the Secretary the individual acting as the Secretary, determines that--

(i) the holder of an approved application under subsection (b) is exporting a drug from the United States to an importer,

(ii) such importer is exporting the drug to a country which is not listed under subsection (b)(4), and

(iii) such export presents an imminent hazard to the public health in such country,

the Secretary shall immediately prohibit the export of the drug to such importer, give the person exporting the drug from the United States prompt notice of the determination, and afford such person an opportunity for an expedited hearing.

(B) The authority conferred by subparagraph (A) shall not be delegated by the Secretary. A determination by the Secretary under subparagraph (A) may not be stayed pending final action by a reviewing court.

(4)(A) If the Secretary, or in the absence of the Secretary the individual acting as the Secretary, determines that the holder of an approved application under subsection (b) is exporting a drug to a country which is not listed under subsection (b)(4) and that the export of the drug presents an imminent hazard, the Secretary shall immediately prohibit the export of the drug to such country, give the holder prompt notice of the determination, and afford the holder an opportunity for an expedited hearing.

(B) The authority conferred by subparagraph (A) shall not be delegated by the Secretary. A determination by the Secretary under subparagraph (A) may not be stayed pending final action by a reviewing court.

(5) If the Secretary receives credible evidence that the holder of an application approved under subsection (b) is exporting a drug to a country which is not listed under subsection (b)(4), the Secretary shall give the holder 60 days to provide information to the Secretary respecting such evidence and shall provide the holder an opportunity for an informal hearing on such evidence. Upon the expiration of such 60 days the Secretary shall prohibit the export of such drug to such country if the Secretary determines that the holder of the application is exporting the drug to a country which is not listed under subsection (b)(4).

(6) If the Secretary receives credible evidence that an importer is exporting a drug to a country which is not listed under subsection (b)(4), the Secretary shall notify the holder of the application authorizing the export of such drug of such evidence and shall require the holder to investigate the export by such importer and to report to the Secretary within 14 days of the receipt of such notice the findings of the holder. If the Secretary determines that the importer has exported a drug to such a country, the Secretary shall prohibit such holder from exporting such drug to the im-

porter unless the Secretary determines that the export by the importer was unintentional.

(f)(1) A drug (including a biological product) which is to be used in the prevention or treatment of a tropical disease may, upon approval of an application submitted under paragraph (2), be exported if--

(A) the Secretary finds, based on credible scientific evidence, including clinical investigations, that the drug is safe and effective in the country to which it is to be exported in the prevention or treatment of a tropical disease in such country,

(B) the drug is manufactured, processed, packaged, and held in conformity with current good manufacturing practice and is not adulterated under paragraph (a)(1), (a)(2)(A), (a)(3), (c), or (d) of section 501,

(C) the outside of the shipping package is labeled with the following statement: "This drug may be sold or offered for sale only in the following countries: ", the blank space being filled with a list of the countries to which export of the drug is authorized under this subsection,

(D) the drug is not the subject of a notice by the Secretary or the Secretary of Agriculture of a determination that the manufacture of the drug in the United States for export to a country is contrary to the public health and safety of the United States, and

(E) the requirements of subparagraphs (A) through (D) of section 801(d)(1) have been met.

(2) Any person may apply to have a drug exported under paragraph (1). The application shall--

(A) describe the drug to be exported,

(B) list each country to which the drug is to be exported,

(C) contain a certification by the applicant that the drug will not be exported to a country for which the Secretary cannot make a finding described in paragraph (1)(A),

(D) identify the establishments in which the drug is manufactured, and

(E) demonstrate to the Secretary that the drug meets the requirements of paragraph (1).

(3) The holder of an approved application for the export of a drug under this subsection shall report to the Secretary--

(A) the receipt of any information indicating that the drug is being or may have been exported from a country for which the Secretary made a finding under paragraph (1)(A) to a country for which the Secretary cannot make such a finding, and

(B) the receipt of any information indicating any adverse reactions to such drug.

(4)(A) If the Secretary determines that--

(i) a drug for which an application is approved under paragraph (2) does not continue to meet the requirements of paragraph (1),

(ii) the holder of such application has not made the report required by paragraph (3), or

(iii) the manufacture of such drug in the United States for export is contrary to the public health and safety of the United States and an application for the export of such drug has been approved under paragraph (2),

then, before taking action against the holder of an application for which a determination was made under clause (i), (ii), or (iii), the Secretary shall notify the holder in writing of the determination and provide the holder 30 days to take such action as may be required so that the Secretary would be unable to make such determination. When the Secretary takes action against such holder because of such a determination, the Secretary shall provide the holder a written statement specifying the reasons for such determination and provide the person, on request, an opportunity for an informal hearing with respect to such determination.

(B) If at any time the Secretary, or in the absence of the Secretary the individual acting as the Secretary, determines that--

(i) the holder of an approved application under paragraph (2) is exporting a drug from the United States to an importer,

(ii) such importer is exporting the drug to a country for which the Secretary cannot make a finding under paragraph (1)(A), and

(iii) such export presents an imminent hazard to the public health in such country,

the Secretary shall immediately prohibit the export of the drug to such importer, give the person exporting the drug from the United States prompt notice of the determination, and afford such person an opportunity for an expedited hearing. A determination by the

Secretary under this subparagraph may not be stayed pending final action by a reviewing court. The authority conferred by this subparagraph shall not be delegated by the Secretary.

(C) If the Secretary, or in the absence of the Secretary the individual acting as the Secretary, determines that the holder of an approved application under paragraph (2) is exporting a drug to a country for which the Secretary cannot make a finding under paragraph (1)(A), and that the export of the drug presents an imminent hazard, the Secretary shall immediately prohibit the export of the drug to such country, give the holder prompt notice of the determination, and afford the holder an opportunity for an expedited hearing. A determination by the Secretary under this subparagraph may not be stayed pending final action by a reviewing court. The authority conferred by this subparagraph shall not be delegated by the Secretary.

(D) If the Secretary receives credible evidence that the holder of an application approved under paragraph (2) is exporting a drug to a country for which the Secretary cannot make a finding under paragraph (1)-(A), the Secretary shall give the holder 60 days to provide information to the Secretary respecting such evidence and shall provide the holder an opportunity for an informal hearing on such evidence. Upon the expiration of such 60 days the Secretary shall prohibit the export of such drug to such country if the Secretary determines the holder is exporting the drug to a country for which the Secretary cannot make a finding under paragraph (1)(A).

(E) If the Secretary receives credible evidence that an importer is exporting a drug to a country for which the Secretary cannot make a finding under paragraph (1)(A), the Secretary shall notify the holder of the application authorizing the export of such drug of such evidence and shall require the holder to investigate the export by such importer and to report to the Secretary within 14 days of the receipt of such notice the findings of the holder. If the Secretary determines that the importer has exported a drug to such a country, the Secretary shall prohibit such holder from exporting such drug to the importer unless the Secretary determines that the export by the importer was unintentional.

(g) For purposes of this section--

(1) a reference to the Secretary shall in the case of a biological product which is required to be licensed under the Virus- Serum Toxin Act be considered to be a reference to the Secretary of Agriculture, and

(2) a reference in paragraph (3), (4), (5), or (6) of subsection (e) and in subparagraph (B), (C), (D), or (E) of subsection (f)(4) to the holder of an application shall be considered a reference to any person which is under common control with holder, is controlled by the holder, controls the holder, is owned by the holder, or owns the holder.

SEC. 803 [21 U.S.C. 383] Office of International Relations

(a) There is established in the Department of Health and Human Services an Office of International Relations.

(b) In carrying out the functions of the office under subsection (a), the Secretary may enter into agreements with foreign countries to facilitate commerce in devices between the United States and such countries consistent with the requirements of this Act. In such agreements, the Secretary shall encourage the mutual recognition of--

(1) good manufacturing practice regulations promulgated under section 520(f), and

(2) other regulations and testing protocols as the Secretary determines to be appropriate.

CHAPTER. IX--Miscellaneous

SEC. 903 [21 U.S.C. 393] Food and Drug Administration

(a) In General. There is established in the Department of Health and Human Services the Food and Drug Administration (hereinafter in this section referred to as the "Administration").

(b) Commissioner.

(1) Appointment. There shall be in the Administration a Commissioner of Food and Drugs (hereinafter in this section referred to as the "Commissioner") who shall be ap-

pointed by the President by and with the advice and consent of the Senate.

(2) General powers. The Secretary, through the Commissioner, shall be responsible for executing this Act and for--

(A) providing overall direction to the Food and Drug Administration and establishing and implementing general policies respecting the management and operation of programs and activities of the Food and Drug Administration;

(B) coordinating and overseeing the operation of all administrative entities within the Administration;

(C) research relating to foods, drugs, cosmetics, and devices in carrying out this Act;

(D) conducting educational and public information programs relating to the responsibilities of the Food and Drug Administration; and

(E) performing such other functions as the Secretary may prescribe.

(c) Technical and Scientific Review Groups. The Secretary through the Commissioner of Food and Drugs may, without regard to the provisions of title 5, United States Code, governing appointments in the competitive service and without regard to the provisions of chapter 51 and subchapter III of chapter 53 of such title relating to classification and General Schedule pay rates, establish such technical and scientific review groups as are needed to carry out the functions of the Administration, including functions under the Federal Food, Drug, and Cosmetic Act, and appoint and pay the members of such groups, except that officers and employees of the United States shall not receive additional compensation for service as members of such groups.

SEC. 903 [21 U.S.C. 394] Scientific Review Groups

Without regard to the provisions of title 5, United States Code, governing appointments in the competitive service and without regard to the provisions of chapter 51 and subchapter III of chapter 53 of such title relating to classification and General Schedule pay rates, the Commissioner of Food and Drugs may--

(1) establish such technical and scientific review groups as are needed to carry out the functions of the Food and Drug Administration (including functions prescribed under this Act); and

(2) appoint and pay the members of such groups, except that officers and employees of the United States shall not receive additional compensation for service as members of such groups.

GOOD MANUFACTURING PRACTICES

FDA's Current Good Manufacturing Practice (CGMP) regulations for the manufacturing, processing, packing, or holding of drugs (21 CFR 210) and for finished pharmaceuticals (21 CFR 211) were first proposed in February 1976 to update the drug GMPs in light of new technology and to adopt specific requirements to assure the quality of drug products. They were finalized in September 1978. Revisions that have been made to the drug GMPs since the publication of the *Validation Compliance Annual: 1995* are as follows:

PART 211-CURRENT GOOD MANUFACTURING PRACTICE FOR FINISHED PHARMACEUTICALS

§211.1 Scope.

* * * * *

(d) The Director of the Center for Drug Evaluation and Research or the Director of the Office of Compliance, Center for Drug Evaluation and Research, may approve an exception or alternative to any application of this part to the manufacture of positron emission tomography (PET) radiopharmaceuticals. Requests for such exceptions or alternatives should ordinarily be made in writing. However, in certain circumstances, such requests may be made orally and permission may be granted orally. Oral requests and oral approvals must be followed by written requests and written approvals. Approval of a request for an exception or alternative must be obtained from either specified Director prior to the use of any affected PET radiopharmaceutical.

(1) A request for an exception or alternative is required to contain one of the following:

(i) An explanation, with supporting data as necessary, why compliance with a particular requirement of this part is unnecessary or cannot be achieved;

(ii) A description, with supporting data as necessary, of alternative procedures or controls that satisfy the purpose of the requirement; or

(iii) Other information justifying an exception or alternative.

(2) The Director may approve a request for an exception or alternative if the Director finds one of the following:

(i) The requestor's compliance with the requirement is unnecessary to provide suitable assurance that the drug meets the requirements of the act as to safety, and has the identity and strength and meets the quality and purity characteristics that it purports or is represented to possess, or compliance with the requirement cannot be achieved;

(ii) The requestor's alternative procedures or controls satisfy the purpose of the requirement; or

(iii) The requestor's submission otherwise justifies an exception or alternative.

(3) The Director may withdraw approval of an exception or alternative if the Director finds, on the basis of new information, that the criteria for approval in paragraph (d)(2) of this section are no longer met. Withdrawal of approval shall be accomplished by providing written notice of such withdrawal, and the reasons for the withdrawal, to the original requestor.

PART 211–CURRENT GOOD MANUFACTURING PRACTICE FOR FINISHED PHARMACEUTICALS

§211.42 *Design and construction features.*

* * * * *

(c) * * * There shall be separate or defined areas or such other control systems for the firm's operations as are necessary to prevent contamination or mixups during the course of the following procedures:

* * * * *

§211.68 *Automatic, mechanical, and electronic equipment.*

* * * * *

(b) * * * The degree and frequency of input/output verification shall be based on the complexity and reliability of the computer or related system. * * *

§211.137 *Expiration dating.*

* * * * *

(g) New drug products for investigational use are exempt from the requirements of this section, provided that they meet appropriate standards or specifications as demonstrated by stability studies during their use in clinical investigations. Where new drug products for investigational use are to be reconstituted at the time of dispensing, their labeling shall bear expiration information for the reconstituted drug product.

* * * * *

§211.170 *Reserve samples.*

* * * * *

(b) * * * Except for those for drug products described in paragraph (b)(2) of this section, reserve samples from representative sample lots or batches selected by acceptable statistical procedures shall be examined visually at least once a year for evidence of deterioration unless visual examination would affect the integrity of the reserve sample. * * *

* * * * *

§*211.180 General requirements.*

* * * * *

(e) * * *

(1) A review of a representative number of batches, whether approved or rejected, and, where applicable, records associated with the batch.

* * * * *

FDA CGMPs for Medical Devices (21 CFR 820) were proposed in March 1977 to assure that medical devices are safe and effective by defining the methods used in, and the facilities and controls used for, the manufacture, packing, storage, and installation of medical devices. They were finalized in July 1978. A major revision to the medical device GMPs was proposed in November 1993. This revision replaced current quality assurance *PROGRAM* requirements with quality *SYSTEM* requirements and ensures that the GMP requirements for medical devices are compatible with specifications for quality systsms that are contained in the international ISO 9001 standards. In July 1995, the FDA developed a "working draft" of the medical device GMPs from the comments to the November 1993 proposal. Additional changes were made after a public workshop on the medical device GMPs in August 1995. While the "working draft" is still only a draft, the entire proposed regulation is presented below.

WORKING DRAFT OF THE CGMP FINAL RULE - REGULATION

Subpart A - General Provisions

§*820.1 Scope.*

(a) Applicability.

(1) Current good manufacturing practice (CGMP) requirements are set forth in this quality system regulation. The requirements in this part govern the methods used in, and the facilities and controls used for, the design, purchasing, manufacture, packaging, labeling, storage, installation, and servicing of all finished devices intended for human use. The regulations in this part are intended to ensure that finished devices will be safe and effective and otherwise in compliance with the Federal Food, Drug, and Cosmetic Act (the act). This part establishes basic requirements applicable to manufacturers of finished medical devices. With respect to class I devices, design controls apply only to those devices listed in §820.30(a)(2). The regulations in this part do not apply to manufacturers of components or parts of finished devices when such components or parts are not intended specifically for use as part of a medical device, but such manufacturers are encouraged to use appropriate provisions of this regulation as guidelines. Manufacturers of human blood and blood components are not subject to this part, but are subject to part 606 of this chapter.

(2) The provisions of this part shall be applicable to any finished device and component described above, as defined in this part, intended for human use, that is manufactured, imported, or offered for import in any State or Territory of the United States, the District of Columbia, or the Commonwealth of Puerto Rico.

(b) Limitations. The quality system regulation in this part supplements regulations in other parts of this chapter except where explicitly stated otherwise. In the event that it is impossible to comply with all applicable regulations, both in this part and in other parts of this chapter, the regulations specifically applicable to the device in question shall supersede any other regulations.

(c) Authority. Part 820 is established and

promulgated under authority of sections 501, 502, 510, 513, 514, 515, 518, 519, 520, 522, 701, 704, 801, and 803 of the act (21 U.S.C. 351, 352, 360, 360c, 360d, 360e, 360h, 360i, 360j, 360l, 371, 374, 381, and 383). The failure to comply with any applicable provision in this part renders the device adulterated under section 501(h) of the act. Such a device, as well as any person responsible for the failure to comply, is subject to regulatory action under sections 301, 302, 303, 304, and 801 of the act.

(d) Foreign manufacturers. If a manufacturer who offers devices for import into the United States refuses to permit or allow the completion of an FDA inspection of the foreign facility for the purpose of determining compliance with this part, it shall appear for purposes of section 801(a) of the act, that the methods used in, and the facilities and controls used for, the design, purchasing, manufacture, packaging, labeling, storage, installation, or servicing of any devices produced at such facility that are offered for import into the United States do not conform to the requirements of section 520(f) of the act and this part and that the devices manufactured at that facility are adulterated under section 501(h) of the act.

(e) Exemptions or variances.

(1) Any person who wishes to petition for an exemption or variance from any device quality system requirement is subject to the requirements of section 520(f)(2) of the act. Petitions for an exemption or variance shall be submitted according to the procedures set forth in §10.30 of this chapter, the Food and Drug Administration's administrative procedures. Guidance is available from the Center for Devices and Radiological Health, Division of Small Manufacturers Assistance, Regulatory Assistance Branch (HFZ-220), 1350 Piccard Drive Rockville, MD 20850, telephone 1-800-638-2041, or 1-301-443-6597, FAX 301-443-8818.

(2) FDA may initiate and grant a variance from any device quality system requirement where the agency determines that such variance is in the best interest of the public health. Such variance will remain in effect only so long as there remains a public need for the device and the device would not likely be made sufficiently available without the variance.

§820.3 Definitions.

(a) Act means the Federal Food, Drug, and Cosmetic Act, as amended (secs. 201-903, 52 Stat. 1040 et seq., as amended (21 U.S.C. 321-394)). All definitions in section 201 of the act shall apply to these regulations.

(b) Complaint means any written, electronic, or oral communication that alleges deficiencies related to the identity, quality, durability, reliability, safety, effectiveness, or performance of a device.

(c) Component means any raw material, substance, piece, part, software, firmware, packaging, labeling, or assembly which is intended to be included as part of the finished, packaged, and labeled device.

(d) Control number means any distinctive symbols, such as a distinctive combination of letters or numbers, or both, from which the complete history of the purchasing, manufacturing, packaging, labeling, and distribution of a unit, lot, or batch of finished devices can be determined.

(e) Design History File means a compilation of records which describe the complete design history of a finished device.

(f) Design input means the physical and performance requirements of a device that are used as a basis for device design.

(g) Design output means the results of a design effort at each design phase and at the end of the total design effort. The total finished design output consists of the device, its packaging and labeling, the associated specifications and drawings, and the production and quality assurance specifications and procedures. The finished design output will be the basis for the device master record.

(h) Design review means a documented, comprehensive, systematic examination of a design to evaluate the adequacy of the design requirements, to evaluate the capability of the design to meet these requirements, and to identify problems and propose the development of solutions.

(i) Device history record means a compilation of records containing the complete production history of a finished device.

(j) Device master record (DMR) means a compilation of records containing all the procedures and specifications related to a specific finished device, as required by this part.

(k) End of life means an established time to failure period, determined by the original device manufacturer, based upon reliability data and analysis to characterize nonrepairable product.

(l) Establish means define, document (written or electronic), and implement.

(m) Finished device means any device or accessory to any device that is suitable for use or capable of functioning, whether or not it is packaged, labeled, or sterilized.

(n) Lot or batch means one or more components or finished devices that consist of a single type, model, class, size, composition, and software version that are manufactured under essentially the same conditions and that are intended to have uniform characteristics and quality within specified limits.

(o) Management with executive responsibility means those senior employees of a manufacturer who have the authority to establish or make changes to the manufacturer's quality policy and quality system.

(p) Manufacturer means any person who designs, manufactures, fabricates, assembles, or processes a finished device or components or parts intended specifically for use as part of a finished medical device, and includes contract sterilizers, specification developers, repackers, relabelers, refurbishers, servicers, and initial distributors of imported devices.

(q) Manufacturing material means any material or substance used in, or used to facilitate, a manufacturing process, or a naturally occurring substance, that is not intended by the manufacturer to be included in the finished device, including cleaning agents, mold-release agents, lubricating oils, and sterilant residues, or other byproducts of the manufacturing process.

(r) Nonconformity means the nonfulfillment of a specified requirement.

(s) Product means components, manufacturing materials, in-process devices, finished devices, and returned devices.

(t) Quality means the totality of features and characteristics that bear on the ability of a device to satisfy fitness-for-use, including safety and performance.

(u) Quality audit means an established systematic, independent, examination of a manufacturer's quality system that is performed at defined intervals and at sufficient frequency to ensure that both quality system activities and the results of such activities comply with specified quality system procedures, that these procedures are implemented effectively, and that these procedures are suitable to achieve quality system objectives.

(v) Quality policy means the overall quality intentions and direction of an organization with respect to quality, as formally expressed by management with executive responsibility.

(w) Quality system means the organizational structure, responsibilities, procedures, processes, and resources for implementing quality management.

(x) Record means any written or electronic document, including specifications, procedures, protocols, standards, methods, instructions, plans, files, forms, notes, reviews, analyses, data, and reports.

(y) Refurbisher means any person who processes, conditions, renovates, or restores a finished device which has been previously distributed, and has reached its established end-of-life or is considered to be nonrepairable.

(z) Reprocessing means all or part of a manufacturing operation which is intended to correct nonconformance in a component or finished device before distribution.

(aa) Servicing means maintenance or repair of a finished device after distribution for purposes of returning it to its safety and performance specifications established by the original finished device manufacturer and to meet its original intended use, prior to the device's established end-of-life or before it is considered to be nonrepairable.

(bb) Specification means any requirement with which a product, process, service, or other activity must conform.

(cc) Validation means establishing and documenting evidence which provides a high degree of assurance that a process will consistently produce a result or product meeting its predetermined specifications.

(dd) Verification means confirmation by examination and provision of objective evidence that specified requirements related to a product or process have been met.

§*820.5 Quality system.*

Each manufacturer shall establish and maintain a quality system that is appropriate to the specific medical device manufactured and meets the requirements of this part. Each manufacturer shall:

(a) Establish effective quality system instructions and procedures in accordance with the requirements of this part; and,

(b) Maintain the established quality system instructions and procedures effectively.

Subpart B - Quality System Requirements

§*820.20 Management responsibility.*

(a) Quality policy. Management with executive responsibility shall establish its policy and objectives for, and commitment to, quality. Management with executive responsibility shall ensure that the quality policy is understood, implemented, and maintained at all levels of the organization.

(b) Organization. Each manufacturer shall establish and maintain an adequate organizational structure to ensure that devices are produced in accordance with the requirements of this part.

(1) Responsibility and authority. Each manufacturer shall establish the appropriate responsibility, authority, and interrelation of all personnel who manage, perform, and verify work affecting quality, and provide the independence and authority necessary to perform these tasks.

(2) Resources. Each manufacturer shall provide adequate resources, including the assignment of trained personnel, for management, performance of work, and verification activities, including internal quality audits, to carry out the requirements of this part.

(3) Managementrepresentative.Management with executive responsibility shall appoint, and document such appointment of, a member of management who, irrespective of other responsibilities, shall have established authority over and responsibility for:

(i) Ensuring that quality system requirements are established and maintained in accordance with this part; and

(ii) Reporting on the performance of the quality system to management with executive responsibility for review.

(c) Management review. Management with executive responsibility shall review the suitability and effectiveness of the quality system at defined intervals and at sufficient frequency according to established procedures to ensure that the quality system satisfies the requirements of this part and the manufacturer's established quality policy and objectives. The results of quality system reviews shall be documented.

§*820.22 Quality audit.*

Each manufacturer shall conduct quality audits to verify that the quality system is in compliance with the established quality system requirements and to determine the effectiveness of the quality system. Quality audits shall be conducted in accordance with established audit procedures by individuals who do not have direct responsibility for the matters being audited. A report of the results of each quality audit, and reaudit(s) where taken, shall be made and such reports shall be reviewed by management having responsibility for the matters audited. Follow-up corrective action, including a reaudit of deficient matters, shall be taken when necessary. The dates on which the audit and reaudit were performed shall be documented.

§*820.25 Personnel.*

(a) General. Each manufacturer shall employ sufficient personnel with the necessary education, background, training, and experience to ensure that all activities required by this part are correctly performed.

(b) Training. Each manufacturer shall ensure that all personnel are trained to adequately perform their assigned responsibilities. Training shall be conducted in accordance with established procedures to ensure that employees have a thorough understanding of their current job functions and with the CGMP requirements applicable to their job functions. As part of their training, employees shall be made aware of device defects which may occur from the improper performance of their specific jobs. Personnel who perform verification activities shall be made aware of defects and errors that may be encountered as part of their verification functions. Employee training shall be documented.

Subpart C--Design Controls

§*820.30 Design controls.*

(a) General.

(1) Each manufacturer of any class III, or class II device, and the class I devices listed in paragraph (a)(2) of this section, shall establish and maintain procedures to control and verify the design of the device in order to ensure that specified design requirements are met.

(2) The following class I devices are subject to design controls:

(i) Devices automated with computer soft ware; and

(ii) The devices listed in the chart below.

Section	Device
868.6810	Catheter, Tracheobronchial Suction
878.4460	Glove, Surgeon's
880.6760	Restraint, Protective
892.5650	System, Applicator, Radionuclide, Manual
892.5740	Source, Radionuclide Teletherapy

(b) Design and development planning. Each manufacturer shall establish and maintain plans that describe or reference the design and development activities and define responsibility for the implementation. The design and development activities shall be assigned. The plan shall include and describe the interfaces with different groups or activities. The plans shall be reviewed, updated, and approved as design and development evolves.

(c) Design input. Each manufacturer shall establish and maintain procedures to ensure that the design requirements relating to a device are appropriate and address the intended use of the device, including the needs of the user and patient. The design input requirements shall be documented and shall be reviewed and approved by a designated individual(s). The approval including the date and signature of the individual(s) approving the requirements, shall be documented.

(d) Design output. Each manufacturer shall establish and maintain procedures for defining and documenting design output in terms that allow an adequate evaluation of conformance to design input requirements. Design output procedures shall ensure that design output meets the design input requirements. Design output procedures shall contain or make reference to acceptance criteria and shall ensure that those design outputs that are essential for the proper functioning are identified. Design output documents shall be reviewed and approved before release. The approval, including the approval date and the signature of the individual(s) approving release, shall be documented.

(e) Design review. Each manufacturer shall establish and maintain procedures to ensure that formal documented reviews of the design results are planned and conducted at appropriate stages of the device's design development. Each manufacturer shall assign an individual(s), who does not have direct responsibility for the design development, to participate in the design reviews. The procedures shall ensure that participants at each design review include representatives of all functions concerned with the design stage being reviewed, as well as specialists where appropriate. The results of a design review shall be documented in the design history file.

(f) Design verification and validation. Each manufacturer shall establish and maintain procedures for verifying and validating the device design. Design verification shall confirm that design output meets the design input requirements. Design validation shall ensure that devices conform to defined user needs and intended uses. Design validation shall be performed under defined operating conditions and on the initial production units, lots, or batches. Design validation, and verification where appropriate, shall include testing of production units under actual or simulated use conditions before distribution. The results of the design verification and validation, including identification of the design, method(s), the date, and the individual(s) performing the verification and validation shall be documented in the design history file. Design verification and validation shall include software validation, and an analysis of available information to identify potential sources of harm and estimate their probable rate of occurrence and degree of severity.

(g) Design transfer. Each manufacturer shall establish and maintain procedures to ensure that the design basis for a device and its components is correctly translated into production specifications.

(h) Design changes. Each manufacturer shall establish and maintain procedures for the identification, documentation, validation, or verification where appropriate, review, and approval of design changes.

(i) Design history file. Each manufacturer shall establish and maintain a design history file for each design. The design history file shall contain or reference all records necessary to demonstrate that the design was developed in accordance with the approved design plan and the requirements of this part.

Subpart D--Document Controls

§820.40 Document controls.

Each manufacturer shall establish and maintain procedures to control all documents that are required to be established, maintained, and removed under this part. The procedures shall provide for the following:

(a) Document approval and distribution. Each manufacturer shall designate an individual(s) to review and approve all documents established to meet the requirements of this part for adequacy prior to issuance. The approval, including the approval date and signature of the individual(s) approving the document, shall be documented. Documents established to meet the requirements of this part shall be available at all locations for which they are designated, used, or otherwise necessary, and all obsolete documents shall be promptly removed from all points of use or otherwise prevented from unintended use.

(b) Document changes. Changes to documents shall be reviewed and approved by individuals in the same functions/organizations that performed the original review and approval, unless specifically designated otherwise. Approved changes shall be communicated to the appropriate personnel in a timely manner. Each manufacturer shall maintain records of changes to documents. Change records shall include a description of the change, identification of the affected documents, the signature of the approving individual(s), the approval date, and when the change becomes effective.

Subpart E--Purchasing Controls

§820.50 Purchasing controls.

Each manufacturer shall establish and maintain procedures to ensure that all purchased or otherwise received product and services conform to specified requirements.

(a) Evaluation of suppliers, contractors, and consultants. Each manufacturer shall establish and maintain evaluation criteria for suppliers, contractors, and consultants that specify the requirements, including quality requirements, that must be met. Each manufacturer shall:

(1) Evaluate and select potential suppliers, contractors, and consultants on the basis of their ability to meet specified requirements, including quality requirements. Records of the evaluation results shall be maintained.

(2) Define the type and extent of control to be exercised over the product, services, suppliers, contractors, and consultants based on

the evaluation results.

(3) Establish and maintain records of acceptable suppliers, contractors, and consultants.

(b) Purchasing data. Each manufacturer shall establish and maintain purchasing data that clearly describe or reference the specified requirements, including quality requirements, for purchased product and services. Purchasing documents shall include, where possible, an agreement that the suppliers and contractors agree to notify the manufacturer of any changes in the product or service so that manufacturers may determine whether the change may affect the quality of a finished device. Each manufacturer shall review and approve purchasing data for adequacy of the specified requirements prior to release. The approval, including the approval date and signature of the individual(s) approving the data, shall be documented.

Subpart F--Identification and Traceability

§820.60 Identification.

Each manufacturer shall establish and maintain procedures for identifying product during all stages of receipt, production, distribution, and installation to prevent mixups.

§820.65 Traceability.

Each manufacturer shall establish and maintain procedures for identifying each unit, batch, or lot of finished devices and components with a control number where necessary to ensure the protection of the public health. The procedures shall facilitate corrective action. Such identification shall be recorded in the device history record.

Subpart G--Production and Process Controls

§820.70 Production and process controls.

(a) General. Each manufacturer shall design, conduct, control, and monitor production processes to ensure that a device conforms to its specifications. Where deviations from device specifications could occur as a result of the manufacturing process, the manufacturer shall establish and maintain process control procedures that describe any process controls necessary to ensure conformance to specifications. Process controls shall include:

(1) Documented instructions, standard operating procedures (SOP's), and methods that define and control the manner of production;

(2) Monitoring and control of process parameters and component and device characteristics during production;

(3) Compliance with applied reference standards or codes and process control procedures;

(4) The approval of processes and process equipment; and,

(5) Criteria for workmanship which shall be expressed in documented standards or by means of representative samples.

(b) Production and process changes. Each manufacturer shall establish and maintain procedures for changes to a specification, method, process, or procedure. Such changes shall be validated before approval and implementation, unless inspection and test fully verifies the results of the changes. The results of the validation, or verification, shall be documented. Approved changes shall be communicated to the appropriate personnel in a timely manner.

(c) Environmental control. Each manufacturer shall establish and maintain requirements for the environment to which product is exposed. Where environmental conditions could have an adverse effect on a device's fitness for use, these environmental conditions shall be controlled, and procedures for such controls shall be established and maintained. Any environmental control system shall be periodically inspected to verify that the system is adequate and functioning properly. Results of such inspections shall be documented and reviewed.

(d) Personnel. Each manufacturer shall establish and maintain requirements for health, cleanliness, personnel practices, and clothing of personnel if contact between such personnel and product or environment could adversely affect the quality of the product. The manufacturer shall ensure that all personnel who are required to work temporarily under special environmental conditions are appropriately trained or supervised by a trained individual.

(e) Contamination control. Each manufacturer shall establish and maintain procedures to prevent contamination of equipment or product by substances that could adversely affect device safety or effectiveness.

(f) Buildings. Buildings shall be of suitable design and contain sufficient space to perform necessary operations, prevent mixups, and assure orderly handling.

(g) Equipment. Each manufacturer shall ensure that all equipment used in the manufacturing process meets specified requirements and is appropriately designed, constructed, placed, and installed to facilitate maintenance, adjustment, cleaning, and use.

(1) Maintenance schedule. Each manufacturer shall establish and maintain schedules for the maintenance, adjustment, and cleaning of equipment to ensure that manufacturing specifications are met. Records shall be maintained documenting the date when scheduled maintenance activities were performed and the individual(s) performing the maintenance activity.

(2) Inspection. Each manufacturer shall conduct periodic inspections in accordance with established procedures to ensure adherence to applicable equipment maintenance schedules. The inspections, including the date and individual conducting the inspections, shall be documented.

(3) Adjustment. Each manufacturer shall ensure that any inherent limitations or allowable tolerances are visibly posted on or near equipment requiring periodic adjustments or are readily available to personnel performing these adjustments.

(h) Manufacturing material. Each manufacturer shall establish and maintain procedures for the use and removal of manufacturing material to ensure that such material is removed from the device or limited to a specified amount that does not adversely affect the device's quality. The removal of such manufacturing material shall be documented.

(i) Automated processes. When computers are used as part of production, the quality system, or automated data processing systems, the manufacturer shall validate computer software for its intended use according to an established protocol. The results shall be documented. All software changes shall be validated before approval and issuance. The results shall be documented.

§820.75 Process validation.

(a) Where the results of a process cannot be fully verified by subsequent inspection and test, the process shall be validated according to established procedures. Records shall be made of the validation activities and results, including the date and individual(s) performing the validation.

(b) Each manufacturer shall establish and maintain procedures for continuous monitoring and control of process parameters to ensure that the specified requirements are met.

(c) Each manufacturer shall ensure that validated processes are performed by qualified individual(s).

(d) Records shall be maintained for the validated processes to include monitoring and control methods and data, signature of the individual(s) performing the process, the date performed, and where appropriate, the major equipment used.

Subpart H--Acceptance Activities

§820.80 Receiving, in-process, and finished device acceptance.

(a) General. Each manufacturer shall establish and maintain the acceptance activities necessary to ensure that specified requirements are met. Acceptance activities include inspections, tests, and other verification activities.

(b) Receiving acceptance activities. Each manufacturer shall establish and maintain procedures for acceptance of incoming product. Incoming product shall be inspected or otherwise verified as conforming to specified requirements. Acceptance and rejection shall be documented.

(c) In-process acceptance activities. Each manufacturer shall establish and maintain acceptance procedures to ensure that specified requirements for in-process product are met. Such procedures shall ensure that in-process product is held until the required inspection and tests or other verification activities have been completed, or necessary approvals are received, and are documented.

(d) Final acceptance activities. Each manufacturer shall establish and maintain procedures for finished device acceptance to ensure that each production run, lot, or batch of finished devices meets specified requirements. Finished devices shall be held in quarantine or otherwise adequately controlled until released. Finished devices shall not be released for distribution until the required activities specified in the DMR are completed, the associated data and documentation is reviewed, and release is authorized by the signature of a designated individual(s). Such authorization shall be dated.

(e) Acceptance records. Each manufacturer shall maintain records of the results of acceptance activities required by this part. These records shall include the acceptance criteria, acceptance activities performed, dates performed, results, the signature of the individual(s) conducting the acceptance activities and, where appropriate, equipment used. These records shall be part of the device history record.

§820.84 Inspection, measuring, and test equipment.

Each manufacturer shall ensure that all measurement and test equipment, including mechanical, automated, or electronic inspection and test equipment, is suitable for its intended purposes and is capable of producing valid results. Each manufacturer shall establish and maintain procedures to ensure that equipment is routinely calibrated, inspected, checked, and maintained. Records documenting these activities shall be maintained.

(a) Calibration. Calibration procedures shall include specific directions and limits for accuracy and precision. There shall be provisions for remedial action when accuracy and precision limits are not met.

(b) Calibration standards. Calibration standards used for measurement equipment shall be traceable to national or international standards. If national or international standards are not practical or available, the manufacturer shall use an independent reproducible standard. If no applicable standard exists, the manufacturer shall establish and maintain an in-house standard.

(c) Calibration records. Each manufacturer shall ensure that records of calibration dates, the individual performing each calibration, and the next calibration date are maintained. These records shall be displayed on or near each piece of equipment or shall be readily available to the personnel using such equipment and the individuals responsible for calibrating the equipment.

(d) Maintenance. Maintenance procedures shall include provisions for handling, preservation, and storage of inspection, measuring, test equipment, and test software so that their accuracy and fitness-for-use are maintained.

§820.86 Acceptance status.

Each manufacturer shall identify by suitable means the acceptance status of product, to indicate the conformance or nonconformance of these items with respect to acceptance criteria. The identification of acceptance status shall be maintained throughout component acceptance, manufacturing, packaging, labeling, installation, and servicing of the product to ensure that only products which have passed the required acceptance activities are distributed, used, or installed.

Subpart I--Nonconforming Product

§820.90 Nonconforming product.

(a) Control of nonconforming product. Each manufacturer shall establish and maintain procedures to ensure that product that does not conform to specified requirements is not used or distributed. The procedures shall provide for the identification, documentation, evaluation, segregation, and disposition of nonconforming product. The procedures shall provide for the investigation of nonconformances and notification of the persons or organizations responsible for the nonconformance.

(b) Nonconformity review and disposition.

(1) Each manufacturer shall establish and maintain procedures that define the responsi-

bility for review and the authority for the disposition of nonconforming product. The procedures shall set forth the review and disposition process. Nonconforming disposition shall be documented to include the justification for any concession and the signature of the individual(s) authorizing the concession.

(2) Each manufacturer shall establish and maintain procedures for the reprocessing, to include retesting and reevaluation of the nonconforming product after reprocessing, to ensure that it meets its original, or subsequently modified and approved, specifications. Reprocessed product shall be clearly identified during reprocessing, and shall be subjected to reevaluation. The reprocessing and reevaluation results shall be recorded in the device history record. When there is reprocessing of a product, a determination of the effect of the reprocessing upon the product shall be made and documented.

Subpart J--Corrective and Preventive Action

§820.100 Corrective and preventive action.

(a) Each manufacturer shall establish and maintain procedures for implementing corrective and preventive action. The procedures shall include requirements for:

(1) Analyzing processes, work operations, concessions, quality audit reports, quality records, service records, complaints, returned product, and other sources of quality data to identify existing and potential causes of nonconforming product, or other quality problems. Appropriate statistical methodology shall be employed to detect recurring quality problems;

(2) Investigating the cause of nonconformities relating to product, process, and quality system;

(3) Identifying action needed to correct the cause and prevent recurrence of nonconforming product and other quality problems;

(4) Verifying or validating the adequacy of the corrective and preventive action to ensure that the corrective and preventive action does not adversely affect the finished device and that such action is effective;

(5) Implementing and recording changes in methods and procedures needed as a result of the identification of quality problems, and corrective and preventive action;

(6) Ensuring that information related to nonconforming product or quality problems is disseminated to those directly responsible for assuring the quality of such product or the absence of such problems; and

(7) Confirming that relevant information on actions taken is submitted for management review.

(b) All activities required under this section, and their results, shall be documented.

Subpart K--Handling, Storage, Distribution, and Installation

§820.120 Handling.

Each manufacturer shall establish and maintain procedures to ensure that mixups, damage, deterioration, or other adverse effects to product do not occur during handling.

§820.122 Storage.

(a) Each manufacturer shall establish and maintain procedures for the control of storage areas and stock rooms for product to prevent mixups, damage, deterioration, or other adverse effects pending use or distribution and to ensure that all obsolete, rejected, or deteriorated product is not used or distributed.

(b) Each manufacturer shall establish and maintain procedures that describe the methods for authorizing receipt from and dispatch to such designated areas. When the quality of product deteriorates over time, it shall be stored in a manner to facilitate proper stock rotation, and its condition shall be assessed at appropriate intervals.

§820.124 Distribution.

(a) Each manufacturer shall establish and maintain procedures for control and distribution of finished devices to ensure that only those devices approved for release are distributed. Where a device's fitness-for-use or quality deteriorates over time, the procedures shall ensure that expired devices or devices deteriorated beyond acceptable fitness for use are not distributed.

(b) Each manufacturer shall maintain distribution records which include or refer to the location of:

(1) The name and address of the initial consignee;

(2) The identification and quantity of devices shipped, the date shipped; and

(3) Any control number(s) used for traceability, if required by §820.65.

§*820.126 Installation.*

Each manufacturer of a device requiring installation shall establish and maintain adequate instructions and procedures for proper device installation. Instructions and procedures shall include directions for ensuring proper performance of the installation and that the device will perform as intended after installation. The manufacturer shall ensure that the installation instructions and inspection procedures are distributed with the device or otherwise available to the person(s) installing the device. The person installing the device shall ensure that the installation was performed in accordance with the manufacturer's instructions and procedures and shall record the inspection results to demonstrate proper installation. The results of the installation inspection shall made available to FDA upon request.

Subpart L--Packaging and Labeling Control

§*820.160 Device packaging.*

Each manufacturer shall ensure that device packaging and shipping containers are designed and constructed to protect the device from alteration or damage during the customary conditions of processing, storage, handling, and distribution.

§*820.162 Device labeling.*

Each manufacturer shall establish and maintain procedures to maintain labeling integrity and to prevent labeling mixups.

(a) Label integrity. Each manufacturer shall ensure that labels are printed and, where applicable, applied so as to remain legible and affixed to the device during the customary conditions of processing, storage, handling, distribution, and use.

(b) Labeling inspection. Labeling shall not be released for storage or use until a designated individual(s) has examined the labeling for accuracy including, where applicable, the correct expiration date, control number, storage instructions, handling instructions, and additional processing instructions. The release, including the date and signature of the individual(s) performing the examination, shall be documented in the device history record.

(c) Labeling storage. Each manufacturer shall store and maintain labeling in a manner that provides proper identification and is designed to prevent mixups.

(d) Labeling operations. Each manufacturer shall control labeling and packaging operations to prevent labeling mixups.

(e) Control number. Where a control number is required by §820.65, that control number shall be on the device itself or its label.

Subpart M--Records

§*820.180 General requirements.*

All records required by this part shall be maintained at the manufacturing establishment or other location that is reasonably accessible to responsible officials of the manufacturer and to employees of the Food and Drug Administration designated to perform inspections. Such records, including those not stored at the inspected establishment, shall be made readily available for review and copying by FDA employee(s). Such records shall be legible and shall be stored to minimize deterioration and to prevent loss. Those records stored in automated data processing systems shall be backed up.

(a) Confidentiality. Records deemed confidential by the manufacturer may be marked to aid the Food and Drug Administration in determining whether information may be disclosed under the public information regulation in part 20 of this chapter.

(b) Record retention period. All records

required by this part shall be retained for a period of time equivalent to the design and expected life of the device, but in no case less than 2 years from the date of release for commercial distribution by the manufacturer.

(c) Exceptions. This section does not apply to the reports required by §820.20(c) and §820.22, and supplier audit reports used to meet the requirements of §820.50(a), but does apply to procedures established under these subsections. Upon request of a designated employee of the Food and Drug Administration, an employee in management with executive responsibility shall certify in writing that the management reviews and quality audits required under this part, and supplier audits where applicable, have been performed and documented, the dates on which they were performed, and that any required corrective action has been taken.

§820.181 Device master record.

Each manufacturer shall maintain device master records (DMRs). Each manufacturer shall ensure that each DMR is prepared, dated, and approved with the signature of the qualified individual(s) designated by the manufacturer. The DMR for each type of device shall include, or refer to the location of, the following information:

(a) Device specifications including appropriate drawings, composition, formulation, component specifications, software specifications, and software source code for customized software; (b) Production process specifications including the appropriate equipment specifications, production methods, production procedures, and production environment specifications;

(c) Quality assurance procedures and specifications including quality assurance checks used, and the quality assurance apparatus used;

(d) Packaging and labeling specifications, including methods and processes used; and

(e) Installation, maintenance, and servicing procedures and methods.

§820.184 Device history record.

Each manufacturer shall maintain device history records (DHRs). Each manufacturer shall establish and maintain procedures to ensure that DHRs for each batch, lot, or unit are maintained to demonstrate that the device is manufactured in accordance with the DMR and the requirements of this part. The DHR shall include, or refer to the location of, the following information:

(a) The dates of manufacture;

(b) The quantity manufactured;

(c) The quantity released for distribution;

(d) The specific label and labeling used for each production unit; and

(e) Any device identification(s) and control number(s) used.

§820.186 Quality system records.

Each manufacturer shall maintain quality system records to demonstrate conformance to specified requirements and effective operation of the quality system as defined in §820.5, to include or refer to the location of:

(a) Documentation of activities that establish the objectives and requirements for quality, and the application of quality system elements;

(b) Documentation of the responsibilities, authorities, and interrelationships of personnel who manage, perform, verify, or review work affecting quality;

(c) The quality system procedures and instructions; and

(d) An outline of the structure of the documentation used in the quality system, where appropriate.

§820.198 Complaint files.

(a) Each manufacturer shall maintain complaint files. Each manufacturer shall establish and maintain procedures for receiving, reviewing, and evaluating complaints by a formally designated unit. Such procedures shall ensure that:

(1) All complaints are processed in a uniform and timely manner;

(2) Oral complaints are documented upon receipt; and

(3) Complaints are evaluated to determine whether the complaint represents an event

which is required to be reported to the Food and Drug Administration under part 803 of this chapter, Medical Device Reporting.

(b) Each manufacturer shall review and evaluate all complaints to determine whether an investigation is necessary. When no investigation is made, the unit shall maintain a record that includes the reason no investigation was made and the name of the individual responsible for the decision not to investigate.

(c) Any complaint involving the possible failure of a device, labeling, or packaging to meet any of its specifications shall be reviewed, evaluated, and investigated.

(d) Any complaint pertaining to death, injury, or any hazard to safety shall be i immediately reviewed, evaluated, and investigated by a designated individual(s) and shall be maintained in a separate portion of the complaint files or clearly and visibly identified as pertaining to a death, injury, or hazard to safety. Such investigations shall also include a determination of whether there was an actual device failure to perform pursuant to specifications, whether the device was being used to treat or diagnose a patient, and the relationship, if any, of the device to the reported incident or adverse event.

(e) When an investigation is made under this subpart, a written record of each investigation shall be maintained by the formally designated unit identified in paragraph (a) of this section. The record of investigation shall include:

(1) The name of the device;

(2) The date the complaint was received;

(3) Any device identification(s) and control number(s) used;

(4) The name, address, and phone number of the complainant;

(5) The nature and details of the complaint;

(6) The dates and results of the investigation;

(7) Any corrective action taken; and

(8) Any reply to the complainant.

(f) When the manufacturer's formally designated complaint unit is located at a site separate from the actual manufacturing establishment, the complaint and the record of investigation shall be concurrently maintained at the actual manufacturing establishment in a file designated for device complaints.

(g) If a manufacturer's formally designated complaint unit is located outside of the United States, records required under this section shall be concurrently maintained in the United States at either:

(1) A location in the United States where the manufacturer's records are regularly kept; or

(2) The location of the agent designated by the manufacturer under §803.26(g)(2) of this chapter [as proposed in the Federal Register of November 26, 1991 (56 FR 60024)].

Subpart N--Servicing

§820.200 Servicing.

Each original manufacturer (including a refurbisher) shall establish and maintain instructions and procedures to ensure that finished devices that are serviced meet safety and performance specifications for the original intended use of the device(s). The instructions and procedures shall include directions for ensuring that the device(s) will perform as intended after servicing. Such manufacturer shall ensure that the device's safety and performance specifications, to include the device's end-of-life date or period, accompany the device at the time of the initial sale or are otherwise made available to the person(s) servicing the device. Procedures for servicing shall include provisions for determining if service requests represent an event which must be reported to the Food and Drug Administration under the requirements of part 803 of this chapter.

(a) Service reports. Each person that services a device shall establish and maintain procedures to ensure that service reports are maintained and identify the device serviced, including any device identification(s) and control number(s) used, the date of service, the service performed, and individual(s) servicing the device. Service reports shall be recorded and made available to FDA upon request. Such reports shall demonstrate that the finished device serviced meets the manufacturer's safety and performance specifications. A copy of all service reports shall be forwarded to the original manufacturer.

(b) Service report evaluation. Each original

manufacturer shall analyze service reports with appropriate statistical methodology in accordance with §820.100; however, when a service report involves a death, injury, or hazard to safety, the report shall automatically be considered a complaint and shall be investigated in accordance with the requirements of §820.198.

Subpart O--Statistical Techniques

§820.250 Statistical techniques.

(a) Where appropriate, each manufacturer shall establish and maintain procedures for identifying valid statistical techniques required for establishing, controlling, and verifying the acceptability of process capability and product characteristics.

(b) Sampling plans, when used, shall be written and based on a valid statistical rationale. Each manufacturer shall establish and maintain procedures to ensure that sampling methods are adequate for their intended use and are regularly reviewed.

FDA's regulations for Good Clinical Practices (GCPs) are found primarily in 21 CFR 312 (Regulations for Investigational New Drug Applications) and 21 CFR 314 (Regulations for New Drug Applications. These regulations define FDA's requirements for reviewing drug applications and the responsibilities of drug manufacturers. Since the publication of the *Validation Compliance Annual: 1995*, the FDA has proposed the following changes to these regulations:

PART 312 -- INVESTIGATIONAL NEW DRUG APPLICATION

§312.23 IND content and format.

(a) * * *

(3) * * *

(v) A description of the safety monitoring and evaluation program developed by the sponsor in order to evaluate safety data reported by investigators.

* * * * *

(6) * * *

(iii) * * *

(*h*) A description of any adverse clinical or laboratory outcomes in the study that are to be reported by the investigators to the sponsor immediately. This would ordinarily include any death, any life-threatening even, any serious event that might reflect potential toxicity, as suggested by preclinical data, laboratory values that exceed specified limits, or any markedly abnormal laboratory value. The identified events and abnormal laboratory values are to include those that focus attention on toxicity that may target the same organs and body systems as the underlying disease or concomitant medications for the disease. The events are to be reported to the sponsor even if they are potentially attributable to the patient's underlying disease or to other medications the patient may have received. This section of the protocol shall include instructions for the investigator encountering such an event, such as reporting requirements, a remeasurement or challenge procedure, or discontinuation of the study drug.

(*i*) Sponsors should consider use of a formal control group (for example, placebo, active, documented historical) in studies that focus on safety when the underlying disease is likely to produce serious adverse events that might be confused with drug toxicity.

(*j*) The sponsor shall estimate the expected incidence of deaths and serious adverse experiences in the study population that may arise from the underlying disease or from medications used to treat the underlying disease. Deaths or serious adverse experiences that exceed these estimates would create a presumption that the events are associated with the use of the investigational drug.

(*k*) The sponsor shall determine and include in each protocol an appropriate followup period and appropriate followup procedures based on the preclinical data, experience with other members of the drug class, the drug's mechanism of action, and prior human experience. The sponsor shall include a brief description of the rationale used in selecting the followup period and procedures. The intensity of the followup may change with time; e.g., a full evaluation might be scheduled for 2 weeks after the end of drug dosing, with a telephone followup at a later time. Ordinarily, in Phase 1 and 2 studies, there should be at least telephone followup for 3 months after drug dosing is completed, but alternative timeframes and procedures can be proposed by the sponsor. In some cases, available information may dictate followup periods longer than 3 months.

* * * * *

§312.32 IND safety reports.

(a) * * *

Serious adverse experience * * * A serious adverse drug experience means an experience occurring at any dose that is fatal or life-threatening, results in permanent or significant disability/incapacity, requires or prolongs inpatient hospitalization, necessitates medical or surgical intervention to preclude permanent impairment of a body function or permanent damage to a body structure, or is a congenital anomaly.

* * * * *

(c) *IND safety reports* -- (1) *Written reports.* (i) The sponsor shall notify FDA and all participating investigators in a written IND safety report of any adverse experience associated with use of the drug that is both serious and unexpected. This includes notification of any death or other serious adverse experience that exceeds the estimate of the expected incidence of deaths and serious adverse experiences required under §312.23(a)(6)(iii)(*j*). Such notification shall be made as soon as possible and in no event later than 15 calendar days after the sponsor's initial receipt of the information. Each written notification may be submitted on FDA Form 3500A or in a narrative form and shall bear prominent identification of its contents, i.e., "IND Safety Report." Each written notification to FDA shall be transmitted to the FDA division of the Center for Drug Evaluation and Research that has responsibility for review of the IND. If FDA determines that insufficient data were submitted on FDA Form 3500A, the agency may require further narrative data to be submitted.

* * * * *

(2) *Telephone safety reports.* The sponsor shall also notify FDA by telephone, either orally or by facsimile transmission, of any unexpected fatal or life-threatening experience associated with the use of the drug no later than 7 calendar days after the sponsor's initial receipt of the information. * * *

* * * * *

(d) * * *

(3) If the results of a sponsor's investigation show that an adverse experience not initially determined to be reportable under paragraph (c) of this section is so reportable, the sponsor shall report such experience in a written safety report as soon as possible after the determination is made, but in no event longer than 15 calendar days.

* * * * *

§312.23 Annual and semiannual reports.

(a) *Annual reports.* A sponsor shall within 60 days of the anniversary date that the IND went into effect, submit a brief report of the progress of the investigation that includes:

(1) *Individual study information.* A brief summary of the status of each study in progress and each study completed during the previous year. The summary is required to include the following information for each study:

(i) The title of the study (with any appropriate study identifiers such as protocol number), its purpose, a brief statement identifying the patient population, and a statement as to whether the study is completed.

(ii) The total number of subjects initially planned for inclusion in the study, the number entered into the study to date, the number whose participation in the study was completed as planned, and the number who dropped out of the study for any reason.

(iii) If the study has been completed, or if interim results are known, a brief description of any available study results.

(2) *Summary information.* Information obtained during the previous year's clinical and nonclinical investigations, including:

(i) A narrative or tabular summary showing the most frequent and most serious adverse experiences by body system.

(ii) A summary of all IND safety reports submitted during the past year.

(iii) A list of subjects who died during participation in the investigation, with the cause of death for each subject.

(iv) A list of subjects who dropped out during the course of the investigation in association with any adverse experience, whether or not thought to be drug related.

(v) A brief description of what, if anything, was obtained that is pertinent to an understanding of the drug's actions, including, for example, information from controlled trials, and information about bioavailability.

(vi) A list of the preclinical studies (including animal studies) completed or in progress during the past year and a summary of the major preclinical findings.

(vii) A summary of any significant manufacturing or microbiological changes made during the past year.

(3) A description of the general investigational plan for the coming year to replace that submitted 1 year earlier. The general investigational plan shall contain the information required under §312.23(a)(3)(iv).

(4) If the investigator brochure has been revised, a description of the revision and a copy of the new brochure.

(5) A description of any significant Phase 1 protocol modifications made during the previous year and not previously reported to the IND in a protocol amendment.

(6) A brief summary of significant foreign marketing developments with the drug during the past year, such as approval of marketing in any country or withdrawal or suspension from marketing in any country.

(7) If desired by the sponsor, a log of any outstanding business with respect to the IND for which the sponsor requests or expects a reply, comment, or meeting.

(b) *Semiannual reports.* A sponsor shall submit a report of the progress of the investigation with respect to safety issues for the 6-month period following the day the IND goes into effect, and for each 6-month period thereafter, until the end of the followup period specified in the protocol. The report shall be due within 60 days of the end of the reporting period. The semiannual safety report that is due during the same period as the annual report required under paragraph (a) of this section shall be submitted with the annual report. These semiannual reports shall include:

(1) A summary and analysis of all deaths, all serious adverse experiences, and all study discontinuations resulting from an adverse experience that occurred during the study or within the prescribed followup period, whether the deaths or adverse experiences were expected or unexpected and whether or not there is thought to be a possibility that the death or adverse experience or study discontinuation was caused by the drug. This summary shall include data not only from studies conducted under the IND, but also data from all premarketing studies of the drug conducted worldwide, with an analysis of all unexpected deaths, serious adverse experiences, and study discontinuations thought to be related to the study drug from foreign postmarketing clinical trials and from foreign postmarketing spontaneous or required reporting systems. For purposes of this section, serious adverse events shall include laboratory changes identified in the study protocol as reportable events or that result in discontinuation. The sponsor shall present in the summary both the data that accumulated during the reporting period and cumulatively. The sponsor shall also submit an analysis of the data that assumes that the investigational drug is responsible for the deaths, serious adverse experiences, and study discontinuations, and refute, as feasible, this presumption with appropriate data and evaluations. The expected incidence of deaths and serious adverse experiences in the study population that may arise from the underlying disease or from medications used to treat the underlying disease that was estimated in the protocol should be considered in this evaluation.

(2) All available autopsy reports and termi-

nal medical reports concerning all deaths reported in this summary, with a discussion of any inconsistencies between autopsy and medical reports and the cause of death reported to FDA by the sponsor.

(3) At the request of the sponsor, or on its own initiative, FDA may modify the requirements of paragraph (b) of this section. A sponsor requesting such a modification should submit to the division responsible for review of the IND a written request for modification and justification for such modification. FDA shall issue a written response to the sponsor either granting or denying, in whole or in part, the request for modification.

§312.37 Special safety summary and final clinical study report.

(a) *Special safety summary.* Upon request of FDA, a sponsor shall prepare and submit special summaries of safety data regarding the investigational drug. These summaries may include safety data available to the sponsor from previous studies of the drug and of closely related drugs identified in consultation with FDA. Examples of types of events that may be requested to be summarized include, among others, deaths, serious adverse experiences, study discontinuations for safety reasons, patients who reach or exceed safety endpoints as defined in the protocol, and any unusual or extreme changes in study subjects. The special safety summary shall be submitted within 30 days after a request by the agency unless the sponsor demonstrates that extraordinary circumstances warrant a later date and the agency has agreed to that later date.

(b) *Final clinical study report.* Upon request by FDA, a sponsor shall submit a final report on a clinical study. The final report shall be submitted within 90 calendar days after a request by the agency unless the sponsor demonstrates that extraordinary circumstances warrant a later date and the agency has agreed to that later date.

§312.42 Clinical holds and requests for modification.

* * * * *

(b) * * *

(1) * * *

(v) The sponsor has failed to submit a special safety summary or final clinical study report, as required by §312.37, for the drug that is the subject of the investigation. This provision applies to special safety summaries and final clinical study reports from other investigations on the same drug and special safety summaries and final clinical study reports requested by FDA for investigations on closely related drugs conducted by the sponsor.

* * * * *

§312.44 Termination.

* * * * *

(b) * * *

(1) * * *

(viii) The sponsor fails to submit an accurate and timely annual or semiannual safety report of the investigations in accordance with §312.33.

* * * * *

§312.56 Review of ongoing investigations.

* * * * *

(c) Before the initiation of clinical studies, the sponsor shall develop a safety monitoring and evaluation program to evaluate safety data reported by the investigator(s). The sponsor shall review and evaluate the evidence relating to the safety and effectiveness of the drug as it is obtained from the investigator(s). The sponsor shall make such reports to FDA regarding information relevant to the safety of the drug as required under §§312.32 and 312.37. The sponsor shall make annual and semiannual safety reports in accordance with §312.33.

* * * * *

§312.64 Investigator reports.

* * * * *

(b) * * * An investigator shall report to the sponsor all adverse clinical and laboratory

outcomes that are required to be reported under the protocol for the study. These reports shall be made within the time period specified within the protocol.
* * * * *

21 CFR PART 314 -- APPLICATIONS FOR FDA APPROVAL TO MARKET A NEW DRUG OR AN ANTIBIOTIC DRUG

§314.50 Content and format of an application.

* * * * *

(h) *Patent information.* The application is required to contain the patent information described under §314.53.

(i) *Patent certification -- Contents.* A 505 (b)(2) application is required to contain the following:

(i) *Patents claiming drug, drug product, or method of use.* (A) Except as provided in paragraph (i)(2) of this section, a certification with respect to each patent issued by the United States Patent and Trademark Office that, in the opinion of the applicant and to the best of its knowledge, claims a drug (the drug product or drug substance that is a component of the drug product) on which investigations that are relied upon by the applicant for approval of its application were conducted or that claims an approved use for such drug and for which information is required to be filed under section 505(b) and (c) of the act and §314.53. For each such patent, the applicant shall provide the patent number and certify, in its opinion and to the best of its knowledge, one of the following circumstances:

(*1*) That the patent information has not been submitted to FDA. The applicant shall entitle such a certification "Paragraph I Certification";

(*2*) That the patent has expired. The applicant shall entitle such a certification "Paragraph II Certification";

(*3*) The date on which the patent will expire. The applicant shall entitle such a certification "Paragraph III Certification"; or

(*4*) That the patent is invalid, unenforceable, or will not be infringed by the manufacture, use, or sale of the drug product for which the application is submitted. The applicant shall entitle such a certification "Paragraph IV Certification". This certification shall be submitted in the following form:

I (*name of applicant*), certify that Patent No. ____________ (*is invalid, unenforceable, or will not be infringed by the manufacture, use or sale of*) (*name of proposed drug product*) for which this application is submitted.

The certification shall be accompanied by a statement that the applicant will comply with the requirements under §314.52(a) with respect to providing a notice to each owner of the patent or their representatives and to the holder of the approved application for the drug product which is claimed by the patent or a use of which is claimed by the patent and with the requirements under §314.52(c) with respect to the content of the notice.

(B) If the drug on which investigations that are relied upon by the applicant were conducted is itself a licensed generic drug of a patented drug first approved under section 505(b) of the act, the appropriate patent certification under this section with respect to each patent that claims the first-approved patented drug or that claims an approved use for such a drug.

(ii) *No relevant patents.* If, in the opinion of the applicant and to the best of its knowledge, there are no patents described in paragraph (i)(1)(i) of this section, a certification in the following form:

In the opinion and to the best knowledge of (*name of applicant*), there are no patents that claim the drug or drugs on which investigations that are relied upon in this application were conducted or that claim a use of such drug or drugs.

(iii) *Method of use patent.* (A) If information that is submitted under section 505(b) or (c) of the act and §314.53 is for a method of use patent, and the labeling for the drug product for which the applicant is seeking approval does not include any indications that are covered by the use patent, a statement explaining that the method of use patent does not claim any of the proposed indications.

(B) If the labeling of the drug product for which the applicant is seeking approval includes an indication that, according to the patent information submitted under section 505(b) or (c) of the act and §314.53 or in the opinion of the applicant, is claimed by a use patent, the applicant shall submit an applicable certification under paragraph (i)(1)(i) of this section.

(2) *Method of manufacturing patent.* An applicant is not required to make a certification with respect to any patent that claims only a method of manufacturing the drug product for which the applicant is seeking approval.

(3) *Licensing agreements.* If a 505(b)(2) application is for a drug or method of using a drug claimed by a patent and the applicant has a licensing agreement with the patent owner, the applicant shall submit a certification under paragraph (i)(1)(i)(A)(4) of this section ("Paragraph IV Certification") as to that patent and a statement that it has been granted a patent license. If the patent owner consents to an immediate effective date upon approval of the 505(b)(2) application, the application shall contain a written statement from the patent owner that it has a licensing agreement with the applicant and that it consents to an immediate effective date.

(4) *Late submission of patent information.* If a patent described in paragraph (i)(1)(i)(A) of this section is issued and the holder of the approved application for the patented drug does not submit the required information on the patent within 30 days of issuance of the patent, an applicant who submitted a 505(b)(2) application that, before the submission of the patent information, contained an appropriate patent certification is not required to submit an amended certification. An applicant whose 505(b)(2) application is filed after a late submission of patent information or whose 505(b)(2) application was previously filed but did not contain an appropriate patent certification at the time of the patent submission shall submit a certification under paragraph (i)(1)(i) or (i)(1)(ii) of this section or a statement under paragraph (i)(1)(iii) of this section as to that patent.

(5) *Disputed patent information.* If an applicant disputes the accuracy or relevance of patent information submitted to FDA, the applicant may seek a confirmation of the correctness of the patent information in accordance with the procedures under §314.53(f). Unless the patent information is withdrawn or changed, the applicant must submit an appropriate certification for each relevant patent.

(6) *Amended certification.* A certification submitted under paragraphs (i)(1)(i) through (i)(1)(iii) of this section may be amended at any time before the effective date of the approval of the application. An applicant shall submit an amended certification as an amendment to a pending application or by letter to an approved application. If an applicant with a pending application voluntarily makes a patent certification for an untimely filed patent, the applicant may withdraw the patent certification for the untimely filed patent. Once an amendment or letter for the change in certification has been submitted, the application will no longer be considered to be one containing the prior certification.

(i) *After finding of infringement.* An applicant who has submitted a certification under paragraph (i)(1(i)(A)(4) of this section and is sued for patent infringement within 45 days of the receipt of notice sent under the §314.52 shall amend the certification if a final judgment in the action is entered finding the patent to be infringed unless the final judgement also finds the patent to be invalid. In the amended certification, the applicant shall certify under paragraph (i)(1)(i)(A)(*3*) of this section that the patent will expire on a specific date.

(ii) *After removal of a patent from the list.* If a patent is removed from the list, any applicant with a pending application(including a tentatively approved application with a delayed effective date) who has made a certification with respect to such patent shall amend its certification. The applicant shall certify under paragraph (i)(1)(ii) of this section that no patents described in paragraph (i)(1)(i) of this section claim the drug or, if other relevant patents claim the drug, shall amend the certification to refer only to those relevant patents. In the amendment, the applicant shall state the reason for the change in certification (that the patent is or has been removed from the list). A patent that is the subject of a lawsuit under §314.107(c) shall not be removed from the list until FDA determines either that no delay in effective dates of approval is required under that section as a result of the lawsuit, that the patent has expired, or that any such period of delay in effective dates of approval is ended. An applicant shall submit an amended certification as an amendment to a pending application. Once an amendment for the change has been submitted, the application will no longer be considered to be one containing a certification under paragraph (i)(1)(i)(A)(4) of this section.

(iii) *Other amendments.* (A) Except as provided in paragraphs (i)(4) and (i)(6)(iii)(B) of this section, an applicant shall amend a submitted certification if, at any time before the effective date of the approval of the application, the applicant learns that the submitted certification is no longer accurate.

(B) An applicant is not required to amend a submitted certification when information on an otherwise applicable patent is submitted after the effective date of approval for the 505(b)(2) application.

(j) *Claimed exclusivity.* A new drug product, upon approval, may be entitled to a period of marketing exclusivity under the provisions of §314.108. If an applicant believes its drug product is entitled to a period of exclusivity, it shall submit with the new drug application prior to approval the following information:

(1) A statement that the applicant is claiming exclusivity.

(2) A reference to the appropriate paragraph under §314.108 that supports its claim.

(3) If the applicant claims exclusivity under §314.108(b)(2), information to show that, to the best of its knowledge or belief, a drug has not previously been approved under section 505(b) of the act containing any active moiety in the drug for which the applicant is seeking approval.

(4) If the applicant claims exclusivity under §314.108(b)(4) or (b)(5), the following information to show that the application contains "new clinical investigations" that are "essential to approval of the application or supplement" and were "conducted or sponsored by the applicant:"

(i) *"New clinical investigations."* A certification that to the best of the applicant's knowledge each of the clinical investigations included in the application meets the definition of "new clinical investigation" set forth in §314.108(a).

(ii) *"Essential to approval."* A list of all published studies or publicly available reports of clinical investigations known to the applicant through a literature search that are relevant to the conditions for which the applicant is seeking approval, a certification that the applicant has thoroughly searched the scientific literature and, to the best of the applicant's knowledge, the list is complete and accurate and, in the applicant's opinion, such published studies or publicly available reports do not provide a sufficient basis for the approval of the conditions for which the applicant is seeking approval without reference to the new clinical investigation(s) in the application, and an explanation as to why the studies or reports are insufficient.

(iii) *"Conducted or sponsored by."* If the applicant was the sponsor named in the Form FDA-1571 for an investigational new drug application (IND) under which the new clinical investigation(s) that is essential to the approval of its application was conducted, identification of the IND by number. If the applicant was not the sponsor of the IND under which the clinical investigation(s) was conducted, a certification that the applicant or its predecessor in interest provided substantial support for the clinical investigation(s) that is essential to the approval of its application, and information supporting the certification. To demonstrate "substantial support," an applicant must either provide a certified statement from a certified public accountant that the applicant provided 50 percent or more of the cost of conducting the study or provide an explanation of why FDA should consider the applicant to have conducted or sponsored the study if the applicant's financial contribution to the study is less than 50 percent or the applicant did not sponsor the investigational new drug. A predecessor in interest is an entity, e.g., a corporation, that the applicant has taken over, merged with, or purchased, or from which the applicant has purchased all rights to the drug. Purchase of nonexclusive rights to a clinical investigation after it is completed is not sufficient to satisfy this definition.

* * * * *

§314.52 Notice of Certification of invalidity or noninfringement of a patent.

(a) *Notice of certification.* For each patent which claims the drug or drugs on which investigations that are relied upon by the applicant for approval of its application were conducted or which claims a use for such drug or drugs and which the applicant certifies under §314.50(i)(1)(i)(A)(*4*) that a patent is invalid, unenforceable, or will not be infringed, the applicant shall send notice of such certification by registered or certified mail, return receipt requested to each of the following persons:

(1) Each owner of the patent that is the subject of the certification or the representative designated by the owner to receive the notice. The name and address of the patent owner or its representative may be obtained from the United States Patent and Trademark Office; and

(2) The holder of the approved application under section 505(b) of the act for each drug product which is claimed by the patent or a use of which is claimed by the patent and for which the applicant is seeking approval, or, if the application holder does not reside or maintain a place of business within the United States, the application holder's attorney, agent, or other authorized official. The name and address of the application holder or its attorney, agent, or authorized official may be obtained from the Division of Drug Information Resources (HFD-80), Center for Drug Evaluation and Research, Food and Drug Administration, 5600 Fishers Lane, Rockville, MD 20857.

(3) This paragraph does not apply to a use patent that claims no uses for which the applicant is seeking approval.

(b) *Sending the notice.* The applicant shall send the notice required by paragraph (a) of this section when it receives from FDA an acknowledgement letter stating that its application has been filed. At the same time, the applicant shall amend its application to include a statement certifying that the notice has been provided to each person identified under paragraph (a) of this section and that the notice met the content requirement under paragraph (c) of this section.

(c) *Content of a notice.* In the notice, the applicant shall cite section 505(b)(3)(B) of the act and shall include, but not be limited to, the following information:

(1) A statement that a 505(b)(2) application submitted by the applicant has been filed by FDA.

(2) The application number.

(3) The established name, if any, as defined in section 502(e)(3) of the act, of the proposed drug product.

(4) The active ingredient, strength, and dosage form of the proposed drug product.

(5) The patent number and expiration date, as submitted to the agency or as known to the applicant, of each patent alleged to be invalid, unenforceable, or not infringed.

(6) A detailed statement of the factual and legal basis of the applicant's opinion that the patent is not valid, unenforceable, or will not be infringed. The applicant shall include in the detailed statement:

(i) For each claim of a patent alleged not to be infringed, a full and detailed explanation of why the claim is not infringed.

(ii) For each claim of a patent alleged to be invalid or unenforceable, a full and detailed explanation of the grounds supporting the allegation.

(7) If the applicant does not reside or have a place of business in the United States, the name and address of an agent in the United States authorized to accept service of process for the applicant.

(d) *Amendment to an application.* If an application is amended to include the certification described in §314.50(i), the applicant shall send the notice required by paragraph (a) of this section at the same time that the amendment to the application is submitted to FDA.

(e) *Documentation of receipt of notice.* The

applicant shall amend its application to document receipt of the notice required under paragraph (a) of this section by each person provided the notice. The applicant shall include a copy of the return receipt or other similar evidence of the date the notification was received. FDA will accept as adequate documentation of the date of receipt a return receipt or a letter acknowledging receipt by the person provided the notice. An applicant may rely on another form of documentation only if FDA has agreed to such documentation in advance. A copy of the notice itself need not be submitted to the agency.

(f) *Approval.* If the requirements of this section are met, the agency will presume the notice to be complete and sufficient, and it will count the day following the date of receipt of the notice by the patent owner or its representative and by the approved application holder as the first day of the 45-day period provided for in section 505(c)(3)(C) of the act. FDA may, if the applicant amends its application with a written statement that a later date should be used, count from such later date.

§314.53 Submission of patent information.

(a) *Who must submit patent information.* This section applies to any applicant who submits to FDA a new drug application or an amendment to it under section 505(b) of the act and §314.50 or a supplement to an approved application under §314.70, except as provided in paragraph (d)(2) of this section.

(b) *Patents for which information must be submitted.* An applicant described in paragraph (a) of this section shall submit information on each patent that claims the drug or a method of using the drug that is the subject of the new drug application or amendment or supplement to it and with respect to which a claim of patent infringement could reasonably be asserted if a person not licensed by the owner of the patent engaged in the manufacture, use, or sale of the drug product. For purposes of this part, such patents consist of drug substance (ingredient) patents, drug product (formulation and composition) patents, and method of use patents. Process patents are not covered by this section and information on process patents may not be submitted to FDA. For patents that claim a drug substance or drug product, the applicant shall submit information only on those patents that claim a drug product that is the subject of a pending or approved application, or that claim a drug substance that is a component of such a product. For patents that claim a method of use, the applicant shall submit information only on those patents that claim indications or other conditions of use of a pending or approved application.

(c) *Reporting requirements* -- (1) *General requirements.* An applicant described in paragraph (a) of this section shall submit the following information for each patent described in paragraph (b) of this section:

(i) Patent number and the date on which the patent will expire.

(ii) Type of patent, i.e., drug, drug product, or method of use.

(iii) Name of the patent owner.

(iv) If the patent owner or applicant does not reside or have a place of business within the United States, the name of an agent (representative) of the patent owner or applicant who resides or maintains a place of business within the United States authorized to receive notice of patent certification under section 505(b)(3) and (j)(2)(B) of the act and §§314.52 and 314.95.

(2) *Formulation, composition, or method of use patents* -- (i) *Original declaration.* For each formulation, composition, or method of use patent, in addition to the patent information described in paragraph (c)(1) of this section the applicant shall submit the following declaration:

The undersigned declares that Patent No. _______ covers the formulation, composition, and/or method of use of (*name of drug product*). This product is (*currently approved under section 505 of the Federal Food, Drug, and Cosmetic Act*) [or] (*the subject of this application for which approval is being sought*):

____

(ii) *Amendment of patent information upon approval.* Within 30 days after the date of approval of its application, if the application contained a declaration required under paragraph (c)(2)(i)of this section, the applicant shall by letter amend the declaration to identify each patent that claims the formulation, composition, or the specific indications or other conditions of use that have been approved.

(3) *No relevant patents.* If the applicant believes that there are no patents which claim the drug or the drug product or which claim a method of using the drug product and with respect to which a claim of patent infringement could reasonably be asserted if a person not licensed by the owner of the patent engaged in the manufacture, use, or sale of the drug product, it shall so declare.

(4) *Authorized signature.* The declarations required by this section shall be signed by the applicant or patent owner, or the applicant's or patent owner's attorney, agent (representative), or other authorized official.

(d) *When and where to submit patent information* -- (1) *Original application.* An applicant shall submit with its original application submitted under this part, including an application described in section 505(b)(2) of the act, the information described in paragraph(c) of this section on each drug (ingredient), drug product (formulation and composition), and method of use patent issued before the application is filed with FDA and for which patent information is required to be submitted under this section. If a patent is issued after the application is filed with FDA, but before the application is approved, the applicant shall, within 30 days of the date of issuance of the patent, submit the required patent information in an amendment to the application under §314.60.

(2) *Supplements.* (i) An applicant shall submit patent information required under paragraph (c) of this section for a patent that claims the drug, drug product, or method of use for which approval is sought in any of the following supplements:

(A) To change the formulation;

(B) To add a new indication or other condition of use, including a change in route of administration;

(C) To change the strength;

(D) To make any other patented change regarding the drug, drug product, or any method of use.

(ii) If the applicant submits a supplement for one of the changes listed under paragraph (d)(2)(i) of this section and existing patents for which information has already ben submitted to FDA claim the changed product, the applicant shall submit a certification with the supplement identifying the patents that claim the changed product.

(iii) If the applicant submits a supplement for one of the changes listed under paragraph (d)(2)(i) of this section and no patents, including previously submitted patents, claim the changed product, it shall so certify.

(iv) The applicant shall comply with the requirements for amendment of formulation or composition and method of use patent information under paragraphs (c)(2)(ii) and (d)(3) of this section.

(3) *Patent information deadline.* If a patent is issued for a drug, drug product, or method of use after an application is approved, the applicant shall submit to FDA the required patent information within 30 days of the date of issuance of the patent.

(4) *Copies.* The applicant shall submit two copies of each submission of patent information, an archival copy and a copy for the chemistry, manufacturing, and controls section of the review copy, to the Central Document Room, Center for Drug Evaluation and Research, Food and Drug Administration, Park Bldg., rm. 2-14, 12420 Parklawn Dr., Rockville, MD 20857. The applicant shall submit thee patent information by letter separate from, but at the same time as, submission of the supplement.

(5) *Submission date.* Patent information shall be considered to be submitted to FDA as of the date the information is received by the Central Document Room.

(6) *Identification.* Each submission of patent information, except information submitted with an original application, and its mailing cover shall bear prominent identification as to its contents, i.e., "Patent Information," or, if

submitted after approval of an application, "Time Sensitive Patent Information."

(e) *Public disclosure of patent information.* FDA will publish in the list the patent number and expiration date of each patent that is required to be, and is, submitted to FDA by an applicant, and for each use patent, the approved indications or other conditions of use covered by a patent. FDA will publish such patent information upon approval of the application, or, if the patent information is submitted by the applicant after approval of an application as provided under paragraph (d)(2) of this section, as soon as possible after the submission to the agency of the patent information. Patent information submitted by the last working day of a month will be published in that month's supplement to the list. Patent information received by the agency between monthly publication of supplements to the list will be placed on public display in FDA's Freedom of Information Staff. A request for copies of the file shall be sent in writing to the Freedom of Information Staff (HFI-35), Food and Drug Administration, rm. 12A-16, 5600 Fishers Lane, Rockville, MD 20857.

(f) *Correction of patent information errors.* If any person disputes the accuracy or relevance of patent information submitted to the agency under this section and published by FDA in the list, or believes that an applicant has failed to submit required patent information, that person must first notify the agency in writing stating the grounds for disagreement. Such notification should be directed to the Drug Information Services Branch (HFD-84), Center for Drug Evaluation and Research, Food and Drug Administration, 5600 Fishers Lane, Rockville, MD 20857. The agency will then request of the applicable new drug application holder that the correctness of the patent information or omission of patent information be confirmed. Unless the application holder withdraws or amends its patent information in response to FDA's request, the agency will not change the patent information in the list. If the new drug application holder does not change the patent information submitted to FDA, a 505(b)(2) application or an abbreviated new drug application under section 505(j) of the act submitted for a drug that is claimed by a patent for which information has been submitted must, despite any disagreement as to the correctness of the patent information, contain an appropriate certification for each listed patent.

§314.54 Procedure for submission of an application requiring investigations for approval of a new indication for, or other change from, a listed drug.

(a) * * *

(1) * * *

(vii) If the applicant believes the change for which it is seeking approval is entitled to a period of exclusivity, the information required under §314.50(j).

* * * * *

§314.70 Supplements and other changes to an approved application.

* * * * *

(f) *Claimed exclusivity.* If an applicant claims exclusivity under §314.108 upon approval of a supplemental application for a change to its previously approved drug product, the applicant shall include with its supplemental application the information required under §314.50(j).

§314.94 Content and format of an abbreviated application.

(a) * * *

(12) *Patent certification* -- (i) *Patents claiming drug, drug product, or method of use.* (A) Except as provided in paragraph (a)(12(iv) of this section, a certification with respect to each patent issued by the United States Patent and Trademark Office that, in the opinion of the applicant and to the best of its knowledge, claims the reference listed drug or that claims a use of such listed drug for which the applicant is seeking approval under section 505(j) of the act and for which information is required to be filed under section 505(b) and (c) of the act and §314.53 for each such patent, the applicant shall provide the patent number and certify, in its opinion and to the best of its

knowledge, one of the following circumstances:

(*1*) That the patent information has not been submitted to FDA. The applicant shall entitle such a certification "Paragraph I Certification";

(*2*) That the patent has expired. The applicant shall entitle such a certification "Paragraph II Certification";

(*3*) The date on which the patent will expire. The applicant shall entitle such a certification "Paragraph III Certification"; or

(*4*) That the patent is invalid, unenforceable, or will not be infringed by the manufacture, use, or sale of the drug product for which the abbreviated application is submitted. The applicant shall entitle such a certification "Paragraph IV Certification". This certification shall be submitted in the following form:

I (*name of applicant*), certify that Patent No. ____________ (*is invalid, unenforceable, or will not be infringed by the manufacture, use or sale of*) (*name of proposed drug product*) for which this application is submitted.

The certification shall be accompanied by a statement that the applicant will comply with the requirements under §314.95(a) with respect to providing a notice to each owner of the patent or their representatives and to the holder of the approved application for the listed drug and with the requirements under §314.95(c) with respect to the content of the notice.

(B) If the abbreviated new drug application refers to a listed drug that is itself a licensed generic product of a patented drug first approved under section 505(b) of the act, the appropriate patent certification under paragraph (a)(12)(i) of this section with respect to each patent that claims the first-approved patented drug or that claims a use for such drug.

(ii) *No relevant patents*. If, in the opinion of the applicant and to the best of its knowledge, there are no patents described in paragraph (a)(12)(i) of this section, a certification in the following form:

In the opinion and to the best knowledge of (*name of applicant*), there are no patents that claim the listed drug referred to in this application or that claim a use of the listed drug.

(iii) *Method of use patent*. (A) If patent information is submitted under section 505(b) or (c) of the act and §314.53 fora patent claiming a method of using the listed drug, and the labeling for the drug product for which the applicant is seeking approval does not include any indications that are covered by the use patent, a statement explaining that the method of use patent does not claim any of the proposed indications.

(B) If the labeling of the drug product for which the applicant is seeking approval includes an indication that, according to the patent information submitted under section 505(b) or (c) of the act and §314.53 or in the opinion of the applicant, is claimed by a use patent, an applicable certification under paragraph (a)(12)(i) of this section.

(iv) *Method of manufacturing patent*. An applicant is not required to make a certification with respect to any patent that claims only a method of manufacturing the listed drug.

(v) *Licensing agreements*. If the abbreviated new drug application is for a drug or method of using a drug claimed by a patent and the applicant has a licensing agreement with the patent owner, a certification under paragraph (a)(12)(i)(A)(*4*) of this section ("Paragraph IV Certification") as to that patent and a statement that it has been granted a patent license.

(vi) *Late submission of patent information*. If a patent on the listed drug is issued and the holder of the approved application for the listed drug does not submit the required information on the patent within 30 days of issuance of the patent, an applicant who submitted an abbreviated new drug application for that drug that contained an appropriate patent certification before the submission of the patent information is not required to submit an amended certification. An applicant whose abbreviated new drug application is submitted after a late submission of patent information, or whose pending abbreviated application was previously submitted but did not contain an appropriate patent certification at the time of the patent submission, shall submit a certification under paragraph (a)(12)(i) of this section or a statement under paragraph (a)(12)(iii) of this section as to that patent.

(vii) *Disputed patent information*. If an applicant disputes the accuracy or relevance of patent information submitted to FDA, the applicant may seek a confirmation of the correctness of the patent information in accordance with the procedures under §314.53(f). Unless the patent information is withdrawn or changed, the applicant shall submit an appropriate certification for each relevant patent.

(viii) *Amended certifications*. A certification submitted under paragraphs (a)(12)(i) through (a)(12)(iii) of this section may be amended at any time before the effective date of the approval of the application. However, an applicant who has submitted a paragraph IV patent certification may not change it to a paragraph III certification if a patent infringement suit has been filed against another paragraph IV applicant unless the agency has determined that no applicant is entitled to 180-day exclusivity or the patent expires before the lawsuit is resolved or expires after the suit is resolved but before the end of the 180-day exclusivity period. If an applicant with a pending application voluntarily makes a patent certification for an untimely filed patent, the applicant may withdraw the patent certification for the untimely filed patent. An applicant shall submit an amended certification by letter or as an amendment to a pending application or by letter to an approved application. Once an amendment or letter is submitted, the application will no longer be considered to contain the prior certification.

(A) *After finding of infringement*. An applicant who has submitted a certification under paragraph (a)(12)(i)(A)(*4*) of this section and is sued for patent infringement within 45 days of the receipt of notice sent under §314.95 shall amend the certification if a final judgment in the action against the applicant is entered finding the patent to be infringed. In the amended certification, the applicant shall certify under paragraph (a)(12)(A)(*3*) of this section that the patent will expire on a specific date. Once an amendment or letter for the change has been submitted, the application will no longer be considered to be one containing a certification under paragraph (a)(12)(i)(A)(*4*) of this section. If a final judgment finds the patent to be invalid and infringed, an amended certification is not required.

(B) *After removal of a patent from the list*. If a patent is removed from the list, any applicant with a pending application (including a tentatively approved application with a delayed effective date) who has made a certification with respect to such patent shall amend its certification. The applicant shall certify under paragraph (a)(12)(ii) of this section that no patents described in paragraph (a)(12)(i) of this section claim the drug or, if other relevant patents claim the drug, shall amend the certification to refer only to those relevant patents. In the amendment, the applicant shall state the reason for the change in certification (that the patent is or has been removed from the list). A patent that is the subject of a lawsuit under §314.107(c) shall not be removed from the list until FDA determines either that no delay in effective dates of approval is required under that section as a result of the lawsuit, that the patent has expired, or that any such period of delay in effective dates of approval is ended. An applicant shall submit an amended certification. Once an amendment or letter for the change has been submitted, the application will no longer be considered to be one containing a certification under paragraph (a)(12)(i)(A)(*4*) of this section.

(C) *Other amendments*. (*1*) Except as provided in paragraphs (a)(12)(vi) and (a)(12)(viii)(C)(*2*) of this section, an applicant shall amend a submitted certification if, at any time before the effective date of the approval of the application, the applicant learns that the submitted certification is no longer accurate.

(*2*) An applicant is not required to amend a submitted certification when information on a patent on the listed drug is submitted after the effective date of approval of the abbreviated application.

* * * * *

§314.95 Notice of certification of invalidity or

noninfringement of a patent.

(a) *Notice of certification.* For each patent that claims the listed drug or that claims a use for such listed drug for which the applicant is seeking approval and that the applicant certifies under §314.94(a)(12) is invalid, unenforceable, or will not be infringed, the applicant shall send notice of such certification by registered or certified mail, return receipt requested to each of the following persons:

(1) Each owner or the patent which is the subject of the certification or the representative designated by the owner to receive the notice. The name and address of the patent owner or its representative may be obtained from the United States Patent and Trademark Office; and

(2) The holder of the approved application under section 505(b) of the act for the listed drug that is claimed by the patent and for which the applicant is seeking approval, or, if the application holder does not reside or maintain a place of business within the United States, the application holder's attorney, agent, or other authorized official. The name and address of the application holder or its attorney, agent, or authorized official may be obtained from the Division of Drug Information Resources (HFD-80), Center for Drug Evaluation and Research, Food and Drug Administration, 5600 Fishers Lane, Rockville, MD 20857.

(3) This paragraph does not apply to a use patent that claims no uses for which the applicant is seeking approval.

(b) *Sending the notice.* The applicant shall send the notice required by paragraph (a) of this section when it receives from FDA an acknowledgement letter stating that its abbreviated new drug application is sufficiently complete to permit a substantive review. At the same time, the applicant shall amend its abbreviated new drug application to include a statement certifying that the notice has been provided to each person identified under paragraph (a) of this section and that the notice met the content requirement under paragraph (c) of this section.

(c) *Content of a notice.* In the notice, the applicant shall cite section 505(j)(2)(B)(ii) of the act and shall include, but not be limited to, the following information:

(1) A statement that FDA has received an abbreviated new drug application submitted by the applicant containing any required bioavailability or bioequivalence data or information.

(2) The abbreviated application number.

(3) The established name, if any, as defined in section 502(e)(3) of the act, of the proposed drug product.

(4) The active ingredient, strength, and dosage form of the proposed drug product.

(5) The patent number and expiration date, as submitted to the agency or as known to the applicant, of each patent alleged to be invalid, unenforceable, or not infringed.

(6) A detailed statement of the factual and legal basis of the applicant's opinion that the patent is not valid, unenforceable, or will not be infringed. The applicant shall include in the detailed statement:

(i) For each claim of a patent alleged not to be infringed, a full and detailed explanation of why the claim is not infringed.

(ii) For each claim of a patent alleged to be invalid or unenforceable, a full and detailed explanation of the grounds supporting the allegation.

(7) If the applicant does not reside or have a place of business in the United States, the name and address of an agent in the United States authorized to accept service of process for the applicant.

(d) *Amendment to an abbreviated application.* If an abbreviated application is amended to include the certification described in §314.94(a)(12)(i)(A)(*4*), the applicant shall send the notice required by paragraph (a) of this section at the same time that the amendment to the abbreviated application is submitted to FDA.

(e) *Documentation of receipt of notice.* The applicant shall amend its abbreviated application to document receipt of the notice required under paragraph (a) of this section by each person provided the notice. The applicant shall include a copy of the return receipt or other similar evidence of the date the notification was received. FDA will accept as adequate documentation of the date of receipt a return receipt or a letter acknowledging receipt by the person provided the notice. An applicant may rely on another form of documentation only if FDA has agreed to such documentation in advance. A copy of the notice itself need not be submitted to the agency.

(f) *Approval.* If the requirements of this section are met, FDA will presume the notice to be complete and sufficient, and it will count the day following the date of receipt of the notice by the patent owner or its representative and by the approved application holder as the first day of the 45-day period provided for in section 505(j)(4)(B)(iii) of the act. FDA may, if the applicant provides a written statement to FDA that a later date should be used, count from such later date.

§314.101 Filing an application and an abbreviated antibiotic application and receiving an abbreviated new drug application.

* * * * *

(e) The agency will refuse to file an application or abbreviated antibiotic application or will consider an abbreviated new drug application not to have been received if any of the following applies:

(1) The drug product is subject to licensing by FDA under the Public Health Service Act (42 U.S.C. 201 *et seq.*) and subchapter F of this chapter.

(2) In the case of a 505(b)(2) application or an abbreviated new drug application, the drug product contains the same active moiety as a drug that:

(i) was approved after september 24, 1984, in an application under section 505(b) of the act, and

(ii) Is entitled to a 5-year period of exclusivity under section 505(c)(3)(D)(ii) and (j)(4)(D)(ii) of the act and §314.108(b)(2), unless the 5-year exclusivity period has elapsed or unless 4 years of the 5-year period have elapsed and the application or abbreviated application contains a certification of patent invalidity or noninfringement described in §314.50(i)(1)(i)(A)(*4*) or §314.94(a)(12)(i)(A)(*4*).

* * * * *

§314.107 Effective date of approval of a 505(b)(2) application or abbreviated new drug application under section 505(j) of the act.

(a) *General.* A drug product may be introduced or delivered for introduction into interstate commerce when approval of the application or abbreviated application for the drug product becomes effective. Except as provided in this section, approval of an application or abbreviated application for a drug product becomes effective on the date FDA issues an approval letter under §314.105 for the application or abbreviated application.

(b) *Effect of patent on the listed drug.* If approval of an abbreviated new drug application submitted under section 505(j) of the act or of a 505(b)(2) application is granted, that approval will become effective in accordance with the following:

(1) *Date of approval letter.* Except as provided in paragraphs (b)(3), (b)(4), and (c) of this section, approval will become effective on the date FDA issues an approval letter under §314.105 if the applicant certifies under §314.50(i) or §314.94(a)(12) that:

(i) There are no relevant patents; or

(ii) The applicant is aware of a relevant patent but the patent information required under section 505(b) or (c) of the act has not been submitted to FDA; or

(iii) The relevant patent has expired; or

(iv) The relevant patent is invalid, unenforceable, or will not be infringed.

(2) *Patent expiration.* If the applicant certifies under §314.50(i) or §314.94(a)(12) that the relevant patent will expire on a specified date, approval will become effective on the specified date.

(3) *Disposition of patent litigation.* (i)(A) Except as provided in paragraphs (b)(3)(ii), (b)(3)(iii), and (b)(3)(iv) of this section, if the applicant certifies under §314.50(i) or §314.94(a)(12) that the relevant patent is invalid, unenforceable, or will not be infringed, and the patent owner or its representative or the exclusive patent licensee brings suit for patent infringement within 45 days of receipt by the patent owner of the notice of certification from the applicant under §314.52 or §314.95,

approval may be made effective 30 months after the date of the receipt of the notice of certification by the patent owner or by the exclusive licensee (or their representatives) unless the court has extended or reduced the period because of a failure of either the plaintiff or defendant to cooperate reasonably in expediting the action; or

(B) If the patented drug product qualifies for 5 years of exclusive marketing under §314. 108(b)(2) and the patent owner or its representative or the exclusive patent licensee brings suit for patent infringement during the 1-year period beginning 4 years after the date the patented drug was approved and within 45 days of receipt by the patent owner of the notice of certification, the approval may be made effective at the expiration of the 7 1/2 years from the date of approval of the application for the patented drug product.

(ii) If before the expiration of the 30-month period, or 7 1/2 years where applicable, the court issues a final order that the patent is invalid, unenforceable, or not infringed, approval may be made effective on the date the court enters judgment;

(iii) If before the expiration of the 30-month period, or 7 1/2 years where applicable, the court issues a final order or judgment that the patent has been infringed, approval may be made effective on the date the court determines that the patent will expire or otherwise orders; or

(iv) If before the expiration of the 30-month period, or 7 1/2 years where applicable, the court grants a preliminary injunction prohibiting the applicant from engaging in the commercial manufacture or sale of the drug product until the court decides the issues of patent validity and infringement, and if the court later decides that the patent is invalid, unenforceable, or not infringed, approval may be made effective on t he date the court enters a final order or judgment that the patent is invalid, unenforceable, or not infringed.

(v) In order for an approval to be made effective under paragraph (b)(3) of this section, the applicant must receive an approval letter from the agency indicating that the application has received final approval. Tentative approval of an application does not constitute "approval" of an application and cannot, absent a final approval letter from the agency, result in an effective approval under paragraph (b)(3) of this section.

(4) *Multiple certifications*. If the applicant has submitted certifications under §314.50(i) or §314.94(a)(12) for more than one patent, the date of approval will be calculated for each certification, and the approval will become effective on the last applicable date.

(c) *Subsequent abbreviated new drug application submission*.

(1) If an abbreviated new drug application contains a certification that a relevant patent is invalid, unenforceable, or will not be infringed and the application is for a generic copy of the same listed drug for which one or more substantially complete abbreviated new drug applications were previously submitted containing a certification that the same patent was invalid, unenforceable, or would not be infringed and the applicant submitting the first application has successfully defended against a suit for patent infringement brought within 45 days of the patent owner's receipt of notice submitted under §314.95, approval of the subsequent abbreviated new drug application will be made effective no sooner than 180 days from whichever of the following dates is earlier:

(i) The date the applicant submitting the first application first commences commercial marketing of its drug product; or

(ii) The date of a decision of the court holding the relevant patent invalid, unenforceable, or not infringed.

(2) For purposes of paragraph (c)(1) of this section, the "applicant submitting the first application" is the applicant that submits an application that is both substantially complete and contains a certification that the patent was invalid, unenforceable, or not infringed prior to the submission of any other application for the same listed drug that is both substantially complete and contains the same certification. A "substantially complete" application must contain the results of any required bioequivalence studies, or, if applicable, a request for a waiver of such studies.

(3) For purposes of paragraph (c)(1) of this section, if FDA concludes that the applicant

submitting the first application is not actively pursuing approval of its abbreviated application, FDA will make the approval of subsequent abbreviated applications immediately effective if they are otherwise eligible for an immediately effective approval.

(4) For purposes of paragraph (c)(1)(i) of this section, the applicant submitting the first application shall, if sued for patent infringement, notify FDA of the date that it commences commercial marketing of its drug product. Commercial marketing commences with the first date of introduction or delivery for introduction into interstate commerce outside the control of the manufacturer of a drug product, except for investigational use under part 312 of this chapter, but does not include transfer of the drug product for reasons other than sale within the control of the manufacturer or application holder. If an applicant does not promptly notify FDA of such date, the effective date of approval shall be deemed to be the date of the commencement of first commercial marketing.

(d) *Delay due to exclusivity.* The agency will also delay the effective date of the approval of an abbreviated new drug application under section 505(j) of the act or a 505(b)(2) application if delay is required by the exclusivity provisions of §314.108. When the effective date of an application is delayed under both this section and §314.108, the effective date will be the later of the 2 days specified under this section and §314.108.

(e) *Court actions.* (1) References to actions of "the court" in paragraphs (b) and (c) of this section are to the court that enters final judgment from which no appeal can be or has been taken.

(2) For purposes of establishing the effective date of approval based on a court judgment, the following dates shall be deemed to be the date of the final decision of the court on the patent issues:

(i) If the district court enters a decision that the patent is invalid, unenforceable, or not infringed, and the decision is not appealed, the date on which the right to appeal lapses.

(ii) If the district court enters a decision that the patent is invalid, unenforceable, or not infringed, and the decision is appealed, the date of the first decision or order by a higher court holding or affirming the decision of the district court that the patent is invalid, unenforceable, or not infringed.

(iii) If the district court enters a decision that the patent is infringed, and the decision is appealed, the date on which the district court enters a judgment that the patent is invalid, unenforceable, or not infringed pursuant to a mandate issued by a court of appeals.

(iv) The applicant shall submit a copy of the entry of the order or judgment to the Office of Generic Drugs (HFD-600), or to the appropriate division in the Office of Drug Evaluation I (HFD-100) or Office of Drug Evaluation II (HFD-500), whichever is applicable, within 10 working days of a final judgment.

(f) *Computation of 45-day time clock.* (1) The 45-day clock described in paragraph (b)(3) of this section begins on the day after the date of receipt of the applicant's notice of certification by the patent owner or its representative, and by the approved application holder. When the 45th day falls on Saturday, Sunday, or a Federal holiday, the 45th day will be the next day that is not a Saturday, Sunday, or a Federal holiday.

(2) The abbreviated new drug applicant or the 505(b)(2) applicant shall notify FDA immediately of the filing of any legal action filed within 45 days of receipt of the notice of certification. If the applicant submitting the abbreviated new drug application or the 505 (b)(2) application or patent owner or its representative does not notify FDA in writing before the expiration of the 45-day time period or the completion of the agency's review of the application, whichever occurs later, that a legal action for patent infringement was filed within 45 days of receipt of the notice of certification, approval of the abbreviated new drug application or the 505(b)(2) application will be made effective immediately upon expiration of the 45 days or upon completion of the agency's review and approval of the application, whichever is later. The notification to FDA of the legal action shall include:

(i) The abbreviated new drug application or 505(b)(2) application number.

(ii) The name of the abbreviated new drug or 505(b)(2) application applicant.

(iii) The established name of the drug product or, if no established name exists, the name(s) of the active ingredient(s), the drug product's strength, and dosage form.

(iv) A certification that an action for patent infringement identified by number, has been filed in an appropriate court on a specified date.

The applicant of an abbreviated new drug application shall send the notification to FDA's Office of Generic Drugs (HFD-600). A 505(b)(2) applicant shall send the notification to the appropriate division in the Center for Drug Evaluation and Research reviewing the application. A patent owner or its representative may also notify FDA of the filing of any legal action for patent infringement. The notice should contain the information and be sent to the offices or divisions described in this paragraph.

(3) If the patent owner or approved application holder who is an exclusive patent licensee waives its opportunity to file a legal action for patent infringement within 45 days of a receipt of the notice of certification and the patent owner or approved application holder who is an exclusive patent licensee submits to FDA a valid waiver before the 45 days elapse, approval of the abbreviated new drug application or the 505(b)(2) application will be made effective upon completion of the agency's review and approval of the application. FDA will only accept a waiver in the following form:

(*Name of patent owner or exclusive patent licensee*) has received notice from (*name of applicant*) under (*section 505(b)(3) or 505(j)(2)(B) of the act*) and does not intend to file an action for patent infringement against (*name of applicant*) concerning the drug (*name of drug*) before (*date on which 45 days elapses*). (*Name of patent owner or exclusive patent licensee*) waives the opportunity provided by (*section 505(c)(3)(C) or 505(j)(B)(iii) of the act*) and does not object to FDA's approval of (*name of applicant*)'s (*505(b)(2) or abbreviated new drug application*) for (*name of drug*) with an immediate effective date on or after the date of this letter.

§314.108 New drug product exclusivity.

(a) *Definitions*. The following definitions of terms apply to this section:

Active moiety means the molecule or ion, excluding those appended portions of the molecule that cause the drug to be an ester, salt (including a salt with hydrogen or coordination bonds), or other noncovalent derivative (such as a complex, chelate, or clathrate) of the molecule, responsible for the physiological or pharmacological action of the drug substance.

Approved under section 505(b) means an application submitted under section 505(b) and approved on or after October 10, 1962, or an application that was "deemed approved" under section 107(c)(2) of Pub. L. 87-781.

Clinical investigation means any experiment other than a bioavailability study in which a drug is administered or dispensed to, or used on, human subjects.

Conducted or sponsored by the applicant with regard to an investigation means that before or during the investigation, the applicant was named in Form FDA-1571 filed with FDA as the sponsor of the investigational new drug application under which the investigation was conducted, or the applicant or the applicant's predecessor in interest, provided substantial support for the investigation. To demonstrate "substantial support," an applicant must either provide a certified statement from a certified public accountant that the applicant provided 50 percent or more of the cost of conducting the study or provide an explanation why FDA should consider the applicant to have conducted or sponsored the study if the applicant's financial contribution to the study is less than 50 percent or the applicant did not sponsor the investigational new drug. A predecessor in interest is an entity, e.g., a corporation, that the applicant has taken over, merged with, or purchased, or from which the applicant has purchased all rights to the drug. Purchase of nonexclusive rights to a clinical investigation after it is completed is not sufficient to satisfy this definition.

Date of approval means the date on the

letter from FDA stating that the new drug application is approved, whether or not final printed labeling or other materials must yet be submitted as long as approval of such labeling or materials is not expressly required. "Date of approval" refers only to a final approval and not to a tentative approval that may become effective at a later date.

Essential to approval means, with regard to an investigation, that there are no other data available that could support approval of the application.

FDA means the Food and Drug Administration.

New chemical entity means a drug that contains no active moiety that has been approved by FDA in any other application submitted under section 505(b) of the act.

New clinical investigation means an investigation in humans the results of which have not been relied on by FDA to demonstrate substantial evidence of effectiveness of a previously approved drug product for any indication or of safety for a new patient population and do not duplicate the results of another investigation that was relied on by the agency to demonstrate the effectiveness or safety in a new patient population of a previously approved drug product. For purposes of this section, data from a clinical investigation previously submitted for use in the comprehensive evaluation of the safety of a drug product but not to support the effectiveness of the drug product would be considered new.

(b) *Submission of and effective date of approval of an abbreviated new drug application submitted under section 505(j) of the act or a 505(b)(2) application.* (1) [Reserved]

(2) If a drug product that contains a new chemical entity was approved after September 24, 1984, in an application submitted under section 505(b) of the act, no person may submit a 505(b)(2) application or abbreviated new drug application under section 505(j) of the act for a drug product that contains the same active moiety as in the new chemical entity for a period of 5 years from the date of approval of the first approved new drug application, except that the 505(b)(2) application or abbreviated application may be submitted after 4 years if it contains a certification of patent invalidity or noninfringement described in §314.50(i)(1)(i)(A)(*4*) or §314.94(a)(12)(i)(A)(*4*).

(3) The approval of a 505(b)(2) application or abbreviated application described in paragraph (b)(2) of this section will become effective as provided in §314.107(b)(1) or (b)(2), unless the owner of a patent that claims the drug, the patent owner's representative, or exclusive licensee brings suit for patent infringement against the applicant during the 1-year period beginning 48 months after the date of approval of the new drug application for the new chemical entity and within 45 days after receipt of the notice described at §314.52 or §314.95, in which case, approval of the 505(b)(2) application or abbreviated application will be made effective as provided in §314.107(b)(3).

(4) If an application:

(i) Was submitted under section 505(b) of the act;

(ii) Was approved after September 24, 1984;

(iii) Was for a drug product that contains an active moiety that has been previously approved in another application under section 505(b) of the act; and

(iv) Contained reports of new clinical investigations (other than bioavailability studies) conducted or sponsored by the applicant that were essential to approval of the application, the agency will not make effective for a period of 3 years after the date of approval of the application the approval of a 505(b)(2) application or an abbreviated new drug application for the conditions of approval of the original application, or an abbreviated new drug application submitted pursuant to an approved petition under section 505(j)(2)(C) of the act that relies on the information supporting the conditions of approval of an original new drug application.

(5) If a supplemental application:

(i) Was approved after September 24, 1984; and

(ii) Contained reports of new clinical investigations (other than bioavailability studies)that were conducted or sponsored by the applicant that were essential to approval of the supplemental application, the agency will not make effective fora period of 3 years after the date of approval of the supplemental application the approval of a 505(b)(2) application or an abbreviated new drug application for a change, or an abbreviated new drug application submitted pursuant to an approved petition under section 505(j)(2)(C) of the act that relies on the information supporting a change approved in the supplemental new drug application.

PART 314 -- APPLICATIONS FOR FDA APPROVAL TO MARKET A NEW DRUG OR AN ANTIBIOTIC DRUG

§314.80 Postmarketing reporting of adverse drug experiences.

(a) * * *

* * * * *

Data lock-point means the date designated as the cutoff date for data to be incorporated into a specific postmarketing adverse drug experience periodic report. Data available to the applicant after this date will not be incorporated into the report, unless it represents important information.

Disability means a substantial disruption of a person's ability to carry out normal life functions.

* * * * *

International birth date means the date that the first regulatory authority in the world approved the human drug product for marketing.

Life-threatening means that the patient was, in the view of the initial reporter, at immediate risk of death from the adverse experience as it occurred. It does not include an adverse experience that, had it occurred in a more serious form, might have caused death. For example, product-induced hepatitis that resolved without evidence of hepatic failure would not be considered life-threatening even though hepatitis of a more severe nature can be fatal. Similarly, an allergic reaction resulting in angioedema of the face would not be life-threatening, even though angioedema of the larynx, allergic bronchospasm, or anaphylaxis can be fatal.

Serious means an adverse drug experience occurring at any dose that is fatal or life-threatening, results in the persistent or significant disability/incapacity, requires or prolongs inpatient hospitalization, necessitates medical or surgical intervention to preclude permanent impairment of a body function or permanent damage to a body structure, or is a congenital anomaly.

* * * * *

(b) * * * Applicants should not resubmit to FDA adverse product experience reports forwarded to the applicant by FDA; however, they should submit all followup information to FDA. Applicants shall also develop written procedures for the surveillance, receipt, evaluation, and reporting of adverse drug experiences.

(c) *Reporting requirements* The applicant shall report to FDA adverse drug experience information, as described in this section. The applicant shall submit two copies of each report described in this section to the Central Document Room, Park Bldg., rm. 2-14, 12420 Parklawn Dr., Rockville, MD 20857. FDA may waive the requirement for the second copy in appropriate instances.

(1)(i) *Postmarketing 15-day "Alert reports".* The applicant shall report each adverse drug experience that is both serious and unexpected as soon as possible but in any case within 15 calendar days of initial receipt of the information. These reports shall be submitted on FDA Form 3500A.

(ii) *Postmarketing Fifteen-day "Alert reports" -- followup.* The applicant shall promptly investigate all adverse drug experiences that are the subject of these postmarketing 15-day Alert reports and shall submit followup reports within 15 calendar days of receipt of new information or as requested by FDA. If additional information is not obtainable, records should be maintained of the unsuccessful steps taken to seek additional information. These postmarketing 15-day Alert reports and followups to them shall be submitted under separate cover and may not be included, except for summary or tabular purposes, in a postmarketing adverse drug experience periodic report.

(iii) *Increased frequency report.* The applicant shall review periodically (at least as often as the periodic reporting cycle) the frequency of reports of adverse drug experiences that are both serious and expected and reports of therapeutic failure (lack of effect), regardless of source, and report any significant increase in frequency as soon as possible but in any case within 15 calendar days of determining that a significant increase in frequency exists. Upon written notice, FDA may require that applicants review the frequency of reports of serious, expected adverse drug experiences at intervals different than the periodic reporting cycle. Reports of a significant increase in frequency are required to be submitted in narrative form (including the time period on which the increased frequency is based, the method of analysis, and the interpretation of the results), rather than using FDA Form 3500A. 15-day Alert reports based on increased frequency are required to be submitted under separate cover and may not be included, except for summary purposes, in a periodic report.

(iv) *Submission of reports.* The requirements of paragraphs (c)(1)(i), (c)(1)(ii), and (c)(1)(iii) of this section, concerning the submission of postmarketing 15-day Alert reports, shall also apply to any person (other than the applicant) whose name appears on the label of an approved drug product as a manufacturer, packer, or distributor. However, in order to avoid unnecessary duplication in the submission to FDA of reports required by paragraphs (c)(1)(i), (c)(1)(ii), and (c)(1)(iii) of this section, obligations of a nonapplicant may be met by submission of all reports of serious adverse drug experiences to the applicant. If a nonapplicant elects to submit adverse drug experience reports to the applicant rather than to FDA, it shall submit each report to the applicant within 3 calendar days of its receipt by the nonapplicant, and the applicant shall then comply with the requirements of this section. Under this circumstance, the nonapplicant shall maintain a record of this action which shall include:

(A) A copy of the drug experience report.

(B) Date the report was received by the nonapplicant.

(C) Date the report was submitted to the applicant.

(D) Name and address of the applicant.

(v) *Report identification.* Each report submitted under this paragraph shall bear prominent identification as to its contents, i.e., "15-day Alert report," "15-day Alert report -- followup," or "Increased frequency report."

(2) *Periodic adverse drug experience reports.* (i) The applicant shall report every 6 months each adverse drug experience not reported under paragraphs (c)(1)(i) and (c)(1)(ii) of this section. The periodic reporting term shall be based upon the international birth date of the human drug product. The first 6-month anniversary of the international birth date after the application is approved in the United States is the data lock-point for the first periodic reporting term. Each subsequent 6-month anniversary of the international birth date is the data lock-point for subsequent periodic reporting terms for that particular product. Periodic reports shall be submitted to FDA within 45 days after the data lock-point. Upon written notice, FDA may require that the applicant submit reports under this section at times other than those stated. An applicant who wishes to submit periodic reports at different intervals must submit to FDA a request for a waiver under §314.90. Followup information to adverse drug experiences submitted initially in a periodic report may be submitted in the next periodic report. If the applicant does not receive any adverse experience reports during the reporting period, the applicant shall, in place of a periodic report, send a copy of the current approved U.S. labeling and a letter identifying the product, the application number, and the reporting period, stating that no adverse drug experience reports were received.

(ii) *Reports.* Each periodic report shall contain:

(A) *Title page, table of contents, and introduction.* The introduction shall be a summary of the periodic report with page references to detailed data and information.

(B) *Applicant's core safety data sheet.* The applicant's core safety data sheet shall be a document prepared by the applicant that contains all relevant safety information, including adverse drug experiences, which the applicant believes should be listed for the drug in all countries where the drug is marketed. It may be used by the applicant as the reference document by which an adverse drug experience is judged to be expected or unexpected for purposes of this postmarketing periodic report. For all other determinations of whether an adverse drug experience is expected or unexpected, the definition in paragraph (a) of this section shall apply.

(*1*) FDA recognizes that the postmarketing periodic report may be submitted by the applicant to multiple countries and the product may have different approved labels in the different countries. The use of the applicant's core safety data sheet as the reference document for determining whether an adverse drug experience is unexpected or not may result in some overreporting of unexpected adverse events that actually are expected by the U.S. approved product label. This is because the approved label for the United States may have more safety information included in it than the manufacturer's core safety data sheet. If an adverse event is not listed in the U.S. label, but is in the manufacturer's core safety data sheet, this shall be clearly noted in the "Overall safety evaluation" (see paragraph (c)(2)(ii)(H) of this section). This section also shall highlight clearly any changes and the reasons for the changes in the applicant's core safety data sheet since the previous postmarketing periodic report.

(*2*) An applicant may also use the approved U.S. label as the reference by which expected and unexpected adverse drug experiences are determined for the postmarketing periodic report. If an applicant chooses to use the approved U.S. label for this purpose, it shall clearly be stated in this section of the report.

(C) *The product's marketing status.* This section shall contain a table containing a chronological history of the marketing status of the product worldwide (all regulatory decisions affecting the product and all market launches) from the date it was first approved through its current status. Approvals or applications voluntarily withdrawn for safety reasons shall be included at a minimum. The product shall be listed by chemical (or proper name) and brand name(s).

(D) *Regulatory actions for safety reasons.* This section shall contain a narrative identifying the reasons for significant regulatory authority or manufacturer-initiated actions taken anywhere in the world, or to be taken imminently, for safety reasons during the reporting period. This includes, for example, application withdrawal or license suspension or failure to renew, distribution restrictions, clinical trial suspension, labeling changes due to significant safety concerns, dosage modifications, or pharmaceutical changes.

(E) *Patient exposure.* This section shall include the product's domestic and foreign marketing distribution data during the reporting period. This shall be used to calculate the extent of patient exposure. The method used by the manufacturer to estimate patient exposure shall always be described and shall include the total number of dosage units of each dosage form and strength or potency (e.g., 100,000/ 5-milligram tablets, 50,000/10-milliliter vials).

(F) *Individual case histories.* This section shall consist of individual case reports of adverse drug experiences thought possibly associated with the use of the drug that are:

(*1*) Serious, unexpected reports from published or unpublished clinical studies where the applicant has concluded that there is a reasonable possibility that the drug caused the adverse experience;

(*2*) Serious, expected or unexpected spontaneous adverse drug experience reports and nonserious, unexpected spontaneous adverse drug experience reports (causality always assumed in spontaneous reports) received directly by the applicant from the initial reporter or received by the applicant from a drug regulatory authority, both U.S. or foreign; and

(*3*) Serious, expected or unexpected, individual published case histories and nonserious, unexpected individual published case histories.

(*4*) All of these reports in paragraphs (c)(2)(ii)(F)(*1*) through (c)(2)(ii)(F)(*3*) of this section shall be presented in line listing format with the following 10 columns: country, source, age, gender, dose, duration of treatment (prior to event), time to onset, description of reaction (as reported), outcomes (e.g., fatal, resolved), other comments (e.g., manufacturer's report number). This format is consistent with that suggested by CIOMS. In addition, a tabular summary of the number of adverse events by body system may be included. This section shall end with an analysis by the reporter, in narrative form, of the cases submitted. The applicant shall also attach to the end of the postmarketing periodic report a completed FDA Form 3500A for all U.S. spontaneous reports of adverse drug experiences except those reported under paragraphs (c)(1)(i) and (c)(1)(ii) of this section, or those sent by FDA to the applicant.

(G) *Safety studies.* This section shall contain an analysis and full critical discussion of all toxicological, clinical, and epidemiological studies containing important safety information.

(H) *Overall safety evaluation.* This section shall contain a critical analysis and full discussion of the safety information provided in the periodic report as it pertains to serious unexpected reactions, increased frequencies of known toxicity, reactions listed in the manufacturer's core safety data sheet but not included in the U.S. label, drug interactions, overdose, drug abuse, experiences during pregnancy or lactation, chronic treatment, pediatric or geriatric treatment, and new safety issues. The applicant shall indicate when any significant information has not been obtained. The evaluation shall indicate whether the safety profile of the product remains consistent with cumulative experience to date and with the previous manufacturer's core safety data sheet. The evaluation shall specify any action recommended and the reasons for such recommendations.

(I) *Other information.* This section shall include important information received after the data lock-point.

(J) *FDA Form 3500A.* An FDA Form 3500A shall be used for each spontaneous U.S. adverse drug experience not reported under paragraphs (c)(1)(i) and (c)(1)(ii) of this section.

(K) *Location of adverse experience records.* The current addresses where all adverse experience reports and records are maintained.

(d) * * *

(1) * * * The 15-day reporting requirements in paragraph (c)(1)(iii) of this section (i.e., a significant increase in frequency of a serious, expected adverse drug experience, or of a therapeutic failure) apply only to the reports found in scientific and medical journals either as case reports or as the result of a formal clinical trial. * * *

* * * * *

(f) *Reporting FDA Form 3500A.* (1) Except as provided in paragraphs (c)(1)(iii) and (f)(3) of this section, the applicant shall complete FDA Form 3500A for each report of an adverse drug experience.

* * * * *

(3) * * *

(ii) The format is agreed to in advance by MedWatch: The FDA Medical Products Reporting Program.

(4) Ten copies or fewer of FDA Form 3500A and/or a copy of the instructions for completing the form may be obtained from the Division of Epidemiology and Surveillance (HFD-730), Center for Drug Evaluation and Research, Food and Drug Administration, 5600 Fishers Lane, Rockville, MD 20857. More than 10 copies of the form may be obtained by writing to the Consolidated Forms and Publications Distribution Center, Washington Commerce Center 3222 Hubbard R., Landover, MD 20785.

Copies of this guideline may be obtained from the CDER Executive Secretariat Staff (HFD-8), Center for Drug Evaluation and Research, Food and Drug Administration, 7500 Standish Pl., Rockville, MD 20857.

* * * * *

(l) * * * For purposes of this provision, the term "applicant" also includes any person reporting under paragraph (c)(1)(iv) of this section.

* * * * *

Chapter 4

EPA REGULATIONS

THE EPA APPROACH

Under the authority of section 4 of the Toxic Substances Control Act (TSCA), the EPA evaluates laboratory data submitted to the agency regarding tests of the health effects of chemical substances and mixtures. The EPA also, under authority of the Federal Insecticide, Fungicide and Rodenticide Act (FIFRA), evaluates laboratory test data relating to hazards to humans arising from the use of a pesticide product when the agency evaluates pesticide registration applications.

The EPA was aware of problems, uncovered by the FDA in the mid-1970s, relating to unacceptable laboratory practices. The EPA responded by forming the Toxicology Auditing Program in the agency's Office of Pesticide Programs. The EPA also held public hearings to solicit comments on how appropriate the agency's approach was to data quality assurance for pesticide testing.

GOOD LABORATORY PRACTICES

Like the FDA, the EPA determined that the promulgation of Good Laboratory Practice regulations would most effectively handle the problem of compliance with adequate control standards, and the agency published proposed health effects standards for testing under TSCA on May 9, 1979. Proposed GLP regulations applicable to laboratory studies submitted to the EPA in compliance with FIFRA were published on April 18, 1980. Supplemental GLP standards for the development of data on physical, chemical, persistence and ecological effects of chemical substances for which the EPA requires testing under section 4 of TSCA were published November 21, 1980. The EPA's FIFRA and TSCA GLP regulations were both issued in final form on November 29, 1983. No revisions have been made to the EPA's GLP regulations since the publication of the *Validation Compliance Annual: 1995.*

GOOD AUTOMATED LABORATORY PRACTICES

In many laboratories, computers are replacing many manual operations: they manage operations, interface with equipment, and generate reports, for example. The proliferation of automated data collection, however, has resulted in new problems of corruption, loss, and inappropriate modification in data provided to the EPA. The GALPs are EPA's answer to the need for standardized laboratory data management principles. The lack of such principles was underscored after an investigation by EPA's Office of Inspector General, and an EPA survey of automated laboratory practices disclosed that the integrity of computer-resident data was at risk in many laboratories providing data to EPA because of gaps in system security, data validation, and basic documentation. At the same time, laboratory staffs expressed the need for -- and were frustrated with their inability to obtain -- adequate EPA guidance to protect integrity of computer-resident data. These laboratories wanted a single source of guidance for automated operations.

The EPA determined that data management procedures should be standardized in laboratories supporting EPA programs, and that the Agency itself should assume the responsibility to establish those standards. The GALPs and the GALP *Implementation Guidance* were published in draft form on December 28, 1990. A final edition was released in August 1995. The intent of these documents is to enable laboratories that provide data to have a clear understanding of what EPA considers to be adequate controls to assure data integrity.

Good Automated Laboratory Practices

August 10, 1995

Principles and Guidance to
Regulations for Ensuring Data Integrity
In Automated Laboratory Operations

1995 Edition

Scientific Systems Staff
Office of Information Resources
Management
U.S. Environmental Protection Agency
Research Triangle Park
North Carolina 27711

Preface

Most EPA regulatory and research programs have regulations or requirements by contract clause that govern the conduct of laboratory studies. The GALPs *do not* supersede any existing requirements or regulations of EPA's organizations, nor do they augment them. Some of the GALP provisions guide EPA staff and its agents (contractors or grantees) to existing EPA requirements such as the System Life Cycle Management, Chapter 17 of *Information Resources Management Policy Manual.*

The GALPs are developed from essential principles inherent to sustaining challenges to the reliability of data. These include traceability, accountability, standardized procedures, adequate resources, and , importantly,t he availability of documentation that supports conformance with these principles. Each GALP provision embraces at least one of these principles.

The intended objective of the GALPs is to provide EPA organizations with a set of benchmarks to examine in light of their needs and established requirements or regulations. If an organization then determines that changes or additions to their own requirements or regulations are needed, it is the responsibility f that organization to amend their requirements or regulations.

The GALPs have been constructed to address realities of 1995. They may be modified over time to reflect changes in U.S. laws such as the congressionally-mandatedComputer Security Act, requirements by the Office of Management and budget, and others. They may also be modified over time to address advances in automated data management technologies.

Executive Summary

This document describes, benchmarks, Good Automated Laboratory Practices (GALPs), for assuring the reliability of laboratory data. The GALPs are principles and guidelines to regulations for laboratories that use or are planning to use a wide range of automated data collection and management systems. The GALPs are EPA's response to mounting evidence of corruption, loss, and inappropriate modifications of computerized laboratory data by EPA contractors.

The GALPs area union of Federal regulations, policies, and guidance documents. Several of the GALP provisions are embodied in EPA's Good Laboratory Practice Standards (GLPs). The GLPs are regulations that govern the management and conduct of most nonclinical laboratory studies submitted to EPA's Office of Toxic Substances and its Office of Pesticide Programs.

Several GALPs are contained in EPA's Information Resource Management (IRM) policies. These policies prescribe methodologies and practices for using automated data processing hardware and software. The IRM policies are directed to EPA staff and its agents (contractors and grantees) and generally implement broader Federal mandates such as the congressionally-mandated Computer Security Act of 1987, the Office of Management and Budget Circular A-130, and others. Most of these are also specifically required by EPA Acquisition Regulations.

This document is divided into two sections. The first chapter formally establishes the GALPs, describes the purpose they serve, provides background information bout studies that led to their development, and explains their scope and applicability. The second chapter provides laboratories with additional explanations of each provision and other relevant information to assist laboratory staff in implementing each applicable provision.

Acknowledgments

This document culminates a six year program by EPA's Office of Information Resources Management (OIRM). Numerous experts in national and international laboratory standards, laboratory automation experts, senior managers and technical staff in government and private companies provided invaluable support.

Mr. Mickey Cline and Dr. Walter Shackelford, both of OIRM, identified the need for the program, ensured that resources were provided, offered many valuable suggestions that helped to focus the program, and provided encouragement when obstacles seemed insurmountable. Without their support this program likely would not be completed.

Ms. Lynn Laubisch's (Durham, NC) contribution to the publication of this document far exceeded her title, "Micro Graphics Specialist." She is responsible for transforming what could have been a dull, monotonous and probably difficult-to-follow publication into refreshing, easy to read "text book" that enables complex concepts to be easily accessible to a diverse readership. While a cursory review of the document demonstrates her skill in page layout, font selection, and icon and diagram creation, a careful reading of the text is indicative of her oversight in helping to eliminate convoluted sentences and make the text easily readable.

Ms. Stephanie Taublee, Mr. David Brodish both of Research Triangle Institute (RTI), and Ms. Terrie Baker, formerly of RTI, deserve most of the credit for the areas of quality assurance (QA) the GALPs embrace and explain. Their professional QA experience, dedication, determination and commitment to doing the right thing on time, and their ability to examine highly charged and sensitive issues from several angles were essential/

Mr. Keith McLaurin of Technology Planning and Management Corporation (TPMC), Mr. Don Weyel, formerly of TPMC, and Mr. Bill Hampton, a Consultant to TPMC, instilled a wealth of the discipline of Computer Science to the GALPs. Their knowledge and experience in automated system design and development, computing and communication technologies, and the evolving specialized area of computer security enabled issues related to current computing environments, system life cycle and a myriad of intricate factors affecting computing security to be thoroughly and accurately explained in the document.

Mr. Dexter Goldman of Goldman and Associates enthusiastically supported this program from its inception. His extensive experience in EPA's *Good Laboratory Practice Standards* is reflected in many areas of the document. His critical review of earlier drafts was essential. He identified and recommended numerous changes not noted by other reviewers that, though subtle, had profound impact.

Dr. Sandy Weinberg of Weinberg, Sax and Spelton Associates deserves much of the credit for getting this program started in the right direction. He afforded the program with an unparalleled wealth of experience in assisting laboratories in complying with national laboratory standards, auditing laboratory operations, and translating national and international guidelines into laboratory operating standards.

Rick Johnson Voice:(919) 541-1132
EPA (MD-34)
RTP, NC 27711 FAX: (919) 541-1383
Internet: johnson.rick@epamail.epa.gov

Chapter 1 -- GALP Overview

1. PURPOSE

Most of the health and environmental data EPA uses in its regulatory programs are analyzed in and reported by laboratories. Increasingly, these laboratories employ laboratory information management systems (LIMS) to acquire, record, manipulate, store, and archive their data (see **2.c Applicable Systems**). Though many benchmarks are scattered across EPA's regulatory programs, EPA has no consistent set of standards for the use of LIMS that promote integrity of laboratory data.

The purpose of the Good Automated Laboratory Practices (GALPs) is to establish a uniform set of procedures to assure that all LIMS data used by EPA are reliable and credible.

2. SCOPE AND APPLICABILITY

a. Organizations

The GALPs are applicable to all EPA organizations, personnel, or agents (contractors and grantees) of EPA who collect, analyze, process, or maintain laboratory data for EPA. These organizations include the Agency's Regional Laboratories, and laboratories submitting data through contracts or grants with EPA, including the Superfund Contract Laboratory Program (CLP). Other organizations who wish to improve assurance of the integrity of laboratory data where LIMS are used are encouraged to review and implement applicable GALP provisions (see also **6. RESPONSIBILITIES**).

b. Relation to Other Regulations and Requirements

Federal regulations, EPA directives, policies, and its contract requirements govern the activities performed by laboratories that submit data to the Agency. Various laboratories are involved in the collection and analysis of environmental data and not all laboratories are subject to the same set of regulations and requirements. EPA's Contract Laboratory Program sets requirements by explicit clauses and clauses incorporated by reference in their governing contracts. Similarly, laboratories that submit studies in support of the registration or re-registration of pesticides under the Federal Insecticide, Fungicide, and Rodenticide Act (FIFRA) are subject to the Good Laboratory Practice (GLP) Standards [40 Code of Federal Regulations (CFR) Part 160. Federal Register Vol. 54, No. 158, August 17, 1989]. Laboratories that submit studies required by the test rules and negotiated testing agreement section of the Toxic Substances Control Act (TSCA) are subject to t he GLP regulations at 40 CFR Part 792.

The GALPs include many of the GLP requirements for managing the conduct of studies. The GALPs supplement the GLPs with Federal and EPA policies that address automated hardware, software development and operation, electronic transfer, and systems security. These are collectively referred to by the term Information Resources Management (IRM) policies. Thus the GALPs integrate GLP practices and procedures with IRM practices and procedures, to ensure the integrity of data that are entered, stored, and manipulated by the LIMS.

c. Applicable Systems

The GALPs use the acronym LIMS, laboratory information management system, to describe the automated laboratory systems that collect and manage data discussed in this Directive. There is a limitless range of possible configurations of automated data collection and processing equipment, communication components, types of operating system software, database management systems, and application software that can constitute a LIMS. The GALPs are directed to *most* configurations that are involved with entering, recording, manipulating, modifying, and retrieving data.

Not all automated laboratory systems are LIMS. Automated laboratory systems that record data but do not allow changes to the data are not LIMS. For example, an instrument that measures weights and produces or maintains a readout of the weight is not a LIMS, if the true reading cannot be altered by a person prior to recording.

The ability to effect changes to original observations or measurements is the factor in determining whether the automated laboratory system is a LIMS. If data entering automated laboratory systems can be manipulated or changed in any way by the actions of a person prior to being recorded, then that automated laboratory system is a LIMS.

3. DOCUMENT ORGANIZATION

This document is organized into two chapters. This first Chapter, GALP OVERVIEW, describes basic facts about the GALPs, including the purpose they serve, the scope, applicability and organization of this directive, the policy the GALPs implement, authorities and references supporting the GALPs, responsibilities of organizations, background information, the GALP provisions, definitions of terms, lists of acronyms, and sources for Federal information resources management publications referenced in the GALP.

Chapter 2, GALP IMPLEMENTATION ASSISTANCE, provides additional information about each GALP provision. It is intended to assist in the successful application of each GALP provision. See the introduction to Chapter 2 for additional discussion.

4. POLICY

It is EPA policy to implement and comply with all applicable information management laws mandated by Congress, all requirements issued by the Office of Management and Budget (OMB), all federal Information Resource Management Regulations (FIRMR) issued by the General Services Administration (GSA), and all Information Processing Regulations issued by the National Institute of Science and Technology (NIST).

It is also EPA policy that data collected, analyzed, processed, or maintained to support health and environmental effects studies be of sufficient accuracy and integrity to support effective environmental management.

EPA recognizes that absolute data integrity is not possible and that reliability and defensibility are determined by adherence to principles and practices that contribute to improving integrity. The GALPs balance risk against cost, incorporating existing Federal and EPA policies.

5. AUTHORITIES AND REFERENCES

a. Authorities

(1) Computer Security Act of 1987, Public Law 100-235
(2) EPA Information Resources Manage-

ment Policy Manual, Chapter 17 and Chapter 18, September 1994

(3) EPA Information Security Manual, December 1989

(4) EPA Operations and Maintenance Manual, April 1990

(5) Federal Information Processing Standards (FIPS) Publication 31: Guidelines for Automatic Data Processing Physical Security and Risk Management, June 1974

(6) Federal Information Processing Standards (FIPS) Publication 65: Guidelines for Automatic Data Processing Risk Analysis, August 1979

(7) Federal Information Processing Standards (FIPS) Publication 73: Guidelines for Security of Computer Applications, June 1980

(8) Federal Insecticide, Fungicide and Rodenticide (FIFRA); Good Laboratory Practice Standards. 40 CFR Part 160. Federal Register Vol. 54, No. 158, August 17, 1989

(9) Office of Management and Budget (OMB) Circular A-130, Management of Federal Information Resources, as Amended, April 29, 1992 (*this Circular may be subject to revision*)

(10) Office of Management and Budget (OMB) Bulletin 90-08, Guidance for Preparation of Security Plans for Federal Computer Systems that Contain Sensitive Information, July 1990

(11) Toxic Substances Control Act (TSCA); Good Laboratory Practice Standards. 40 CFR Part 792. Federal Register Vol. 54, No. 158, August 17, 1989.

b. References

(1) Automated Laboratory Standards: Current Automated Laboratory Data Management Practices, EPA/OIRM (Final, June 1990)

(2) Automated Laboratory Standards: Evaluation of Good Laboratory Practices for EPA Programs, EPA/ OIRM (Draft, June 1990)

(3) Automated Laboratory Standards: Survey of Automated Technology, EPA/OIRM (Final, June 1990)

(4) Automated Laboratory Standards: Evaluation of the Use of Automated Financial System Procedures, EPA/ OIRM (Final, June 1990)

(5) Automated Laboratory Standards: Evaluation of the Standards and Procedures Used in Automated Clinical Laboratories, EPA/OIRM (Draft, May 1990)

(6) National Institute of Science and Technology (NIST) Special Publication 500-166, Computer Viruses and Related Threats; A Management Guide (August 1989)

(7) U.S. Department of Commerce National Bureau of Standards (NBS) Special Publication 500-101, Care and Handling of Computer Magnetic Storage Media (June 1983)

6. RESPONSIBILITIES

a. The Office of Information Resources Management (OIRM) shall:

(1) be responsible for developing, establishing, providing, and maintaining the GALPs.

(2) provide guidance and technical assistance, where feasible and appropriate in implementing and improving the provisions of the GALPs.

b. Each "Primary Organization Head" (defined by EPA Order 1000.24 as the Deputy Administrator, Assistant Administrators, Regional Administrators, the Inspector General, and the General Counsel) is responsible for:

(1) complying with all applicable Federal and EPA rules and regulations affecting the collection, analysis, processing, storage, or maintenance of LIMS data. These are in-dicated in each GALP provision by the use of underlined lettering, such as <u>EPA Information Security Manual</u>.

(2) reviewing the GALPs and taking the

necessary measures to implement appropriate provisions provided in the GALPs that will improve the integrity of LIMS data.

7. BACKGROUND

a. EPA relies heavily on laboratory data to accomplish its mission. The accuracy and integrity of these data are essential to EPA's ability to effectively formulate policy, make decisions, and take action on issues involving public health and the environment. Laboratory data are therefore critical Agency assets and must be managed and protected as such.

b. The computer is increasingly replacing and augmenting many manual operations in the laboratory. Much of the laboratory data now submitted to EPA have been created, collected, processed, managed, or in other ways manipulated by LIMS.

c. Laboratory data are exposed to potential loss and misuse from a variety of accidental and deliberate causes. Cases involving the corruption, loss, and inappropriate modification of computerized laboratory data provided to EPA have resulted in debarments, suspensions, fines, and criminal prosecution.

d. EPA's OIRM conducted several studies to assess the automated data management practices employed by laboratories to ensure data integrity. Principal findings and recommendations of these studies included:

(1) The integrity of computer-resident data is at risk in many laboratories providing scientific and technical data to EPA. Inadequate system security, data verification, standardized procedures, designation of responsibility, and documentation are to a large extent responsible for these risks.

(2) EPA has no Agencywide policy for laboratories that collect and manage LIMS data. The laboratories that provide data to EPA are subject to differing regulations, policies, and contract requirements for the conduct of studies and management and operation of the laboratory.

(3) In many cases, the requirements that a laboratory must follow in conducting a study are vague or ambiguous regarding the special concerns and issues related to LIMS. For example, FIFRA and TSCA GLPs refer to "recorded data from automated instruments"; however, standards or guidance for performing LIMS risk assessments and LIMS software development and modification are not directly addressed in the GLPs.

(4) EPA has no definitive guidelines to aid the Agency's inspectors and auditors when they inspect laboratories that use LIMS in the conduct of a study.

(5) The need for Agencywide standards and guidance is recognized and acknowledged by the laboratory community and LIMS vendors.

(6) Data management practices should be standardized for all laboratories supporting EPA programs and the Agency should assume the responsibility for establishing these standards. The guidance and training provided to the Agency's inspectors and auditors should also be augmented accordingly.

e. In response to the findings of these studies, OIRM initiated the development of the GALP. The first draft of the GALP was issued in December 1990. Since that time, over one thousand copies of the draft GALP document have been distributed to EPA regional and program offices, other Federal agencies, industry, associations, and private citizens and groups.

f. OIRM received over 600 individual comments on the first draft of the GALP document. OIRM additionally contracted for the review of the document by subject-area experts in the fields of laboratory data systems, laboratory management, systems security, telecommunications, systems development, quality assurance, and information resource management. Document comments received from all sources were reviewed and evaluated by OIRM in the development of this final version of the GALP.

8. GOOD AUTOMATED LABORATORY PRACTICES

8.1 LABORATORY MANAGEMENT

When LIMS Raw Data (see **8.4.1**) are collected, analyzed, processed, or maintained, laboratory management shall:

8.1.1. ensure that personnel clearly understand the function(s) they are to perform on the LIMS.

8.1.2 ensure that a Quality Assurance Unit (QAU) monitors LIMS activities as described in **8.3**.

8.1.3 ensure that personnel, resources, and facilities are adequate and available as scheduled.

8.1.4 receive reports of QAU inspections of the LIMS (see **8.3.3**) and audits of LIMS Raw Data (see 8.3.5) and ensure that corrective actions are promptly taken in response to any deficiencies.

8.1.5 approve the standard operating procedures (SOPs) setting forth the methods that assure LIMS Raw Data integrity, ensure that any deviations from SOPs and applicable GALP provisions are appropriately documented and that corrective actions are taken and documented, and approve subsequent changes to SOPs (see **8.11**).

8.1.6 assure that each applicable GALP provision is followed. With the exception of **8.1, 8.2,** and **8.3**, laboratory management may delegate GALP implementation and compliance to one or more responsible persons.

8.2 PERSONNEL

When LIMS Raw Data are collected, analyzed, processed, or maintained, laboratory management shall ensure that all LIMS support staff and users:

8.2.1 have adequate education, training, and experience to perform assigned LIMS functions.

8.2.2 have a current summary of their training, experience, and job description, including their knowledge relevant to LIMS design and operation, maintained at the facility.

8.2.3 are of sufficient number for timely and proper operation of the LIMS.

8.3 QUALITY ASSURANCE UNIT

When LIMS Raw Data are collected, analyzed, processed, or maintained, laboratory management shall designate a Quality Assurance Unit (QAU) to monitor LIMS functions and procedures. The QAU shall:

8.3.1 be entirely separate from and independent of LIMS personnel, and shall report directly to laboratory management.

8.3.2 have immediate access to the LIMS data, SOPs and other records pertaining to t he operation and maintenance of the LIMS.

8.3.3 inspect the LIMS at intervals adequate to ensure the integrity of the LIMS Raw Data (see **8.3.5**); prepare inspection reports that include a description of the LIMS operation inspected, the dates of the inspection, the person performing the inspection, findings and problems observed, action recommended and taken to resolve existing problems, and any scheduled dates for reinspection; and report to laboratory management any problems that may affect data integrity.

8.3.4 determine that no deviations from

approved SOPs were made without proper authorization (see **8.1.5**) and sufficient documentation.

8.3.5 periodically audit the LIMS Raw Data to ensure their integrity.

8.3.6 ensure that the responsibilities and procedures applicable to the QAU, the records maintained by the QAU, and the method of indexing such records are documented and are maintained.

8.4 LIMS RAW DATA

Laboratory management shall ensure that:

8.4.1 LIMS Raw Data (LRD) and LRD storage media on which they reside (see **9. DEFINITIONS** LIMS Raw Data and LIMS Raw Data storage Media) are identified and documented. This documentation shall be included in the laboratory's SOPs.

8.4.2 the individual(s) responsible for entering and recording LIMS Raw Data is (are) uniquely identified when the data are recorded, and the time(s) and date(s) are documented.

8.4.3 the instrument transmitting LIMS Raw Data is uniquely identified when the data are recorded, and the time and date are documented.

8.4.4 procedures and practices to verify the accuracy of LIMS Raw Data are documented and included in the laboratory's SOPs, and managed as described in **8.11**.

8.4.5 procedures and practices for making changes to LIMS Raw Data are documented and provide evidence of change, preserve the original recorded documentation (see **8.4.3**), are dated, indicate the reason for the change, identify the person who made the change, and, if different, the person who authorized the change. These procedures shall be included in the laboratory's SOPs, and managed as described in **8.11**.

8.5 SOFTWARE

When software is used to collect, analyze, process, or maintain LIMS Raw Data, laboratory management shall ensure that:

8.5.1 SOPs are established, approved, and managed as described in **8.11** for:

8.5.1.1 development methodologies that are based on the size and nature of software being developed. EPA and its agents shall comply with <u>*EPA Information Resources Management Policy Manual, Chapter 17*</u>.

8.5.1.2 testing and quality assurance methods to ensure that all LIMS software accurately performs its intended functions, including: acceptance criteria, tests to be used, personnel responsible for conducting the tests, documentation of test results, and test review and approval.

8.5.1.3 change control methods that include instructions for requesting, testing, approving, documenting, and implementing changes. When indicated, change control methods shall also include reporting and evaluating problems, as well as implementing corrective actions.

8.5.1.4 version control methods that document the LIMS software version currently used.

8.5.1.5 maintaining a historical file of software, software operating procedures (manuals), software changes, and software version numbers.

8.5.2 documentation is established and maintained to demonstrate the validity of software used in the LIMS:

8.5.2.1 for existing and commercially-available LIMS, minimum documentation shall include, but not be limited to: a description of the software and functional requirements; listing of all algorithms and formulas; and, as they occur, testing and quality assurance, installation and operation, maintenance/enhancement, and retirement.

8.5.2.2 for new LIMS development or modifications of existing LIMS, documentation shall cover all phases of the generic software life cycle. EPA laboratories and those of its agents (contractors and grantees) shall comply with the documentation requirements specified in *EPA Information Resources Management Policy Manual, Chapter 17*.

8.5.3 all documentation specified in **8.5.2** is readily available in the facility where the software is used, and the SOPs specified in **8.5.1** are readily available in the laboratory areas where procedures are performed.

8.5.4 a historical file of software and the documentation specified in **8.5.2** are retained according to procedures outlined in **8.9**.

8.6 SECURITY

Laboratory management shall ensure that security practices to assure the integrity of LIMS data are adequate. EPA laboratories and those of its agents (contractors and grantees) shall comply with EPA's Information Security Policy.

8.7 HARDWARE

When LIMS Raw Data are collected, analyzed, processed, or maintained, laboratory management shall ensure that LIMS hardware and communications components are:

8.7.1 of adequate design and capacity, and a description is documented and maintained.

8.7.2 installed and operated in accordance with manufacturer's recommendations and, at installation, undergo acceptance testing that conforms to acceptance criteria. SOPs shall be established and maintained to define the acceptance criteria, testing, documentation, and approval required for changes to LIMS hardware and communications components.

8.7.3 adequately tested, inspected, and maintained. SOPs for and documentation of these routine operations shall be maintained. Documentation of non-routine maintenance shall also include a description of the problem, the corrective action, acceptance testing criteria, and the acceptance testing performed to ensure that the LIMS hardware and communications components have been adequately repaired.

8.8 COMPREHENSIVE TESTING

When LIMS Raw Data are collected, analyzed, processed, or maintained, laboratory management shall ensure that comprehensive testing of LIMS performance is conducted, at least once every 24 months or more frequently as a result of software (see **8.5.2**) or hardware (see **8.7.2**) changes or modifications. These tests shall be documented and the documentation shall be retained and available for inspection or audit.

8.9 RECORDS RETENTION

Laboratory management shall ensure that retention of LIMS Raw Data, documentation, and records pertaining to the LIMS comply with EPA contract, statute, or regulation; and SOPs for retention are documented, maintained, and managed as described in **8.11**.

8.10 FACILITIES

When LIMS Raw Data are collected, analyzed, processed, or maintained, laboratory management shall ensure that:

8.10.1 the environmental conditions of the facility housing the LIMS are regulated to protect against LIMS Raw Data loss.

8.10.2 environmentally adequate storage capability for retention of LIMS Raw Data, LIMS Raw Data storage media, documentation, and records pertaining to the LIMS are provided.

8.11 STANDARD OPERATING PROCEDURES

Laboratory management shall ensure that:

8.11.1 SOPs include, but are not limited to, those specified in **8.4.1, 8.4.4, 8.4.5, 8.5.1.1** through **8.5.1.5, 8.7.2, 8.7.3,** and **8.9**. Each current SOP shall be readily available where the procedure is performed.

8.11.2 SOPs are periodically reviewed at a frequency adequate to ensure that they accurately describe the current procedures.

8.11.3 SOPs are authorized and changed in accordance with **8.1.5.**

8.11.4 a historical file of SOPs is maintained.

9. DEFINITIONS

The definitions below generally come from existing Federal and EPA information management publications. While broader or narrower definitions, published in other authoritative sources, could have been used, those below were selected because they are more focused on the environment of laboratory data management.

Acceptance testing Formal testing conducted to determine whether or not a system satisfies its acceptance criteria and to enable the customer to determine whether or not to accept the system. *FIPS Publication 101, June 1983.*

Assurance A measure of confidence that the security features and architecture of [a LIMS] accurately mediate and enforce the security policy. Modified from *EPA Risk Analysis Guideline (Draft) March 1992.*

Audit A qualitative and quantitative evaluation of the documentation and procedures associated with the LIMS to verify that resulting LIMS Raw Data are of acceptable quality. Modified from *EPA Quality Assurance Management Staff, January 6, 1994.*

Change control Management and implementation methodologies associated with increasing or correcting system capabilities, a partial system redesign, or determining software obsolescence. *EPA Operations and Maintenance Manual, April 1990.*

Commercially-available software Software that is available through lease or purchase in the commercial market. Software that is furnished as part of the [LIMS] system but that is separately priced is included. *EPA Information Resources Management Policy Manual, Chapter 17, September 1994.*

Data A representation of facts, concepts, information, or instructions suitable for communication, interpretation, or processing by humans [or a LIMS]. *EPA Risk Analysis Guideline (Draft) March 1992.*

Design (software life cycle) The stage that specifies the automated and manual functions and procedures, the computer programs, and data storage techniques that meet the requirements identified and the security and control techniques that assure the integrity of the system. *EPA Information Resources Management Policy Manual, Chapter 17, September 1994.*

Documentation The process of gathering written or electronic information describing, defining, specifying, reporting, or certifying activities, requirements, procedures, or results. Modified from *ASME NQA-1, Quality Assurance Program Requirements for Nuclear Facilities, 1989 edition as cited in ANSI/ASQC E4-1994.*

Facility The premises and operational unit(s) that are necessary for operating a LIMS. Modified from *Organization for Economic Cooperation and development Series on Principles of Good Laboratory Practice and Compliance Monitoring Number 1: The*

OECD Principles of Good Laboratory Practice. Environment Monograph No. 45 (1992).

Hardware Physical equipment such as the computer and its related peripheral devices, tape drives, disk drives, printers, etc. *EPA Information Resources Management Policy Manual, Chapter 17, September 1994.*

Information Any communication or reception of knowledge such as facts, data or opinions, including numerical, graphic, or narrative forms, whether oral or maintained in any medium, including computerized databases (e.g., floppy disk and hard disk), papers, microform (microfiche or microfilm), or magnetic tape. *EPA Risk Analysis Guideline (Draft) March 1992.*

Initiation (software life cycle) A request for the development of a system to meet a need for information or to solve a problem for the individual making the request. *EPA Information Resources Management Policy Manual, Chapter 17, September 1994.*

Inspect To measure, examine, test or gauge one or more characteristics of an entity and compare the results with specified requirements in order to establish whether conformance is achieved for each characteristic. Modified from *ANSI/ASQC 34-1994 Specifications and Guidelines for Quality Systems for Environmental Data Collection and Environmental Technology Programs, January 3, 1995.*

Installation and operation (software life cycle) Incorporation and continuing use of the new system by the organization. *EPA Information Resources Management Policy Manual, Chapter 17, September 1994.*

Integrity Sound, unimpaired or perfect condition. That computer security characteristic that ensures that computer resources operate correctly and that the data in the databases are correct. This characteristic protects against deliberate or inadvertent unauthorized manipulation of the system and ensures and maintains the security of entities of a computer system under all conditions. Integrity is concerned with protecting information from corruption. *EPA Risk Analysis Guideline (Draft) March 1992.*

Laboratory Information Management System (LIMS) See **2.c APPLICABLE SYSTEMS**.

Laboratory management Those individuals directly responsible and accountable for planning, implementing, and assessing work, and for the overall operation of a facility. Modified from *ANSI/ASQC 34-1994 Specifications and Guidelines for Quality Systems for Environmental Data Collection and Environmental Technology Programs, January 3, 1995.*

LIMS Raw Data (LRD) Original observations recorded by the LIMS that are needed to verify, calculate, or derive data that are or may be reported.

LIMS Raw Data (LRD) storage media The media to which LIMS Raw Data are first recorded.

Maintenance/enhancement (software life cycle) Resolving problems not detected during testing, improving the performance of the product and modifying the system to meet changing requirements. (Full-scale enhancements require full life cycle analysis). *EPA Information Resources Management Policy Manual, Chapter 17, September 1994.*

Original observations The first occurrence of human-readable information.

Programming (software life cycle) Coding of the program modules that implement the design. *EPA Information Resources Management Policy Manual, Chapter 17, September 1994.*

Quality Assurance Unit Any person or organizational element designated by laboratory management to monitor the LIMS functions and procedures. Modified from *EPA GLPs, August 17, 1989.*

Records All books, papers, maps, photographs, machine-readable materials, or other documentary materials, regardless of physical form or characteristics, made or received by an agency of the United States Government under Federal law or in connection with the transaction of public business and preserved or appropriate for preservation by that agency or its legitimate successor as evidence of the organization, functions, policies, decisions, procedures, operations, or other activities of the government or because of the informational value of the data in them. Library and museum material made or acquired and preserved solely for reference or exhibition purposes, extra copies of documents preserved only for convenience of reference, and stocks of publications and of processed documents are not included. *44 U.S.C. 3301.*

Requirements analysis (software life cycle) Determination of what is required to automate the function(s) identified by the organization. *EPA Information Resources Management Policy Manual, Chapter 17, September 1994.*

Retirement (software life cycle) The stage which ends use of the system. *EPA Information Resources Management Policy Manual, Chapter 17, September 1994.*

Security The set of laws, rules, and practices that regulate how an organization manages, protects, and distributes sensitive data *EPA Risk Analysis Guideline (Draft) March 16, 1992.*

Software Computer programs, procedures, rules and associated documentation pertaining to the operation of a computer system. *EPA Information Resources Management Policy Manual, Chapter 17, September 1994.*

Software life cycle The period of time beginning when a software product is conceived and ending when the product no longer performs the function for which it was designed. The software life cycle is typically broken into phases such as initiation, requirements analysis, design, programming, testing and quality assurance, installation and operation, maintenance, and retirement. *EPA Information Resources Management Policy Manual, Chapter 17, September 1994.*

Software version control Management of changes or revisions to a specific baseline software module or application. Software version control provides a mechanism to control changes and to return to any previous revision of the application or module.

Standard Operating Procedures (SOPs) Documentation setting forth methods of operation that laboratory management is satisfied are adequate to insure the quality and integrity of LIMS Raw Data. Modified from *EPA GLPs, August 17, 1989.*

Testing The examination of the behavior of a program by executing the program on sample data sets. *EPA Information Resources Management Policy Manual, Chapter 17, September 1994.*

Testing and quality assurance (software life cycle) Ensuring that the system works as intended and that it meets applicable organization standards of performance, reliability, integrity and security. *EPA Information Resources Management Policy Manual, Chapter 17, September 1994.*

Validity A state or quality of software that provides conformation that the particular requirements for a specific intended use are fulfilled. In design and development, validity concerns the process of examining a product or result to determine conformance to user needs. Modified from *ISO 8402: 1994, Quality Management and Quality Assurance* as cited in *ANSI/ASQC E4-1994.*

Verify To review, inspect, test, check, audit, or otherwise establish and document whether or not LIMS Raw Data are accurate. Modified from *FIPS Publication 101, June 1983.*

10. LIST OF ACRONYMS

CFR	Code of Federal Regulations
CLP	Contract Laboratory Program
EPA	Environmental Protection Agency
FIFRA	Federal Insecticide, Fungicide, and Rodenticide Act
FIPS	Federal Information Processing Standard
FIRMR	Federal Information Resource Management Regulation
GALP	Good Automated Laboratory Practice
GLP	Good Laboratory Practice
GSA	General Services Administration
IRM	Information Resources Management

LIMS	Laboratory Information Management System
LRD	LIMS Raw Data
NIST	National Institute of Science and Technology
OIRM	Office of Information Resources Management
OMB	Office of Management and Budget
QAU	Quality Assurance Unit
SOP	Standard Operating Procedure
TSCA	Toxic Substances Control Act

11. SOURCES

Copies of the Federal information resources management publications referenced in the GALP can be ordered via mail, telephone, or the Internet.

Computer Security Act of 1987

This is a Federal regulation and should be available in local public libraries.

The Internet World Wide Web address is:
http://www.first.org/secplcy/csa_87.txt

Office of Management and Budget (OMB) publications

Office of Management and Budget
Assistant Director of Administration
OMB Publications
725 17th Street, NW
Washington, D.C. 20503

telephone: (202) 395-7332 (then press 2)

The Internet addresses for OMB publications are:
World Wide Web: http://www2.infoseek.com/ Titles?qt=OMB

Gopher: gopher: //pula.financenet.gov:70/11/docs/central/omb

EPA Publications

U.S. Environmental Protection Agency
OARM/FMSD
Publication Distribution Section
Mailcode 3204
401 M St., SW
Washington, D.C. 20460
telephone: (202)260-5797

For References 1 through 5 on page 1-6 (Automated Laboratory Standards), contact:

Rick Johnson Voice: (919)541-1132
EPA (MD-34) Fax: (919)541-1383
RTP, NC 27711
Internet: johnson.rick@epamail.epa.gov

The Internet addresses for EPA IRM documents are:
World Wide Web: http?//www.epa.gov/docs/IRMPolicy.html

Gopher: gopher: //gopher.epa.gov:70/11/Initiatives/IRM.Policy

National Institute of Standards and Technology (NIST) and National Bureau of Standards (NBS) publications

National Technical Information Service
U.S. Department of Commerce
5285 Port Royal Road
Springfield, VA 22161
(703)487-4650

The Internet World Wide Web address for NIST is:
http://www.ncsl.nist.gov

The Internet World Wide Web address for FIPS Publications is:
http://www.ncsl.nist.gov/fips/

Chapter 5

EUROPEAN UNION REGULATIONS

OVERVIEW

The European Union (EU), formerly the European Community (EC), provides guidelines for medical devices and pharmaceuticals through directives issued by the Commission of the European Communities. In the *Validation Compliance Annual: 1995*, we presented three directives which were major factors in complying with systems validation requirements in the EU: the directive on good manufacturing practices for medicinal products, the active implantable medical device directive, and the general medical device directive. The Commission's guideline on Good Clinical Practices, also found in *Validation Compliance Annual: 1995*, provides additional guidance on system validation.

In April 1994, the European Commission prepared a working document on guidelines relating to the application of medical devices. That document was not a part of the previous edition of this work but is presented here in its entirety.

Early in 1995, the European Union prepared a draft proposal for a directive on *in vitro* diagnostic medical devices. That draft proposal is also presented below.

EUROPEAN COMMISSION

WORKING DOCUMENT

GUIDELINES RELATING TO THE APPLICATION OF: *THE COUNCIL DIRECTIVE 90/385/EEC ON ACTIVE IMPLANTABLE MEDICAL DEVICES *THE COUNCIL DIRECTIVE 93/42/EEC ON MEDICAL DEVICES

Definition of "medical devices"
Definition of "accessory"
Definition of "manufacturer"

Brussels, 21 April, 1994

Note

The present first draft for a Guidelines Document has been elaborated taking account of various comments and questions addressed to the Commission by Member States and parties concerned. It forms a support for the subjects to be discussed and requires further elaboration subsequently to such discussions.

The various parts of the guidelines will at the present stage of preparation be the subject of separate circulation and discussion.

1. FIELD OF APPLICATION -- DEFINITIONS

1. Directive 93/42/EEC on medical devices

1.1 Definition of "medical devices"

a) devices -- accessory

The definition of the term "medical device" together with the definition of "accessory" is determinant for the delimitation of the field of application of Directive 93/42/EEC. A slight difference exists between the definition in article 1(2)(a) of Directive 93/42/EEC and in article 1(2) of Directive 90/385/EEC.

Following the latter directive, accessories are by definition medical devices, whilst following Directive 93/42/EEC, a distinction is made between "devices" and "accessories". Therefore within the meaning of Directive 93/42/EEC, accessories are products in their own right and, although being treated as devices (article 1(1)) do not follow, as a general rule, the classification of related devices in conjunction with which they are used.

Accessories are therefore following Directive 93/42/EEC to be classified in their own right.

b) medical purpose

Medical devices are defined as articles which are intended to be used for a medical purpose. The medical purpose is assigned to a product by the manufacturer. The manufacturer determines through the label, the instruction for use and the promotional material related to a given device its specific medical purpose. As the directive aims essentially the protection of patients and users, the medical purpose relates in general to finished products regardless of whether they are intended to be used alone or in combination. This means that the protection ensured by the directive becomes valid for products having a a stage of manufacture, where they are supplied to the final user.

Following this concept, raw materials, components or intermediate products are as such normally not medical devices. Such raw materials may need to present properties or characteristics which are determinant for the safety and quality of finished devices. It is therefore the responsibility of the manufacturer of finished devices to select and control by adequate means his raw materials or intermediate products (see Annex I, section 7.1; Annex II, section 3.2; Annex V, section 3.2 of Directive 93/42/EEC).

Spare parts supplied for replacement of existing components of a device, the conformity of which has already been established, are not medical devices. If spare parts, however, change significantly the characteristics or performances of a device with regard to its already established conformity, such spare parts are to be considered as devices in their own right.

c) customizing

The concept of "finished device" does not imply that a device when reaching the final user is already in a state ready for use. Prior to use further preparatory processing, preparation, configuration, installation, assembling, adaptation or fitting to the needs of the user or patient may be required. Examples:

-- sterilisation of medical devices supplied non-sterile
-- assembling of systems
-- configuration of electronic equipment
-- preparation of a dental filling
-- fitting of contact lenses
-- adaptation of prosthesis to the needs of the patient

The aforementioned activities are normally not manufacturers activities if they are carried out by the final user as part of the use or preparation for use. In this context a distinction needs to be made between a typical professional activity performed by a healthcare professional and processing and assembling activities done by a specialist for such processing. In the latter case relevant activities may become proper manufacturing or assembling activities relevant within the meaning of articles 11 and 12.

A particular consideration in this context needs to be given for the manufacture of custom-made devices. Custom-made devices (such as dental appliances, prosthesis, hearing-aid inserts) are in most cases one-off devices and the Directive subjects their manufacture to the procedure of article 11(6) in conjunction with Annex VIII. In these cases intermediate products specifically intended for this kind of custom-made devices may be considered also as medical devices. This applies essentially to dental alloys, dental ceramics, modular components for prosthesis, if the intended purpose of such products is specifically related to medical devices.

d) medical-toiletry purpose

The definition of "medical device" should be understood to include products intended to be used principally for a medical use. Therefore products intended to have a toiletry or cosmetic purpose are not medical devices even though they may be used for prevention of a disease. Examples for products for which a medical purpose can normally not be established:

-- tooth brushes, dental sticks, dental floss;
-- baby diapers, hygiene tampons;
-- contact lenses without corrective function intended to provide another colour to the eyes;
-- bleaching products for teeth [needs to be further discussed with regard to products intended specifically for application by dentists]
-- instruments for tattooing.

Examples for products where depending on the circumstances, a medical purpose can be established:

-- incontinence products.

e) aids for handicapped persons

In the case of equipment intended for alleviation of or compensation for a handicap, three must be a direct link between the corrective function and the person concerned. Therefore following equipment are not medical devices:

-- acoustic signals at traffic lights,
-- special water tapes, toilet equipment for handicapped.

f) software

Following distinction can be made: software influencing the proper functioning of a device and software used in combination with non-medical equipment.

Software related to the functioning of a medical device may be part of a device or a device in its own right if it is placed on the market separately from the related device.

In the case of software intended for use with multipurpose informatic equipment a distinction has to be made between software providing for a proper diagnostic of therapeutic tool and software for handling general patient-related data. Only in the first case may a medical purpose be determined. Examples for medical devices:

-- calculation of anatomical sites of the body,
-- image enhancing software intended for diagnostic purpose.
-- software for programming a medical device.

There is no medical purpose in the case of software used for administration of general patient data.

g) multipurpose products

Products with a multiple purpose which may be used occasionally in a medical environment are normally not medical devices, unless a specific medical intended purpose is assigned to them. Examples:

-- multipurpose PC, printer, scanner,...
-- magnetoscope, screen.

1.2 Definition of "accessory"

The question whether a product is a "device" or a "accessory" has not practical consequence. Following article 1(1) of Directive 93/42/EEC, "accessories shall be treated as medical devices in their own right". Therefore the main question is whether a product with a rather remote link to a medical use can still be considered as "accessory" (article 1(2)(b) and as a matter of consequence is covered by the directive.

The definition of "accessory" requires that the accessory is specifically intended by the manufacturer of the accessory to be used together with a device. The intended use of the accessory must be such as to enable a device being used in accordance with its intended use. Therefore a product can only become an accessory to a medical device if the manufacturer of such a product establishes an intended use in conjunction with one or several medical devices.

Examples for accessories depending on defined circumstances of device-related use:

-- sterilizers for use in a medical environment,
-- pouches for packaging re-sterilised medical devices,
-- specific battery chargers for battery driven electromedical devices,
-- contact lens care products, disinfectants specifically intended for invasive medical devices,
-- special water treatment device for use in conjunction with dialyzing machines,

-- gas cylinders/pressure release devices for use in conjunction with anaesthesia machines.

1.3 Definition of "manufacturer"

Users in-house manufacturing

The Directive defines a manufacturer as the natural or legal person responsible for defined manufacturing activities related to a device with a view to its being placed on the market under the manufacturer's own name. The reason for this link with the placing on the market is that the directive aims to subject to its protection requirements the transaction of a device from the sphere of a manufacturer towards the public. The directive does not provide any specific provisions for the case where a device is manufactured by the user (for example, a hospital) without being transferred to another person. The decision to which extent such in-house manufacturing activities by hospitals are subjected to legal requirements, belongs therefore to the national legislator. This relates however exclusively to such in-house manufacturing activities where a device remains within the users, but not to cases where, for example, an hospital produces orthopaedic devices for use with patients.

PROPOSAL FOR A DIRECTIVE OF THE EUROPEAN PARLIAMENT AND THE COUNCIL ON IN VITRO DIAGNOSTIC MEDICAL DEVICES

95/0013 (COD)

THE EUROPEAN PARLIAMENT AND THE COUNCIL OF THE EUROPEAN UNION,

Having regard to the Treaty establishing the European Community, and in particular Article 100a thereof,

Having regard to the proposal from the Commission,

Having regard to the opinion of the Economic and Social Committee,

Acting in accordance with the procedure referred to in Article 189B of the Treaty establishing the European Community,

Whereas measures should be adopted for the progressive establishment of the internal market; whereas the internal market is an area without internal frontiers in which the free movement of goods, persons, services and capital is ensured;

Whereas the content and scope of the laws, regulations and administrative provisions in force in the Member States with regard to the safety, health protection and performance characteristics and authorization procedures for in vitro diagnostic medical devices are different; whereas the existence of such disparities constitute barriers to trade and the need to establish harmonised rules was confirmed by a comparative survey of national legislations carried out on behalf of the Commission;

Whereas the harmonization of national legislations is the only means of removing these barriers to free trade and preventing new barriers; whereas this objective cannot be achieved in a satisfactory manner at another level by the individual Member States; whereas this Directive only lays down necessary and sufficient requirements for the free circulation of the in vitro diagnostic medical devices to which it is applicable;

Whereas the harmonized provisions must be distinguished from the measures adopted by the Member States to manage the funding of public health and sickness insurance schemes relating directly or indirectly to such devices; whereas, therefore, the provisions do not affect the ability of the Member States to implement the abovementioned measures provided Community law is complied with;

Whereas in vitro diagnostic medical devices should provide patients, users and third parties with a high level of protection and attain the performance levels attributed to them by the manufacturer; whereas, therefore, the maintenance or improvement of the level of protection attained in the Member States is one of the essential objectives of this Directive;

Whereas, in accordance with the principles set out in the Council resolution of 7 May 1985 concerning a new approach to technical harmonization and standardization, rules regarding the design and manufacture of relevant products must be confined to the provisions required to meet the essential requirements; whereas, because they are essential, such requirements should replace the corresponding national provisions; whereas the essential requirements, including requirements to minimize and reduce risks, should be applied with discretion, taking into account the technology and practice at the time of design and of technical and economic considerations compatible with a high level of protection of health and safety;

Whereas the major part of medical devices are covered by Council Directive 90/385/EEC relating to active implantable medical devices and Council Directive 93/42/EEC relating to all other medical devices with the exclusion of in vitro diagnostic medical devices; the present Directive shall extend the harmonization to in vitro diagnostic medical devices; whereas, in the interest of uniform Community rules, this Directive is based largely on the provisions of Directives 90/385/EEC and 93/42/EEC;

Whereas certified international reference materials and materials used for external quality assessment schemes are not covered by this Directive, calibrators and control materials needed by the user to establish or verify performances of devices are in vitro diagnostic medical devices;

Whereas reagents produced within user laboratories and which are not subject to commercial transactions, taking account of the principle of subsidiarity, are not included in this Directive;

Whereas the electromagnetic compatibility aspects form an integral part of the essential requirements of this Directive, the Council Directive 89/336/EEC of 3 May 1989 relating to electromagnetic compatibility shall not apply;

Whereas, in order to ease the task of proving conformity with the essential requirements and to enable conformity to be verified, it is desirable to have harmonized standards in respect of the prevention of risks associated with the design, manufacture and packaging of medical devices; whereas such harmonized standards are drawn up by private-law bodies and should retain their status as non-mandatory texts; whereas, to this end, the European Committee for Standardization (CEN) and the European Committee for Electrotechnical Standardization (CENELEC) are recognized as the competent bodies for the adoption of harmonized standards in accordance with the general guidelines on cooperation between the Commission and these two bodies signed on 13 November 1984;

Whereas, for the purpose of this Directive, a harmonized standard is a technical specification (European standard or Harmonization Document) adopted, on a mandate from the Commission, by either or both of these bodies in accordance with Council Directive 83/189/-EEC of 28 March 1983 laying down a procedure for the provision of information in the field of technical standards and regulations, and pursuant to the abovementioned general guidelines;

Whereas, in Decision 90/683/EEC of 13 December 1990 concerning the modules for the various phases of the conformity assessment procedures which are intended to be used in the technical harmonization directives, the Council has laid down harmonized conformity assessment procedures; whereas the details added to these modules are justified by the nature of the verification required for in vitro diagnostic medical devices and by the need for consistency with the previous directives on medical devices;

Whereas it is necessary, essentially for the purpose of the conformity assessment procedures, to group in vitro diagnostic medical devices into two product classes; whereas, since the large majority of such devices do not constitute a direct risk to patients and are used by competently trained professionals, and the results obtained can often be confirmed by other means, the conformity assessment procedures can be carried out, as a general rule, under the sole responsibility of the manufacturer; whereas, taking account of existing national regulations and of notifications received following the procedure of Directive 83/189/EEC, the intervention of notified bodies is needed only for defined devices, the correct performance of which is essential to medical practice and the failure of which can cause a serious risk to health;

Whereas the list of in vitro diagnostic medical devices to be subjected to third party conformity assessment needs updating, taking account of technological progress and of evolution in the domain of protection of health; whereas such updating measures must be taken in line with the procedure IIIa as laid down in the Council Decision 87/373/EEC; whereas a system of adverse incident reporting (vigilance) constitutes a useful tool for the surveillance of the market, including the performance of new devices; whereas information obtained from vigilance as well as from external quality assessment schemes becomes useful for decision-making on classification of devices;

Whereas medical devices should, as a general rule, bear the CE marking to indicate their conformity with the provisions of this Directive to enable them to move freely within the Community and to be put into service in accordance with their intended purpose;

Whereas manufacturers shall have the possibility, when the intervention of a notified body is required, to choose amongst those bodies published by the Commission; whereas the Member States do not have an obligation to designate such notified bodies, they must ensure however that bodies designated as notified bodies comply with the assessment criteria laid down in this directive;

Whereas the competent authorities in charge of market surveillance must be able, particularly in emergencies, to contact the manufacturer or his authorized representative established in the Community; whereas cooperation and exchange of information between Member States are necessary in view of uniform application of this Directive, in particular for the purpose of market surveillance;

Whereas this Directive includes in vitro diagnostic medical devices incorporating substances derived from the human body, it does not affect national regulations relating to the ethics of use of such substances; whereas for the purpose of overall consistency between medical devices directives, the Directive 93/42/EEC needs to be amended accordingly;

HAVE ADOPTED THIS DIRECTIVE :

Article 1

Scope, Definitions

1. This Directive shall apply to in vitro diagnostic medical devices and their accessories. For the purposes of this Directive, accessories shall be treated as in vitro diagnostic medical devices in their own right. Both in vitro diagnostic medical devices and accessories shall hereinafter be termed devices.

2. For the purpose of this Directive, the following definitions shall apply :

(a) "medical device" means any instrument, apparatus, appliance, material or other article, whether used alone or in combination, including the software necessary for its proper application intended by the manufacturer to be used for human beings for the purpose of :

- diagnosis, prevention, monitoring, treatment or alleviation of disease,

- diagnosis, monitoring, treatment, alleviation or compensation for an injury or handicap,

- investigation, replacement or modification of the anatomy or of a physiological process,

- control of conception

and which does not achieve its principal intended action in or on the human body by pharmacological, immunological or metabolic means, but which may be assisted in its function by such means;

(b) "in vitro diagnostic medical device" means any medical device which is a reagent, reagent product, calibrator, control material, kit, instrument, apparatus, equipment or system whether used alone or in combination, intended by the manufacturer to be used in vitro for the examination of specimens including blood and tissue donations, derived from the human body, solely or principally for the purpose of providing information concerning a physiological state, state of health or disease or congenital abnormality or to determine the safety and compatibility with potential recipients.

For the purpose of this directive, a specimen receptacle, whether evacuated or not, specifically intended by its manufacturer to contain a specimen for the purpose of in vitro diagnostic examination is considered to be a device; Products for general laboratory use are not devices unless such products, in view of their characteristics, are specifically intended by their manufacturer to be used for in vitro diagnostic examination;

(c) "accessory" means an article which, whilst not being a device in the meaning of paragraph 2(b), is intended specifically by its manufacturer to be used together with a device to enable this device to be used in accordance with its intended purpose;

(d) "Device for self-testing" means any device intended by the manufacturer to be used in the home environment;

(e) "Device for performance evaluation" means any device intended by the manufacturer to be subject to one or more performance evaluation studies in clinical laboratories or in other appropriate environments outside his own premises;

(f) "manufacturer" means the natural or legal person with responsibility for the design, manufacture, packaging and labelling of a device before it is placed on the market under his own name, regardless of whether these operations are carried out by that person himself or on his behalf by a third party.

The obligations of this Directive to be met by manufacturers also apply to the natural or legal person who assembles, packages, processes, fully refurbishes and/or labels one or more ready-made products and/or assigns to them their intended purpose as devices with a view to their being placed on the market under his own name;

(g) "authorized representative" means the natural or legal person established within the Community who, being explicitely designated by the manufacturer, acts and may be addressed by authorities and bodies in the Community in substitution for the manufacturer with regard to the latter's obligations in accordance with this Directive.

(h) "intended purpose" means the use for which the device is intended according to the data supplied by the manufacturer on the labelling, in the instructions for use and/or in promotional materials;

(i) "placing on the market" means the first making available in return for payment or free of charge of a device with a view to distribution and/or use on the Community market, regardless of whether it is new or fully refurbished;

(j) "putting into service" means the stage at which a device is ready for use on the Community market for the first time for its intended purpose;

(k) "calibrator", "control material" mean any substance, material or article intended by its manufacturer to establish and/or verify performance characteristics of a device in conjunction with the use of that device;

3. Where a device incorporates tissues or substances of human origin, this directive shall not affect national regulations relating to the ethics of the collection of tissues or substances of human origin as well as any regulations relating to the ethics governing distribution of given types of devices of such origin.

4. This Directive shall not apply to devices manufactured and used only within the same institution and on the premises of their manufacture. This does not affect the right of Member States to subject such activities to appropriate protection requirements.

5. This Directive is a specific directive within the meaning of Article 2(2) of the Directive 89/336/EEC.

Article 2

Placing on the market and putting into service

Member States shall take all necessary steps to ensure that devices may be placed on the market and put into service only if they comply with the requirements laid down in this Directive when properly installed, maintained and used in accordance with their intended purpose.

Article 3

Essential requirements

Devices must meet the essential requirements set out in Annex 1 which apply to them, taking account of the intended purpose of the devices concerned.

Article 4

Free movement

1. Member States shall not create any obstacle to the placing on the market or the putting into service within their territory of devices bearing the CE-marking provided for in article 14 or labelled as devices for performance evaluation if these devices have been the subject of an assessment of their conformity in accordance with the provisions of article 9.

2. At trade fairs, exhibitions, demonstrations, Member States shall not create any obstacle to the showing of devices which do not conform to this Directive, provided that a visible sign clearly indicates that such devices cannot be marketed or put into service until they have been made to comply.

3. Member States may require, when a device reaches the final user, information, as referred to in Annex 1 point 13, to be in their national language(s) or in other Community language(s) to the extent that it is needed for safe and correct use of the device. In the application of this provision Member States shall take into account the principle of proportionality and in particular whether information can be provided by recognised symbols or codes and whether the device is intended for use by trained professionals.

4. Where the devices are subject to other Directives concerning other aspects and which also provide for the affixing of the CE marking, the latter shall indicate that the devices also fulfil the provisions of the other Direc-

tives. However, should one or more of these directives allow the manufacturer, during a transitional period, to choose which arrangements to apply, the CE marking shall indicate that the devices fulfil the provisions only of those directives applied by the manufacturer. In this case, the particulars of these directives, as published in the Official Journal of the European Communities, must be given in the documents, notices or instructions required by the directives and accompanying such devices.

Article 5

Reference to standards

1. Member States shall presume compliance with the essential requirements referred to in Article 3 in respect of devices which are in conformity with the relevant national standards which transpose the harmonized standards the reference of which have been published in the Official Journal of the European Communities; Member States shall publish the references of such national standards.

2. If the Member State or the Commission considers that the harmonized standards do not entirely meet the essential requirements referred to in Article 3, the measures to be taken with regard to these standards and the publication referred to in paragraph 1 of this Article shall be adopted by the procedure defined in Article 6(2).

Article 6

Committee on Standards and Technical Regulations

1. The Commission shall be assisted by the Committee set up by Article 5 of Directive 83/189/EEC.

2. The representative of the Commission shall submit to the Committee a draft of the measures to be taken. The Committee shall deliver its opinion on the draft within a time limit which the chairman may lay down according to the urgency of the matter, if necessary by taking a vote. The opinion shall be recorded in the minutes; in addition, each Member State shall have the right to ask to have its position recorded in the minutes. The Commission shall take the utmost account of the opinion delivered by the Committee. It shall inform the Committee of the manner in which its opinion has been taken into account.

Article 7

Committee on Medical Devices

1. The Commission shall be assisted by the Committee set up by Article 6(2) of Directive 90/385/EEC.

2. The representative of the Commission shall submit to the Committee a draft of the measures to be taken. The Committee shall deliver its opinion on the draft within a time limit which the chairman may lay down according to the urgency of the matter. The opinion shall be delivered by the majority laid down in Article 148(2) of the Treaty in the case of decisions which the Council is required to adopt on a proposal from the Commission. The votes of the representatives of the Member States within the Committee shall be weighted in the manner set out in that Article. The chairman shall not vote.

The Commission shall adopt the measures envisaged if they are in accordance with the opinion of the Committee.

If the measures envisaged are not in accordance with the opinion of the Committee, or if no opinion is delivered, the Commission shall, without delay, submit to the Council a proposal relating to the measures to be taken. The Council shall act by a qualified majority.

If, on the expiry of a period of three months from the date of referal to the Council, the Council has not acted, the proposed measures shall be adopted by the Commission.

3. The Committee referred to in paragraph 1 may examine any question connected with the implementation of this Directive.

Article 8

Safeguard clause

1. Where a Member State ascertains that the devices referred to in Article 4(1), when correctly installed, maintained and used for their intended purpose may compromise the health and/or safety of patients, users or, where applicable, other persons, and the safety of property, it shall take all appropriate interim measures to withdraw such devices from the market or prohibit or restrict their being placed on the market or put into service. The Member State shall immediately inform the Commission of any such measures, indicating the reasons for its decision and, in particular, whether non-compliance with this Directive is due to:

(a) failure to meet the essential requirements referred to in Article 3;

(b) incorrect application of the standards referred to in Article 5, in so far as it is claimed that the standards have been applied;

c) shortcomings in the standards themselves.

2. The Commission shall enter into consultation with the parties concerned as soon as possible. Where, after such consultation, the Commission finds that:

- the measures are justified, it shall immediately so inform the Member State which took the initiative and the other Member States; where the decision referred to in paragraph 1 is attributed to shortcomings in the standards, the Commission shall, after consulting the parties concerned, bring the matter before the committee referred to in Article 6(1) within two months if the Member State which has taken the decision intends to maintain it and shall initiate the procedures referred to in Article 6;
- the measures are unjustified, it shall immediately so inform the Member State which took the initiative and the manufacturer or his authorized representative established within the Community.

3. Where a non-complying device bears the CE marking, the competent Member State shall take appropriate action against whomsoever has affixed the marking and shall inform the Commission and the other Member States thereof.

4. The Commission shall ensure that the Member States are kept informed of the progress and outcome of this procedure.

Article 9

Conformity assessment procedures

1. For all devices other than those covered by Annex 2 and devices for performance evaluation, the manufacturer shall, in order to affix the CE marking, follow the procedure referred to in Annex 3 and draw up the EC declaration of conformity required before placing the devices on the market.

For all devices for self-testing the manufacturer shall, prior to the drawing up of the aforementioned declaration of conformity, fulfil the supplementary requirements set out in Annex 3 point 6. Instead of applying this procedure, the manufacturer may also follow the procedure referred to in paragraph 2.

2. For all devices covered by Annex 2 other than those intended for performance evaluation, the manufacturer shall, in order to affix the CE-marking either :

(a) follow the procedure relating to the EC declaration of conformity set out in Annex 4 (full quality assurance) or

(b) follow the procedure relating to EC type examination set out in Annex 5 coupled with:

(i) the procedure relating to EC verification set out in Annex 6, or

(ii) the procedure relating to the EC declaration of conformity set out in Annex 7 (production quality assurance)

3. In the case of devices for performance evaluation, the manufacturer shall follow the procedure referred to in Annex 8 and draw up the statement set out in that Annex before placing such devices on the market.

4. During the conformity assessment procedure for a device, the manufacturer and, if involved, the notified body shall take account of the results of any assessment and verification operations which, where appropriate, have been carried out in accordance with this Directive at an intermediate stage of manufacture.

5. The manufacturer may instruct his authorized representative established in the Community to initiate the procedures provided for in Annexes 3, 5, 6 and 8.

6. The manufacturer must keep the declaration of conformity, the technical documentation referred to in Annexes 3 to 8, as well as the decisions, reports and certificates established by notified bodies, and make it available to the national authorities for inspection purposes for a period ending five years after the last product has been manufactured. Where the manufacturer is not established in the Community, the obligation to make the aforementioned documentation available on request applies to his authorized representative established within the Community.

7. Where the conformity assessment procedure involves the intervention of a notified body, the manufacturer, or his authorized representative established in the Community, may apply to a body of his choice within the framework of the tasks for which the body has been notified.

8. The notified body may require, where duly justified, any information or data, which is necessary for establishing and maintaining the attestation of conformity in view of the chosen procedure.

9. Decisions taken by the notified bodies in accordance with Annexes 3, 4 and 5 shall be valid for a maximum of five years and may be extended on application, made at a time agreed in the contract signed by both parties, for further periods of five years.

10. The records and correspondence relating to the procedures referred to in paragraphs 1 to 3 shall be in one of the official languages of the European Community and in the case of involvement of a notified body in a Community language acceptable to the notified body.

11. By derogation from paragraphs 1 to 3, the competent authorities may authorize, on duly justified request, the placing on the market and putting into service, within the territory of the Member State concerned, of individual devices for which the procedures referred to in paragraphs 1 to 3 have not been carried out and the use of which is in the interest of protection of health.

Article 10

Registration of manufacturers

1. Any manufacturer, who places devices on the market under his own name, shall notify the competent authorities of the Member State in which he has his registered place of business of the address of the registered place of business, the categories of devices as defined in terms of common characteristics of technology and/or analytes and of any significant change thereto. The before mentioned notification shall also include any new device as referred to in Article 11(4).

2. Where a manufacturer, who places devices on the market under his own name, does not have a registered place of business in a Member State, he shall designate an authorized representative established in the Community. The authorized representative shall inform the competent authorities of the Member State, in

which he has his registered place of business, of all particulars as referred to in paragraph 1.

3. The Member States shall on request inform the other Member States and the Commission of the details referred to in paragraphs 1 and 2. The procedures implementing the present Article shall be adopted in accordance with the procedure referred to in Article 7(2).

Article 11

Vigilance

1. Member States shall take the necessary steps to ensure that any information brought to their knowledge, in accordance with the provisions of this Directive, regarding the incidents mentioned below involving devices bearing CE marking is recorded and evaluated centrally:

(a) any malfunction, failure or deterioration in the characteristics and/or performance of a device, as well as any inadequacy in the labelling or the instructions for use which, directly or indirectly, might lead to or might have led to the death of a patient or user or to a serious deterioration in his state of health;

(b) any technical or medical reason in relation to the characteristics or performance of a device for the reasons referred to in subparagraph (a), leading to systematic recall of devices of the same type by the manufacturer.

2. Where a Member State requires medical practitioners, the medical institutions or the organizers of external quality assessment schemes to inform the competent authorities of any incidents referred to in paragraph 1, it shall take the necessary steps to ensure that the manufacturer of the device concerned, or his authorized representative established in the Community, is also informed of the incident.

3. After carrying out an assessment, if possible together with the manufacturer, Member States shall, without prejudice to Article 8, immediately inform the Commission and the other Member States of the incidents referred to in paragraph 1 for which relevant measures have been taken or are contemplated.

4. Where, in the context of notification referred to in Article 10, a device notified, bearing a CE marking, is a "new" product, the manufacturer shall indicate this fact on his notification. The competent authority so notified may at any time within the following two years and on justified grounds, require the manufacturer to submit a report relating to the experience with the device subsequent to its being placed on the market. For this purpose, a device is "new" if :

(a) such a device, for the relevant analyte or other parameter, has not been continuously available on the Community market during the previous three years,

(b) the procedure involves analytical technology not used in connection with a given analyte or other parameter on the Community market continuously during the previous three years.

5. The Member States shall on request inform the other Member States of the details referred to in paragraphs 1 and 4. The procedures implementing the present Article shall be adopted in accordance with the procedure referred to in Article 7(2).

Article 12

Modification of Annex 2, derogation clause

1. Where a Member State considers that :

(a) the list of devices covered by Annex 2 should be modified or extended, or

(b) the conformity of a device or a category of devices should be established by way of derogation from the provisions of Article 9, by applying one or several alternative given procedures taken from amongst those referred to in Article 9, it shall submit a duly substantiated request to the Commission and ask it to take the necessary measures. The measures shall be adopted in accordance with Article 7(2).

2. When a measure is to be taken in accordance with paragraph 1, due consideration shall be given :

(a) to any relevant information available from the vigilance procedures and from external quality assessment schemes as referred to in Article 11

(b) to the following criteria

(i) whether total reliance has to be put on the result obtained with a given device, having a direct impact on the subsequent medical action, and

(ii) whether an action taken on the basis of an incorrect result obtained using a given device could prove to be hazardous to the patient or third party as a consequence of false positive or false negative results, and

(iii) whether the involvement of a notified body is appropriate and justified.

3. The Commission shall inform the Member States of the measures taken and, where appropriate, publish these measures in the Official Journal of the European Communities.

Article 13

Notified bodies

1. Member States shall notify the Commission and other Member States of the bodies which they have designated for carrying out the tasks pertaining to the procedures referred to in Article 9 and the specific tasks for which the bodies have been designated. The Commission shall assign identification numbers to these bodies, hereinafter referred to as "notified bodies".

The Commission shall publish a list of the notified bodies, together with the identification numbers it has allocated to them and the tasks for which they have been notified, in the Official Journal of the European Communities. It shall ensure that the list is kept up to date.

2. Member States shall apply the criteria set out in Annex 9 for the designation of bodies. Bodies that meet the criteria laid down in the national standards which transpose the relevant harmonised standards shall be presumed to meet the relevant criteria.

3. A Member State that has notified a body shall withdraw that notification if it finds that the body no longer meets the criteria referred to in paragraph 2. It shall immediately inform the other Member States and the Commission of any withdrawal of notification for the aforementioned or any other reason.

4. The notified body and the manufacturer, or its authorized representative established in the Community, shall lay down, by common accord, the time limits for completion of the assessment and verification operations referred to in Annexes 3 to 7.

5. The notified body must make available to the other notified bodies and the competent authority, on request, all relevant information concerning approvals and certificates issued, refused or withdrawn.

Article 14

CE-marking

1. Devices, other than devices for performance evaluation, considered to meet the essential requirements referred to in Article 3 must bear the CE marking of conformity when they are placed on the market.

2. The CE-marking of conformity, as shown in Annex 10, must appear in a visible, legible and indelible form on the device, where

practicable and appropriate, and on the instructions for use. Where applicable the CE marking of conformity must also appear on the sales packaging. The CE marking shall be accompanied by the identification number of the notified body responsible for implementation of the procedures set out in Annexes 4, 6 and 7.

3. It is prohibited to affix marks or inscriptions which are likely to mislead third parties with regard to the meaning or the graphics of the CE marking. Any other mark may be affixed to the device, to the packaging or to the instruction leaflet accompanying the device provided that the visibility and legibility of the CE marking is not thereby reduced.

Article 15

Unduly affixed CE marking

Without prejudice to Article 8 :

(a) where a Member State establishes that the CE marking has been affixed unduly, the manufacturer or his authorized representative established within the Community shall be obliged to end the infringement under conditions imposed by the Member State

(b) here non-compliance continues, the Member State must take all appropriate measures to restrict or prohibit the placing on the market of the product in question or to ensure that it is withdrawn from the market, in accordance with the procedure in Article 8.

Article 16

Decisions in respect of refusal or restriction

1. Any decision taken pursuant to this Directive :

(a) to refuse or restrict the placing on the market or the putting into service of a device or

(b) to withdraw devices from the market

shall state the exact grounds on which it is based. Such decisions shall be notified without delay to the party concerned, who shall at the same time be informed of the remedies available to him under the national law in force in the Member State in question and of the time limits to which such remedies are subject.

2. In the event of a decision as referred to in paragraph 1, the manufacturer, or his authorized representative established in the Community, shall have an opportunity to put forward his viewpoint in advance, unless such consultation is not possible because of the urgency of the measure to be taken.

Article 17

Confidentiality

Without prejudice to the existing national provisions and practices on medical secrets, Member States shall ensure that all the parties involved in the application of this Directive are bound to observe confidentiality with regard to information obtained in carrying out their tasks. This does not affect the obligations of Member States and notified bodies with regard to mutual information and the dissemination of warnings, nor the obligations of the persons concerned to provide information under criminal law.

Article 18

Cooperation between Member States

Member States shall take appropriate measures to ensure that competent authorities charged with the implementation of the present Directive cooperate with each other and convey to each other the information necessary to ensure compliance with the provisions of this Directive.

Article 19

Amendment of Directives

1. Directive 93/42/EEC is hereby amended as follows:

a) Article 1(2) c) shall read as follows:

"(c) in vitro diagnostic medical device means any medical device which is a reagent, reagent product, calibrator, control material, kit, instrument, apparatus, equipment or system whether used alone or in combination, intended by the manufacturer to be used in vitro for the examination of specimens including blood and tissue donations, derived from the human body, solely or principally for the purpose of providing information concerning a physiological state, state of health or disease or congenital abnormality or to determine the safety and compatibility with potential recipients".

b) Article 1(5)f shall read as follows:

"(f) transplants or tissues or cells of human origin, unless a device is manufactured utilizing tissues or substances derived from such tissues which are non viable or rendered non viable. In this case, the directive shall not affect national regulations relating to the ethics of the collection of tissues or substances of human origin, as well as any regulations relating to the ethics governing distribution of given types of devices of such origin."

c) Annex I, paragraph 8.2, shall read as follows:

"8.2. Where a device is manufactured utilizing human tissues or substances derived from human tissues, the use of such tissues or substances must be subject to the relevant validated selection and screening procedures, including traceability as appropriate in relation to the inherent risk.

Tissues of animal origin must originate from animals that have been subjected to veterinary controls and surveillance adapted to the intended use of the tissues.

Processing, preservation, testing and handling of tissues, cells and substances of human or animal origin must be carried out so as to provide optimal security. In particular safety with regard to viruses and other transmissible agents must be addressed as appropriate by implementation of validated methods of elimination or inactivation of such viruses or other transmissible agents in the course of the manufacturing process, taking into account the sourcing and control methods applied."

d) Annex I paragraph 13.3 shall be completed by the following subparagraph :

"n) in the case of devices incorporating tissues of human origin or substances derived from such tissues, a statement indicating that the device incorporates tissue or substances derived from tissue of human origin as appropriate"

e) In annex II paragraph 3.2c) and annex III paragraph 3 the following indent shall be included as third last indent:

"in the case of devices incorporating tissues of human or animal origin, information on the selection and origin"

f) Annex IX, section III, paragraph 4.5: the following sub-paragraph shall be added:

"All devices manufactured utilizing human tissues or substances derived from such tissues are in class III."

2. In Directive 89/392/EEC, the text of Article 1(3), second indent "machinery for medical use, used in direct contact with patients" is replaced by the following text :

"- medical devices"

Article 20

Implementation, transitional provisions

1. Member States shall adopt and publish the laws, regulations and administrative provisions necessary to comply with this Directive not later than 1 April 1998. They shall immediately inform the Commission thereof.

The Standing Committee referred to in Article 7 may assume its tasks from the date of notification of this Directive. The Member States may take the measures referred to in Article 13 on notification of this Directive.

When Member States adopt these provisions, these shall contain a reference to this Directive or shall be accompanied by such a reference at the time of their official publication. The procedures for such reference shall be adopted by Member States. Member States shall apply these provisions with effect from1 July 1998.

2. Member States shall communicate to the Commission the texts of the provisions of national law which they adopt in the field covered by this Directive.

3. Member States shall take the necessary action to ensure that the notified bodies which are responsible pursuant to Article 9 for conformity assessment take account of any relevant information regarding the characteristics and performance of such devices, including in particular the results of any relevant tests and verification already carried out under pre-existing national law, regulations or administrative provisions in respect of such devices.

4. Member States shall accept during a period of four years following adoption of the present Directive the placing on the market and putting into service of devices which conform to the rules in force in their territory on the date of adoption of this Directive.

Article 21

This Directive shall enter into force the twentieth day after its publication.

Article 22

This Directive is addressed to the Member States.

Annex 1

ESSENTIAL REQUIREMENTS

I. GENERAL REQUIREMENTS

1. The devices must be designed and manufactured in such a way that, when used under the conditions and for the purposes intended, they will not compromise, directly or indirectly, the clinical condition or the safety of the patients, or the safety and health of users and, where applicable, other persons, and the safety of property. Any risks, which may be associated with their use, must be acceptable when weighed against the benefits to the patient and are compatible with a high level of protection of health and safety.

2. The solutions adopted by the manufacturer for the design and construction of the devices must conform to safety principles, taking account of the generally acknowledged state of the art. In selecting the most appropriate solutions, the manufacturer must apply the following principles in the following order :

- eliminate or reduce risks as far as possible (inherently safe design and construction);
- where appropriate take adequate protection measures in relation to risks that cannot be eliminated;
- inform users of the residual risks due to any shortcomings of the protection measures.

3. The devices must be designed and manufactured in such a way that they are suitable for the purposes referred to in Article 1(2)(b), as specified by the manufacturer, taking account of the generally acknowledged state of the art. They must achieve the performances in particular in terms of analytical sensitivity, specificity, accuracy, repeatability, reproducibility and limits of detection stated by the manufacturer.

The traceability of values assigned to calibrators and/or control materials must be assured through available reference measurement procedures and/or available reference materials of higher order.

4. The characteristics and performances referred to in Sections 1 and 3 must not be adversely affected to such a degree that the health or the safety of the patient or the user and, where applicable, of other persons, are compromised during the lifetime of the device, as indicated by the manufacturer, when the device is subject to the stresses which can occur during normal conditions of use. When no lifetime is stated, the same applies for the lifetime reasonably to be expected of a device of that kind, having regard to the intended purpose and the anticipated use of the device.

5. The devices must be designed, manufactured and packed in such a way that their characteristics and performances during their intended use will not be adversely affected in the storage and transport conditions (temperature, humidity, ...) taking account of the instruction and information provided by the manufacturer.

II. REQUIREMENTS REGARDING DESIGN AND MANUFACTURING

6. Chemical and physical properties

6.1. The devices must be designed and manufactured in such a way as to achieve the characteristics and performances referred to in Section I on the "General Requirements". Attention must be paid to the possibility of impairment of analytical performance due to incompatibility between the materials used and the samples (such as biological tissues, cells, body fluids and micro-organisms) intended to be used with the device, taking account of its intended purpose.

6.2. The devices must be designed, manufactured and packed in such a way as to reduce as far as possible the risk posed by leakage products, contaminants and residues to the persons involved in the transport, storage and use of the devices, taking account of the intended purpose of the products.

7. Infection and microbial contamination

7.1. The devices must be designed in such a way as to eliminate or reduce as far as possible the risk of infection to the user or other persons. The design must allow easy handling and, where necessary, reduce as far as possible contamination of and leakage from the device during use and, in the case of specimen receptacles, the risk of contamination of the specimen. The manufacturing processes must be appropriate for these purposes.

7.2. Where a device incorporates biological substances, the risks of infection must be reduced as far as possible by selecting appropriate donors, appropriate substances and using appropriate inactivation, conservation, test and control procedures.

7.3. Devices labelled either as "STERILE", or, as having a special microbiological state must be designed, manufactured and packed in an appropriate pack, according to procedures suitable to ensure that they remain in the microbiological state indicated on the label when placed on the market, under the storage and transport conditions as laid down by the manufacturer, until the protective packaging is damaged or opened.

7.4. Devices labelled either as "STERILE", or, as having a special microbiological state must have been processed by an appropriate, validated method.

7.5. Packaging systems for devices other than those referred to in section 7.3 must keep the product without deterioration at the level of cleanliness, if any, as indicated by the manufacturer and, if the devices are to be sterilized prior to use, reduce as far as possible the risk of microbial contamination.

Steps shall be taken to reduce microbial contamination during selection and handling of raw materials, manufacture, storage and distribution where the performance of the device can be adversely affected by such contamination.

8. Construction and environmental properties

8.1. If the device is intended for use in combination with other devices or equipment, the whole combination, including the connection system must be safe and must not impair the specified performances of the devices. Any restrictions on use must be indicated on the label and/or in the instructions for use.

8.2. Devices must be designed and manufactured in such a way as to remove or reduce as far as possible :

- the risk of injury, in connection with their physical features, including aspects of volume x pressure, dimensional and where appropriate ergonomic features;
- risks connected with reasonably foreseeable external influences, such as magnetic fields, external electrical effects, electrostatic discharge, pressure,temperature or variations in pressure or acceleration.

Devices must be designed and manufactured in such a way as to provide an adequate level of intrinsic immunity from electromagnetic disturbance to enable them to operate in accordance with their intended purpose.

8.3. Devices must be designed and manufactured in such a way as to reduce as far as possible the risks of fire or explosion during normal use. Particular attention must be paid to devices whose intended use includes exposure to or use in association with flammable substances or substances which could cause combustion.

8.4. Devices must be designed and manufactured in such a way as to facilitate the management of safe waste disposal.

8.5. The measuring, monitoring or display scale (including colour change and other visual indicators) must be designed and manufactured in line with ergonomic principles, taking account of the intended purpose of the device.

9. Devices which are instruments or apparatus with a measuring function

9.1. Devices which are instruments or apparatus having a primary analytical measuring function must be designed and manufactured in such a way as to provide adequate stability and accuracy of measurement within appropriate accuracy limits, taking into account the intended purpose of the device and of available and appropriate reference measurement procedures and materials. The accuracy limits have to be specified by the manufacturer.

9.2. When values are expressed numerically, they must be given in legal units conforming to the provisions of Council Directive 80/181/EEC.

10. Protection against radiation

10.1 Devices shall be designed and manufactured in such a way that exposure of users and other persons to the emitted radiation is reduced as far as possible.

10.2 When devices are intended to emit potentially hazardous, visible and/or invisible radiation, they must be :

- designed and manufactured in such a way as to ensure that the quality and quantity of radiation emitted can be adjusted and controlled;
- fitted, where practicable, with visual displays and/or audible warnings of such emissions.

10.3 The operating instructions for devices emitting radiation must give detailed information as to the nature of the emitted radiation, means of protecting the user, and on ways of avoiding misuse and of eliminating the risks inherent in installation.

11. Requirements for medical devices connected to or equipped with an energy source

11.1 Devices incorporating electronic programmable systems must be designed to ensure the repeatability, reliability and performance of these systems according to the intended use.

11.2 Devices must be designed and manufactured in such a way as to minimize the risks of creating electromagnetic fields which could impair the operation of other devices or equipment in the usual environment.

11.3 Devices must be designed and manufactured in such a way as to avoid, as far as possible, the risk of accidental electric shocks during normal use and in single fault condition, provided the devices are installed and maintained correctly.

11.4 Protection against mechanical and thermal risks

11.4.1 Devices must be designed and manufactured in such a way as to protect the user against mechanical risks connected with, for example, resistance, stability and moving parts. Annex 1, sections 1.3 and 1.4 of Directive 89/392/EEC shall apply, were appropriate.

11.4.2 Devices must be designed and manufactured in such a way as to reduce to the lowest possible level the risks arising from vibration generated by the devices, taking account of technical progress and of the means available for limiting vibrations, particularly at source, unless the vibrations are part of the specified performance.

11.4.3 Devices must be designed and manufactured in such a way as to reduce to the lowest possible level the risks arising from the noise emitted, taking account of technical progress and of the means available to reduce noise, particularly at source, unless the noise emitted is part of the specified performance.

11.4.4 Terminals and connectors to electricity, gas or hydraulic and pneumatic energy supplies which the user has to handle must be designed and manufactured in such a way as to minimize all possible risks.

11.4.5 Accessible parts of the devices (excluding the parts or areas intended to supply heat or reach given temperatures) and their surroundings must not attain potentially dangerous temperatures under normal use.

12. Requirements for devices for self-testing

Devices for self-testing must be designed and manufactured in such a way that they perform appropriately for their intended purpose taking into account the skills and the means available to users and the influence resulting from variation that can reasonably be anticipated in users' technique and environment. The information and instructions provided by the manufacturer should be easily understood and applied by the user.

12.1 Devices for self-testing must be designed and manufactured in such a way as to reduce as far as practicable the risk of user's error in handling and in interpretation of the result.

12.2 Devices for self-testing must, where reasonably possible, include a method of user control, i.e. a procedure by that, at the time of use, the product will perform as intended.

13. Information supplied by the manufacturer

13.1 Each device must be accompanied by the information needed to use it properly, taking account of the training and knowledge of the potential users and to identify the manufacturer.

This information comprises the data on the label and in the instructions for use.

As far as practicable and appropriate, the information needed to use the device properly must be set out on the device itself and/or on the packaging or, where appropriate, on the sales packaging. If individual full labelling of each unit is not practicable, the information must be set out on the packaging and/or in the instructions for use supplied with one or more devices.

Instructions for use must accompany or be included in the packaging of one or more devices.

By way of exception, no such instructions for use are needed for a device if it can be used properly and safely without them.

13.2 Where appropriate, this information to be supplied should take the form of symbols. Any symbol and identification colour used must conform to the harmonized standards. In areas for which no standards exist, the symbols and colours as used must be described in the documentation supplied with the device.

13.3 In the case of devices containing a substance or being a preparation which may be considered as being dangerous, taking account of the nature and quantity of its constituents and the form under which they are present, relevant danger symbols and labelling requirements of Directive 67/548/EEC and Directive 88/379/EEC shall apply. Where there is insufficient space to put all information on the device itself or on its label, the relevant danger symbols shall be put on the label and the other information required by those Directives shall be given in the instructions for use.

The provisions of the aforementioned directives on the safety data sheet shall apply, unless all relevant information as appropriate is already made available by the instructions for use.

13.4 The label must bear the following particulars which may take the form of symbols as appropriate:

a) the name or trade name and address of the manufacturer. For devices imported into the Community, in view of their distribution in the Community, the label, or the outer packaging, or instructions for use, shall contain in addition the name and address of the authorized representative of the manufacturer established within the Community;

b) the details strictly necessary for the user to identify the device and the contents of the packaging.

c) where appropriate, the word "STERILE" or a statement indicating any special microbiological state or state of cleanliness;

d) where appropriate, the batch code, preceded by the word "LOT", or the serial number;

e) where appropriate, an indication of the date by which the device or part of it should be used, expressed as the year, the month and, where relevant, the day;

f) in cases of devices for performance evaluation, the words "for performance evaluation only";

g) where appropriate, a statement indicating the in vitro nature of the device;

h) any special storage and/or handling conditions;

i) where applicable any special operating instructions;

j) appropriate warnings and/or precautions to take;

k) if the device is intended for self-testing, this fact should be clearly stated.

l) In the case of devices incorporating tissues of human origin or substances derived from such tissues, a statement indicating that the device incorporates tissues or substances derived from tissue of human origin.

13.5 If the intended purpose of the device is not obvious to the user, the manufacturer must clearly state the intended purpose in the instructions for use and, if appropriate, on the label.

13.6 Wherever reasonable and practicable, the devices and detachable components must be identified, where appropriate in terms of batches, to allow all appropriate action to detect any potential risk posed by the devices and detachable components.

13.7 Where appropriate, the instructions for use must contain the following particulars :

a) the details referred to in Section 13.4, with exception of points d) and e);

b) composition of the reagent product by nature and amount or concentration of the active ingredient(s) of the reagent(s) or kit as well as a statement that the device contains other ingredients influencing the measurement;

c) the storage conditions and shelf life following the first opening of the immediate container, together with the storage conditions and stability of working reagents;

d) the performances referred to in Section 3;

e) a statement of any special materials required including information necessary to enable those special materials to be identified for proper use;

f) the type of specimen to be used, any special conditions of collection, pre-treatment and, if necessary, storage conditions;

g) a detailed description of the procedure to be followed in using the device;

h) the measurement procedure to be followed with the device including as appropriate :
- the principle of the method;
- the specific analytical performance characteristics (e.g. analytical sensitivity, specificity, accuracy, repeatability, reproducibility and limits of detection), limitations of the method and information about the use of available reference measurement procedures and materials by the user;
- the details of any further procedure or handling needed before the device can be used (for example reconstitution,

incubation, dilution, instrument checks, etc);
- the indication whether any particular training is required;

i) the mathematical approach upon which the calculation of the analytical result is made;

j) measures to be taken in the event of changes in the analytical performance of the device;

k) information appropriate to users on :
- internal quality control including specific validation procedures,
- the traceability of the calibration of the device;

l) the reference intervals for the quantities being determined;

m) if the device must be used in combination with or installed with or connected to other medical devices or equipment in order to operate as required for its intended purpose, sufficient details of its characteristics to identify the correct devices or equipment to use in order to obtain a safe and proper combination;

n) all the information needed to verify whether the device is properly installed and can operate correctly and safely, plus details of the nature and frequency of the maintenance and calibration needed to ensure that the device operates properly and safely; information about safe waste disposal;

o) the necessary instructions in the event of damage to the sterile protective packaging and details of appropriate methods of resterilization or decontamination;

p) if the device is reusable, information on the appropriate processes to allow reuse, including cleaning, disinfection, packaging and resterilization or decontamination, and any restriction on the number of reuses;

q) precautions to be taken as regards exposure, in reasonably foreseeable environmental conditions, to magnetic fields, external influences, electrostatic discharge, pressure or variations in pressure, acceleration, thermal ignition sources, etc.;

r) precautions to be taken against any special, unusual risks related to the use or disposal of the device including special protective measures; where the device includes substances of human or animal origin, attention shall be drawn to their potential infectious nature;

s) specifications for devices for self-testing :
- the results need to be expressed and presented in a way that is readily understood by a non-professional person; information needs to be provided with advice to the user on action to be taken (in case of positive, negative or indeterminate result) and on the possibility of false positive or false negative result;
- specific particulars may be omitted provided that the other information supplied by the manufacturer is sufficient

to enable the user to know how to use the device and to understand the result(s) produced by the device;
- the information provided must include a statement clearly directing that the user should not take any decision of medical relevance without first consulting his or her medical practitioner;

t) date of issue or last revision of the instructions for use.

Annex 2

LIST OF DEVICES REFERRED TO IN ARTICLE 9(2)

1. Reagents and reagent products for blood grouping (A B O system and Rh/D)

2. Reagents and reagent products for the detection in human specimens of markers of HIV infection, Hepatitis B and C .

Annex 3

EC DECLARATION OF CONFORMITY

1. The EC declaration of conformity is the procedure whereby the manufacturer or his authorized representative established in the Community who fulfils the obligations imposed by Sections 2 to 5 and additionally, in the case of devices for self-testing, the obligations imposed by Section 6, ensures and declares that the products concerned meet the provisions of this Directive which apply to them. The manufacturer must affix the CE marking in accordance with Article 14.

2. The manufacturer must prepare the technical documentation described in Section 3 and ensure that the manufacturing process follows the principles of quality assurance as set out in section 4.

3. The technical documentation must allow assessment of the conformity of the product with the requirements of the Directive. It must include in particular :

- a general description of the product, including any variants planned;
- the documentation of the quality system;
- design information, including performance characteristics and limitation, methods of manufacture and, in the case of instruments, design drawings, diagrams of components, sub--assemblies, circuits, etc.;
- the descriptions and explanations necessary to understand the abovementioned characteristics, drawings and diagrams and the operation of the product;
- the results of the risk analysis, where appropriate, and a list of the standards referred to in Article 5, applied in full or in part, and descriptions of the solutions adopted to meet the essential requirements of the Directive if the standards referred to in Article 5 have not been applied in full;
- in the case of sterile products or products with a special microbiological state or state of cleanliness, a description of the procedures used;
- the results of the design calculations and of the inspections carried out, etc...
- if the device is to be combined with other device(s) in order to operate as intended, proof must be provided that it conforms to the essential requirements when combined with any such device(s) having the characteristics specified by the manufacturer;
- the test reports;
- adequate performance evaluation data, supported by a reference measurement system (when available), originated from studies in a clinical or other appropriate environment or resulting from available literature; this applies in particular to biological materials, culture media and devices using a new technology;
- the labels and instructions for use.

4. The manufacturer shall take necessary measures to ensure that the manufacturing process follows the principles of quality assurance as appropriate for the products manufactured.

The system shall address :

- the organizational structure and responsibilities,
- the manufacturing processes and systematic quality control of production;
- the means to monitor the performance of the quality system;

5. The manufacturer shall institute and keep up to date a systematic procedure to review experience gained from devices in the post-production phase and to implement appropriate means to apply any necessary corrective actions, taking account of the nature and risks in relation to the product. He shall notify the competent authorities of the following incidents immediately on learning of them :

(i) any malfunction, failure or deterioration in the characteristics and/or performance of a device, as well as any inadequacy in the labelling or the instructions for use which, directly or indirectly, might lead to or might have led to the death of a patient or user or to a serious deterioration in his state of health;

(ii) any technical or medical reason connected with the characteristics or the performance of a device for the reasons referred to in subparagraph (i) above leading to systematic recall of devices of the same type by the manufacturer.

6. For devices for self-testing the manufacturer shall lodge an application for examination of the design with a notified body.

6.1 The application shall enable the design of the device to be understood and shall enable conformity with the design-related requirements of the directive to be assessed.

It shall include :

- test reports including, where appropriate, results of studies carried out with lay persons;
- data showing the handling suitability of the device in view of its intended purpose for self-testing;
- the information to be provided with the device on its label and its instructions for use.

6.2 The notified body shall examine the application and, if the design conforms to the relevant provisions of this Directive shall issue the applicant with an EC design-examination certificate. The notified body may require theapplication to be completed by further tests or proof to allow assessment of conformity with the design-related requirements of the Directive. The certificate shall contain the conclusions of the examination, the conditions of validity, the data needed for identification of the approved design and, where appropriate, a description of the intended purpose of the product.

6.3 The applicant shall inform the notified body which issued the EC designed-examination certificate of any significant change made to the approved design. Changes to the approved design must receive further approval from the notified body which issued the EC design-examination certificate wherever the changes could affect conformity with the essential requirements of the Directive or with the conditions prescribed for use of the product. This additional approval shall take the form of a supplement to the EC design-examination certificate.

Annex 4

EC DECLARATION OF CONFORMITY (FULL QUALITY ASSURANCE SYSTEM)

1. The manufacturer must ensure application of the quality system approved for the design, manufacture and final inspection of the products concerned, as specified in Section 3, and is subject to audit as laid down in section 3.3 and to the EC surveillance as specified in Section 4.

2. The declaration of conformity is the procedure whereby the manufacturer who fulfils the obligations imposed by Section 1 ensures and declares that the products concerned meet the provisions of this Directive which apply to them.

The manufacturer shall affix the CE marking in accordance with Article 14 and shall draw up a declaration of conformity covering the products concerned.

3.1 The manufacturer must lodge an application for assessment of his quality system with a notified body.

The application must include:

- the name and address of the manufacturer and any additional manufacturing site covered by the quality system;
- adequate information on the product or product category covered by the procedure;
- a written declaration that no such application has been lodged with any other notified body for the same product-related quality system;
- the documentation on the quality system;
- an undertaking by the manufacturer to fulfil the obligations imposed by the quality system approved;
- an undertaking by the manufacturer to keep the approved quality system adequate and efficacious;
- an undertaking by the manufacturer to institute and keep up to date a systematic procedure to review experience gained from devices in the post production phase and to implement appropriate means to apply any necessary corrective action and notification as referred to in Annex 3, section 5.

3.2 Application of the quality system must ensure that the products conform to the provisions of this Directive which apply to them at every stage, from design to final inspection. All the elements, requirements and provisions adopted by the manufacturer for his quality system must be documented in a systematic and orderly manner in the form of written policies and procedures, such as quality programmes, quality plans, quality manuals and quality records.

It shall include in particular an adequate description of:

(a) the manufacturer's quality objectives;

(b) the organization of the business and in particular:
- the organizational structures, the responsibilities of the managerial staff and their organizational authority where quality of design and manufacture of the products is concerned;
- the methods of monitoring the efficient operation of the quality system and in particular its ability to achieve the desired quality of design and of product, including control of products which fail to conform;

(c) the procedures for monitoring and verifying the design of the products and in particular:
- a general description of the product, including any variants planned;
- all documentation referred to in Annex 3,

section 3, indents 3 to 11;
- in the case of devices for self-testing, the information referred to in Annex 3, section 6.1;
- the techniques used to control and verify the design and the processes and systematic measures which will be used when the products are being designed;

(d) the inspection and quality assurance techniques at the manufacturing stage and in particular:
- the processes and procedures which will be used, particularly as regards sterilization and purchasing;
- the product identification procedures drawn up and kept up to date from drawings, specifications or other relevant documents at every stage of manufacture;

(e) the appropriate tests and trials which will be carried out before, during and after manufacture, the frequency with which they will take place, and the test equipment used; it must be possible to trace back the calibration of the test equipment adequately.

3.3 The notified body must audit the quality system to determine whether it meets the requirements referred to in Section 3.2. It must presume that quality systems which implement the relevant harmonized standards conform to these requirements.

The assessment team must include at least one member with past experience of assessments of the technology concerned.

The assessment procedure must include an inspection on the manufacturer's premises and, in duly substantiated cases, on the premises of the manufacturer's suppliers and/or subcontractors to inspect the manufacturing processes.

The decision shall be notified to the manufacturer. It must contain the conclusions of the inspection and a reasoned assessment.

3.4 The manufacturer must inform the notified body which approved the quality system of any plan for substantial changes to the quality system or the product-range covered.

The notified body must assess the changes proposed and verify whether after these changes the quality system still meets the requirements referred to in Section 3.2. It must notify the manufacturer of its decision. This decision must contain the conclusions of the inspection and a reasoned assessment.

4. Surveillance

4.1 The aim of surveillance is to ensure that the manufacturer duly fulfils the obligations imposed by the approved quality system.

4.2 The manufacturer must authorize the notified body to carry out all the necessary inspections and supply it with all relevant information, in particular:

- the documentation on the quality system;
- the data stipulated in the part of the quality system relating to design, such as the results

of analyses, calculation, tests, etc.;
- the data stipulated in the part of the quality system relating to manufacture, such as inspection reports and test data, calibration data, qualification reports of the personnel concerned, etc.

4.3 The notified body must periodically carry out appropriate inspections and assessments to make sure that the manufacturer applies the approved quality system and must supply the manufacturer with an assessment report.

4.4 In addition, the notified body may pay unannounced visits to the manufacturer. At the time of such visits, the notified body may, where necessary, carry out or ask for tests in order to check that the quality system is working properly. It must provide the manufacturer with an inspection report and, if a test has been carried out, with a test report.

Annex 5

EC TYPE EXAMINATION

1. EC type examination is the part of the procedure whereby a notified body ascertains and certifies that a representative sample of the production envisaged fulfils the relevant provisions of this Directive.

2. The application for EC type examination shall be lodged by the manufacturer or by his authorized representative established in the Community with a notified body.

The application shall include:

- the name and address of the manufacturer and the name and address of the authorized representative if the application is lodged by the representative;
- the documentation described in Section 3 needed to assess the conformity of the representative sample of the production in question, hereinafter referred to as the "type", with the requirements of this Directive. The applicant shall make a "type" available to the notified body. The notified body may request other samples as necessary;
- a written declaration that no application has been lodged with any other notified body for the same type.

3. The documentation must allow an understanding of the design, the manufacture and the performances of the product. The documentation shall contain the following items in particular:

- a general description of the type, including any variants planned;
- all documentation referred to in Annex 3, section 3, indents 3 to 5, 7, 10 and 11.

4. The notified body shall:

4.1 examine and assess the documentation and verify that the type has been manufactured in conformity with that documentation; it shall also record the items designed in conformity with the applicable provisions of the standards referred to in Article 5, as well as the items not designed on the basis of the relevant provisions of the abovementioned standards;

4.2 perform or have performed appropriate examinations and the tests necessary to verify whether the solutions adopted by the manufacturer meet the essential requirements of this Directive if the standards referred to in Article 5 have not been applied; if the device is to be combined with other device(s) in order to operate as intended, proof must be provided that it conforms to the essential requirements when combined with any such device(s) having the characteristics specified by the manufacturer;

4.3 carry out or ask for the appropriate examinations and the tests necessary to verify whether, if the manufacturer has chosen to apply the relevant standards, these have actually been applied;

4.4 agree with the applicant on the place where the necessary examinations and tests will be carried out.

5. If the type conforms to the provisions of this Directive, the notified body shall issue the applicant with an EC type examination certificate. The certificate shall contain the name and address of the manufacturer, the conclusions of the examination, the conditions of validity and the data needed for identification of the type approved. The relevant parts of the documentation shall be annexed to the certificate and a copy shall be kept by the notified body.

6. The applicant shall inform the notified body which issued the EC type examination certificate of any significant change made to the approved product.

Changes to the approved product must receive further approval from the notified body which issued the EC type-examination certificate wherever the changes may affect conformity with the essential requirements or with the conditions prescribed for use of the product. This new approval shall, where appropriate, take the form of a supplement to the initial EC type examination certificate.

7. Administrative provisions

Other notified bodies may obtain a copy of the EC type examination certificates and/or the supplements thereto. The annexes to the certificates must be made available to the other notified bodies on reasoned application, after the manufacturer has been informed.

Annex 6

EC VERIFICATION

1. EC verification is the procedure whereby the manufacturer or his authorized representative established in the Community ensures and declares that the products which have been subject to the procedure set out in Section 4 conform to the type described in the EC type examination certificate and meet the requirements of this Directive which apply to them.

2.1 The manufacturer must take all the measures necessary to ensure that the manufacturing process produces products which conform to the type described in the EC type examination certificate and to the requirements of the Directive which apply to them. Before the start of manufacture, the manufacturer must prepare documents defining the manufacturing process, in particular as regards sterilization and the suitability of starting materials, where necessary, and define the necessary testing procedures according to the state of the art. All the routine, pre established provisions must be implemented to ensure homogeneous production and conformity of the products with the type described in the EC type examination certificate and with the requirements of this Directive which apply to them. The manufacturer must affix the CE marking in accordance with Article 14 and draw up a declaration of conformity covering the products concerned.

2.2 To the extent that for certain aspects the final testing according to section 6.3 is not appropriate, adequate in process testing, monitoring and control methods shall be established by the manufacturer with the approval of the notified body. The provisions of Annex 4 section 4 shall apply accordingly in relation to the above mentioned approved procedures.

3. The manufacturer must undertake to institute and keep up to date a systematic procedure to review experience gained from devices in the post-production phase and to implement appropriate means to apply any necessary corrective and notification action as referred to in Annex 3 section 5.

4. The notified body must carry out the appropriate examinations and tests in order to verify the conformity of the product with the requirements of the Directive either by examining and testing every product as specified in section 5 or by examining and testing products on a statistical basis as specified in section 6, as the manufacturer decides. In as far as the conduct of examinations and tests on a statistical basis is not appropriate, examinations and tests may be carried out on a random basis provided that such procedure in conjunction with the measures taken in accordance with section 2.2 ensures an equivalent level of conformity.

5. Verification by examination and testing of every product

5.1 Every product is examined individually and the appropriate tests defined in the relevant standard(s) referred to in Article 5 or equivalent tests must be carried out in order to verify the conformity of the products with the EC type described in the type examination certificate and with the requirements of the Directive which apply to them.

5.2 The notified body must affix, or have affixed, its identification number to each approved product and must draw up a written certificate of conformity relating to the tests carried out.

6. Statistical verification

6.1 The manufacturer must present the manufactured products in the form of homogeneous batches.

6.2 A random sample is taken from each batch. The products which make up the sample are examined individually and the appropriate tests defined in the relevant standard(s) referred to in Article 5 or equivalent tests must be carried out to verify, where appropriate, the conformity of the products with the type described in the EC type examination certificate and with the requirements of the Directive which apply to them in order to determine whether to accept or reject the batch.

6.3 Statistical control of products will be based on attributes, entailing a sampling system ensuring a limit quality corresponding to a probability of acceptance of 5%, with a non conformity percentage of between 3 and 7%. The sampling method will be established by the harmonized standards referred to in Article 5, taking account of the specific nature of the product categories in question.

6.4 If the batch is accepted, the notified body affixes, or has affixed its identification number to each product and draws up a written certificate of conformity relating to the tests carried out. All products in the batch may be put on the market except any in the sample which failed to conform.

If a batch is rejected the competent notified body must take appropriate measures to prevent the batch from being placed on the market. In the event of frequent rejection of batches, the notified body may suspend the statistical verification.

The manufacturer may, on the responsibility of the notified body, affix the notified body's identification number during the manufacturing process.

Annex 7

EC DECLARATION OF CONFORMITY (PRODUCTION QUALITY ASSURANCE)

1. The manufacturer must ensure application of the quality system approved for the manufacture of the products concerned and carry out the final inspection, as specified in Section 3, and is subject to the EC surveillance referred to in Section 4.

2. The declaration of conformity is the part of the procedure whereby the manufacturer who fulfils the obligations imposed by Section 1 ensures and declares that the products concerned conform to the type described in the EC type examination certificate and meet the provisions of this Directive which apply to them.

The manufacturer must affix the CE marking in accordance with Article 14 and draw up a declaration of conformity.

3. Quality system

3.1 The manufacturer must lodge an application for assessment of his quality system with a notified body. The application must include:

- all documentation and undertakings referred to in Annex 4 section 3.1 and
- the technical documentation on the types approved and a copy of the EC type-examination certificates.

3.2 Application of the quality system must ensure that the products conform to the type described in the EC type-examination certificate.

All the elements, requirements and provisions adopted by the manufacturer for his quality system must be documented in a systematic and orderly manner in the form of written policy statements and procedures. This quality system documentation must permit uniform interpretation of the quality policy and procedures such as quality programmes, plans, manuals and records.

It must include in particular an adequate description of:

(a) the manufacturer's quality objectives;

(b) the organization of the business and in particular:
- the organizational structures, the responsibilities of the managerial staff and their organizational authority where manufacture of the products is concerned;
- the methods of monitoring the efficient operation of the quality system and in particular its ability to achieve the desired quality of product, including control of products which fail to conform;

(c) the inspection and quality assurance techniques at the manufacturing stage and in particular:

- the processes and procedures which will be used, particularly as regards sterilization and purchasing;
- the product identification procedures drawn up and kept up to date from drawings, specifications or other relevant documents at every stage of manufacture;

(d) the appropriate tests and trials to be carried out before, during and after manufacture, the frequency with which they will take place, and the test equipment used; it must be possible adequately to trace back the calibration of the test equipment.

3.3 The notified body must audit the quality system to determine whether it meets the requirements referred to in Section 3.2. It must presume that quality systems which implement the relevant harmonized standards conform to these requirements.

The assessment team must include at least one member with past experience of assessments of the technology concerned. The assessment procedure must include an inspection on the manufacturer's premises and, in duly substantiated cases, on the premises of the manufacturer's suppliers and/or subcontractors to inspect the manufacturing processes. The decision must be notified to the manufacturer after the final inspection and contain the conclusions of the inspection and a reasoned assessment.

3.4 The manufacturer shall inform the notified body which approved the quality system of any plan for substantial changes to the quality system.

The notified body must assess the changes proposed and verify whether after these changes the quality system still meets the requirements referred to in Section 3.2. It must notify the manufacturer of its decision. This decision must contain the conclusions of the inspection and a reasoned assessment.

4. Surveillance

The provision of Annex 4, section 4, shall apply.

Annex 8

STATEMENT AND PROCEDURES CONCERNING DEVICES FOR PERFORMANCE EVALUATION

1. For devices for performance evaluation the manufacturer or his authorized representative established in the Community shall draw up the statement containing the information stipulated in Section 2 of this Annex and ensure that the relevant provisions of this Directive are met.

2. The statement shall contain the following information:

- the list of laboratories or other institutions taking part in the performance evaluation study;
- in the case of devices for self--testing, the place, starting date and scheduled duration for the investigations and the number of lay-persons involved.
- a statement that the device in question conforms to the requirements of the Directive, apart from the aspects covered by the investigation and apart from those specifically itemised in the statement, and that every precaution has been taken to protect the health and safety of the patient, user and third parties.

3. The manufacturer shall also undertake to keep available for the competent national authorities the documentation allowing an understanding of the design, manufacture and performances of the product, including the expected performances, so as to allow assessment of conformity with the requirements of this Directive. This documentation must be kept for a period ending at least five years after the end of the performance evaluation.

The manufacturer shall take all the measures necessary for the manufacturing process to ensure that the products manufactured conform to the documentation mentioned in the first paragraph.

Annex 9

CRITERIA FOR THE DESIGNATION OF NOTIFIED BODIES

1. The notified body, its Director and the assessment and verification staff shall not be the designer, manufacturer, supplier, installer or user of the devices which they inspect, nor the authorized representative of any of these persons. They may not be directly involved in the design, construction, marketing or maintenance of the devices, nor represent the parties engaged in these activities. This in no way precludes the possibility of exchanges of technical information between the manufacturer and the body.

2. The notified body and its staff must carry out the assessment and verification operations with the highest degree of professional integrity and the requisite competence in the field of medical devices and must be free from all pressures and inducements, particularly financial, which might influence their judgment or the results of the inspection, especially from persons or groups of persons with an interest in the results of the verifications.

Should the notified body subcontract specific tasks connected with the establishment and verification of the facts, it must first ensure that the subcontractor meets the provisions of the Directive. The notified body shall keep at the disposal of the national authorities the relevant documents assessing the subcontractor's qualifications and the work carried out by the subcontractor under this Directive.

3. The notified body must be able to carry out all the tasks assigned to such bodies by one of Annexes 3 to 7 and for which it has been notified, whether these tasks are carried out by the body itself or on its responsibility. In particular, it must have the necessary staff and possess the facilities needed to perform properly the technical and administrative tasks entailed in assessment and verification. It must also have access to the equipment necessary for the verifications required.

4. The notified body must have:

- sound vocational training covering all the assessment and verification operations for which the body has been designated;
- satisfactory knowledge of the rules on the inspections which they carry out and adequate experience of such inspections;
- the ability required to draw up the certificates, records and reports to demonstrate that the inspections have been carried out.

5. The impartiality of the inspection staff must be guaranteed. Their remuneration must not depend on the number of inspections carried out, nor on the results of the inspections.

6. The body must take out civil liability insurance, unless liability is assumed by the State under domestic legislation or the Member State itself carries out the inspections directly.

7. The staff of the inspection body are bound to observe professional secrecy with regard to all information gained in the course of their duties (except vis-vis the competent administrative authorities of the State in which their activities are carried out) under this Directive or any provision of national law putting it into effect.

Annex 10

CE MARKING OF CONFORMITY

The CE conformity marking shall consist of the initials "CE" taking the following form :

- If the marking is reduced or enlarged the proportions given in the above graduated drawing must be respected.

- The various components of the CE marking must have substantially the same vertical dimension, which may not be less than 5 mm. This minimum dimension may be waived for small-scale devices.

Chapter 6

SITUATIONAL REVIEW BY INDUSTRY

PHARMACEUTICAL INDUSTRY

Background

The Pharmaceutical Industry is almost totally dependent upon computer systems. Computerized laboratory devices test compounds and report to Laboratory Information Management Systems. Clinical studies are computer designed, results are computer recorded, and statistical analyses are computer calculated. Computer Aided Manufacturing Systems use robotic devices to move inventory into computerized manufacturing systems, while other computers test final product for stability and purity.

This reliance is in part a result of the need for complex regulatory documentation of all aspects of the process. But it is also a result of financial pressures to maximize production efficiency in a production process ideally suited to automation. In effect, technology makes it possible; costs make it desirable, and regulation makes it appropriate. The result is the ideal climate for automation.

As production methods grow in complexity and sophistication, and as OSHA and public health safety requirements for workers continue to tighten, pharmaceutical manufacturing will become increasingly reliant on robotic devices for production, testing, and distribution. That reliance will necessitate even greater systems controls, in much the same way that safety concerns have increased regulatory scrutiny of systems in the blood processing industry.

That move toward automation has generated increasing needs to assure control of the computers involved. As a result, the pharmaceutical industry is highly oriented toward validation, the documentation of those controls. All automated aspects of research, laboratory analysis, production, and distribution are subject to validation guidelines. Good Manufacturing Practice (GMP) and Good Laboratory Practice (GLP) inspections focus on a number of areas, but almost always include computer validation as one of their major themes.

History

Demands for evidence of computer validation were first expressed in 1983 and 1984, as an extension of process validation requirements. Early process equipment was generally electric or manual, but as electronic and automated pieces of equipment were introduced most approaches simply "tested around" the computer components. That is, a new mixer now controlled by a computerized programmable logic controller (PLC) was validated by comparing input to output and expectations, ignoring the controller in the middle. By the mid 1980s, though, this "test around" procedure had lost favor in financial auditing, and was being seriously questioned elsewhere. "Black box" testing could tell if a system component was functional at present, but without understanding the inner workings of that system or software there was no certainty of continued performance and no ability to develop appropriate controls to prevent or detect likely future problems.

In 1987, an FDA spokesperson provided the Drug Information Association with a possible timetable, later proven impractical. All systems, it was announced, should be thoroughly tested and validated by 1992, providing a five year window for compliance. Unfortunately, the key parameters and guidelines were not defined until late in 1989, and in 1992 clear definitions of systems to be included were still emerging. It was not until 1994 that the Center for Biologics Evaluation and Research and the Center for Devices and Radiological Health announced a mechanism for registering compliance (using the 510(k) medical device registration, identifying software as a medical device). In 1995, most pharmaceutical companies were still in the process of defining the appropriate validation standards for their operations.

Present

The pharmaceutical industry has been struggling to understand, define, and comply with validation requirements for more than ten years. Currently, acceptance of the need to validate all major computer systems is widespread. Most major pharmaceutical companies have established a policy or Standard Operating Procedure (SOP) for validation that includes a definition of systems to validate; a categorization of those systems with a multi-tier approach to the extend of evidence necessary, based largely upon uniqueness and importance of the systems; and methodology for accomplishing that validation. The standard validation methodology of Dr. Sandy Weinberg, first issued in 1989, has become the general basis for these policies, and for the FDA approach to auditing the results.

The most common validation model that has emerged depends upon the formation of ad hoc task groups with representatives from the user community, the technical support group, and the organization's quality assurance group. These task groups develop SOPs, design and conduct acceptance tests, develop disaster recovery and security plans, and document system and user controls. An outside audit team or internal audit group then periodically reviews the validation effort and certifies compliance to the company and the regulatory authorities.

While all relevant systems have not yet been fully validated, most companies have in place a plan for completing all validations within the next few years. In addition, the validation of new systems as they are introduced is rapidly diminishing the number of non-validated systems currently in use in the pharmaceutical industry.

Future

In the *Federal Register* of June 3, 1994, the FDA announced that as part of a review of regulations for blood establishments and blood products, it would also review other regulations applicable to blood products, 21 CFR 210 and 211 (the Current Good Manufacturing Practice regulations in the manufacture, processing, packing and holding of human and veterinary drugs). At a pharmaceutical industry meeting in March 1995, FDA staff indicated that the agency was also preparing to develop Good Manufacturing Practice regulations specifically for the Bulk Pharmaceutical Chemicals (BCPs) that are used to manfuacture drugs; these chemicals are currently regulated under the GMPs for finished pharmaceuticals. Any changes in the GMP would probably follow changes already made to the Medical Device Good Manufacturing Practices, and increase the requirements for system validation.

Whatever the outcome of proposed revisions to the drug GMPs, the future of pharmaceutical validation will include a continuation of the current plans for validation of all relevant systems and the periodic (generally annual) reaudit of validated systems. But this classic approach to validation, critical as it is, is rapidly being supplemented by an important development in the software industry: the inclusion of validation modules in LIMS, MRP, CAM, and other major systems.

An algorithm has been developed, and is currently being integrated into major software, for the industry to perform a continuous statistical co-validation monitoring on all variable data streams. This process involves the detection of unexpected changes in datapoints, in the relationships between data elements, and in the relationships between current and previous elements. The calculations are performed thousands of times per second, and flag discrepancies for investigation and possible diagnostic interpretation. A cybernetic feature allows the system to continuously "learn" from prior functionality, and to feedback that experience to current control situations.

Unlike other automated tools that simply check for consistency in software design, this algorithm identifies problems in the application of the software to specific pharmaceutical regulatory functions. When coupled with a classic validation study to develop appropriate control and use documentation, the validation algorithm provides a high level of confidence in the results of any controlled system.

In the future, this and similar algorithms will provide all software and software driven devices, including medical devices, blood processing systems, and biologics systems, with the ability to supplement classic validation efforts with continuous monitoring of functionality. This combination will meet the original regulatory intent, and will provide the necessary and continuous controls of systems that the high dependency of the pharmaceutical industry requires.

Summary

The Pharmaceutical industry has paid particular attention to the need to develop and document controls for computer systems in use, largely in response to a high dependency upon automation and to the regulatory attention that dependency has generated. Currently, most major companies have developed and are implementing a strategic plan for classic validation of all relevant systems, including internal testing, SOP and user documentation, and security. In the future, an algorithm built into major software systems will supplement these classic validation studies with a continuous monitoring of data streams, using sophisticated statistical analysis to diagnose possible system performance and control problems.

BLOOD PROCESSING INDUSTRY

Background

It is in the Blood Processing industry that regulatory concerns regarding the validation of computer systems have been most clearly focused and most rigorously pursued. This intense scrutiny is a result of an interaction of two vectors. First, the AIDS crisis has heightened public awareness of the potential danger an untested blood supply represents as a carrier of pathogens. Since there is yet to be developed any processing method that effectively sterilizes blood, undetected disease is passed from donor to recipient unimpeded.

Second, since most of the screening tests utilize sophisticated, computerized equipment to detect contaminations, and since virtually all blood processing centers rely upon computer systems to manage the process of evaluating and labeling blood products, the dependency of the blood processing industry upon computer systems is extremely high. In most blood processing centers, the decision to label as safe for transfusion or to destroy as potentially dangerous is placed in the "hands" of a computer system, well beyond conventional human controls.

In most environments, blood donors are initially screened for self-reported disease risks, and a donor profile (health history, donation history, and personal identification) is created. Blood is then "harvested" in bags for processing and in tubes for testing.

The tubes are tested for type and for a series of disease conditions, including hepatitises, syphilis, and AIDS. Simultaneously, the bags are separated into blood products, including packed cells, plasma, and whole blood.

When the testing is completed, the results of the test are married to the created products, which are then labeled for safe use or destroyed.

To accomplish these steps, a computer program with a series of software "switches" is utilized. A donation is assigned a unique number, and all switches are set as "off." The donor profile is evaluated, and the initial switch is left as "off" if a disqualifying factor is identified, or turned "on" if the profile is appropriate.

The donor identification is checked against an extensive donor deferral file, seeking a match with a donor whose blood may have been previously disqualified. If no match is found, the next switch is turned "on;" a match results in a disqualifying "off." Each additional test establishes the disposition profile of the donation, eventually resulting in a file in which all significant "offs" have been changed (safe for use) or in which one or more disqualifications remain, forcing destruction (or, in some circumstances, other appropriate disposition). Finally, the unique donation number is matched to the number on the products, and a label is generated. For all practical purposes, human personnel are not involved in the entire process.

In such an environment, non-conventional secondary human controls, in the form of the computer system controls that validation demands and documents, are critical.

History

In the early 1980s, FDA investigators observed an increased reliance upon computer systems in the processing, testing, and distribution of human blood products. As scattered but highly publicized incidents of inappropriate release of possibly AIDS-contaminated blood reached the public later in the decade, numerous industry conferences attempted to move the industry toward a more extensive and rigorous control over computer systems. The result was a widespread industry shakedown, as small blood processing centers consolidated to reach cost effective levels that would permit the extensive validation controls necessary in such a vulnerable setting.

The remaining centers have been struggling to comply ever since, working to improve systems while developing new alternatives and organizing their operations around new control principles. This reorganization is somewhat complicated by the variety of blood processing operations in place in the United States. Some, like the American Red Cross, are not-for-profit organizations with interests and goals extending well beyond blood processing. These groups generally have highly involved public boards as well as professional staffs, adding to the difficulty of change and of long-term planning for new solutions. Other blood processing centers are for-profit companies, organized much as pharmaceutical companies. These for-profit companies are responsive to regulatory compliance issues and generally can afford the cost of innovation, but they have a historical adversarial relationship with the FDA, and hence await clear and unambiguous guidelines before proceeding. Those guidelines have been long delayed in deference to the planning cycles of the not-for-profits. Finally, a number of small hospitals operate their own blood processing centers. These units fall under joint regulation of the FDA and the voluntary hospital regulatory associations, and hence are often ignored by both groups.

This mix has led to chaos, only now leading to a clear regulatory position and stiff enforcement.

Present

At the Food and Drug Law Institute's annual Education Conference in December 1993, Food and Drug Commissioner Dr. David Kessler delineated his agency's current emphasis on vigilance and scrutiny of the blood processing industry. He has committed the FDA to hold the industry to rigorous Good Manufacturing Practice standards, and announced that the agency would revoke licenses or seek court injunctions against blood centers to ensure compliance with those standards. In addition, he emphasized that CBER would ensure compliance with the "Draft Guideline for the Validation of Blood Establishment Computer Systems" issued in September 1993.

In March of 1994, the FDA sent the following letter to all software vendors selling to the blood processing industry:

To: Blood Establishment Computer Software Manufacturers

Dear Sir/Madam:

The purpose of this letter is to advise you that the Food and Drug Administration (FDA) considers software products intended for use in the manufacture of blood and blood components or for the maintenance of data that personnel use in making decisions regarding the suitability of donors and the release of blood or blood components for transfusion or further manufacture to be devices under section 201(h) of the Federal Food, Drug, and Cosmetic Act (the Act) [21 U.S.C. §321(h)].

These software products are designed to receive and store data used by blood establishments during the manufacturing process, from determining donor suitability through component processing, testing, and labeling to product release. They are designed to receive and store data regarding blood donor status, including donors' answers to health history questions and the results of laboratory tests, including blood grouping and typing, hepatitis, and antibody to the human immunodeficiency virus (anti-HIV). Blood establishment personnel later access and use the data to determine whether donors are suitable and whether blood or blood components are free from disease-causing agents transmissible by blood, such as hepatitis and HIV. In addition, the data are used to label blood and blood components prior to release for use in hospitals and other health care facilities or for further manufacturing. Because they aid in the prevention of disease (e.g., hepatitis, HIV, etc.) by identifying unsuitable donors and preventing release of unsuitable blood and blood components for transfusion or for further manufacturing use, these software products meet the definition of device under the Act.

Facilities that manufacture and distribute these software products are subject to the device provisions of the Act and FDA's device regulations, including establishment registration, product listing, premarket notification or approval, Current Good Manufacturing Practices (CGMP), and adverse event reporting. FDA's CGMP regulations for devices appear at Title 21, Code of Federal Regulations (CFR), Part 820 and the MDR regulations at 21 CFR, Part 803.

According to FDA's information, your facility manufactures software intended for use in the manufacture of blood and blood components. Consequently, you are required under the Act to register your establishment and list your devices. In addition, your manufacturing operations are required to be in compliance with CGMP for devices, and you must report adverse events and other problems as required by FDA's Medical Device Reporting (MDR) regulations.

We are enclosing a registration package for your convenience. We will forward a device listing package to you in the near future. When completing the device listing form (Form FDA 2892), identify your product as Software, Blood-Bank, Stand Alone Products, Product Code 75MMH. The registration form should be submitted within 60 days of receipt of this letter if you intend to continue to distribute software products to blood establishments for use in manufacturing.

In addition, you are required to submit a premarket notification or application for premarket approval for each of your devices unless you can demonstrate that you commercially distributed the devices in interstate commerce prior to May 28, 1976 (the date of enactment of the Medical Device Amendments) and have continued to do so without any significant changes to these devices. If you claim such preamendment status for any product(s), please complete only the registration form, and send it to: Center for Devices and Radiological Health (HFZ-300), 2098 Gaither Road, Rockville, MD, 20850.

If you do not intend to submit a premarket submission, and intend to submit proof of the claimed preamendment distribution, this information should be sent to: Center for Biologics Evaluation and Research (CBER), Division of Blood Applications (HFM-370), 1401 Rockville Pike, Rockville, MD, 20852-1448. Finally, if you do not currently manufacture software products for blood establishments, please advise CBER promptly.

When we forward the product listing information to you, we will include guidance on how to prepare your premarket submissions. If you have questions about the content or format of a premarket submission once you have reviewed our guidance, CBER staff are available to help answer such questions. Premarket submissions should be submitted to CBER no later than March 31, 1995.

In the interim, FDA will continue to conduct inspections of blood establishment software vendors. These inspections will include, among other things, a review of standards for software development, testing, validation, and quality assurance. The primary focus of these inspections will be to assess compliance with the CGMP regulations for devices (21 CFR, Part 820).

The agency will also review and assess your procedures for investigating reports of product problems and defects and for implementing and evaluating corrective actions. We will also review and assess your procedures for notifying your customers and the agency when you take corrective actions.

Please be advised that during this interim period if a manufacturer of software products for blood establishments is not making good faith efforts to comply with the Act and FDA's regulations as stated above, the agency will not hesitate to take appropriate steps to bring the firm into compliance.

If you have questions concerning: (1) the preparation of the establishment registration and device listing notification, contact Center for Devices and Radiological Health, Division of Small Manufacturers Assistance (HFZ-220), at 301-443-6597, or (2) guidance for premarket submissions, contact Center for Biologics Evaluation and Research, Division of Blood Applications (HFM-370), at 301-594-2012.

Sincerely yours,
Kathryn C. Zoon, Ph.D.
Director
Center for Biologics Evaluation
and Research

On February 10, 1995, Dr. Zoon sent the following letter to the same software manufacturers:

To: Blood Establishment Computer Software Manufacturers

Dear Sir/Madam:

The purpose of this letter is to notify you of the revised schedule for compliance with the various provisions of the Federal Food, Drug, and Cosmetic Act for premarket submissions for blood establishment computer software products regulated as medical devices. The schedule has been developed after careful evaluation of the concerns expressed by the software manufacturers, the impact of the regulatory initiative on blood establishments, and the public health significance of assuring the safety and quality of this software.

In a letter dated March 31, 1994, the FDA stated that the agency considers software products intended for use in the manufacture of blood and blood components or for the maintenance of data that personnel use in making decisions regarding the suitability of donors and the release of blood or blood components for transfusion or further manufacture are devices under section 201(h) of the Federal Food, drug, and Cosmetic Act (the Act) [21 U.S.C. §321(h)]. This initiative was also described in a Federal Register (FR) notice dated August 31, 1994 (59 FR 44991)

As a medical device manufacturer, you are currently required under the Act to register your establishment and list your devices. In addition, your manufacturing operations are required to be in compliance with CGMP for devices, and you must report adverse events and other problems as required by FDA's Medical Device Reporting (MDR) regulations.... These and other specific points relating to establishment inspections noted in the March 31, 1994, letter and the August 31, 1994, Federal Register notice remain unchanged.

In these documents, FDA stated that manufacturers of blood establishment computer software would be required to submit to the Center for Biologics Evaluation and Research (CBER) a premarket notification or application for premarket approval for each of their devices no later than March 31, 1995. The agency received numerous responses from organizations representing both software manufacturers and blood establishments. The principal concern expressed in these responses related to the requirements for premarket clearance or approval for blood establishment computer software products. The concerns included, but were not limited to,the difficulty of expeditious compliance with the requirement for premarket clearance or approval, the need for additional, detailed guidance to be used in the preparation of premarket submissions for these specific software products, and additional time needed to remove software from use by blood establishments in situations where a software manufacturer does not intend to seek premarket clearance or approval for the particular product.

After careful evaluation of the needs expressed by the software manufacturers and the impact of this regulatory initiative on blood establishments, the FDA has concluded that a one year extension of the March 31, 1995, deadline is warranted. Therefore, premarket submissions should be submitted to CBER no later than March 31, 1996. The extension period for premarket submissions, does not, however, affect other responsibilities of the computer software manufacturers and distributors who are subject to the device provisions of the Act and implementing regulations as previously stated.

To effectively implement this important and complex regulatory program, the agency intends to work with industry to clarify the expectations concerning premarket submissions through issuance of guidance. We also plan to have a continuing dialogue with affected establishments and industry organizations.

Also, within this extension period, it is the FDA's expectation that vendors and blood establishments will cooperatively conduct all transitions from software products for which premarket clearance or approval will not be sought to software products for which premarket clearance or approval is being actively pursued. These transitions should also be conducted in an orderly and effective manner so that they have minimal impact on the blood establishment's operations as they relate to the identity, safety, purity, and quality of blood products. These transitions should also be completed by March 31, 1996.....

Sincerely,
Kathryn C. Zoon, Ph.D.
Director
Center for Biologics Evaluation and Research[1]

These letters have explicitly declared software used in the processing and testing of blood (and, by extension, in other FDA regulatory areas) to be medical devices, subject to the Food, Drug, and Cosmetic Act and the Safe Medical Devices Act of 1990's requirements for establishment registration, product listing, adverse event reporting, premarket approval or premarket notification, and the documentation of software validation. The letters also announced that the FDA would conduct inspections of vendors supplying software to the blood processing industry, and that these inspections "will include, among other things, a review of your standards for software development, testing, validation, and quality assurance. Sixty days after the release of the first letter, the FDA began visiting software vendors to enforce those validation guidelines.

Concurrently, the FDA increased the frequency and extent of its visitations to blood processing centers, asking for clear evidence of system controls and validation documentation. Centers unable to comply have been cited for violations of the Good Manufacturing Practice (GMP) regulations, regardless of their ownership or organizational status.

This two-prong thrust, pressuring both the vendors and users of systems, has provided a clear message to the blood processing industry. While full compliance will no doubt require some time as well as expense and effort, there is now industry certainty that only systems that can be documented as validated can be sold, supported, or used. After almost ten years of less-than-subtle FDA signals, the message has finally been received.

---

[1]Letters from Dr. Kathryn C. Zoon, Director of the Center for Biologics Evaluation and Research, to Blood Establishment Computer Software Manufacturers, March 31, 1994, February 10, 1995.

Future

In the *Federal Register* of June 3, 1994, the FDA announced that as part of a review of regulations for blood establishments and blood products, it would also review other regulations applicable to blood products, 21 CFR 606 (the Current Good Manufacturing Practice regulations for blood and blood components). Any changes in the GMP would probably follow changes already made to the Medical Device Good Manufacturing Practices, and increase the requirements for system validation. A July 1995 FDA "Talk Paper" discussed FDA's attitude toward new methodologies applicable to blood products:

> FDA has provided guidance covering the computer software that governs much of what happens to blood products in many modern blood centers -- including the labeling and quarantining of products. And FDA is proposing regulations that will require registered blood banks, as well as licensed facilities, to report to FDA any problems ("errors and accidents") within 72 hours, so that appropriate follow-up measures can occur promptly.[2]

The increased cost of well designed, controlled, and validated systems will probably result in a further consolidation in the blood processing industry. Small centers, processing ten thousand or fewer bags of blood a year, will have great difficulty demonstrating cost efficiency. Even the largest facilities are likely to have management problems as they try and balance community service priorities, volunteer involvement, and strict federal requirements.

Vendors, too, are likely to decrease in number. Several vendors have already re-evaluated their opportunities, unwilling to invest in one market segment when their same software can be sold in other segments without the overhead and liability that blood regulation implies. Those vendors whose products were equally suited to blood processing and other non-regulated applications are simply withdrawing from the market.

In the not-for-profit environment, discussions are still proceeding. The American Red Cross has publicly announced a plan to adapt a single system for all of its processing sites, while simultaneously reducing the number of those sites. Interim plans, during the development and phase-in of that adapted system, are still hazy.

Several large, independent blood processors have encouraged the formation of a remote teleconnected central blood systems organization that is offering validated and controlled computerized services to blood facilities all over the country. This "time share" concept may well prove the ideal trend of the future, providing the tight controls necessary in a regulated and critical environment but placing those controls in the hands of experienced and focused computer professionals rather than medical personnel.

---

[2]Jim O'Hara and Lawrence Bachorik, "The Safety of Blood and Blood Products." FDA Talk Paper T95-36, July 12, 1995.

Summary

The Blood Processing industry is still in transition. It is perhaps poetically appropriate that the validation "leading edge - bleeding edge" is in blood; regardless of the pun potential, the high reliance upon systems and high vulnerability of this industry has caused a sharp regulatory focus. Organizational complexities related to the mixed status of blood processing centers have delayed the implementation and enforcement of the guidelines resulting from that focus, but official patience has been exhausted. Today, and tomorrow, blood processing centers will utilize compliance-validated software selected from 510(k) (medical device Premarket Notification Submission) approved systems, or will be cited for GMP violations.

BIOLOGICS MANUFACTURING

Background

In the regulated pharmaceutical industry, biologics are commonly defined as any substances which are obtained from animal products or other biological sources for use in the treatment or prevention of disease. They are governed by the Good Manufacturing Practice (GMP) Regulations which have been in existence, inclusive of various modifications, since 1963. The current GMP regulations were initially published in 1978 and include subsequent revisions.

As the use of computers dramatically increased to support the demand for automation in the manufacturing of biologic products, those recognized principles and standards used in conjunction with process validation formed the basis for demonstrating evidence of control in the use of computer systems.

History

The regulation of biologics in the U.S. is mandated in the *Code of Federal Regulations* (CFR), Title 21 as follows: Part 210, Current Good Manufacturing Practice in Manufacturing, Processing, Packing, or Holding of Drugs, Subpart A -- General Provisions and Part 211 Current Good Manufacturing Practice for Finished Pharmaceuticals, Subpart A - General Provisions.

It is apparent that the discipline of process validation became the basis for providing evidence of control with respect to computer systems involved in the manufacture of biologic products. In many cases, the procedures to confirm processing integrity and to provide assurance of computer system functionality and inherent resident data were applied from previously acknowledged standards, practices, and techniques. These methods were used to regulate the use of process control equipment in production operations.

The Division of Field Investigations, Office of Regional Operations, Office of Regulatory Affairs and the Division of Manufacturing and Product Quality, Office of Compliance, Center for Drug Evaluation and Research originally issued the *Guide to Inspection of Bulk Pharmaceutical Chemicals* in April 1984. These reference materials and training aids were intended for use by field investigators; they were revised in February 1987 to include changes submitted by the Pharmaceutical Research and Manufacturers Association (PhRMA, formerly the PMA). In May 1987, the FDA published the *General Principles of Process Validation* to cover drugs and medical devices. These principles provided those practices generally accepted by the FDA, along with guidance to assist manufacturers in fulfilling process validation requirements contained in 21 CFR Parts 210 and 211.

Before being marketed in the United States, new biological products must be approved by the FDA. Based on the Safe Medical Devices Act of 1990, the FDA implemented a mechanism for extending regulation to the registration of products that incorporate drug, device, and device/biologic combinations. As part of the approval process, manufacturers of biological products are required to submit to the Center for Biologics Evaluation and Research (CBER) adequate data concerning the safety and efficacy of its products. These submissions often include reports, analyses, tabulations, and case reports.

In July 1990, CBER published a number of considerations with respect to computer-assisted submissions for license applications. These materials provided biologic product and establishment license manufacturers with information regarding computer-assisted submissions. In order to evaluate such computer assisted submissions, CBER instituted a program to acquire practical experience. Procedures were developed leading to generalized acceptance of portions or complete license application submitted to the center on computer-readable media. By using computer technology, it was envisioned that efficiency improvements in the transmission, storage, retrieval, and analysis of data submitted to CBER as part of the development, licensing, and marketing of a biological products could be realized.

CBER issued an updated *Guide to Inspection of Bulk Pharmaceutical Chemicals* in September 1991. Revisions were included that were sought by a number of federal regulatory and industry association sources. The relevant stipulation regarding validation process and control procedures in this document is the following:

- An important factor in the assurance of product quality includes the adequate design and control of manufacturing process. Each step of the manufacturing process must be controlled to the extent necessary to assure that the product meets established specifications. The concept of validation is a key element in assuring that these quality assurance goals are met.

Many guidelines on process validation were made applicable to bulk pharmaceutical chemicals (BPCs) to assure that they were manufactured in accordance with GMPs. Sections in the *Guide to Inspection of Bulk Pharmaceutical Chemicals* became an integral standard practice for computer systems supporting biologics manufacturing. The areas of validation consideration included:

- A formal process change system.
- Standard operating procedures.
- Independence of quality assurance to have responsibility and authority for final approval of process changes.
- Performance and recording of tests and results.

- Calibration and maintenance of equipment.
- Independence of an operational function having the responsibility and authority to reject quality assurance activities including procedure approvals, investigation of product failures, process change approvals, and product record reviews.
- Procedures to assure that the correct label is applied.
- Formalization of a written testing program.
- Determination of a test sampling.
- Documentation of the manufacturing process, including a written description of production records.

As a practical matter, when computers and other sophisticated equipment are employed in the biologic manufacturing process emphasis focuses on:

- Systems and procedures that demonstrate that equipment is in fact performing as intended.
- Checking the calibration of the equipment at appropriate intervals.
- Retention of suitable backup systems.
- Assurance that changes in the program are clearly documented and are made only by authorized personnel.

The objective of the *Biotechnology Inspection Guide, Reference Materials and Training Aids*, issued by the FDA in November 1991, was to provide a basis to determine whether a biologic manufacturer was operating in a state of control and was in compliance with federal regulations and requirements. One important aspect of an inspection was identification of system defect or failure. The inspection approach of reviewing standard operating procedures to assure adequacy of controls became the criterion for performing such regulatory review and audit of biologic manufacturers.

Present

In December 1993, Dr. David Kessler, Commissioner of the FDA, stated that CBER would emphasize compliance with computer system validation guidelines issued earlier in the year. In March 1994, CBER Director Dr. Kathryn Zoon stated in a letter to software vendors that computer system software would be treated as a medical device, a significant statement of policy.

The first draft of the *Application of the Medical Device GMPs to Computerized Devices and Manufacturing Processes, Medical Device GMP Guidance For FDA Investigators* was issued in November 1990 by the FDA Office of Compliance and Surveillance, Division of Compliance Programs. This document outlines requirements which apply to the control of computerized manufacturing and quality assurance systems. It relates to manufacturers who utilize automated systems for manufacturing, quality assurance, and/or record keeping purposes. It also refers to manufacturers of medical devices that are driven or controlled by software. This document is intended to provide assistance to FDA investigators and to supplement FDA document 84-4191, *Medical*

Device GMP Guidance for FDA Investigators. As such, these materials were designed to supplement FDA issued compliance policy statements and references.

Future

The use of computer systems has integrated into every area of the manufacture of biologics. The reliance on computer systems for automation will continue to increase as computer integrated manufacturing (CIM), electronic batch record (EBR), and automated document management systems are used to their full potential and capability by the pharmaceutical and medical device industries. Predictions can be made that even the notorious Title 21 CFR Part 211 Subpart J Section §211.186 calling for "full signature, handwritten" will be modified to enable acceptance of electronic signatures.

As mandated by Section 351 of the Public Health Service Act; the Federal Food Drug, and Cosmetic Act, Section 502; and 21 CFR Part 600 (licensing of the manufacture of biological products) can only be processed when a completed "Application For Establishment License For Manufacture of Biological Products" form is received. Section nine of the updated Department of Health and Human Services, Public Health Service, Food and Drug Administration, FDA Form 3210 (5/94 Form Approved: OMB No. 0910-0124, Expiration Date: April 30, 1997) specifically requires information relating to computer systems used in the biologic manufacturing process. Questions requiring response are:

A. Describe all processing steps which are computer controlled. Describe each system used in these steps. Include a schematic diagram of the system and the functional statement.
B. Describe any computer system used to control the tracking and/or status of raw materials, in-process materials, or final product.
C. Describe the validation and security of the hardware and software of each system.
D. Indicate where the software was developed and describe procedures for program update.

Section fourteen of the same form provides calibration and validation requirements:

A. For all major systems (except those where validation has been previously described in another section) such as waste disposal, clean-in-place (CIP), steam-in-place (SIP), compressed air, etc., please provide a description of the validation studies performed including the results. Include a table of all validations done and the related protocols and a summary of the results for each major system. Include:

   1. Installation Qualification
   2. Operation Qualification
   3. Performance Qualification
   4. Revalidation

B. For major pieces of equipment (autoclaves, dry heat ovens, lyophilizers, hoods, etc; reference chromatographic columns to Product License Application (PLA) if appropriate) please provide a description of the validation studies performed including the results. Include a table of all validations done and the related protocols and a summary of the results for each piece of equipment. Include:

1. Installation Qualification
2. Operation Qualification
3. Performance Qualification
4. Revalidation

C. For critical processes such as cleaning of product contact parts and major equipment in product contact, disinfection, product changeover, etc., please describe the validation studies that were performed and the results of each.

Section fifteen describes the records requirements:

A. Describe the relationship between the master production record and each batch production record.
B. Explain preparation of records and sign off authority within the organization.
C. How are records assembled, stored and accessed?
D. Who reviews the records for each lot prior to release of the lot? Describe the method of documenting such reviews.
E. How long are records kept?
F. Describe the records used for distribution. How do these records allow for efficient recall?
G. Describe the complaint file and adverse reaction reporting procedures.
H. Describe audit procedures and when and how trend analyses are performed.

In addition, applicants must provide an overview of the GMP training program for biologics establishment employees.

Summary

Certainly, there is increasing pressure to provide evidence of control for the development, installation, and utilization of computer systems. This pressure has produced an appropriate response in the pharmaceutical, medical device, and biologics industries regulated by the FDA with the result being regulation, field inspection, and impending threat of regulatory response. Since computer systems are increasingly used by manufacturers of biologic products in their production operations, the connectivity between computer systems, regulation, and economic realities becomes inseparable. Clearly, validation becomes a solution to protect the best interests of all parties concerned with biologic manufacture.

Validation reviews of computer systems correlate directly to providing evidence of control, with the ultimate goal of producing a total quality product. The strategic validation protocol, tactical validation plan, and the entire validation process serve as guideposts leading biologic manufacturers away from defect detection and towards defect prevention.

MEDICAL DEVICE INDUSTRY

Background

The Medical Device Industry encompasses a wide range of apparatuses, from the most sophisticated computerized imaging systems through the lowest, technologically mundane bandages. In this universe of devices, validation has had a significant impact in three ways:

- in the approval and control of computerized devices;
- in the control of computerized systems involved in the manufacture and distribution of all medical devices; and
- in the classification of software itself as a medical device.

Medical devices that utilize computers as either control or data systems must submit evidence of the validation of the computer elements of the product as a part of the Pre-Market Approval (PMA) or Premarket Notification (510(k) submission) process. The validation of software, and evidence of the controls built into the code, use, or design is an important review element, and it seems that the scrutiny of that validation evidence is most rigorous when the device is a direct (no professional human intervention) delivery system of medical treatment, when the computer elements are utilized in a critical decision-making capacity, or when the computer serves as the safety or protective mechanism for the device.

Systems involved in the testing, manufacture, labeling, recall, or distribution of medical devices are required to be supported with appropriate levels of validation evidence. The depth of that evidence -- and the rigor with which the requirement is enforced -- is dependent largely upon the risks or hazards associated with the final product and with the contribution made to the safety of that product by the computer system. A system that counts and boxes band-aids may be subject to only minimal controls; a system that checks the sterility of those band-aids would be, appropriately, subject to much more rigorous validation testing.

In recent years, federal regulators have begun to discuss the classification of computer software as a medical device when that software is utilized in the production or testing of regulated products, including biologics, other medical devices, and pharmaceuticals. The theory underlying this classification derives from the broad definition of a medical device as anything utilized in providing medical care, triggering FDA Commissioner David Kessler's famous, if melodramatic, comment that a pencil, used by a physician to record data on a patient's chart, is a medical device subject to FDA regulations. While such a statement is clearly an exaggeration intended for effect, the principle is sound.

Current regulation does give the FDA broad interpretative powers, and the extension through those powers to the regulation of software is not absurd. In situations in which that software is deeply ingrained in decision-making processes involving human health and safety, either through diagnostic testing, labeling, manufacture, or clinical analysis, the FDA's regulatory responsibility is clear and appropriate.

History

Validation has been a criteria for medical device registration since the early 1960s. In fact, registration of a medical device is ultimately nothing more than a validation process -- a third-party review of the design, control, and performance aspects of a (medical) system. In the early 1970s, the concept of validation was formalized as an element of process review in the production of medical devices (and pharmaceuticals). Subsequent decades have expanded the concept in depth to the review of software used to control devices and their manufacture.

The key element in the validation of medical device software is the hazard analysis, first introduced in the early 1990s. Evaluating the potential danger (in degree and likelihood) of a system failure, varying levels of control may be acceptable, and hence acceptably tested and documented. The same manufacturing process might be used to test batteries in internal pacemakers and in external electronic thermometers. But a failure of the testing software in the pacemaker case may prove fatal, or may require emergency surgery, while a failure of that software in the thermometer may result in an inoperative diagnostic instrument. Under the hazard analysis umbrella, proof of control is more critical, and would require higher standards of evidence, in the first of these two cases.

With a reorganization of the Center for Devices and Radiological Health (CDRH) at FDA in 1993, and subsequent higher reliance upon PMAs and the 510(k) registration, validation has taken on a much greater importance than was previously assigned. Field investigators are checking the manufacturing of devices, including the validation evidence in support of systems involved in that manufacturing process, while registration officers are examining and reviewing validation evidence with a high level of scrutiny. In many ways, validation concerns, particularly in the review of systems, have emerged as the main issues in the regulation of medical devices.

Present

Currently, the FDA is informing medical device manufacturers of the need to file validation evidence as a part of the 510(k) submission process. Spot inspections of medical devices production facilities utilize the Good Manufacturing Practices (GMPs) as a guideline, and include the requirements for validation of software used in the manufacturing and tracking process as an emphasis of those inspections. In situations in which a computer is the main vehicle for potential product recalls, validation of that computer system is given special scrutiny.

Since other regulatory divisions of FDA are now calling for registration of software used by their regulatees (see the section on the Blood Processing Industry, above), CDRH is expecting an increase in the number of software registrations over the next five years. The Center for Biologics Evaluation and Research (CDRH), which regulates the blood processing industry, has been particularly active in this area. GMP inspections of software vendors have been held, and citations of GMP violations have been issued. A notification program informing all software vendors of their registration responsibilities has been introduced, and those registrations are now under receipt and review.

In a December 1994 speech to the Food and Drug Law Institute, the FDA's Dr. Kessler announced that the agency had significantly reduced its backlog of 510(k) applications, and was planning on proposing more low-risk devices for exemptions from the necessity of submitting 510(k) applictions. Rather than reduce the importance of validation, however, this means that for those products that must contiue to submit 510(k)s, validation of processes and software will become more and more important in proving safety and efficacy. In April 1995, CDRH wrote to medical device manufacturers about the changes the FDA was making to its program linking class I and class II medical device 510(k) applications with GMP requirements, changes which wree publihsed later in the year as part of the FDA's Standard Compliance Program 7382.830 for medical device GMPs.

Future

There is always a lag period between an announced FDA interpretation of a regulation and the widespread enforcement of that interpretation. During that period, the FDA and industry publicize the interpretation and the methods of documenting compliance.

When the new interpretation impacts upon an industry segment not previously regulated, that lag time is necessarily lengthened, since affected organizations are not generally monitoring the communication channels normally used by FDA to publicize new requirements. Because the software industry has not been subject to any previous regulation, extension of the GMPs and registration requirements to software used in the manufacture or testing of regulated products will be a slow and gradual process. By 1996 virtually all blood processing software will be registered as medical devices; by 1998, most software used by the pharmaceutical and biologics industries will similarly be in compliance.

To achieve that compliance, some software companies will develop and document their internal testing and control procedures, and register that evidence of compliance with FDA. Still other software companies will withdraw from the market, rather than risk operating in a now-regulated environment. These companies will eventually be replaced by companies with FDA regulatory experience, introducing new and compliant products.

All of this shift, re-introduction, and registration will be managed and defined through the definition of software as a medical device, subject to the PMA and 510(k) requirements.

Summary

Medical devices utilizing software will continue to be regulated; software used in the manufacture of medical devices will be subject to the validation requirements of the GMPs. The real change in medical device validation is an outgrowth of the definition of virtually all software used in the pharmaceutical and biologics industries for manufacture, testing, clinical studies, laboratory management, labeling, tracking, and design as medical devices. This "software as a medical device" position is forcing 510(k) registration, which includes a clear definition of, and requirement for, the validation of the registered software. This will become even more important as the CDRH moves forward on establishing design controls as the cornerstone of good manufacturing practice regulations.

VETERINARY PRODUCTS

Background

The majority of drug regulation responsibility belongs to the FDA. New drugs -- both prescription and over-the-counter -- must be approved by the FDA before being marketed. The FDA is responsible for ensuring the safety and effectiveness of drugs and overseeing their manufacture. These responsibilities are carried out under the Federal Food, Drug and Cosmetic Act. Veterinary products -- such as feeds, pet foods, and veterinary drugs and devices -- come under FDA's jurisdiction. However, the U.S. Department of Agriculture's Animal and Plant Health Inspection Service tests and licenses all animal vaccines and serums.

History

Because of the lower standards used in the 1940s and 1950s, and the subsequent need for improving these standards, the 1962 amendments to the Food, Drug and Cosmetic Act required firms to establish that their products were effective as well as safe.

In response to the need for a systematic approach to evaluating the effectiveness and safety of new veterinary drugs, the New Animal Drug Application (NADA) Approval Process was adopted. The NADA assembles data to provide the Center for Veterinary Medicine (CVM) reviewers a complete rationale for approval of veterinary drugs by defining the ingredients, the manufacturing procedures and quality control checks, the environmental impact of the manufacturing process and use of the drug, the safety and effectiveness of the drug, and a summary for publication which relieves the FDA from the burden of defending decisions made behind closed doors.

The process of approval is a dynamic one, affected by scientific and technical developments, personalities of the regulatory staff, social trends and -- very importantly -- by court decisions in cases where disputes between the FDA and the regulated industry or the public have been decided. In reality, the "approval process," such as it is, never ends, because commitments made in the original NADA and subsequent supplemental filings require continued surveillance and evaluation by regulatory personnel.

Present

Eleven major sections comprise the NADA. Each section assists different FDA personnel to analyze the legal and scientific grounds for ultimate acceptance of a given product.

- *Section One* identifies the submission as an original or supplemental application and includes the applicant's name and address, the active ingredient's generic name, and the proposed product brand name.
- *Section Two*, the Table of Contents and Summary, organizes the NADA in order to introduce the active ingredients in the product and provide a summary of the pharmacological rationale upon which to base the drug review.
- *Section Three*, Label Copy, is the sponsor's proposal for product labeling which would permit the intended user (whether lay person or veterinarian) to determine conditions for which the drug is to be used, the proper route of administration and dosage level, and the frequency of use.
- *Section Four*, Components and Composition. The information provided in this section allows the FDA reviewer to check the batch formula quantitatively against the composition of the final dosage unit and facilitates a quick review of all ingredients, active and inactive, including those removed during the manufacturing process itself, such as water removed during a drying process.
- *Section Five*, the Manufacturing Section, provides crucial information concerning the quality and reliability of manufacturing procedures used in drug formulation and packaging. The sponsor must submit a certificate of compliance with the Good Manufacturing Practice (GMP) regulations codified under 21 CFR 210 and 211. This is usually a letter from a responsible company official affirming the sponsor's compliance with the regulations.
- *Section Six*, the Samples Section, states that samples of the drug components and/or the finished drug product may be requested.
- *Section Seven*, the Analytical Methods for Residues Section, requires the firm to specify a method for determining the quantity and identity of the drug, its impurities, and its metabolites present in edible tissue of treated food animals.
- *Section Eight*, the Safety and Effectiveness Section, comprises evidence demonstrating that the drug product is safe and effective when administered according to label directions. The data in this section include information regarding laboratory trials, chronic and acute toxicity tests, and clinical or field observations of the drug.
- *Section Nine*, Good Laboratory Practice Compliance, applies to each non-clinical laboratory study contained in the application. In this regard, the NADA should contain either a statement that each study was conducted in compliance with the Good Laboratory Practice regulations set forth in 21 CFR 58, or, if the study was not conducted in compliance with those regulations, a statement that describes in detail all differences between the practices used in the study and those required in the regulations.
- *Section Ten*, the Environmental Assessment (EA), details the probable and potential environmental impact of all aspects of the manufacture, distribution, administra-

tion, metabolism and excretion of the product. The preparation of an EA is an involved process, but in many cases some portion of the requirement can be waived under an exemption.

- *Section Eleven*, the Freedom of Information (FOI) Summary, completes the NADA. The purpose of this section is to provide as much information to the general public concerning the approval of new animal drugs without jeopardizing the applicant's proprietary rights. Failure to submit an adequate FOI Summary constitutes grounds for refusal to approve a pending NADA.

What happens after the separate NADA sections are assembled is even more important to the future of the new animal drug product. Once the sponsoring firm has satisfied itself that the NADA is adequate, the file is forwarded in triplicate to CVM. The three copies are logged into the Document Control Unit, a large storage and dispersal library, where they are assigned an NADA number. This six-digit number permanently identifies the newly-submitted file. The NADA is then routed to the Office of New Animal Drug Evaluation, where it is given to the proper CVM statisticians and others who are involved in various aspects of the review process, depending on the nature of the drug and whether it is intended for use in food-producing species. This process takes about 180 days. If acceptable, the NADA is deemed approved when either a regulation to that effect is officially published in the *Federal Register* or when the sponsor firm receives an approval letter from the FDA.

One or more of the 425-plus discrete chemical entities approved for veterinary use by the CVM is incorporated into each of the 1,350 approved NADAs which exist today. Many others have emerged long enough to be published as approved and then have been withdrawn. In reality, the "approval Process" never ends, because commitments made in the original NADA and subsequent supplemental filings require continued surveillance and evaluation by regulatory personnel.

Future

Currently, any documentation in support of the validation of computer equipment critical to the manufacture or handling of a veterinary drug product is submitted as part of the response in Section Five (the Manufacturing Section) of the NADA. This situation may change in the near future, however. If the proposed changes to 21 CFR 210 and 211 (see discussion above, in the Pharmaceutical Industry section) include significantly increased requirements for validation of computer systems, this will also affect veterinary product manufacturers.

Although there are currently no requirements for medical devices manufactured for veterinary use to comply with medical device GMPs, this situation, too, may be undergoing changes at the FDA. A recent (4/7/94) warning letter to a veterinary device manufacturer recommended "that veterinary devices be manufactured in a state of the art facility using the device GMP as guidance."

ENVIRONMENTAL LABORATORY

Background

Laboratories that provide data to the Environmental Protection Agency (EPA) are governed by the jurisdiction, requirements, and specifications of the EPA program(s) which they support. Consequently, standards for data management procedures concerning automated equipment, system and application software, and associated operating environments which maintain these program(s) are included under the scope of the agency's authority. The objective of this chapter is to present the background and status of validation as it relates to environmental laboratories regulated by the EPA.

History

Certain EPA programs adopted and required laboratories to follow Good Laboratory Practice (GLP) regulations to provide confidence regarding the integrity of laboratory-generated data produced through manual operations, procedures, and practices. Over the years, as the GLP guidelines were implemented, they became accepted by agency and independent contract laboratories as well as by private industry. These groups submitted information and data to the EPA. Based on the results of proven field experience, these standards became, and continue to represent, the expectations and requirements for laboratory management and performance. The EPA's authority to impose controls over laboratories dealing with the agency was initially stipulated in the following federal legislation and published documents:

- Computer Security Act of 1987. Public Law 100-235, January 8, 1988.
- Environmental Protection Agency System Design and Development Guidance. OIRM 87-02, June 1989.
- Environmental Protection Agency Data Standards for Electronic Transmission of Laboratory Measurement Results (EPA Order 2180.2,12/10/87).
- Environmental Protection Agency Security Manual for Personal Computers, December 1989.
- Toxic Substances Control Act (TSCA); Good Laboratory Practices. 40 CFR part 792.
- Federal Fungicide, Insecticide and Rodenticide Act (FIFRA); Good Laboratory Practices. 40 CFR Part 160.
- Findings of EPA's Electronic Reporting Standards Workgroup.

Present

As the use of computer power increased and replaced manual laboratory operations, the EPA had neither the standards to guide laboratories nor the means to provide direction for agency auditors and inspectors. Findings from investigations by the EPA's Office of Inspector General and Office of Information Resources Management (OIRM) raised concerns regarding the integrity of data received by the EPA and highlighted the need for standard data-management practices. The end-result culminated in The United States

Environmental Protection Agency, Office of Administration and Resources Management to publish and issue the draft *Good Automated Laboratory Practices* on December 28, 1990. This two-part document quickly became known by the acronym GALPs.

The GALPs is a guidance document providing minimum standards, practices, and procedures for Laboratory Information Management (LIMS) and Laboratory Data (LDS) computer systems used to collect, analyze, process, transmit, report, summarize, maintain, and/or manipulate data for EPA health and/or environmental programs. They provide specifications regarding the demonstration of evidence of control in 14 areas of validation consideration: Personnel; Laboratory Management, Responsible Person, Quality Assurance, Facilities, Equipment, Security, Standard Operating Procedures, Software, Data Entry, Raw Data, Records and Archives, Reporting, and Comprehensive Ongoing Testing. Recommendations are presented to ensure data integrity in automated laboratory operations and an implementation guidance serves as a tool for assuring successful and cost-effective compliance with the GALPs.

The topical areas of coverage of GALP validation acceptance criteria fall into distinct categories. When taken as a composite, they define requirements that would enable laboratories to demonstrate control over activities which develop, maintain, and support computerized environments. Standards in the GALPs are based upon a foundation of the following six concepts for computer systems which should provide:

- A method of assuring the integrity of all data.
- Confidence that the formulae and decision algorithms employed by a system are accurate and appropriate.
- An audit trail, which allows the tracking of data entry and modification to the responsible individual.
- A consistent and appropriate change control procedure capable of tracking system operation and application software.
- A requirement for establishment of and compliance to standard operating procedures.
- For alternate plans of recovery operation in the event of system outage or failure.

Together, these concepts specify the evidence of control defined by the GALPs.

Recognizing that there is an ever-growing availability of a wide variety of computer system design, technology, laboratory purpose, and application functionality, the GALPs were formulated as general guidelines to enable the evolution of alternative control strategies which could not have been anticipated by the original guideline standards. Therefore, the GALPs are intended to serve as guideposts in the interpretation of the standards and as evaluation criteria in considering proposed and warranted forthcoming equivalencies. The August 1995 edition of EPA's *Good Automated Laboratory Practices*, the draft GALPs will serve as the most comprehensively organized set of validation standards for computer systems used by environmental laboratories.

Future

Regarding implementation, adoption of the GALPs is dependent solely on the requirements of the EPA programs under which participating laboratories fall. Furthermore, compliance to the GALPs is also contingent upon consideration of the jurisdiction and by the specifications of the governing EPA program. In those instances where the GALPs are not requirements of a particular EPA program, compliance is voluntary.

Summary

Since laboratories are relying increasingly upon computer systems to provide a wide range of operational services, situations have occurred in which the manual GLPs provide little or ambiguous guidance. To avoid confusion and potential problems, and to deal with the newly arising problems associated with computerized systems, the Environmental Protection Agency developed a series of supplemental guidelines called the Good Automated Laboratory Practices (GALPs). In essence, the GALPs provide the elements to be included in a validation strategic protocol and tactical program. These supplemental guidelines are intended to provide clarification of expectations of performance and control for laboratories electing to utilize computer systems.

The "bottom line" issue in expanding GLP concepts to encompass the GALPs in environmental laboratories using computer systems is that of control. Effective management and confidence in these environments cannot be assured unless the design, use, and operation of systems are based on a consistent set of standards.

Chapter 7

SITUATIONAL REVIEW IN THE U.S.A.

OVERVIEW

There are two kinds of industries in the United States -- those that are regulated, and wish they weren't; and those that aren't regulated, and wish they were. Regulation adds costly overhead and removes critical management decisions out of the hands of company officers. But in a highly litigious society, regulation can be a valuable tool in limiting liability and in defining the extent of protective actions that are necessary and appropriate. In effect, regulation can serve as an operationalization of industry standards that provide an umbrella of protection for a variety of lawsuits.

Yet regulated industries complain loudly about the interference and restrictions that regulation represents. In the pharmaceutical industry, for example, PhRMA (Pharmaceutical Researchers and Manufacturers of America, formerly the PMA), launched a 1993 advertising campaign to counter claims of unfair profits by blaming the Food and Drug Administration for the high cost of pharmaceuticals.

The FDA, on the other hand, considers the pharmaceutical industry to be "self-regulated." This "self-regulation" designation is common in the United States. With very few exceptions (largely in the setting of interstate rail transportation rates), the regulatory agencies have asked their industries to set standards of performance. Rather than directly regulating, the agencies simply evaluate and randomly review that self-regulation.

Self-regulation may be necessary as the complexity and number of regulated companies exceeds the financial capacity of the regulatory agency to conduct appropriate evaluations, but this approach has produced two generic criticism of United States regulatory agencies.

These criticisms are widespread, although stating them explicitly may surprise the critics themselves.

First, the United States regulatory agencies -- the Food and Drug Administration, the Environmental Protection Agency, the Nuclear Regulatory Commission, the Department of Energy, and other agencies -- have been criticized as being insufficiently bureaucratic. The critics, of course, would deny this characterization. Stripped of its pejorative meaning, however, bureaucracy refers to a clear, explicit, standardized code of requirements applied uniformly. The Social Security Administration, for example, is highly bureaucratic. When that agency is functioning as designed, an applicant in any part of the United States would receive the same disposition on an application of disability benefits regardless of the state of application, the reviewer, or the date of application. Employees of the SSA have no discretionary power; the rules they follow are explicit, rigid, and uniform.

In direct contrast, the FDA uses vague standards of conduct enforced by investigators with wide latitude. The Good Manufacturing Practices, Good Laboratory Practices, and Good Clinical Practices are frameworks, rather than blueprints. In inspections and application reviews, FDA professionals are expected to use their education, training, and experience to apply these general guidelines to specific situations that vary greatly. There is no assumption that two investigators will come to identical conclusions, or that the same investigator will remain inflexible in a variety of environments.

Such flexibility allows for a variety of industry solutions to common problems, permits variation of interpretation in situations of unique circumstances, and allows a "learning process" for both regulators and the regulated. But this flexibility leads to frustration for managers who want clear directives and explicitly defined requirements.

Whether the current situation is good or bad is a matter for debate; the direct relationship between bureaucratic explicitness and inflexibility of application is established. Where United States regulatory agencies become more specific, they are less able to deal with unusual situations and to respond to changing conditions. In the validation arena, flexibility is critical; the frustrating lack of explicit clarification is inevitable. New technologies emerge much more rapidly than new regulation. A guideline can take five years from drafting to publication, including the legally-required comment processes. A memorandum detailing requirements for testing a software feature or hardware device is likely to be obsolete by the time it is published. The result would likely be the kind of anti-effectiveness requirements that haunt OSHA -- requirements that call for obsolete types of fire detectors, for example.

The second regulatory criticism flows directly from the first: That many of the employees of these agencies are inadequately qualified for their decision-making roles. In the absence of explicit requirements, judgment becomes all the more crucial. Too often, that judgement is based upon too little information.

Eroding federal employee benefits, lucrative opportunities in private industry, and decreased budgets for participation in conventions and training programs have all contributed to this problem. No doubt, most regulatory professionals are well motivated, but access to the latest technologies, and to information about those technologies, may be limited. With possible exceptions in the Department of Defense, government is not the cutting edge of innovation.

In the validation area, this problem has been accelerated. Few regulators have first-class, technical backgrounds. All too often, the field investigator is forced to review complex technologies with inadequate support, training, and background. An investigator hired as a chemist or biologist, for example, may find himself or herself in a computer validation review for FDA, EPA, or DOD, staring at indecipherable code or examining test protocols well beyond understanding.

Increasingly, regulators are demanding third-party evidence in support of validation efforts, relying upon outside experts to evaluate the systems at issue. But the criticism is well founded, and it is acknowledged by regulators who, appropriately, see increased training and support as the appropriate remedy.

Criticism of lack of regulatory support and lack of bureaucratic standardization are widespread and pervasive. The simple solution of firm and inflexible standardization of requirements is, like most simple solutions, inappropriate. Both industry and the public mount significant pressures for standardization and clarity of definition, citing rising costs and lower profits as rationales. Weaknesses in regulatory expertise seem to reinforce the argument. Yet standardized approaches to widely varying and non-standardized problems is a recipe for chaos. Despite pressures to speed approval, lower expenses, and clarify requirements, regulatory agencies must jealously protect their flexibility if they are to be responsive in a technologically innovative world. Bureaucracies, staffed by well-directed but inflexible personnel, handle large case loads rapidly and efficiently. But the well-deserved, negative reputation of bureaucracies is an inevitable outgrowth of a lack of flexibility in a changing environment. In the validation area, bureaucratic inflexibility would be disastrous.

THE FOOD AND DRUG ADMINISTRATION

The FDA is facing a number of major problems. Despite the increased review staff in the Center for Drug Evaluation and Research made possible by the user-fee program, budget limitations are preventing real growth in the number of investigators, even as the number of regulated companies increases every year. Pressures to accelerate the review process are counterbalanced by increased consumer demands (and their law suits) to assure increasingly far-reaching levels of product safety. Simultaneously, the range of products falling under FDA regulation is in an upward spiral.

Regulation of cosmetics increases as treatment claims are used to enhance products. Dentifrices, for example, are laced with chemical additives, forcing reclassification as Over-the-Counter (OTC) drugs. Shampoos and soaps bear bacteria-fighting claims that require clinical studies, review, and evaluation. Medical devices, vitamins, and so-called New Age treatments require further review within already stretched budgets.

Faced with these mounting pressures, the FDA has subtly adopted four coping tactics:

- User Fees
- Deferred Costs
- Cooperative Agreements
- Selective Targeting

Together, these four tactics are proving successful, even as they rewrite the nature of FDA regulation.

User Fees

The most overt of the new tactics, the user-fee program calls for direct agency billing of regulated industries for identified services. In effect, preferential scheduling is available through a complex fee formula. The user fee concept, first introduced for the pharmaceutical industry in 1992 -- and currently being looked at for medical devices and biologics regulation as well -- carries the promise of accelerated action industry-wide, although that promise has been somewhat subjugated as demand has increased in excess of the services the FDA is legally required to provide.

Deferred Costs

Deferred costs are indirectly billed to regulated industries that, under this tactic, are required to provide reports to the FDA which, previously, FDA would have generated itself. For example, in the early 1980s, FDA investigators routinely made validation inspections and cited companies for failures in design, manufacturing, or delivery. Today, companies are required to conduct their own validation studies, generally through the services of outside, independent consulting companies, and to submit those validation reports to the FDA. FDA investigators then review these reports, rather than the system or process. The costs of the investigation have been deferred from the agency to the regulated company.

Cooperative Agreements

Still in their infancy, cooperative agreements allow the FDA to accept investigations by other international agencies. Some agreements with Swiss, EU, and Swedish agencies have eased the process for companies importing certain kinds of products, although this trend is still slow and cumbersome. While future agreements will permit further cooperation, counter pressures to restrict trade and favor domestic companies may offset any gains this tactic offers.

Selective Targeting

Selective targeting opens FDA to the criticism that its technique for randomly investigating companies is, in reality, far from random. Instead, anecdotal evidence seems to indicate that regulators tend to concentrate on a singly, high-profile company for an in-depth investigation of a particular issue. Presumably, the publicity generated by that investigation pressures the rest of the industry into compliance. While this tactic may raise some fairness issues, it is an effective method of regulating (albeit by press pressure) in an era of budgetary limitations.

These tactics, and the pressures to which they respond, combine to define an agency under great industry and legislative pressure, yet expanding its power base even as that expansion threatens its effectiveness. Under several current health care reform proposals, the FDA will increase its interest in hospitals and medical treatment. Recent revelations suggest that regulation of tobacco will shift from the Bureau of Alcohol, Tobacco, and

Firearms to the FDA. And an increasingly vocal vitamin therapy movement will drive increased FDA activity in that area.

Yet as the FDA increases the width of its focus, it will have to use the diffusion tactics described above to maintain itself within the inevitable budgetary limitations of a United States federal deficit environment. The current (1995-96) climate in Congress which is looking for drastic FDA reform on the premise that the agency "takes too long" in approving new drugs and devices, has led the agency to look inwardly and embark on an ambitious process of internal reform across all the agency's centers. This internal reform, an attempt to balance the FDA's mission of protecting the public health while eliminating unnecessary regulatory burdens, promises to bring a significant change in the way the agency deals with the regulated industries. Part of the result of the pressure toward reform will be greater FDA concern about regulated products, but a shallower depth of regulation and an increased reliance on outside assistance, particularly felt in the validation area.

OTHER U.S. REGULATORY AGENCIES

The Environmental Protection Agency first approached the problems of system validation in the early 1990s, particularly in dealing with its own contract testing laboratories. These private laboratories perform compliance testing and forward results to a variety of EPA databases. Concern for the integrity of that data led to the development of the Good Automated Laboratory Practices (GALPs), a series of guidelines for controlling computer systems.

Today, the EPA is increasingly relying upon computerized data, and it is in the process of designing a series of common databases that can collect, combine, and analyze water, air, and pesticide data from a variety of sources. These computer databases require careful control, bringing validation issues to the forefront of EPA planning.

Unfortunately, the structure of the EPA makes the establishment of a cohesive validation policy a near impossibility. The agency is actually a collection of semi-autonomous programs, each responsible internally for its own standards and policies. The Superfund Program, for example, has different priorities and procedures from Contract Laboratories, which shares little in common with Pesticides, Air Quality, or other EPA programs. While the GALPs stand as guidelines for all EPA activity areas, enforcement is fragmented.

A recently-introduced legislative initiative, tentatively entitled the Laboratory Accreditation Act, is designed to establish uniform standards for all testing facilities utilized by the agency and by industries reporting results to the EPA. The system validation requirements contained within those standards should represent significant progress in establishing uniform system controls in all EPA establishments.

The Department of Energy is actively involved in a massive effort to clean up nuclear weapons production facilities which were originally established in the early 1950s. That effort is beginning with the creation of measurement laboratories, which will be used to assess progress throughout the multi-year program.

In order to assure uniformity and reliability of measurements at the DOE laboratories, the department has decided to incorporate the GALPs into its design standards. Energy laboratories will be required to meet system validation standards for all LIMS and related data collection and analysis software.

The Nuclear Regulatory Commission regularly inspects nuclear power facilities throughout the United States. While most of the current facilities were originally designed with manual analog gauges, the current industry trend is to replace these original systems with computerized measurement and control devices. In this increasingly automated environment, the NRC has request, and is incorporating, the System Validation Standards of Weinberg, Spelton & Sax, Inc. (the basis for the EPA's GALPs) into its internal guidelines. In much the same way, regulators from the department of Agriculture and the Department of Defense are incorporating the standards developed for the pharmaceutical and health-care industries into the validation requirements for laboratory and control systems under their authority.

STATE OF THE ART

A variety of new guidelines, regulations, and standards are currently in release or preparation, further defining and narrowing the responsibilities of system vendors, users, and developers. These directives are both a result of increasing regulatory emphasis upon validation in field investigations or NDA/IND reviews and have been anticipated by investigators to cybernetically increase that scrutiny.

The new clarifications are not only potentially valuable as interpretations and clarifications of necessarily vague requirements. These new interpretations are critical for the development of specific, in-house standard operating procedures and policies. The interpretive guidelines serve as the template for organizational policy review and as the assurance that those internal policies are compliant with regulatory interpretations.

Four new guidelines are emerging:

- The latest revision of the ISO 9000 series
- The finalized Good Automated Laboratory Practices
- Clarifications of requirements for OTC dental products
- Medical device (diagnostics) requirements

ISO 9000

The ISO 9000 quality guidelines have never had the impact on the pharmaceutical-related industries that they have commanded in other economic sectors. The major U.S. automobile manufacturers seek ISO 9000 compliance from all of their suppliers. Heavy manufacturing industries use ISO 9000 as a baseline for performance assurances. In agriculture, ISO certification lends tacit credibility to seed quality and product purity.

But to heavily regulated industries, ISO 9000 represents a sub-level standard. Because GMP and GLP requirements are more rigorous than ISO 9000 guidelines, certification seems like additional paperwork without additional value or credibility. While this position is sound, there is a subtle shift in the wind that is making ISO 9000 an increasingly valuable tool in the pharmaceutical industries.

FDA has had a long-standing concern with the production environment. While products, not sites, are licensed, the GMPs prescribe site conditions -- cleanliness, organization, control -- as requisites of product approval. Attempts to reconcile differences between the GMPs and ISO 9000 have stumbled over this site vs. product inconsistency, but the realization that the ISO guidelines can help to clarify a significant portion of the GMPs has given a new impetus to harmonization.

Perhaps coincidentally, the fact that the sections of the GMPs that ISO clarifies are transferable from product to product -- that is, the same quality facility may support a number of licensed products -- makes ISO certification cost effective, where it otherwise would not be. An ISO 9000 review costs upwards of $30,000 to $50,000. That investment would not survive a cost/benefit analysis for a single product line. However, if a facility is used to manufacture 10 products, the $5,000 investment (1/10 the cost of the review) is a realistic and cost-effective method of assuring facility compliance with GMP requirements.

For facilities investing in new equipment, of course, ISO 9000 certification of suppliers provides the same baseline quality assurances that any manufacturer seeks. A coater that comes from a registered company may not be any better than a non-registered piece of equipment, but the assurances of quality are established.

ISO 9000 guidelines are currently in final revision, with new versions expected in late 1995 or early 1996. Reviews of pre-release drafts suggests that this set of revisions is clarificational rather than substantive, at least as the guidelines affect the pharmaceutical industries. In either their current or the soon-to-be-released form, the ISO 9000 guidelines represent a reasonable and cost-effective tool for self-assessing compliance with site controls of GMP facilities.

GALPs

The EPA began developing the GALPs almost five years ago. These guidelines, providing a practical interpretation of Good Laboratory Practices for laboratories relying increasingly upon the use of computerized instruments, data management systems, and reporting systems, have been available in draft form for almost three years. In the interim, the guidelines have been reviewed, field tested, and found to be useful as both a management tool for the control of laboratories and as a regulatory tool for the assessment of that control.

The final GALP guidelines were released in August 1995. The revision is not significantly different from the original draft, which has been used by investigators from Superfund, DOE, CLP, and other programs. A number of FDA field people have even been seen using the GALPs as a guideline, despite official FDA refusal to acknowledge the EPA document.

Most importantly, however, the GALPs have proven to be useful for laboratories assessing, organizing, and managing their operations. Although a non-regulated environment would, no doubt, dispense with a number of the documentary requirements the GALPs suggest, the Good Automated Laboratory Practices serve an important function in defining areas of concern and recommending methods of control that are fundamental to any automated scientific arena.

One FDA-regulated, pre-clinical animal laboratory used the GALPs as the defining document for developing their own internal Standard Operating Procedures, cutting their start-up time by an estimated 80%. A non-regulated university research laboratory is using the GALPs as an internal template for quality control. And a major energy (oil and natural gas) company has adopted the GALPs as the basis for a corporate-wide, system validation policy.

With final publication and wide distribution, the Good Automated Laboratory Practices will serve as a valuable guideline in a variety of settings, and as an invaluable self-assessment tool for laboratory managers in regulated environments.

OTC -- Dental

The focus of regulatory interest in validation is constantly broadening, partly in response to real or perceived problems with data and systems, and partly in a "what else" philosophy that expands investigations as more mainstream concerns are routinely satisfied. When investigators find good evidence of organization and control, for example, they begin to devote more of their on-site time and energy to the factors that precede and assure that control.

One area of new examination by the validation microscope involves that most cross-related of product categories -- dental hygiene tools, dentifrices, and washes. While classification of the ethical drug products is clear, the OTC products vary widely, falling under medical device (toothbrushes), OTC drug (most dentifrices), and even cosmetic (some polishes) categories.

Some clarification is now available, through a notice of preliminary 21 CFR 356 regulations published in the *Federal Register* on February 9, 1994 (Vol. 59, No. 27, p. 6084) and a final OTC dental monograph published in November 1995. While dealing with a variety of important dental topics, the notice includes an indication of the widening of validation requirements to the manufacturers of OTC dentifrice products. By extension, the same policy will ultimately include other boarder-area products, including anti-bacterial soaps and related detergents.

Simultaneously, the major manufacturers of these product lines have reported increased FDA correspondence and visitations. Clearly, the FDA is following through on its commitment to extend validation requirements to all sectors of its regulated industries; the OTC dentifrice areas are just the latest focus group.

There is good news for these companies, incidentally. Validation compliance has been consistently found to be cost effective. While the regulatory agenda is critical to focusing attention, and to leveling the cost playing field for all competitors, validation can be justified purely on a coast analysis basis. Uncontrolled, inadequately documented, poorly designed, or mismanaged computer systems can represent a huge expense sink. Validation requirements are, of course, documentation heavy. But that documentation establishes and assures the proper designs and controls. And while everyone may gripe about the need for carefully-written SOPs and well-organized logs, flow diagrams, and support materials, those gripes turn to sighs of relief when a programmer resigns to pursue a different challenge, a system manager is off on vacation during a critical incident, or a vendor decides to no longer support an important software program.

Medical Devices

While the organizational shakedown in the FDA's medical devices offices is still not complete, an important validation trend is emerging. International consensus has been reached: software, particularly software used in control and decision making with the pharmaceutical industries, out to be regulated as a medical device.

The concept is an expansion of regulation of diagnostic devices. A blood analyzer, for example, is clearly a medical device. It assists in classification and treatment; it is decision based; and its functionality has an associated human health risk factor. All of these elements are important in the definition of medical devices, although the FDA tends to err on the side of a very loose classification. As a top FDA spokesperson half-jokingly stated in a recent address: "If a physician uses a pencil to write a diagnosis on a patient chart, that pencil is a medical device subject to appropriate registration and regulation."

In past years, the FDA field investigators responsible for biologics -- particularly blood -- have been most aggressive in demanding documentation of software as medical devices. Biologics people have visited software companies involved in the design and coding of data management systems and of tracking systems used in blood banks, requesting evidence of compliance with medical device GMPs. By widespread statement and recent -- if thus far, narrow -- action, other FDA groups have expanded the same concept to other critical software systems, including Laboratory Information Management Systems (LIMS), Material Requirements Planning (MRP) systems, lot tracking software, and CIM (manufacturing) systems. In investigator training programs and directives, investigators in at least two of the three largest FDA districts have been instructed to treat all software used in laboratory and manufacturing decision making as medical devices.

This trend is not limited to the United States. At a recent conference in Toronto, a spokesperson for the Canadian Health Protection Bureau (HPB) stated that the Canadian FDA-equivalent would consider all laboratory computer systems to be subject to medical-device regulation. In subsequent conversation, he indicated that his organization's policy had been harmonized to coincide with emergent interpretations by the UK's Department of Health and the Australian Ministry of Health. In a not unrelated announcement, the UK DOH has indicated that they had harmonized their guidelines with BGA in Germany to form the core of the European Union policy.

Should a software company be concerned? Probably not, in the short run. Regulators are unlikely to visit sites of otherwise unregulated industries for many years. Some spot checking, of course, will continue in the interim, but the regulatory concentration will focus on the users rather than developers of software.

On the other hand, anticipating the future regulations, and registering validation evidence for software under the 510(k) provision has three immediate advantages. First, registration can be advertised, providing an inexpensive quality assurance to potential customers. The marketing advantage is as yet unmeasurable, but persuasive. Second, under a ruling by the First Circuit Court of Appeals last year, registration of medical devices provides some kinds of liability immunity. In a litigious society, that advantage can be significant, with saved insurance costs well exceeding registration and validation expenses. Finally, the trend has been to continually tighten the requirements for (510(k) registration, particularly through increasing requests for clinical trials. Arguably, if registration is inevitable, better now rather than later.

SUMMARY

Validation regulation is subject to continual refinement, both in response to emergent clarification of guidelines and to an extension of the environments to which those guidelines apply. While the four areas of change summarized above -- ISO 9000, GALPs, Dental OTC products, and Medical Devices -- are dramatic and significant, they are not unique. Each month new documents are released; every quarter new policies are developed; every year new requirements are announced.

Validation is the foundation of regulation, for it is the evidence upon which all other analyses builds. A toxicology review examines data stored and summarized in an electronic database; a product sampling relies upon the systems that select that sample, the devices that conduct that analysis, and the software that statistically reviews and interprets the results. The validation of the systems underlying the industry provides not just the tools for all other reviews, but the calibration and honing of the edges on those tools.

In such an environment, a clear understanding of validation guidelines, principles, and requirements is critical to maintain a compliant and efficient operation. And that understanding is dependent largely upon a continual updating of the ever-emerging body of validation information.

Chapter 8

INTERNATIONAL SITUATIONAL REVIEW

OVERVIEW

The international situation with regards to validation standards and compliance is in a state of flux. ISO standards are undergoing major revisions. European Union directives are taking effect in only portions of Europe. Asian nations are developing their own standards, or borrowing from everyone else. And former Soviet-influenced portions of the world are both growing markets and suppliers of pharmaceuticals, medical devices, and biological products. The most important strides on the international level, however, are being taken by efforts at international harmonization.

ISO 9000

The ISO 9000 series of voluntary international standards for quality management and quality control were first issued by the International Organization for Standardization in 1987. Since their first publication, these standards have become internationally recognized for their importance in developing quality management practices in a large number of industries.

There are four basic standards in the ISO 9000 series:

- ISO 9001 -- *Quality Systems -- Model for Quality Assurance in Design/Development, Production, Installation and Servicing.* This is the most comprehensive of the ISO 9000 standards. It deals with all the requirements in ISO 9002 and ISO 9003, and also covers design, development, and servicing.

- ISO 9002 -- *Quality Systems -- Model for Quality Assurance in Production and Installation.* This standard addresses installation, procurement and production aspects.

- ISO 9003 -- *Quality Systems -- Model for Quality Assurance in Final Inspection and Test.* Although this is the least comprehensive of the ISO 9000 standards (covering only final inspection and testing in the production process), it is important to system validation because it provides some guidance in applying ISO 9001 standards to software development.

- ISO 9004 -- *Quality Management and Quality System Elements -- Guidelines.* This standard provides guidelines for specific industry application.

The ISO is currently rewriting the ISO 9000 series standards. The revision was planned to be released in the Fall of 1994, but their release has been delayed until early in 1995. Although the revision will supposedly have more specific systems validation guidance, elements of the currently-applied standards are still useful for quality management of automated equipment and software.

Elements of ISO 9001

4.1 Management Responsibility

- Quality Policy
  - Defined and documented by management
  - Includes objectives
  - Appears at all levels of the organization
- Worker authority and responsibility
  - For all whose work affects quality
  - Includes those who verify quality
- Verification of resources and personnel
  - Verification requirements identified
  - Adequate resources assigned
  - Includes those who audit the quality system
  - Independent of the line organization
- Management review
  - Representative responsible for ISO 9000 requirements assigned
  - Quality system and audits reviewed and documented

4.2 Quality System

- Documented procedures
- Effectively implemented
  - Quality plan
  - Quality manual
  - Clarification of standards of acceptance
  - Quality records planned

4.3 Contract Review

- Requirements adequately defined
- Differences with tender offer resolved
- Capabilities exist
- Documentation of review

4.4 Design Control

- Planned responsibilities for all design and development activities
- Interfaces between groups documented
- Documented design input requirements
    - -- Meet design input requirements
    - -- Embody acceptance criteria
    - -- Conform to regulatory requirements
    - -- Identify design characteristics crucial to safe and proper functioning of the product
- Documented design verification
- Design reviews
- Qualification tests and demonstrations
- Alternative calculations
- Comparison with any available proven design(s)
- Design changes
    - -- Documented
    - -- Reviewed
    - -- Approved

4.5 Document Control

- Approval and issue
    - -- Procedures exist
    - -- Documents reviewed and approved for adequacy
    - -- Documents available where needed
    - -- Removal of obsolete documents
    - -- Listing maintained of current versions of documents
- Documentation changes
    - -- Reviewed and approved by originators
    - -- Documents reissued whenever changes become confusing

4.6 Purchasing Control

- Subcontractors
    - -- Acceptable ones listed
    - -- Previous performance recorded
- Purchasing data
    - -- Clear and complete definition of product

  - Reviewed and approved before issuance
- Verification of purchased product
  - As necessary
  - At supplier's site or upon receipt
  - Not a substitute for supplier's verification

4.7 Purchaser-Supplied Product

- Procedures in place for storage, maintenance, and verification

4.8 Product Identification and Traceability

- Where appropriate, procedures exist
- For traced products
  - During all stages of production, deliver, and installation
  - Individual items traceable as necessary

4.9 Process Control

- Work instruction documented if lack would impair quality
- Approval of processes and equipment as appropriate
- Workmanship standards are written or in samples
- Special processes are documented
- Appropriate records exist for qualified processes, equipment, and personnel

4.10 Inspection and Testing

- Incoming Product
  - Verification procedures exist
  - Product held until release
  - Product used in emergency will be positively identified
- In-process product
  - Procedures exist to inspect, test, and identify
  - Product held until release
  - Product released in emergency will be positively identified
- Final product
  - Procedures exist to inspect and test
  - Require passing of all previous tests
  - Product held until release
  - No emergency product release permitted
- Identification of nonconforming product
- Test records maintained

4.11 Inspection Test Equipment

- Calibration procedures established and checked
- Control, calibrate, and maintain
- Mark equipment to show calibration status
- Calibration records maintained
- Whenever out of calibration, document recent testing
- Safeguard from in-validating adjustments
- Testing hardware and software
  - -- Proved capable of verifying product acceptability
  - -- Procedures in place for rechecking
  - -- Records for evidence of control

4.12 Inspection Test Status

- Identification of conformance or nonconformance to performed tests
- Inspection authority responsible for release identified
- Maintained throughout production, shipping, and installation

4.13 Control of Nonconforming Product

- Procedures to prevent inadvertent use
- Review procedures documented with responsibility defined
- Disposition
  - -- Authority defined
  - -- Actions documented
  - -- Product reinspected when repaired and/or reworked

4.14 Corrective Action

- Procedures documented
  - -- Investigate nonconforming product for cause and corrective action
  - -- Analyze all operations, exceptions, and complaints to eliminate
  - -- Ensure that corrective actions are taken and are effective
  - -- Implement necessary changes in procedures

4.15 Handling, Storage, Packaging, and Delivery

- Procedures documented
- Secure storage areas and stock rooms
- Periodically reassess stored product for deterioration
- Quality protected after final inspection(s) and test(s)

4.16 Quality Records

- Procedures to control records for their full life cycle
- Subcontractor records included
- Keyed to the product involved
- Retention records recorded
- Safeguarded

4.17 Internal Quality Audits

- Procedures documented
- Comprehensive, planned, documented
- Results given to responsible auditee
- Timely corrective action taken by auditee

4.18 Training

- Procedures for
  - -- Identifying training needs
  - -- Providing training
  - -- Qualifying those performing specific assigned tasks
  - -- Maintaining training records

4.19 Servicing

- Procedures for performance and verifying requirements are met

4.20 Statistical Techniques

- Procedures for identifying adequacy for verifying
  - -- Process capability
  - -- Product characteristics

The chart on the following page compares these elements as they are included ISO 9001, ISO 9002, and ISO 9003:

Requirement	ISO 9001	ISO 9002	ISO 9003
Management Responsibility	■	●	○
Quality System	■	■	○
Contract Review	■	■	
Design Control	■		
Document Control	■	■	○
Purchasing	■	■	
Purchaser-Supplied Product	■	■	
Product Identification and Traceability	■	■	○
Process Control	■	■	
Inspection and Testing	■	■	○
Inspection Test Equipment	■	■	○
Inspection Test Status	■	■	○
Control of Nonconforming Product	■	■	○
Corrective Action	■	■	
Handling, Storage, Packaging and Delivery	■	■	○
Quality Records	■	■	○
Internal Quality Audits	■	●	
Training	■	●	○
Servicing	■		
Statistical Techniques	■	■	○

● Requirements less stringent than in ISO 9001
○ Requirements less stringent than in ISO 9002

THE WORLD VIEW

Many nations have not attempted to develop their own regulations regarding systems validation in the manufacturing of drugs and medical devices, but have instead relied on world-wide or regional Good Manufacturing Practice requirements -- primarily those developed by the World Health Organization (WHO) -- or regional requirements such as those established by the by the Pharmaceutical Inspection Convention in Europe or the Association of South-East Asian Nations (ASEAN).

The World Health Organization's Good Manufacturing Practice requirements do not specifically address system validation. The PIC's "Guide to Good Manufacturing Practice for Pharmaceutical Products," however, generally defines validation as an assurance that manufacturing procedures, processes, equipment, material, and systems lead to the results expected of them. Under PIC guidelines, validation responsibility is vested in a firm's quality control unit. In relation to documentation, the integrity of data processed by automated equipment must be assured through detailed procedures to assure their accuracy, security, and change control.

LOCAL REQUIREMENTS

There is a three-tier system at work in the world when it comes to GMP, GLP, and GCP requirements. Locally-issued regulations, if there are any, take precedence. If there are no local regulations, regional requirements, such as those issued by the PIC or ASEAN, are followed. In those areas where there are neither local nor regional requirements, international requirements, such as those of the WHO, are applied.

Many national Good Manufacturing Practice (GMP), Good Laboratory Practice (GLP), and Good Clinical Practice (GCP) standards, guidelines, and regulations do not specifically address systems validation requirements. However, such requirements can generally be considered necessary under other aspects of these requirements. For example, requirements for adequate personnel training and experience in manufacturing or laboratory equipment, equipment operation and maintenance, quality assurance/quality control activities, documentation of procedures, and record retention would all include appropriate actions when automated equipment is used in place of manual operations.

The following discussion will examine some of the more significant, specific national requirements for system validation and their application in GMPs, GLPs, and GCPs.

Japan

The Koseisho is the Japanese government's Ministry of Health and Welfare responsible for drug and medical device registration, with responsibilities similar to those of the FDA in the United States. Japan is in the forefront of establishing guidelines for computer system validation in GLPs and GMPs for pharmaceutical products and medical devices.

Japan's "GLP Inspection of Computer System" covers eight basic points:

1. Computer Hardware and Computer Room

Computers must be located in an appropriate, controlled environment, where authorized access can be regulated and data adequately archived.

2. Development of Computer Systems

   Computer systems must be developed with appropriate design and with adequate functioning capacity.

3. Software Developed in-house

   Written development and programming standards must be approved and followed. Functional specifications must be met. Testing documents must be reviewed, approved, and retained. Validation testing must be followed and documented.

4. Vendor-Supplied Systems

   System specifications must be documented. Vendors must provide confirmation that systems were developed, reviewed, and approved appropriately. Vendors must have established change-control procedures. Vendors must supply development documentation as a validation package in order to evaluate the system's reliability.

5. Existing Systems

   The system's historical documentation must be reviewed to accomplish retrospective validation. Testing must be reviewed and approved to confirm the system's reliability. Documentation must be maintained as a retrospective validation package.

6. Operation and Maintenance

   Computer systems must be operated and maintained to perform their required functions through supervision by a responsible person and established operation and maintenance procedures.

7. GLP Compliance of the Data Directly Recorded in Computers

   Data processing must comply with Japan's general GLP standard. Adequate data change control procedures must be in place and followed.

8. Computer System Inspections by the Quality Assurance Unit

   Quality assurance unit must inspect the system to assure that its design complies with the GLP standard, and that the system was developed and is operated to function as designed. QA units must have SOPs for inspecting systems, must inspect all developmental phases of the system, and must periodically inspect the system's operation, maintenance, and data reliability.

The Japanese Ministry of Health and Welfare issued a "Guideline on Control of Computerized Systems in Drug Manufacturing" in 1993 for use by pharmaceutical firms governed by GMP regulations. The guideline applies to systems used for manufacturing process control and management, production control (storage and inventory), manufacturing documentation, and quality control. The Guideline covers three areas:

1. System Development

   Responsible individuals must be designated to supervise each step of system development. Development must be documented in a System Development Manual which includes development procedures, evaluation procedures, and approval procedures. A development schedule must be established to describe system development objectives, timetables, staffing, and equipment selection. System engineering and program development must be controlled to ensure development of appropriate system specifications. System development tests, installation tests, and operation tests must be conducted according to established procedures and adequately evaluated and approved.

2. Control of System Operation

   Adequate operation procedures must be established for hardware operation, system inspection and maintenance, disaster recovery, and security control. A "system alteration standard code" must be prepared to describe procedures for alteration or amendment of system operation. System hardware must be operated according to appropriate SOPs. Security control procedures must be established to prevent system use by unauthorized individuals. Continued system reliability is ensured by periodic self-inspection.

3. System Documentation

   Pharmaceutical manufacturers must retain documentation relating to system development and operation.

In October 1993, a Study Group on Policies for Medical Devices submitted a "Proposal for Reclassification and Regulation of Medical Devices." One of the recommendations was to expand the scope of medical devices and include computer software in the category of regulated medical device because of the necessity to standardize its quality, effectiveness and safety.

United Kingdom

In 1989, the GLP Compliance Programme of the U.K.'s Department of Health (DOH) issued "The Application of GLP Principles to Computer Systems" as an Advisory Leaflet to accompany its 1989 GLP Compliance Programme. The guideline outlined how DOH inspectors would examine computer systems in laboratories conducting health and environmental safety studies. Inspectors would focus on three areas:

1. Identification and Functional Definition of Systems

   Inspectors must require definitions for hardware used in data processing, data input, and data output. Information would be gathered on operation management, system users, and QA system monitoring.

2. Control Procedures

   Inspectors will check procedures for developing, installing, maintaining, and documenting systems and examine data security, change control, archiving, and disaster recovery procedures.

3. Evaluation of Effectiveness of Specific Functions

   Inspectors will examine function documentation, methods used by laboratories to test system function, and observe specific system operations.

The DOH's "Guide to Good Pharmaceutical Manufacturing Practices" contains a section on Electronic Data Processing to regulate computer systems used in drug manufacturing, storage, and distribution. Specific requirements relating to such systems include:

1. Personnel must have appropriate expertise and training for system use.
2. System descriptions must be maintained and kept up to date.
3. Before placing a system on line it must be tested and its functions confirmed.
4. Adequate, periodic operational checks must be performed.
5. Change control procedures should be well defined.
6. Only authorized users should enter data into a system, and a procedure for system security should be defined.
7. Independent checks should be made on entry of critical data.
8. Archived data should be accessible and checked for accuracy when equipment or software programs are changed.
9. Disaster recovery procedures should be defined and tested.
10. Formal agreements should be made with system vendors to define their responsibilities.

Australia

Australia's Good Manufacturing Practices for pharmaceutical products also contains a section specifically addressing automated processes and procedures. These requirements include:

1. Computer systems must be designed and installed by personnel with appropriate training and expertise; systems must be tested before being put on line.
2. A specifications document must be prepared detailing system objectives, the types of data entered, testing procedures, operations, and maintenance.
3. System users must be appropriately trained and have access to written SOPs.
4. Change-control procedures must be defined.
5. System operation must be periodically checked and corrective actions taken when necessary.
6. Security procedures must be in place to allow access by authorized system users.
7. When critical data is entered, data entry will be checked by a second individual.
8. Written disaster recovery procedures must be tested, and any recovery actions must be documented.
9. Audit trails must exist to confirm the accurate and reliable transfer of data to or from manufacturing or monitoring equipment.

France

France's Good Pharmaceutical Manufacturing Practices contains four specific requirements listed for electronic data processing:

1. Preliminary Study

   Preliminary studies must be undertaken to determine what kind of automated equipment is used and its location. Personnel from the firm's departments which will be concerned with data processing (such as quality control and production), must undertake the study. These individuals, and personnel who will use the system, must have appropriate training.

2. System Implementation

   Systems descriptions, specifying their functions, must be kept up to date. System validation procedures must be implemented to assure reliability of function.

3. System Operations

   Data entry or modification, and any other access to a system must be authorized. A specific security procedure must be established, as well as a validated disaster recovery procedure. Modifications to the system must be authorized, verified, validated, and documented.

4. Batch Records

   Data relating to the manufacture of batches must be verified by responsible individual(s) and must be maintained.

The Netherlands

The Good Manufacturing Practice guidelines of the Netherlands contain a section on Electronic Data Processing which includes the following requirements:

1. Hardware should be tested for reliability, and the test results recorded.
2. Documentation must be maintained for systems used.
3. Software must be validated, and validated again after any modification.
4. There must be a specific procedure for data entry and modification, and data entry must be verified.
5. Operation of computer/equipment combinations must be validated.
6. To prevent data loss due to equipment malfunction, copies of electronic data must be retained in another medium.

INTERNATIONAL HARMONIZATION

In his keynote address to the Food and Drug Law Institute in December 1994, FDA Commissioner David Kessler noted that his agency had been actively participating in the International Conference on Harmonization of Technical Requirements of Pharmaceuticals for Human Use (ICH) in efforts to reduce the necessity of drug and device manufacturers to comply with differing and sometimes incompatible standards and regulations throughout the international marketplace.

The ICH is primarily concerned with the harmonization of technical requirements in for drug product registration among three regions -- the European Union, Japan, and the United States. Tripartite harmonization initiatives are developed by the ICH with input from industry and regulatory representatives.

In 1994, the FDA published four final ICH guidelines. In 1995, Dr. Kessler stated, the agency would publish seven more and continue the development process on fourteen more that were at various stages of the drafting process.

In its efforts at internal reform as well, the FDA noted that a major goal for the agency was the expansion of its efforts to achieve international harmonization. This effort would be made in three basic areas in the area of drugs, biologics, and medical devices:

1. Beginning work on new harmonization topics relating to testing drugs, biologics, and medical devices and standards for the electronic transfer of information; the FDA would issue harmonized standards as agency guidelines.
2. The agency would increase the pace of its work on harmonizing GMP, GLP, and GCP standards and inspections.
3. New harmonization initiatives with North American Free Trade Agreement (NAFTA) partners.

The following five ICH guidelines are of specific concern regarding validation issues:

- Final ICH Guideline on Clinical Safety Data Management: Definitions and Standards for Expedited Reporting, published in the *Federal Register* on March 1, 1995.
- Final ICH Guideline on Validation of Analytical Procedures: Definitions and Terminology, published in the *Federal Register* on March 1, 1995
- Draft Guideline on Good Clinical Practice, published in the *Federal Register* for public comment on August 17, 1995.
- Draft ICH Guideline on Structure and Content of Clinical Study Reports, published in the *Federal Register* for public comment on August 23, 1995.
- FDA's Policy on the Development and Use of Standards With Respect to International Harmonization of Regulatory Requirements and Guidelines, published in the *Federal Register* on October 11, 1995.

Clinical Safety Data Management: Definitions and Standards for Expedited Reporting

I. Introduction

It is important to harmonize the way to gather and, if necessary, to take action on important clinical safety information arising during clinical development. Thus, agreed definitions and terminology, as well as procedures, will ensure uniform Good Clinical Practice standards in this area. The initiatives already undertaken for marketed medicines through the CIOMS-1 and CIOMS-2 Working Groups on expedited (alert) reports and periodic safety update reporting, respectively, are important precedents and models. However, there are special circumstances involving medicinal products under development, especially in the early stages and before any marketing experience is available. Conversely, it must be recognized that a medicinal product will be under various stages of development and/or marketing in different countries, and safety data from marketing experience will ordinarily be of interest to regulators in countries where the medicinal product is still under investigational only (Phase 1, 2, or 3) status. For this reason, it is both practical and well-advised to regard premarketing and postmarketing clinical safety reporting concepts and practices as interdependent, while recognizing that responsibility for clinical safety within regulatory bodies and companies may reside with different departments, depending on the status of the product (investigational versus marketed).

There are two issues within the broad subject of clinical safety data management that are appropriate for harmonization at this time:

(1) The development of standard definitions and terminology for key aspects of clinical safety reporting, and

(2) The appropriate mechanism for handling expedited (rapid) reporting, in the investigational (i.e., preapproval) phase.

The provisions of this guideline should be used in conjunction with other ICH Good Clinical Practice guidelines.

II. Definitions and Terminology Associated with Clinical Safety Experience

A. Basic Terms

Definitions for the terms adverse event (or experience), adverse reaction, and unexpected adverse reaction have previously been agreed to by consensus of the more than 30 Collaborating Centers of the WHO International Drug Monitoring Centre (Uppsala, Sweden). (Edwards, I. R., et al., "Harmonisation in Pharmacovigilance," Drug Safety, 10(2): 93-102, 1994.) Although those definitions can pertain to situations involving clinical investigations, some minor modifications are necessary, especially to accommodate the preapproval, development environment.

The following definitions, with input from the WHO Collaborative Centre, have been agreed.

1. Adverse Event (or Adverse Experience).

Any untoward medical occurrence in a patient or clinical investigation subject administered a pharmaceutical product and which does not necessarily have to have a causal relationship with this treatment.

An adverse event (AE) can therefore be any unfavorable and unintended sign (including an abnormal laboratory finding, for example), symptom, or disease temporally associated with the use of a medicinal product, whether or not considered related to the medicinal product.

2. Adverse Drug Reaction (ADR).

In the preapproval clinical experience with a new medicinal product or its new usages, particularly as the therapeutic dose(s) may not be established:

All noxious and unintended responses to a medicinal product related to any dose should be considered adverse drug reactions.

The phrase "responses to medicinal products" means that a causal relationship between a medicinal product and an adverse event is at least a reasonable possibility, i.e., the relationship cannot be ruled out.

Regarding marketed medicinal products, a well-accepted definition of an adverse drug reaction in the postmarketing setting is found in WHO Technical Report 498 (1972) and reads as follows:

"A response to a drug which is noxious and unintended and which occurs at doses normally used in man for prophylaxis, diagnosis, or therapy of disease or for modification of physiological function."

The old term "side effect" has been used in various ways in the past, usually to describe negative (unfavorable) effects, but also positive (favorable) effects. It is recommended that this term no longer be used and particularly should not be regarded as synonymous with adverse event or adverse reaction.

3. Unexpected Adverse Drug Reaction

An adverse reaction, the nature or severity of which is not consistent with the applicable product information (e.g., Investigator's Brochure for an unapproved investigational medicinal product). See III. C.

B. Serious Adverse Event Or Adverse Drug Reaction

During clinical investigations, adverse events may occur which, if suspected to be medicinal product-related (adverse drug reactions), might be significant enough to lead to important changes in the way the medicinal product is developed (e.g., change in dose, population, needed monitoring, consent forms). This is particularly true for reactions which, in their most severe forms, threaten life or function. Such reactions should be reported promptly to regulators.

Therefore, special medical or administrative criteria are needed to define reactions that, either due to their nature ("serious") or due to the significant, unexpected information they provide, justify expedited reporting.

To ensure that no confusion or misunderstanding exist of the difference between the terms "serious" and "severe," which are not synonymous, the following note of clarification is provided:

The term "severe" is often used to describe the intensity (severity) of a specific event (as in mild, moderate, or severe myocardial infarction); the event itself, however, may be of relatively minor medical significance (such as severe headache). This is not the same as "serious," which is based on patient/event outcome or action criteria usually associated with events that pose a threat to a patient's life or functioning. Seriousness (not severity) serves as a guide for defining regulatory reporting obligations.

After reviewing the various regulatory and other definitions in use or under discussion elsewhere, the following definition is believed to encompass the spirit and meaning of them all:

A serious adverse event (experience) or reaction is any untoward medical occurrence that at any dose:

** Results in death,
** Is life-threatening,
(NOTE: The term "life-threatening" in the definition of "serious" refers to an event in which the patient was at risk of death at the time of the event; it does not refer to an event which hypothetically might have caused death if it were more severe.)
** Requires inpatient hospitalization or prolongation of existing hospitalization,
** Results in persistent or significant disability/incapacity, or
** Is a congenital anomaly/birth defect.

Medical and scientific judgment should be exercised in deciding whether expedited reporting is appropriate in other situations, such as important medical events that may not be immediately life-threatening or result in death or hospitalization but may jeopardize the patient or may require intervention to prevent one of the other outcomes listed in the definition above. These should also usually be considered serious.

Examples of such events are intensive treatment in an emergency room or at home for allergic bronchospasm; blood dyscrasias or convulsions that do not result in hospitalization; or development of drug dependency or drug abuse.

C. Expectedness of an Adverse Drug Reaction

The purpose of expedited reporting is to make regulators, investigators, and other appropriate people aware of new, important information on serious reactions. Therefore, such reporting will generally involve events previously unobserved or undocumented, and a guideline is needed on how to define an event as "unexpected" or "expected" (expected/unexpected from the perspective of previously observed, not on the basis of what might be anticipated from the pharmacological properties of a medicinal product).

As stated in the definition (II.A.3.), an "unexpected" adverse reaction is one, the nature or severity of which is not consistent with information in the relevant source document(s). Until source documents are amended, expedited reporting is required for additional occurrences of the reaction.

The following documents or circumstances will be used to determine whether an adverse event/reaction is expected:

1. For a medicinal product not yet approved for marketing in a country, a company's Investigator's Brochure will serve as the source document in that country. See III.F. and ICH Guideline for the Investigator's Brochure.

2. Reports which add significant information on specificity or severity of a known, already documented serious ADR constitute unexpected events. For example, an event more specific or more severe than described in the Investigator's Brochure would be considered "unexpected." Specific examples would be (a) acute renal failure as a labeled ADR with a subsequent new report of interstitial nephritis and (b) hepatitis with a first report of fulminant hepatitis.

III. Standards for Expedited Reporting

A. What Should Be Reported?

1. Single Cases of Serious, Unexpected ADR's

All ADR's that are both serious and unexpected are subject to expedited reporting. This applies to reports from spontaneous sources and from any type of clinical or epidemiological investigation, independent of design or purpose. It also applies to cases not reported directly to a sponsor or manufacturer (for example, those found in regulatory authority generated ADR registries or in publications). The source of a report (investigation, spontaneous, other) should always be specified.

Expedited reporting of reactions that are serious but expected will ordinarily be inappropriate. Expedited reporting is also inappropriate for serious events from clinical investigations that are considered not related to study product, whether the event is expected or not. Similarly, nonserious adverse reactions, whether expected or not, will ordinarily not be subject to expedited reporting.

Information obtained by a sponsor or manufacturer on serious, unexpected reports from any source should be submitted on an expedited basis to appropriate regulatory authorities if the minimum criteria for expedited reporting can be met. See section III.B. Causality assessment is required for clinical investigation cases. All cases judged by either the reporting health care professional or the sponsor as having a reasonable suspected causal relationship to the medicinal product qualify as ADR's. For purposes of reporting, adverse event reports associated with marketed drugs (spontaneous reports) usually imply causality.

Many terms and scales are in use to describe the degree of causality (attributability) between a medicinal product and an event, such as certainly, definitely, probably, possibly, or likely related or not related. Phrases such as "plausible relationship," "suspected causality," or "causal relationship cannot be ruled out" are also invoked to describe cause and effect. However, there is currently no standard international nomenclature.

The expression "reasonable causal relationship" is meant to convey in general that there are facts (evidence) or arguments to suggest a causal relationship.

2. Other Observations

There are situations in addition to single case reports of "serious" adverse events or reactions that may necessitate rapid communication to regulatory authorities; appropriate medical and scientific judgment should be applied for each situation. In general, information that might materially influence the benefit-risk assessment of a medicinal product or that would be sufficient to consider changes in medicinal product administration or in the overall conduct of a clinical investigation represents such situations. Examples include:

a. For an "expected, serious ADR, an increase in the rate of occurrence which is judged to be clinically important.

b. A significant hazard to the patient population, such as lack of efficacy with a medicinal product used in treating life-threatening disease.

c. A major safety finding from a newly completed animal study (such as carcinogenicity).

B. Reporting Time Frames

1. Fatal Or Life-Threatening Unexpected ADR's

Certain ADR's may be sufficiently alarming so as to require very rapid notification to regulators in countries where the medicinal product or indication, formulation, or population for the medicinal product are still not approved for marketing, because such reports may lead to consideration of suspension of, or other limitations to, a clinical investigation program. Fatal or life-threatening, unexpected ADR's occurring in clinical investigations qualify for very rapid reporting. Regulatory agencies should be notified (e.g., by telephone, facsimile transmission, or in writing) as soon as possible but no later than 7 calendar days after first knowledge by the sponsor that a case qualifies, followed by as complete a report as possible within 8 additional calendar days. This report should include an assessment of the importance and implication of the findings, including relevant previous experience with the same or similar medicinal products.

2. All Other Serious, Unexpected ADR's

Serious, unexpected reactions (ADR's) that are not fatal or life-threatening should be filed as soon as possible but no later than 15 calendar days after first knowledge by the sponsor that the case meets the minimum criteria for expedited reporting.

3. Minimum Criteria for Reporting

Information for final description and evaluation of a case report may not be available within the required timeframes for reporting outlined above. Nevertheless, for regulatory purposes, initial reports should be submitted within the prescribed time as long as the following minimum criteria are met:

An identifiable patient; a suspect medicinal product; an identifiable reporting source; and an event or outcome that can be identified as serious and unexpected, and for which, in clinical investigation cases, there is a reasonable suspected causal relationship. Followup information should be actively sought and submitted as it becomes available.

C. How To Report

The CIOMS-I form has been a widely accepted standard for expedited adverse event reporting. However, no matter what the form or format used, it is important that certain basic information/data elements, when available, be included with any expedited report, whether in a tabular or narrative presentation. The listing in Attachment 1 addresses those data elements regarded as desirable; if all are not available at the time of expedited reporting, efforts should be made to obtain them. See III.B.

All reports must be sent to those regulators or other official parties requiring them (as appropriate for the local situation) in countries where the drug is under development.

D. Managing Blinded Therapy Cases

When the sponsor and investigator are blinded to individual patient treatment (as in a double-blind study), the occurrence of a serious event requires a decision on whether to open (break) the code for the specific patient. If the investigator breaks the blind, then it is assumed the sponsor will also know the assigned treatment for that patient. Although it is advantageous to retain the blind for all patients prior to final study analysis, when a serious adverse reaction is judged reportable on an expedited basis, it is recommended that the blind be broken only for that specific patient by the sponsor even if the investigator has not broken the blind. It is also recommended that, when possible and appropriate, the blind be maintained for those persons, such as biometrics personnel, responsible for analysis and interpretation of results at the study's conclusion.

There are several disadvantages to maintaining the blind under the circumstances described which outweigh the advantages. By retaining the blind, placebo and comparator (usually a marketed product) cases are filed unnecessarily. When the blind is eventually opened, which may be many weeks or months after reporting to regulators, it must be ensured that company and regulatory data bases are revised. If the event is serious, new, and possibly related to the medicinal product, then if the Investigator's Brochure is updated, notifying relevant parties of the new information in a blinded fashion is inappropriate and possibly misleading. Moreover, breaking the blind for a single patient usually has little or no significant implications for the conduct of the clinical investigation or on the analysis of the final clinical investigation data.

However, when a fatal or other "serious" outcome is the primary efficacy endpoint in a clinical investigation, the integrity of the clinical investigation may be compromised if the blind is broken. Under these and similar circumstances, it may be appropriate to reach agreement with regulatory authorities in advance concerning serious events that would be treated as disease-related and not subject to routine expedited reporting.

E. Miscellaneous Issues

1. Reactions Associated With Active Comparator or Placebo Treatment

It is the sponsor's responsibility to decide whether active comparator drug reactions should be reported to the other manufacturer and/or directly to appropriate regulatory agencies. Sponsors should report such events to either the manufacturer of the active control or to appropriate regulatory agencies. Events associated with placebo will usually not satisfy the criteria for an ADR and, therefore, for expedited reporting.

2. Products With More Than One Presentation or Use

To avoid ambiguities and uncertainties, an ADR that qualifies for expedited reporting with one presentation of a product (e.g., a dosage form, formulation, delivery system) or product use (e.g., for an indication or population) should be reported or referenced to regulatory filings across other product presentations and uses.

It is not uncommon that more than one dosage form, formulation, or delivery system (oral, IM, IV, topical, etc.) of the pharmacologically active compound(s) is under study or marketed; for these different presentations there may be some marked differences in the clinical safety profile. The same may apply for a given product used in different indications or populations (single dose versus chronic administration, for example). Thus, "expectedness" may be product or product-use specific, and separate Investigator's Brochures may be used accordingly. However, such documents are expected to cover ADR information that applies to all affected product presentations and uses. When relevant, separate discussions of pertinent product-specific or use-specific safety information will also be included.

It is recommended that any adverse drug reactions that qualify for expedited reporting observed with one product dosage form or use be cross referenced to regulatory records for all other dosage forms and uses for that product. This may result in a certain amount of overreporting or unnecessary reporting in obvious situations (for example, a report of phlebitis on IV injection sent to authorities in a country where only an oral dosage form is studied or marketed). However, underreporting is completely avoided.

3. Poststudy Events

Although such information is not routinely sought or collected by the sponsor, serious adverse events that occurred after the patient had completed a clinical study (including any protocol required posttreatment followup) will possibly be reported by an investigator to the sponsor. Such cases should be regarded for expedited reporting purposes as though they were study reports. Therefore, a causality assessment and determination of expectedness are needed for a decision on whether or not expedited reporting is required.

F. Informing Investigators and Ethics Committees/Institutional Review Boards of New Safety Information

International standards regarding such communication are discussed within the ICH GCP Guidelines, including the addendum on "Guideline for the Investigator's Brochure." In general, the sponsor of a study should amend the Investigator's Brochure as needed, and in accord with any local regulatory requirements, so as to keep the description of safety information updated.

Attachment 1

Key Data Elements for Inclusion in Expedited Reports of Serious Adverse Drug Reactions

The following list of items has its foundation in several established precedents, including those of CIOMS-I, the WHO International Drug Monitoring Centre, and various regulatory authority forms and guidelines. Some items may not be relevant depending on the circumstances. The minimum information required for expedited reporting purposes is: an identifiable patient, the name of a suspect medicinal product, an identifiable reporting source, and an event or outcome that can be identified as serious and unexpected and for which, in clinical investigation cases, there is a reasonable suspected causal relationship. Attempts should be made to obtain followup information on as many other listed items pertinent to the case.

1. Patient Details:

** Initials,
** Other relevant identifier (clinical investigation number, for example),
** Gender,
** Age and/or date of birth,
** Weight,
** Height.

2. Suspected Medicinal Product(s):

** Brand name as reported,
** International Nonproprietary Name (INN),
** Batch number,
** Indication(s) for which suspect medicinal product was prescribed or tested,
** Dosage form and strength,
** Daily dose and regimen (specify units--e.g., mg, mL, mg/kg)
** Route of administration,
** Starting date and time of day,
** Stopping date and time, or duration of treatment.

3. Other Treatment(s):

** For concomitant medicinal products (including nonprescription/OTC medicinal products) and nonmedicinal product therapies, provide the same information as for the suspected product.

4. Details Of Suspected Adverse Drug Reaction(s)

** Full description of reaction(s) including body site and severity, as well as the criterion (or criteria) for regarding the report as serious should be given. In addition to a description of the reported signs and symptoms, whenever possible, attempts should be made to establish a specific diagnosis for the reaction.

** Start date (and time) of onset of reaction,

** Stop date (and time) or duration of reaction,

** Dechallenge and rechallenge information,

** Setting (e.g., hospital, out-patient clinic, home, nursing home),

** Outcome: Information on recovery and any sequelae; what specific tests and/or treatment may have been required and their results; for a fatal outcome, cause of death and a comment on its possible relationship to the suspected reaction should be provided. Any autopsy or other post-mortem findings (including a coroner's report) should also be provided when available. Other information: anything relevant to facilitate assessment of the case, such as medical history, including allergy, drug or alcohol abuse; family history; findings from special investigations.

5. Details on Reporter of Event (Suspected ADR):

** Name,

** Address,

** Telephone number,

** Profession (specialty).

6. Administrative and Sponsor/Company Details:

** Source of report: was it spontaneous, from a clinical investigation (provide details), from the literature (provide copy), other?

** Date event report was first received by sponsor/manufacturer,

** Country in which event occurred,

** Type of report filed to authorities: initial or followup (first, second, etc.).

** Name and address of sponsor/manufacturer/company,

** Name, address, telephone number, and FAX number of contact person in reporting company or institution,

** Identifying regulatory code or number for marketing authorization dossier or clinical investigation process for the suspected product (for example, IND or CTX number, NDA number).

** Sponsor/manufacturer's identification number for the case. (This number should be the same for the initial and followup reports on the same case.)

Text on Validation of Analytical Procedures

1. Introduction

This document presents a discussion of the characteristics for consideration during the validation of the analytical procedures included as part of registration applications submitted within the European Union, Japan, and the United States. This document does not necessarily seek to cover the testing that may be required for registration in, or export to, other areas of the world. Furthermore, this text presentation serves as a collection of terms, and their definitions, and is not intended to provide direction on how to accomplish validation. These terms and definitions are meant to bridge the differences that often exist between various compendia and regulators of the European Union, Japan, and the United States.

The objective of validation of an analytical procedure is to demonstrate that it is suitable for its intended purpose. A tabular summation of the characteristics applicable to identification, control of impurities, and assay procedures is included. Other analytical procedures may be considered in future additions to this document.

2. Types of Analytical Procedures to be Validated

The discussion of the validation of analytical procedures is directed to the four most common types of analytical procedures:

** Identification tests.

** Quantitative tests for impurities' content.

** Limit tests for the control of impurities.

** Quantitative tests of the active moiety in samples of drug substance or drug product or other selected component(s) in the drug product.

Although there are many other analytical procedures, such as dissolution testing for drug products or particle size determination for drug substance, these have not been addressed in the initial text on validation of analytical procedures. Validation of these additional analytical procedures is equally important to those listed herein and may be addressed in subsequent documents.

A brief description of the types of tests considered in this document is provided below.

** Identification tests are intended to ensure the identity of an analyte in a sample. This is normally achieved by comparison of a property of the sample (e.g., spectrum, chromatographic behavior, chemical reactivity, etc.) to that of a reference standard.

** Testing for impurities can be either a quantitative test or a limit test for the impurity in a sample. Either test is intended to accurately reflect the purity characteristics of the sample. Different validation characteristics are required for a quantitative test than for a limit test.

** Assay procedures are intended to measure the analyte present in a given sample. In the context of this document, the assay represents a quantitative measurement of the major component(s) in the drug substance. For the drug product, similar validation characteristics also apply when assaying for the active or other selected component(s). The same validation characteristics may also apply to assays associated with other analytical procedures (e.g., dissolution).

The objective of the analytical procedure should be clearly understood since this will govern the validation characteristics which need to be evaluated. Typical validation characteristics which should be considered are listed below:

Accuracy;
Precision: Repeatability; Intermediate precision;
Specificity;
Detection limit;
Quantitation limit;
Linearity;
Range.

Each of these validation characteristics is defined in the attached Glossary. The table lists those validation characteristics regarded as the most important for the validation of different types of analytical procedures. This list should be considered typical for the analytical procedures cited but occasional exceptions should be dealt with on a case-by-case basis. It should be noted that robustness is not listed in the table but should be considered at an appropriate stage in the development of the analytical procedure.

Furthermore revalidation may be necessary in the following circumstances:
** Changes in the synthesis of the drug substance;
** Changes in the composition of the finished product;
** Changes in the analytical procedure.

The degree of revalidation required depends on the nature of the changes. Certain other changes may require validation as well.

Glossary

1. Analytical Procedure

The analytical procedure refers to the way of performing the analysis. It should describe in detail the steps necessary to perform each analytical test. This may include but is not limited to: The sample, the reference standard and the reagents preparations, use of the apparatus, generation of the calibration curve, use of the formulae for the calculation, etc.

2. Specificity

Specificity is the ability to assess unequivocally the analyte in the presence of components which may be expected to be present. Typically these might include impurities, degradants, matrix, etc.

Lack of specificity of an individual analytical procedure may be compensated by other supporting analytical procedure(s).

This definition has the following implications:

Identification: To ensure the identity of an analyte.

Purity Tests: To ensure that all the analytical procedures performed allow an accurate statement of the content of impurities of an analyte, i.e., related substances test, heavy metals, residual solvents content, etc.

Assay (content or potency): To provide an exact result which allows an accurate statement on the content or potency of the analyte in a sample.

3. Accuracy

The accuracy of an analytical procedure expresses the closeness of agreement between the value which is accepted either as a conventional true value or an accepted reference value and the value found.

This is sometimes termed trueness.

4. Precision

The precision of an analytical procedure expresses the closeness of agreement (degree of scatter) between a series of measurements obtained from multiple sampling of the same homogeneous sample under the prescribed conditions. Precision may be considered at three levels: Repeatability, intermediate precision and reproducibility.

Precision should be investigated using homogeneous, authentic samples. However, if it is not possible to obtain a homogeneous sample it may be investigated using artificially prepared samples or a sample solution.

The precision of an analytical procedure is usually expressed as the variance, standard deviation, or coefficient of variation of a series of measurements.

4.1. Repeatability

Repeatability expresses the precision under the same operating conditions over a short interval of time. Repeatability is also termed intra-assay precision.

4.2. Intermediate precision

Intermediate precision expresses within laboratories' variations: Different days, different analysts, different equipment, etc.

4.3. Reproducibility

Reproducibility expresses the precision between laboratories (collaborative studies, usually applied to standardization of methodology).

5. Detection Limit

The detection limit of an individual analytical procedure is the lowest amount of analyte in a sample which can be detected but not necessarily quantitated as an exact value.

6. Quantitation Limit

The quantitation limit of an individual analytical procedure is the lowest amount of analyte in a sample which can be quantitatively determined with suitable precision and accuracy. The quantitation limit is a parameter of quantitative assays for low levels of compounds in sample matrices, and is used particularly for the determination of impurities and/or degradation products.

7. Linearity

The linearity of an analytical procedure is its ability (within a given range) to obtain test results which are directly proportional to the concentration (amount) of analyte in the sample.

8. Range

The range of an analytical procedure is the interval between the upper and lower concentration (amounts) of analyte in the sample (including these concentrations) for which it has been demonstrated that the analytical procedure has a suitable level of precision, accuracy, and linearity.

9. Robustness

The robustness of an analytical procedure is a measure of its capacity to remain unaffected by small, but deliberate, variations in method parameters and provides an indication of its reliability during normal usage.

Guideline for Good Clinical Practice

Introduction

Good clinical practice (GCP) is an international ethical and scientific quality standard for designing, conducting, recording, and reporting trials that involve the participation of human subjects. This standard has its origin in the Declaration of Helsinki. Compliance with this standard provides public assurance that the rights, well-being, and confidentiality of trial subjects are protected and that the clinical trial data are credible.

The objective of this ICH GCP Guideline is to provide a unified standard for the European Union (EU), Japan, and the United States to facilitate the mutual acceptance of clinical data by the regulatory authorities in these jurisdictions.

The guideline was developed with consideration of the current good clinical practices of the European Union, Japan, and the United States, as well as those of Australia, Canada, the Nordic countries, and the World Health Organization (WHO).

This guideline should be followed when generating clinical data that are intended to be submitted to regulatory authorities.

The principles established in this guideline should also be applied to other investigations that involve therapeutic intervention in, or observation of, human subjects.

1. Glossary

1.1 Adverse Drug Reaction (ADR)

All noxious and unintended responses to a medicinal product (i.e., where the relationship between an adverse event and product cannot be ruled out) related to any dose.

1.2 Adverse Event (AE)

Any untoward medical occurrence in a patient or clinical investigation subject administered a pharmaceutical product and which does not necessarily have to have a causal relationship with this treatment. An adverse event (AE) can therefore be any unfavorable and unintended sign (including an abnormal laboratory finding), symptom, or disease temporally associated with the use of a medicinal (investigational) product, whether or not related to the medicinal (investigational) product.

1.3 Amendment (to the protocol)

See Protocol Amendment.

1.4 Applicable Regulatory Requirement(s)

Any law(s) and regulation(s) addressing the conduct of clinical trials.

1.5 Audit

A systematic and independent examination to determine whether trial-related activities were conducted and analyzed according to the protocol, sponsor's standard operating procedures (SOP's), good clinical practice (GCP), the applicable regulatory requirement(s), and whether the trial reports accurately reflect the procedures carried out, and the data collected.

1.6 Audit Certificate

A written statement by the sponsor's auditor documenting/confirming that an audit of clinical trial-related activities has been conducted.

1.7 Audit Report

A written evaluation by the sponsor's auditor of the accuracy of the audited trial data, and the adherence of the trial to the protocol, the sponsor's SOP's, to GCP, and to applicable regulatory requirement(s).

1.8 Audit Trail

Traceable documents or proof for audit of GCP and/or applicable regulatory requirement(s), which include the essential documents, and allow reconstruction of the course of events.

1.9 Case Report Form (CRF)

A printed, optical, or electronic document designed to record all of the protocol required information that is to be reported to the sponsor on each trial subject.

1.10 Clinical Trial

Any systematic study in human subjects intended to discover or verify the clinical, pharmacological and/or other pharmacodynamic effects of investigational products, and/or to identify any adverse reactions to investigational products, and/or to study absorption, distribution, metabolism, and excretion of these products with the object of ascertaining their safety and/or efficacy.

1.11 Clinical Trial/Study Report

A written description of a trial/study of any therapeutic, prophylactic, or diagnostic agent conducted in human subjects, in which the clinical and statistical description, presentations, and analysis are fully integrated into a single report.

1.12 Comparator (Product)

An investigational or marketed product, or placebo, used as a reference in a clinical trial.

1.13 Compliance (in relation to trials)

Adherence to all trial-related requirements, GCP requirements, and applicable regulatory requirements.

1.14 Confidentiality

Maintenance of secrecy of the sponsor's proprietary information and the subject's source data.

1.15 Contract

A written, dated, and signed agreement between two or more involved parties that sets out any arrangements on delegation and distribution of tasks and obligations and, if appropriate, on financial matters. The protocol may act as the basis of a contract.

1.16 Coordinating Investigator

An investigator assigned the responsibility for the coordination of investigators at different centers participating in a multicenter trial.

1.17 Contract Research Organization (CRO)

A person or an organization (commercial, academic, or other) contracted by the sponsor to perform one or more of a sponsor's trial-related responsibilities.

1.18 Direct Access

Direct access refers to permission for domestic and foreign regulatory authorities, and the sponsor's auditors and monitors, to see, review, analyze, verify, and reproduce any records and reports that are important to evaluation of the clinical trial. When accessing trial-related documents, the regulatory authorities, and the sponsor's monitors and auditors, will take all reasonable precautions within the constraints of the applicable regulatory requirement(s) to maintain the confidentiality of subjects' identities.

1.19 Documentation

All records, in any form (including, but not limited to, written, electronic, magnetic, and optical records, and scans, x-rays, and EKG's) that describe or record the methods, conduct, and/or results of a trial, the factors affecting a trial, and the actions taken.

1.20 Essential Documents

Documents which individually and collectively permit evaluation of the conduct of a study and the quality of the data produced.

1.21 Good Clinical Practice (GCP)

A standard for the design, conduct, performance, monitoring, auditing, recording, analyses, and reporting of clinical trials that provides assurance that the data and reported results are credible and accurate, and that the rights, integrity, and confidentiality of trial subjects are protected.

1.22 Identification Code

A unique identifier assigned by the investigator to each trial subject to protect the subject's identity and used in lieu of subjects' names when the investigator reports on adverse events and other trial data.

1.23 Impartial Witness

A person, who is independent of the trial, who cannot be unfairly influenced by people involved with the trial, who participates in the informed consent process, and who documents (by signing and dating the written informed consent form) that the subject freely gave informed consent to participate in the trial.

1.24 Independent Ethics Committee (IEC)

An independent body (a review board or a committee, institutional, regional, national, or supranational), constituted of medical/scientific professionals and nonmedical/nonscientific members, whose responsibility it is to ensure the protection of the rights and well-being of human subjects involved in a trial and to provide public assurance of that protection, by, among other things, reviewing and approving/providing favorable opinion, the trial protocols and amendment(s), and the methods and material to be used in obtaining and documenting informed consent of the trial subjects.

1.25 Informed Consent

A subject's voluntary confirmation of willingness to participate in a particular trial, after having been informed of all aspects of the trial that are relevant to the subject's decision to participate. Informed consent is documented by means of a written informed consent form that contains relevant information about the trial and that is signed and dated by the subject or the subject's legally acceptable representative.

1.26 Inspection

The act by regulatory authorities of conducting an official review of documents, facilities, records, and any other resources deemed by them to be related to the clinical trial that may be located at the site of the trial, at the sponsor's facilities, at CRO's, or at other establishments deemed appropriate by such authorities.

1.27 Institution (medical)

Any public or private entity or agency or medical or dental facility where clinical trials are conducted.

1.28 Institutional Review Board (IRB)

See Independent Ethics Committee (IEC).

1.29 Interim Clinical Trial/Study Report

A report of intermediate results and their evaluation based on analyses performed during the course of a trial.

1.30 Investigational Product

A pharmaceutical form of an active ingredient or placebo being tested or used as a reference in a clinical trial, including a product with a marketing authorization when used or assembled other than as authorized, i.e., in a way different from the approved form, or when used in a clinical setting other than the one approved, or when used to gain further information about the approved use.

1.31 Investigator

A person responsible for the conduct of the clinical trial at a trial site. In the event that a trial is conducted by a team of individuals at a trial site, the investigator is the responsible leader of the team and may be called the principal investigator. See also Subinvestigator.

1.32 Investigator/Institution

Expression meaning "the investigator or institution, if required by the applicable laws and regulations."

1.33 Investigator's Brochure

A compilation of the clinical and nonclinical data on the investigational product that are relevant to its study in human subjects.

1.34 Legally Acceptable Representative

An individual or juridical or other body authorized under applicable law to consent, on behalf of a prospective subject, to the subject's participation in the clinical trial.

1.35 Monitoring

The act of overseeing the progress of a clinical trial, and of ensuring that it is conducted, recorded, and reported in accordance with the protocol, GCP, and the applicable regulatory requirement(s). A person, designated by the sponsor, who is responsible for the monitoring is a monitor.

1.36 Monitoring Report

A written report from the monitor to the sponsor after each site visit and after all trial-related communications (audit trail concept). Reports should include findings and any actions taken.

1.37 Multicenter Trial

A clinical trial conducted according to one single protocol but at different centers (institutions), and therefore, carried out by more than one investigator.

1.38 Nonclinical Study

Biomedical studies not performed on human subjects.

1.39 Opinion (in relation to Independent Ethics Committee)

A professional judgment and/or advice provided by an Independent Ethics Committee (IEC).

1.40 Original Medical Record

See Source Documents.

1.41 Protocol

A document that provides the background, rationale, and objective(s) of the trial and describes its design, methodology, and organization, including statistical considerations.

1.42 Protocol Amendment

A written description of the change(s) to a protocol.

1.43 Quality Assurance (QA)

All those planned and systematic actions that are established to ensure that the trial is performed and the data are generated, documented (recorded), and reported in compliance with GCP and the applicable regulatory requirement(s).

1.44 Quality Control (QC)

The operational techniques and activities undertaken within the quality assurance system to verify that the requirements for quality of the trial have been fulfilled. Quality control activities are undertaken by all members of the investigational team, including the staff of the sponsor or contract research organization (CRO), involved with planning, conducting, monitoring, data processing/management, documenting (recording), analyzing, evaluating, and reporting a trial with the objective of performing the trial in compliance with the protocol, GCP, and applicable regulatory requirement(s) and drawing conclusions from reliable data.

1.45 Randomization

The process of assigning trial subjects to treatment or control groups using a procedure by which only chance (unbiased) determines the assignments (random allocation).

1.46 Serious Adverse Event (SAE)

Any untoward medical occurrence that at any dose:

--Results in death,
--Is life-threatening,
--Requires inpatient hospitalization or prolongation of existing hospitalization,
--Results in persistent or significant disability/incapacity, or
--Results in a congenital anomaly/birth defect.

1.47 Source Data

All information in original records, and certified copies of original records of clinical findings, observations, or other activities in a clinical trial necessary for the reconstruction and evaluation of the trial. Source data are contained in source documents (original records or certified copies).

1.48 Source Documents

Original documents and records (e.g., laboratory notes, memoranda, subjects' diaries or evaluation checklists, pharmacy dispensing records, recorded data from automated instruments, copies or transcriptions certified after verification as being accurate copies, microfiches, photographic negatives, microfilm or magnetic media, x-rays, and subject files) kept at the pharmacy, the laboratories, and at medico-technical departments involved in the clinical trial.

1.49 Sponsor

An individual, a company, an institution, or an organization which takes responsibility for the initiation, management, and/or financing of a clinical trial.

1.50 Sponsor-Investigator

An individual who both initiates and conducts, alone or with others, a clinical trial, and under whose immediate direction the investigational product is administered to, dispensed to, or used by a subject. The term does not include any person other than an individual (e.g., it does not include a corporation or an agency). The obligations of a sponsor-investigator include both those of a sponsor and those of an investigator.

1.51 Standard Operating Procedures (SOP's)

Detailed, written instructions to achieve uniformity of the performance of a specific function.

1.52 Subinvestigator

Any individual member of the clinical trial team designated and supervised by the investigator at a trial site. See also Investigator.

1.53 Subject

An individual who participates in a clinical trial, either as a recipient of the investigational product(s) or as a control.

1.54 Trial Site

The location(s) where trial-related activities are actually conducted.

1.55 Unexpected Adverse Drug Reaction

An adverse reaction, the nature or severity of which is not consistent with the applicable product information (e.g., Investigator's Brochure for an unapproved investigational product or package insert for an approved product).

1.56 Vulnerable Subjects

Individuals whose willingness to volunteer in a clinical trial may be unduly influenced by the expectation, whether justified or not, of benefits associated with participation, or of a retaliatory response from senior members of the hierarchy or institution in case of refusal to participate. Examples are members of a group with a hierarchical structure, such as medical, pharmacy, dental, and nursing students, subordinate hospital and laboratory personnel, employees of the pharmaceutical industry, members of the armed forces, and persons kept in detention. Other vulnerable subjects include patients with incurable diseases, persons in nursing homes, unemployed or impoverished persons, patients in emergency rooms, ethnic minority groups, homeless persons, nomads, refugees, children, and those incapable of giving consent.

1.57 Well-being (of the trial subjects)

The physical and mental integrity of the subjects participating in a clinical trial.

2. The Principles of ICH GCP

2.1 Clinical trials should be conducted in accordance with the ethical principles that have their origin in the Declaration of Helsinki, and that are consistent with GCP and the applicable regulatory requirement(s).

2.2 Before a trial is initiated, foreseeable risks and inconveniences should be weighed against the anticipated benefit for the individual trial subject and society. A trial should be initiated and continued only if the anticipated benefits justify the risks.

2.3 The rights, safety, and well-being of the trial subjects are the most important considerations and should prevail over interests of science and society.

2.4 The available nonclinical and clinical information on an investigational product should be adequate to support the proposed clinical trial.

2.5 Clinical trials should be scientifically sound, and described in a clear, detailed protocol.

2.6 A trial should be conducted in compliance with the protocol and amendment(s) that have received prior institutional review board (IRB)/independent ethics committee (IEC) approval/favorable opinion.

2.7 The medical care given to, and medical decisions made for, subjects should always be the responsibility of a qualified physician or, when appropriate, of a qualified dentist.

2.8 Each individual involved in conducting a trial should be qualified by education, training, and experience to perform his or her respective task(s).

2.9 Freely given informed consent should be obtained from every subject prior to clinical trial participation.

2.10 All clinical trial information should be recorded, handled, and stored in a way that allows its accurate reporting, interpretation, and verification.

2.11 The confidentiality of records that could identify subjects should be protected, respecting the privacy and confidentiality rules in accordance with the applicable regulatory requirement(s).

2.12 Investigational products should be manufactured, handled, and stored in accordance with applicable good manufacturing practice (GMP). They should be used in accordance with the approved protocol and amendment(s).

2.13 Systems with procedures that assure the quality of every aspect of the trial should be implemented.

3. Institutional Review Board/Independent Ethics Committee (IRB/IEC)

3.1 Responsibilities

3.1.1 An IRB/IEC should safeguard the rights, safety, and well-being of all trial subjects. Special attention should be paid to trials that may include vulnerable subjects.

3.1.2 To fulfill its responsibilities, the following documents should be submitted to the IRB/IEC: Trial protocols, protocol amendment(s), written informed consent forms, consent form updates, subject recruitment procedures (e.g., advertisements), written information to be provided to subjects, Investigator's Brochure (IB), safety reports, documents related to payments and compensation available to subjects, and any other documents that the IRB/IEC may require.

The IRB/IEC should review a proposed clinical trial within a reasonable time limit and document their views in writing, clearly identifying the trial, the documents reviewed, and the dates for the following:

Approval/favorable opinion;
Modifications required prior to approval/favorable opinion;
Disapproval/negative opinion; and
Suspension of any prior approval/favorable opinion.

3.1.3 The IRB/IEC should consider the qualifications of the investigator for the proposed trial as documented by a current curriculum vitae.

3.1.4 The IRB/IEC should conduct continuing review of each ongoing trial at intervals appropriate to the degree of risk to human subjects, but at least once per year.

3.1.5 The IRB/IEC should require that more information than is outlined in paragraph 4.8.10 for written informed consent forms, and for written information to be provided to subjects, be given to subjects when, in the judgment of the IRB/IEC, the additional information is necessary to protect the rights and/or well-being of the subjects.

3.1.6 The IRB/IEC should review both the amount and method of payment to subjects to assure that neither present problems of coercion or undue influence on the trial subjects.

3.1.7 The IRB/IEC should ensure that all information regarding payment, including amounts and schedule of payment to trial subjects, is set forth in the written informed consent form. Such payments should not be wholly contingent upon the trial subject's completion of the trial. Prorated payment should be specified in the written informed consent form.

3.1.8 Except where national law does not permit, the IRB/IEC should evaluate whether remuneration is coercive by reviewing the extent to which investigators/institutions may be rewarded/compensated for participation (see 4.4.2, 5.9).

3.2 Functions and Operations

3.2.1 The IRB/IEC should consist of a reasonable number of members, who collectively have the qualifications and experience to review properly the science and ethics of the proposed trial. It is recommended that the IRB/IEC should include:

(a) At least five members.
(b) At least one member whose primary concern is in a nonscientific area.
(c) All members independent of the investigator and the sponsor.
(d) At least one member who is independent of the trial site.

A list of IRB/IEC members and their qualifications should be maintained.

3.2.2 The IRB/IEC should perform its functions according to written operating procedures, should maintain written records of its activities and minutes of its meetings, and should comply with GCP and with the applicable regulatory requirement(s).

3.2.3 An IRB/IEC should make its decisions at announced meetings where at least a quorum, as stipulated in its written operating procedures, is present.

3.2.4 Only members who participate in the IRB/IEC review and discussion should vote/provide their opinion and advice.

3.2.5 The investigator may provide information on any aspect of the trial, but at the time the IRB/IEC makes a decision, the investigator should not participate in the vote/opinion.

3.2.6 An IRB/IEC may invite nonmembers who have expertise in special areas for assistance.

3.3 Procedures

The IRB/IEC should establish, document in writing, and follow its procedures that include:

3.3.1 Determining its composition (names and qualifications of the members and the authority under which the committee is established).

3.3.2 Scheduling, notifying its members of, and conducting its meetings.

3.3.3 Conducting initial and continuing review of trials.

3.3.4 Determining the frequency of continuing review.

3.3.5 Providing expedited review and approval/favorable opinion of trials involving no more than minimal risk or of change(s) not increasing the risk for a trial for which approval/favorable opinion was provided by the IRB/IEC.

3.3.6 Prohibiting admission of subjects to a trial before the IRB/IEC issues its written approval/favorable opinion of the trial.

3.3.7 Prohibiting initiation of changes in the protocol without prior IRB/ IEC approval/favorable opinion of an appropriate amendment, except when necessary to eliminate immediate hazards to the human subjects or when the change involves logistical or administrative aspects of the trial (e.g., change in monitor, phone number).

3.3.8 Providing that the investigator promptly reports to the IRB/IEC:

(a) Changes in a trial implemented to eliminate immediate hazards to the trial subjects.

(b) Changes affecting significantly the conduct of the trial, and/or increasing the risk to subjects.

(c) All serious and unexpected adverse drug reactions (ADR's).

(d) New information that may affect adversely the safety of the subjects or the conduct of the trial.

3.3.9 Ensuring that the IRB/IEC promptly notify in writing the investigator/institution concerning:

(a) Its trial-related decisions.

(b) The reasons for its decisions.

(c) Procedures for appeal of its decisions.

3.4 Records

Where required by applicable regulation, the IRB/IEC should retain all relevant records (e.g., written procedures, membership lists, submitted documents, minutes of meetings, and correspondence) for a period of at least 3 years after completion of the trial and make them available upon request from the regulatory authority(ies).

Written procedures and membership lists should be made available upon request from the investigator(s), and/or the sponsor(s), except where national law does not permit.

4. INVESTIGATOR

4.1 Investigator's Qualifications and Agreements

4.1.1 The investigator(s) should be qualified by education, training, and experience to assume responsibility for the proper conduct of the trial, should meet all the qualifications required by the applicable regulatory requirements, and should provide evidence of such qualifications through up-to-date curriculum vitae and other credentials.

4.1.2 The investigator should be thoroughly familiar with the appropriate use of the investigational product(s), as described in the protocol/ amendment(s), in the current Investigator's Brochure, in other information sources provided by the sponsor, and in the available literature.

4.1.3 The investigator should be aware of, and should comply with, GCP and the applicable regulations.

4.1.4 The investigator/institution should permit monitoring and auditing by the sponsor, and inspecting by the appropriate regulatory authorities.

4.1.5 The investigator should maintain a written record of appropriately qualified persons, delegated to assume specified investigator trial responsibilities.

4.2 Adequate Resources

4.2.1 The investigator should be able to demonstrate (e.g., based on retrospective data) a potential for recruiting the required number of suitable subjects within the agreed recruitment period.

4.2.2 The investigator should have sufficient time to properly conduct and complete the trial within the agreed trial period.

4.2.3 To conduct the trial properly and safely, the investigator should have available an adequate number of qualified staff and adequate facilities for the foreseen duration of the trial.

4.2.4 The investigator should ensure that all persons assisting with the trial are adequately informed about the protocol/amendment(s), the investigational product(s), and their trial-related responsibilities.

4.3 Medical Care of Trial Subjects

4.3.1 A qualified physician (or dentist, when appropriate), who is an investigator or a subinvestigator for the trial, should be responsible for all trial-related medical (or dental) decisions.

4.3.2 For the duration of a subject's participation in a trial, the investigator should ensure that adequate medical care (or dental care, when appropriate) is made available to the subject for trial-related illness(es)/adverse event(s). The investigator should inform the subject when medical care is needed for intercurrent illness(es) of which the investigator becomes aware. Following a subject's participation in a trial, the investigator should ensure that adequate medical care is provided for any adverse events, including clinically significant laboratory values, related to the trial.

4.3.3 It is recommended that the investigator inform the subject's primary physician, when there is one, with the subject's consent, about the participation in the trial.

4.3.4 Although a subject is not obliged to give his/her reason(s) for withdrawing prematurely from a trial, the investigator should make a reasonable effort to ascertain the reason(s), while fully respecting the subject's rights.

4.4 Communication With IRB/IEC

4.4.1 Before initiating a trial, the investigator should have written and dated approval/favorable opinion from the IRB/IEC for the trial protocol/ amendment(s), written informed consent form, consent form updates, subject recruitment procedures (e.g., advertisements), and written information to be provided to subjects.

4.4.2 Except where national law does not permit, the investigator also should obtain approval/favorable opinion from the IRB/IEC for the trial-related costs and payments (see 3.1.8, 5.9, 5.11).

4.4.3 As part of the investigator's written application to the IRB/IEC, the investigator should provide the IRB/IEC with a current copy of the Investigator's Brochure. If the updated Investigator's Brochure contains important updates on safety information and clinical trial analyses, the investigator should supply a copy to the IRB/IEC.

4.4.4 During the trial the investigator should provide to the IRB/IEC all documents subject to review (see 3.1.2, 3.1.5, 3.1.7, 3.3.8, 4.10, 4.11).

4.5 Compliance with Protocol/Amendment(s)

4.5.1 The investigator/institution should conduct the trial in compliance with the protocol/amendment(s) agreed to by the sponsor, and the IRB/IEC and, if required, by the appropriate authority(ies). The investigator/institution and the sponsor should sign the protocol/amendment(s), or an alternative contract, to confirm agreement.

4.5.2 The investigator should not implement any changes in a trial without agreement by the sponsor and prior review or expedited review by the IRB/IEC, and without documented approval/favorable opinion of an appropriate amendment, except when the change involves logistical or administrative aspects of the trial.

4.5.3 The investigator, or person designated by the investigator, should document and explain any deviation from the approved protocol/amendment(s).

4.5.4 The investigator may implement a deviation, where necessary, to eliminate an immediate hazard(s) to trial subjects without prior IRB/IEC approval/favorable opinion.

As soon as possible, the implemented deviation, the reasons, and, if appropriate, the proposed protocol amendment(s) should be submitted: (1) to the IRB/IEC for review and approval/favorable opinion, (2) to the sponsor for agreement, and, if required, (3) to the appropriate authority(ies).

4.6 Investigational Product(s)

4.6.1 Responsibility for investigational product(s) accountability at the trial site(s) rests with the investigator.

4.6.2 Where allowed/required, the investigator may/should assign some/all of the investigator's duties for investigational product(s) accountability at the trial site(s) to an appropriate pharmacist.

4.6.3 The investigator should ensure that the investigational product(s) are used only in accordance with the approved protocol/amendment(s).

4.6.4 The investigator, or a person designated by the investigator, should clearly explain the correct use of the investigational product(s) to each subject and should check, at each subject visit, that the subject is using the product(s) properly.

4.6.5 The investigator and/or pharmacist should maintain records of the product's delivery to the trial site, the inventory at the site, the use by each subject, and the return to the sponsor or alternative disposition of unused product(s). These records should include dates, quantities, batch/serial numbers, expiration dates (if applicable), and the unique code numbers assigned to the investigational product(s) and trial subjects. Investigators should maintain records that document adequately that the subjects were provided the doses specified by the protocol/amendment(s) and reconcile all investigational product(s) received from the sponsor.

4.6.6 The investigational product(s) should be properly and safely stored in accordance with applicable regulatory requirement(s).

4.7 Randomization Procedures

The investigator should follow the trial's randomization procedures, if any, and should ensure that the code is broken only in accordance with the protocol. If the trial is blinded, the investigator should document and explain any unblinding of the investigational product(s) promptly to the sponsor.

4.8 Informed Consent of Trial Subjects

4.8.1 In obtaining and documenting informed consent, the investigator should comply with the applicable regulatory requirement(s), and should adhere to GCP and to the ethical principles that have their origin in the Declaration of Helsinki. Prior to the beginning of the trial, the investigator should have the IRB/IEC's written approval/favorable opinion of the written informed consent form and written information to be provided to subjects.

4.8.2 The written informed consent form and written information to be provided to subjects should be revised whenever new information becomes available that may be relevant to the subject. Any revised written informed consent form, and written trial information should receive the IRB/IEC's approval/favorable opinion in advance of use.

4.8.3 Neither the investigator, nor the trial staff, should coerce or unduly influence a subject to participate or to continue to participate in a trial.

4.8.4 None of the oral and written information concerning the trial, including the written informed consent form, should contain any language that causes the subject or the subject's legally acceptable representative to waive or to appear to waive any legal rights, or that releases or appears to release the investigator, the institution, the sponsor, or their agents from liability for negligence.

4.8.5 The investigator, or a person designated by the investigator, should fully inform the subject or, if the subject is unable to provide informed consent, the subject's legally acceptable representative, of all pertinent aspects of the trial including the written information as approved/received favorable opinion from the IRB/IEC.

4.8.6 The language used in the oral and written information about the trial, including the written informed consent form, should be nontechnical and should be understandable to the subject or the subject's legally acceptable representative.

4.8.7 Before informed consent is given, the investigator, or a person designated by the investigator, should provide the subject or the subject's legally acceptable representative ample time to decide and opportunity to inquire about details of the trial. All questions about the trial should be answered to the satisfaction of the subject or the subject's legally acceptable representative.

4.8.8 Prior to participation in the trial, the written informed consent form should be signed and personally dated by the subject, or the subject's legally acceptable representative, and by the person who conducted the informed consent discussion.

4.8.9 If the subject or the subject's legally acceptable representative is unable to read, an impartial witness should be present during the entire informed consent discussion. After the written informed consent form is read to the subject and orally consented to, and signed and personally dated by the subject, if capable of doing so, or the subject's legally acceptable representative, the witness should also sign and personally date the consent form, attesting that informed consent was freely given by the subject or the subject's legally acceptable representative.

4.8.10 Both the informed consent discussion and the written informed consent form should include clear explanations of the following:

(a) The trial involves research.

(b) The purpose of the trial.

(c) The trial treatment(s).

(d) The trial procedures to be followed, including all invasive procedures.

(e) The subject's responsibilities.

(f) Those trial features that are experimental.

(g) The reasonably foreseeable risks or inconveniences to the subject.

(h) The reasonably expected benefits. When there is no intended clinical benefit to the subject, the subject should be made aware of this.

(i) The alternative procedure(s) or course(s) of treatment that may be available to the subject, and their potential benefits and risks.

(j) The compensation and/or treatment available to the subject in the event of trial-related injury.

(k) The anticipated prorated payment, if any, to the subject for participating in the trial.

(l) The anticipated expenses, if any, to the subject for participating in the trial.

(m) That the subject's participation in the trial is voluntary and the subject may refuse to participate or withdraw from the trial, at any time, without penalty or loss of benefits to which the subject is otherwise entitled.

(n) That the monitor(s), the auditor(s), the IRB/IEC, and the regulatory authority(ies) will be granted direct access to the subject's original medical records for verification of clinical trial procedures and/or data without violating the confidentiality of the subject to the extent permitted by the applicable laws and regulations and that, by signing a written informed consent form, the subject or the subject's legally acceptable representative is authorizing such access.

(o) That confidentiality of records that identify the subject will be maintained and will not be made publicly available to the extent permitted by the applicable laws and/or regulations.

(p) That the subject or the subject's legally acceptable representative will be informed in a timely manner if information becomes available that may be relevant to the subject's participation in the trial.

(q) The person(s) to contact for further information regarding the trial and the rights of trial subjects, and whom to contact in the event of trial-related injury.

(r) The circumstances under which the subject's participation in the trial may be terminated without the subject's consent.

(s) The expected duration of the subject's participation in the trial.

(t) The approximate number of subjects involved in the trial.

4.8.11 The subject or the subject's legally acceptable representative should receive a copy of the signed and dated written informed consent form and signed and dated consent form updates, and a copy of the written trial information to be provided to subjects and its amendment(s).

4.8.12 To the extent that a minor child is capable of understanding and granting informed consent, the minor should give consent and, in addition to the minor's legally acceptable representative, the minor should sign and personally date the written informed consent form if the minor is capable of doing so.

4.8.13 In a nontherapeutic trial (i.e., when there is no anticipated direct clinical benefit to the subject), consent should always be given by the subject and the written consent form should be signed and personally dated by the subject.

4.8.14 Where prior consent of the trial subject is not possible, the protocol submitted to the IRB/IEC may provide that such consent need not be obtained and that only the consent of the subject's legally acceptable representative, if present, should be solicited. Absence of the subject's legally acceptable representative should require other measures that are described in the protocol to ensure compliance with applicable regulatory requirements. The trial subject or the subject's legally acceptable representative should be informed as soon as possible and consent should be requested.

4.9 Records and Reports

4.9.1 The source documents should support the data reported on the CRF, should identify the trial, and should document the dates of the subject's participation.

4.9.2 The investigator should ensure the accuracy, completeness, legibility, and timeliness of the data reported to the sponsor in the CRF's and in all required reports.

4.9.3 Any correction to a CRF should not obscure the original entry; this applies to both written and electronic corrections (see 5.18.5(n)). Sponsors should provide guidance to investigators or the investigators' designated representatives on making such corrections. Sponsors should have written procedures to assure that changes made by sponsor's designated representatives in CRF's are necessary, documented by an audit trail, and endorsed by the investigator.

4.9.4 The investigator/institution should maintain the trial documents as specified in the ICH Guideline for "Essential Documents for the Conduct of a Clinical Trial," and as required by the applicable regulatory requirements. The investigator/institution should take measures to prevent accidental or premature destruction of records, including the subject's original medical records and accountability records for the investigational product(s).

4.9.5 Where there are no applicable regulatory requirements that require the retention of the essential documents for at least 2 years after the last marketing application approval or 2 years after formal discontinuation of the clinical development of the investigational product, the investigator/institution should prevent accidental or premature destruction of records for the period specified in the agreement with the sponsor, which should be at least as long as indicated above.

4.9.6 Upon request of the monitor, auditor, IRB/IEC, or regulatory authority, the investigator/institution should make available for direct access all requested trial-related records.

4.10 Progress Reports

4.10.1 The investigator should routinely submit written summaries of the status of the trial to the IRB/IEC. These reports should be submitted annually or more frequently if required by the IRB/IEC.

4.10.2 The investigator should promptly provide written reports to the sponsor and the IRB/IEC (see 3.3.8) on any problems, changes, or occurrences affecting the conduct of the trial, and/or increasing the risk to subjects.

4.11 Safety Reporting

4.11.1 All serious adverse events (SAE's) should be reported immediately to the sponsor unless the protocol or other document (e.g., Investigator's Brochure) identifies certain expected SAE's as not needing immediate reporting. This immediate reporting should be followed promptly by detailed, written reports, which should identify subjects by unique code numbers assigned to the trial subjects instead of by the subjects' names, personal identification numbers, and/or addresses. The investigator should also comply with local regulatory requirements related to the reporting of unexpected serious adverse drug reactions to the regulatory authority(ies) and/or the IRB/IEC.

4.11.2 In addition, those adverse events and/or laboratory abnormalities, which are identified in the protocol/amendment(s) as critical to safety evaluations, should be reported to the sponsor according to the reporting requirements and within the time periods that are specified by the sponsor in the protocol/amendments.

4.11.3 For all reported deaths, the investigator should supply the sponsor and, where required, the IRB/IEC with relevant information including autopsy reports and terminal medical reports.

4.12 Trial Termination or Suspension

4.12.1 If the trial is terminated or suspended by the investigator without prior agreement of the sponsor, the investigator should inform the institution where applicable and the investigator/institution should provide promptly to the sponsor and the IRB/IEC a detailed explanation of the termination/suspension.

4.12.2 If the trial is terminated/suspended by the sponsor, the investigator should provide promptly to the IRB/IEC and the institution, where required, a detailed explanation of the termination/suspension.

4.12.3 If the IRB/IEC terminates or suspends a trial, the investigator should immediately notify the sponsor and provide the sponsor with a detailed written explanation of the termination/suspension.

5. SPONSOR

5.1 Quality Assurance and Quality Control

5.1.1 The sponsor is responsible for implementing and maintaining quality assurance and quality control systems with written SOP's to ensure that trials are conducted and data are generated, documented (recorded), and reported in compliance with the protocol(s), GCP, and the applicable regulatory requirements.

5.1.2 The sponsor is responsible for securing agreement from all involved parties to ensure the availability of all trial-related sites and source data/documents, and reports for the purpose of monitoring and auditing by the sponsor, and inspecting by domestic and foreign regulatory authorities.

5.1.3 Quality control should be applied to each stage of data handling to ensure that all data are reliable and have been processed correctly.

5.1.4 Agreements, made by the sponsor with the investigator/institution and any other parties involved with the clinical trial, should be in writing, as part of the protocol or a written contract, to assure the quality of the trial-related activities.

5.2 Contract Research Organization (CRO)

5.2.1 A sponsor may transfer any or all of the sponsor's trial-related responsibilities to a CRO. However, the ultimate responsibility for the quality and integrity of the trial data should always reside with the sponsor. The sponsor should ensure that the CRO has and implements quality assurance and quality control.

5.2.2 Each trial-related responsibility that is transferred to and assumed by a CRO should be specified in a written contract.

5.2.3 All trial-related responsibilities not specifically transferred to and assumed by a CRO should be considered to be retained by the sponsor.

5.2.4 All references to a sponsor in this guideline also apply to a CRO to the extent that a CRO has assumed the trial-related responsibilities of a sponsor.

5.3 Medical Expertise

The sponsor should designate appropriately qualified medical personnel, who are readily available to advise on trial-related questions or problems. If necessary, outside consultant(s) may be appointed for this purpose.

5.4 Trial Design

5.4.1 The sponsor should utilize qualified experts (e.g., biostatisticians, clinical pharmacologists, and physicians) as appropriate, throughout all stages of the trial process, from designing the protocol and CRF's and planning the analysis to analyzing and preparing interim and final clinical trial reports.

5.4.2 See Chapter 6 "Clinical Trial Protocol and Protocol Amendment(s)," and the ICH Guideline for "Structure and Content of Clinical Study Reports," and other appropriate ICH guidance on trial design and protocols.

5.5 Trial Management, Data Handling, and Recordkeeping

5.5.1 The sponsor should utilize appropriately qualified individuals to supervise the overall conduct of the trial, to handle the data, to verify the data, to conduct the statistical analyses, and to prepare the trial reports.

5.5.2 When using electronic trial data handling and/or remote electronic trial data systems, the sponsor should:

(a) Ensure that only validated data processing systems are used.

(b) Maintain SOP's for using these systems.

(c) Ensure that the systems are designed to allow data changes without any deletion of entered data (i.e., maintain an audit trail).

(d) Maintain a security system that prevents unauthorized access to the data.

(e) Maintain a list of the individuals who are authorized to make data changes.

(f) Maintain adequate backup of the data.

(g) Safeguard the blinding, if any.

5.5.3 The sponsor should ensure the greatest possible accuracy when processing data. If data are transformed during processing, it should always be possible to compare the original data and observations with the processed data.

5.5.4 The sponsor should use an unambiguous subject code that enables identification of all the data reported for each subject.

5.5.5 The sponsor, or other owners of the data, should retain all of the sponsor-specific essential documents pertaining to the trial. (See ICH Guideline for "Essential Documents for the Conduct of a Clinical Trial.")

5.5.6 The sponsor should retain all sponsor-specific essential documents in conformance with the applicable regulatory requirement(s) of the country(ies) where the product is approved, and/or where the sponsor intends to apply for approval(s).

5.5.7 If the sponsor discontinues the clinical development of an investigational product for any or all indications, routes of administration, or dosage forms, the sponsor should maintain all sponsor-specific essential documents for at least 2 years after formal discontinuation or in conformance with the applicable regulatory requirement(s).

5.5.8 If the sponsor discontinues the clinical development of an investigational product, the sponsor should notify all the trial investigators and all the appropriate regulatory authorities.

5.5.9 Any transfer of ownership of the data should be reported to the appropriate authority(ies), as required by the applicable regulatory requirement(s).

5.5.10 Where there are no applicable regulatory requirement(s) that require the investigator/institution to retain the trial-related essential documents for at least 2 years after the last marketing application approval or 2 years after formal discontinuation of the clinical development of the investigational product for any or all indications, routes of administration, or dosage forms, it is recommended that the sponsor agree with the investigator/institution to prevent accidental or premature destruction of the essential documents, for a period, which should be at least as long as indicated above.

5.6 Investigator Selection

5.6.1 The sponsor is responsible for selecting the investigator(s)/institution(s). The sponsor should select the investigator(s) who is/are qualified by training and experience to conduct the trial(s) and have adequate resources (see 4.1, 4.2). If coordinating committees and/or selection of coordinating investigator(s) are appropriate in multicenter trials, the organization and/or selection is the sponsor's responsibility.

5.6.2 Before entering an agreement with an investigator/institution to conduct a trial, the sponsor should provide the investigator(s)/institution(s) with the protocol and an up-to-date Investigator's Brochure, and should provide sufficient time for the investigator/institution to review the protocol and the information provided.

5.6.3 The sponsor should obtain the investigator's/institution's agreement to conduct the trial in compliance with the agreed to and/or approved protocol/amendment(s), and with GCP and the applicable regulatory requirement(s), and to accept procedures for data recording/reporting, monitoring, auditing, and inspecting. The sponsor and the investigator/institution should sign the protocol/amendment(s), or an alternative document, to confirm this agreement.

5.7 Allocation of Responsibilities

Prior to initiating a trial, the sponsor should define, establish, and allocate all trial-related responsibilities to either the sponsor, investigator(s), and/or other parties.

5.8 Compensation to Subjects and Investigators

5.8.1 If required by the applicable regulatory requirement(s), the sponsor should provide insurance or should indemnify (legal and financial coverage) the investigator/institution against claims arising from the trial, except for claims that arise from malpractice and/or negligence.

5.8.2 The sponsor's policies and procedures should address the costs of treatment of trial subjects in the event of trial-related injuries in accordance with the applicable regulatory requirement(s).

5.8.3 When trial subjects receive compensation, the method and manner should comply with applicable regulatory requirement(s).

5.9 Financing

Except where national law does not permit, the financial aspects of the trial should be documented as an agreement between the sponsor and the investigator/institution, and should be reviewed by the IRB/IEC and by the appropriate authority(ies) to evaluate if remuneration is not coercive (see 3.1.8, 4.4.2).

5.10 Notification/Submission to Regulatory Authority(ies)

Before initiating the clinical trial(s), the sponsor (or the sponsor and the investigator), if required by the applicable regulatory requirement(s) should submit any required application(s) to the appropriate authority(ies) for review, acceptance, and/or permission (as required by the applicable regulatory requirement(s)) to begin the trial(s). Any notification/submission should be dated and contain sufficient information to identify the protocol/amendment(s).

5.11 Confirmation of Review by IRB/IEC

5.11.1 The sponsor should obtain from each investigator:

(a) The name and address of the investigator's IRB/IEC.

(b) Statement from the IRB/IEC that it complies with GCP and the applicable laws and regulations.

(c) Documented IRB/IEC approval/favorable opinion and, where required, a current copy of protocol/amendment(s), written informed consent forms, and written information to be provided to subjects, subject recruiting procedures, documents related to payments and compensation available to the subjects, and any other documents that the IRB/IEC may require.

5.11.2 If the IRB/IEC conditions its approval/favorable opinion of the protocol and/or amendment(s), written informed consent form, written information to be provided to subjects and other procedures upon modifications, the sponsor should obtain from the investigator:

(a) A copy of the modification(s) as approved/received favorable opinion by the IRB/IEC.

(b) Documentation that the modification(s) was/were approved/received favorable opinion by the IRB/IEC and the date of the IRB/IEC's approval/favorable opinion.

5.11.3 The sponsor should obtain documentation and dates of any reapprovals/re-evaluations with favorable opinion, and of any withdrawals or suspensions of approval/favorable opinion.

5.12 Information on Investigational Product(s)

5.12.1 When planning trials, the sponsor should ensure the availability of sufficient safety and efficacy data for the product(s), including the available data from investigations and/or marketing worldwide. Sufficient safety and efficacy data from nonclinical studies and/or clinical trials should be available to justify human exposure by the route, at the dosages, and for the duration proposed to be studied during the trial and should be appropriate to the phase, type, and target population of the proposed trial.

5.12.2 The sponsor should update the Investigator's Brochure as significant new information becomes available. (See ICH Guideline for "Investigator's Brochure.")

5.13 Manufacturing, Packaging, Labeling, and Coding Investigational Product(s)

5.13.1 The sponsor should ensure that the investigational product(s) (including active comparators and placebo, if applicable) is characterized as appropriate to the stage of development of the product(s), manufactured in accordance with any applicable GMP, and coded and labeled in a manner that protects the blinding, if applicable. In addition, the labeling should comply with applicable regulatory requirement(s).

5.13.2 The sponsor should determine, for the investigational product(s), acceptable storage temperatures, storage times, reconstitution fluids and procedures, and devices for product infusion, if any. The sponsor should inform all involved parties (e.g., monitors, investigators, pharmacists, storage managers) of these determinations.

5.13.3 The investigational product(s) should be packaged to prevent contamination and unacceptable deterioration during transport and storage.

5.13.4 In blinded trials, the coding system for the investigational product(s) should include a mechanism that permits rapid identification of the product(s) in case of a medical emergency, but does not permit undetectable breaks of the blinding.

5.13.5 If significant formulation changes are made in the investigational or comparator product during the course of the clinical development/trial, the results of additional studies (e.g., stability, comparative dissolution rate, comparative bioavailability) demonstrating that these changes would not be expected to alter the pharmacokinetic profile or other clinical characteristics of the product should be available prior to the use of the new formulation in the clinical trial.

5.14 Supplying and Handling Investigational Product(s)

5.14.1 The sponsor is responsible for supplying the investigator(s) with the investigational product(s).

5.14.2 The sponsor should not supply an investigator with the investigational product(s) until the sponsor obtains documentation of all required approvals (e.g., IRB/IEC and authorities).

5.14.3 The sponsor should ensure that written procedures include the requirements that the investigator/institution follow for the handling and storage of investigational product(s) for the trial and documentation thereof. The procedures should address adequate and safe receipt, handling, storage, dispensing, retrieval of unused product from subjects, and return of unused investigational product(s) to the sponsor (or alternative disposition if authorized by the sponsor and in compliance with the applicable regulatory requirement(s)).

5.14.4 The sponsor should:

(a) Ensure timely delivery of investigational product(s) to the investigator(s).

(b) Maintain records that document shipment, delivery, receipt, disposition, return, and destruction of the investigational product(s). (See ICH Guideline for "Essential Documents for the Conduct of a Clinical Trial.")

(c) Maintain a system for retrieving investigational products and documenting this retrieval (e.g., for deficient product recall, reclaim after trial completion, expired product reclaim).

(d) Maintain a system for the disposition of unused investigational product(s) and documenting this disposition.

5.14.5 Where GMP does not apply to investigational product(s), the sponsor should:

(a) Take steps to ensure that the investigational product(s) are stable over the period of use.

(b) Maintain sufficient quantities of batch samples of the investigational product(s) in order to reconfirm specifications, if it appears necessary, and maintain records of batch sample analyses and characteristics. To the extent stability permits, batch samples should be retained either until the statistical analyses are complete or as required by the applicable regulatory requirement(s), whichever represents the longer retention period.

5.15 Record Access

5.15.1 The sponsor should ensure that it is specified in the protocol that the investigator(s)/institution(s) provide direct access to source data/ documents for trial-related monitoring, audits, IRB/IEC review, and regulatory inspection.

5.15.2 The sponsor should verify that each subject has consented, in writing, to direct access to his/her original medical records for trial-related monitoring, audit, IRB/IEC review, and regulatory inspection.

5.16 Safety Information

5.16.1 The sponsor is responsible for the ongoing safety evaluation of the investigational product(s).

5.16.2 The sponsor should promptly notify all concerned investigator(s)/institution(s) and the regulatory authority(ies) of findings that could affect adversely the safety of subjects, impact the conduct of the trial, or alter the IRB/IEC's approval/favorable opinion to continue the trial.

5.17 Adverse Drug Reaction Reporting

5.17.1 The sponsor should expedite the reporting to all concerned investigator(s)/institutions(s), to the IRB(s)/IEC(s), where required, and to the regulatory authority(ies) of all adverse drug reactions (ADR's) that are both serious and unexpected.

5.17.2 Such expedited reports should comply with the applicable regulatory requirement(s) and with the ICH Guideline for "Clinical Safety Data Management: Definitions and Standards for Expedited Reporting."

5.17.3 The sponsor should submit to the regulatory authority(ies) all safety updates and periodic reports, as required by applicable regulatory requirement(s).

5.18 Monitoring

5.18.1 Purpose

The purposes of trial monitoring are to verify that:

(a) The rights and well-being of human subjects are protected.

(b) The reported trial data are accurate, complete, and verifiable from source documents.

(c) The conduct of the trial is in compliance with the currently approved protocol/amendment(s), with GCP, and with the applicable regulatory requirement(s).

5.18.2 Selection and Qualifications of Monitors.

(a) Monitors should be appointed by the sponsor.

(b) Monitors should be appropriately trained, and should have the scientific and/or clinical knowledge needed to monitor the trial adequately.

(c) Monitors should be thoroughly familiar with the investigational product(s), the protocol/amendment(s), written informed consent form, written information to be provided to subjects, GCP, and the applicable regulatory requirement(s).

5.18.3 Number of Monitors

The sponsor should ensure that the trials are appropriately monitored at the trial sites.

The number of monitors needed to ensure adequate monitoring of a trial depends primarily on the complexity of the trial, the number and locations of trial sites, and the number of trial subjects.

5.18.4 Monitoring Schedule

The monitor(s) should visit the trial site(s) before, during, and after the trial. The on-site visits should be frequent enough to monitor the trial adequately.

5.18.5 Monitor's Responsibilities

The monitor(s) should ensure that the trial is conducted and documented properly by:

(a) Acting as the main line of communication between the sponsor and the investigator.

(b) Verifying that the investigator has adequate qualifications and resources (see 4.1, 4.2, 5.6) and remain adequate throughout the trial period, that facilities, including laboratories, equipment, and staff, are adequate to safely and properly conduct the trial and remain adequate throughout the trial period.

(c) Verifying, for the investigational product(s):

(i) That storage times and conditions are acceptable, and that supplies are sufficient throughout the trial.

(ii) That the investigational product(s) are supplied only to subjects who are eligible to receive it and at the protocol specified dose(s).

(iii) That subjects are provided with necessary instruction on properly using, handling, storing, and returning the investigational product(s).

(iv) That the receipt, use, and return of the investigational product(s) at the trial sites are controlled and documented adequately.

(v) That the disposition of unused investigational product(s) at the trial sites complies with applicable regulatory requirement(s) and is in accordance with the sponsor.

(d) Verifying that the investigator follows the approved protocol and all approved amendment(s), if any.

(e) Verifying that written, informed consent was obtained before each subject's participation in the trial.

(f) Ensuring that the investigator receives the current Investigator's Brochure, all documents, and all trial supplies needed to conduct the trial properly and to comply with the applicable regulatory requirement(s).

(g) Ensuring that the investigator and the investigator's trial staff are adequately informed about the trial.

(h) Verifying that the investigator and the investigator's trial staff are performing the specified trial functions, as per written agreement between the sponsor and the investigator/institution, and have not delegated these functions to unauthorized or unapproved individuals.

(i) Verifying that the investigator is enrolling only eligible subjects.

(j) Reporting the subject recruitment rate.

(k) Verifying that accurate, complete, and current source documents and trial records are maintained.

(l) Verifying that the investigator provides all the required reports, notifications, applications, and submissions, and that these documents are accurate, complete, timely, legible, dated, and identify the trial.

(m) Checking the accuracy and completeness of the CRF entries against the subjects' source documents and other trial-related records. The monitor specifically should verify that:

(i) The data required by the protocol are reported accurately on the CRF's and are consistent with the source documents.

(ii) Any dose and/or therapy modifications are well documented for each of the trial subjects.

(iii) Concomitant medications and intercurrent illnesses are reported in accordance with the protocol on the CRF's.

(iv) Visits that the subjects fail to make, tests that are not conducted, and examinations that are not performed are clearly reported as such on the CRF's.

(v) All withdrawals and dropouts are reported and explained on the CRF's.

(n) Informing the investigator of any CRF entry error, omission, or illegibility. The monitor should ensure that appropriate corrections, additions, or deletions are made, dated, explained (if necessary), and initialed by the investigator or by a member of the investigator's trial staff who is authorized to initial CRF changes for the investigator. This authorization should be documented.

(o) Determining whether all adverse events (AE's) are appropriately reported within the time periods required by GCP, the protocol, the IRB/IEC, the sponsor, and the applicable regulatory requirement(s).

(p) Determining whether the investigator is maintaining the essential documents. (See Guideline for "Essential Documents for the Conduct of a Clinical Trial.")

(q) Communicating significant deviations to the investigator and taking appropriate action to prevent recurrence of the detected deviations.

5.18.6 Monitoring Procedures

The monitor should follow established written Standard Operating Procedures.

5.18.7 Monitoring Report

(a) The monitor should submit a written report to the sponsor after each trial-site visit or trial-related communication.

(b) Reports should include the time, date, site, name of the monitor, and name of the investigator or other individual(s) contacted.

(c) Reports should include the monitor's statements concerning the significant findings/facts, deviations and deficiencies, conclusions, actions taken or to be taken and/or actions recommended to secure compliance.

(d) The review and followup of the monitoring report with the sponsor should be documented.

5.19 Audit

If or when, as part of implementing quality assurance, the sponsors performing audits should consider:

5.19.1 Purpose

The purpose of a sponsor's audit, which is independent of and separate from routine monitoring or quality control functions, should be to evaluate trial conduct, protocol compliance, and GCP compliance.

5.19.2 Selection and Qualification of Auditors

(a) The sponsor should appoint individuals who are independent of the trial to conduct audits.

(b) The sponsor should ensure that the auditors are qualified by training and experience to conduct audits properly.

5.19.3 Auditing Procedures

(a) The sponsor should ensure that the auditing of clinical trials/systems is conducted in accordance with the sponsor's written procedures on what to audit, how to audit, the frequency of audits, and the form and content of audit reports.

(b) The sponsor's audit plan and procedures should be guided by the importance of the trial to submissions to regulatory authorities, the number of subjects in the trial, the type and complexity of the trial, the level of risks to the trial subjects, and any identified problem(s).

(c) The observations and findings of the auditor(s) should be documented in an audit report.

(d) The audit reports should not routinely be made available for inspection, but should be made available for inspection upon request by the regulatory authority(ies).

(e) When required by applicable law or regulation, the sponsor should provide an audit certificate.

5.20 Noncompliance

5.20.1 Noncompliance with the protocol, GCP, and/or applicable regulatory requirement(s) by an investigator/institution, or by member(s) of the sponsor's staff should lead to prompt action by the sponsor to secure compliance.

5.20.2 If monitoring and/or auditing identified serious and/or persistent noncompliance of an investigator, the sponsor should terminate the investigator's participation in the trial. When an investigator's participation is terminated because of noncompliance, the sponsor should notify promptly the responsible IRB(s)/IEC(s) and the appropriate authority(ies).

5.21 Premature Termination of a Trial

If a trial is prematurely terminated, the sponsor should promptly inform the investigators/institutions, the IRB(s)/IEC(s), and the appropriate regulatory authority(ies) of the termination and the reason(s) for the termination.

5.22 Clinical Trial Reports

Whether the trial is completed or prematurely terminated, the sponsor should ensure that the clinical trial reports are prepared and provided to the regulatory agency(ies) as required by the applicable regulatory requirement(s). The sponsor should also ensure that the clinical trial reports meet the standards of the ICH guideline entitled, "Structure and Content of Clinical Study Reports."

5.23 Multicenter Trials

For multicenter trials, the sponsor should ensure that:

5.23.1 All investigators conduct the trial in strict compliance with the same protocol, or with well-documented amended protocols that are agreed to by the sponsor and approved/provided a favorable opinion by the IRB/IEC and authorities, when required, for specific sites.

5.23.2 The CRF's are designed to capture the required data at all multicenter trial sites, with exceptions for those investigators who are collecting additional data.

5.23.3 The responsibilities of a coordinating investigator and the other participating investigators are documented prior to the start of the trial.

5.23.4 All investigators are given sufficient instructions on how to follow the protocol and to comply with a uniform set of standards for assessment of the clinical and laboratory findings and for completing the CRF's.

5.23.5 Communication between investigators is possible.

6. CLINICAL TRIAL PROTOCOL AND PROTOCOL AMENDMENT(S)

The contents of a trial protocol should generally include the following topics:

6.1 General Information

6.1.1 Protocol title, protocol identifying number, and date. Any amendment(s) should also bear the amendment number(s) and date(s).

6.1.2 Name and address of the sponsor and monitor (if other than the sponsor).

6.1.3 Name and title of the person(s) authorized to sign the protocol and the protocol amendment(s) for the sponsor.

6.1.4 Name, title, address, and telephone number(s) of the sponsor's medical expert (or dentist when appropriate) for the trial.

6.1.5 Name and title of the investigator(s) who is (are) responsible for conducting the trial, and the address and phone number(s) of the trial site(s).

6.1.6 Name, title, address, and telephone number(s) of the qualified physician (or dentist, if applicable) who is responsible for all trial-site related medical (or dental) decisions (if other than investigator).

6.1.7 Name and address of the clinical laboratory and other medical and/or technical department(s) involved in the trial.

6.1.8 Site specific information may be provided on separate protocol page(s), if not addressed in a separate agreement.

6.2 Background Information

6.2.1 Name and description of the investigational product(s).

6.2.2 A summary of clinically significant findings from nonclinical studies and clinical trials that are relevant to the trial.

6.2.3 Summary of the known and potential risks and benefits, if any, to human subjects.

6.2.4 Description of and justification for the route of administration, dosage,. dosage regimen, and treatment period(s).

6.2.5 A statement that the trial will be conducted in compliance with GCP and the applicable regulatory requirement(s).

6.2.6 Description of the population to be studied.

6.2.7 References to literature and data that are relevant to the trial, and that provide background for the trial.

6.3 Trial Objectives and Purpose

A detailed description of the objectives and the purpose of the trial.

6.4 Trial Design

(NOTE: The scientific integrity of the trial and the credibility of the data from the trial depend substantially on the trial design). A description of the trial design, including:

6.4.1 A specific statement of the primary endpoints and the secondary endpoints, if any, to be measured during the trial.

6.4.2 A description of the type/design of trial to be conducted (e.g., double-blind, placebo-controlled, parallel design) and a schematic diagram of trial design, procedures, and stages.

6.4.3 A description of the measures taken to minimize/avoid bias, including:

(a) Randomization.
(b) Blinding.

6.4.4 A description of the dosage and dosage regimen of the trial treatment(s).

6.4.5 The expected duration of subject participation, and a description of the sequence and duration of all trial periods, including followup, if any.

6.4.6 A description of the "stopping rules" or "discontinuation criteria" for individual subjects, parts of trial, and entire trial. Criteria for removing subjects from the trial should be outlined.

6.4.7 Investigational product accountability procedures.

6.4.8 Maintenance of trial treatment randomization codes and procedures for breaking codes.

6.5 Selection and Withdrawal of Subjects

6.5.1 Subject inclusion criteria.

6.5.2 Exclusion criteria.

6.5.3 Withdrawal criteria and procedures specifying:

(a) When and how to withdraw subjects from the trial.
(b) The type and timing of the data to be collected for withdrawn subjects.

6.6 Treatment of Subjects

6.6.1 The treatment(s) to be administered, including the name of the product(s), the dose(s), the dosing schedule(s), the route/mode of administration, and the treatment period(s) for the product(s).

6.6.2 Medication(s) permitted (including rescue medication) and not permitted before and/or during the trial.

6.6.3 Procedures for monitoring subject compliance.

6.7 Assessment of Efficacy

6.7.1 Specification of the efficacy parameters.

6.7.2 Methods and timing for assessing, recording, and analyzing of efficacy parameters.

6.8 Assessment of Safety

6.8.1 Specification of safety parameters.

6.8.2 Procedures for eliciting reports of and recording and reporting adverse events.

6.8.3 The duration of the followup period(s) after adverse events.

6.9 Statistics

6.9.1 A description of the statistical methods to be employed, including timing of any planned interim analysis(ses).

6.9.2 The number of subjects planned to be included. In multicenter trials, the numbers of subjects projected for each trial site should be specified. Reason for choice of sample size, including reflections on (or calculations of) the power of the trial and clinical justification.

6.9.3 The level of significance to be used.

6.9.4 Criteria for the termination of the trial.

6.9.5 Procedure for accounting for missing, unused, and spurious data.

6.9.6 Procedures for reporting any deviation(s) from the original statistical plan.

(Any deviation(s) from the original statistical plan should be described and justified in protocol amendment(s) and/or in the final report, as appropriate.)

6.10 Quality Control and Quality Assurance Procedures

Monitoring and audit procedures, if not addressed in a separate agreement.

6.11 Ethics

Description of ethical considerations relating to the trial.

6.12 Data Handling and Recordkeeping

6.13 Financing and Insurance

Financing and insurance, if not addressed in a separate agreement.

6.14 Publication Policy

Publication policy, if not addressed in a separate agreement.

6.15 Supplements

(Note: Since the protocol and the clinical trial/study report are closely related, further relevant information can be found in the ICH Guideline for "Structure and Content of Clinical Study Reports.")

Structure and Content of Clinical Study Reports

Introduction to the Guideline

The objective of this guideline is to allow the compilation of a single worldwide core clinical study report acceptable to all regulatory authorities. The regulatory authority specific additions will consist of modules to be considered as appendices, available upon request.

The clinical study report described in this guideline is an "integrated" full report of an individual study of any therapeutic, prophylactic, or diagnostic agent (referred to herein as drug or treatment) conducted in patients, in which the clinical and statistical description, presentations, and analyses are integrated into a single report, incorporating tables and figures into the main text of the report, or at the end of the text, and with appendices containing the protocol, sample case report forms, investigator information, trial material information, technical statistical documentation, related publications, patient data listings, and technical statistical details such as derivations, computations, analyses, and computer output. The integrated full report of a study should not be derived by simply joining a separate clinical and statistical report. Although this guideline is mainly aimed at efficacy and safety trials, the basic principles and structure described can be applied to other kinds of trials, such as clinical pharmacology studies.

The guideline is intended to assist sponsors in the development of a report that is complete, free from ambiguity, well organized, and easy to review. The report should provide a clear explanation of how the critical design features of the study were chosen and enough information on the plan, methods, and conduct of the study so that there is no ambiguity in how the study was carried out. The report with its appendices should also provide enough individual patient data, including the demographic and baseline data, and details of analytical methods, to allow replication of the critical analyses when authorities wish to do so. It is also particularly important that all analyses, tables, and figures carry, in text or as part of the table, clear identification of the set of patients from which they were generated.

Depending on the regulatory authority's review policy, abbreviated reports using summarized data or with some sections deleted may be acceptable for uncontrolled studies or other studies not designed to establish efficacy (but a controlled safety study should be reported in full), for seriously flawed or aborted studies, or for controlled studies which examine conditions clearly unrelated to those for which a claim is made. However, a full description of safety aspects should be included in these cases. If an abbreviated report is submitted, there should be enough detail of design and results to allow the regulatory authority to determine whether a full report is needed. If there is any question regarding whether the reports are needed, it may be useful to consult the regulatory authority.

In presenting the detailed description of how the study was carried out, it may be possible simply to restate the description in the initial protocol. Often, however, it is possible to present the methodology of the study more concisely in a separate document. In each section describing the design and conduct of the study, it is particularly important to clarify features of the study that are not well-described in the protocol and identify ways in which the study as conducted differed from the protocol, and to discuss the statistical methods and analyses used to account for these deviations from the planned protocol.

The full integrated report of the individual study should include the most detailed discussion of individual adverse events or laboratory abnormalities, but these should usually also be reexamined as part of an overall safety analysis of all available data in any application.

The report should describe demographic characteristics of the study population and, where the study is large enough to permit this, present data for demographic (e.g., age, sex, race, weight) and other (e.g., renal or hepatic function) subgroups so that possible differences in efficacy or safety can be identified. Usually, however, subgroup responses should be examined in the larger data base used in the overall analysis.

The data listings requested as part of the report (usually in an appendix) are those needed to support critical analyses, not the more comprehensive archival data listings required by some regulatory authorities. Data listings which are part of the report should be readily usable by the reviewer. Thus, although it may be desirable to include many variables in a single listing to limit size, this should not be at the expense of clarity. An excess of data should not be allowed to lead, for example, to overuse of symbols instead of words or easily understood abbreviations or to too small displays. In this case, it is preferable to produce several listings.

Data should be presented in the report at different levels of detail: overall summary figures, and tables for important demographic, efficacy, and safety variables may be placed in the text to illustrate important points; other summary figures, tables, and listings for demographic, efficacy, and safety variables should be provided in section 14; individual patient data for specified groups of patients should be provided as listings in Appendix 16.2; and all individual patient data (U.S. archival listings) should be provided in Appendix 16.4.

In any table, figure, or data listing, estimated or derived values, if used, should be identified in a conspicuous fashion. Detailed explanations should be provided as to how such values were estimated or derived and what underlying assumptions were made.

The guidance provided below is detailed and is intended to notify the applicant of virtually all of the information that should routinely be provided so that postsubmission requests for further data clarification and analyses can be reduced as much as possible. Nonetheless, specific requirements for data presentation and/or analysis may depend on specific situations, may evolve over time, may vary from drug class to drug class, may differ among regions, and cannot be described in general terms. It is therefore important to refer to specific clinical guidelines and to discuss data presentation and analyses with the reviewing authority, whenever possible. Detailed written guidance on statistical approaches is available from some authorities.

Each report should consider all of the topics described (unless clearly not relevant) although the specific sequence and grouping of topics may be changed if alternatives are more logical for a particular study. Some data in the appendices are specific requirements of individual regulatory authorities and should be submitted as appropriate. The numbering should then be adapted accordingly.

In the case of very large trials, some of the provisions of this guideline may be impractical or inappropriate. When planning and when reporting such trials, contact with regulatory authorities to discuss an appropriate report format is encouraged.

The provisions of this guideline should be used in conjunction with other ICH guidelines.

Structure and Content of Clinical Study Reports

1. Title Page

The title page should contain the following information:

- Report title
- Name of drug
- Indication studied
- If not apparent from the title, a brief (one to two sentences) description giving design (parallel, cross-over, blinding, randomized) comparison (placebo, active, dose/response), duration, dose, and patient population
- Name of the sponsor
- Protocol identification
- Drug development phase
- Study initiation date (first patient enrolled, or any other verifiable definition)
- Date of early study termination, if any
- Study completion date (last patient completed)
- Name and affiliation of signatory (principal) investigator(s) or responsible medical officer
- Name of company signatory (the person responsible for the study report within the company.) The name, telephone number, and fax number of the company contact persons for questions arising during review of the study report should be indicated on this page or in the letter of application.
- Statement indicating whether the study was performed in compliance with good clinical practices (GCP's), including the archiving of essential documents
- Date of the report (identify any earlier reports from the same study by title and date)

2. Synopsis

A brief synopsis (maximum: three pages) that summarizes the study should be provided (see Annex I of the guideline for an example). The synopsis should include numerical data to illustrate results, not just text or p-values.

3. Table of Contents for the Individual Clinical Study Report

The table of contents should include:
- The page number or other locating information of each section, including summary tables, figures, and graphs;
- A list and the locations of appendices, tabulations, and any case report forms provided.

4. List of Abbreviations and Definitions of Terms

A list of the abbreviations and definitions of unusual or specialized terms or measurements units used in the report should be provided. In addition, abbreviated terms should be spelled out and the abbreviation indicated at first appearance in the text.

5. Ethics

5.1 Independent Ethics Committee (IEC) or Institutional Review Board (IRB)

It should be confirmed that the study and any amendments were reviewed by an independent ethics committee or institutional review board. A list of all IEC's or IRB's consulted should be given in Appendix 16.1.3 and, if required by the regulatory authority, the name of the committee Chair should be provided.

5.2 Ethical Conduct of the Study

It should be confirmed that the study was conducted in accordance with the ethical principles that have their origins in the Declaration of Helsinki.

5.3 Patient Information and Consent

How and when informed consent was obtained in relation to patient enrollment (e.g., at allocation, prescreening) should be described.

Representative written information for the patient (if any) and a sample patient consent form should be provided in Appendix 16.1.3.

6. Investigators and Study Administrative Structure

The administrative structure of the study (e.g., principal investigator, coordinating investigator, controller, steering committee, administration, monitoring and evaluation committees, institutions, statistician, central laboratory facilities, clinical research organization, clinical trial supply management) should be described briefly.

A list of the investigators with their affiliations, their role in the study, and their qualifications (curriculum vitae or equivalent), and a similar list for other persons whose participation materially affected the conduct of the study should be provided in Appendix 16.1.4. In the case of very large trials with many investigators, the above information may be abbreviated to consist of general statements of qualifications for persons carrying out particular roles in the study with only the name, degree, and institutional affiliation and roles of each investigator or other participant.

The listing should include:

(a) Investigators.

(b) Any person carrying out important study observations, such as a nurse, physician's assistant, clinical psychologist, clinical pharmacist, or house staff physician. It is not necessary to include in this list a person with only an occasional role, e.g., an on-call physician who dealt with a possible adverse effect or a temporary substitute for any of the above.

(c) The author(s) of the report, including the responsible biostatistician(s).

Where signatures of the principal investigators are required by regulatory authorities, these should be included in Appendix 16.1.5 (see Annex II for a sample form). Where these are not required, the signature of the responsible medical officer should be provided in Appendix 16.1.5.

7. Introduction

The introduction should contain a brief statement (maximum: one page) placing the study in the context of the drug's development, relating the critical features of the study (e.g., rationale and aims, target population, treatment, duration, primary endpoints) to that development. Any guidelines which were followed in the development of the protocol or any other agreements/meetings between the Company and regulatory authorities which are relevant to the particular study should be identified or described.

8. Study Objectives

A statement describing the overall purpose(s) of the study should be provided.

9. Investigational Plan

9.1 Overall Study Design and Plan - Description

The overall study plan and design (configuration) of the study (e.g., parallel, cross-over) should be described briefly but clearly using charts and diagram as needed. If other studies used a very similar protocol, it may be useful to note this and describe any important differences. The actual protocol and any changes should be included as Appendix 16.1.1 and a sample case report form (unique pages only; i.e., it is not necessary to include identical pages from forms for different evaluations or visits) as Appendix 16.1.2. If any of the information in this section comes from sources other than the protocol, these should be identified.

The information provided should include:

- Treatments studied (specific drugs, doses, and procedures).
- Patient population studied and the number of patients to be included.
- Level and method of blinding/masking (e.g., open, double-blind, single-blind, blinded evaluators, and unblinded patients and/or investigators).
- Kind of control(s) (e.g., placebo, no treatment, active drug, dose-response, historical) and study configuration (parallel, cross-over).
- Method of assignment to treatment (randomization, stratification).
- Sequence and duration of all study periods, including prerandomization and post-treatment periods, therapy withdrawal periods, and single- and double-blind

treatment periods. The timing of randomization should be specified. It is usually helpful to display the design graphically with a flow chart which includes timing of assessments (see Annexes IIIa and IIIb for an example).

- Any safety, data monitoring, or special steering or evaluation committees.
- Any interim analyses.

9.2 Discussion of Study Design, Including the Choice of Control Groups

The specific control chosen and the study design used should be discussed, as necessary. Examples of design issues meriting discussion follow.

Generally, the control (comparison) groups that are recognized are placebo concurrent control, no treatment concurrent control, active treatment concurrent control, dose comparison concurrent control, and historical control. In addition to the type of control, other critical design features that may need discussion are use of a cross-over design and selection of patients with particular prior history, such as response or nonresponse to a specific drug or member of a drug class. If randomization was not used, it is important to explain how other techniques, if any, guarded against systematic selection bias.

Known or potential problems associated with the study design or control group chosen should be discussed in light of the specific disease and therapies being studied. For a cross-over design, for example, there should be consideration, among other things, of the likelihood of spontaneous change in the disease and of carry-over effects of treatment during the study.

If efficacy was to be demonstrated by showing equivalence, i.e., the absence of a specified degree of inferiority of the new treatment compared to an established treatment, problems associated with such study designs should be addressed. Specifically there should be provided a basis for considering the study capable of distinguishing active from inactive therapy. Support may be provided by an analysis of previous studies similar to the present study with respect to important design characteristics (e.g., patient selection, study endpoints, duration, dose of active control, concomitant therapy) showing a consistent ability to demonstrate superiority of the active control to placebo. How to assess the ability of the present study to distinguish effective from ineffective therapy should also be discussed. For example, it may be possible to identify a treatment response (based on past studies) that would clearly distinguish between the treated population and an untreated group. Such a response could be the change of a measure from baseline or some other specified outcome like healing rate or survival rate. Attainment of such a response would support the expectation that the study could have distinguished the active drug from an inactive drug. There should also be a discussion of the degree of inferiority of the therapy (often referred to as the delta value) the study was intended to show was not exceeded.

The limitations of historical controls are well known (e.g., difficulty of assuring comparability of treated groups, inability to blind investigators to treatment, change in therapy/disease, difference due to placebo effect) and deserve particular attention.

Other specific features of the design may also deserve discussion, including presence or absence of washout periods and the duration of the treatment period, especially for a chronic illness. The rationale for dose and dose-interval selection should be explained, if it is not obvious. For example, once daily dosing with a short half-life drug whose effect is closely related in time to blood level is not usually effective; if the study design uses such dosing, this should be explained, e.g., by pointing to pharmacodynamic evidence that effect is prolonged compared to blood levels. The procedures used to seek evidence of "escape" from drug effect at the end of the dose-interval, such as measurements of effect just prior to dosing, should be described. Similarly, in a parallel design dose-response study, the choice of doses should be explained.

9.3 Selection of Study Population

9.3.1 Inclusion Criteria

The patient population and the selection criteria used to enter the patients into the study should be described, and the suitability of the population for the purposes of the study discussed. Specific diagnostic criteria used, as well as specific disease requirements (e.g., disease of a particular severity or duration, results of a particular test or physical examination, particular features of clinical history, such as failure or success on prior therapy, or other potential prognostic factors, and any age, sex, or ethnic factors) should be presented.

Screening criteria and any additional criteria for randomization or entry into the drug treatment part of the trial should be described. If there is reason to believe that there were additional entry criteria, not defined in the protocol, the implications of these should be discussed. For example, some investigators may have excluded, or entered into other studies, patients who were particularly ill or who had particular baseline characteristics.

9.3.2 Exclusion Criteria

The criteria for exclusion at entry into the study should be specified and the rationale (e.g., safety concerns, administrative reasons, or lack of suitability for the trial) provided. The impact of exclusions on the generalisability of the study should be discussed in the study report or in an overview of safety and effectiveness.

9.3.3 Removal of Patients From Therapy or Assessment

The predetermined reasons for removing patients from therapy or assessment observation, if any, should be described, as should the nature and duration of any planned followup observations in those patients.

9.4 Treatments

9.4.1 Treatments Administered

The precise treatments or diagnostic agents to be administered (active drug(s), placebo(s), procedures) in each arm of the study, and for each period of the study, should be fully described by name of drug(s), marketing formulation, route and mode of administration, dose, and dosage schedule.

9.4.2 Identity of Investigational Product(s)

In the text of the report, a brief description of the investigational product(s) (formulation, batch number) should be given.

In Appendix 16.1.6 of the report, the batch/serial number(s) of treatment(s) administered and a description of their formulation, content, external appearance (manufacturing and packaging batch records) should be provided. If more than one batch of test drug was used, patients receiving each batch should be identified.

The source of placebos and active control drugs should be provided. Any modification of active control drugs from their usual commercial state should be noted, and the steps taken to assure that their bioavailability was unaltered should be described.

For long-duration trials of investigational products with limited shelf- lives or incomplete stability data, the logistics of resupply of the materials should be described. Any use of test materials past their expiry date should be noted, and patients receiving them identified. If there were specific storage requirements, these should also be described.

9.4.3 Method of Assigning Patients to Treatment Groups

The specific methods used to assign patients to treatment groups, e.g., centralized allocation, allocation within sites, adaptive allocation (that is, assignment on the basis of earlier assignment or outcome), should be described in the text of the report, including any stratification or blocking procedures. Any unusual features should be explained.

A detailed description of the randomization method, including how it was executed, should be given in Appendix 16.1.7 with references cited if necessary. A table exhibiting the randomization codes, patient identifier, and treatment assigned should also be presented in the appendix. For a multicenter study, the information should be given by center. The method of generating random numbers should be explained.

For a historically controlled trial, it is important to explain how the particular control was selected and what other historical experiences were examined, if any, and how their results compared to the control used.

9.4.4 Selection of Doses in the Study

The doses or dose ranges used in the study should be given and the basis for choosing them described (e.g., prior experience in humans, animal data).

9.4.5 Selection and Timing of Dose for Each Patient

Procedures for selecting each patient's dose of test drug and control agent should be described. These procedures can vary from simple random assignment to a selected fixed drug/dose regimen, to some specified titration procedure, to more elaborate response-determined selection procedures, e.g., where dose is titrated upward at intervals until intolerance or some specified endpoint is achieved. Procedures for back-titration, if any, should also be described.

The timing (time of day, interval) of dosing and the relation of dosing to meals should be described, and if it was not specified, this should be noted.

Any specific instructions to patients about when or how to take the dose(s) should be described.

9.4.6 Blinding

A description of the specific procedures used to carry out blinding should be provided (e.g., how bottles were labeled, double dummy techniques), including the circumstances in which the blind would be broken for an individual or all patients, the procedures used, and who had access to patient codes. If the study allowed for some investigators to remain unblinded (e.g., to allow them to adjust medication), the means of shielding other investigators should be explained. Measures taken to ensure that drug and placebo were indistinguishable and evidence that they were indistinguishable, should be described, as should the appearance, shape, smell, and taste of the test material. Measures to prevent unblinding by laboratory measurements, if used, should be described.

If blinding was considered unnecessary to reduce bias for some or all of the observations, this should be explained; e.g., use of a random-zero sphygmomanometer eliminates possible observer bias in reading blood pressure and Holter tapes are often read by automated systems that are presumably immune to observer bias. If blinding was considered desirable, but not feasible, the reasons and implications should be discussed. Sometimes blinding is attempted but is known to be imperfect because of obvious drug effects in at least some patients (dry mouth, bradycardia, fever, injection site reactions, changes in laboratory data). Such problems or potential problems should be identified and if there were any attempts to assess the magnitude of the problem or manage it (e.g., by having some end point measurements carried out by people shielded from information that might reveal treatment assignment), they should be described.

A description of any packaging and labeling techniques used for blinding, e.g., double dummy techniques, special labels that reveal blind-breakage, sealed code lists/envelopes, etc., should be provided in Appendix 16.1.7.

9.4.7 Prior and Concomitant Therapy

Which drugs or procedures were allowed before and during the study, whether and how their use was recorded, and any other specific rules and procedures related to permitted or forbidden concomitant therapy should be described. How allowed concomitant therapy might affect the outcome due either to drug- drug interaction or to direct effects on the study endpoints should be discussed, and how the independent effects of concomitant and study therapies could be ascertained should be explained.

9.4.8 Treatment Compliance

The measures taken to ensure and document treatment compliance should be described, e.g., drug accountability, diary cards, blood, urine, or other body fluid drug level measurements, or medication event monitoring.

9.5 Efficacy and Safety Variables

9.5.1 Efficacy and Safety Measurements Assessed and Flow Chart

The specific efficacy and safety variables to be assessed and laboratory tests to be conducted, their schedule (days of study, time of day, relation to meals, and the timing of critical measures in relation to test drug administration, e.g., just prior to next dose, 2 hours after dose), the methods for measuring them, and the persons responsible for the measurements should be described. If there were changes in personnel carrying out critical measurements, these should be reported.

It is usually helpful to display graphically in a flow chart (see Annex III of the guideline) the frequency and timing of efficacy and safety measurements; visit numbers and times should be shown, or, alternatively, times alone can be used (visit numbers alone are more difficult to interpret). Any specific instructions (e.g., guidance or use of a diary) to the patients should also be noted.

Any definitions used to characterize outcome (e.g., criteria for determining occurrence of acute myocardial infarction, designation of the location of the infarction, characterization of a stroke as thrombotic or hemorrhagic, distinction between TIA and stroke, assignment of cause of death) should be explained in full. Any techniques used to standardize or compare results of laboratory tests or other clinical measurements (e.g., ECG, chest X-ray) should also be described. This is particularly important in multicenter studies.

If anyone other than the investigator was responsible for evaluation of clinical outcomes (e.g., the sponsor or an external committee to review X- rays or ECG's or to determine whether the patient had a stroke, acute infarction, or sudden death) the person or group should be identified. The procedures, including means of maintaining blindness, and centralizing readings and measurements, should be described fully.

The means of obtaining adverse event data should be described (volunteered, checklist, questioning), as should any specifically planned followup procedures for adverse events or any planned rechallenge procedure.

Any rating of adverse events by the investigator, sponsor, or external group (e.g., severity rating, likelihood of drug causation) should be described and criteria for such ratings, if any, given. If efficacy or safety was to be assessed in terms of categorical ratings or numerical scores, the criteria used for point assignment (e.g., definitions of point scores) should be provided. For multicenter studies, how methods were standardized should be indicated.

9.5.2 Appropriateness of Measurements

If any of the efficacy or safety assessments were not standard, i.e., widely used and generally recognized as reliable, accurate, and relevant (able to discriminate between effective and ineffective agents), its reliability, accuracy, and relevance should be documented. It may be helpful to describe alternatives considered but rejected.

If a surrogate end point (a laboratory measurement or physical measurement or sign that is not a direct measure of clinical benefit) was used as a study end point, this should be justified, e.g., by reference to clinical data publications, guidelines, or previous actions by regulatory authorities.

9.5.3 Primary Efficacy Variable(s)

The primary measurements and end points used to determine efficacy should be clearly specified. Although the critical efficacy measurements may seem obvious, when there are multiple variables, or when variables are measured repeatedly, the protocol should identify the primary ones, with an explanation of why they were chosen, or designate the pattern of significant findings or other method of combining information that would be interpreted as supporting efficacy. If the protocol did not identify the primary variables, the study report should explain how these critical variables were selected (e.g., by reference to publications, guidelines, or previous actions by regulatory authorities) and when they were identified (i.e., before or after the study was completed and unblinded). If an efficacy threshold was defined in the protocol, this should be described.

9.5.4 Drug Concentration Measurements

Any drug concentrations to be measured, and the sample collection times and periods in relation to the timing of drug administration, should be described. Any relation of drug administration and sampling to ingestion of food, posture, and the possible effects of concomitant medication/alcohol/ caffeine/nicotine should also be addressed. The biological sample measured and the method of measurement used should be described, referring to published and/or internal assay validation documentation for methodological details. Where other factors are believed important in assessing pharmacokinetics (e.g., soluble circulating receptors, renal or hepatic function), the timing and plans to measure these factors should also be specified.

9.6 Data Quality Assurance - Audit Certificate

The quality assurance and quality control systems implemented to assure the quality of the data should be described in brief. If none were used, this should be stated.

Any steps taken at the investigation site or centrally to ensure the use of standard terminology and the collection of accurate, consistent, complete, and reliable data, such as training sessions, monitoring of investigators by sponsor personnel, instruction manuals, data verification, cross-checking, use of a central laboratory for certain tests, centralized ECG reading, or data audits, should be described. It should be noted whether investigator meetings or other steps were taken to prepare investigators and standardize performance.

If the sponsor used an independent internal or external auditing procedure, it should be mentioned here and described in Appendix 16.1.8 and audit certificates (if available; see Annexes IVa and IVb for descriptions) provided in the same appendix.

9.7 Statistical Methods Planned in The Protocol and Determination of Sample Size

9.7.1 Statistical and Analytical Plans

The statistical analyses planned in the protocol and any changes made before outcome results were available should be described. In this section emphasis should be on which analyses, comparisons, and statistical tests were planned, not on which ones were actually used. If critical measurements were made more than once, the particular measurements (e.g., average of several measurements over the entire study, values at particular times, values only from study completers, or last on-therapy value) planned as the basis for comparison of drug and control should be specified. Similarly, if more than one analytical approach is plausible, e.g., changes from baseline response, slope analysis, life table analysis, the planned approach should be identified. Also, whether the primary analysis is to include adjustment for covariates should be specified.

If there were any planned reasons for excluding from analysis patients for whom data are available, these should be described. If there were any subgroups whose results were to be examined separately, these should be identified. If categorical responses (global scales, severity scores, responses of a certain size) were to be used in analyzing responses, they should be clearly defined.

Planned monitoring of the results of the study should be described. If there was a data monitoring committee, either within or outside the sponsor's control, its composition and operating procedures should be described and procedures to maintain study blinding should be given. The frequency and nature of any planned interim analysis, any specified circumstances in which the study would be terminated, and any statistical adjustments to be employed because of interim analyses should be described.

9.7.2 Determination of Sample Size

The planned sample size and the basis for it, such as statistical considerations or practical limitations, should be provided. Formulae for sample size and power calculation should be given together with their derivations or source of reference. Estimates used in the formulae should be given and explanations provided as to how they were obtained. For a positive control study intended to show that a new therapy is at least as effective as the standard therapy, the sample size determination should specify a "delta value" (a difference between treatments that would be considered unacceptably large) and therefore the difference the study is designed to be able to exclude.

9.8 Changes in the Conduct of the Study or Planned Analyses

Any change in the conduct of the study or planned analyses (e.g., dropping a treatment group, changing the entry criteria or drug dosages, adjusting the sample size, etc.) instituted after the start of the study should be described. The time(s) and reason(s) for the change(s), the procedure used to decide on the change(s), the person(s) or group(s) responsible for the change(s), and the nature and content of the data available (and to whom they were available) when the change was made should also be described, whether the change was documented as a formal protocol amendment or not. (Personnel changes need not be included.) Any possible implications of the change(s) for the interpretation of the study should be discussed briefly in this section and more fully in other appropriate sections of the report. In every section of the report, a clear distinction between conditions (procedures) planned in the protocol and amendments or additions should be made.

10. Study Patients

10.1 Disposition of Patients

There should be a clear accounting of all patients who entered the study, using figures or tables in the text of the report. The numbers of patients who were screened, were randomized, and who entered and completed each phase of the study (or each week/month of the study) should be provided, as well as the reasons for all postrandomization discontinuations, grouped by treatment and by major reason (e.g., lost to followup, adverse event, poor compliance). In some cases, it may also be relevant to provide a breakdown of the reasons for excluding patients during screening, if this could help clarify the appropriate patient population for eventual drug use. A flow chart is often helpful (see Annexes Va and Vb of the guideline). Whether patients are followed for the duration of the study, even if drug is discontinued, should be made clear.

In Appendix 16.2.1, there should also be a listing of all patients discontinued from the study after enrollment, broken down by center and treatment group, giving a patient identifier, the specific reason for discontinuation, the treatment (drug and dose), cumulative dose (where appropriate), and the duration of treatment before discontinuation. Whether or not the blind for the patient was broken at the time of discontinuation should be noted. It may also be useful to include other information, such as critical demographic data (e.g., age, sex, race), concomitant medication, and the major response variable(s) at termination. See Annex VI for an example of such a listing.

10.2 Protocol Deviations

All important deviations related to study inclusion or exclusion criteria, conduct of the trial, patient management, or patient assessment should be described.

In the body of the text, protocol deviations should be appropriately summarized by center and grouped into different categories, such as:

- Those who entered the study even though they did not satisfy the entry criteria.
- Those who developed withdrawal criteria during the study but were not withdrawn.
- Those who received the wrong treatment or incorrect dose.
- Those who received an excluded concomitant treatment.

In Appendix 16.2.2, individual patients with these protocol deviations should be listed, broken down by center for multicenter studies.

11. Efficacy Evaluation

11.1 Data Sets Analyzed

Exactly which patients were included in each efficacy analysis should be precisely defined, e.g., all patients receiving any drug, all patients with any efficacy observation or with a certain minimum number of observations, only patients completing the trial, all patients with an observation during a particular time window, only patients with a specified degree of compliance, etc. It should be clear, if not defined in the study protocol, when (relative to study unblinding), and how inclusion/exclusion criteria for the data sets analyzed were developed. As a general rule, even if the applicant's proposed primary analysis is based on a reduced subset of the patients with data, there should also be for any trial intended to establish efficacy an additional analysis using all randomized (or otherwise entered) patients with any on-treatment data.

There should be a tabular listing of all patients, visits, and observations excluded from the efficacy analysis provided in Appendix 16.2.3 (see Annex VII of the guideline for an example). The reasons for exclusions should also be analyzed for the whole treatment group over time (see Annex VIII of the guideline for an example).

11.2 Demographic and Other Baseline Characteristics

Group data for the critical demographic and baseline characteristics of the patients, as well as other factors arising during the study that could affect response, should be presented in this section and comparability of the treatment groups for all relevant characteristics should be displayed by use of tables or graphs in section 14.1. The data for the patient sample included in the "all patients with data" analysis should be given first. This can then be followed by data on other groups used in principal analyses, such as the "per-protocol" analysis or other analyses, e.g., groups defined by compliance, concomitant disease/therapy, or demographic/baseline characteristics. When such groups are used, data for the complementary excluded group should also be shown. In a multicenter study, comparability should be assessed by center, and centers should be compared.

A diagram showing the relationship between the entire sample and any other analysis groups should be provided.

The critical variables will depend on the specific nature of the disease but will usually include:

** Demographic variables
- Age
- Sex
- Race

** Disease factors
- Specific entry criteria (if not uniform), duration, stage, and severity of disease and other clinical classifications and subgroupings in common usage or of known prognostic significance.
- Baseline values for critical clinical measurements carried out during the study or identified as important indicators of prognosis or response to therapy.
- Concomitant illness at trial initiation such as renal disease, diabetes, heart failure.

- Relevant previous illness.
- Relevant previous treatment for illness treated in the study.
- Concomitant treatment maintained, even if the dose was changed during the study, including oral contraceptive and hormone replacement therapy; treatments stopped at entry into the study period (or changed at study initiation).

** Other factors that might affect response to therapy (e.g., weight, renin status, antibody levels, metabolic status).

** Other possibly relevant variables (e.g., smoking, alcohol intake, special diets) and, for women, menstrual status and date of last menstrual period, if pertinent for the study.

In addition to tables and graphs giving group data for these baseline variables, relevant individual patient demographic and baseline data, including laboratory values, and all concomitant medication for all individual patients randomized (broken down by treatment and by center for multicenter studies) should be presented in by-patient tabular listings in Appendix 16.2.4. Although some regulatory authorities will require all baseline data to be presented elsewhere in tabular listings, the appendix to the study report should be limited to only the most relevant data, generally the variables listed above.

11.3 Measurements of Treatment Compliance

Any measurements of compliance of individual patients with the treatment regimen under study and drug concentrations in body fluids should be summarized, analyzed by treatment group and time interval, and tabulated in Appendix 16.2.5.

11.4 Efficacy Results and Tabulations of Individual Patient Data

11.4.1 Analysis of Efficacy

Treatment groups should be compared for all critical measures of efficacy (primary and secondary end points; pharmacodynamics), as well as benefit/risk assessment measured in each patient where these are utilized. In general, the results of all analyses contemplated in the protocol and an analysis including all patients with on-study data should be performed in studies intended to establish efficacy.

Analyses based on continuous variables (e.g., mean blood pressure or depression scale score) and categorical responses (e.g., cure of an infection) can be equally valid; ordinarily both should be presented if both were planned and are available. If categories are newly created (i.e., not in the statistical plan), the basis for them should be explained. Even if one variable receives primary attention (e.g., in a blood pressure study, supine blood pressure at week x), other reasonable measures (e.g., standing blood pressure and blood pressures at other particular times) should be assessed, at least briefly. In addition, the time course of response should be described, if possible. For a multicenter study, data display and analysis of individual centers should be included to give a clear picture of the results at each site, especially the larger sites.

If any critical measurements or assessments of efficacy or safety outcomes were made by more than one party (e.g., both the investigator and an expert committee may offer an opinion on whether a patient had an acute infarction), overall differences between the ratings should be shown, and each patient having disparate assessments should be identified. The assessments used should be clear in all analyses.

In many cases, efficacy and safety end points are difficult to distinguish (e.g., stroke in a thrombolytic trial, deaths in a fatal disease study). Many of the principles addressed below should be adopted for critical safety measures as well.

11.4.2 Statistical/Analytical Issues

The statistical analysis used should be described for clinical and statistical reviewers in the text of the report, with detailed documentation of statistical methods (see section Annex IX) presented in Appendix 16.1.9. Important features of the analysis, including the particular tests used, adjustments made for demographic or baseline measurements or concomitant therapy, handling of dropouts and missing data, adjustments for multiple comparisons, special analyses of multicenter studies, and adjustments for interim analyses, should be discussed. Any changes in the analysis made after blind-breaking should be identified.

In addition to the general discussion, the following specific issues should be addressed (unless not applicable):

11.4.2.1 Adjustments for Covariates

Selection of, and adjustments for, demographic or baseline measurements, concomitant therapy, or any other covariate or prognostic factor should be explained in the report, and methods of adjustment, results of analyses, and supportive information (e.g., ANCOVA or Cox regression output) should be included in the detailed documentation of statistical methods. If the covariates or methods used in these analyses differed from those planned in the protocol, the differences should be explained and where possible and relevant, the results of planned analyses should also be presented. Although not part of the individual study report, comparisons of covariate adjustments and prognostic factors across individual studies may be an informative analysis in a summary of clinical efficacy data.

11.4.2.2 Handling of Dropouts or Missing Data

There are several factors that may affect dropout rates. These include the duration of the study, the nature of the disease, the efficacy and toxicity of the drug under study, and other factors that are not therapy related. Ignoring the patients who dropped out of the study and drawing conclusions based only on patients who completed the study can be misleading. A large number of dropouts, however, even if included in an analysis, may introduce bias, particularly if there are more early dropouts in one treatment group or the reasons for dropping out are treatment or outcome related. Although the effects of early dropouts, and sometimes even the direction of bias, can be difficult to determine, possible effects should be explored as fully as possible. It may be helpful to examine the observed cases at various time points or, if dropouts were very frequent, to concentrate on analyses at time points when most of the patients were still under observation and when the full effect of the drug was realized. It may also be helpful to examine modeling approaches to the evaluation of such incomplete data sets.

The results of a clinical trial should be assessed not only for the subset of patients who completed the study, but also for the entire patient population as randomized (the intent-to-treat analysis) or at least for all those with any on-study measurements. Several factors should be considered and compared for the treatment groups in analyzing the effects of dropouts: the reasons for the dropouts, the time to dropout, and the proportion of dropouts among treatment groups at various time points. Procedures for dealing with missing data, e.g., use of estimated or derived data, should be described. Detailed explanation should be provided as to how such estimations or derivations were done and what underlying assumptions were made.

11.4.2.3 Interim Analyses and Data Monitoring

The process of examining and analyzing data accumulating in a clinical trial, either formally or informally, can introduce bias and/or increase type I error. Therefore, all interim analyses, formal or informal, preplanned or ad hoc, by any study participant, sponsor staff member, or data monitoring group should be described in full, even if the treatment groups were not identified. The need for statistical adjustment because of such analyses should be addressed. Any operating instructions or procedures used for such analyses should be described. The minutes of meetings of any data monitoring group and any data reports reviewed at those meetings, particularly a meeting that led to a change in the protocol or early termination of the study, may be helpful and should be provided in Appendix 16.1.9. Data monitoring without code-breaking should also be described, even if this kind of monitoring is considered to cause no increase in type I error.

11.4.2.4 Multicenter Studies

A multicenter study is a single study under a common protocol, involving several centers (e.g., clinics, practices, hospitals) where the data collected are intended to be analyzed as a whole (as opposed to a post-hoc decision to combine data or results from separate studies). Individual center results should be presented, however, and statistical tests for homogeneity across centers, i.e., for detecting treatment-by center interaction, should be provided, if feasible. The significance level used to declare the significance of a given test for treatment-by-center interaction should be considered in light of the sample sizes involved. Any extreme or opposite results among centers should be noted and discussed, considering such possibilities as differences in study conduct, patient characteristics, or clinical settings. As mentioned in previous sections, demographic, baseline, and postbaseline data, as well as efficacy data, should be presented by center, even though the combined analysis is the primary one.

11.4.2.5 Multiple Comparisons

False positive findings increase in number as the number of significance tests (number of comparisons) performed increases. If there was more than one primary endpoint (outcome variable), or if there were multiple treatment groups, or subsets of the patient population being examined, the statistical analysis should reflect awareness of this and either explain the statistical adjustment used for type I error criteria or give reasons why it was considered unnecessary.

11.4.2.6 Use of an "Efficacy Subset" of Patients

Particular attention should be devoted to the effects of dropping patients with available data from analyses because of poor compliance, missed visits, ineligibility, or any other reason. As noted above, an analysis using all available data should be carried out for all studies intended to establish efficacy, even if it is not the analysis proposed as the primary analysis by the applicant.

11.4.2.7 Active-Control Studies Intended to Show Equivalence

If an active control study is intended to show equivalence (i.e., lack of a difference greater than a specified size) between a new drug and a control, the analysis should show the confidence interval for the comparison between the two agents for critical end points and the relation of that interval to the prespecified degree of inferiority that would be considered unacceptable. (See 9.2, for important considerations when using the active control equivalence design.)

11.4.2.8 Examination of Subgroups

If the size of the study permits, important demographic or baseline value- defined subgroups should be examined for unusually large or small responses and the results presented, e.g., comparison of effects by age, sex, or race, by severity or prognostic groups, by history of prior treatment with a drug of the same class. If these analyses were not carried out because the study was too small, it should be noted. These analyses are not intended to "salvage" an otherwise nonsupportive study but may suggest hypotheses worth examining in other studies or be helpful in refining, for example, labeling information, patient selection, or dose selection. Where there is a prior hypothesis of a differential effect in a particular subgroup, this hypothesis and its assessment should be part of the planned statistical analysis.

11.4.3 Tabulation of Individual Response Data

In addition to tables and graphs representing group data, individual response data and other relevant study information should be presented in tables. Some regulatory authorities may require all individual data in archival case report tabulations. What needs to be included in the report will vary from study to study and from one drug class to another and the applicant should decide, if possible after consultation with the regulatory authority, what to include in an appendix to the study report. The study report should indicate what material is included as an appendix, what is in the more extensive archival case report tabulations, if required by the regulatory authority, and what is available on request.

For a controlled study in which critical efficacy measurements or assessments (e.g., blood or urine cultures, pulmonary function tests, angina frequency, or global evaluations) are repeated at intervals, the data listings accompanying the report should include, for each patient, a patient identifier, all measured or observed values of critical measurements, including baseline measurements, with notation of the time during the study (e.g., days on therapy and time of day, if relevant) when the measurements were made, the drug/dose at the time (if useful, given as milligrams per kilogram (mg/kg)), any measurements of compliance, and any concomitant medications at the time of, or close to the time of, measurement or assessment. If, aside from repeated assessments, the study included some overall responder versus nonresponder evaluation(s) (bacteriologic cure or failure), it should also be included. In addition to critical measurements, the tabulation should note whether the patient was included in the efficacy evaluation (and which evaluation, if more than one), provide patient compliance information, if collected, and a reference to the location of the case report form, if included. Critical baseline information such as age, sex, weight, disease being treated (if more than one in study), and disease stage or severity, is also helpful. The baseline values for critical measurements would ordinarily be included as zero time values for each efficacy measurement.

The tabulation described should usually be included in Appendix 16.2.6 of the study report, rather than in the more extensive case report tabulations required by some regulatory authorities, because it represents the basic efficacy data supporting summary tables. Such a thorough tabulation can be unwieldy for review purposes, however, and more targeted displays should be developed as well. For example, if there are many measurements reported, tabulations of the most critical measurements for each patient (e.g., the blood pressure value at certain visits might be more important than others) will be useful in providing an overview of each individual's results in a study, with each patient's response summarized on a single line or small number of lines.

11.4.4 Drug Dose, Drug Concentration, and Relationships to Response

When the dose in each patient can vary, the actual doses received by patients should be shown and individual patient's doses should be tabulated. Although studies not designed as dose-response studies may have limited ability to contribute dose-response information, the available data should be examined for whatever information they can yield. In examining the dose response, it may be helpful to calculate dose as mg/kg body weight or milligrams per meter squared (mg/m2) body surface.

Drug concentration information, if available, should also be tabulated (Appendix 16.2.5), analyzed in pharmacokinetic terms and, if possible, related to response.

Further guidance on the design and analysis of studies exploring dose- response or concentration response can be found in the ICH Guideline "Dose-Response Information to Support Drug Registration."

11.4.5 Drug-Drug and Drug Disease Interactions

Any apparent relationship between response and concomitant therapy and between response and past and/or concurrent illness should be described.

11.4.6 By-Patient Displays

While individual patient data ordinarily can be displayed in tabular listings, it has on occasion been helpful to construct individual patient profiles in other formats, such as graphic displays. These might, for example, show the value of (a) particular parameter(s) over time, the drug dose over the same period, and the times of particular events (e.g., an adverse event or change in concomitant therapy). Where group mean data represent the principal analyses, this kind of "case report extract" may offer little advantage; it may be helpful, however, if overall evaluation of individual responses is a critical part of the analysis.

11.4.7 Efficacy Conclusions

The important conclusions concerning efficacy should be concisely described, considering primary and secondary endpoints, prespecified and alternative statistical approaches, and results of exploratory analyses.

12. Safety Evaluation

Analysis of safety-related data can be considered at three levels. First, the amount of exposure (dose, duration) should be examined to determine the extent to which safety can be assessed from the study. Second, the more common adverse events, e.g., laboratory test changes, should be identified, classified in some reasonable way, compared for treatment groups, and analyzed, as appropriate, for that may affect the frequency of adverse reactions, such as time dependence, relation to demographic characteristics, relation to dose or drug concentration. Finally, serious adverse events and other significant adverse events should be identified, usually by close examination of patients who left the study prematurely because of an adverse event, whether or not identified as drug related, or who died.

The ICH Guideline on Clinical Safety Data Management, Definitions and Standards for Expedited Reporting defines serious adverse events as follows: A "serious adverse event" (experience) or reaction is any untoward medical occurrence that at any dose: results in death, is life-threatening, requires inpatient or prolongation of existing hospitalization, results in persistent or significant disability/incapacity, or is a congenital anomaly/birth defect.

For the purpose of this guideline, "other significant adverse events" are marked hematological and other laboratory abnormalities and any medical events that led to an intervention, including withdrawal of drug treatment, dose reduction, or significant additional concomitant therapy.

In the following sections, three kinds of analysis and display are suggested:

(1) Summarized data, often using tables and graphical presentations presented in the main body of the report;

(2) Listings of individual patient data; and

(3) Narrative statements of events of particular interest.

In all tabulations and analyses, events associated with both test drug and control treatment should be displayed.

12.1 Extent of Exposure

The extent of exposure to study drugs (and to active control and placebo) should be characterized according to the number of patients exposed, the duration of exposure, and the dose to which they were exposed.

** Duration: Duration of exposure to any dose can be expressed as a median or mean, but it is also helpful to describe the number of patients exposed for specified periods of time, such as for 1 day or less, 2 days to 1 week, more than 1 week to 1 month, more than 1 month to 6 months. The numbers exposed to drug for the various durations should also be broken down into age, sex, and racial subgroups, and any other pertinent subgroups, such as disease (if more than one is represented), disease severity, concurrent illness.

** Dose: The mean or median dose used and the number of patients exposed to specified daily dose levels should be given; the daily dose levels used could be the maximum dose for each patient, the dose with longest exposure for each patient, or the mean daily dose. It is often useful to provide combined dose-duration information, such as the numbers exposed for a given duration (e.g., at least 1 month) to the most common dose, the highest dose, the maximum recommended dose. In some cases, cumulative dose might be pertinent. Dosage may be given as the actual daily dose or on a mg/kg or mg/m2 basis as appropriate. The numbers of patients exposed to various doses should be broken down into age, sex, and racial subgroups, and any other pertinent subgroups.

** Drug concentration: If available, drug concentration data (e.g., concentration at the time of an event, maximum plasma concentration, area under curve) may be helpful in individual patients for correlation with adverse events or changes in laboratory variables. (Appendix 16.2.5.)

It is assumed that all patients entered into treatment who received at least one dose of the treatment are included in the safety analysis; if that is not so, an explanation should be provided.

12.2 Adverse Events (AE's)

12.2.1 Brief Summary of Adverse Events

The overall adverse event experience in the study should be described in a brief narrative, supported by the following more detailed tabulations and analyses. In these tabulations and analyses, events associated with both the test drug and control treatment should be displayed.

12.2.2 Display of Adverse Events

All adverse events (including events likely to be related to the underlying disease or likely to represent concomitant illness, unless there is a prior agreement with the regulatory authority to consider specified events as disease related) should be displayed in summary tables (section 14.3.1). The tables should include changes in vital signs and any laboratory changes that were considered serious adverse events or that resulted in withdrawal of treatment.

In most cases, it will also be useful to describe in such tables, "treatment emergent signs and symptoms" (TESS; those not seen at baseline, and those which worsened even if present at baseline).

The tables should list each adverse event, the number of patients in each treatment group in whom the event occurred, and the rate of occurrence. Adverse events should be grouped by body system. Each event may then be divided into defined severity categories (e.g., mild, moderate, severe) if these were used. The tables may also divide the adverse events into those considered at least possibly related to drug use and those considered not related, or use some other causality (e.g., unrelated or possibly, probably, or definitely related). For any such categorization, the categories should be defined and the person(s) responsible for classification of each event identified. Even when such a causality assessment is used, the tables should include all adverse events, whether or not considered drug related, including events thought to represent intercurrent illnesses. Subsequent analyses of the study or of the overall safety data base may help to distinguish between adverse events that are, or are not, considered drug related. So that it is possible to analyze and evaluate the data in these tables, it is important to identify each patient having each adverse event. An example of such a tabular presentation is shown below.

In addition to these complete tables provided in 14.3.1, an additional summary table comparing treatment and control groups, without the patient identifying numbers limited to relatively common adverse events (e.g., those in at least 1 percent of the treated group), should be provided in the body of the report.

In presenting adverse events, it is important both to display the original terms used by the investigator and to attempt to group related events (i.e., events that probably represent the same phenomena) so that the true occurrence rate is not obscured. One way to do this is with a standard adverse reaction dictionary.

12.2.3 Analysis of Adverse Events

The basic display of adverse event rates described in section 12.2.2 (and located in section 14.3.1) of the report should be used to compare rates in treatment and control groups. For this analysis it may be helpful to combine the event severity categories and the causality categories, leading to a simpler side-by-side comparison of treatment groups. In addition, if study size and design permit, it may be useful to examine the more common adverse events that seem to be drug related for relationship to dosage and to mg/kg or mg/m2 dose, to dose regimen, to duration of treatment, to total dose, to demographic characteristics, such as age, sex, race, to other baseline features, such as renal status to effectiveness outcomes, and to drug concentration, if data are available.

It may also be useful to examine time of onset and duration of adverse events. A variety of additional analyses may be suggested by the study results or by the pharmacology of the drug.

It is not intended that every adverse event be subjected to rigorous statistical evaluation. It may be apparent from initial display and inspection of the data that a significant relation to demographic or other baseline features is not present. If the studies are small and if the number of events is relatively small, it may be sufficient to limit analyses to a comparison of treatment and control.

Under certain circumstances, life table or similar analyses may be more useful than reporting of crude adverse event rates.

12.2.4 Listing of Adverse Events by Patient

All adverse events for each patient, including the same event on several occasions, should be listed in Appendix 16.2.7, giving both preferred term and the original term used by the investigator. The listing should be by investigator and by treatment group and should include:

- Patient identifier
- Age, race, sex, weight (height, if relevant)
- Location of case report forms (CRF's), if provided.
- Preferred term
- The adverse event -
    - reported term
    - Duration of the adverse event
    - Severity (e.g., mild, moderate, severe)
    - Seriousness (serious/nonserious)
    - Action taken (none, dose reduced, treatment stopped, specific treatment instituted, etc.)
    - Outcome (e.g., CIOMS format)
    - Causality assessment (e.g., related/not related). How this was determined should be described in the table or elsewhere.
    - Date of onset or date of clinic visit at which the event was discovered
    - Timing of onset of the adverse event in relation to last dose of study drug (when applicable)
    - Study treatment at time of event or most recent study treatment taken
    - Study drug dose in absolute amount, mg/kg or mg/m2 at time of event
    - Drug concentration (if known)
    - Duration of study drug treatment
    - Other drug treatment during study

Any abbreviations and codes should be clearly explained at the beginning of the listing or, preferably, on each page.

12.3 Deaths, Other Serious Adverse Events, and Other Significant Adverse Events

Deaths, other serious adverse events, and other significant adverse events (see definition in section 12) deserve special attention.

12.3.1 Listing of Deaths, Other Serious Adverse Events, and Other Significant Adverse Events

Listings, containing the same information as called for in section 12.2.4 above, should be provided for the following events.

12.3.1.1 Deaths

All deaths during the study, including the posttreatment followup period, and deaths that resulted from a process that began during the study, should be listed by patient in section 14.3.2.

12.3.1.2 Other Serious Adverse Events

All serious adverse events (other than death but including the serious adverse events temporally associated with or preceding the deaths) should be listed in section 14.3.2. The listing should include laboratory abnormalities, abnormal vital signs, and abnormal physical observations that were considered serious adverse events.

12.3.1.3 Other Significant Adverse Events

Marked hematological and other laboratory abnormalities (other than those meeting the definition of serious) and any events that led to an intervention, including withdrawal of drug treatment, dose reduction, or significant additional concomitant therapy, other than those reported as serious adverse events, should be listed in section 14.3.2.

12.3.2 Narratives of Deaths, Other Serious Adverse Events, and Certain Other Significant Adverse Events

There should be brief narratives describing each death, each other serious adverse event, and those of the other significant adverse events that are of special interest because of clinical importance. These narratives can be placed either in the text of the report or in section 14.3.3, depending on their number. Events that were clearly unrelated to the study drug may be omitted or described very briefly. In general, the narrative should describe:

- The nature and intensity of event, the clinical course leading up to the event, with an indication of timing relevant to drug administration; relevant laboratory measurements whether the drug was stopped, and when; countermeasures; post mortem findings; investigator's opinion on causality, and sponsor's opinion on causality, if appropriate.

In addition, the following information should be included:

- Patient identifier
- Age and sex of patient; general clinical condition of patient, if appropriate
- Disease being treated (if the same for all patients this is not required) with duration of current episode of illness
- Relevant concomitant/previous illnesses with details of occurrence/duration
- Relevant concomitant/previous medication with details of dosage
- Study drug administered, drug dose, if this varied among patients, and length of time administered

12.3.3 Analysis and Discussion of Deaths, Other Serious Adverse Events, and Other Significant Adverse Events

The significance of the deaths, other serious adverse events, and other significant adverse events leading to withdrawal, dose reduction, or institution of concomitant therapy should be assessed with respect to the safety of the drug. Particular attention should be paid to whether any of these events may represent a previously unsuspected important adverse effect of the drug. For serious adverse events that appear of particular importance, it may be useful to use life table or similar analyses to show their relation to time on drug and to assess their risk over time.

12.4 Clinical Laboratory Evaluation

12.4.1 Listing of Individual Laboratory Measurements by Patient (16.2.8) and Each Abnormal Laboratory Value (14.3.4)

When required by regulatory authorities, the results of all safety-related laboratory tests should be available in tabular listings, using a display similar to the following, where each row represents a patient visit at which a laboratory study was done, with patients grouped by investigator (if more than one) and treatment group, and columns include critical demographic data, drug dose data, and the results of the laboratory tests. As not all tests can be displayed in a single table, they should be grouped logically (e.g., hematological tests, liver chemistries, electrolytes, urinalysis). Abnormal values should be identified, e.g., by underlining, bracketing. These listings should be submitted as part of the registration/marketing application, when this is required, or may be available on request.

For all regulatory authorities, there should be a by-patient listing of all abnormal laboratory values in section 14.3.4, using the format described above. For laboratory abnormalities of special interest (abnormal laboratory values of potential clinical importance), it may also be useful to provide additional data, such as normal values before and after the abnormal value, and values of related laboratory tests. In some cases, it may be desirable to exclude certain abnormal values from further analysis. For example, single, nonreplicated, small abnormalities of some tests (e.g., uric acid or electrolytes) or occasional low values of some tests (e.g., transaminase, alkaline phosphatase, blood urea nitrogen (BUN)) can probably be defined as clinically insignificant and excluded. Any such decisions should be clearly explained, however, and the complete list of values provided (or available to authorities on request) should identify every abnormal value.

12.4.2 Evaluation of Each Laboratory Parameter

The necessary evaluation of laboratory values must in part be determined by the results seen, but, in general, the following analyses should be provided. For each analysis, comparison of the treatment and control groups should be carried out, as appropriate, and as compatible with study size. In addition, normal laboratory ranges should be given for each analysis.

12.4.2.1 Laboratory Values Over Time

For each parameter at each time over the course of the study (e.g., at each visit) the following should be described: the group mean or median values, the range of values, and the number of patients with abnormal values, or with abnormal values that are of a certain size (e.g., twice the upper limit of normal, 5 times the upper limit; choices should be explained). Graphs may be used.

12.4.2.2 Individual Patient Changes

An analysis of individual patient changes by treatment group should be given. A variety of approaches may be used, including:

I. "Shift tables" - These tables show the number of patients who are low, normal, or high at baseline and then at selected time intervals.

II. Tables showing the number or fraction of patients who had a change in parameter of a predetermined size at selected time intervals. For example, for BUN, it might be decided that a change of more than 10 mg per deciliter BUN should be noted. For this parameter, the number of patients having a change less than this or greater than this would be shown for one or more visits, usually grouping patients separately depending on baseline BUN (normal or elevated). The possible advantage of this display, compared to the usual shift table, is that changes of a certain size are noted, even if the value is not abnormal.

III. A graph comparing the initial value and the on-treatment values of a laboratory measurement for each patient by locating the point defined by the initial value on the abscissa and a subsequent value on the ordinate. If no changes occur, the point representing each patient will be located on the 45 deg. line. A general shift to higher values will show a clustering of points above the 45 deg. line. As this display can show only a single time point for a single treatment, interpretation requires a time series of these plots for treatment and control groups. This kind of display identifies outliers readily (it is useful to include patient identifiers for the outliers).

12.4.2.3 Individual Clinically Significant Abnormalities

Clinically significant changes (defined by the applicant) should be discussed. A narrative of each patient whose laboratory abnormality was considered a serious adverse event and, in certain cases, considered another significant event should be provided under section 12.3.2 or 14.3.3. When toxicity grading scales are used (e.g., World Health Organization, National Cancer Institute), changes graded as severe should be discussed regardless of seriousness. An analysis of the clinically significant changes, together with a recapitulation of discontinuations due to laboratory measurements, should be provided for each parameter. The significance of the changes and likely relation to the treatment should be assessed, e.g., by analysis of such features as relationship to dose, relationship to drug concentration, disappearance on continued therapy, positive dechallenge, positive rechallenge, and the nature of concomitant therapy.

12.5 Vital Signs, Physical Findings, and Other Observations Related to Safety

Vital signs, other physical findings, and other observations related to safety should be analyzed and presented in a way similar to laboratory variables. If there is evidence of a drug effect, any dose-response or drug concentration-response relationship or relationship to patient variables (e.g., disease, demographics, concomitant therapy) should be identified and the clinical relevance of the observation described. Particular attention should be given to changes not evaluated as efficacy variables and to those considered to be adverse events.

12.6 Safety Conclusions

The overall safety evaluation of the study drug(s) should be reviewed, with particular attention to events resulting in changes of dose or need for concomitant medication, serious adverse events, events resulting in withdrawal, and deaths. Any patients or patient groups at increased risk should be identified and particular attention paid to potentially vulnerable patients who may be present in small numbers, e.g., children, pregnant women, frail elderly, people with marked abnormalities of drug metabolism or excretion. The implication of the safety evaluation for the possible uses of the drug should be described.

13. Overall Conclusions

The efficacy and safety results of the study and the relationship of risks and benefit should be briefly summarized, referring to the tables, figures, and sections above as needed. The presentation should not simply repeat the description of results nor introduce new results.

The conclusions should clearly identify any new or unexpected findings, comment on their significance, and discuss any potential problems such as inconsistencies between related measures. The clinical relevance and importance of the results should also be discussed in the light of other existing data. Any specific benefits or special precautions required for individual subjects or at-risk groups and any implications for the conduct of future studies should be identified.

14. Tables, Figures, and Graphs Referred to but not Included in the Text

Figures should be used to visually summarize the important results, or to clarify results that are not easily understood from tables.

Important demographic, efficacy, and safety data should be presented in summary figures or tables in the text of the report. However, if these become obtrusive because of size or number they should be presented here, cross- referenced to the text, along with supportive, or additional, figures, tables, or listings.

The following information may be incorporated in this section:

14.1 Demographic Data
Summary figures and tables
14.2 Efficacy Data
Summary figures and tables
14.3 Safety Data
Summary figures and tables
14.3.1 Displays of Adverse Events
14.3.2 Listings of Deaths, Other Serious and Significant Adverse Events
14.3.3 Narratives of Deaths, Other Serious and Certain Other Significant Adverse Events
14.3.4 Abnormal Laboratory Value Listing (each patient)

15. Reference List

A list of articles from the literature pertinent to the evaluation of the study should be provided. Copies of important publications should be attached in an appendix (16.1.10 and 16.1.11). References should be given in accordance with the internationally accepted standards of the 1979 Vancouver Declaration on "Uniform Requirements for Manuscripts Submitted to Biomedical Journals" or the system used in "Chemical Abstracts."

16. Appendices

This section should be prefaced by a full list of all appendices available for the study report. Where permitted by the regulatory authority, some of the following appendices should not be submitted with the report but should be provided only on request.

The applicant should therefore clearly indicate those appendices that are submitted with the report. N.B. To have appendices available on request, they should be finalized by the time of filing of the submission.

16.1 Study Information
16.1.1 Protocol and protocol amendments
16.1.2 Sample case report form (unique pages only)
16.1.3 List of IEC's or IRB's (plus the name of the committee Chair if required by the regulatory authority) - Representative written information for patient and sample consent forms

16.1.4 List and description of investigators and other important participants in the study, including brief (1 page) curriculum vitaes or equivalent summaries of training and experience relevant to the performance of the clinical study

16.1.5 Signatures of principal investigator(s) (or coordinating investigators) or responsible medical officer, depending on the regulatory authority's requirement

16.1.6 Analytical documentation--batch certificate for the investigational product(s)

16.1.7 Randomization scheme and codes (patient identification and treatment assigned)

16.1.8 Audit certificates (if available) (see Annex IVa and IVb of the guideline)

16.1.9 Documentation of statistical methods

16.1.10 Publications based on the study

16.1.11 Important publications referenced in the report

16.2. Patient Data Listings

16.2.1 Discontinued patients

16.2.2 Protocol deviations

16.2.3 Patients excluded from the efficacy analysis

16.2.4 Demographic data

16.2.5 Compliance and/or Drug Concentration Data (if available)

16.2.6 Individual Efficacy Response data

16.2.7 Adverse event listings (each patient)

16.2.8. Listing of individual laboratory measurements by patient, when required by regulatory authorities

16.3 Case Report Forms

16.3.1 CRF's of serious adverse events and withdrawals for adverse events

16.3.2 Other CRF's submitted

16.4. Individual Patient Data Listings (U.S. Archival Listings)

ANNEX IV a

AUDIT CERTIFICATES (if available)

The audit certificate is a document indicating:

- The audit (type, identification number)
- The audited system, clinical trial, or organization
- The audit dates
- The date of release of the audit report

The Audit Certificate should be prepared and used in accordance with standard operating procedures. To be valid the Audit Certificate should be signed by the responsible auditor(s) or the head of the independent audit organization as defined in the standard operating procedures.

Audit Certificates may be added to the clinical study report.

Since it is recognized that not every clinical study will be subject to an independent audit, Audit Certificates for systems audits can demonstrate the independent assessment of the systems or part of the systems used to ensure that the trial is performed and the data are generated in compliance with good clinical practice.

Audit Certificates pertaining to audits of systems implemented and existing during the planning, conduct, data analysis, and reporting of the reported clinical trial may be added to the clinical study report together with the certificate for the audit of the reported study, if such an audit was conducted.

The Audit Certificate should be retained in accordance with any regulatory requirements.

ANNEX IX

Guidance for Section 11.3.2--Statistical/Analytical Issues

A. Statistical Considerations

Details of the statistical analysis performed on each primary efficacy measure should be presented in an appendix. Details reported should include at least the following information:

(a) The statistical model underlying the analysis. This should be presented precisely and completely, using references if necessary.

(b) A statement of the clinical claim tested in precise statistical terms, e.g., in terms of null and alternative hypotheses.

(c) When statistically reasonable and appropriate, the power against specific clinically meaningful alternatives for those tests that fail to reject the null hypothesis to indicate whether the study size was adequate.

(d) The statistical methods applied to estimate effects, construct confidence intervals, etc. Literature references should be included where appropriate.

(e) The assumptions underlying the statistical methods. It should be shown, insofar as statistically reasonable, that the data satisfy crucial assumptions, especially when necessary to confirm the validity of an inference. When extensive statistical analyses have been performed by the applicant, the extent to which the analyses were planned prior to the availability of data should be considered, and, if they were not, how bias was avoided in choosing the particular analysis used as a basis for conclusions should be described. This is particularly important in the case of any subgroup analyses, because if such analyses are not preplanned, they will ordinarily not provide an adequate basis for definitive conclusions.

(i) In the event data transformation was performed, a rationale for the choice of data transformation along with interpretation of the estimates of treatment effects based on transformed data should be provided.

(ii) A discussion of the appropriateness of the choice of statistical procedure and the validity of statistical conclusions will guide the regulatory authority's statistical reviewer in determining whether reanalysis of data is needed.

(f) The test statistic, the sampling distribution of the test statistic under the null hypothesis, the value of the test statistic, significance level (i.e., p-value), and intermediate summary data, in a format that enables the regulatory authority's statistical reviewer to verify the results of the analysis quickly and easily. The p-values should be designated as one- or two-tailed. The rationale for using a one-tailed test should be provided.

For example, the documentation of a two-sample t-test should consist of the value of the t-statistic, the associated degrees of freedom, the p-value, the two sample sizes, mean and variance for each of the samples, and the pooled estimate of variance. The documentation of multi-center studies analysed by analysis of variance techniques should include, at a minimum, an analysis of variance table with terms for centers, treatments, their interaction, error, and total. For crossover designs, the documentation should include information regarding sequences, patients with sequences, baselines at the start of each period, washouts and length of washouts, dropouts during each period, treatments, periods, treatment by period interaction error, and total. For each source of variation, aside from the total, the table should contain the degrees of freedom, the sum of squares, the mean square, the appropriate F-test, the p-value, and the expected mean square. Generally, it is recommended that regression type sums of squares be provided in addition to any other analyses.

Intermediate summary data should display the demographic data and response data, averaged or otherwise summarized, for each center-by-treatment combination (or other design characteristic such as sequence) at each observation time.

B. Format and Specifications for Submission of Data Requested by Regulatory Authority's Statistical Reviewers

In the report of each controlled clinical study, there should be data listings (tabulations) of patient data utilized by the sponsor for statistical analyses and tables supporting conclusions and major findings. These data listings are necessary for the regulatory authority's statistical review, and the sponsor may be asked to supply these patient data listings in a computer-readable form.

International Harmonization of Regulatory Requirements and Guidelines

I. Background

The purpose of this document is to articulate FDA's policy on development and use of standards with respect to international harmonization of regulatory requirements and guidelines. As used throughout this document, the term "standards" includes what are commonly referred to as "consensus standards," "voluntary standards," and "industry standards." Also, FDA sometimes accepts standards and makes them mandatory regulatory requirements. Although the draft policy focuses on international harmonization and international standards, its principles are applicable as well to domestic standards activities in which FDA participates.

A. Statutory Mandates for FDA-Regulated Products

FDA is the principal regulatory agency within the Public Health Service (PHS). The agency protects the public health by, among other things, implementing statutory provisions designed to ensure that food is safe and otherwise not adulterated or misbranded; that human and veterinary drugs, human biological products, and medical devices are safe and effective; that cosmetics are safe; and that electronic product radiation is properly controlled. FDA-regulated products must be truthfully and accurately labeled and in compliance with all applicable laws and regulations. The statutory mandates for safeguarding the public health in these product sectors are prescribed in several statutes, notably in the Federal Food, Drug, and Cosmetic Act; the Public Health Service Act; and the Fair Packaging and Labeling Act.

B. International Harmonization of Regulatory Requirements and Guidelines

In recent decades, great changes in the world economy, together with expanded working relationships of regulatory agencies around the globe, have resulted in increased interest in international harmonization of regulatory requirements. Increased international commerce, opportunities to enhance public health through cooperative endeavors, and scarcity of government resources for regulation have resulted in efforts by the regulatory agencies of different nations to work together on standards and harmonize their regulatory requirements. Such harmonization enhances public health protection and improves government efficiencies by reducing both unwarranted contradictory regulatory requirements and redundant applications of similar requirements by multiple regulatory bodies. Harmonization facilitates cooperation in regulatory activities.

Harmonization of FDA's regulatory requirements and guidelines with those of other countries was recently embraced as a pillar of the President's and Vice President's National Performance Review. In Reinventing Drug and Device Regulation (April 1995), international harmonization was identified as a high priority initiative across FDA programs. Recognizing the considerable synergy between its domestic policy and its international policy priorities, FDA is sharpening and focusing its planning for enhanced alignment of FDA and international standards.

In 1992, an FDA Task Force on International Harmonization had provided a broad assessment of the goals, scope, and direction of FDA's international activities. These activities were found to comprise a wide variety of efforts by FDA to retain and strengthen its public health safeguards, while striving toward common ground with its foreign government counterparts on product standards, criteria for the assessment of test data, and enforcement procedures. The task force's recommendations for the agency included an overall FDA policy on international harmonization, which is to encourage the initiation and support of efforts, consistent with the agency's goals and principles, that will further the international harmonization of standards and policies for the regulation of products for which FDA has authority. Soon thereafter, FDA's strategic plan began to recognize standards as the premier focus of the agency's international activities.

1. Goals

FDA's goals in participating in international harmonization are:

** To safeguard U.S. public health,

** To assure that consumer protection standards and requirements are met,

** To facilitate the availability of safe and effective products,

**To develop and utilize product standards and other requirements more effectively, and

** To minimize or eliminate inconsistent standards internationally.

2. General Principles

FDA participation in international harmonization efforts should be guided by the following general principles:

** The harmonization activity should be consistent with U.S. Government policies and procedures and should promote U.S. interests with foreign countries.

** The harmonization activity should further FDA's mission to protect the public health by, among other things, ensuring that food is safe and otherwise not adulterated or misbranded; that human and veterinary drugs, human biological products, and medical devices are safe and effective as required by law and are not adulterated or misbranded; that cosmetics are not adulterated or misbranded; that electronic product radiation is properly controlled; and that all of these products are labeled truthfully and informatively.

** FDA's input into international standard setting activities should be open to public scrutiny and should provide the opportunity for the consideration of views of all parties concerned.

** FDA should accept, where legally permissible, the equivalent standards, compliance activities, and enforcement programs of other countries, provided that FDA is satisfied such standards, activities, and programs meet FDA's level of public health protection.

** Scientific and regulatory information and knowledge should be exchanged with foreign government officials, to the extent possible within legal constraints, to expedite the approval of products and protect public health.

Thus, the agency's primary goal in all of its international harmonization activities is to preserve and enhance its ability to accomplish its public health mission. Global harmonization is also approached with the aim of enhancing regulatory effectiveness, by providing more consumer protection with scarce government resources, and increasing worldwide consumer access to safe, effective, and high quality products.

C. Other Obligations and Policies

1. International Agreements

The U.S. Government is a party to international trade agreements. In the United States, such trade agreements become effective only after implementing legislation is signed into law. FDA has participated in recent international trade negotiations to ensure that FDA's requirements are preserved and that regulatory practices can remain focused on fulfilling the agency's mission to protect the public health while being supportive of emerging, broader U.S. Government obligations and policies. In addition, FDA continues to be involved in work of the World Trade Organization (WTO) as well as the North American Free Trade Agreement (NAFTA) committees on sanitary and phytosanitary measures, and on technical barriers to trade, in order to foster international harmonization of regulatory requirements and to facilitate consultation on trade issues. Recently FDA has begun to be involved in other regional activities, e.g., the Forum on Asia Pacific Economic Cooperation (APEC), work on initial steps toward a Free Trade Area of the Americas (FTAA), and work towards a Transatlantic Area that strengthens our ties with Europe.

The principal international trade agreement is the General Agreement on Tariffs and Trade (GATT), which entered into force on January 1, 1948. GATT has since been amended several times following negotiation sessions known as rounds.

The GATT Agreement on Technical Barriers to Trade (TBT), popularly known as the Standards Code, was negotiated during the Tokyo Round of the GATT in the 1970's and entered into force on January 1, 1980. As part of a general effort to reduce unnecessary nontariff barriers to trade, the TBT agreement was intended to promote use by countries of standards, technical regulations, and conformity assessment procedures that are based on work done by international standards bodies. The implementing legislation for the TBT agreement, provided in the Trade Agreements Act of 1979 as amended in 1994 (Pub. L. 103- 465; 19 U.S.C 2531-2582), has thus provided additional authority for FDA's international standards activity. To assure that harmonization does not result in lowering safety or quality standards for U.S. consumers, this law contains the safeguard that:

"* * * No standard-related activity of any private person, Federal agency, or State agency shall be deemed to constitute an unnecessary obstacle to the foreign commerce of the United States if the demonstrable purpose of the standards-related activity is to achieve a legitimate domestic objective including, but not limited to, the protection of legitimate health or safety, essential security, environmental, or consumer interests and if such activity does not operate to exclude imported products which fully meet the objectives of such activity."

The most recent GATT round, the Uruguay Round, was concluded on December 15, 1993, and was formally signed at the Marrakech Ministerial Meeting on April 15, 1994. The new WTO will administer the new GATT and other Uruguay Round agreements, and every country that is a member of the WTO will be required to adhere to all of these agreements. On December 8, 1994, Pub. L. 103-465 was enacted in the United States to approve and implement the Uruguay Round agreements. This law included updating changes in the Trade Agreements Act that reaffirmed the duty of Federal agencies to participate in international standards activities, subject to available resources.

One of the agreements of the Uruguay Round administered by the WTO is the new agreement on TBT, which is similar in many respects to the 1980 TBT agreement. As with the 1980 TBT agreement, the purpose of the new TBT agreement is to ensure that product standards, technical regulations, and related procedures do not create unnecessary obstacles to trade. The new TBT agreement ensures, and clearly states, that each country has the right to establish and maintain technical regulations for the protection of human, animal, and plant life and health and the environment, and for prevention against deceptive practices.

In the new TBT agreement, the term "standard" is defined as:

"[A] document approved by a recognized body, that provides, for common and repeated use, rules, guidelines or characteristics for products or related processes and production methods, with which compliance is not mandatory [emphasis added]. It may also include or deal exclusively with terminology, symbols, packaging, marking or labelling requirements as they apply to a product, process or production method."

Also, "technical regulation" is defined as:

"[A] document which lays down product characteristics or their related processes and production methods, including applicable administrative provisions, with which compliance is mandatory [emphasis added]. It may also include or deal exclusively with terminology, symbols, packaging, marking, or labelling requirements as they apply to a product, process or production method."

Thus, in the language of the new TBT agreement, when a government acts to accept a voluntary standard to make it mandatory, the resulting document is a technical regulation. A measure used to ascertain compliance with a standard or technical regulation is a conformity assessment procedure.

The new TBT agreement continues and strengthens the reference to international standards found in the 1980 TBT agreement. Specifically, the agreement states that, where technical regulations are required and relevant international standards exist or their completion is imminent, WTO-member countries shall use them, or the relevant parts of them, as a basis for their technical regulations, except when such international standards or relevant parts would be an ineffective or inappropriate means for the fulfillment of the legitimate objectives pursued. Further, the agreement states that, with a view towards harmonizing technical regulations on as wide a basis as possible, WTO-member countries shall play a full part within the limits of their resources in the preparation by appropriate international standards bodies of international standards for products for which they either have adopted or expect to adopt technical regulations.

Another agreement of the Uruguay Round administered by the WTO is the Agreement on the Application of Sanitary and Phytosanitary Measures (SPS agreement). This agreement pertains to those measures intended: (1) To protect animal or plant life or health within a territory from risks arising from the entry, establishment, or spread of pests, diseases, disease-carrying organisms, or disease-causing organisms; (2) to protect human or animal life or health within a territory from risks arising from additives, contaminants, toxins, or disease-causing organisms in foods, beverages, or feedstuffs; (3) to protect human life or health within a territory from risks arising from diseases carried by animals, plants, or products thereof, or from entry, establishment, or spread of pests; or (4) to prevent or limit other damage within a territory from the entry, establishment, or spread of pests.

In order to harmonize SPS measures on as wide a basis as possible, the SPS agreement encourages Members to base their SPS measures on international standards, guidelines, or recommendations. Thus, the SPS agreement, like the new TBT agreement, encourages use of international standards. The SPS agreement refers specifically to standards established by the Codex Alimentarius Commission, as discussed below.

NAFTA also contains TBT and SPS agreements similar to those in the new WTO agreements.

2. Internal U.S. Government Policy

The United States Office of Management and Budget (OMB), in its revision to OMB Circular No. A-119 (58 FR 57643, October 26, 1993), provides policy on Federal use of standards and agency participation in voluntary standards bodies and standards-developing groups:

"It is the policy of the Federal Government in its procurement and regulatory activities to:

a. Rely on voluntary standards, both domestic and international, whenever feasible and consistent with the law and regulation pursuant to law;

b. Participate in voluntary standards bodies when such participation is in the public interest and is compatible with agencies' missions, authorities, priorities, and budget resources; and

c. Coordinate agency participation in voluntary standards bodies so that: (1) The most effective use is made of agency resources and representatives; and (2) the views expressed by such representatives are in the public interest and, as a minimum, do not conflict with the interests and established views of the agencies."

OMB Circular No. A-119 also establishes additional policy guidance and responsibilities for U.S. Government agencies. It is applicable to all executive agency participation in voluntary standards activities, domestic and international, but not to activities carried out pursuant to treaties and international standardization agreements.

The term "standard," as defined in OMB Circular No. A-119, means:

"* * * a prescribed set of rules, conditions, or requirements concerned with the definition of terms; classification of components; delineation of procedures; specification of dimensions, materials, performance, design, or operations; measurement of quality and quantity in describing materials, products, systems, services, or practices; or descriptions of fit and measurement of size."

The circular defines "voluntary standards" as:

"* * * established generally by private sector bodies, both domestic and international, and are available for use by any person or organization, private or governmental. The term voluntary standard includes what are commonly referred to as "industry standards" as well as "consensus standards," but does not include professional standards of personal conduct, institutional codes of ethics, private standards of individual firms, or standards mandated by law, such as those contained in the United States Pharmacopeia and the National Formulary, as referenced in 21 U.S.C. 351."

These definitions in OMB Circular No. A-119 conform to common usage and are consistent with the usage of these terms throughout this policy document. It should be noted that, under the TBT, "standards" are considered to be nonmandatory (i.e., voluntary) unless promulgated into mandatory technical regulations.

II. Standards Programs and Practices Within FDA

A. Purpose of FDA Involvement in Standards

The central purpose of FDA involvement in the development and use of standards is to assist the agency in fulfilling its public health, regulatory mission. The agency intends to participate in the development of standards, domestic or international, and adopt or use standards when such action will enhance its ability to protect consumers and the effectiveness or efficiency of its regulatory efforts. In doing so, FDA recognizes that standards often serve as useful adjuncts to agency regulatory controls and that economies of time and human resources are often realized in solving problems when consensus-building activities are undertaken and conducted in open, public arenas. The working together of FDA staff with other professionals outside the agency in standards bodies effectively multiplies the technical resources available to FDA. Further, standards bodies generally have in place procedures for periodically reviewing and updating completed standards, thus extending the resource-multiplier effect, as well as keeping the solutions current with the state of knowledge. The economy of effort translates into monetary savings to the agency, regulated industries, and ultimately consumers. Further, using standards, especially international ones, is a means to facilitate the harmonization of FDA regulatory requirements with those of foreign governments, to better serve domestic and global public health.

Another benefit of participating in the development of standards at both domestic and international levels is that in sharing technical information with technical groups and professionals outside FDA, staff members have opportunities to learn of other viewpoints on an issue, to establish scientific leadership, and to remain informed of state-of-the-art science and technology.

B. Past and Present Activities

FDA has been involved in standards activities for many years, and on January 25, 1977, the agency promulgated a final regulation, now found at 21 CFR 10.95 (Sec. 10.95), covering the participation by FDA employees in standards-setting activities outside the agency. This regulation encourages FDA participation in standard-setting activities that are in the public interest and specifies the circumstances under which FDA employees can participate in various types of standards bodies.

Standards activities of multilateral organizations such as the World Health Organization (WHO) and the Organization for Economic Cooperation and Development (OECD, an international organization with 25 member countries with advanced industrial economies) are often important to FDA and frequently involve multiple product types. For example, OECD is developing Genetic Toxicology Test Guidelines that are of interest to all FDA Centers. Similarly, guidelines developed under the International Programme on Chemical Safety of the WHO relate to chemicals that may be in a wide variety of FDA- regulated products, such as food additives, pesticides, drugs, animal drugs, biologics, and devices. The United States Pharmacopeia is a national standard setting body in which FDA officials actively participate.

The principal standards organizations that are not connected with a treaty are the International Organization for Standardization (ISO) and the International Electrotechnical Commission (IEC). Private organizations and government agencies, including FDA, participate in ISO and IEC activities through the American National Standards Institute (ANSI). ANSI represents the United States in the ISO and IEC and coordinates much of the standards development activity in the United States. As discussed below, FDA is active in many ISO, IEC, ANSI, and standards development organization activities. For example, FDA is represented on the Board of Directors of ANSI and on several of its committees and working groups.

1. Foods and Veterinary Medicine

FDA's Center for Food Safety and Applied Nutrition (CFSAN) and Center for Veterinary Medicine (CVM) actively participate in the development of international standards by the Codex Alimentarius Commission (Codex). Codex is an international organization formed in 1962 to facilitate world trade in foods and to promote consumer protection. It is a subsidiary of two United Nations components, the Food and Agriculture Organization (FAO) and the WHO. Codex standards cover food commodity standards (similar to FDA standards of identity), food additives, food contaminants, and residues of veterinary drugs in food. FDA officials chair two Codex committees, the Food Hygiene Committee and the Residues of Veterinary Drugs in Foods Committee, and participate in many others. Through its involvement, FDA has been influential in the establishment of many Codex standards. FDA's procedures for reviewing Codex standards for purposes of regulation are codified in 21 CFR 130.6 and 564.6.

A provision of the United States implementing legislation for the Uruguay Round Agreements, Pub. L. 103-465, requires the President to designate an agency to inform the public, through a notice published in the Federal Register each year by June 1, of certain Codex Alimentarius standard-setting activities. The President, pursuant to Proclamation No. 6780 of March 23, 1995 (60 FR 15845, March 27, 1995), designated the Department of Agriculture to have this responsibility, and the first such notice of Codex activities was published in the Federal Register of May 23, 1995 (60 FR 27250).

In 1988, the governments of the United States and Canada entered into the Canada-United States Free Trade Agreement (now largely superseded by NAFTA). Since then, officials from CFSAN and CVM have participated in technical working groups responsible for implementation of the chapter of the agreement that deals with agriculture, food, beverage, and related goods (the CUSFTA Technical Working Groups).

Officials from CFSAN and CVM also participate in the development of standards by such domestic and international groups as the Food Chemicals Codex (FCC), AOAC International (previously, the Association of Official Analytical Chemists), expert committees of the WHO, FAO, ISO, and other international consensus standards bodies. Standards developed by these organizations are used by industry, both in the United States and abroad. These standards provide industry with guidance for food grade materials and processes, and thus help elevate the quality of food and food chemicals in domestic and international trade.

CFSAN has adopted many FCC and American Society for Testing and Materials (ASTM) standards and AOAC methods, incorporating them into regulations for both food additives and generally recognized as safe food ingredients. CFSAN also refers industry to relevant FCC, Codex, or ASTM standards when discussing particular issues related to good manufacturing practices. CFSAN accepts many AOAC and equivalent methods for use by laboratories in assaying food and in testing for contaminants in food.

CVM accepts many AOAC and equivalent methods for use by laboratories in testing for drug residues in animal tissues. CVM has adopted the consumption estimates used by the FAO/WHO Joint Expert Committee on Food Additives in the development of standards for drug residues in animal tissues.

CVM is also an active participant in a new harmonization effort under the auspices of the Office of International Epizooties (OIE). This activity is known as the International Cooperation on Harmonisation of Technical Requirements for Registration of Veterinary Medicinal Products (VICH).

2. Biologics and Drugs

There has been active international standard setting for biological products for more than 50 years. Officials from FDA's Center for Biologics Evaluation and Research (CBER) serve as experts or members of a variety of international committees that perform standard-setting functions. Activities have encompassed collaborative studies to establish international units of measure and to develop internationally accepted standards for control of biologics, including WHO standards. Efforts have been directed to many kinds of biological products, including vaccines, human blood and plasma products, blood testing reagents, and allergenic extracts, and have extended to biotechnology-derived growth factors, cytokines, and monoclonal antibody products.

FDA's Center for Drug Evaluation and Research (CDER), CBER, and the National Center for Toxicological Research (NCTR) actively participate in the International Conference on Harmonisation of Technical Requirements for Registration of Pharmaceuticals for Human Use (ICH). This ongoing project, begun in 1989, has been undertaken by governmental agencies responsible for regulation of drugs and by industry trade organizations from the European Union (EU), Japan, and the United States. Specifically, ICH is sponsored jointly by the Commission of the European Communities (CEC), the Japanese Ministry of Health and Welfare (MHW), FDA, the European Federation of Pharmaceutical Industries' Associations (EFPIA), the Japan Pharmaceutical Manufacturers Association (JPMA), and the Pharmaceutical Research and Manufacturers

Association (PhRMA) of the United States. In addition, the International Federation of Pharmaceutical Manufacturers Associations (IFPMA) participates as an umbrella organization for the pharmaceutical industry and provides the secretariat function for ICH. ICH operates under the direction of the ICH Steering Committee, which is comprised of representatives of these organizations. Official observer status has been given to WHO, the European Free Trade Area (EFTA), and the Health Protection Branch of Canada.

The purposes of ICH are to: (1) Provide a forum for a dialogue between regulatory agencies and the pharmaceutical industry on differences in the technical requirements for product registration (i.e., requirements for product marketing) in the EU, Japan, and the United States; (2) identify areas where modifications in technical requirements or greater mutual acceptance of research and development procedures could lead to more efficient use of human, animal, and material resources without compromising safety, quality, and efficacy; and (3) make recommendations of practical ways to achieve greater harmonization in the interpretation and application of technical guidelines and requirements for registration. The work products of ICH, created in working groups of experts from the regulatory agencies and industry, consist of a series of consensus guidance documents. These guidance documents, after successive ICH steps of review and acceptance, including an opportunity for public review and comment in the respective jurisdictions, are forwarded to the regulatory agencies with the expectation that they will be formally adopted by the agencies.

Officials from both CBER and CDER also participate in a consensus standard setting activity sponsored by the Council for International Organizations of Medical Sciences (CIOMS) that is aimed at standardizing medical definitions and adverse experience reporting.

3. Medical Devices and Radiation-Emitting Products

FDA's Center for Devices and Radiological Health (CDRH) has had extensive involvement with standards in its regulation of medical devices, as well as electronic products that emit radiation. The development of standards to solve problems related to medical devices involves many groups outside FDA. The interaction between CDRH and the manufacturing and health care communities that frequently occurs during the standards development process provides knowledge and insight into the use of products, problems, and the effectiveness of solutions. Frequently, the public discussion of the problem that occurs in the consensus-building process results in the manufacturers and the users of the subject medical device implementing the solution before a standard is formally completed. Thus, CDRH has encouraged participation in the development of standards as a useful adjunct to regulatory controls. CDRH's approach to use and participation in the development of consensus standards was described in a letter dated June 29, 1993, to all interested parties from the Director of CDRH. (This policy did not apply to mandatory performance standards, i.e., technical regulations, for class II medical devices as specified under the Medical Device Amendments of 1976 (Pub. L. 94- 295). The Safe Medical Devices Act of 1990, SMDA (Pub. L. 101-629), puts the promulgation of mandatory standards at the discretion of the agency.)

Over 200 completed consensus standards and selected sections of additional draft standards that are not yet complete have been incorporated into guidance documents for applications for conducting clinical trials with investigational devices and applications for permitting devices to be marketed. These guidance documents are widely disseminated by CDRH to all interested parties. Other standards used by CDRH, or which CDRH has helped to develop, concern measurement or test methods, or support good manufacturing practices and quality assurance.

A new ISO Committee, Technical Committee 210 (TC-210) is developing harmonized standards in these areas. Also, CDRH is an active participant in the Global Harmonization Task Force, in cooperation with officials from Canada, the EU, Japan, and other countries.

CDRH has published a notice of a working draft of a final rule to revise the current good manufacturing practice regulations for medical devices (60 FR 37856, July 24, 1995), in part to ensure that they are compatible with specifications for quality systems contained in an international quality standard developed by ISO, namely ISO 9001 "Quality Systems Part 1. Specification for Design/Development, Production, Installation, and Servicing." This standard (ISO 9001) is becoming widely recognized by medical device regulatory authorities worldwide and is finding application in many other industry sectors as well. CDRH officials, working with counterpart foreign government officials, are pursuing in step-wise fashion the harmonization of quality system inspection procedures and enforcement.

The process of harmonizing regulatory requirements is facilitated by using an international standard as a basis. Such harmonization is not only recognized public policy, but for medical devices, it is explicitly encouraged by provisions of SMDA (Pub. L. 101-629), which states, in part, that FDA "* * * may enter into agreements with foreign countries to facilitate commerce in devices between the United States and [foreign] countries consistent with the requirements of this Act." 21 U.S.C. 383.

In a recent (April 1995) program review, CDRH reported that in 1994, 192 Center staff members served as primary and alternate liaison representatives on 440 committees and subcommittees in 38 standards developing organizations (domestic and foreign). CDRH actively reviewed 286 draft standards; of these, 134 were with nine international standards organizations. The experience CDRH has acquired over the years has provided the foundation for the standards policy it announced on June 29, 1993. The essential features of that policy are reflected in the FDA policy presented in section IV. below.

4. Regulatory Affairs

FDA's Office of Regulatory Affairs (ORA) is increasingly active in international standards activities relevant to quality control and conforming assessment, including activities relevant to ISO--9001 and laboratory regulation.

III. Response to Comments

In response to its request for comments on the draft international harmonization policy on standards, FDA received comments from ten organizations (standards setting organizations, trade and professional associations, a manufacturer, and a consumer organization). A discussion of the comments received and the agency's responses follow.

1. In general, the comments supported the agency's proposed international harmonization policy on standards. For example, one comment stated that the policy demonstrated the agency's commitment to the international standards development process as well as international harmonization. Another comment pointed out that the policy will better enable the agency to establish agreements with other global regulatory bodies, and ultimately permit FDA to carry out its mandate to protect the public health in a more efficient and cost effective manner. Other comments stated that the harmonization of regulatory requirements and supportive standards could benefit U.S. companies engaged in international trade. In addition, one of these comments pointed out that standards reflect technology, and that the first priority with regard to standards should always be to develop standards that represent the best available technological solutions, adding that 'harmony' and 'consistency' can be achieved through the general acceptance of excellent technological solutions. FDA agrees with these comments.

A. Potential for Lowered Standards

2. Two comments stated that harmonization has the potential to result in lowered standards, with potential adverse effects on public health protection. One of the comments expressed concern that FDA was subordinating the public interest in favor of voluntary standards bodies and standards developing groups in a manner that is inconsistent with the vital tasks assigned to the agency to protect health. The second comment argued that the first priority with regard to standards should be to develop standards that represent the best available technological solutions, and that FDA should not support international standards that reflect inferior or compromised technological solutions that become obstacles, rather than benefits, to U.S. industry. Another comment, while agreeing that international standards should be adopted as national standards whenever possible, stated that international standards may sometimes not meet the needs of our health care community, adding that some may contain safety standards only and no performance parameters, and that the international standards may also be inconsistent with our country's codes and regulations.

FDA wishes to reassure those who commented that FDA's participation in international harmonization activities is intended to safeguard the U.S. public health and to assure that consumer protection standards and requirements are met. Indeed, a central principle that guides the agency's international harmonization activities is that the activities should further FDA's mission to protect the public health. In addition, international agreements to which the U.S. Government is a party have provisions that ensure that harmonization activities will not result in lowered standards. For example, the WTO Agreement on SPS provides that each country may determine its appropriate level of protection; therefore, the encouragement to use international standards as the basis for technical regulations will not result in "downward harmonization." Safeguards have been built into the TBT agreement and U.S. implementing legislation that protect the ability of each country to establish requirements necessary to fulfill a legitimate objective. As stated in section I.C.1. above, the implementing legislation for the new TBT agreement, which provides additional authority for FDA's international standards activities, provides further assurance that such harmonization would not result in lowering safety or quality standards for U.S. consumers. Thus, the agency does not agree that harmonization will result in inferior standards. Furthermore, FDA's participation in standards development, consistent with Sec. 10.95 and OMB Circular No. A-119, and FDA's use of standards in its regulatory programs, will be dependent not only on the substantive aspects of the standards for ensuring the safety, effectiveness, and quality of products, but also on the development process for the standard. The standard itself must also comply with all applicable statutes, regulations, and policies.

B. Regulatory Issues

3. One comment stated that any time a voluntary standard is used in a regulation, the scope of that standard needs to be unambiguously determined. The comment used a hypothetical case of a voluntary standard intended to be used in the home environment being inappropriately extended to the hospital environment. The comment argued that if the regulatory need exceeds the scope of the existing voluntary standard, it may indicate the need for yet another voluntary standard that addresses the additional scope, or else provisions may need to be added to the existing voluntary standard so as to accommodate the broader scope in which the standard is to be applied.

The agency agrees that when a voluntary standard is used in a regulation, the scope of that standard should be unambiguously expressed. As stated above, a basic tenet in the development and adoption of a standard is that it contributes to safer, more effective, and higher quality products. Inappropriate application of a standard (voluntary or as part of a technical regulation) would run counter to this notion.

4. Three comments questioned the need to make voluntary standards mandatory regulations. One of the comments stated that if the agency uses a voluntary standard as a "referee" standard, which means that the agency uses the voluntary standard as a frame of reference for determining safety and efficacy, there should be no need for the agency to go through the procedures required for creating a regulation. The second comment stated that if technical standards are based on state-of-the-art science and are revised as needed to incorporate advances, they should be voluntary standards as opposed to regulatory requirements. The third comment asserted that existence of a standard does not warrant a regulation and FDA should avoid unnecessary regulations.

The agency agrees that it should avoid unnecessary regulations but notes that there are times when it finds it is necessary to propose and promulgate regulations for the efficient enforcement of the laws it administers. Voluntary standards that will serve agencies' purposes and are consistent with applicable laws and regulations can be adopted and used by Federal agencies. This principle is stated in both FDA's policy and in section 7(a) of OMB Circular No. A-119 on Reliance on Voluntary Standards. Thus, when appropriate, FDA will adopt voluntary standards by referencing them in the regulations it promulgates. In all other instances, these standards will remain voluntary.

As stated above, the purpose of FDA's involvement in the development and use of standards is to assist the agency in fulfilling its public health and regulatory missions. Thus, the agency intends to participate in the development of domestic and international standards, and to adopt or use standards, when such action will enhance its ability to protect consumers and the effectiveness or efficiency of its regulatory efforts.

C. Transparency

5. Four comments addressed the need for transparency during the development of standards, in determining "official" use of a standard, or when standards are used in a regulation or adopted as regulations. The comments asserted that, if voluntary standards are incorporated into guidance documents and compliance policy guides and serve as the bases for mandatory standards and other regulations promulgated by FDA, ample opportunity should be provided for interested parties to comment through the established procedures of notice and comment rulemaking. One comment further stated that the policy should include a statement of assurance that FDA will engage potentially affected parties whenever it intends formal inclusion of a voluntary standard in an FDA document or process.

The agency agrees that the development of standards should be conducted in an open fashion. Under Sec. 10.95, one of the criteria for FDA participation in standards-setting bodies is that the group or organization responsible for the standard-setting activity must have a procedure by which an interested person will have an opportunity to provide information and views on the activity and standards involved, and that the information and views will be considered. This is why FDA clearly states in its policy (section IV. below) that one of the factors for FDA's participation in standards development and use is the transparency of the process, i.e., the process must be open to public scrutiny and provide the opportunity for the consideration of views of all parties concerned.

With regard to transparency when standards are used in a regulation or adopted as regulations, under the Administrative Procedure Act (APA), an agency that issues, amends, or revokes a regulation, whether on its own initiative or when petitioned by an interested person, must act in an open manner with adequate time provided for comment from interested parties, which will be considered before a final regulation is promulgated. FDA's rule on promulgation of regulations, found in 21 CFR 10.40, is explicit with respect to the need for transparency of the process and opportunity for participation in the process by interested persons. Other procedural regulations govern guidelines and similar documents (21 CFR 10.90), and interested persons may use correspondence or meetings (21 CFR 10.65), petitions (21 CFR 10.30), or reviews by supervisors (21 CFR 10.75) to raise issues and present views about other nonbinding guidance documents, which provide industry with useful information about recommended or alternative ways to comply with requirements. In fact, FDA has increasingly used public meetings to elicit and share information with regard to its guidance documents and it currently is reviewing the procedures it uses to develop guidance documents to ensure sufficient transparency in the process.

Thus, with regard to the comment that this policy should include a statement of assurance that FDA will engage potentially interested parties whenever it intends formal inclusion of a voluntary standard in an FDA document or process, the agency finds that it is not necessary to do so as part of this document because there are established mechanisms under the APA and the agency's administrative practices and procedure regulations for obtaining and considering views of interested persons. Of course, FDA is not foreclosing future consideration of additional mechanisms toward this end.

D. Comments on Specific Issues

6. One comment suggested the alternative language, "The standard contributes to safe and effective products that meet consumers' requirements for quality," instead of "The standard contributes to safer, more effective, and higher quality products" (section III.A.1., proposed policy) and "The standard, if adhered to, would help ensure the safety, effectiveness, or quality of products" (section III.D.1., proposed policy). The alternative language was offered to simplify future negotiations and to allow the agency to participate more fully in standards development and promulgation. The comment also questioned the use of the terms 'safer' and 'more effective' in section III.A.1. of the proposed policy (see above) because it is not clear what the measures for 'safer' and 'more effective' are. The comment further stated that the term "higher quality" is relative, leaving open to question who determines higher quality. Finally, the comment added that an international standard could conceivably result in requirements for the same degree of safety, effectiveness, and quality as those required by FDA.

The agency is revising the policy in a manner similar to that suggested by the comment. FDA agrees that an international standard could indeed result in requirements for the same degree of safety, effectiveness, and quality as required by FDA. In fact, one of FDA's guiding principles in its international harmonization activities is that FDA should accept, where legally permissible, equivalent standards of other countries provided such standards meet FDA's goals to facilitate the availability of safe, effective, and properly labeled products. The agency further agrees that the alternative language would allow the agency more flexibility to participate in standards development, without compromising public health, and is therefore amending the policy accordingly.

7. One comment supported FDA's intent to develop standards on the basis of sound scientific and technical information. The comment added that the use of sound scientific and technical information will permit the development of food regulations and standards that cannot be misconstrued as unreasonable trade barriers. However, the comment cautioned that a decision on participation in standards development should be based on the purpose of the standard, not whether the standard is based on sound scientific and technical information.

The agency agrees that while all standards should be based on sound scientific and technical information, not all scientifically sound standards will serve purposes justifying the agency's participation in developing them. FDA's regulation on participation in outside standard-setting activities states that not only will the activity be based upon consideration of sound scientific and technological information, but it also will be designed to protect the public against unsafe, ineffective, or deceptive products and practices (21 CFR 10.95(d)(5)(i)). In addition, OMB Circular No. A-119 states that it is the policy of the Federal Government to participate in voluntary standards bodies when such participation is in the public interest and is compatible with agencies' missions, authorities, priorities, and budget resources. The OMB Circular adds that the providing of agency support to a voluntary standards activity should be limited to that which is clearly in furtherance of an agency's mission and responsibility. These directives are adequately reflected in the policy.

8. One comment suggested that proposed section III.A.4., "The development of an international standard that achieves the agency's public health objectives is generally, but

not always, given a higher priority than the development of a domestic standard," be deleted because it is made clear in other parts of the draft policy that FDA complies with U.S. obligations under the GATT, other international trade agreements, and OMB Circular No. A-119.

The agency agrees that the draft policy document does make clear that FDA complies with U.S. obligations under international agreements and the OMB Circular. However, the agency does not agree that proposed section III.A.4. should be deleted. FDA's belief that the development of an international standard that achieves the agency's public health objectives is generally (but not always) given a higher priority than the development of a domestic standard, is an important factor on which agency participation in standards development is based and merits being clearly delineated. This is more so because proposed section III (section VI. of this document), FDA Policy on Standards, is intended to stand on its own as the agency's policy on harmonization of standards and, therefore, needs to be as complete as possible.

FDA emphasizes that there are three routes to development of a harmonized international standard, all of which are favored under the FDA policy: (1) The U.S. voluntary standards community or an agency, such as FDA, develops a U.S. standard and takes it to an international forum so it can be made an international standard; (2) a standard already developed in an international forum (or by another country or a regional standards body) is adopted as a U.S. voluntary or regulatory standard; or (3) a new international standard is developed, "from scratch," in an international forum. Which of these routes is followed in the particular case will vary with the facts of that case. While starting a standards activity in an international forum offers many efficiencies in avoiding duplication of effort, there will continue to be times when it makes sense first to develop a domestic standard (voluntary or regulatory) and then to take it, as appropriate, to an international forum.

9. One comment asserted that the intent of proposed factors III.A.6. and III.D.6., which state: "Wherever appropriate for the product, the standard stresses product performance rather than product design, but where necessary, covers all factors required to ensure safety, effectiveness, and quality," was not clear. The comment added that inspection can be used to prevent poor quality products from being consumed but that safety cannot be inspected into a product. The comment stated further that safety must be designed into products during development, subsequent manufacturing, packaging, and transport. Further, the comment stated that product performance or product functionality issues with regard to safety are the primary focus in the development of food regulations, and therefore, the comment recommended alternative language to that in the proposed policy: "Wherever appropriate for the product, the standard stresses product safety, performance, and functionality, but where necessary, covers all factors required to ensure safety and effectiveness, including product and process design, and process performance."

The agency believes that the suggested language is helpful in capturing FDA's intentions in formulating these factors as the basis for participation in standards development, and use of standards in its regulatory programs. Therefore, the agency is making editorial changes in the factors along the lines suggested in the comment.

E. Other Comments

10. One comment recommended that FDA review and revise current U.S. guidelines for toxicity testing of food additives as outlined in the Toxicological Principles for the Safety Assessment of Direct Food Additives and Color Additives Used in Food (Redbook I), as well as the proposed guidelines set forth in the revised draft, Redbook II, and harmonize with those recommended by the OECD. The comment added that this will allow more universal acceptance of results performed throughout the world and will minimize the need to repeat expensive testing to meet different testing standards in different countries.

The agency has stated that standards activities of organizations such as OECD are often important to FDA, and that the development of international standards, and harmonization with international standards if they achieve the agency's public health objectives, will in most instances be given a high priority.

The agency announced that the draft Redbook II was available (March 29, 1993, 58 FR 16536) and solicited comments on the draft revised guidelines. Redbook II is being finalized in light of comments received by the agency, including a comment that the guidelines should be harmonized with those of the OECD; the final revised Redbook II has yet to be issued. The agency notes that, in revising the guidelines in the Redbook, it took into account the fact that differences among guidelines can result in unnecessary duplication of effort and inefficient use of scarce testing resources. The agency also wants to make clear that the Redbook is designed to provide guidance. Strict adherence to Redbook guidelines is not a requirement for toxicological studies conducted to establish the safety of an additive.

11. One comment indicated concern that a long standing participation by the United States Public Health Service in the 3-A Sanitary Standards (for dairy and related food industries) is not mentioned in the text of the draft policy. The comment also stated that it is necessary that, as international agreements anticipate trade and importation of equipment, compliance with 3-A Sanitary Standards should be applied by reference to assure receipt of acceptable equipment.

The domestic and international standards-setting organizations or bodies listed under section II.B. of this document are those in which FDA has been or is most actively involved in developing standards. The listing is not meant to be exhaustive nor is it meant to list all standards setting bodies in which FDA has an interest.

12. One comment urged FDA to reference voluntary standards rather than adopting and publishing standards, to maintain appropriate support for standards development. The comment argued that referencing rather than publishing the text of voluntary standards as regulations or guidance protects the standards organizations' copyrights which provide the financial support for national and international programs.

FDA uses standards in the manner described in OMB Circular No. A-119, which states that while voluntary standards adopted by Federal agencies should be referenced along with their dates of issuance and sources of availability in appropriate publications, regulatory orders, and in related in-house documents, such adoption should take into account any applicable requirements of copyright law and other similar restrictions.

13. One comment advised that the value of standards is that they are the consensus product of all technology experts, not just the consensus of experts from government.

Therefore, care should be exercised that government participation in voluntary standards organizations and its use of voluntary standards does not lead to an appearance that voluntary standards organizations are unduly directed or influenced by government.

The agency is sensitive to the need for balanced participation in voluntary standards bodies and works within OMB's guidelines regarding policy to be followed by executive agencies in working with voluntary standards bodies. OMB Circular No. A-119 states that agency representatives serving as members of standard-developing groups should participate actively and on a basis of equality with private sector representatives but that, in doing so, agency representatives should not seek to dominate such groups. In addition, the number of individual agency participants in a given voluntary standards activity should be kept to the minimum required for effective presentation of the various program, technical, or other concerns of Federal agencies. Finally, while the circular encourages agency representatives to participate in the policy-making process of voluntary standards bodies, particularly in matters such as establishing priorities, developing procedures for preparing, reviewing and approving standards, and creating standards-developing groups, it also states that in order to maintain the private, nongovernmental nature of such bodies, agency representatives should refrain from decisionmaking involvement in the internal day-to-day management of such bodies.

F. Conclusion

Therefore, after considering the comments received, FDA is issuing this statement of policy.

IV. FDA Policy on Standards Note /1/

Note /1/ This policy document does not create or confer any rights, privileges, or benefits, for or on any person, nor does it operate to bind FDA in any way.

It is the intent of this policy to enable FDA to: (1) Continue to participate in international standards activities that assist it in implementing statutory provisions for safeguarding the public health, (2) increase its efforts to harmonize its regulatory requirements with those of foreign governments, including setting new standards that better serve public health, and (3) respond to laws and policies such as the Trade Agreements Act and OMB Circular No. A-119 that encourage agencies to use international standards that provide the desired degree of protection. Accordingly, it is the policy of FDA, concerning the development and use of standards, that:

A. FDA participation in standards development will be based on the extent to which the development activity and expected standard conform to certain factors, with consideration also being given to the resources available in FDA to devote to the effort and expected efficiencies to be gained as a result of the effort; the factors are as follows:

1. The standard stresses product safety and effectiveness and therefore contributes to safe, effective, and high quality products; when necessary, the standard also covers all factors required to ensure safety and effectiveness, including product and process design, and process performance;

2. The standard is based on sound scientific and technical information and permits revision on the basis of new information;

3. The development process for the standard is transparent (i.e., open to public scrutiny), complies with applicable statutes, regulations, and policies, specifically including Sec. 10.95 and OMB Circular A-119, and is consistent with the codes of ethics that must be followed by FDA employees;

4. The development of an international standard that achieves the agency's public health objectives is generally, but not always, given a higher priority than the development of a domestic standard; and

5. The development of a horizontal standard which applies to multiple types of products is generally, but not always, given higher priority than the development of a vertical standard which applies to a limited range of types of products.

B. FDA is not bound to use standards developed with FDA participation. For example, the agency will not use a standard when, in the judgment of FDA, doing so will compromise the public health.

C. The uses of final (and selected draft or proposed) standards, or selected relevant parts, will include, where appropriate: (1) Incorporating such standards into guidance documents for nonclinical testing, applications for conducting clinical trials with investigational products, and applications for permitting products to be marketed; (2) conducting reviews of such applications; (3) incorporating such standards into compliance policy guides; (4) conducting reviews of test protocols used by firms as part of good manufacturing practices; (5) conducting reviews of study protocols submitted by firms as required for postmarket surveillance studies or programs; (6) serving as the basis for mandatory standards or other regulations promulgated by FDA; and (7) serving as the basis for reference (e.g., evaluation criteria) in a memorandum of understanding with other government agencies.

D. The use of a standard in the regulatory programs of FDA is dependent upon the following factors:

1. The standard stresses product safety and effectiveness and therefore, if adhered to, would help ensure the safety, effectiveness, or quality of products; when necessary, the standard also covers all factors required to ensure safety and effectiveness, including product and process design, and process performance;

2. The standard is based on sound scientific and technical information and is current;

3. The development process for the standard was transparent (i.e., open to public scrutiny), was consistent with the codes of ethics that must be followed by FDA employees, and the standard is not in conflict with any statute, regulation, or policy under which FDA operates;

4. Where a relevant international standard exists or completion is imminent, it will generally be used in preference to a domestic standard, except when the international standard would be, in FDA's judgment, insufficiently protective, ineffective, or otherwise inappropriate; and

5. Where a relevant horizontal standard which applies to multiple types of products exists or its completion is imminent, it will generally be used in preference to a vertical standard, which applies to a limited range of types of products, except when such horizontal standard would be insufficiently protective, ineffective or otherwise inappropriate.

E. FDA employees will comply with agency regulations (Sec. 10.95) covering participation in standard setting activities outside the agency.

Chapter 9

DRUG INSPECTIONS

OVERVIEW

Although validation concepts had been used in analytical test methods, the real "validation story," as it has been called,[1] began in the mid-1970s when the FDA published its new Good Manufacturing Practice (GMP) regulations. Beginning with sterilization validation, the FDA later began looking at validation of aseptic manufacturing processes, and by the early 1980s non-aseptic processes had been included for FDA scrutiny.

In 1986, the FDA issued its "Guidelines on General Principles of Process Validation, still an important document for FDA-regulated industries to consider when applying validation principles and practices in manufacturing processes.

FDA began considering computer system validation as the pharmaceutical industry turned to automated production methods in the 1980s. Today, system validation is an increasingly important aspect of FDA inspections of pharmaceutical facilities, and new guidelines are being prepared by the agency to help the industry comply with newly-emerging regulations.

The document included in this edition of the *Validation Compliance Annual* is:

1. FDA's Computer Assisted New Drug Application (CANDA) Manual

---

[1]See Kenneth G. Chapman, "A History of Validation in the United States, Part I," *Pharmaceutical Technology*, Vol. 15 (October 1991): 82-96; Kenneth G. Chapman, "A History of Validation in the United States -- Part II, Validation of Computer-Related Systems," *Pharmaceutical Technology*, Vol. 15 (November 1991):54-68.

CANDA Guidance Manual

Acknowledgement

The guidance in this document is the result of the considerable efforts of the Pharmaceutical Manufacturers Association (PMA), its member firms, and the Food and Drug Administration (FDA), in conjunction with the support and agreement of the Health Protection Branch (HPB) of Health and Welfare Canada.

The information contained in the following chapters has been gleaned from the many efforts of the different PMA/FDA task forces. The task force members would like to express our gratitude to everyone who contributed to this effort. In addition, we want to express appreciation to our colleagues who read the works-in-progress and raised questions and supplied answers that improved the final product.

Preface

The Center for Drug Evaluation and Research (CDER) within the Food and Drug Administration (FDA) has been experimenting with computer assisted new drug applications (CANDAs) for the past 6 years. In a Federal Register notice published September 15, 1988 (See Appendix I-A), FDA provided basic guidance to drug sponsors and other interested parties on the factors that the Agency considers in accepting computer assisted new drug applications. This guidance is still applicable today.

Over the years, additional guidelines and information addressing the specific needs of the various FDA review disciplines have evolved. Because no one source existed for all these guidances, the Pharmaceutical Manufacturers Association (PMA) and FDA decided to compile all this information into one comprehensive reference guide.

What we have learned from the early CANDAs and the information included in this guide can be used as a framework for future CANDA submissions. it is clear that a pharmaceutical company may do more or less than is outlined in the guidelines. However, we hope that each drug sponsor preparing a CANDA will give careful consideration to what is contained here. It is also hoped that this information will be useful to drug sponsors in planning and preparing their CANDA submissions. Each new drug application sponsor must, of course, ensure that the application fulfills all statutory and regulatory requirements.

The purpose of this guidance manual is to increase awareness concerning computerization and to encourage industry to computerize their NDA submissions so that we may achieve the goal of virtually all submissions being computerized by 1995.

The subject of CANDAs is dynamic. Rapidly developing technologies and the proliferation of new information and ideas will require that the guidance manual be updated regularly to maintain the work's usefulness.

A person may also choose to use alternate procedures even though they are not provided for in the guide. If a person chooses to depart from the practices and procedures set forth in the guide, that person may wish to discuss the matter further with the agency to prevent an expenditure of money and effort on activities that may later be determined to be unacceptable by FDA. This guideline does not bind the agency, and it does not create or confer any rights, privileges, or benefits for or on any persons.

CHAPTER I

ADMINISTRATIVE/TECHNICAL REQUIREMENTS FOR THE U.S. FOOD AND DRUG ADMINISTRATION

A. Introduction and Background

History

The Food and Drug Administration (FDA) has been receiving data in an automated fashion for more than two decades. The Division of Biometrics pioneered the original submission of data in order to avoid the task of rekeying the data from the application to analyze it. When the Division of Biometrics initiated these efforts, their statisticians were among the few experienced in computers staff in the agency. Several years ago, several sponsors and a third party developed the early computer assisted new drug applications and initiated experiments with Center review divisions. These early efforts took place at a time when there was a limited amount of equipment in the agency and relatively few staff that were experienced in using. At that point, the agency and the Pharmaceutical Manufacturers Association (PMA) initiated joint working groups to explore the feasibility of the computer assisted new drug application and its possible impacts on both the agency and sponsors.

There has been dramatic change since that time so that now nearly all members of the staff of the Center for Drug Evaluation and Research (Center) have a microcomputer or terminal on their desk and staff are rapidly becoming experienced in computers.

One of the significant benefits of the availability of automated communications is the reduction in the turnaround time between an agency request for information and the applicant's response. Submissions still require a large amount of paper. In the future it might be possible to require that only a relatively small amount of paper, if any, actually be submitted, with the bulk of the submission, including both the required and the frequently requested case reports, submitted in an automated form, e.g., optical disk.

The Center has received more than 60 CANDAs to date. This does not include the submissions of data on tape and diskette that the Center has accepted for some time. Many of the reviews involving CANDAs have been completed. The agency has not only committed to continuing to explore the use of automated technology via the CANDA program but is both educating and encouraging its own staff and urging the industry to accelerate their efforts in this area so that by FY95, or earlier, virtually all submissions to the agency will either be full CANDAs or have major automated components.

The Center has seen the CANDAs improve in quality since the inception of the program and with the availability of new technology.

The agency is continuing to explore with the industry automated methods to further increase the efficiency and effectiveness of the review process. Although a number of CANDAs have been received, the agency is not in a position to conclude that one technical approach is superior to another. However, the agency's experience with CANDAs does allow it to conclude that data from case report forms and tabular listings can be effectively retrieved and reviewed using computer technology and that a significant decrease in the amount of paper submitted to the agency is possible. It is also clear that submission of text material in an automated format is of significant benefit.

FEDERAL REGISTER Notice

To further enhance the process initiated with the PMA and to assure that all interested parties were aware of the efforts and developments, a FEDERAL REGISTER notice was published on September 15, 1988 (see Appendix I-A) which announced that the agency believed that the increased use of computers may improve the efficiency of the drug review process. The notice was intended to provide basic guidance to drug sponsors and other interested parties on the factors that the agency considers in accepting computer assisted new drug applications and the topics that the sponsors should be prepared to discuss with the agency.

The notice made clear that it applied not only to new drug applications (NDAs) and antibiotic applications, but that the policies applied to investigational new drugs (INDs) and abbreviated new drug applications (ANDAs) as well.

The guidance discusses resolution of potential problems in filing a CANDA with respect to logistical problems in the installation of computer equipment, consultation with the reviewing divisions on the development of CANDAs, the characteristics of the proposed application including the methods and data files to be used, computer hardware and software requirements, plans for reviewer orientation and training, and other factors relating to the submission of applications using computer technology.

Computer Assisted New Drug Application

The CANDA can take any number of various forms including direct access by reviewers to the applicant's mainframe, direct access to a third party mainframe, review of the application on a microcomputer in the reviewer's office, review of the application on an agency mainframe, or review of the application from an optical disk. The CANDA can also be a combination of these methods. A CANDA might also include an electronic mail component or the use of facsimile with the applicant to speed communications regarding a submission.

Minimum Alternative

The agency is in the process now of defining what it considers to be a minimum alternative for submission of a CANDA. The agency realizes that many of the CANDAs developed to date are quite sophisticated and expensive to develop and believes that there is an alternative possible that would not entail a large expense to the sponsors. For example, if the text material were submitted as a work processing document and the data in a standard ASCII format, that in itself would significantly aid the Center in completing its review efforts.

B. Interaction with FDA

Office of Management

The Center strongly urges drug sponsors to explore the use of CANDAs. Drug sponsors and other organizations considering the use of CANDAs should contact Robert A. Bell, Director, Office of Management (address in Appendix I-B).

A drug sponsor intending to submit a CANDA is encouraged to discuss its plans with the Center's Office of Management to identify and resolve potential problems involved in filing a CANDA. In particular, sponsors should be prepared to discuss possible logistical and security problems in the installation of computer equipment, if additional equipment is required, or current reviewer equipment could be upgraded.

The Center urges that the sponsor meet with the involved review division well in advance of submission of the CANDA to discuss both the sponsor's planned approach and the needs of the reviewers. Center participants in such a preliminary discussion will ordinarily include management of the responsible review division, review personnel that will have responsibility for review of the application, as well as statisticians from the Division of Biometrics. A list of FAX numbers within the Center can be found in Appendix I-C. In addition, a member of the Division of Information Systems Design may be present to assist in resolution of technical questions. CANDAs may be developed by the sponsor or a third party. In its initial discussion with the agency, the sponsor should identify the developer of the CANDA. Because of limited resources, reviewing divisions will provide detailed consultation on the development of CANDAs only in the course of their ordinary regulatory discussions with sponsors. The reviewing divisions are not in a position to offer assistance directly to third party vendors who are not currently working with a sponsor involved in obtaining FDA marketing approval of a drug or drugs (i.e., those whose intention is to develop and market data processing methodology for use by drug sponsors). Such third party vendors should contact Mr. Bell for information on CANDAs.

A drug sponsor should be prepared to discuss the following additional aspects of its proposed CANDA with the Center:

o General characteristics of the proposed application, including the methods and data files to be used. The Center is interested in the methods used to capture and store data.

o The Center prefers that the equipment already on the reviewer's desks be used to the extent possible. If the specific equipment is not adequate for the CANDA to be submitted, the agency will attempt to provide the equipment if it can be done in a timely manner or the agency will work with the sponsor to either upgrade the existing equipment, add peripheral equipment or, if necessary, accept the loan of hardware and software required for the specific CANDA.

If the sponsor is going to loan equipment to the agency, it is imperative that it discusses this with Mr. David M. Moss, Director, Division of Information Systems Design, Office of Management (address in Appendix I-B). He will consult with the sponsor regarding the equipment planned for the loan. In some cases, planned configurations may not be acceptable to the agency due to security, environmental, or space factors. Mr. Moss will work with the sponsor to ensure that the equipment is properly identified as loaned equipment and that this is entered in agency records. It is the sponsor's responsibility to assure that any loaned hardware and software are returned at the completion of the review. The sponsor should also confirm that not agency information is on the system before they remove it.

To assure that both the agency and the sponsor avoid any impropriety or appearance of impropriety, the Center will need a full description of any hardware and software being loaned including its name, configuration, serial numbers, the drug application number, the name of drug, the name(s) of the reviewer(s) to whom the equipment is being loaned and a statement to the effect that all hardware and software will be removed when the review is complete or the hardware and software are no longer needed. This information must be provided to Mr. Moss. A sample letter addressing issues that must be covered is included in Appendix I-D.

It is essential that the sponsor observe all software copyright restrictions. The agency will follow these explicitly. The agency will assume that all provisions of copyright agreements and licenses are being followed when software is loaned to the agency.

o In selecting software, a sponsor should consider the overall requirements of analysis and data retrieval in the review process, the availability of software at FDA, and the ease of operation for the reviewer. The sponsor should clearly identify what program language and/or data base system is being used.

- Orientation and training should include instructing reviewers in the use of the computer hardware, data file content and structure, and retrieval routines. Sponsors should also be prepared to provide continuing technical support and advice to reviewing personnel throughout the course of the review. The agency and sponsors have found that this interaction is beneficial and prevents delays for technical reasons.

- Although all of the review divisions do accept CANDAs, the filing of a CANDA is conditional on the reviewing division's willingness to accept the application in the CANDA format used by the sponsor. For this reason, it is important that the acceptability of the proposed CANDA be established, to the degree it is possible, well in advance of the submission, before substantial effort has been expended.

- The submission of a CANDA will not influence the priority given to an application. The Center has written policies regarding the factors that affect the priority given to the review of an application. In general, review priority is based on the Center's classification of a new drug product under the IND/NDA classification system (see Center for Drug Evaluation and Research Staff Manual Guide 4820.3, Drug Classification and Priority Review Policy).

- Currently, the submission of a CANDA will generally not affect the required submission of an application in paper form. Future development of CANDA procedures should result in the reduction of the amount of paper required to be filed, especially with respect to case report forms and tabular listings. At the current time, the majority of the case reports need only be submitted in requested by the reviewer. These nonrequired case reports form a large portion of the volume of a submission. In many cases, reviewers are willing to accept true copies of the case reports on optical disks if available. This must be discussed with the individual reviewers involved.

- It is critical that the electronic submission and the hardcopy be identical. If any data or information are not included in the electronic version, that must be clearly noted and all differences clearly identified.

- Over time, review staff will begin to review only the CANDA and spend little or no time reviewing the hardcopy. There may in fact, as indicated earlier, be a time when no hardcopy is submitted. For these reasons, the Center must consider the CANDA to be an official part of the submission and, therefore, all diskettes, hard disks or copies of information on hard disks, optical disks and other media must remain at the Center upon completion of the review. The Center, however, does not want to keep any loaned equipment; the Center will take responsibility for accessing the data after final action. If the CANDA involves the review of data at a remote site, that data also becomes an official part of the application and procedures will be established to make such a CANDA part of the official submission.

o Once the drug sponsor is prepared to submit a CANDA, it should talk to Mr. Moss to identify and resolve any potential problems involved in filing the CANDA. This includes logistical problems such as electrical needs, security concerns, environmental concerns, communications, leased lines, dedicated circuits, and modems.

Electronic Mail

Electronic mail has been a frequent component of CANDAs submitted to date. The early experience with electronic mail systems showed that they were not used very effectively and their use was minimal. However, now that agency reviewers have become more experienced in computers, are using agency mail systems, and becoming more comfortable with the systems, their use and effectiveness have improved considerably.

Most of the early electronic mail systems put in place required the agency reviewer to access a mail system on the applicant's system. Although this remains an option at the discretion of the reviewers involved and the applicant, it has been found that since reviewers are already accessing the ALL-IN-1 electronic mail system at the agency, it is an extra burden both to learn a new mail system and to log off of the agency system and check the sponsor system. It has also been found that unless there is a fairly large volume of message traffic, the reviewers will access the electronic mail with decreasing frequency over time.

Some sponsors have set up electronic mail systems using a commercial third party system that requires both the sponsor and the reviewers to access the remote facility and in most cases learn new mail systems. Again, this remains an option at the discretion of the reviewers involved and the sponsor but is not recommended.

There are two other alternatives that the agency has found to be preferable for the reviewers and believes are most likely to work effectively to enhance communication between the agency and the sponsors, mail boxes on the agency system, and mail systems using a common third party interface.

The Center has a MicroVAX in place on which it is only running mail systems, both VMSmail and ALL-IN-1. The Center is providing access to the MicroVAX and instructions on its use in order to provide an effective and easy to use electronic messaging system for both its own reviewers and sponsors. The use of the system requires nothing additional for the Center's reviewers. Sponsors must dial up and log onto the MicroVAX. Messages sent between the reviewers and the sponsors will go directly to their mail boxes. In the case of the reviewers, mail will appear in their regular ALL-IN-1 account and they will have no new procedures to follow. The systems are fully protected. Messages can only be read by the senders and recipients of messages.

The agency has established a system using a third party that provides the benefits of the MicroVAX-based system described above to sponsors. The differences from the reviewers' standpoint are fairly minimal. There are charges by the third party to use this system, and when sending a message, the reviewer must input a bit more information. However, the reviewers still use ALL-IN-1 just as they did before. The advantage to the sponsor over the previously described agency MicroVAX-based system is that they can now use their own company mail systems without having to log onto the agency system. There is a charge by the third party. For both the reviewers and the sponsors, messages are sent from their own systems and received in their own accounts.

One of these two latter methods is recommended to assure effective electronic mail communications. In both of the latter methods, Mr. Moss can arrange for the establishment of the necessary accounts.

Another vehicle for communications with the Center is via BITNET. A Center for Drug Evaluation and Research node has been established on the VAXcluster. Center staff can receive and send BITNET messages from their ALL-IN-1 accounts.

An additional method of exchanging information is through the use of facsimile. All of the reviewing divisions have facsimile equipment. The facsimile numbers for all of the review divisions are given in the appendix that describes each review division's computing environment.

The agency has not set uniform procedures to be followed in electronic communications via electronic mail or use of facsimile. Some divisions require that a sponsor submit copies of all transmissions of both the sponsor and the agency as part of a formal submission on a monthly basis. Some division require that all electronic messages be precleared before sending. In some divisions, consumer safety officers serve as an intermediary for all electronic communications. Because the procedures do vary depending on the division involved, it is important that the division's requirements are determined before such communications are initiated.

To minimize the security risks to both the Center and to sponsors, the Center will not allow executable data files to be sent via electronic mail. Diskettes or other media can be used to get executable data files to the Center rapidly via overnight mail or other suitable vehicles.

C. Hardware and Software Common in CDER

o VAXcluster

The VAXcluster is used to support the Center's ORACLE information systems as well as electronic mail, word processing, and application system development and testing. The Division of Drug Analysis, located in St. Louis, has a VAX which is used to support all the ORACLE information systems currently under development in DDA, electronic mail, word processing, sample inventory, administrative systems, and application system development and testing.

At this time, we do not envision using the VAXcluster for CANDAs with the current configuration although such use might be possible when sufficient capacity is available to accommodate them. This is only precluded by current budget restraints.

- Fee-for-Service Data Centers

  The Parklawn Computer Center (PCC) is a fee for service center which currently has an IBM ES/9021-520 and an IBM 3090-400J. PCC is used by the Center for large reports, to back up all major databases, for statistical analysis, for publications, and for pharmacokinetic analysis. The Center also uses the Division of Computer Research and Technology at the NIH to a much lesser extent for some statistical and scientific analysis. The Center also uses a CRAY for pharmacokinetic analysis.

  Although PCC is not currently being used for CANDAs, the facility is available for this use.

- Microcomputers

  The Center for Drug Evaluation and Research has a large number of microcomputers. Many of the microcomputers are laptops that are used on trips, for work at home, and for emergency backup. Several of them are connected to laboratory equipment. Several are specialized microcomputers which are shared by a number of users in the Center.

  The Center's goal is to have a microcomputer or terminal for every employee who can use one.

  The microcomputers are used for CANDAs, word processing, statistical analysis, pharmacokinetics analysis, electronic mail, modeling, graphics, telecommunications, data entry, databases, spreadsheets, systems development, inventories, and laboratory data capture and analysis.

Chapter XII describes generally the mix of microcomputers and software in the review divisions. Future updates will provide more detailed information on each division.

D. The Future of CANDAs

To date the agency has had over 60 CANDAs submitted. The agency believes that the computer does help in the review process. The Center is in the process of providing the necessary hardware, software, and training to every reviewer although there is still a way to go as can be seen by the current mix of hardware available in the Center. A long-term goal is a commitment for every reviewer in the Center to become experienced in computers.

CANDAs are clearly here to stay and all sponsors are strongly encouraged to both use them and submit them. As you will see in Chapter III on the early evaluation of CANDAs, the computer systems developed for the agency are easiest to use and most effective when they have been developed and used by the scientific staff in the sponsor company. The company should develop a computer application for the collection of data that parallels the development of the NDA submission. In doing this, the company encourages its own professional staff to be a part of the development. The outcome will be a computer system which is not only beneficial to the company but to FDA as well. We believe that there are several benefits to be derived.

o First, the quality of the submission will improve. Jonathan Cook, formerly an analyst in FDA's Office of Planning and Evaluation and now a member of the management team in the Center's Office of Generic Drugs, published in Pharmaceutical Executive, December 1990, an article titled, "Plain and Simple Advice for Improving NDAs." This article outlines the NDA submission deficiencies found from a review of several NDAs. We believe that all of the types of deficiencies identified in the article can be corrected through the proper use of computer assisted submissions.

o Second, based on an evaluation of the early CANDAs, we believe that there will be a decrease in the overall approval time. In five of the early CANDAs that have gone through to final action, the automated application came in at the same time as the paper copy of the NDA. In these five cases, the average time to action was 17 months instead of the usual average of approximately 30 months. We are in the process of analyzing each of these applications to see if we can understand exactly what caused this significant difference.

o Third, we expect a significant reduction in paper. The Center is looking at, and questioning, the requirements for NDA submissions. If the case report forms were submitted on optical disk, there would be no need for this material to come in hard copy when requested. But, more significantly, we are beginning to look at such things as summary tables. If we had a CANDA, we would not need summary tables because we could build our own. Thus, the Center is beginning to review the entire application to see how computer tools can assist and improve the process.

o Fourth, the development of new submission technologies continues to evolve. The submission technology of 1985 would have utilized something analogous to the IBM AT. Today, microcomputer technologies are using 386 and 486 level platforms to speed overall processing in addition to architectures that transfer information more quickly. There are coprocessors to speed math and statistical computations. We have the options of better graphics, spreadsheets, database management, word processing, and desktop publishing. Storage, too, has changed significantly.

We now have diskettes that hold over three times the amount of data of a 1985 diskette, and can be carried in a shirt pocket, or fixed disks that store data in multiples of 10 from the capabilities of 1985. With the development and refinements with Windows, DOS, and the OS/2 operating systems, microcomputers can perform multiple tasks simultaneously.

With optical disk storage and readability, optical character recognition, CD-ROM, and similar technologies, it is possible to eliminate a huge amount of typing and copying that used to be necessary.

Monitoring CANDA Progress

To monitor the progress of CANDA development and use, the Center is in the process of developing a database that will include demographic data regarding the application, including the type of CANDA utilized, its features, and an indication of its degree of success in meeting the desired goals.

To keep this data current, the agency asks that companies be willing to participate in case studies of their CANDAs. FDA staff will review expectations, benefits of features, and observed weaknesses of CANDAs with both he reviewing staff at FDA and with company staff.

When a CANDA is initiated, we would like to send out a questionnaire to the company to get a clear idea of expectations. When the CANDA is completed, we will ask both FDA staff and company staff to complete another short questionnaire. The results of these questionnaires will create a case study of each CANDA. Periodically, the Center will use the data from these case studies to develop issue papers and to update this guidance.

One final word on CANDAs, we are very concerned about data security and integrity. Chapter V discusses this in some detail. We are particularly concerned about the proliferation of viruses and their impact. Although the Center has anti-virus software installed on all of its equipment, it is critical that sponsors check their automated submissions before submitting them to avoid possible contamination. If you have any concerns in this area, please discuss them with us. If a CANDA is submitted for or on microcomputers which do not contain anti-virus software, the Center will load appropriate anti-virus software to protect the Center, the sponsor and the integrity of the CANDA.

Though this introductory chapter contains a lot of information, it is still considered a brief overview of the process. The Center strongly encourages discussion between sponsors creating and submitting CANDAs and the Center. We hope the following chapters will smooth the way for that to happen.

APPENDIX I-A Federal Register Notice on Submitting CANDAs

FOOD AND DRUG ADMINISTRATION

(Docket No. 88N-0197)

SUBMISSION OF DRUG APPLICATIONS TO THE FOOD AND DRUG ADMINISTRATION USING COMPUTER TECHNOLOGY

AGENCY: Food and Drug Administration

ACTION: Notice

SUMMARY: The Food and Drug Administration (FDA) believes the increased use of computers may improve the efficiency of the drug application review process. This notice is intended to facilitate the use of computers in the drug approval process by giving the drug sponsors and other interested parties guidance on the factors FDA considers in accepting computer assisted new drug applications (CANDAs). This notice also identifies topics drug sponsors should be prepared to discuss with FDA before submitting CANDAs. Although this notice deals primarily with the use of computer technology in new drug applications (NDAs) and antibiotic applications, the policies described also apply to investigational new drug applications (INDs) and abbreviated new drug applications (ANDAs).

DATE: Comments should be submitted on or before March 14, 1989.

ADDRESS: Written comments to the Dockets Management Branch (HFA-305), Food and Drug Administration, Rm 4-62, 5600 Fishers Lane, Rockville, MD 20857.

FOR FURTHER INFORMATION CONTACT:
About the submission and contents of CANDAs:

Robert A. Bell
Center for Drug Evaluation and Research (HFD-10)
Food and Drug Administration
5600 Fishers Lane, Rockville, MD 20857.
301-443-0845

For information about this notice contact:

Wayne H. Mitchell
Center for Drug Evaluation and Research (HFD-362)
Food and Drug Administration
5600 Fishers Lane
Rockville, MD 20857
301-295-8046

SUPPLEMENTARY INFORMATION: FDA must approve any new drug before it may be marketed in the United States. As part of the approval process, the sponsor of a new drug must submit to FDA's Center for Drug Evaluation and Research (CDER) large quantities of data about the safety and efficacy of its product. The original application for a new drug ordinarily includes thousands of pages of reports, analyses, tabulations, and case reports. These data are carefully reviewed by CDER.

A continuing goal of FDA is to improve the speed and quality of the drug review program. The agency views computer technology as a promising means of making the new drug application review process more efficient. Reflecting this assessment. FDA's long-range plans commit CDER to continue to explore the use of computer technology to enhance the timeliness, effectiveness, and efficiency of the drug review process and reduce burdensome, nonessential hard-copy handling. This notice is intended to provide guidance to those considering the submission of a computer assisted new drug application.

I. AGENCY EXPERIENCE WITH CANDAs

CDER has received a limited number of CANDAs to date. These applications have been developed both by drug manufacturers and by third party vendors and consultants. They reflect a variety of strategies and technical approaches, using different software and hardware. Data have been located both at remote sites, accessed by telecommunications, and at FDA.

CDER's experience with CANDAs has not been sufficiently extensive to permit it to draw conclusions about the degree to which CANDAs will ultimately improve the efficiency of the drug approval process, nor to allow conclusions about the superiority of one technical approach to another. The agency's experience does, however, allow CDER to conclude that data from case report forms and tabular listings can be effectively retrieved and reviewed using computer technology and that further exploration of the use of CANDAs should be encouraged. The agency will continue to cooperate with interested sponsors and urges sponsors to explore new systems and technologies. CDER is also interested in promoting the exchange of information on CANDAs that will not infringe on property rights.

II. GUIDANCE ON SUBMISSIONS OF CANDAs

CANDAs can be defined as any method using computer technology to improve the transmission, storage, retrieval, and analysis of data submitted to FDA as part of the drug development and marketing approval process. Computer technology may be applied to INDs, NDAs, ANDAs, and antibiotic applications.

CDER has developed the following guidance for sponsors of drug applications:

1. CDER encourages drug sponsors to explore the use of CANDAs. Drug sponsors considering the use of CANDAs should contact Robert A. Bell, Director, Office of Management (address above).

2. A drug sponsor intending to submit a CANDA is encouraged to discuss its plans with CDER's Office of Management to identify and resolve potential problems involved in filing a CANDA. In particular, sponsors should be prepared to discuss possible logistical problems involved in the installation of computer equipment. If a drug sponsor proposes to meet with FDA to discuss an application for a specific drug product, CDER participants in such a preliminary discussion will ordinarily include a scientific personnel from the review division that will have responsibility for review of the application. Other reviewers in CDER having experience with CANDAs may also be consulted.

3. CANDAs may be developed by the sponsor or a third party. In its initial discussion with FDA, the sponsor should identify the developer of the CANDA. Because of limited resources, reviewing divisions will provide detailed consultation on the development of CANDAs only in the course of their ordinary regulatory discussions with sponsors. The reviewing divisions are not in a position to offer assistance directly to third party vendors who are not currently working with a sponsor involved in obtaining FDA marketing approval of a drug or drugs (i.e., those whose intention is to develop and market data processing methodology for use by drug sponsors). Such third party vendors should contact Robert A. Bell (address above) for information on CANDAs.

4. A drug sponsor should be prepared to discuss the following additional aspects of its proposed CANDA with CDER:

   a. General characteristics of the proposed application including the methods and data files to be used. CDER is interest in the method used to store data.

   b. Computer hardware requirements. Computers used at FDA may not be compatible or available for a specific requirement. If FDA's equipment is not available, the sponsor should make appropriate arrangements for the agency to gain access to the data.

   c. Computer software considerations. In selecting software, a sponsor should consider the overall requirements of analysis and data retrieval in the review process, the availability of software at FDA, and the ease of operation for the reviewer. The sponsor should state what program language, and/or data base system, is being used.

   d. Plans for CDER reviewer orientation and training. Orientation and training should include instructing reviewers in the use of the computer hardware, data file content and structure, and retrieval routines. Sponsors should also be prepared to provide continuing technical support and advice to reviewing personnel.

5. Filing of a CANDA is conditional on the reviewing division's willingness to accept the application in the CANDA format used by the sponsor. For this reason it is important that the acceptability of the proposed CANDA be established to the degree it is possible, well in advance of the submission, before substantial effort has been expended.

6. The submission of a CANDA will not influence the priority given to an application. CDER has written policies regarding the factors that affect the priority given to the review of an application. In general, review priority is based on CDER's classification of anew drug product under the IND/NDA classification system (see Food and Drug Administration, "Staff Manual Guide: Center for Drugs and Biologics," Guide No. 4820.3).

7. Currently, the submission of a CANDA will generally not affect the required submission of an application in paper form. Future development of CANDA procedures should result in the reduction of the amount of paper required to be filed, especially with respect to case report forms and tabular listings.

III. COMMENTS

Interested persons are encouraged to submit comments and suggestions regarding CANDAs. FDA requests comments that will help it develop an overall CANDA policy, rather than comments that propose adoption of specific equipment, systems, computer programs, and services. However, FDA is also interested in learning about new computer hardware, systems, and programs that have a particular applicability to the drug approval process.

Interested persons may, on or before March 14, 1989, submit written comments and other communications related to this notice to the Dockets Management Branch (address above). Two copies of any comments are to be submitted, except that individuals may submit one copy. Comments are to be identified with the docket number found in the heading of this document. Received comments may be seen in the Dockets Management Branch between 9 a.m.and 4 p.m., Monday through Friday.

This notice is not an offer to buy nor a solicitation of offers to sell any product or service. Non-FDA participants in current experiments are utilizing their own resources, and this arrangement is expected to continue for the foreseeable future.

Dated: September 4, 1988
Frank Young, Commissioner of Food and Drugs.

FR/Vol. 53, No. 179/Thursday, September 15, 1988/Notices E.
[FR Doc. 88-21053 Filed 9-14-88; 8:45 a.m.]

APPENDIX I-B Office of Management Contacts

Robert A. Bell
Director, Office of Management
Center for Drug Evaluation and Research
Food and Drug Administration, HFD-10
5600 Fishers Lane
Rockville, MD 20857
(301) 443-0845

David M. Moss
Director, Division of Information Systems Design
Office of Management
Center for Drug Evaluation and Research
Food and Drug Administration, HFD-70
5600 Fishers Lane
Rockville, MD 20857
(301) 443-2836
FAX NUMBER: (301) 227-6183

Technical Questions can also be addressed by the following:

Gregory V. Brolund
Chief, Information Systems Branch I
Division of Information Systems Design
(301) 443-1066

Michael L. Buster
Chief, Systems and Network Section, ISB I
Division of Information Systems Design
(301) 443-1066

Kenneth Edmunds, Jr.
Chief, Information Systems Branch II
Division of Information Systems Design
(301) 443-5368

APPENDIX I-C Center for Drug Evaluation and Research FAX Numbers

OFFICE OFTHE DIRECTOR

- o PKLWN, 13B-45
  Office of the Director -- Immediate Office
  Don Carrington
  443-2894
  FAX NUMBER:443-2763

- o PKLWN, 9B-45
  Pilot Drug Evaluation Staff
  Sandra Barnes
  443-3741
  FAX NUMBER: 443-0754

OFFICE OF MANAGEMENT

- o PKLWN, 13b-45
  Office ofthe Director -- Immediate Office
  Don Carrington
  443-2894
  FAX NUMBER:443-2763

- o PKLWN, 8B-11
  Division of Information Systems Design
  Ladeana Frye
  443-5368
  FAX NUMBER: 227-6183

OFFICE OF DRUG EVALUATION I (ODE I)

- o PKLWN, 14B-45
  ODE I -- Immediate Office
  Linda Carter/Jane Peterson
  443-4330
  FAX NUMBER: 443-1653

- o PKLWN, 16B-30
  Division of Cariorenal Drug Products
  Natalia Morgenstern
  443-4730
  FAX NUMBER: 443-9283

- o PKLWN, 10B-45
  Division of Neuropharmacological Drug Products
  Mike Bernstein
  443-3800
  FAX NUMBER: 443-9280

- o PKLWN, 17B-45
  Division of Oncologyand Pulmonary Drug Products
  Richard Potter
  443-4260
  FAX NUMBER: 443-9284

- o PKLWN, 18B-08
  Division of Medical Imaging, Surgical andDental Drug Products
  Warren Rumble
  443-3500
  FAX NUMBER: 443-9281

- o PKLWN, 10-90
  Division of Gastrointenstinal and Coagulation Drug Products
  Thomas Hassall
  443-0487
  FAX NUMBER: 443-9285

OFFICE OF DRUG EVALUATION II (ODE II)

- o PKLWN, 13B-28
  ODE II -- Immediate Office
  Zulema Collins
  443-2544
  FAX NUMBER: 443-4467

- o PKLWN,12B-03
  Division of Anti-infective Drug Products
  James Bona
  443-6797
  FAX NUMBER: 443-5803

- o NRC, 200
  Division ofAnti-viral Drug Products
  Julie Basore
  443-9550
  FAX NUMBER: 443-9292

- o PKLWN, 14B-04
  Division of Metabolism and Endocrine Drug Products
  John Short
  443-3490
  FAX NUMBER:443-9282

OFFICE OF EPIDEMIOLOGY AND BIOSTATISTICS (OEB)

- o PKLWN, 15B-18
  OEB -- Immediate Office
  Charles Anello
  443-4227
  FAX NUMBER: 443-5161

APPENDIX I-D Sample Letter for Loaning Equipment/Software to FDA

Mr. David M. Moss
Director, Division of Information Systems Design
Center for Drug Evaluation and Research
Food and Drug Administration
5600 Fishers Lane, 8B-11
Rockville, MD 20857

Dear Sir:

In our previous discussions, it has become apparent that it is necxessary for us to loan the Center for Drug Evaluation and Research (CDER) equipment and software to facilitate thereview of our computer assisted new drug application.

As requested, we are providing the following information regarding this loan.

1. Identify application by name and number.

2. Aspecific listing of the hardware and/or software that will be loaned to CDER, including serial numbers where applicable.

3. Anticipated date of arrival.

4. The name of the reviewer(s) that will be using the equipment along with the name(s) of their division(s)and the location(s) to which you will be delivering the equipment/software.

5. Details on why this loan is necessary.

All software provided is in full compliance with the licensing agreements for that software.

We will arrange to take the equipment back as soon as the review is completed or the equipment is no longer needed for the review of this specific application.

We understand that if we have not loaded anti-virus software on this equipment, you will do that.

We further understand that although the loaned equipment and software is a temporary loan to help in facilitating the review, the data -- and possibly the media it is on, such as optical disk, diskette orhard disk -- is considered to be an official aprt of the NDA submission and, as such, will be retained by the Agency for a possible audit, archive or other such purposes.

Any differences between the automated submission and the hardcopy submission have been clearly identified. Other than the specifics identified, our computer assisted submission is identical in content to the hardcopy.

Sincerely yours,

Vice President for Regulatory Affairs

cc: Chief, Document Management and Reporting Branch, HFD-53
Consumer Safety Officer, review division

CHAPTER II

INTERIM ADMINISTRATION/TECHNICAL REQUIREMENTS OF CANADIAN DRUGS DIRECTORATE FOR COMPUTER ASSISTED NEW DRUG SUBMISSIONS (CANDS)

July 31, 1991

A. Introduction and Background

The Health Protection Branch (HPB) of Health and Welfare Canada is currently evaluating its requirements for the computer aided review of drug submissions. This present chapter in the CANDA GUIDANCE MANUAL represents an interim document that reflects short term requirements and conditions as HPB. As the requirements definition and feasibility studies are completed, elements of a definitive set of guidelines will be prepared. Guidelines describing acceptable format and content for the preparation and filing of paper New Drug Submissions (NDS) are available (by request to appropriate Bureau in the Drugs Directorate -- see Appendix A for mailing addresses).

As noted in the FDA's introduction, the goals of this manual are not limited to providing guidance for the further development of computer assisted new drug submissions (CANDS), in which the usual paper submission is aided by a computer assisted query of the submission, but may lead the way to establish a framework for the eventual replacement of the paper submission with a complete electronic new drug submission (ENDS) in which all information required for a complete evaluation of a drug is provided in a computer-accessible format.

History

The Branch received its first computer assisted new drug submission (CANDS) in 1986. However, it was not until January, 1988, that members of HPB and the Pharmaceutical Manufacturers Association of Canada (PMAC) met to explore the potential benefits and technical feasibility of machine readable drug submissions. In December, 1989, the Joint HPB/PMAC WORKING GROUP on CANDS was established.

To date, HPB has accepted approximately 15 CANDS projects. Most CANDS were adaptations of CANDA systems that had also been submitted to the FDA.

The main difference between NDS and NDA as well as CANDS and CANDA systems is primarily related to the style of presentation rather than in the type of information required for the evaluation process. The overall objectives of the evaluation processes are very similar.

Administrative Requirements

Interaction with the Drugs Directorate, Health Protection Branch:

- o Drug sponsors and other organizations considering the use of CANDS should contact the appropriate Division Chief in the Bureau to which the submission will be directed (addresses in Appendix II-A) well in advance of the preparation of the CANDS to discuss the sponsor's planned approach.

- o CANDS may be developed by the sponsor or jointly with a third party consultants. However, the sponsor assumes full responsibility for all technical and training support required for their particular CANDS.

- o A drug sponsor should be prepared to discuss the general characteristics of the proposed CANDS including the structure and content of all data files.

- o At present, HPB prefers that sponsors provide all hardware and software required for the CANDS.

- o All components of the CANDS will be logged to the submission file and will include a full description of all hardware and software.

- o Sponsors will recover all CANDS components logged to the submission at the completion of the review.

- o HPB will assume that all provisions of copyright agreements and licenses are being followed when a CANDS is submitted.

- o In selecting software, a sponsor should consider the overall requirements of analysis and data retrieval in the review process, the availability of software at HPB, and the ease of operation for the reviewer. The sponsor should clearly identify what program language, and/or database system is being used.

- o Orientation and training should include instructing reviewers in the use of the computer hardware and software, data file content and structure, and retrieval routines. Sponsors should also be prepared to provide continuing hardware and software support throughout the full period of the review.

- o The filing of a CANDS is conditional on the reviewing division's willingness to accept the submission in the CANDS format used by the sponsor.

- o The filing of a CANDS will not influence the priority given to a submission. HPB has written policies regarding the factors affecting the priority given to the review of a submission.

- o The submission of a CANDS will not affect the required filing of a submission in paper form.

- o The actual filing of a CANDS should not be made until HPB notifies the sponsor that the review is scheduled to begin.

- o Sponsors should take all reasonable precautions to ensure that all CANDS components are computer virus free. In the case of a contamination submission, HPB reserves the right to return the CANDS to the sponsor until decontamination has been verified.

B. Technical Requirements

Interim Approach

System configurations should be limited to the IBM PC-Compatible Family of Computers (although UNIX based system configurations could be considered under certain circumstances).

HPB recognizes that drug companies are in the process of adapting modern information technology to the drug development and submission preparation processes. CANDS systems can be technically demanding and expensive to develop. HPB recognizes that CANDS development is a joint initiative with industry and is committed to an ongoing dialogue with industry through its Joint Working Group(s) with industry.

Interim Design Considerations

Two reports on CANDS related activities are available for distribution to interested parties and individuals.

The *Report of the Joint HPB/PMAC Working Group on Computer Assisted New Drug Submissions (CANDS) Demonstration Day* documents the results of the questionnaires filled by HPB personnel during the CANDS demonstrations of September 1991.

The Report of the Requirements Definition Study for Computer Assisted New Drug Submissions (CANDS) for the Drugs Directorate is the interim report of the analysis conducted by an external consulting firm engaged by the Drugs Directorate. This firm's mandate was to provide High Level User Requirements and to make a preliminary assessment of the impacts to the Directorate of computerization of the drug review process.

Written requests for information about the Directorate's CANDS related activities or the above named reports can be addressed to Dr. Jacques Messier, Director, Bureau of Human Prescription Drugs, Health Protection Branch, 355 River Road, Vanier, Ontario, K1A 1B8.

C. Hardware and Software at the Drugs Directorate

Three computer platforms are used for specific purposes in the Drugs Directorate as follows:

1. Microcomputers

Each drug evaluator in the Drugs Directorate is equipped with a desktop 386-based microprocessor (mainly Olivetti), IBM-compatible, with a minimum of 4 Mb of RAM, 7 expansion slots, a 1.44 Mb 3.5 inch floppy disk, VGA colour monitor, mouse, hard disk with minimum capacity of 80 Mb, DOS 3.3 or better, Word Perfect 5.1 and Windows 3.1. Also, a few laptops (mainly Toshiba) are available.

Most workstations are equipped with a printer but only a few have modems; some workstations have 6 MB or better of RAM. A few workstations are equipped with other peripheral equipment such as scanners, laser printers, etc.

There are also other commercial software packages used across the Directorate. The more common ones are: Windows Express, MS Excel, Lotus 1-2-3, RapidFile, dBase III Plus, SAS/PC, LeMenu,Harvard Graphics, MS Works, Norton Utilities, and Word Perfect Library. The more specialized packages include TOPKAT, PCNONLIN, and BIOPAK amongst others. A number of communications software (mainly terminal emulators) are used, such as CEO Connection, Procomm or Procomm Plus, Smarterm 400, and Minisoft.

The bulk of these workstations are not connected to any host computer and are operating in single-user mode. There are no LANs in place in the Drugs Directorate. The Branch (HPB) is developing guidelines and standards for LAN implementation.

2. Minicomputers

The Department has adopted a *de facto* standard for minicomputer equipment i.e. DATA GENERAL. The Directorate is the owner of 1 Data General - MV20000 Model I for the exclusive use by the Bureau of Dangerous Drugs (BDD) for application development and production.

The Drugs Directorate also has access to 2 HPB minicomputer facilities; the first one is a production machine and is reserved for Oracle-based application production and was scheduled to house the newly-developed submissions tracking system. This minicomputer is an MV40000-HA and is currently shared with the Field Operations Directorate of HPB. The second minicomputer (MV40000) accessible to the Drugs Directorate is a development facility for Oracle-based applications. This equipment is shared currently with the Field Operations Directorate. Other Directorates in HPB may, in the future, process their own applications on these machines.

These minicomputer systems are not proposed to house any CANDS-related information. The Directorate also uses two HP-3000 minicomputers. One of these is an HP-3000 Series 30 used exclusively by the Bureau of Dangerous Drugs and will be completely phased out in the next 6 to 12 months due to its very limited processing power and storage capabilities. The other HP minicomputer is an HP-3000 series 70 and is owned by Environmental Health Directorate and houses only one application of the Directorate, that is, the former submissions tracking system. Currently, these two computers are not intended to house any CANDS-related information.

Access to these computer systems is done through the HC Data Communications Network except for the BDD minicomputer which is restricted to local users in the Jackson Building.

3. Mainframe

The Directorate also has access to a departmental IBM-compatible mainframe (AMDAHL-5880) under a facility management contract with STM systems Corp. This computer system is used mainly to process batch-oriented applications, written in the 1970s, such as the Drug Notification System, the Drug Abuse System and the Drug Adverse Reactions Reporting Program. There are a number of department-wide applications, such as the Departmental Financial System, the Personnel Management Information System and the Correspondence Tracking System, which are processed on this computer system as well as other Directorate and Branch applications. Most of the Drugs Directorate applications, residing on this mainframe, are planned to be redesigned and redeveloped in the next 2 or 3 years.

Data Communications Network at Health and Welfare Canada

The HC Data Network is a wide-area network (WAN) based on Gandalf real-time data switching technology that provides switching and connection capabilities for workstations to access both local and remote computing resources. These nodes are linked using various communicating services provided by common carriers and by HC owned wiring distribution systems. The common carrier services include Telecom Canada's Dataroute, Datapac and local leased lines. The standards for HC-owned intra-building wiring is derived from the Northern Telecom's Integrated Building Distribution System with 4-pair twisted copper wires at each termination point using an RJ45 plug. The inter-building communications in the Tunney's Pasture complex use a combination of fibre-optic cables and twisted wires. Any workstation connected to the network using the standard RS-232C asynchronous protocol, either hard-wired or dialing-in from remote locations, is capable of accessing any computing resources connected to the network on a need-to-know basis. The personal workstation at each termination point is effectively used as a multi-purpose workstation. The speed of communication is 9600 baud for hard-wired workstations and 1200, 24,00 or 9600 baud for dial-in access.

Electronic Mail

Electronic mail is used in the Drugs Directorate and is available to assist communications between sponsor personnel and Drugs Directorate reviewers for Computer-Assisted New Drug Submissions.

There are 3 different types of electronic mail in the Drugs Directorate:

1. Comprehensive Electronic Office (CEO) is the name of Data General's office automation software and is an integrated package comprised of electronic message, document transfer, list processing, calendaring, work processing and a reminder facility. It has been adopted by Health and Welfare Canada as its department-wide Office Communication system and is the recommended approach for E-Mail communication between the drug sponsors and Drug Directorate personnel. There are over 1800 employees connected to the CEO system in Health and Welfare Canada with approximately 100 users in the Drugs Directorate. These are mainly division Chiefs and Bureau Directors. Any reviewer needing a CEO account has to meet the following requirements:

    i) Complete a request to the CEO Implementation Project.
    ii) Acquire terminal emulation software (preferably CEO Connection).
    iii) Be connected to the HWC Data Communications Network or have a modem and a dedicated telephone line. Training and support is provided through a Hotline service. The CEO system is not intended to carry particularly sensitive data.

2. DISNET, Drug Information Systems NETwork, is used as an E-Mail by a limited number of employees in the Drugs Directorate. DISNET is mainly used to communicate with external regulatory bodies around the world. DISNET is a computerized mail and conference system open for persons working in the field of drugs and is a service of SWEDIS Development Centre in Uppsala, Sweden which provides administration and support to users of the network.

3. ONET, Ontario regional NETwork is another recent addition to the E-Mail systems available to Drugs Directorate employees on a need-to-know basis. ONET is part of CANET, which, in turn, is part of the world-wide INTERNET system, an international scientific electronic mail service which connects many national and international organizations and institutions. Currently only a few staff members of the Directorate are connected to ONET.

APPENDIX II-A Addresses in the Canadian Drug Directorate

Director, Bureau of Human Prescription Drugs
Place Vanier, Tower B
355 River Road
Vanier, Ontario
K1A 1B8. (613) 991-0107

Aids & Viral Diseases Division
Dr. S. Khan
Acting Chief, (613) 993-3744

Central Nervous System Division
Dr. C. Petersen
Acting Chief, (613) 993-3203

Endocrinology, Metabolism & Allergy Division
Dr. P. Roufail
Acting Chief, (613) 993-3231

Cardio-Vascular Division
Dr. T. Uscinowicz
Acting Chief, (613) 993-3104

Infection & Immunology Division
Dr. E. Gadd
Acting Chief, (613) 993-3230

Gastro-Enterology, Hematology, Oncology Division
Dr. A. Klein
Chief, (613) 993-1552

Pharmaceutical Evaluation Division
Mr. P. Jeffs
Chief, (613) 993-3743

A/Administrative Officer
Ms. J. Richmyre, (613) 993-3101

Director, Bureau of Drug Research
Sir Frederick G. Banting Building
Tunney's Pasture
Ottawa, Ontario, K1A 0L2

Computer Services Section
Mr. J. Raby
Head, (613) 954-6488

Biostatistics & Epidemiology Division
Dr. G. Jarvis
Chief, (613) 954-6481

Biostatistics Section
Mr. R. Li
(613) 954-6485

Director, Bureau of Non-Prescription Drugs
Place Vanier, Tower A
333 River Road
Vanier, Ontario, K1A OL1

Drug Evaluation Division
Dr. B. Gillespie
Acting Chief, (613) 954-6456

Pharmaceutical Assessment & Cosmetics Division
Dr. N. Pound, Chief, (613) 954-1254

Administrative Officer
Ms. A. Bouchard, (613) 954-6495

Director, Bureau of Pharmaceutical Surveillance
Finance Annex Building
Tunney's Pasture
Ottawa, Ontario, K1A OL2

Biopharmaceutics Evaluation Division
Dr. W. Nitchuk
Acting Chief, (613) 957-1489

Administrative Officer
Mrs. P. Barrette
(613) 957-1834

Director, Bureau of Biologics
Virus Building
Tunney's Pasture
Ottawa, Ontario, K1A OL2

Mrs. D. Krepps
Chief, Compliance Division
(613) 957-0363

Administrative Officer
Mr. B. Mallet, (613) 957-8069

Director, Bureau of Veterinary Drugs
Main Statistics Canada Building
Tunney's Pasture
Ottawa, Ontario, K1A OL2

Manufacturing and Control Division
Mr. W. A. McWilliam
(613) 957-3865

Mr. M. J. Crocker
Manufacturing & Compliance Division
(613) 957-3869

Administrative Officer
Mrs. P. Potvin
(613) 957-3824

CHAPTER III

PMA/FDA CANDA Survey Results on CANDAs Submitted 1985-1988

Summary Report

A. Preface

This Summary Report was prepared by the PMA-FDA Evaluation Task Group and reflects the experience from 23 CANDAs submitted to FDA from 1985-1988.

This document is intended to provide a high level management overview regarding the viability of these CANDAs and their effects on the NDA review process, from both FDA and sponsoring company viewpoints.

B. Overview

A CANDA is a project sponsors and FDA interact to design, develop, implement, and use a computer system for the FDA review of new drug submission data. The survey results, even for these early CANDAs, clearly indicate that CANDAs have value and have a very promising future in facilitating the review of new drug applications.

The results from personal interviews with participating sponsor company and FDA personnel gathered in this comprehensive survey of early CANDAs reveal that a strong commitment to computer assisted NDA review exists throughout the pharmaceutical industry and FDA.

C. Background

The concept of a computer system that could facilitate the drug review process began to take shape in 1985. Because of the limited experience of both FDA reviewers and pharmaceutical firms in computerizing the NDA, PMA and FDA agreed to encourage individual firms to make pilot submissions of portions of NDA data in a variety of computerized formats. It was hoped that this experimentation would allow a variety of options to be explored without the limitations of prescribed technology or preconceived planning.

In order to evaluate systematically the impact of the CANDA experiment, 23 CANDAs, sponsored by 13 firms, and submitted to FDA during the period of 1985-1988, were targeted for study through a personal interview survey. All firms agreed to participate, and their representatives openly discussed their CANDA experiences with interviewers from Statistical Consultants, Inc. (SCI). FDA agency reviewers were equally open in their discussions. This evaluation process could not have been successful without this high level of cooperation by everyone who was involved in the CANDA experiment.

D. Results

To measure the effectiveness of these CANDAs, the Evaluation Task Group developed a questionnaire for use by the inte viewers. It was designed primarily to identify both sponsor's and reviewer's objectives for CANDAs and, secondarily, to identify those system features that were beneficial to the review process. The subjective responses to the questionnaires form the basis of this evaluation.

OVERALL RESULTS OF THIS EVALUATION
CANDAs HAVE:

- clearly facilitated FDA reviewer's access to information
- clearly improved communications between sponsor and FDA: most sponsors felt that CANDAs satisfied FDA requests and helped focus the NDA review process
- clearly enabled sponsor companies to have a better understanding of the review process
- clearly been successful in helping FDA experiment with computer technology
- generally enhanced the quality of the NDA review
- to some degree, reduced time to approval
- to some degree, enhanced the quality of submission provided to FDA
- to some degree, reduced the need for hard copy

Specific results of these interviews are characterized in Appendices III-B and III-C.

Appendix III-B, Table 1, displays the results for objectives common to both FDA and Sponsor (i.e., what percent of these objectives were met respectively).

Appendix III-B, Table 2, displays the results for individual objectives of FDA and individual objectives of sponsor companies (i.e., what percent of these objectives were met respectively).

Appendix III-C, displays the results of the evaluation of individual features.

E. Other Key Findings

1. Sponsor medical and statistical personnel should be involved in the development and testing of a CANDA to ensure its utility as an aid to the FDA review.

2. CANDA capabilities have the potential to improve the quality of the NDA submission and to facilitate its preparation.

F. Conclusion

While it is clear that these early CANDAs have done well in achieving their objectives, much work remains to be done in maximizing the potential of CANDAs for both the FDA review process and inhouse company use in developing submissions.

CANDAs have represented a significant achievement in aiding both FDA reviewers and sponsor companies in their difficult task of ensuring that safe and effective drugs are available in a timely manner.

APPENDIX III-A PMA/FDA Survey Sponsorship and Methodology Survey Sponsorship

This survey was sponsored by the PMA-FDA Evaluation Task Group, a subcommittee of the PMA Board-appointed working group for CANDAs. The Evaluation Task Group is charged to measure the effectiveness of CANDA systems and to report its findings to the PMA Board Committee, PMA member companies, and FDA. SCI was engaged by this Task Group for the purposes of conducting and reporting on this initial survey.

Survey Methodology

For each of the 23 CANDAs, sponsor company personnel and FDA personnel were interviewed by SCI consultants. Sponsor company interviews were conducted with regulatory, clinical, statistical, and information systems personnel at each company participating in a CANDA. FDA interviews were conducted with the medical and statistical reviewers at FDA for each of these respective CANDAs. The results of these interviews were recorded onto the questionnaire by SCI and then collected into the SCI Final Report. The objective and features were summarized into this Summary Report by the Evaluation Task Group.

APPENDIX III-B PMA/FDA CANDA Survey Results of Evaluated Objectives

How to Read Tables 1 and 2

Tables 1 and 2 describe the objectives of FDA and sponsor companies for the 23 CANDAs.

Not all respondents that had a particular objective indicated as to whether that objective was satisfied or not; therefore the tables reflect a range of percent success.

For example in Table 1, the objective of 'Facilitate Access to Information':

- In 19 of 23 CANDAs, the sponsor company had this objective.
- Nineteen companies responded, one company did not respond.
- Of the 18 responses, 17 indicated the objective had been met: therefore, 17 ÷ 18 = 94%.
- If you assume the objective was not met for the non-responder; then 17 ÷19 = 89% had met their objective.
- Therefore, the range of succes as reported is: 89% ÷ 94% had this particular objective met.

Table 1

Common Objective Summary

Objectives	Measure of Success*			Conclusion
	Sponsor	*FDA clinical/statistical*		
Facilitate Access	89 - 94%	62 - 77%	100%	Clearly Achieved
Enhance Quality of Submission	89 - 100%	44 - 80%	100%	Sponsor Clearly Achieved; FDA Partially Achieved
Reduce Time to Approval	61%	14 - 25%	33 - 100%	Promising but Objective difficult to Measure
Reduce Hardcopy	100%	50 - 75%	33%	Promising but inconclusive
Experiment with Computer Technology	100%	85 - 100%	100%	Clearly Achieved

* Percent of Objective Met

Underlined numbers are percent of objective met of those that responded.

Non-underlined numbers assume the objective was not met for those that did not respond.

Table 2

Individual Objective Summary

Objective	Measure of Success*	Conclusion
Sponsor		
Improve Communications with FDA	100%	Clearly Achieved
Better Understand FDA Review Process	100%	Clearly Achieved
Satisfy FDA Request	100%	Clearly Achieved
Focus the NDA Review	85%	Clearly Achieved
FDA		
Enhance the Quality of Review	Clinical 27 - 43% Statistical 71 - 83%	Promising but Inconclusive for Clinical Review; Clearly Achieved for Statistical Review

* Percent of Objective Met

Underlined numbers are percent of objective met of those that responded.

Non-underlined numbers assume the objective was not met for those that did not respond.

APPENDIX III-C

System Features Usefulness to FDA

Feature	Summary Comment on Usefulness
Training Support	
Online Help	Not used very Effectively
Systems Manuals	Not very Effective - 'Poorly Prepared'
User Manuals	More useful than System Manuals
User Interface	
Menu Selection	Clearly Useful
Command Driven	Second Choice to Menu Driven
Review Capabilities	
Online Index to CANDA	Useful to Medical Reviewers
Online Index to Hardcopy NDA	Very Useful to Medical, Essential for Cross Referencing
CRF Image	Very Useful to Medical Reviewers
CRF Tabulations	Very Useful to Medical Reviewers
Review Capabilities	
Ad Hoc Query	Very Useful to Medical and Statistical
Report Formatting	Very Useful to Medical
Graphics	Not Used Effectively (Data Import Problem)
Data Manipulation	Very Useful to Medical and Statistical
Construct Tables	Medical Found Tables Difficult
Spreadsheet	Inconclusive
Statistical Analysis	Very Useful to Medical and Statistical
Full Text Search	A Key Feature for Medical
Image Retrieval	Very Useful to Medical
Review Capabilities	
SAS Datasets	A 'MUST' for Statistical Reviewer
ASCII Files	Useful for Medical Reviewer
Derived Datasets	Difficult for Medical - A 'MUST for Statistical
Reviewer Storage	Not Useful
Electronic Mail	Not Used Effectively
Word Processing	A Key Feature if Integrated
Audit Trail	Not Useful

CHAPTER IV

CANDA TERMS & FEATURES

Prepared by the CANDA Terms and Features Task Group

A. Introduction

The CANDA Terms and Features Task Group was established to develop a set of definitions for commonly used terms and features applicable in the development and implementation of any CANDA submissions to facilitate communications between the sponsor and FDA. The attached report conveys the results of its efforts. Because CANDAs are new and are still rapidly evolving, the report attempts to describe current practice and usage, rather than provide a finished catalogue of features. Additional terms and features may be added at a later date.

Each of the many CANDAs submitted to FDA by various sponsors was tailored for the specific product under review, and to a greater or lesser degree for the specific requirements of the reviewer. In addition, each of the sponsors had its own guiding concepts of systems design and its own data processing procedures. In consequence, the CANDAs have differed in both structure and basic terminology. To improve communications between sponsors and FDA, the Task Group has focused on producing an accurate and comprehensive description of current practice and usage. The Task Group attempted to provide a single definition for each term where possible. Because of the inherent variety across CANDA systems and companies, however, the glossary often provides multiple definitions or conveys the need for users of a term to seek clarification.

B. CANDA Terms and Features: Definitions

<u>DATA</u>

<u>Data Review</u>

CRF Browse/Review
This is the means by which the reviewer may examine the data on case report forms (CRFs). In some cases, derived and/or summarized data are also available.

by patient
Ability to review a patient's case report form data.

by variable (parameter)
Ability to review specific variables for patients across time.

Patient Profile
Selected data for a single patient as opposed to a full CRF or data listing.

Patient (Key ID, Subject ID)
A means of uniquely identifying patients within a CANDA.
Patient Header (Screen header)
The common information that appears on a screen for a given patient (e.g., patient ID., patient demographics, treatment group and dose, study visit number, days on therapy, and any other clinical information).

Data Terms

Original Data
Data collected in the study, such as CRF, central laboratory data, etc. These data have met company standards for completeness and accuracy.

Derived (Computed) Data
Values for an individual patient that are calculated or selected from original data.

Summary Statistics
Values for a group of patients that are calculated using mathematical or statistical functions.

Formatting
The process of preparing data chosen for reporting. This includes the preparation of headings, footings, subheadings, subfootings for whole reports, sections, pages, columns, or other subdivisions of data.

Data Subsetting
Dividing a set of data into subsets according to the value of one or more variables.

Data Merging/Combining
Combining two ore more sets of data into one set.

Data Pooling
Combining data across studies.

Concatenating
Refers to linking of data sets either horizontally or vertically and therefore requires clarification.

Data Analysis and Reporting

Data/Programs Behind Statistical Tables
Data and/or programs which would allow one to reproduce tables in the hard copy NDA. This may include all patients in a study along with their evaluability status.

Report/Output
Any formatted output of data, such as listings or summaries for presentation or storage. Data can be sent to the screen, printer, or stored for later access.

Reports

summary
Summary statistics, such as total, mean, variance, etc. to describe characteristics of a set of data.

listing
Formatted display of patient data.

table
Original or derived patient data, formatted in rows and columns. The terminology used to describe tables varies with the user and with whether the table in question contains only original data, only derived data, or a mixture of the two. Care must be exercised to avoid confusion.

cross tabulation
Derived data arrayed in rows and columns.

frequency table (cross tabulation, contingency table)
Table or counts meeting criteria for specific variables/parameters

Types of Reports

supplied report
A previously generated report that exists in the system that the reviewer can retrieve.

preprogrammed report
A report whose format has been predefined, but the reviewer is permitted to select data subsets.

ad hoc report
A report generated at the time of the review.

custom report
Can refer to either a preprogrammed or an ad hoc report and must therefore be clarified.

Tools for Generating Reports

function
A mathematical or logical expression that allows one to compute, transform, select, or otherwise manipulate variables.

query
Identify data that meet one or more search criteria, such as: treatment group x, with headache.

Inventory by Study Event
A count or summary of unique occurrences of a study event. The summary can be by patient, time point, etc. Study events are as defined in the study protocol or as reported by the patient.

TEXT

Summaries (Synopses)

Protocol Summary
A brief description of the salient features of the protocol (i.e., objectives, design, patient population, clinical and lab assessments) to provide information regarding the study plan.

Integrated Overall Summary of Effectiveness Data
There may be two parts to this option. The sponsor may want to include the entire text of the "Integrated Summary of Effectiveness Data" or parts thereof to view and search electronically (see the FDA's Guideline for the Format and Content of the Clinical and Statistical Sections of New Drug Applications (July 1988) for a complete description of this summary). The sponsor may also launch the ability to view preprogrammed reports, generate custom reports across studies, or examine the original data from individual studies, etc., from this section.

Integrated Overall Summary of Safety Information
There may be two parts to this option. The sponsor may want to include the entire text of the "Integrated Summary of Safety Information" or parts thereof to view and search electronically (see the FDA's Guideline for the Format and Content of the Clinical and Statistical Sections of New Drug Applications (July 1988) for a complete description of this summary). The sponsor may also launch the ability to view preprogrammed reports, generate custom reports across studies, or examine the original data from individual studies, etc., from this section.

Summary and Conclusions

There may be two parts to this option. The sponsor may want to include the entire text of the "Summary and Conclusions" or parts thereof to view and search electronically (see the FDA's Guideline for the Format and Content of the Clinical and Statistical Sections of New Drug Applications (July 1988) for a complete description of this summary). The sponsor may also launch the cross-reference of all pertinent materials from this section.

Terms such as "Applications Summary," "Overall Summary of clinical Pharmacology," "Overall Summary of Controlled Studies," "Overall Summary of Uncontrolled Studies," and "Integrated Summary of Risks and Benefits" are also defined in the Guideline for the Format and Content of the Clinical and Statistical Sections of New Drug Applications (July 1988).

Key Word Search
Locate text that matches key word search criteria.

Word Processing
An application that facilitates the entry and modification of information for the production of documents such as summary bases of approval and other reports.

SYSTEMS: UTILITIES

Utility Functions

Electronic Mail (E-Mail)
A facility to create, edit, save, transmit, and receive text/data electronically.

Facsimile (Fax)
A facility to transmit and receive images of documents using telecommunications equipment.

Help
A system feature providing on-line assistance upon request.

Notepad
A system feature that provides the ability to compose and store comments, tables, graphs, etc., with appropriate identifiers. These can be retrieved for subsequent review.

Annotate
The ability to add notes to data listings and summary tables for future reference; related to specific pages, and generally more specific than notepad comments.

Audit Trail
A record of changes to a data base that includes the old value, the new value, the date/time of change, and the person making the change.

Save
The ability to store for later retrieval information such as reports or commands used to generate reports or data subsets.

Library
A collection of data sets, programs, reports, etc., created by the user.

Spreadsheet
An application that facilitates the manipulation of rows and columns of data.

Activity Log
A record of user CANDA activities that may include data access, manipulation, and report generation. It may include sufficient information to reproduce those activities.

Backup
An electronic copy of computer files, systems, and software that will permit recovery from system failure, or the process of creating such a copy.

Downloading/Uploading Data
Electronic transfer of data from one computer to another.

Data Base Updates

The correction or addition of data to the data base that forms the basis of the CANDA, such as the inclusion of a new study or additional safety data.

Menu

A list of features, functions, or options provided by the system.

Menu-Driven System

A system organized through a set of menus that provide access to various functions.

Menu Access

The ability to choose the desired feature, function, or option by using a menu.

Documentation

User Manual
A nontechnical guide explaining the features and use of the system.

Systems Manual
The documentation of the technical aspects of the system.

Reference Manual
This term may refer to either or both of the above-defined manuals. It should therefore be clarified when used.

Indices

Index to CANDA
This section of the CANDA, immediately and easily available to the reviewer, provides an overview and guide to the system and serves as a table of contents and index. It gives the reviewer a general idea of what and how much information is available and provides a starting point for accessing that information.

Index to NDA
This is an outline of the NDA document based on the various sections as required by FDA. It identifies what information is in each section and where that information can be located in the hard copy submission.

Cross Reference
An index linking specific clinical/statistical data or reports in the CANDA to specific sections of the hard copy NDA; also sued to link summary to detail data.

TERMS

Local Computer

The computer that resides with the reviewer. The computer with which the reviewer is interacting. This can be either a microcomputer, a minicomputer, or a mainframe, and can be either a single-user or multi-user computer.

Remote Computer

Any computer other than the local computer. This will most commonly refer to a computer at the company or a third party, or to the FDA mainframe.

GRAPHICS

Pictorial displays of data, such as bar charts, scatter plots, and line graphs.

CHAPTER V

GUIDELINES FOR CANDA SYSTEM SECURITY AND INTEGRITY

Prepared by the Pharmaceutical Manufacturers Association
Working Group on CANDAs

A. Introduction

CANDA systems have become common supplements to new drug applications. These systems provide FDA officials and drug sponsors with tools to facilitate the drug review and approval process. While the systems provide opportunities for timeliness and efficiency improvements in the NDA process, safeguards must be established to assure the security and integrity of NDA information accessible via the CANDAs.

A CANDA system should enhance the NDA review without changing the fundamental process. Most, if not all, data provided in a CANDA system is included in the hardcopy NDA, and the confidential and secure nature of this data is clearly understood by all parties. A CANDA system can facilitate accessing and manipulating data. Providing this flexibility, however, also requires controls over the processes involved to assure their ongoing proper operation. Components of a CANDA project that should be reviewed for appropriate controls include the data, software, telecommunications, hardware, and FDA interactions required for the system.

This document provides general guidance for controls that should be considered for all CANDA projects. Though all items will not be necessary for every project, please review all controls and identify those required for your CANDA work.

Definitions

The following terms are used in this document:

<u>CANDA</u>
Computer Assisted NDA. Any automated system to improve the transmission, storage, retrieval, and/or analysis of information submitted to FDA as part of the drug development and marketing approval process.

<u>FDA</u>
Food and Drug Administration. The specific group within FDA that handles new drug applications is the Center for Drug Evaluation and Research (CDER), headquartered in Rockville, Maryland.

Hardcopy NDA
The paper volumes, as specified by the Federal regulations, required for a new drug application. An NDA usually includes text, graphics, tables, forms, and other forms of data.

Sponsor
The company that submits the NDA to FDA for review.

CANDA Alternatives

Many different technical approaches have been used in CANDA projects to date, and new approaches will be implemented as the process is refined and new technologies develop. System implementations have included:

- o Implementation of the entire CANDA system, both hardware and software, in the FDA reviewers office (e.g., install a microcomputer with all software and data on a hard disk).

- o Installation of remote access equipment (e.g., a terminal and communications equipment) at the FDA site, and maintain the CANDA software and data on computer hardware located at the sponsor (or third party) site.

- o Submission of data and/or software on electronic media for use on FDA computer equipment.

Regardless of approach, however, the security and integrity controls identified in this document should be considered by the project team.

Responsibilities

The sponsor firm is responsible for establishing and implementing appropriate CANDA system security and integrity controls, including providing necessary information to FDA reviewers for proper operation. In reviewing these control guidelines, sponsors should identify the individual(s) responsible for implementing necessary actions. Groups that may play roles in a CANDA project include:

- o Senior Management -- provide leadership and assure adequate resources
- o Regulatory Affairs -- responsible for submission (installation) of system, FDA interaction, assure adequate system safeguards are functional
- o Disciplines -- interact with FDA counterparts during development and support phases, provide expert input into system design, validate functions; disciplines may include:
  - -- Biopharmaceutics
  - -- Chemistry, Manufacturing, and Controls
  - -- Medical
  - -- Toxicology and Pharmacology

- o Biostatistics -- system integrity, interact with FDA counterparts
- o Information Systems -- system development and support, including implementation of necessary hardware and software security and integrity controls.

B. Data Security and Integrity

The data stored in the CANDA system represents the NDA itself. Many different components of the NDA could be stored as data -- text from the NDA, document images, laboratory or patient records, or tables and graphs. Data also can be generated by the reviewer and stored in the CANDA during the course of the review process.

Security and integrity procedures for NDA documents should be well established within a pharmaceutical company and with FDA, and those procedures are fully applicable to CANDA system data. In addition, steps must be taken to assure consistency between the NDA hardcopy and CANDA data.

Controls

- o Are appropriate security procedures to protect data in place at the sponsor company and FDA sites? (Example: is access to the CANDA system and/or CANDA data limited to authorized staff?)

- o Are CANDA data sets identical to those used to create the hardcopy NDA? Were the data validated after they were placed in the CANDA data base?

- o Are the CANDA data validated periodically to assure that unauthorized changes have not been made to the data? Are data sets adequately protected so that an FDA reviewer cannot make accidental changes to sponsor data?

- o Are data change control procedures in place? Do they address changes to electronic and hardcopy data? How are corrections/changes to CFR data handled within the CANDA system? Are data audit reports produced by the CANDA system, tracing data changes?

- o Are procedures established to appropriately secure data generated by the FDA reviewer?

- o Is there a formal schedule for making backups of data? Are data backups stored in a secure area? Are data restore/recovery plans in place?

- o Is an electronic data submission to FDA handled using similar procedures as those for hardcopy submissions (e.g., submission coordinated by and archival copies retained by regulatory affairs)? Are FDA staff members informed of the confidential and secure nature of the CANDA data and system?

- o Are there security procedures to address and properly control data transfers (e.g., downloading) to other computers?

C. Software System Security and Integrity

Application and system software are sets of computer programs that provide the system functions, allowing access to data or creation of new data. Most system software packages -- e.g., operating systems, data base management systems, and word processors -- are purchased from vendors. Application systems can be purchased or custom developed by a programming staff.

Adequate controls must be in place for both the development and operation of a CANDA system. A logical approach to system development must be used, such as a Systems Development Methodology. Primary input for system functions must come from the appropriate FDA and sponsor groups, typically including the FDA reviewer(s), regulatory, and discipline (e.g., CMC, metabolism, medical, statistics) specialists.

Although specific validation requirements are not available from FDA, a general requirement for all CANDA systems is to develop documented evidence that the system performs as intended.

Controls

- o Are security functions, such as password protection and controlled data access, built into the CANDA system?

- o Does the system prohibit the FDA user from modifying the NDA data and/or CANDA system programs?

- o Can reports in the hardcopy NDA be replicated using CANDA functions? Are terminology and functions consistent between the CANDA and NDA?

- o Did the FDA users provide input into the system design? Does the CANDA meet the FDA's needs?

- o Is an adequate testing plan in place to prove that the CANDA operates correctly before the system is installed at FDA? Were the internal testing and test results approved by the sponsor management?

- o What plans are in place for user training and system support/enhancements?

- o Is a change control process in place? Is a change log maintained (for audit and control purposes)?

- o Does the CANDA have sufficient hardcopy and on-line user documentation and help information? Do the FDA users know how to use the documentation?

- o Is technical system documentation available and kept up to date?

- o For CANDAs that require systems operations support (e.g., CANDAs that operate on a host computer), is a secure operations process in place?

- o Are appropriate anti-virus software and procedures in place? Have FDA users been informed of potential virus problems and how to avoid them?

- o Are "end user computing" guidelines provided to FDA users?

D. Telecommunications Security and Integrity

Telecommunications links between the sponsor, FDA, and/or a third party site may be either a continually required or periodic component of a CANDA system. For systems utilizing a remote host computer (e.g., operate on a mainframe computer at the sponsor site), an ongoing telecommunications link is necessary to use the CANDA. Other types of CANDA systems may use telecommunications periodically for data exchange, electronic mail, and/or program maintenance.

Transmitting data over telephone lines presents both security and integrity concerns. Proper procedures must be in place to prevent unauthorized access or inaccurate data transmission.

Controls

- o Are procedures established to ensure appropriate protection from unauthorized alteration, destructions, or disclosure of data, or other acts of fraud?

- o What procedures are in place to ensure accurate transmission of data via telecommunications? Are error correcting capabilities included in telecommunications exchanges?

- o Is the telecommunications network reliable? Is a backup plan established?

- o Have the following factors been considered for the telecommunications network?
  - -- security features such as encryption devices, dial-back modems, and location and/or device ID dependent communications?
  - -- secure physical location of telecommunications devices and lines?
  - -- leased telephone lines?
  - -- public versus private networks?
  - -- password protection?
  - -- security violation monitoring and management reporting?

E. Hardware Security and Integrity

Hardware security and integrity on CANDA projects require proper operations procedures by FDA and sponsor firm. Hardware required for a CANDA system could be installed at FDA site only (e.g., a microcomputer), or both in FDA's and sponsor firm's offices (e.g., workstations at FDA connected to a remote host computer).

Proper hardware operations include established end-user and computer operations procedures.

Controls

- o Are adequate hardware security devices provided at the FDA site? Can the hardware be physically secured in an FDA reviewer's office?

- o Are security procedures established for CANDA hardware at the sponsor site? Are computer operators informed on the sensitivity of the project?

- o Is an identical system available for backup, testing, and/or disaster recovery? Is a disaster recovery plan in place?

- o Is timely hardware maintenance available at the FDA site? Are procedures established to replenish supplies?

- o If multiple CANDA systems are operating on the same machine, are appropriate security safeguards in place to allow access to authorized staff only?

- o Have all hardware installations been coordinated through FDA's Center for Drug Evaluation and Research, Office of Management? Are plans in place to remove all sponsor-owned equipment following completion of the NDA review? Are all installation and removal plans documented, with copies sent to FDA Office of Management?

F. FDA Interaction on CANDA Projects

All CANDA projects require extensive work with FDA staff. In addition to the standard interaction involved with an NDA review, a CANDA project requires discussions between FDA reviewers and the sponsor on t he development, installation, and support of the system.

The CANDA system also may provide an electronic means for communication, via electronic mail or through sharing CANDA-based data.

Controls

- o Is a formal development methodology in place to guide the system design, development, and support process?
- o Are procedures established for system documentation, software, and data submissions to FDA?
- o Are CANDA training sessions and installations coordinated by regulatory affairs staff?
- o Have agreements been established with FDA for confidentiality of FDA-generated data files on the CANDA system (e.g., is the sponsor prohibited from reviewing all files on the CANDA created by the FDA user)?
- o Has FDA been informed of the system's ability to record user activity (i.e., functions used, data accessed, keystrokes entered)? Has an agreement been established on access to such data?
- o Are FDA interaction roles clearly established within the sponsor company?
- o If a third party is involved with CANDA work, is their role in FDA interactions clearly defined and monitored?
- o Are all FDA interactions documented?

G. Other Topics

Controls

- o Have FDA/ and or PMA guidelines on computer system validation been reviewed as part of the CANDA development process?
- o Are responsibilities for CANDA development and support and for FDA interaction clearly established within the sponsor company?

CHAPTER VI

FDA MODEL CANDA

(THE MODEL IS CURRENTLY BEING DEVELOPED)

CHAPTER VII

A. GENERAL MEDICAL CANDA CONSIDERATIONS

A. Introduction

Unlike other scientific disciplines involved in the NDA review process, a high degree of variability exists in the medical review process based on drug class and specific characteristics of each individual drug. In addition, the Center for Drug Evaluation and Research includes nine drug product divisions, and within each division a different procedure may be used to review different types of drugs. Due to this variation, specific CANDA requirements applicable for all medical reviews cannot be defined at this time.

The appendices of this document include some guidelines for individual drug product divisions. However, following are general topics and statements on medical CANDAs that should be considered for each project.

NDA Text Access

The hardcopy NDA presents text reports describing and summarizing clinical characteristics of the study drug. The reviewer may wish to review parts of this information periodically during the review period. A text search function can help locate topics of interest. Access and display of text and/or graphics also can be helpful when the reviewer is creating his/her review document.

Patient Data Access

Large volumes of data are collected during clinical studies, and create the basis for drug efficacy and safety claims. Easy access to data, presented for a single patient, a subset or the entire patient population, can facilitate the review process. Data can be presented as a listing (tabulation) or as a patient record (single screen for one patient). Consider how the data should be presented, e.g., by study, investigator, patient, visit, or occurrence, based on how a medical reviewer may look at the data.

Patient Data Analysis

Patient data may be analyzed using many different functions. Graphs, tables, descriptive statistics, inferential statistics, and other approaches are commonly used within an NDA. CANDA systems that include some of these tools can help the reviewer validate, reanalyze or create new analyses of the data. Analysis approaches for safety data are relatively standard. Efficacy analysis, however, varies significantly based on drug type.

Document Creation

The end result of the medical review is a review document, passed on to supervisory levels for additional review. Document processing tools, including a word processor and graphics tools, can help expedite the medical review document creation process.

Other Considerations

The CANDA system should be easy to use and easy to understand, utilizing, if possible, computing tools that are familiar to the medical reviewer. The data and analysis results must be consistent with those presented in the hardcopy NDA. CANDA submissions should be handled following the same procedures used for NDA submissions and updates.

Drug Product Division Guidance

A PMA/FDA Task Group is currently interviewing reviewers within the various drug product divisions. The appendices of this chapter will eventually contain some guidelines for the individual drug product divisions. This information will be developed and incorporated into this chapter as the interviews ar completed. It is hoped that this information will assist companies in designing their CANDAs to meet the needs of the various drug product reviewing divisions.

B. Therapeutic Area Analysis

These analyses are being developed to outline the NDA clinical review process within the review divisions of the Center for Drug Evaluation and Research. Though reviews may vary by drug, drug class, or reviewer, these analyses will cover general activities required for most NDA reviews. Presently, included in this guidance manual are the Therapeutic Area Analyses for the Division of Anti-Infective Drug Products for the review of anti-infective and dermatologic drug products. As the analyses are completed for other review divisions and drug types they will be published in later versions of the guidance manual.

Working Group on Computer Assisted NDA Reviews
FDA/PMA Therapeutic Area Analysis

Anti-Infective Drug Products Division

Introduction

This document outlines the NDA clinical review process within the Anti-Infective Drug Products Division, and identifies useful functions for a Computer Assisted NDA

(CANDA) system. Though review activities may vary by study drug, reviewer, or drug class, this process covers general activities required for most NDA reviews. It is intended as a general guideline for development of CANDA systems submitted to the division. Prior to development and submission of a CANDA system (9 months-1 year), the division highly recommends discussions between the sponsor and the review team to agree on the detailed design of system functions.

Since the medical review is the critical path activity in most NDA reviews, the CANDA system features are focused on the needs of the medical reviewer. Other review disciplines also can benefit from automation support, but are not addressed in this document.

The Anti-Infective medical review activities are grouped into six categories:

-- NDA Overview
-- General Clinical Review
-- Efficacy Review
-- Safety Review
-- Create Review Document
-- Other Activities

CANDA functions are identified for each category, though some functions may be applicable to many of the activities listed. A CANDA system must be easy to understand, follow a logical approach to an NDA review, and be easy to use for a medical reviewer with little computer experience.

<u>Overview</u>

Before the Medical Officer begins reviewing the data sections of the NDA, the Reviewer performs an overall review of the submission. This is primarily done to get an overall understanding of the compound, the submission, and to determine whether there is sufficient data included in the submission to suport the claims. Activities can also include comparing the compound in the submission to other approved drugs in the same classifiction. Reviewers approach this overview differently, but essentially all elements are eventually reviewed.

-- Summary - NDA Summary (Item 2) is read to get an understanding of the drug and application including the number of claims being requested.

- The Index (Item 1) is read to determine the layout, contents and information that the reviewer will need to know and access.

- Medical Integrated Summaries (Item 8)

- Data tables reviewed in the volumes are reviewed for facts not interpretations.

-- Label - Since the label is a coordinated outlook of the NDA, it contains data on stability, dose information, chemistry, toxicology, and microbiology. Therefore, the entire review group (Medical, CMC, Biopharmaceutics, Tox/Pharm, Microbiology) examines information contained on the package insert information. To do this they review:

-- Package Insert
-- Annotated Package Insert
-- References for key labeling statements

-- Completeness review

- This review ensures that sufficient data has been submitted to warrant a full review (see 45 day meeting).

- Other components of the submission are reviewed to get an overall understanding of the drug. This normally includes toxicology/nonclinical pharmacology, microbiology and pharmacokinetics.

- This also may include reviewing:
  -- Clinical Pharmacology data (Cmax, Cmin, AUC, t1/2 + tissue levels...) and comparing these parameters to other drugs in the same class.
  -- Microbiology (MIC90) to see if there is enough data to support the pathogen and then comparing this information to other drugs in the same class.
  -- Animal Pharmacology to understand the potential adverse events that can occur in humans.

- The medical officer is responsible for compiling the information from all of the different areas and writing the overall review document.

-- 45 day meeting

- Within forty five (45) days after the submission has been filed by the sponsor, the Medical Officer meets with representatives and division management from other reviewing disciplines to determine whether there is sufficient information in the submission to warrant a full review (i.e., file the application).

Functions required
-- Access to sponsor's documents.
-- Access to sponsor's labeling information.

General Clinical Review

In this section of the review, the Medical Officer begins reviewing each of the protocols to determine the types of studies and indications included in the submission. After identifying the pivotal and safety studies, the Reviewer begins the process of reviewing patient level data from a clinical view of "What happened to the patient". The outcome of this step is general understanding by the Medical Officer of the patients included in the studies. Activities in this step include:

-- Individual Protocol

  -- Each study summary is read

  -- Determination of:
    -- Key protocols
    -- Dosing schedule
    -- Study schedule
    -- Methods section
    -- Control group

  -- Determine evaluability criteria
    -- Review each protocol
    -- Evaluate sponsor criteria.

  -- Determine patient population based on sponsor's evaluability criteria.
  -- Determine the number of evaluable patients vs. enrolled.
    -- Identify reasons why patients were disqualified from the efficacy analysis.
    -- List differences between the reviewer's and sponsor's evaluability criteria.

  -- Review of the overall safety for the protocol.

  -- Summary of patients who discontinued due to an adverse event.

-- Validate Patient Data

  -- Patient by patient data review and validation.
    -- Review of line listing of patient data
    -- Reviews key variables:
      -- Age, diagnosis, type of organism/pathogen, patient assessment, follow-up, clinical outcome, adverse events, lab, and lab abnormalities.

    -- Review of key identifiers:
      -- Symptom scores

Functions required:
-- Access to protocol information and study summaries and reports.
-- Access to case report form data in line listing format.
-- The capability to:
   -- Select/subset, sort, graph, and tabulate patient level data.
   -- Ability to create new patient fields/variables (reclassification).
-- Data extraction to other computing tools.
-- Access to complete patient data (CRF - Image).

<u>Efficacy Review</u>

An important component of the Anti-Infective review is the review of each patient's efficacy data. This review confirms whether the patient is evaluable, determines the total population of evaluable patients and re-analyzes the claim with the patient population determined by the Reviewer. This section is completed for each indication requested in the claim. Specific activities for this review section include:

-- Evaluability assessment and validation of efficacy data (by indication):
   -- Determine evaluable patients (Bacteriological or clinical outcome depending on the infection being studied)
   -- Validate Patient data.
   -- Patient by patient data review and validation.
   -- Review of line listing of patient data.
   -- Review key variables:
      -- Age, diagnosis, type of organism, patient assessment and time of patient assessment, follow-up clinical outcome, and safety.
   -- A review of key identifiers:
      -- Demographic information
      -- Disease and disease characteristics
      -- Clinical signs and/or symptom scores
      -- Bacteriology including methodology, culture site, serology, and organism if applicable.

   -- A review of other important parameters:
      -- Chest x-ray, lab tests, scans,

   -- A review of evaluable and non evaluable patients for:
      -- Clinical efficacy (clinical cure)
         -- Microbiology efficacy (eradication rate of pathogen)

         -- Clinical Outcome (Agree/Disagree)
         -- Review data on clinical evaluability and compare with the sponsor.
         -- Cross reference micro and clinical results for patient consistency.
      (Same for microbiological evaluability)

-- Test Claim

- -- Retest claim with new population
  - -- Re-tabulation of the results in support of the claim based on the reviewer's patient populations.
- -- Compare to sponsor's results
  - -- Review results vs. sponsor's and identify, document, and explain differences.
- -- Biostatistics review:
  - -- If different results are found between the reviewer and sponsor, the reviewer may send this additional information to the Biostatistics reviewer and ask for a re-analysis of the information.

Functions required:
-- Access to patient data.
- -- Access to case report form data in line listing format (meet with reviewer prior to CANDA formatting as to which data parameters need to be in the line listings).
- -- Case report forms (image)
- -- Access to sponsor's efficacy analysis results.
- -- The capability to:
  - -- Select/subset data and patients
  - -- Subset non evaluable and evaluable patients from the entire database.
  - -- Subset and select data based on key variables
  - -- Create new patient fields/variables (reclassification)
  - -- Sort,
  - -- Graph,
  - -- Tabulate results (re-do basic statistical analyses)
  - -- Produce listings and tables
- -- Ability to re-create the sponsor's analysis.
- -- Ability to apply analysis to reviewer identified population.
- -- Access to sponsor's efficacy analysis results.
- -- Data extraction to other computing tools.
- -- Access to sponsor's document.

<u>Safety Review</u>

The safety data is normally reviewed on an overall, and study by study basis (with no patients being excluded). Specific activities within this section include:

-- Adverse Event Review

- -- Adverse events by body system:
  - -- Each body system is reviewed for events.
  - -- Non serious adverse events are also reviewed.

-- Trend analysis:
  -- The adverse events are reviewed for trends and other events that could be related to the indication and treatment.
  -- Time relatedness (i.e. 1 dose vs. multiple dose)
  -- Etiology, relatedness

-- Lab Review

  -- Lab results vs. baseline
    -- Lab values are reviewed for changes from baseline and for trends.

  -- Normal ranges:
    -- Lab test results which are outside the normal ranges are also reviewed along with checking the results after the normal ranges are modified (for example: 2 times the upper limit of normal for specific tests).

  -- Lab outliers are also reviewed.

-- Deaths and Discontinuations
  -- Case Report Forms are reviewed and summarized for each death and discontinuation.

  -- Summary of patients who discontinued (by regimen, reason, and adverse event summary).

  -- Narrative reports of each patient are also reviewed.

-- Other Safety

  -- Pre Clinical results

  -- Drug-Drug interactions

  -- "Cluster review within an indication"
    -- A study by study review of adverse events as well as labs outliers, trends and other events which may have been related to the indication.

Functions required:
-- Access to case report form safety data.
-- Access to sponsor's summaries of safety data by body system.
-- Access to sponsor's narrative reports.
-- Access to complete patient data (CRF).
-- Access to the patient data including adverse events, labs, and medications.
  -- Access to sponsor's lab data results including listings and tables of lab abnormalities, outliers, values outside normal range, changes from baseline.
  -- Ability to create new patient fields/variables (modifications of lab normals).

- -- Access to case report form safety data.
- -- Access to sponsor's summaries of safety data by body system.
- -- Data extraction to other computing tools (for trend analysis).
- -- The capability to:
  - -- Select/subset data and patients
  - -- Sort
  - -- Graph
  - -- Tabulate results
  - -- Produce listings and tables
- -- Access to complete patient data (CRF).
- -- Access to the sponsor's documents.

Create Review Document

The Medical Officer is responsible for writing the review document for the submission. Throughout the review, the Medical Officer is compiling information and tables from the submission. Some of this information is taken from the sponsor's documents and from the Reviewer's assessment. In addition, the Medical Officer is also responsible for summarizing the recommendations of the other reviewing disciplines. Activities in this section occur during each of the other sections and include the following specific activities.

- -- Access sponsor documents

  - -- Tables, data listings, and other sections of the sponsor's document are cut and pasted into the review document (If applicable or for comparative purposes)

- -- Write document

  - -- Summary of Volume 1 of the NDA.

  - -- Discussion of key issues: clinical cure, microbiology, stability, dose...).

  - -- Summary of the sections reviewed by the other FDA departments (Biopharmaceutics, Microbiology, CMC, Toxicology, and Statistics).

  - -- Summary of each study reviewed by indication.

  - -- Blood level information

  - -- Introductions of each area are written including the pharmacokinetics and labeling sections.

  - -- Information from the review and references to outside documents are also integrated into the document.

-- Discussion of overall efficacy by indication including a discussion of evaluable and non-evaluable patients.

-- Explanation of variations from sponsor's results.

-- Recommended indications

-- Recommended dose and duration of therapy for each indication.

Functions required:
-- Word processing tool to create the document.
-- Access to the sponsor's documents.
-- Capability to copy the tables, listings, and summaries from the sponsor's documents into the Medical Officer's Review Document.

Other

Communication between the sponsor, reviewing division and Medical Officer was mentioned as one of the key components of the review process. Meetings before the NDA is submitted are primarily to discuss the information and format of the key variables to be included in the line listings. These formats are modified and fine tuned during the review process.

-- Pre NDA meeting
  -- Because of the importance of the line listings, the Reviewers stated that they need to meet with the company to discuss the format of the listings, and why they need them. (Should be around 9 months to 1 year prior to the NDA submission)

-- Sponsor/Reviewer communications
  -- Throughout the review, interactions between the sponsor and reviewer are needed to clarify the information included in the line listings, discuss additional or modifications to the line listings already provided, and to answer specific reviewer questions pertaining to the NDA.

-- Advisory Committee meeting
  -- As determined by the division.

Working Group on Computer Assisted NDA Reviews
FDA/PMA Therapeutic Area Analysis

Dermatologic Drug Products Division

Introduction

This document outlines the NDA clinical review process within the Dermatologic Drug Products Division, and identifies useful functions for a Computer Assisted NDA (CANDA) system. Though review activities may vary by study drug, reviewer, or drug class, this process covers general activities required for most NDA reviews. It is intended as a general guideline for development of CANDA systems submitted to the division. Prior to development and submission of a CANDA system (9 months-1 year), the division highly recommends discussions between the sponsor and the review team to agree on the detailed design of system functions.

Since the medical review is the critical path activity in most NDA reviews, the CANDA system features are focused on the needs of the medical reviewer. Other review disciplines also can benefit from automation support, but are not addressed in this document.

The Dermatology medical review activities are grouped into six categories:

-- NDA Overview
-- General Clinical Review
-- Efficacy Review
-- Safety Review
-- Create Review Document
-- Other Activities

CANDA functions are identified for each category, though some functions may be applicable to many of the activities listed. A CANDA system must be easy to understand, follow a logical approach to an NDA review, and be easy to use for a medical reviewer with little computer experience.

Overview

Before the Medical Officer begins reviewing the data sections of the NDA, the Reviewer performs an overall review of the submission. This is primarily done to get an overall understanding of the compound, the submission, and to determine whether there is sufficient data included in the submission to support the claims. Activities can also include comparing the compound in the submission to other approved drugs in the same classification. Reviewers approach this overview differently, but essentially all elements are eventually reviewed.

-- Summary - NDA Summary is read to get an understanding of the drug and application including the number of claims being requested.

- The Index is read to determine the layout, contents and information that the reviewer will need to know and access.

- Medical Integrated Summaries

-- Label - Since the label is a coordinated outlook of the NDA, it contains data on stability, dose information, chemistry, toxicology, and microbiology. Therefore, the entire review group (Medical, CMC, Biopharmaceutics, Tox/Pharm) examines information contained on the package insert information. To do this they review:

-- Package Insert
-- Annotated Package Insert
-- References for key labeling statements

-- Completeness review

- This review ensures that sufficient data has been submitted to warrant a full review (see 45 day meeting).

- Other components of the submission are reviewed to get an overall understanding of the drug. This normally includes toxicology/nonclinical pharmacology, microbiology and pharmacokinetics.

- The medical officer is responsible for compiling the information from all of the different areas and writing the overall review document.

-- 45 day meeting

- Within forty five (45) days after the submission has been filed by the sponsor, the Medical Officer meets with representatives from other reviewing disciplines and division management to determine whether there is sufficient information in the submission to warrant a full review (i.e., file the application).

Functions required
-- Access to sponsor's documents.
-- Access to sponsor's labeling information.

General Clinical Review

In this section of the review, the Medical Officer begins reviewing each of the protocols to determine the types of studies and indications included in the submission. After identifying the pivotal and safety studies, the Reviewer begins the process of reviewing patient level data from a clinical view of "What happened to the patient". The outcome of this step is general understanding by the Medical Officer of the patients included in the studies. Activities in this step include:

- -- Review Clinical Program
  - -- Determination of:
    - -- Key protocols for safety and clinical efficacy:
      - -- Phase I - safety and non clinical efficacy,
      - -- Phase II - dose ranging studies,
      - -- Phase III - well controlled studies for efficacy.
- -- Individual Protocol
  - -- Each study summary is read
  - -- Determination of:
    - -- Dosing schedule
    - -- Study schedule
    - -- Methods section
    - -- Control group
  - -- Determine evaluability criteria
    - -- Review each protocol
    - -- Evaluate sponsor criteria.
  - -- Determine patient population based on sponsor's evaluability criteria.
  - -- Determine the number of evaluable patients vs. enrolled.
    - -- Identify reasons why patients were disqualified from the efficacy analysis.
    - -- List differences between the reviewer's and sponsor's evaluability criteria.
  - -- Review of the overall safety for the protocol.
    - -- Summary of patients who discontinued due to an adverse event.
    - -- Review of key safety parameters and interactions:
      - -- Contact interactions, culture interactions, photo-toxicity, photocontact, vasoconstriction, and HPA access suppression.
- -- Validate Patient Data
  - -- Patient by patient data review and validation.
    - -- Review of line listing of patient data
    - -- Cross check with Case Record form
    - -- Reviews key variables:
      - -- Age, diagnosis, patient assessment, follow-up, adverse events, lab, and lab abnormalities.

- -- Review of key identifiers:
  - -- clinical signs and symptoms,
  - -- cultures and KOH values (anti-fungal submissions)
- -- Histopathology

Functions required:

-- Access to protocol information and study summaries and reports.
-- Access to case report form data in line listing format.
-- Data extraction to other computing tools.
-- Access to complete patient data (CRF - Image)
-- The capability to:
  -- Select/subset, sort, graph, and tabulate patient level data.
  -- Ability to create new patient fields/variables (reclassification).

Efficacy Review

An important component of the dermatologic review is the review of each patient's efficacy data. This review confirms whether the patient is evaluable, determines the total population of evaluable patients and reanalyses the claim with the patient population determined by the Reviewer. This section is completed for each study and indication requested in the claim. Specific activities for this review section include:

-- Evaluability assessment and validation of efficacy data:

- -- Determine evaluable patients
  - -- Validate Patient data.
  - -- Cross check data on Case Record form.
  - -- Patient by patient data review and validation.
  - -- Review of line listing of patient data.
  - -- Review key variables:
    - -- Age, diagnosis, follow-up clinical outcome, and safety.
  - -- A review of key data points:
    - -- Baseline, end of dose titration, and endpoint.
  - -- A review of key identifiers:
    - -- Demographic information
    - -- Clinical signs
    - -- Cultures and KOH values (anti-fungal submissions)
  - -- Histopathology (if applicable)

  - -- A review of evaluable and non evaluable patients for clinical efficacy (clinical cure rate).

  - -- Clinical Outcome (Agree/Disagree)
    - -- Review data on clinical evaluability and compare with the sponsor.
    - -- Cross reference clinical results for patient consistency.

-- Test Claim

- -- Retest claim with new population
  - -- Re-tabulation of the results in support of the claim based on the reviewer's patient populations

- -- Compare to sponsor's results
  - -- Review results vs. sponsor's and identify, document, and explain differences.

- -- Biostatistics review:
  - -- If different results are found between the reviewer and sponsor, the reviewer may send this additional information to the Biostatistics reviewer and ask for a reanalysis of the information.

Functions required:

- -- Access to patient data.
  - -- Access to case report form data in line listing format (meet with reviewer prior to CANDA formatting as to which data parameters need to be in the line listings).
  - -- Case report forms (image)
  - -- Access to sponsor's efficacy analysis results.
  - -- The capability to:
    - -- Select/subset data and patients
    - -- Subset non evaluable and evaluable patients from the entire database.
    - -- Subset and select data based on key variables
    - -- Create new patient fields/variables (reclassification)
    - -- Sort,
    - -- Graph,
    - -- Tabulate results (re-do basic statistical analyses)
    - -- Produce listings and tables
- -- Ability to apply analysis to reviewer identified population.
- -- Access to sponsor's efficacy analysis results.
- -- Data extraction to other computing tools.
- -- Access to sponsor's document.

<u>Safety Review</u>

The safety review within the Dermatologic Drug Products Division consists of two types of reviews. Besides reviewing the Safety data on an over, and study by study basis (with no patients being excluded, the Phase I and II studies are reviewed for key safety parameters. Specific activities within this section include:

- -- Adverse Event Review
  - -- Adverse events by body system:
    - -- Each body system is reviewed for events.
    - -- Non serious adverse events are also reviewed.

-- Trend analysis:
  -- The adverse events are reviewed for trends and other events that could be related to the indication and treatment.
  -- Time relatedness (i.e. 1 dose vs. multiple dose)
  -- Etiology, relatedness

-- Lab Review
  -- Lab results vs. baseline
    -- Lab values are reviewed for changes from baseline and for trends.

  -- Normal ranges:
    -- Lab test results which are outside the normal ranges are also reviewed along with checking the results after the normal ranges are modified (for example: 2 times the upper limit of normal for specific tests).

  -- Lab outliers are also reviewed.

-- Deaths and Discontinuations
  -- Case Report Forms are reviewed and summarized for each death and discontinuation.

  -- Since deaths and discontinuations are uncommon in these submissions, patient data is reviewed to determine:
    -- Why the patient discontinued:
    -- The progress of the patient in the study (cure/no cure/no change), and
    -- Whether there were any safety or adverse events associated with the discontinuation.

  -- Summary of patients who discontinued (by regimen, reason, and adverse event summary).

  -- Narrative reports of each patient are also reviewed.

-- Other Safety

  -- Pre Clinical results

  -- Review of key safety parameters and interactions:
    -- contact interactions, culture interactions, photo-toxicity, photocontact, vasoconstriction, and HPA access suppression.

  -- Drug-Drug interactions

  -- Special population analyses

-- The Biophamaceutics data is reviewed to determine the amount of drug (if any) that is absorbed into the body.

Functions required:

-- Access to case report form safety data.
-- Access to sponsor's summaries of safety data by body system.
-- Access to sponsor's narrative reports.
-- Access to complete patient data (CRF).
-- Access to the patient raw data listings including adverse events, labs, and medications.
  -- Access to sponsor's lab data results including listings and tables of lab abnormalities, outliers, values outside normal range, changes from baseline.
  -- Ability to create new patient fields/variables (modifications of lab normals).
-- Access to case report form safety data.
-- Access to sponsor's summaries of safety data by body system.
-- Data extraction to other computing tools (for trend analysis).
-- Access to complete patient data (CRF).
-- Access to the sponsor's documents.

-- The capability to:
  -- Select/subset data and patients
  -- Sort
  -- Graph
  -- Tabulate results
  -- Produce listings and tables
-- Access to complete patient data (CRF).
-- Access to the sponsor's documents.

<u>Create Review Document</u>

The Medical Officer is responsible for writing the review document for the submission. Throughout the review, the Medical Officer is compiling information and tables from the submission. Some of this information is taken from the sponsor's documents and from the Reviewer's assessment. In addition, the Medical Officer is also responsible for summarizing the recommendations of the other reviewing disciplines. Activities in this section occur during each of the other sections and include the following specific activities.

-- Access sponsor documents

  -- Tables, data listings, and other sections of the sponsor's document are cut and pasted into the review document (If applicable or for comparative purposes)

-- Write document

-- A summary describing the background of the drug.

-- A section on the safety profile of the drug based on the Phase I and II Safety studies.

-- A section describing the efficacy and safety profile of the drug based on the Phase II and III clinical studies.

-- A study by study discussion describing the method and results (sponsor's and reviewer's) and the Reviewer's conclusion.

-- Explanation of variations from sponsor's results.

-- Overall summary of the submission.

-- Recommended dose and duration of therapy if declared approvable.

Functions required:
-- Word processing tool to create the document.
-- Access to the sponsor's documents.
-- Capability to copy the tables, listings, and summaries from the sponsor's documents into the Medical Officer's Review Document.

<u>Other</u>

Communication between the sponsor, reviewing division and Medical Officer was mentioned as one of the key components of the review process. Meetings before the NDA is submitted are primarily to discuss the information and format of the key variables to be included in the line listings. These formats are modified and fine tuned during the review process.

-- Pre NDA meeting
  -- Because of the importance of the line listings, the Reviewers stated that they need to meet with the company to discuss the format of the listings, and why they need them. (Should be around 9 months to 1 year prior to the NDA submission, even if studies not yet completed)

-- Sponsor/Reviewer communications
  -- Throughout the review, interactions between the sponsor and reviewer are needed to clarify the information included in the line listings, discuss additional or modifications to the line listings already provided, and to answer specific reviewer questions pertaining to the NDA.

-- Advisory Committee meeting
  -- As determined by the division.

CHAPTER VIII

BIOPHARMACEUTIC CANDAs

A. Introduction and Background

An FDA and PMA Biopharmaceutics Task Force has developed a system for exchanging and manipulating data. The File Interchange System (FIS) has two components. The first is the FIS file specifications. The second, also referred to as FIS, is the program which enables the firm to create a FIS file and the FDA reviewer to extract and manipulate the data. Both parts have undergone initial Alpha and Beta testing.

Members of the Division of Biopharmaceutics, Center for Drug Evaluation and Research of FDA and the Drug Metabolism subsection of PMA worked 3 years to develop FIS. It was created to facilitate the use of FIS files as a conduit to exchange biopharmaceutics data between pharmaceutical companies and FDA reviewers using a standardized format. The FIS program has been copyrighted. FDA has applied for permission to use the copyrighted material but, to date, has not received it; therefore, the FIS program is unavailable for distribution. A notice will appear in the Federal Register when the program can be ordered.

Presently, it is designed so that biopharmaceutic data can be developed by a sponsor or hired party without considering interfacing or adapting to the FDA environment in which it will be used. Additionally, it was created to be available, affordable and useable to any company without regard to size, resources or computer sophistication.

The Task Force is aware of the array of computer skills among the people who will be using the system and have designed the system to be user-friendly, able to accept any data, and hardware, software and operating system independent. FIS can, therefore, accept any type of information contained in a file if the construction follows the specifications of the program. We will deal were with an overview of the system for a file which contains data and detailed descriptions of that data without restricting the type of data included. The FIS User's Handbook defines specifically how to construct files. It also tells how to include comments in the file which can provide further descriptive information. Therefore, a file conforming to FIS can be entirely self documenting.

B. System Requirements

The minimum system requirements for creating and using a FIS file are a personal computer of at least AT-class using DOS 3.0 or higher and 640K of RAM.

C. File Construction to Meet FIS Specifications

These are the files that will be crated at the company and eventually transferred to FDA for analysis. These files can be created from initial, manual data entry or exporting existing ASCII files from databases within the company.

There are specific types of records that must be entered in the file to properly structure the data for manipulation when it is imported into FDA files. These are headers, source, comment, logical, field definition, data trailer and volume continuation.

The **header file** is the information identifying the submission. It must contain the submitter's name, phone number, company, creation date and time the file is created.

The **source record** is where the data came from.

The **comment record** allows for comments and is usually located near the beginning of the FIS file. This record's first character must be an asterisk (*). The comment record must be located after the BEG record and the length is limited to 255 characters per line.

The **logical record** contains the sequential logical record number. This should begin with the letters "REC" and be followed by the short name, long name, starting column position, ending record position and the logical record identification. There must be at least one of these records per file and must be associated with at least one field definition record.

The **field definition record** identifies the type of field. This will begin with a field number, which is sequential and non-repeating, and be followed by a short name, long name, the field length, the data type, decimals, non-quantifiable data and missing data identifier and the units of measure. There should be at least one field definition record for each logical record.

The **data record** defines the information in the file. It would begin with the field number, followed by the logical record marker and the data value. The data value must match the associated field definition. For example, if the definition states that it is a numeric field, then a number must be found there.

The **trailer record-end record** could begin with END, the defining REC or FLD. It would then be followed by the logical record or field number, number of times this entity was written to the FIS file, and must be one per logical record and one per field.

The **volume continuation record** begins with VOL, identify the BEG or END, and contains a volume number, field definition count and data occurrence count. This record allows for the FIS file to be contained in multiple diskettes.

D. FIS Program Functions

These are the function files that enable a reviewer to manipulate and/or analyze the data. There are, basically, four: **create, verify, view** and **extract data from FIS** files. They are explained below.

The **create** file allows a range of data entry from initial, manual entry to downloading existing data from micros, minis or mainframe computers. During initial creation of this program file, firms can create the FIS file or define the FIS file structure.

The **verify** file reads in and verifies existing FIS files so that they conform to the FIS file specification.

The **view** file allows the user to view the content of the FIS data files. With this option the headers, comments, logical record and field definitions, individual logical records or the entire file can be seen.

The **extraction** file allows the user to select the output format and output fields, connect multiple logical records and specify extraction criteria. This program file also allows a user to save extraction criteria and data extracted as well as the files that have been viewed and audited.

For example, this function would allow a reviewer to select from a certain bioavailability study file all the subjects who were above a certain age and smoked. Then they would be able to add biopharmaceutics data from a FIS file and review, then save, the combined data as one unit of information. The audit file is a step-by-step tracer of the procedure and will allow both the reviewer and the firm to follow the query. If mistakes were made or the data are not what was wanted, the audit file can be reviewed for historical information or correction.

Appended to this chapter are three sets of information on FIS files. The first, Appendix VIII-A, is a "roadmap" on how to create three types of files. The information on the left of the sheet is the event that the creator wants to happen. The information on the right is the menu and selection that will allow the event to occur.

Second, Appendix VIII-B, a Quick Reference Guide to the complete Users' Manual. And third, Appendix VIII-C, is an index of the complete Users' Manual.

Please use this information as a general guide. To create a biopharmaceutic section for a specific CANDA, a sponsor will still need to call and discuss it with the reviewers. Initial contact should be made to Dr. See-Yan Lam, Food and Drug Administration, HFD-426, 5600 Fishers Lane, Rockville, MD 20857. Dr. Lam's phone number is (301) 443-8098

APPENDIX VIII-A CREATE FIS file from ASCII File(s)
(No template, no comments)

CREATE (main CREATE/VERIFY menu)

FILE(S) (FILE(S) KEYBOARD menu)

Enter name of FIS file
No template files
Enter # of ASCII files
Enter names of ASCII files
Enter header/source information
No comments

RECORDS (RECORDS DATA-INPUT menu)

Enter record definitions

FIELDS (EDIT ADD DELETE FIELDS MENU, Press/to display)
Enter filed definitions
Quit (back to Record menu)

Quit (back to RECORDS DATA-INPUT menu)

DATA-INPUT (RECORDS DATA-INPUT menu)

(FIS file is written)
save individual databases
save template name
enter template name

(FILE(S) KEYBOARD menu)

CREATE FIS file from ASCII File(s)
(No template, with comments)

CREATE (main CREATE/VERIFY menu)

FILE(S) (FILE(S) KEYBOARD menu)

Enter name of FIS file
No template files
Enter # of ASCII files
Enter header/source information
No comments (EDIT COMMENTS THISFILE OTHERFILE menu)

RECORDS (RECORDS DATA-INPUT menu)

Enter record definitions

QUIT (back to Record menu (EDIT ADD DELETE display)
FIELDS menu, Press/to
QUIT (back to RECORDS DATA-INPUT menu)

DATA-INPUT (RECORDS DATA-INPUT menu)

(FIS file is written

save individual databases
save templates
enter template name

(FILE(S) KEYBOARD menu)
CREATE FIS file from ASCII File(s)
(Template (no modifications), with comments)

CREATE (main CREATE/VERIFY menu)

FILE(S) (FILE(S) KEYBOARD menu)

Enter name of FIS file
Use template files
Enter template name
Enter # of ASCII files
Enter names of ASCII files
Enter header/source information
Enter comments (EDIT COMMENTS THISFILE OTHERFILE menu)

OTHERFILE

Enter name of file containing comments

DATA-INPUT (RECORDS DATA-INPUT menu)

(FIS file is written)

save individual databases
do not save templates

(FILE(S) KEYBOARD menu)

APPENDIX VIII-B FIS QUICK REFERENCE GUIDE

Using This Guide

The guide summarized the most commonly used special keys and concepts in the FIS program. In addition, it lists the limitations and restrictions which you are likely to encounter. Please refer to the Users Guide for more complete information.

Special keys are in **boldface**.

Getting Started

To start FIS	Change directory to \FISPRG\ and type FIS
Making Menu Selections	Move your cursor using the arrow keys to the desired menu item and press Enter
or	Press the first letter of the desired menu item.
Escape Key	In general, when you press the **ESC** key; the current operation will be canceled and the previous menu screen will appear. Information previously entered is retained by the FIS program.
To Exit FIS	Repeatedly select the QUIT and EXIT options from the menus presented to you to exit the FIS program. DO NOT REBOOT or TURN OFF your PC while FIS is running as required files may be left in an unusable state.
Restrictions	NO data files may be stored in the \FISPRG\ directory or in a subdirectory of \FISPRG\

CREATE

Limits	Up to 9 ASCII data files may be used when creating a FIS file	
	Up to 15 logical records per FIS file	
	Up to 255 fields per logical record	
Special Keys	**F1**	Displays any available help.
	F2	displays a list of files in the current directory when entering file names. Note that the **Enter** key selects the file to be used from this list.

F7 displays a list of ASCII input files when you are entering their names.

/ displays a menu when none is shown during logical record and field definition.

Restrictions The following SHORT names are NOT ALLOWED for LOGICAL RECORDS or FIELDS

ANSWER	INSERTED
BTEMP	LIST
CHANGED	OUTFLDS
DELETED	PROBLEMS
ENTRY	RECORDS
FIELDS	STRUCT

names beginning with numbers or underscores

SHORT names for LOGICAL RECORDS and FIELDS can be composed of any combination of the characters A-Z and 0-9 within the constraints listed above. LOGICAL RECORD SHORT names CANNOT contain embedded spaces.

Special Terms TEMPLATE -- This refers to a set of 3 files which contain the logical record and field definitions for a FIS file. The files have the name of the FIS file but with following extensions: .NTX, .FLD, .REC. A template can be used instead of manually keying in logical record and field definitions.

VERIFY

Limits Up to 15 logical records per FIS file.

Up to 255 fields per logical record.

Special keys **F1** displays any available help.

F2 displays list of files in the current directory when entering file names. Note that the Entry key selects the file to be used from this list.

Restrictions Data stored in GRAphics or TeXt blocks MUST be in ASCII format to be successfully extracted from the FIS file.

Special Terms SAVINGS DATABASES. These are intermediate files created during a VERIFY session. They are the ACTUAL input used by the VIEW/XTRACT functions. If they have been saved once for a particular FIS file, you do not have to VERIFY it again before you can VIEW/XTRACT.

VIEW/XTRACT

Limits

Large amounts of disk space is needed to view a FIS file. The program will crash if there is insufficient disk space to complete this function.

You may enter up to 8 lines of Boolean selection criteria.

You may join up to 15 logical records.

Special keys

F1 displays any available help.

F2 displays items/signals that you are finished and are ready to proceed (go forward).

CTRL R rotates the columns in a table when you are viewing data.

ENTER selects the logical record/field pair to use when specifying selection criteria.

PgUP scrolls the table of data up and down when you are
PgDN viewing data or results.

Restrictions

Do not give your output file the extension .AUD. It will be overwritten by the audit file.

The audit file has the same name as the output file specified EXCEPT that the extension becomes .AUD.

The audit file is always written unless you request otherwise.

Do not give the SAS program file and SAS data file the same file extensions. One will overwrite the other.

Special Terms

SAVED QUERY. A saved query is a file containing the specifications of a previous extraction. It can be used to extract information from other FIS files which have the same structure. The criteria can not be modified if a saved query is used.

APPENDIX VIII-C Users' Manual Index

Contents

1. INTRODUCTION

2. INSTALLATION NOTES AND GETTING STARTED
- 2.1 Installation Notes 2
  - 2.1.1 System Requirements 2
  - 2.1.2 Installation Procedure 2
    - 2.1.2.1 Diskette Contents 2
    - 2.1.2.2 Using the INSTALL program 2
  - 2.1.3 Known Conflicts 4
- 2.2 Getting Started 5
  - 2.2.1 Conventions 5
  - 2.2.2 Glossary 6

3. USER REFERENCE GUIDE 7
- 3.1 How to Start FIS 7
- 3.2 How to Create a FIS File 10
  - 3.2.1 Creating a FIS File from ASCII file(s) 12
  - 3.2.2 Entering Headers & Comments 18
  - 3.2.3 Defining Logical Records and Fields 24
    - 3.2.3.1 ASCII File Data Entry 37
    - 3.2.3.2 Keyboard Input 41
  - 3.2.4 Entering Data 41
    - 3.2.4.1 ASCII File Data Entry 41
    - 3.2.4.2 Manual Data Entry via Keyboard 41
  - 3.2.5 Writing the FIS File 49
- 3.3 How to Verify a FIS File 49
  - 3.3.1 Multivolume FIS Files 53
  - 3.3.2 Embedded Test Records 57
  - 3.3.3 Embedded Graphics Records 58
- 3.4 VIEW/XTRACT 61
  - 3.4.1 How to View a FIS File 64
    - 3.4.1.1 Viewing Logical Record/Field definitions 66
    - 3.4.1.2 Viewing Header Records 68
    - 3.4.1.3 Viewing Comments 69
    - 3.4.1.4 Viewing Data by Logical record 70
    - 3.4.1.5 Viewing FIS file 72
  - 3.4.2 How to Xtract Data from a FIS File 73
    - 3.4.2.1 XTRACTing data using a Saved Query 80
    - 3.4.2.2 XTRACTing data by specifying criteria 80
      - 3.4.2.2.1 Joining logical records 80
      - 3.4.2.2.2 Select fields for output 86

3.4.2.2.3 Specifying selection criteria 88

3.5 How to Exit FIS 98

APPENDIX A FIS file Specifications

APPENDIX B DERIVED FIS

APPENDIX C Data Files Used to Create DERIVED FIS

APPENDIX D PC-SAS Output and Audit Files

APPENDIX E DELIMITED ASCII Output and Audit Files

APPENDIX F ASCII Table Output and Audit Files

APPENDIX G SAVED QUERY File and Audit Files

APPENDIX H SAVING DATABASE Files

APPENDIX I ERROR MESSAGES

APPENDIX J HINTS & TIPS

DataStorage

Improved Performance

Use of Audit Files

APPENDIX K Restrictions on Logical Record and Field Names

APPENDIX L Memory Requirement for FIS

CHAPTER IX

CURRENT FDA PERSPECTIVE ON COMPUTER ASSISTED PHARMACOLOGY/TOXICOLOGY REVIEW

The pharmacology/toxicology CANDA guidelines are currently being developed.

CHAPTER X

COMPUTER ASSISTED NEW DRUG APPLICATION (CANDA) RECOMMENDED FEATURES FOR ELECTRONIC SUBMISSION OF CHEMISTRY, MANUFACTURING, AND CONTROLS DATA

Prepared by the Task Group on Chemistry, Manufacturing, and Controls Data
PMA Working Group on Computer-Assisted NDA Reviews

A. Introduction

The CANDA Task Group on Chemistry, Manufacturing, and Controls was established as a subcommittee of the Pharmaceutical Manufacturers Association (PMA) CANDA Working Group to address specific needs relating to the chemistry, manufacturing, and controls (CMC) sections of a new drug application (NDA). Initially, the CMC Task Group met with the FDA Chemists' Task Group on CANDA for general discussions on features of a computerized NDA that would best facilitate their review. Such an approach would address both the requirements of an approvable NDA as well as computer features that enhance the efficiency and thoroughness of their review. Subsequent to these meetings, individual members of the Task Group also met with individual FDA reviewing chemists. In addition, members analyzed possible approaches to successful computerization of the CMC section within their own organizations. The original focus was on original NDAs, although in principle, the concept is applicable, and should be applied to INDs, IND amendments, and drug master files (DMFs). It is anticipated that the guidance will also be relevant for the preparation of abbreviated new drug applications (ANDAs) and abbreviated antibiotic drug applications (AADAs) in computer format.

As a result of these experiences, the CMC Task Group has assembled a set of features that are currently perceived to meet the needs of both FDA reviewing chemists and PMA members. This document provides a discussion of those features for reference. The use of some or all of these features should be discussed with the FDA reviewers prior to the preparation of electronic submissions. The concepts provided are meant as guidelines that are expected to evolve with technological advances and FDA/sponsor experimentation. The overall objective is to facilitate and expedite the evaluation of new drug applications.

An FDA notice giving guidance was published in the Federal Register of September 15, 1988 (53 FR 35912). This notice should be consulted early in the process by those involved in electronic submissions to the agency.

Recommended Features for Electronic Submissions of Chemistry, Manufacturing, and Control Data

B. Retrieval/index and Search

1. **Objective of the feature**

Provide the ability to quickly find complete information concerning a specified topic that was included in an original regulatory submission and subsequent amendments and supplements. The retrieved information is readily displayed in electronic format.

2. **Feasibility/limitations**

The stated objective can be achieved, for example, by the development of multi-level indexing providing access to increasing amount of detail:

- History of regulatory submissions
    - Table(s) of contents
        - Full text and/or keyword index

The ease of implementation of the above is dependent on the software used to create and manage the text, graphics, and data files that are part of the regulatory submission. Many commercial packages are available that can perform the required functions. Some limitations may be currently encountered in the management of graphics.

The history provides a tracking system for all regulatory submissions. It may contain summary information on each submission such as: type (IND, DMF, NDA, amendment, supplement, annual update), date submitted, date approved, brief description of contents, references to related submissions.

The easiest way to give FDA this history is as a text file that includes a tabular presentation of the pertinent information. A more flexible approach is to implement a tracking system using data base management software. The text file or data base would be updated by the sponsor every time a regulatory submission is filed.

The table(s) of contents allows the user to locate both the document page and electronic files. This second feature is important especially when graphics and/or data are not stored with the text information. For maximum usefulness, adequate details, such as listing of all subheadings, figures, tables, and attachments, should be included. Table(s) of contents may be provided to the reviewer in both hard copy and electronic format.

Full text and/or keyword indexing is necessary: This is the key feature and one important advantage that distinguishes an electronic submission from a paper submission. If keyword indexing is provided, a list of the key words must be included with the submission. Full text indexing[2] is preferred and is certainly feasible considering the length of files included in the CMC sections of the regulatory submission. This feature gives the reviewer the ability to find any word, phrase, or number in a variety of combinations by using Boolean type connectors.[3] The result of the search is displayed as retrieved text with, for example, the search strings[4] highlighted. Feature limitations may result from inconsistent use of technical terminology within an application and use of terms unique to the applicant and unfamiliar to the reviewer.

3. **Value to the drug sponsor**

The ability to quickly retrieve complete information on a particular topic allows faster and more consistent responses to FDA inquiries and also facilitates the preparation of amendments and supplements. Providing the FDA reviewer with an electronic index to the submission may also reduce the number of inquiries related to locating information.

4. **Value to the FDA Reviewer**

Various FDA reviewers listed the following benefits that should be realized from the availability of an appropriate electronic index to regulatory submissions:

- o Improve capability to locate specific information and data;
- o Save time in information retrieval when amendments and supplements are received;
- o Reduce query time (the time taken to find answers to questions, whether they are in the NDA submission or require new information from sponsor).

In particular the electronic form of amendments and supplements would be advantageous in the retrieval of information during the many years after approval. Information such as the stability data used to justify the expiry period, method of sterilization in the production of a parenteral, composition of the dosage form,

---

[2]A method that allows the user to search for every word (or every significant word) and number in a file or list of files.

[3]Connector words that designate a relationship between defined entities: OR, AND, and NOT.

[4]A sequence of characters (letters and/or numbers) that together with the Boolean and/or relational operators and/or wildcards or other symbols make up a request intended to find information in a file or list of files.

packaging/labeling controls, and current controls for the substance or product are often difficult to find in the thousands of pages of documentation that make up the NDA several years after approval. In addition, early chemistry volumes are placed in off site storage, making them difficult to obtain. Electronic methods are uniquely suited to the rapid retrieval of such information and would be particularly important in reducing the time necessary for the review of NDA amendments and supplements.

5. Overall contribution of this feature

The availability of multi-level indexing will greatly improve the accessibility of information provided to FDA. The history and table(s) of contents provide valuable keys to the overall organization of regulatory submissions and major topics within a submission. A keyword or full-text electronic index allows rapid access to information concerning a specific item and/or information referred to by particular search strings. Although indexing features can also be incorporated into paper submissions, the critical distinction is that, when provided as software features, they significantly decrease the time required to locate information.

C. Copy/edit

1. Objective of the feature

Provide the reviewer with the ability to reproduce segments of one document in another document, including such elements as word processing, graphs, chemical structures, and notes.

2. Feasibility/limitations

Specific documents require the integration of graphics, diagrams, scientific equations, chemical structures, images, and text. At this time, there is not a generally accepted software package that can accept software from other packages to accomplish this integration.

It is feasible to create both text and image, and to copy these elements into another electronic document. It would be desirable to provide for editing after the text has been copied.

Currently computers are used to create both text and images. However, the chemist may use different software to create different image types, i.e., one to create equations, and one to create chemical structures. The image is then output into a software independent format which can be copied into a word-processed document. But, because the image has been separated from the software that created it, it cannot be edited. One solution would be to move the image intact with all its underlying data. Now when copied, it is editable and computable. However, this would require that the sponsor's system be compatible with the reviewer's system.

3. Value to the drug sponsor

If the drug sponsors can acquire a system that will integrate graphics, diagrams, scientific equations, chemical structures, images, and text, while providing access to the raw data, then scientific documents can be more effectively and efficiently produced.

These documents could be electronically created and incorporated into a file, with full copy and edit capabilities, further reducing time to create them. Subsequent revisions could more easily be made, saving additional time.

4. Value to the FDA reviewer

The reviewer could copy information such as specifications, composition, or an abbreviated manufacturing process for either drug substance or product into the review document. Electronically copying this information would save the time it takes to do it manually.

In addition, this would provide the capability of editing the copied data to more easily produce the review report, saving additional time.

5. Overall contribution of this feature

A system that will allow sponsors to electronically integrate scientific documents, and ill allow FDA reviewers to electronically copy and edit information into their review, will save time. The sponsor will save time in creating the submission, and FDA will save time in the evaluation process.

D. Electronic Mail

1. Objective of the feature

Provide the ability to communicate either messages or documents electronically between two or more internal or remote sites.

2. Feasibility/limitations

Possible difficulties in document transfer or messaging result from the inclusion of multi-format information containing graphics and/or figures, tabulated computer data output, and photocopied literature references, as well as text. Currently, this information is usually provided via facsimile device or by conversion into text, but it is anticipated that, in the near future, computerized transmissions will be commonplace.

It is necessary to have a compatible system with established linkages between the sponsor and FDA or a third party, if they are utilized. Assurance is necessary that data are not altered by conversion into different file formats and that scientific characters and symbols are transmitted accurately across file and printers. The hardware/software itself presents issues of in-house development versus purchased packages, ease of reuse for subsequent filings, and of affordability. Electronic mail sent to a reviewer will be included in the review, however this mail does not constitute an "official" submission until it is sent directly to the appropriate document room for inclusion in the archival file.

Sending it to the document room gets it logged into the tracking system, which makes sure it is acted upon. This should not impact on the time saving associated with the rapid transmission of electronic mail. Assurance that the data remain confidential, retaining their integrity, and are secured, is vital.

3. **Value to the drug sponsor**

Electronic mail has proven to be a valuable asset in the CANDA experience to date. It has provided a mechanism for frequent and rapid contacts between the FDA nd the sponsor to address questions raised during the original application review process and/or in supplemental information previously filed. This ability to communicate easily should certainly enhance the review process.

4. **Value to the FDA Reviewer**

As indicated above, electronic mail has proved to be a valuable asset. From FDA's perspective, it alleviates the problems associated with "telephone tag." Time may also be saved in delivery of responses to FDA's questions. It also enhances person-to-person communication necessary to resolve various issues during the FDA review process.

5. **Overall contribution of this feature**

The most significant feature of electronic mail is the ability to have an interactive communication between the reviewer and technical experts from the sponsor on an on-going basis, resulting in increased effectiveness and efficiency of drug product evaluation.

E. Two-page Display

1. **Objective of the feature**

Provide simultaneous display of two full document pages for ease of comparison on the monitor screen.

2. **Feasibility/limitations**

Two full pages may be displayed on an 18 inch or larger square screen or by the use of two separate screens. The single screen is preferred because of the greater flexibility that it allows. Although large screens exist, software that allows full utilization of large screens to support such applications is less readily available.

3. **Value to the drug sponsor**

This feature is useful during the preparation of a submission for the same reasons it is useful during document review.

4. **Value to the FDA Reviewer**

This feature would:

- o Facilitate comparisons between documents or between sections of one document; and
- o Facilitate (in conjunction with edit/copy feature) preparation of review documents by the reviewer.

Long documents, larger than two to three pages, are difficult to follow on a small screen displaying on third of a page only. Because the information is spread across so many screens, little is visible at any one time, and the reference frame may even change during different passes through the same document. The ability to read one full and fixed page is a necessity for an electronic review. The ability to display a second full page along side the first further enhances the review capability.

One of the key elements of any review is comparisons; how do the different sections of a document agree, how do various stability lots compare, what has been changed in the amendment as compared to the original submission, and what did the referenced section of the DMF contain? Such comparisons are facilitated by simultaneous display of the two pages in question.

In preparing a review document, it is common practice to use in an editorial fashion appropriate sections of text from the document being reviewed. Such cut and paste[5] text editing operations are facilitated by simultaneous display of the page of the original document from which text is being moved along with the page of the review document to which text is to be added.

5. **Overall contribution of this feature**

Electronic reviews will only find favor if their benefits outweigh their drawbacks. The full two-page display and user friendly software will go a long way towards

---

[5]The movement of text or any element of a document from one position to another position in the same document or a different document.

minimizing major potential drawbacks for reviewers while facilitating the work that they must do.

F. Manipulation of Numerical Data

1. Objective of the feature

Provide the capability for the FDA reviewer to perform statistical analysis, graphing, and rearrangement of numerical data.

2. Feasibility/limitations

Stability studies represent the major form of numerical data in chemistry, manufacturing, and controls sections of INDs and NDAs, and the results of such studies are used by the sponsor to determine product shelf life and to demonstrate integrity of the product throughout the proposed shelf life. Methods validation is another category of data in which statistical analysis is useful.

Analysis of the data may be done by any of a number of techniques ranging from the use of statistical functions on a hand calculator, with manual plotting of data, to the use of sophisticated statistical packages; spreadsheet[6] applications are also frequently used. Statistical analysis software is generally available for personal computers. The stability results should be provided in files addressable by the sponsor/FDA software. Statistical analysis can then be performed by the reviewer, either to confirm the sponsor's calculations or to attempt alternative modeling of the data. Graphing capabilities are generally integral to statistical analysis software.

Rearrangement of stability data should also be possible, allowing the reviewer to sort according to a variable, such as package type or storage condition. For this application, the stability data should be presented in a relational database.[7]

3. Value to drug sponsor

The ability to resort stability data could be of value to a drug sponsor by enabling the reviewer to obtain more readily an understanding of the full data set used for stability calculations. Reanalysis of stability data may be useful in situations where the reviewer wants to confirm a sponsor's conclusion. However, it is important that any assumptions and methods of analysis are fully understood by both reviewer and sponsor.

---

[6]An application that facilitates the manipulation of rows and columns of data.

[7]A data-base that can access two or more tables simultaneously by linking records on a common field.

4. **Value to the reviewer**

Reviewers have expressed interest in the ability to resort and reanalyze the data using statistical routines of their own choosing. Presenting the data in a relational data base and providing statistical software would be of great value because this would eliminate the need to re-enter large amounts of data into alternate application programs.

5. **Overall contribution of this feature**

For some sponsors and reviewers statistical software could provide a means for facilitating discussions on stability and expiration-dating. Potentially this is a very valuable feature that is unique to the electronic form. It is a feature that is likely to be increasingly desirable in the future.

G. Archival Data Storage

1. **Objective of the feature**

Provide the reviewer and sponsor with access to the entire contents and chronology of the NDA.

2. **Feasibility/limitations**

Optical disk storage allows the user to browse the pages that have been scanned onto the disk. A computer system and accessories would be required to provide the user with the features recommended and discussed in this report. A typical optical disk system includes an optical disk drive, high resolution monitor, laser printer, computer, and a modem[8] for electronic mail. A 12 inch optical disk can hold approximately 80,000 to 100,000 pages. This is equivalent to about 5,000 floppy disks. Obviously, the storage space required by both the sponsor and the FDA would be decreased significantly as compared to the space required for an equal amount of paper; however, the space needed for the workstation should be considered. The workstation can be a mobile unit and could be maintained in a central storage area when not in use.

An optical disk is durable, and repeated use will not adversely affect the quality of the image. Problems incurred by exposure of magnetic media to magnetic fields or electrical charges will not occur with an optical disk. The information stored on an optical disk cannot be accidentally erased, and it is considered an archival form.

---

[8]Abbreviation for modulator-demodulator, a device that converts computer output to audio tones for transmission and reverses the process on the receiving end.

Location and retrieval of information are considered to be two of the main benefits of a computerized submission. The more extensive the indexing (i.e., key word/phrases, Boolean logic), the easier for the user to locate information.

The extent of the indexing should be evaluated by the sponsor because it can be costly and time consuming.

3. **Value to the drug sponsor**

As indicated above, an optical disk requires much less storage space as compared to paper. Optical disks are also durable and cost effective.

4. **Value to the FDA reviewer**

Optical disk storage provides the reviewer with an exact replication of the paper NDA so the reviewer doesn't have to change his or her normal review process. With the availability of various computer software/hardware additions, the reviewer can perform a variety of tasks much easier than with a paper submission, such as making notes/comments, split screen access to multiple pages, earmarking information, and/or data analysis. Importantly, it is an archival form of storage and, unlike magnetic media, the information will not be changed with time. Just how long the medium will stay archival has not been determined.

Optical disk storage offers the ability to access the entire NDA, assuming the proper indexing has been provided. This is a considerable advantage for the review chemist because, for example, access to the body fluid drug assay method is often required as are other aspects of the NDA in sections other than the chemistry section.

5. **Overall contribution to drug product evaluation**

Optical disk storage provides the sponsor and the reviewer with easy and quick access to the entire NDA. The paper NDA is scanned onto the optical disk and the result is an exact image of the paper NDA accessible and visible through a high resolution monitor. Alternately, digitized information can be recorded directly on the disk. Computer systems provide the capability to do more than just browse the information. The software/hardware provided to the various divisions or reviewers could be tailored to that division/reviewer's needs.

As with any CANDA, the system relies on the sophistication of the software/hardware systems available and provided by the sponsor, and also the reviewer's ability and familiarity with the system.

CHAPTER XI

STATISTICAL CANDA GUIDELINES

A. Introduction and Background

This chapter describes the statistical review of NDAs and the experience of the Division of Biometrics with the implementation of CANDAs. In reviewing an NDA, the statistical reviewer focuses primarily on issues relating to study design and the statistical evidence provided by the sponsor.

In general, the primary requirements of a CANDA for statistical review are:

1. A database of all information collected on randomized patients;
2. Data management, analysis, and graphics program files; and
3. Electronic files of statistical reports.

All elements of a statistical CANDA (the database, the program files, and the text files) must be well documented and carefully linked to the NDA submission.

The content of statistical CANDAs should be discussed before submission with personnel and reviewing officials from the Division of Biometrics. Sponsors should contact Robert T.O'Neill, Ph.D., Food and Drug Administration, Center for Drug Evaluation and Research, HFD-710, 5600 Fishers Lane, Rockville, MD 10857. Dr. O'Neill's phone number is (301) 443-4594.

B. Statistical Reviews

With the completion of Phase III trials, sponsors submit the NDA to FDA for marketing approval. "New Drug and Antibiotic Regulations" published in the *Federal Register* on February 22, 1985 (50 FR 7452), commonly referred to as the NDA Rewrite, established a through review process designed to ensure the approval of drugs scientifically demonstrated to be safe and effective, and the disapproval of those drugs that are not. One of the measures implemented in the Rewrite was the requirement that sponsors submit separate NDA volumes to six different scientific review disciplines, one of which is statistics. This procedure facilitates the parallel review of NDAs by the disciplines involved in making regulatory decisions on the safety and efficacy of new drugs.

The specific format and content of the technical sections submitted to the various disciplines are discussed in their respective guidelines. The format and content of the statistical section of an NDA are discussed in a 1988 publication entitled, Guideline for the Format and Content of the Clinical and Statistical Sections of New Drug Applications.

The Guideline illustrates what medical and statistical reviewers would like to see in the submission and how they would like to see these contents presented. Clinical and statistical sections conforming to this guideline greatly facilitate the review of NDAs by medical officers and Statistical Evaluation and Research Branch (SERB) statisticians.

C. The Institutional Statistical Review Process

It is informative to examine the course of a statistical review after it has been received by the Division. When a new NDA is received by SERB, a group leader assigns the submission to a statistical reviewer and works closely with the reviewer to evaluate the evidence contained in an NDA and to produce the draft review. The draft review, in turn, receives a peer review by the group leader and branch director.

When an individual reviewer receives an assigned NDA, he or she often starts from scratch. He or she begins by reading the cover letter and index to determine the drug class, proposed indication, and the types of studies included in the NDA submission. Following this initial look,the reviewer locates and examines background information on the disease the drug is intended to treat, the properties of this class of drugs, the sponsor's drug developmental program and previous correspondence between the sponsor and FDA for this NDA.

Following this initial review of background materials, the statistical reviewer, generally,meets with the medical reviewer to obtain essential clinical perspective on the disease, the drug, and the studies. From here, he or she examines the integrated summary of effectiveness data, the integrated summary of safety information, and summaries of the controlled studies (detailed descriptions of these documents can be found in the Guideline for the Format and Content of the Clinical and Statistical Sections of New Drug Applications). Based on a review of these materials, supplemented by discussions with the medical officer, the statistical reviewer tentatively identifies those well controlled studies considered to contain primary scientific evidence relating to the efficacy and safety of the new drug. Such studies may or may not agree with the studies identified by the sponsor as "pivotal."

At this point, the reviewer is ready to begin the detailed evaluation of the individual study reports (entitled the "Full Integrated Clinical and Statistical Report of a Controlled Clinical Study" in the guideline), giving priority to the most important studies. From the protocol contained in each of these reports, the reviewer should be able to determine the primary objective of the study, the type of study design, the randomization scheme used, the degree of blinding, dose levels and frequencies, primary efficacy parameters, important safety parameters, and proposed methods of analyses. The reviewer must determine whether the study design can support the proposed indication and the adequacy of the sample size (particularly for active control trials).

In addition, the reviewer examines the randomization scheme and codes and the extent of error in treatment assignments. To the extent possible, he or she also examines the quality of the data: the extent of noncompliance, missing values, dropouts, etc. The reviewer also pays special attention to sources of bias, confounding, etc. If sources of bias or confounding are discovered, the reviewer must, at a later time, attempt to quantify these factors and assess the potential impact on the primary efficacy parameters.

The statistical reviewer must identify all of the data sets used by the sponsor in the analysis of primary and supportive secondary efficacy measures. Data sets that include, for example, evaluable patients, efficacy patients, patients randomized to a treatment, and those used for intent-to-treat analyses, should be identified. The reviewer must know the inclusion/exclusion criteria used in forming each of these data sets. If there are questions concerning inclusion/exclusion criteria, differential dropouts, reasons for dropouts,etc., the reviewer must determine the magnitude of potential biases and assess the impact of these biases on efficacy measures.

The reviewer also determines the types and appropriateness of the analyses performed by the sponsor to demonstrate efficacy and safety statistically. The statistical reviewer conducts reanalysis or alternative analysis, if necessary, on the same or modified data sets.

In the review of evidence relating to safety, the reviewer identifies the data sets used by the sponsor in presenting the safety profile of the drug (for example, a data set of all treated patients). Paralleling his review of the efficacy data, the SERB statistician must judge the appropriateness of the safety analyses and, if necessary, perform reanalyses or alternative analyses on the same or modified sets of data. (Incidentally, based on a recent informal survey of 53 NDAs conducted by SERB,it was reported that in 83% of the cases, reanalyses were performed.)

The statistical reviewer does this for each of the important studies. It should be noted, however, that the depth of this review varies from study to study and from one NDA to another depending on a variety of factors, such as the quality of the data, the conduct of the trial, the completeness of the individual study report, or any outstanding clinical or statistical issues.

After all of the pivotal studies and supportive secondary studies have been reviewed, the statistical reviewer, based on all of the evidence from these studies, assesses the overall safety and effectiveness of the drug for the proposed indication. A statistical review is then written to summarize the findings and to make recommendations based on the statistical evidence.

The statistician's review is then reviewed by his or her group leader. The group leader reviews the report for scientific adequacy and consistency with CDER and agency policy. Differences in opinion between the reviewing statistician and the group leader are discussed. Should the disagreement remain unresolved, the group leader records the basis for his or her disagreement and a recommendation for action by the branch chief.

The branch chief reviews the report and the group leader's recommendations for soundness of the statistical evaluations, conclusions, and recommendations, overall scientific content and standard, adherence to official policies and administrative procedures, and resolves all essential scientific and other issues. The report and all recommendations are then forwarded to the review team leader of the appropriate medical drug review division.

D. The Computer Assisted NDA Review

CANDAs are an on-going effort by the agency to make a more efficient and effective use of the computer as a regulatory review tools. It has two primary goals:

1. The reduction of review time, without sacrificing the quality of review in order to expedite regulatory decisions on new drugs; and

2. The reduction of the paper burden arising from large NDAs by creative applications of computer technology.

The agency has defined a CANDA rather generally as "any submission that involves provision of information to FDA or that permits remote access by FDA in any electronic medium."

CANDA activities to date have concentrated primarily on the design of prototype microcomputer CANDA systems designed to backup the paper submission of large NDAs. Sponsors, working with FDA reviewers, have implemented a variety of systems designed to enhance the ability of FDA scientists to access and review electronically the evidence presented in NDAs. Features of these systems have included, for example:

-- on-line access to a sponsor's mainframe computers;
-- Electronic messaging systems;
-- Access to a sponsor's program and data files on microcomputer data bases;
-- Optical disk systems for review of case report forms;
-- Data base management software for review-oriented subsetting and manipulation of submitted data;
-- Statistics, wordprocessing, and graphics software capable of verifying and reanalyzing sponsor's results and producing reviews.

A great deal has been accomplished to date, but it is clear that more energy must be devoted, if the objectives of CANDAs are to be reached. With an eye towards the future, it is useful to examine our expectations for this project. They include:

1. Conforming to the NDA Rewrite requirements,a user-friendly, enhanced version of the CANDA system, which effectively integrates both the clinical and statistical aspects of the review process for the across-study evaluation of submissions, should be developed.

2. The development of computer systems to access NDA clinical data within and across studies of different designs should require more *a priori* attention to what data are collected, how they will be stored, summarized,retrieved, and used by

industry and regulatory scientists. (This should improve the structure of clinical data systems and the accessibility of any suitable data subset.)

3. NDA Rewrite requirements for tables, data display, efficacy analysis, safety evaluation, and synthesis of effectiveness and safety findings on a new drug should enhance the role of CANDA within the context of both the clinical and statistical review processes.

4. Statistical software should be developed, specifically designed to validate and verify statistical analyses submitted in an NDA.

5. The NDA Rewrite is increasing the intensity of interaction among the primary, secondary, and tertiary statistical reviewers as part of the peer review process. Also, the integrated clinical-statistical guideline is improving interaction among the medical and statistical reviewers. Experimental innovations in the review process, such as "NDA Day," involve intensive interaction among all FDA reviewers of an NDA. Consequently, the CANDA review process will prove more efficient in performing the institutional review of an NDA, thereby reducing review time without sacrificing the quality of statistical reviews.

6. We are in an early, evolutionary stage in the use of microcomputers as review tools. More experience is needed in dealing with the specific concerns of different drug classes. Before we can propose a standardization of NDA data bases or procedures, intensive and extensive interaction between the FDA reviewers and drug sponsors should continue.

   This should result in the development of several strategies and designs appropriate to the drug and disease in the light of clinical and statistical review needs.

CHAPTER XII

COMPUTING ENVIRONMENTS

The following chapter on FDA's reviewing divisions are a combination of information created by the FDA, PAREXEL International Corporation and Mr. Robbin Nighswander. Most of the textual information was written several months ago and may be slightly outdated. For that reason, some of the PAREXEL profiles have been slightly edited for this publication. We sincerely appreciate PAREXEL's permission to use these descriptions.

These sections are reprinted from CANDA: A Regulatory, Technology, and Strategy Report, 1992,with the permission of the publisher, PAREXEL International Corporation, 195 West Street, Waltham, MA 02154.

The information on the Division of Neuropharmalogical Drug Products was adapted from a speech given to the Regulatory Affairs Professional Society's 16th Annual Meeting on September 24,1991, by Mr. Robbin Nighswander, Consumer Safety Officer in this Division.

CANDAs and FDA's DRUG REVIEW DIVISIONS

Like most everything else in thee drug review process, the degree of CANDA use, acceptance, and evolution will be encouraged at the FDA review division level. Although FDA Commissioner David Kessler and other top agency officials are not pushing the agency toward a more uniform CANDA vision, CANDA experimentation and use will be shaped, at least in the near term, by today's CANDA gatekeepers: FDA reviewers themselves.

And that is what makes the question "What does the FDA, want in a CANDA?" so difficult to answer.Unfortunately, there are no real answers to this question, at least not one right answer. For, as Pilot Drug Evaluation Staff Director, John Harter, has stated, "CANDA development today is more like 'skeet shooting' than target practice." An FDA medical reviewer added recently that today's CANDA use is the antithesis of a controlled trial, with the technologies, tools, approaches, reviewers, and reviewer experiences changing so quickly and continually that drawing firm and reliable conclusions about anything important is virtually impossible.

And where does all this leave sponsors hoping to develop a CANDA? In much the same place that a company developing a new drug finds itself in -- that is, to explore, discuss, and negotiate with the individual drug review divisions and the relevant drug reviewers to determine whether a CANDA should be submitted and, if so, what technologies, approaches, and analytical and automated tools should be incorporated to promote the most expeditious approval.

The FDA's two most visible and vocal advocates of CANDA uses -- Dr. Harter and Division of Cardio-Renal Drug Products Director, Raymond Lipicky -- each hold that the CANDA computing environment should be built around the needs,abilities, and preferences of the ultimate CANDA user, the drug reviewer. Unfortunately, for anyone attempting to reach broad characterizations about FDA's CANDA wants, these reviewer needs, abilities, and preferences are often vastly different.

In the profiles below, we attempt to provide some insights into the CANDA philosophies, approaches,and experiences of each of the FDA's nine drug review divisions. In doing so,we attempt to characterize to some degree, each division's CANDA experience, provide an overview of the computing environment within each division, and in some cases, outline divisional CANDA preferences.

The divisional profiles are based on information obtained from literature sources, speeches and presentations given by FDA officials, and informal talks with division officials. In many cases, the perspectives described in these profiles are the personal views, opinions, and experiences of FDA officials and reviewers. But, at least to the extent that the CANDA area is ow being driven by the abilities, perspective, and personal preferences of individual FDA reviewers rather than a divisional or agency-wide standard, this approach is entirely appropriate. In no case is the information presented here intended to represent or suggest an official policy or standard adopted by any single division.

The information presented from division to division is at times uneven, as are the CANDA experiences of each respective review unit. As one would expect, the review divisions with the most CANDA experience were able to provide significantly greater insights regarding CANDA preferences, experiences, and preferences than less experienced divisions.

A. Pilot Drug Evaluation Staff

(The Division of Anesthetic, Analgesic
and Drug Abuse Treatment Drug Products)

CANDA Submissions: 7
1991 CANDA Submissions: 0
CANDA Approvals: 4

As one might expect within a division whose mandate is to experiment with innovative methods to speed drug reviews, the CANDA has been an important vehicle in the Pilot Drug Evaluation Staff's approval strategies. And, while the division is not nearly the FDA's most CANDA-experienced, it is pioneering the use of CANDAs in new ways.

As one of the FDA's two most vocal and visible supporters of the computer-assisted review concept, Division Director Dr. John Harter has characterized the CANDA as "the optimum way to facilitate the review process." (1) Dr. Harter believes that CANDAs can speed drug reviews by quickening the reviewer data analyses and shortening or eliminating reviewer questions/request-sponsor response "recycle" times. Perhaps just as importantly, Dr. Harter has stated that CANDAs can provide sponsors with important insights into how the FDA conducts its reviews, insights that will lead to better submissions and reviewer-sponsor relationships in the future.

True to the "pilot" nature of his division, Dr. Harter hopes to carve a place for the CANDA in the drug development process not just the drug review process. Dr. Harter believes that CANDAs, if they are to reach their highest utility, should be incorporated into a drug's development, allowing FDA reviewers to see study results during clinical testing and to identify potential problems early, allowing them to be addressed during the drug development rather than the NDA stage. Dr. Harter calls this approach CAINDA, which he suggests might be a vehicle that might someday allow the FDA to review drug applications in the 180-day statutory period. Dr. Harter also wants CANDAs to be used during NDA Days, a merging of innovative approaches that he has stated worked well during the division's approval of Ciba-Geigy Voltaren.

Medical reviewers in all three of the division's therapeutic disciplines now have some experience with computer submissions. According to division officials, most of the division's CANDA projects are said to have featured interactive clinical databases, although some optical disk systems have been used. Like many other divisions, the medical reviewers here have found attractive aspects in each type of system -- the ability of optical disks to produce rapid response screen images of original CRFs and the ability to make data queries and reanalysis of data through clinical databases.

As in the other more CANDA-experienced review division, there are some indications of rising CANDA expectations by reviewers. Division officials indicate their ideals in terms of integrated optical disk-database systems and increased functionalities.

Division reviewers are mostly equipped with IBM-clone PCs although a few reviewers have conducted CANDA reviews using the Apple Macintosh. Regarding computer tools and access to equipment, officials put the division in "the middle of the pack" relative to other review units.

Division officials are confident that CANDA systems cut review times -- in both retrieving and accessing documents and in permitting reviewers to perform electronic searches and reanalyses that circumvent the need for more time-consuming request-response cycles with the sponsor. Dr. Harter has stated that, by reducing this recycling time and "thereby sustaining the momentum of the review"one to two years was cut from the division's review of Ciba-Geigy's Voltaren (diclofenac).

Dr. Harter holds that CANDAs will ultimately have profound effects on the nature of drug reviews, and will have serious implications for companies not actively participating in the CANDA age. In comments made during a 1990 interview with the Pharmaceutical Executive, Dr. Harter stated... "Computerized NDAs are going to make it easier for FDA reviewers to explore alternative explanations for research results, which will sometimes lead to identifying problems or issues that were overlooked when the NDA was prepared. To the extent that these issues are identified earlier in the process and are more easily resolved, everyone benefits. With a CANDA, it's easier for a company to see what FDA is doing in a review and why. The companies that have submitted CANDAs have learned a lot about FDA's medical-review process that they didn't know before, and it's improved our relationship with each of them. I would guess that these companies will change some of their procedures for preparing for future NDAs. CANDAs may also affect regular NDAs. As we discover better ways to look at data based on our experience with CANDAs, we will also ask those companies not using CANDAs to generate the same information. The genie is out of the bottle, in a sense, and it's going to be difficult to coax or force him back in. Those who don't have a genie will be at a definite disadvantage." This doesn't necessarily mean all applications have to be full CANDAs. There can be varying degrees of computer assistance from all data available.

About upcoming challenges for sponsors developing CANDAs, Dr. Harter stated in the 1990 interview: "There could be a better integration of CANDAs with word processing and graphics. Also, CANDAs will be more up-to-date, so that we won't have to search through revised paper documents to locate the most recent revisions -- which is often, like looking for a needle in the haystack. There are many technical advances we could put on a wish list, but the biggest change should be to figure out how to support a less limited environment and give people using different data bases, spreadsheets, and programs that ability to work with and share the same information computer systems should be adapted to the user, not the other way around."

B. Office of Drug Evaluation I

Director, Robert J. Temple, M.D.
(301) 443-4330
FAX NUMBER: (301) 443-1653

CANDA Contact: Jane Peterson
(301) 443-4330

Government Owned Microcomputers:
15
AST386
AST486
MacIntosh II
Dell 316

Software:

WordPerfect 5.1
Coterm
dBase
Lotus 1-2-3
Windows
Paradox
Microsoft Word

Approximately 95% of the staff in the Center utilize WordPerfect 5.0, WordPerfect 5.1, or WordPerfect 2.0 (Macintosh) for word processing. Those with version 5.0 are in the process of being upgraded.

There are some users using Microsoft Word, Microsoft Works.

All microcomputers use Coterm. There are a significant number of the staff using Harvard Graphics, Lotus 1-2-3, Windows, Paradox, and dBase. A large number of other software packages are also used.

C. Division of Cardio-Renal Drug Products

CANDAs Received: 15
1991 CANDAs Received: 0
CANDA Approvals: 7

From the beginning of the electronic new drug application (NDA) age, the Division of Cardio-Renal Drug Products has been the FDA's pioneering division in computer submissions. The division participated in the first CANDA experiments -- Abbot's electronic filings for terazosin and cartelol. In the early years, the division handled five of the first 12 CANDAs to obtain agency approval and, even as late as 1989, the division had under its purview as much as half of the CANDAs in the FDA's review pipeline.

The division has a clear computer orientation that starts at the top. Division Director, Dr. Raymond Lipicky, has been, and continues to be, one of the more visible FDA officials on CANDA issues and one of the more vocal proponents of the usee of computer technology in the review and approval of new drugs. Deputy Director, Dr. Robert Fenichel, holds a Ph.D. in Computer Sciences from Harvard University in addition to a medical degree. And, in the division's medical review corps, the unit has at least one well-known "computer jock."

As is true within over review units, most of the division's CANDAs have been clinical in focus. Division officials point out that pharmacologist/toxicologist and chemists were the last within the division to be equipped with computers, an outfitting that was completed roughly a year ago. Most of the division's medical reviewers have participated in at least one CANDA and one reviewer has had two.

As one might expect from the most CANDA-experienced division, the unit has significantly more developed thoughts and perspectives on CANDA issues. Perhaps more than any other, the division has made the transition from having no or limited computer tools and looking to industry for assistance to possessing and achieving a confidence level with computer tools and wanting industry to follow this lead by presenting data in a manner consistent with those tools. Division officials caution that reviewers are much less willing to accept data in technological "black boxes" that require learning new tools.

The CANDA policy implications from this transition, say division officials are that sponsors should consider providing data in the "lowest common denominator" formats -- such as large ASCII files -- so reviewers can dump data into the computer and analytical tools with which they are most comfortable. One suggestion: usee dBase database format, since all database tools can read data in that format, and Lotus 1-2-3 spreadsheet format for the same reason. This general approach, say division officials, should ultimately lead to sponsors being able to prepare CANDAs without knowing which reviewers will be reviewing their CANDA.

Regarding tools, division officials claim that most division reviewers have settled on spreadsheet, database, and word processing tools. Although most reviewers are using WordPerfect as a word processing program, they are said to be using at least three different spreadsheet packages (i.e., Quattro, Excel, and Lotus 1-2-3) and at least two database packages (i.e., dBase and PARADOX). In terms of hardware, division reviewers are working on IBMs, IBM clones, and Apple Macintoshes.

Division reviewers are much less settled regarding graphics processing tools, probably because there is no recognized standard in the field. The division has no high resolution graphics tools.

While division officials stress the lowest common denominator approach to CANDA data submissions, they are also quick to point out that differences between reviewers persist and that some degree of customization is still needed. Some of the division's medical reviewers have considerable statistical sophistication, for instance, and want to see raw data, while other reviewers need pre-programmed menu screens. All medical reviewers, though, want access to the information and an ability to manipulate it in some form. Specifics would need to be worked out with the individual reviewer.

Division officials have much to say about optical disk based systems and interactive databases. Dr. Lipicky has stated that division reviewers like the ability to view screen images of original case report forms, to browse through the documents and that all CRFs should be submitted that way. He has also pointed out that his reviewers want full text, cut and paste, and data analysis capabilities.

Interactive clinical databases seem to contribute more to the primary drug review process, although division officials express concern about response times, which some officials predict will ultimately make remote access databases unpopular with reviewers. Because reviewers are becoming accustomed to rapid response tools, they suggest, response delays associated with off-site databases are becoming increasingly distasteful to reviewers. But Dr. Lipicky says this may not be true for all systems because response speeds vary.

A clear trend toward rising reviewer expectations for CANDAs is also evident in other areas, division officials claim. This means that certain reviewers might be reluctant to use CANDAs that offer less in terms of functionality, querying response time, or user friendliness than their past projects, for example. These expectations, they point out, probably vary greatly between reviewers.

At the same time reviewers are expecting more from CANDAs, division officials report that they have revised their expectations regarding the ultimate impact of CANDAs on the drug approval process. Dr. Lipicky stated in July 1990 that expediting drug reviews through CANDAs should not be the "objective" and, further, that the CANDAs ability to cut drug approval times would "never be assessable" even though it is intuitively obvious that review time could be altered. Because they feel strongly that CANDAs improve the quality of drug reviews, division officials do not believe that these "revised" expectations will reduce the incentive to participate in CANDA projects.

Also factoring into these expectations is the division's realization that CANDAs are not likely to help equally at each level of a drug's review. Obviously, CANDAs have been found to offer the most at the primary review level -- since the primary medical reviewer spends several months sifting through mountains of data, a week invested in start up for a CANDA may be a good time investment provided the CANDA provides advantages and assistance during the review process. Division Director and Deputy Director have a more difficult time seeing the advantage of stand-alone systems at the secondary review level since these reviewers may complete their role in a few to several weeks. Division officials of Office of Drug Evaluation I/II claim that it is "impossible" to see any CANDA advantage at the tertiary review level. Dr. Lipicky does see value in a CANDA system linked to the secondary or tertiary reviewers own microcomputer but it has never been implemented. Such a system could be useful to all involved in the approval process. This is true only for independent stand-alone systems.

D. Division of Neuropharmacological Drug Products

CANDAs Received: 8
1991 CANDA Received: 3
CANDA Approvals: 1

The Division's experience with CANDAs was limited until relatively recently. Early attempts at CANDA submissions were not particularly useful, either because the CANDA itself was of limited value or because the CANDAs were submitted well into the review of the hardcopy NDA, making the electronic submission less of a factor in the review.

The Division of Neuropharmacological Drug Products' recent experience with CANDAs is more extensive than the 8 CANDAs reported, because several of the CANDAs have either been submitted more than once, or they have included both clinical and manufacturing and control components. The Division has also been very active in examining ways to eliminate the need for sponsor supplied equipment at each reviewer's desk and has, on three occasions, utilized the Division's LAN (local area network) to connect to off-site mainframes. This, in addition to eliminating extra equipment in each office, allows for more than one reviewer (or supervisor) to have access to the CANDA.

CANDAs submitted to the Division have included both optical disk components and CANDAs which provide for direct access and analysis of the clinical database of the NDA. CANDAs which allow for *ad hoc* review of the clinical trial safety data, both within and across clinical trials, are by far the most useful to the Division. However, the biggest disadvantage seen so far with these applications has been the relatively long learning curves required for their use. It is strongly urged that vendors increase their efforts to provide logical, user-friendly interfaces to these safety databases.

Optical disk CANDAs have been found to be particularly useful in automating the task review of case reports forms (particularly for indexing and encyclopedic functions). However, use of the optical disk for primary reading of the NDAs has not been found preferable over paper. The future appears to lie in some type of combined/integrated application using the clinical database as platform and optical disk support for the case report forms (although the Division hasn't yet received one).

As yet, the Division has not received efficacy data for review in a CANDA; however, thee Division encourages sponsors to submit efficacy data (formatted for use with Agency statistical review packages) on floppy diskettes.

Features which have been routinely offered to reviewers for CANDAs which have not been found to be particularly useful include electronic mail, file transfer, and the ability to search text of the NDA image (in optical disk applications).

E. Division of Oncology and Pulmonary Drug Products

CANDA Submissions: 3
1991 CANDA Receipts: 1
CANDA Approvals: 0

Like to Division of Medical Imaging, Surgical, and Dental Drug Products, the Division of Oncology and Pulmonary Drug Products may be a division to watch in the CANDA area.

Although the division has received only three CANDAs, it may now be the FDA's most eager to participate in electronic submissions: officials concede that one of the division's highest priorities is to accelerate involvement with computer submissions. This acceleration, promise officials, will be quick and obvious. Division Director Dr. Gregory Burke and the unit's two lead medical reviewers are said to be pushing the agency's CANDA agenda.

The division may also now be the most aggressive in recruiting CANDA filings. Reportedly, the division has suggested to selected firms that they submit their NDAs in an electronic format. Presently, the division is in early negotiations with two firms considering CANDA submissions.

Division officials claim the slow start in the CANDA area is related to a few factors, including a lack of computer equipment and technological capabilities. The point out, however, that computerization has been the division's top priority during the past two years.

A difficult early CANDA experience may also have been a factor. A complicated series of events helped to complicate what might have been the unit's first CANDA experiment, a project based on a computerized clinical database. As part of a 1989 reorganization that brought pulmonary and oncology drugs under the authority of a single division, a

CANDA for a pulmonary drug was shifted to what is now the Division of Oncology and Pulmonary Drug Products. Unfortunately, say division officials now, the new division management had little input in the planing and development of the CANDA, which was submitted just after the reorganization. The result: a computer submission that the reviewing division was ill-prepared to review.

According to officials, both the division's principal therapeutic disciplines -- pulmonary and oncology -- are now CANDA experienced. The division is said to have experience with CANDA based on optical disk systems and interactive clinical databases.

Despite the division's slow start, reviewer computer literacy and accessibility is now characterized as good and improving. Division officials report that computer literacy and receptivity are now divisional hiring criteria for new medical reviewers.

As always, Division officials prefer to discuss each application very early in the developmental process.

F. Division of Medical Imaging, Surgical, and Dental Drug Products

CANDAs Received: 8
1991 CANDAs Received: 3
CANDAs Approved: 3

In May 1991, the Division of Medical Imaging, Surgical, and Dental Drug Products' CANDA receptivity and philosophy changed dramatically. That was when Dr. Wiley Chambers, a medical reviewer within the Division of Anti-Infective Drug Products, was appointed the Acting Director of the Division of Medical Imaging, Surgical, and Dental Drug Products. Dr. Chambers' computer orientation is well-established: he has personally reviewed five CANDAs and one of his three faculty appointments at George Washington University is in computer medicine.

Clearly, the division will have a far different CANDA approach than that evident under former division director Dr. John Palmer. Because the division regulates drugs whose clinical studies typically include fewer patients and less of a paper load compared to other drugs, Dr. Palmer held that his unit had less to gain from CANDA involvement. Not surprisingly, the division did not recruit a single CANDA under Dr. Palmer's tenure.

Although Dr. Chambers promises a more pro-active CANDA approach, he concedes that the division's CANDA preparedness is "low" in terms of both equipment availability and computer literacy. The only CANDA experience the division has was his own -- he reviewed five CANDAs as the FDA's only ophthalmologist (editor's note: the Division of Medical Imaging, Surgical and Dental Drug Products recently hired two other ophthalmologists, each of whom are actively requesting CANDA submissions). Each of those CANDAs involved the submission of clinical/statistical databases filed on diskettes for review on a 386 personal computer.

As part of these CANDAs, Dr. Chambers did not establish CANDA format specifications, but told sponsors to submit their databases in their existing electronic formats. Therefore, the data were filed in a variety of spreadsheets, SAS formats, and database formats.

Dr. Chambers has offered to provide general and technical guidance to any of his reviewers willing to accept a CANDA. All division reviewers are said to be inexperienced with CANDAs. Since Dr. Chambers joined the division, officials report that at least "a few" companies have approached the division about submitted CANDAs.

G. Division of Gastrointestinal and Coagulation Drug Products

Director, Stephen B. Fredd, M.D.
(301) 443-0479
Supervisory CSO, Thomas H. Hassall
FAX NUMBER: (301) 443-9285

CANDA Contact: Annice Bergeris
(301) 443-0479

Government Owned Microcomputers:
28

Software:

Review divisions have a mix of IBM and IBM compatible microcomputers with one or two Macintosh microcomputers in most divisions. Virtually all microcomputers are connected to Ethernet.

The Division of Gastrointestinal and Coagulation Drug Products has all microcomputers connected to a 3COM LAN.

Approximately 95% of the staff in the Center utilize WordPerfect 5.0, WordPerfect 5.1, or WordPerfect 2.0 (Macintosh) for word processing. Those with version 5.0 are in the process of being upgraded.

There are some users using Microsoft Word, Microsoft Works, MultiMate, and some using WPS Plus on the VAXcluster. The latter are primarily using DEC or DEC compatible terminals.

Virtually all microcomputers use Coterm. There are a significant number of the staff using Harvard Graphics, Lotus 1-2-3, Windows, Lotus Agenda, Paradox, Quattro Pro, dBase, PCNONLIN, SAS, Statgraphics, Systat, Statexact, and BMDP. A large number of other software packages are also used for analysis.

H. Office of Drug Evaluation II

Director, James M. Bilstad, M.D.
(301) 443-2544
FAX NUMBER: (301) 443-4467

Government Owned Microcomputers:
18

Software:

Review divisions have a mix of IBM and IBM compatible microcomputers with one or two Macintosh microcomputers in most divisions. Virtually all microcomputers are connected to Ethernet.

Approximately 95% of the staff in the Center utilize WordPerfect 5.0, WordPerfect 5.1, or WordPerfect 2.0 (Macintosh) for word processing. Those with version 5.0 are in the process of being upgraded.

There are some users using Microsoft Word, Microsoft Works, MultiMate, and some using WPS Plus on the VAXcluster. The latter are primarily using DEC or DEC compatible terminals.

Virtually all microcomputers use Coterm. There are a significant number of the staff using Harvard Graphics, Lotus 1-2-3, Windows, Lotus Agenda, Paradox, Quattro Pro, dBase, PCNONLIN, SAS, Statgraphics, Systat, Statexact, and BMDP. A large number of other software packages are also used for analysis.

I. Division of Metabolism and Endocrine Drug Products

CANDAs Received: 4
1991 CANDAs Received: 2
CANDAs Approved: 1

With four CANDA receipts as of August 31, 1991, the Division of Metabolism and Endocrine Drug products has limited experience with electronic filings. There are at least superficial signs pointing to greater involvement: the division's two 1991 CANDA receipts are the most of any review unit.

Despite the division's CANDA receipt count, the unit is experienced in both optical disk systems and interactive clinical databases. Although there is some sentiment that CANDA systems without data querying capabilities have proven to have little utility in the review, others point out that there is a relatively wide range of experiences and opinions in the value of each type of system.

Most of the division's CANDA -experienced medical reviewers are characterized as being enthusiastic about CANDAs and their potential, with a minority finding computer formats to be less than useful.

Division officials are not willing to say that the CANDA experiments have necessarily reduced review times. Reviewers concede that CANDAs do contribute in several ways: (1) allowing the reviewers to see the data in different ways -- that is, permitting the reviewer to manipulate and reanalyze the data, thereby circumventing the need for lengthier interactions with sponsors; (2) allowing reviewers to become "more comfortable" with the data and the resultant decisions; and (3) cutting the time generally needed for the question-response cycle between the division and the sponsor.

The division's computing environment seems relatively advanced, with all reviewers having IBM PCs or IBM clones -- the division recently acquired its first Macintosh. Most reviewers work on WordPerfect word processing software, although a few use selected other packages. The minority of reviewers actively using spreadsheet packages are using Lotus 1-2-3 or Excel. Those reviewers using database tools are functioning on dBase III or dBase IV, with some having limited experiences with other database tools.

As in other divisions, the CANDAs submitted to date have involved the medical/statistical portions of the subject NDAs. All division reviewers are PC-equipped, although officials concede that not all reviewers are positively disposed to CANDAs.

J. Division of Anti-Infective Drug Products

CANDAs Received: 13
1991 CANDAs Received: 1
CANDAs Approved: 4

With 13 CANDA receipts as of August 31, 1991, the Division of Anti-Infective Drug Products is the FDA's second most experienced.

Division officials credit two main factors for the division's high level of CANDA involvement. First, division reviewers are said to have a powerful incentive to computerize: NDAs in the Anti-Infective area tend to be significantly larger than most because of large study sizes -- often 2,000 to 10,000 -- and because many of these NDAs seek approval for multiple indications, all of which must be established in the submission.

Electronic reviews have also served as a useful adjunct to the Government's "flexiplace" program, under which approximately one-third of the division's reviewers work at home two or three days a week. To take advantage of what reviewers characterize as a higher productivity environment, sponsors have installed entire CANDA systems at reviewers' homes, complete with modems and other communications equipment.

Division officials also point out that the division has a crop of younger medical reviewers, many of whom joined the division when the CANDA concept was gaining momentum in the mid-to-late 1980's. They also report that the medical reviewers joining the unit within the last 18 months seem to possess even higher levels of computer sophistication. Given these factors, division officials report that a majority of the unit's clinical reviewers are receptive to CANDAs, and that all medical reviewers are equipped with IBM 386 or IBM 486 personal computers.

The division has received CANDAs employing the full spectrum of automated technologies, from simple WordPerfect files submitted on a diskette, to more recently, a microcomputer-based CANDA integrating text, imaging, and ad hoc database querying systems. According to officials, most projects have featured interactive clinical databases, although a few optical disk systems have been used. The CANDA projects have focused on the clinical area, although one recently submitted application had the microbiology/chemistry, manufacturing, and controls section in electronic format.

As expected from a division more experienced than most in computer projects, division officials have a relatively well developed idea of preferable CANDA approaches. Among the specifications mentioned by division officials:

- o systems that can function in IBM-compatible personal computers;
- o documents in, or compatible with, WordPerfect 5.1, in which 90 percent of the division's reviewers are said to be working;
- o an interactive database system providing reviewer accessibility (and retrievability) to as many pieces of patient data as possible;
- o spreadsheet capability with some basic statistical capability;
- o at a minimum, the study summary and integrated safety summary accessible for cut and paste capability.

There is also some evidence of rising CANDA expectations within the division. The division is now reviewing what is clearly one of the most advanced CANDA, an Upjohn filing that integrates text, imaging, and a powerful ad hoc database query system. This filing is said to be receiving high praise from division reviewers and similar systems may soon be at the top of the division's CANDA wish list.

Division officials are sensitive to the fact that not all sponsors or all CANDAs are necessarily able to fulfill all divisional specifications. They also recognize that several CANDA details -- such as what information each reviewer wants one screen imager -- must be worked out between the sponsor and the reviewer.

Division Director, Mr. Murray Lumpkin, says that the therapeutic area analyses in Chapter Seven of the CANDA Guidance Manual contains more current information.

K. Division of Anti-Viral Drug Products

CANDAs Received: 5
1991 CANDAs Received: 0
CANDA Approvals: 2

Since it evaluates AIDS drugs, the Division of Anti-Viral Drug Products has perhaps the greatest incentive to identify and employ innovative means to expedite drug reviews. Those very important and public responsibilities, however, also dictate that the division remain cautious about accepting CANDA projects that may, due to large up-front time investments, present the possibility of lengthening the review.

Still, division officials characterize themselves as being generally receptive to CANDAs, provided that they are convinced the electronic submission will facilitate the review.

The division is experienced in both optical disk systems and clinical database submissions. As is true in other divisions, these projects have focused on the clinical area. The division has also been active in peripheral activities, such as accepting and revising draft labeling on diskettes and electronic mail hook-ups with sponsors.

The utility of interactive clinical databases, say division officials, is variable, and relates to both the capacity of the computer system and the nature and size of the clinical studies useful for the presentation and analysis of results from an "extremely well-controlled" trial of moderate size, say division officials, but might not be valuable for analyzing a smaller, uncontrolled trial. They also believe that the utility of databases for large trials, such as those run under expanded access protocols, might be useful, provided the system has adequate capacity to provide sufficient responsiveness to adverse experience inquires and other data queries.

Based on division experiences to date, division officials express concern about system functionality and computer documentation for any potential CANDA submission. In one case, officials reported that problems occurred because the medical reviewer was identified late in the CANDA development process, thus limiting the degree of system customization.

Computer literacy within the division is also described as "variable" by officials. All reviewers are equipped with IBM compatibles or Apple Macintosh's, although all medical reviewers are not experienced with computer submissions.

L. Office of Research Resources

Division of Biopharmaceutics
Director, Henry J. Malinowski, Ph.D. (Acting)
(301) 443-1223
FAX NUMBER: (301) 443-4518

CANDA Contact: See Y. Lam, Ph.D.
(301) 443-8098

Government Owned Microcomputers:
53

Software:

Review divisions have a mix of IBM and IBM compatible microcompuers with one or two Macintosh microcomputers in most divisions. Virtually all microcomputers are connected to Ethernet. In addition, the Division of Biopharmaceutics has a SUN SparcStation I+ and a SUN SparcStation II.

Approximately 95% of the staff in the Center utilize WordPerfect 5.0, WordPerfect 5.1, or WordPerfect 2.0 (Macintosh) for word processing. Those with version 5.0 are inthe process of being upgraded.

There are some users using Microsoft Word, Microsoft Works, MultiMate, and some using WPS Plus on the VAXcluster. The latter are primarily those using DEC or DEC compatible terminals.

Virtually all microcomputers use Coterm. There are a significant number of the staff using Harvard Graphics, Lotus 1-2-3, Windows, Lotus Agenda, Paradox, Quattro Pro, dBase, PCNONLIN, SAS, Statgraphics, Systat, Simusolv, NONMEM, Statexact, and BMDP. A large number of other software packages are also used for analysis.

M. Office of Epidemiology and Biostatistics

Division of Biometrics
Director, Robert T. O'Neill, Ph.D.
(301) 443-4594
FAX NUMBER: (301) 443-9279

CANDA Contact: Paul Hepp
(301) 443-1640

Government Owned Microcomputers:
72

Software:

Review divisions have a mix of IBM and IBM compatible microcomputers with one or two Macintosh microcompuers in most divisions. Virtually all microcomputers are connected to Ethernet. In addition, the Division of Biometrics has a SUN SparcStation I.

Approximately 95% of the staff in the Center utilize WordPerfect 5.0, WordPerfect 5.1, or WordPerfect 2.0 (Macintosh) for word processing. Those with version 5.0 are inthe process of being upgraded.

There are some users using Microsoft Word, Microsoft Works, MultiMate, and some using WPS Plus on the VAXcluster. The latter are primarily those using DEC or DEC compatible terminals.

Virtually all microcomputers use Coterm. There are a significant number of the staff using Harvard Graphics, Lotus 1-2-3, Windows, Lotus Agenda, Paradox, Quattro Pro, dBase, PCNONLIN, SAS, Statgraphics, Systat, Statexact, and BMDP. A large number of other software packages are also used for analysis.

N. Office of Generic Drugs

Director, Roger L. Williams, M.D.
(301) 295-8340
FAX NUMBER: (301) 295-8183

CANDA Contact: Jonathan D. Cook
(301) 295-8375

Government Owned Microcomputers:
130

Software:

The Office of Generic Drugs has a MicroVAX based local area network which is connected to the VAXcluster via Ethernet. (The Center is experimenting now in some areas using the VAXcluster as a file server for a local area network. It is expected that most of the Center microcomputers will be on DEC based local area networks in the future.)

All of the staff in OGD utilize WordPerfect 5.1, or WordPerfect 2.0.

Virtually all microcomputers use Coterm. There are a significant number of the staff using Harvard Graphics, Lotus 1-2-3, Windows, Lotus Agenda, Paradox, Quattro Pro, dBase, PCNONLIN, and SAS. A number of other software packages are also used for analysis.

O. Division of Chemistry I

Director, Rashmikant Patel, Ph.D.
(301) 295-8375
FAX NUMBER: (301) 295-8180

CANDA Contact: Dave Holovac
(301) 295-8375

Government Owned Microcomputers:
34

Software:
Review divisions have a mix of IBM and IBM compatible microcomputers with one or two Macintosh microcomputers in most divisions. Virtually all microcomputers are connected to Ethernet.

Approximately 95% of the staff in the Center utilize WordPerfect 5.0, WordPerfect 5.1, or WordPerfect 2.0 (Macintosh) for word processing. Those with version 5.0 are in the process of being upgraded.

There are some users using Microsoft Word, Microsoft Works, MultiMate, and some using WPS Plus on the VAXcluster. The latter are primarily those using DEC or DEC compatible terminals.

Virtually all microcomputers use Coterm. There are a significant number of the staff using Harvard Graphics, Lotus 1-2-3, Windows, Lotus Agenda, Paradox, Quattro Pro, dBase, PCNONLIN, SAS, Statgraphics, Systat, Statexact, and BMDP. A large number of other software packages are also used for analysis.

P. Division of Chemistry II

Director, Michael G. Beatrice
(301) 295-8305
FAX NUMBER: (301) 295-8174

CANDA Contact: Dave Holovac
(301) 295-8375

Frank Holcombe, Jr., Ph.D.

Government Owned Microcomputers:
30

Software:

Review divisions have a mix of IBM and IBM compatible microcomputers in most divisions. Virtually all microcomputers are connected to Ethernet.

Approximately 95% of the staff in the Center utilize WordPerfect 5.0, WordPerfect 5.1, or WordPerfect 2.0 (Macintosh) for word processing.

There are some users using Microsoft Word, Microsoft Works, MultiMate, and some using WPS Plus on the VAXcluster. The latter are primarily those using DEC or DEC compatible terminals.

Virtually all microcomputers use Coterm. There are a significant number of the staff using Harvard Graphics, Lotus 1-2-3, Windows, Lotus Agenda, Paradox, Quattro Pro, dBase, PCNONLIN, SAS, Statgraphics, Systat, Statexact, and BMDP. A large number of other software packages are also used for analysis.

Q. Division of Bioequivalence

Director, Shrikant B. Dighe, Ph.D.
(301) 295-8290
FAX NUMBER: (301) 295-8181

CANDA Contact: Dave Holovac
(301) 295-8375

Government Owned Microcomputers:
35

Software:

Approximately 95% of the staff in the Center utilize WordPerfect 5.0, WordPerfect 5.1, or WordPerfect 2.0 (Macintosh) for word processing. Those with version 5.0 are in the process of being upgraded.

There are some users using Microsoft Word, Microsoft Works, MultiMate, and some using WPS Plus on the VAXcluster. The latter are primarily those using DEC or DEC compatible terminals.

Virtually all microcomputers use Coterm. There are a significant number of the staff using Harvard Graphics, Lotus 1-2-3, Windows, Lotus Agenda, Paradox, Quattro Pro, dBase, PCNONLIN, SAS, Statgraphics, Systat, Statexact, and BMDP. A large number of other software packages are also used for analysis.

Chapter 10

BIOLOGICS INSPECTIONS

OVERVIEW

Biologics and blood products are being increasingly scrutinized by the FDA, primarily because of the seriousness of potential contamination of these products. Validation of manufacturing processes, especially automated processes, therefore, has become a major issue in inspections of biologics and blood establishments, as can be seen in the number of Warning Letters issued to these establishments and other regulatory actions taken by the FDA.

The documents which follow all relate to FDA's regulation and inspection of biologics and blood products. These documents are:

1. Reviewer Guidance, Computer Software
2. Guide to Inspections of Source Plasma Establishments
3. Recommendations to Users of Medical Devices that Test for Infectious Disease Markers by Enzyme Immunoassay (EIA) Test Systems
4. FDA Compliance Program Guide 7342.002 -- INSPECTION OF SOURCE PLASMA ESTABLISHMENTS
5. Points to Consider -- Computer Assisted Submissions for License Applications

Reviewer Guidance, Computer Software

SUMMARY

This device is used for processing blood and blood components. Software controlled hardware, hardware and software interfaces, and peripheral devices are used for data input. Software controlled hardware is used for data storage. Software controlled hardware is used for data output.

LEVEL OF CONCERN

When determining the level of concern, we are concerned about the nature of the device, its indications for use, and the severity of harm that the device could inflict or permit as a result of failure.

PRODUCT CLASS

Class I, II& III [incl. 81MMH Software, Blood Bank, Stand Alone Products] as required.

PRIMARY REFERENCES

21 CFR 807.87 Information required in a premarket notification submission

21 CFR 820 Good Manufacturing Practices for Medical Devices; General

Items I.-V <u>Reviewer Guidance for Computer Controlled Medical Devices Undergoing 510(k) Review</u> August, 1991. FDA, Center for Devices and Radiological Health

PRIMARY AREAS OF CONCERN

I. HAZARD ANALYSIS

A hazards analysis, which provides a description of the hazards presented by the device to the patient, the causes of those hazards, and the methods used to eliminate or mitigate them. This analysis should specifically identify the system hazard, and the components, whose failure could cause those hazards, which are controlled by or interact with software. The analysis should identify this controlling or interacting software, and describe how errors in this software are controlled or mitigated throughout the software development process.

II. FUNCTION REQUIREMENTS AND SYSTEM SPECIFICATIONS

A succinct description of the system and software requirements and specifications, including safety critical functions implemented because of the ongoing hazards analysis and any applicable algorithms.

III. SOFTWARE DESIGN AND DEVELOPMENT

A reasonable detailed description of the software design and development process, that is, written procedures for designing and developing the software to be used in a device, from concept to delivery to the customer. This should include, but is not limited to,a reasonably detailed description of how the following are addressed: specifications/requirements, design, implementation, verification and validation, and maintenance.

IV. VERIFICATION, VALIDATION AND TESTING

A reasonably detailed description of the software verification, validation and testing process, including but not limited to: the techniques and methods used at the module, integration and system level; the testing strategies and methodologies; and, the test acceptance and completion criteria. Include examples and some documentation of the testing results.

V. REVISION CONTROL

A detailed description of revision control procedures.

Reviewer Guidance, Computer Software

SUBMISSION CONTENTS

Item	Description
	21 CFR 807.87 Information required in a premarket notification submission
(a)	Device name, including trade or proprietary name and common or usual name?
(b)	Establishment Registration number?
(c)	Device class determination?
(d)	Has the manufacturer complied with the requirements for performance standards?
(e)	Do the proposed labels, promotional materials, package inserts, etc. include:
	a. The name of the device;
	b. Its intended use;
	c. Directions for use.
(f)	Is there a statement of substantial equivalence with supporting data? If no, must ask for it in review.
(g)	Is there a statement in the submission regarding the safety and effectiveness of the device?
(h)	Other information that should be present in the labeling for these devices are: a. Manufacturing protocols & specifications b. Performance data demonstrating reproducibility of the device;
(i)	Is there a statement in the submission regarding truthfulness and accuracy?
	Reviewer Guidance for Computer Controlled Medical Devices Undergoing a 510(k) Review
	Hazard Analysis
I.	a. a description of the hazards presented by the device to the patient, the causes of those hazards, and the methods used to eliminate or mitigate them.
	b. specifically identify the system hazard, and specifically identify the components, whose failure could cause those hazards, which are controlled by or interact with software.
	c. identify controlling or interacting software, and describe how errors in this software are controlled or mitigated.

Item	Description
II.	Functional Requirements and System Specifications
	a. the system and software requirements, and
	b. the system and software specifications, including
	c. safety and critical functions implemented because of the ongoing hazards analysis and any applicable algorithms.
III.	Software Design and Development written procedures for designing and developing the software including but not limited to, a reasonably detailed description of:
	a. specifications/requirements
	b. design,
	c. implementation,
	d. verification and validation, and
	e. maintenance.
IV.	Verification, Validation and Testing process, including but not limited to:
	a. the techniques and methods used at the module, integration and system level;
	b. the testing strategies and methodologies; and,
	c. the test acceptance and completion criteria.
	d. Examples and some documentation of the testing results.
V.	Revision Control, detailed description of revision control procedures.

GUIDE TO INSPECTIONS OF SOURCE PLASMA ESTABLISHMENTS

December 1994

The Division of Field Investigations
Office of Regional Operations
Office of Regulatory Affairs
Food and Drug Administration

INTRODUCTION

The "Guide[1] to Inspections of Source Plasma Establishments" is a consolidation of information previously provided in the Plasmapheresis Checklist and Report, and the Plasmapheresis Inspection Checklist Instructions, FDA-2722. This guide, which provides the most updated interpretation of certain regulations and guidelines, was prepared by the FDA, Office of Regulatory Affairs and the Center for Biologics Evaluation and Research.

The Plasmapheresis Checklist and Report, and the Plasmapheresis Inspection Checklist Instructions, FDA-2722, are no longer in effect. Consequently, field investigators are no longer required to fill out the checklist during establishment inspections nor submit it with inspectional reports.

The checklist and instruction booklet were last published in November 1990. Since November 1990, CBER has issued a number of memoranda to industry, depicting new recommendations, and modifications to previous guidance, in addition to draft guidelines on quality assurance and validation of computer systems. Although the agency, at the time of this publication, is in the process of revising the regulations on blood and blood products, it will take some time before such revisions are in effect. The current guide provides interim inspectional direction.

This reference provides the most updated interpretation of certain regulations and guidelines. This reference is not intended to be a "How to ...", it is a technical reference and is intended to be used in conjunction with the Inspection Operations Manual (IOM), the Code of Federal Regulations, Title 21 (21 CFR), the Compliance Program for the Inspection of Source Plasma Establishments (CP 7342.002) and the Compliance Policy Guides for biologics (CPG 7134).

In several instances the manual refers to memoranda published by CBER and sent to registered blood establishments. These memoranda should be available at the FDA District Offices; if not, a copy can be obtained from CBER, Division of Inspections and Surveillance, Office of Compliance (HFM-650), (301) 594-1191.

The preparation of products for which there are no published Additional Standards must be described in the establishment's SOP manual and manufactured in accordance

---

[1]Note: This document is reference materials for investigators and other FDA personnel. The document does not bind FDA, and does not confer any rights, privileges, benefits, or immunities for or on any person(s).

with the methods therein. The investigator should offer no advice or recommendation to the manufacturer regarding the preparation of such products. Questions concerning practices which may be hazardous should be addressed to the Division of Inspections and Surveillance (HFM-650) at (301) 594-1191.

GENERAL INFORMATION

Investigators should note the U.S. License number and location number of facilities. This will serve to identify establishments in correspondence, applications and other forms of communications. Registration numbers, essentially central file numbers, are not the same as license numbers.

The establishment should have a validated pink copy of Form FDA-2830, Blood Establishment Registration and Product Listing, for the current calendar year or evidence of having submitted same. If the data on the registration form is not correct, report the corrections to be made, and instruct the establishment to submit in writing the updated information to the Division of Blood Applications (HFM-370), 1401 Rockville Pike, Rockville, MD 20852-1448. If changes in the name and address, Responsible Head, manufacturing procedures, or products have occurred, they should be reported to Division of Blood Applications in order to update the Form FDA-2830.

OPERATIONS

Determine whether Source Plasma is collected for use in manufacturing licensed injectable products, licensed in vitro diagnostic products or unlicensed products.

Determine from whom Source Plasma is being collected: normal donors, immunized donors or disease state donors. Describe what disease states as applicable. Determine what products other than Source Plasma are being collected, for example Source Leukocytes or Therapeutic Exchange Plasma. Source Plasma establishments are required to be licensed and as such they are required to report important changes in manufacturing to CBER. Establishments submit SOPs in reporting these changes. Not all procedures are reviewed by CBER, therefore, if an investigator observes a procedure s/he considers unsafe for the donor or that will affect the safety, purity, or potency of the product, contact the Division of Inspections and Surveillance at (301) 594-1191.

Determine the approximate number of liters of Source Plasma collected annually by manual or automated methods as applicable. Also determine the number of donors processed annually by either manual or automated methods as applicable.

STANDARD OPERATING PROCEDURES

CBER approval letters for SOPs and revisions should be available, either at the plasma center or management's location, for the investigator's review. The firm should have a mechanism for maintaining SOPs and SOPs should be readily available to personnel in each area of manufacturing.

There should be a written procedure to minimize the spread of infectious agents and it should be consistent with current CDC and OSHA recommendations. OSHA published the final rule for the "Occupational Exposure to Bloodborne Pathogens" in the December 6, 1991 Federal Register. Included in the rule are requirements for facilities to develop procedures to ensure the safety of employees with a potential for exposure to biohazardous materials and procedures for medical waste disposal. Precautions that routinely may be followed include wearing gloves and protective clothing and providing vaccination against Hepatitis B. Masks and face or eye coverings may also be worn as the likelihood of exposure to infectious agents increases.

ERRORS AND ACCIDENTS; REPORTABLE INCIDENTS

Refer to CBER's memoranda to registered blood establishments, dated March 20, 1991, titled "Responsibilities of Blood establishments Related to Errors and Accidents in the Manufacture of Blood and Blood Components." When it is apparent that units of plasma were shipped in error, see the following CBER memoranda for requirements for investigation and notification regarding errors related to disease marker testing:

1. "Revised Recommendations for Testing Whole Blood, Blood Components, Source Plasma and Source Leukocytes for Antibody to Hepatitis C Virus Encoded Antigen (Anti-HCV)," dated August 5, 1993.
2. "Revised Recommendations for Testing Whole Blood, Blood Components, Source Plasma and Source Leukocytes for Antibody to Hepatitis C Virus Encoded Antigen (Anti-HCV)," dated April 23, 1992.
3. "Revised Recommendations for the Prevention of HIV Transmission by Blood and Blood Products," dated April 23, 1992.
4. "Clarification of FDA Recommendations for Donor Deferral and Product Distribution Based on the Results of Syphilis Testing," dated December 12, 1991.
5. "Recommendations for the Management of Donors and Units that are initially reactive for Hepatitis B Surface Antigen (HBsAg)," dated December 2, 1987.
6. "Guidance Regarding Post-Donation Information Reports," dated December 10, 1993.
7. "FDA Recommendations Concerning testing for Antibody to Hepatitis B Core Antigen (Anti-HBc)," dated September 10, 1991.

Special attention should be given to any donor fatalities since the last inspection. Fatalities should have been reported to the Office of Compliance at (301) 594-1191. In the event that an investigator becomes aware of a previously unreported fatal donor reaction which occurred since the last inspection, the Director, Division of Inspections and Surveillance, (301) 594-1194, should be notified as soon as possible.

Every adverse reaction experienced by the donor, whether considered insignificant or severe, must be recorded. Donor adverse reactions must be recorded in the donor record file (DRF); a separate adverse reaction file may also be maintained by the firm. The adverse reaction should be satisfactorily explained by the donor center personnel, including measures taken to assist the donor and the resolution of the reaction, and should be reviewed by the licensed physician or physician substitute. Reactions that include urticaria, flushing, wheezing, chest pain, low blood pressure, rapid heart rate, low back pain, loss of consciousness and convulsions, should be evaluated by the physician substitute/physician. Follow-up instructions should also be documented. In those cases where a donor has required prolonged observation, emergency transportation, or hospitalization, the DRF should include appropriate notation and medical approval by the physician (not the physician substitute) for the donor to continue participating in whatever program the donor is qualified.

Wrong red blood cell infusions should be documented. Records relating to a description of the incident must be maintained and made a part of the donor record for each affected donor. Copies of the records associated with unreported incidents and a copy of the EIR should be forwarded to the Biological Product Inspections Branch, (HFM-655).

PLASMAPHERESIS FACILITIES

All rooms and work areas where manufacturing processes, controls, and storage are performed must meet GMP requirements, including orderliness, cleanliness, lack of clutter, good lighting and ventilation, and adequate fly-proofing. Hand washing facilities for phlebotomists should be conveniently located; soap, towels, and hot water should be available. If the sink is not in the same room as the donor collection area, it should be in the immediate proximity. Bathroom facilities should be close enough that donors and personnel may easily reach them if necessary.

In facilities that include laboratories for testing for infectious agents, a separate room for laboratory procedures is ideal; however, if a specific area is designated for infectious agent testing and clearly labeled as such, it may be acceptable. Testing laboratories should observe recommendations as published in the Centers for Disease Control (CDC) and the National Institutes of Health (NIH) publication entitled Biosafety in Microbiological and Biomedical Laboratories. The booklet is available through the Department of Health and Human Services (DHHS). It is publication No. (CDC) 88-8395, 94-95, 2nd Edition, Washington D.C.: U.S. Government Printing Office, 1988.

EQUIPMENT

During the course of an inspection the investigator may observe or review instances where equipment or supplies are either being misused or not functioning as designed. Because of misuse, or lack of adherence to SOPs and/or manufacturing instructions, the use of key equipment or supplies creates circumstances where donor, operator, or product safety may be compromised. It is important, therefore, for the firm's overall use of equipment and supplies be used according to directions, satisfactorily inspected, maintained and operated properly. A record for trends relating to problems attributed to defective soft goods or equipment should be kept. Refer to the Compliance Program for specific instructions regarding documentation of deficiencies relating to the misuse of equipment.

Equipment should be maintained, standardized, and calibrated on a regularly scheduled basis as prescribed in the SOPs and in accordance with the manufacturer's specifications and records of such should be kept.

The standardization and calibration of the microhematocrit centrifuge may be done with a commercially prepared control or by other methods, e.g., duplicate samples tested at multiple intervals.

Bedside and centrifuge room scales. Procedures should include performance checks of all scales used in blood collection, i.e., trip scales and platform scales. Calibration should be done as necessary, with appropriate records maintained. Part of the regular quality control of each automated collection device should include checking the scale with external standard weights to verify the accuracy of the electronic record.

Refractometer. Distilled water should be used to standardize the refractometer to the "zero" point. Certain manufacturer's instructions may specify that a suitable protein based control with a refractometer reading of 6-8 gm/dl be used as a quality control check. Records must be maintained for daily standardization and for calibration in the event that the expected readings are not obtained. Serum or plasma is a very sticky substance, and the surface of the refractometer should be cleaned immediately after use with distilled H_2O on gauze, tissue, cotton, etc. An aqueous disinfectant (1:100 bleach to water) may be used to disinfect the surface of the refractometer once the plasma is removed. If the prism of the refractometer is wiped only with dry material, it is not effectively cleaned. Alcohol should not be used since it precipitates plasma protein and leaves a residue.

Refractometer results should be able to be clearly read. Care should be taken to allow the serum or plasma to flow over the prism, rather than touching the capillary directly on the prism. Extensive scratching of the refractometer prism will result in a "fuzzy" or "blurred" reading. A value for protein concentration will be obtained by looking through the eyepiece and noting where the sharp boundary between dark and light fields crosses the appropriate scale (gm/dl). Sharp contrast can only be verified by looking into the refractometer.

Periodic performance checks of the autoclave are necessary to assure that the time and temperature being recorded are adequate for sterilization. Procedures should provide for calibration before initial use and after repairs. Calibration procedures should provide assurance that the autoclave functions as intended, i.e., sterilization of arm preparation supplies and/or decontamination of biohazardous material. Biologic indicators should be used periodically and temperature controls, such as heat sensitive tape, should be used with each run to verify that the materials are being sterilized. A minimum of 121.5°C (251°F) for 60 minutes by saturated steam at a pressure of 15 atmospheres is recommended for materials contaminated with blood; 20 minutes at the same temperature is recommended for arm preparation supplies.

When the Electronic Devices (e.g. IVAC) are used for blood pressure and pulse, in addition to temperature determinations, the firm should develop a quality control procedure to assure the device is functioning properly. The performance check of the device should be done periodically and should not be limited to the electronic check for the temperature function. The manufacturer's instructions may suggest that the only necessary performance check is the electronic temperature function check; however, performance checks for the accuracy of blood pressure and pulse readings should also be performed.

Only FDA approved donor sets and blood containers (with proper amount of anticoagulant) should be used. Currently approved Whole Blood collection containers with anticoagulant (except heparin) for manual apheresis are manufactured by Cutter, Fenwal, and Terumo. For more recent approvals, contact the Division of Blood Applications, (301) 594-2012. These softgoods should be checked for defects prior to use.

Only FDA approved administration/transfer sets and plasma containers should be in use.

Separated red blood cells may be diluted and resuspended <u>only</u> in 0.85% to 0.9% Sodium Chloride Injection, USP, which can also be used to keep the intravenous line and needle free from clots.

Anticoagulants in use must be CBER approved. The most frequently used anticoagulant for manual apheresis is 4% Sodium Citrate. Automated procedures are approved allowing Sodium Citrate. Heparin is used to produce certain reagents since the plasma is easily defibrinated by commercial defibrinating agents. Other commonly used anticoagulants are citrate dextrose solution (ACD) and citrate phosphate dextrose (CPD). Collection in heparin requires CBER approval as a license or license amendment, and plasma is limited to manufacture of non-injectable products which the container label must clearly state.

Preparation of heparinized blood bags is often done in the establishment and should meet the following criteria:

1. The manufacturer's name and the lot number of the heparin should be recorded.
2. The heparin container may be entered once for preparation of the diluted "stock" heparin solution.
3. Only Sodium Chloride Injection, U.S.P., may be used as a diluent.
4. Diluent heparin "stock" solution should be made up fresh each work day and stored between 1-6°C when not in use.
5. Collection bags prepared with "stock" heparin solution must remain refrigerated until used the same day.

<u>AUTOMATED EQUIPMENT</u> -- For licensure, the operator to device ratio should not exceed 1:6. The ratio at lunch and break periods should not differ from other times. In addition, a fully trained operator should be available as a back-up in the event problems arise.

The amount of plasma to be collected is device specific and based on the device manufacturer's approved nomogram or the FDA abbreviated nomogram, which must also be a part of the establishment's SOP. Refer to CBER memorandum titled "Volume Limits for Automated Collection of Source Plasma," dated November 4, 1992.

Operator training is a vital part of appropriate and safe use of automated plasmapheresis devices. Thee content of the program should include troubleshooting and problem-solving of common problems that occur with the device. Once basic training has been completed and documented, a program for periodic updating and reassessment of operator skills, with appropriate documentation, should be in place.

A log for each device must be maintained. It should contain the identity of the operator on each day of use; the type and number of alarm messages received each day of use; any disposable equipment failures, e.g., leakage and breakage; failures of electronics; donor reactions; and any other problems noted. The maintenance and repair records of each separate electronic/automated device in use should provide evidence of problems or failures and appropriate corrective actions taken.

Before the device is initially placed into use and after repairs, there must be documentation of satisfactory performance of the device. The log should reflect the frequency of such performance checks, results of any problems discovered, and identity of the operator performing the checks. Problems must be satisfactorily resolved as evidenced by supervisory review. A program of periodic preventive maintenance must be written and followed by the establishment and be sufficiently comprehensive that any potential problem is likely to be discovered before a donor is placed in jeopardy. There should be a system to ensure that the maximum time for saline administration and/or anticoagulant set up does not exceed four hours. There is a risk of bacterial contamination once the bag of saline or anticoagulant is entered to connect the solutions to the apheresis set. Four hours is considered the maximum time these solutions should be connected prior to donor collection.

MEDICAL SUPERVISION

Physician substitutes (PS's) have been approved by CBER to perform most of the routine functions of a physician. Refer to CBER memorandum, "Physician Substitutes," dated August 15, 1988.

EXCEPT DURING RBC IMMUNIZATIONS, a physician may be "constructively" on the premises, that is, able to arrive on the premises within 15 minutes of being called. If a physician is not always on the premises, it must be possible for the physician to arrive at the center within 15 minutes.

There should be current, written agreements with ambulance services and emergency medical facilities, as well as explicit instructions regarding when and how to notify the Medical Director and staff, including the physician substitute. The telephone number for emergencies should be available to all staff. The telephone number of a specific emergency care facility or ambulance service may be posted on telephones, or a 911 system may be in effect in the area.

DONOR IDENTIFICATION

Acceptable forms of positive identification include a driver's license with photograph, a military ID, student ID, or any other document with the donor's signature and either a physical description or photograph. Some establishments require two signed documents.

Donors must be checked against the deferral register. Computer records which identify only the donor's name and permanent deferral status are adequate if more detailed information is listed in the DRF. There should be a procedure to keep this information confidential within the center.

Adequate identification of a donor is important to prevent using the same or different names and identification and to prevent cross donation at more than one plasmapheresis center, especially in areas where more than one center is located. The SOP should describe a system to prevent cross donations when applicable.

INFORMED CONSENT

Hazards of the plasmapheresis procedure, both manual and automated, including possible adverse reactions, should be clearly explained to the donor by a qualified, licensed physician or approved physician's substitute. The importance of the donor's active participation in identification of his own red blood cells (for manual procedure only) prior to reinfusion should be stressed. The donor must indicate understanding that two units of whole blood will be removed, one at a time, with reinfusion of the first unit of red blood cells prior to collection of the second unit. The physician's explanation of the hazards should be in simple, non-medical terms but the risks to the donor should not be minimized. Additionally, the discomforts and risks of other systemic reactions including hypotension, convulsions, lightheadedness, possible nausea and vomiting as well as depletion of proteins, including immunoglobulins, and decrease in hemoglobin should be explained to the donor as part of the informed consent. The physician or approved physician substitute must sign the informed consent form to certify that the hazards have been explained by a qualified, licensed physician or approved physician substitute.

The use of automated devices to collect Source Plasma should be explained in such a way that the donor understands the process. The donor should be fully informed regarding possible incomplete collections or "stop and restart" situations using automation. Breakage, leakage, and possible inability to return red blood cells, should also be addressed. A possible reaction to the anticoagulant, i.e., a numbness or "tingling" of the fingers or lips, should also be explained during the informed consent. The hazards of plasmapheresis by manual and automated procedures are not identical. If the same consent form is used for both manual and automated procedures, it must clearly identify the hazards for each procedure.

The hazards of immunization are to be separately discussed. The antigens used have particular risks, and each should be separately identified. In addition, there are general hazards of immunization, such as redness, soreness, possible fever and nausea and vomiting that may accompany any antigen. The possibility of anaphylaxis exists with all immunizing agents and should be part of the discussion of hazards. The donor shall be informed of the entire schedule of immunizations, the criteria for acceptable response, the procedure to be followed if no response occurs, and of the establishment's obligation to provide follow-up for at least one year if red blood cell immunization is done. The donor should be fully informed that s/he may participate in only one immunization program at a time, meaning that until the immunization has clearly failed the donor may not begin a new program. The special hazards of possible hepatitis and AIDS transmission, as well as the risk to a fetus and transplantation and transfusion problems that may occur, should be clearly explained to donors participating in red blood cell immunization programs. Females of child bearing age can not participate in red blood cell immunization programs unless documentation of sterility is obtained from her personal physician.

Each donor should be encouraged to ask questions. All donor questions should be answered fully and completely in a relaxed atmosphere. The interview must be conducted in private in a manner that does not embarrass or unduly pressure the donor to consent. The information on which the donor bases consent should be accurate and understandable to the donor. A donor's decision to refuse consent must be accepted as a matter of fact with no undue pressure to try to alter the decision. The investigator may insist on observing the explanation of the informed consent form by the physician without fear that this is an intrusion on the physician-patient privacy relationship.

All informed consent forms should be CBER approved and, if not revised since 1985, may not reflect currently approved practices. Call Division of Inspections and Surveillance (301) 594-1191 or Division of Blood Applications (301) 594-2012.

INITIAL MEDICAL EXAMINATION

The initial medical examination must be performed by a qualified doctor of medicine or osteopathy currently licensed to practice medicine in the state in which the plasmapheresis center is located or by a CBER approved physician substitute who must be certified or licensed according to state law. The plasmapheresis center should be requested to provide documentation of the CBER approval of <u>all</u> physician substitutes in the establishment.

AIDS educational materials must be presented to each donor at each donation. The AIDS educational materials in use should conform to the latest FDA recommendations or regulations. See CBER memorandum "Revised Recommendations for the Prevention of HIV Transmission by Blood and Blood Products," dated April 23, 1992.

The procedures listed below are the MINIMUM that should be done on the initial physical examination. They physician or approved physician substitute must perform all procedures except the urinalysis. No approval should be signed before the urinalysis results are known. All of the following procedures should be observed with the permission of the physician and the donor, preferably a donor of the same gender as the investigator. If the donor or physician object, remind them of your purpose, that of a federal investigator evaluating compliance with regulations, and gently persuade them of the necessity for your observation. If either persists in objecting, then an assessment of the exam can be determined by questioning the physician and donors and reviewing the records of examinations, or obtaining permission to observe the medical examination from another donor. DO NOT INSIST on observing the physical examination, but DO OBSERVE the explanation of the informed consent, including use of the AIDS educational materials. At some time during observation of the physician with a new donor, privacy should be given for any questions or concerns of a confidential nature the donor may have and for the opportunity for the donor to self-exclude in a confidential manner.

1. Blood pressure (BP) is measured either while the donor is seated or reclining. The cuff should be placed on bare skin, 1-2 inches above the bend of the elbow. The arm should be relaxed, supported by the examiner's hand or arm, when the readings are taken. If the BP is elevated beyond acceptable range, it is permissible to have the donor lie down and relax for 5-10 minutes and retake the pressure. If still elevated, the donor should be temporarily deferred with appropriate medical advice for follow-up, and an appropriate entry should be made in the DRF.
2. Heart and lung sounds should be determined on bare skin, both front and back, and with several intakes of air during the evaluation.

3. Abdominal examination is to determine enlargement of the liver, spleen or lymph nodes. The donor should be relaxed, possibly with knees bent, and the physician should gently but firmly press deeply into the abdomen on both right and left upper abdominal areas. Palpation of the inguinal (groin) areas for lymph node enlargement should also be part of this exam.
4. Neurological examination may consist of reflex assessment using a reflex hammer on knees and possibly elbows, ankles, wrists or other points. Coordination and sensory examinations may also be made, including touch and balance evaluations.
5. Urinalysis for sugar and protein.
6. The lymph node examination should include the neck from the jaw down towards the shoulders, angling forward from the angle of the jaw and nearly straight down from behind the ears. Other areas to evaluate include under the arms, at thee elbows, and the groin region.
7. The skin should be examined for new irregular patches that are reddish to maroon-blue in color and may be slightly raised. These patches can occur in the mouth or nose as well as other skin surfaces.
8. The mouth should be carefully checked for irregular cottony appearing white blotches.
9. It is also important to check the mouth, under the tongue, arms, and legs for needle tracks.

IMMUNIZATION

Report all antigens used and their respective manufacturers. Immunizing agents, such as tetanus, rabies, etc., are licensed products. Red blood cells for immunization currently are not licensed products, however, the source should be approved by CBER in the license application or amendment.

For each antigen used, there should be a current approved SOP. Personnel performing immunizations should be knowledgeable with the SOP. For red blood cell immunizations, see CBER memorandum "Revision of October 7, 1988 Memorandum Concerning Red Blood Cell Immunization Programs," dated December 16, 1992. If the program does not include use of frozen red blood cells for immunization, promptly notify Division of Inspections and Surveillance at (301) 594-1191. Red blood cell immunizations must be evaluated for development of unexpected antibody response and reports of these should be available for FDA review during inspections.

The establishment's physician must evaluate the donor's clinical reaction; for procedures where pretiter is required, an antibody titer should be done prior to immunization and should be compared to the post immunization titer to see if the donor is responding to the antigen. When the approved schedule of immunizations has been completed or interrupted, documentation of the physician's review should appear in the DRF.

A donor may not participate in more than one immunization program concurrently. If the donor has received red blood cell immunization, he/she is excluded from being a normal Source Plasma donor for twelve months after last receipt of red blood cells. Donors participating in other immunization programs may be returned to normal Source

Plasma collection if the donor fails to respond to the immunization program with the desired titer.

Both immediate (hives, localized swelling, etc.) and delayed reactions must be documented with appropriate medical response.

DONOR SUITABILITY

If a trainee is determining donor suitability, he/she should be under close supervision to assure correct performance. NOTE: Donor medical history questions may not be self-administered.

The interview area must offer a degree of privacy so that donors will be comfortable answering questions without fear of being overheard. See Compliance Policy Guide 7134.17.

See the following CBER memoranda for additional information:

1. "Revised Recommendations for Testing Whole Blood, Blood Components, Source Plasma and Source Leukocytes for Antibody to Hepatitis C Virus Encoded Antigen (Anti-HCF)," dated August 5, 1993.
2. "Revised Recommendations for Testing Whole Blood, Blood Components, Source Plasma and Source Leukocytes for Antibody to Hepatitis C Virus Encoded Antigen (Anti-HCF)," dated April 23, 1992.
3. "Revised Recommendations for the Prevention of HIV Transmission by Blood and Blood Products," dated April 23, 1992.
4. "Clarification of FDA Recommendations for Donor Deferral and Product Distribution Based on the Results of Syphilis Testing," dated December 12, 1991.
5. "Recommendations for the Management of Donors and Units that are initially reactive for Hepatitis B Surface Antigen (HBsAg)," dated December 2, 1987.
6. "FDA Recommendations Concerning Testing for Antibody to Hepatitis B Core Antigen (Anti-HBc)," dated September 10, 1991.
7. "Donor Suitability Related to Laboratory Testing for Viral Hepatitis and a History of Viral Hepatitis," dated December 22, 1993.
8. "Deferral of Blood and Plasma Donors based on Medications," dated July 28, 1993.
9. "Exemptions to Permit Persons with a History of Viral Hepatitis Before the Age of Eleven Years to Serve as Donors of Whole Blood and Plasma: Alternative Procedures," dated April 23, 1992.

In addition to the self exclusion provided to the donor in the signed consent statement at each donation, a second self-exclusion opportunity may be offered to the plasma donor in which a private interview conducted by a trained and competent health professional during which the AIDS educational information is presented orally. This second self exclusion may be offered at the initial donation (each time for infrequent plasma donors) and yearly as part of the medical examination.

Determination for suitability includes:

1. Normal temperature (37.5°C (99.6°F) or less). Some plasma centers may have established a low acceptance temperature value, which is usually 97.6°F (36.5°C).

A low temperature is usually of no significance unless the donor has symptoms of viral illness. Mercury-in-glass thermometers cannot be steam sterilized. Use of disposable paper thermometers, plastic covers for glass thermometers, or disposable probes of electronic thermometers are recommended. Although plastic covers may be used for glass thermometers, the thermometers still require disinfection between donors because leaks may occur.

2. Normal blood pressure (90/50 to 180/100). Donor's blood pressure must be determined prior to each donation and should be within normal limits. A systolic range of 90-180 mm and diastolic range of 50-100 mm are considered normal limits. Donors with blood pressures outside this range may be acceptable, but, only with a physician's approval and consistent with a written SOP.
3. Hematocrit (no less than 38%). Hemoglobin must be equal to or greater than 12.5 gm/100mL of blood. If the microhematocrit method is used, a value of 38% is equivalent to 12.5 gm/100 mL.
4. Normal pulse rate (50-100). Recommended normal pulse rate is 50-100 beats per minute (BPM) with regular rhythm. This is a "normal range"; results higher or lower should be reviewed by a physician or the donor should be deferred. Allowances may be made for a lower pulse rate in the athlete, e.g., joggers, but only with physician certification and consistent with a written SOP.
5. Total serum protein (no less than 6.0 gm/100ml). Donors MUST be deferred for quantitative total protein results of less than 6 g/dL or for protein composition test results that are outside the limits established by the testing laboratory, until repeat testing show values within acceptable limits. A donor who fails to appear in time for a four-month test may be plasmapheresed if no more than six months have elapsed and signed approval of the physician or physician substitute is recorded in the DRF. If the donor returns six months, or more, after being deferred, s/he must be treated as a new donor.
6. Weight (at least 110 lbs). Personnel must determine the donor's weight to assure the correct volume of blood is drawn. The donor must be weighed each time he donates. Personnel must read the weight directly from the scale rather than allow the donor to report his weight. In as much as weight loss or gain may be an indicator of disease or an untoward reaction to plasmapheresis, personnel should monitor a donor's weight as part of the donor suitability determination. Additionally, the recommendations to decrease the risk of transmitting AIDS from plasma donors state that the existing cumulative records of each Source Plasma donor's weight should be examined to assure that any weight loss of 10 pounds or more in less than two months is detected. A donor with an unexplained weight loss should be referred promptly to a physician for complete evaluation prior to any further plasma donation. Plasma donated that is still in storage at the plasma center should be quarantined until the reason for the weight loss has been determined and a medical release has been signed in the DRF. If the donor is deferred, plasma from previous units should be quarantined until evaluation is completed.

At the time of each donation, the donor medical history must include questions designed to elicit a history of:

1. Freedom from acute respiratory diseases -- Acute respiratory diseases are colds, influenza, persistent cough, sore throat, sinusitis, or other manifestations of upper respiratory infections and shall b cause for rejections until active symptoms have subsided. Such symptoms may be early indications of a more serious illness.
2. Freedom from diseases other than malaria transmissible by blood transfusion -- The donor must be free of any disease that may be transmissible by a blood transfusion. However, persons with a history of malaria are exempt from the foregoing because it is transmissible only by the cellular elements of blood which are not present in the plasma.
3. Freedom from infectious skin diseases -- A donor with a skin disease should be deferred if it is manifest at the site of phlebotomy. Mild skin disorders such as acne, psoriasis, or poison ivy are not cause for postponement unless prevalent at the phlebotomy area. Donors with boils or other severe skin infections should be deferred. Blue or purple spots (typical of Kaposi's sarcoma) should be referred to the physician.
4. Freedom of the arms and forearms from scars indicative of narcotic addition -- Before a donor is bled, his arms and forearms must be examined for evidence of skin puncture or scars which may indicate addition to narcotics. Such examination may be made by the person determining donor suitability or by the phlebotomist. This should be a close, thorough examination and not a cursory one. Past or present intravenous drug abusers must be permanently rejected.
5. Freedom from close contact with individuals having viral hepatitis within twelve months -- Donors who have had close contact with a person having viral hepatitis are deferred for twelve months after the contact. Health care personnel working in renal dialysis units should be deferred for twelve months after ceasing employment in such areas.
6. Freedom from having received blood or blood products within twelve months -- Donors who have received blood or blood products are deferred for twelve months after receipt of the product, unless specifically immunized by red blood cells in an approved program as defined in 21 CFR 640.66.

All questions should be asked slowly and clearly at each donation so that the donor can hear the questions and have time to respond. Since these questions are asked so often by the personnel who are accustomed to receiving a negative reply, they may be asked in a rapid, meaningless manner. It may be observed that the donor does not bother to answer the questions; yet the screener automatically checks negative responses. If the procedure is perfunctory, the management should be informed.

Screening personnel must defer a donor who is or who seems to be under the influence of alcohol or drugs, or who does not appear to be providing reliable answers to medical history questions. The reason for deferral must be permanently recorded in the DRF.

Some facilities treat donors as infrequent donors on the initial visit, and as regular donors on subsequent donations if the donor returns in less than eight weeks. See CBER memorandum "Requirements for infrequent Plasmapheresis Donors," dated August 27, 1982 and CP 7342.002, Attachment N. Refer to Attachment AA of CP 7342.002, regarding witnessing of donor interviews.

BLOOD COLLECTION

When arm preparation supplies are sterilized by the establishment, SOPs must contain specific information regarding steps to be followed and include directions for acceptable length of time such supplies may be used after the sterile package/container is first opened. If concentrated solutions are used, check that dilutions are properly made and that the proper expiration date/time is followed for dilutions. NOTE: Quaternary ammonia solutions, e.g., Pheneen, are easily neutralized and are NOT ACCEPTABLE for use in storing forceps.

The donor's name may be added to the unique donor bleed number on the container to enable both the phlebotomist and the donor or another person to identify the donor's red blood cells before they are reinfused. The donor's name and/or bed number alone is generally not sufficient identification.

For sterility reasons, a saline container should be used for no more than four hours after entry and for only one donor. If any saline remains in the container after a donor has completed a plasmapheresis procedure, it should be discarded or it may be used for laboratory testing; it must not be used for another donor or for any injection purposes. Generally, the volume of saline reinfused should not exceed the amount of plasma withdrawn. For automated collections, routine replacement of volume with saline is not necessary, although some centers use saline for volume replacement.

Neither the needle itself nor any continuously integral part of the tubing connected with the needle may be used for more than one venipuncture since multiple contacts with skin microbial flora increase the chances of contaminating the blood.

See CBER memorandum "Use of an FDA Cleared or Approved Sterile Connecting Device (STCD) in Blood Bank Practice," dated July 29, 1994.

For manual collection: If the first venipuncture fails, and the donor consents to a second venipuncture, at a minimum a new needle and its integral tubing must be used. If the flow of blood has not reached a connection, only the contaminated portion needs to be discarded.

For automated collection: There may be times when the disposable set must be changed after the venipuncture, or there may be times when a new venipuncture must be performed. A new disposable set may be installed by disconnecting the set from the needle connection and following the manufacturer's instructions. If a new venipuncture is necessary, the disposable set can be disconnected from the needle connection, the new venipuncture performed with a new needle, the disposable set reconnected, and the procedure continued following manufacturer's directions. When it is necessary to change either the venipuncture or the disposable set, this incident, and any resulting red blood cell loss, must be recorded.

There are several ways to do a satisfactory arm preparation. Sufficient duration and vigor of scrubbing are the key factors to removal of superficial microbes. After the venipuncture area is prepared, the vein may be palpated above or below the prepared area; however, the site of needle insertion should not be touched in any manner prior to venipuncture. It is not permissible to put iodine or sterile gauze on the finger to palpate the intended venipuncture site.

For both manual and automated collections: During blood collection or reinfusion of the red blood cells, the phlebotomist should periodically check for slowed or stopped bleeding, or the possibility that the needle slipped out of the vein and the blood is infiltrating the surrounding tissues. The automated device is equipped with warning lights and alarms that will notify the phlebotomist when the venous pressure is high.

Proper technique to assure donor safety, e.g.:

1. <u>Air removed from system</u>
   For manual collection: Before the venipuncture is made, saline should be allowed to flow through the administration set to remove all air from the tubing. This is done by opening the clamps near the saline bag and at the end of the set, allowing the saline to drip into a receptacle until all air is removed from the tubing. The filter chamber should then be half filled with saline. During this procedure, care should be taken not to contaminate the tip of the administration set. For automated collection: Confirm that the startup procedures are consistent with instructions in the operator's manual.
2. <u>Anticoagulant mixed with blood</u>
   The blood should be mixed with anticoagulant to prevent formation of clots. Gentle mechanical mixing throughout the collection is idea; however, if equipment for mixing is not available, periodic manual mixing should be done.
3. <u>Tubing stripped away from donor</u>
   If the flow rate has slowed or stopped, clots may have formed in the tubing. Therefore if stripping of the donor tubing is done to remove the clot, it must <u>always</u> be in a direction <u>away</u> from the donor to preclude forcing clots into the donor's bloodstream.
4. <u>Blood bag tubing sealed</u>
   Hermetic sealing of the blood bag using a metal clamp, white knot, or a dielectric seal must be done immediately after collection to prevent contamination.

For manual collection, monitoring to prevent overbleeding can be done effectively <u>only</u> by weighing each unit of blood. Each time an overbleed is detected, the bedside collection apparatus should be checked and adjusted if necessary before the next unit of blood is collected; it should be noted in the records that a check and/or adjustment was made. The whole blood bag weight records would indicate the possibility of overweight collection by manual methods. Records that show no overweight collections or weights that are identical for all units collected should prompt thorough evaluations of record keeping practices and blood collection system.

Weight of whole blood removed is the total weight less the weight of the blood container and anticoagulant. It is important to know how the weight of the blood container plus anticoagulant is adjusted, i.e., is the scale present to adjust for this weight or is the subtraction made later?

Determination of whether the proper amount has been collected can be calculated as follows:

To collect 500 mL of whole blood, the net weight of the unit should not exceed 526.5 g (500 mL x 1.053 f/mL = 526.5 g whole blood).

For 600 mL of whole blood collected, the net weight should not exceed 631.8 g (600 mL x 1.053 g/mL = 631.8 gm whole blood).

For automated collections, the internal monitor will weigh the collected plasma more accurately than an external scale. Overbleeding may occur when a chair or other object interferes with the plasma bag or bottle hanging on the scale, preventing accurate measurement.

Conversion between manual and automated plasma collection is considered to be a major change in manufacturing methods and must be approved in advance by CBER.

DONOR RECORD FILES (DRF)

A photograph is commonly used for donor identification, however, the regulations permit the use of other methods that provide equal or greater assurances of identifying the donor. Photographs in DRFs should be clear and current.

The DRFs (for donors in an immunization program) should indicate the following: That the donor is only on one immunization program; the type of antigen injected; name and lot number of antigen injected; date of injections; dosage and route of injection; individual who gave injection; description of any untoward reactions in such occurred, and documentation of outcome; record of pre- and post-immunization titers; a signed and dated evaluation by a qualified, licensed physician of the donor's clinical response; that only a qualified, licensed physician selects and schedules the immunizing antigen, and evaluates the donor's clinical response.

For all donors, the DRFs should indicate:

1. proper deferral intervals between: whole blood and plasma donations and plasma donations and past plasma donations when cells were not returned (eight or more weeks must elapse after whole blood donations); donors are not plasmapheresed more frequently than: once in a 2 day period and twice in a 7 day period.

   Collecting from donors in less than 48 hours is acceptable if the donations are two calendar days apart. Twice in 7 days for regular Source Plasma donors; once in eight or more weeks for infrequent Source Plasma donors. Donors with rare transitory antibodies may be plasmapheresed within 8 weeks of Whole Blood donation or after inadvertent loss of a volume of red blood cells that would otherwise require an eight week deferral only if examined and certified by a physician to be acceptable for plasmapheresis. The special characteristics of the antibody and the need for plasmapheresing the donor must be documented.
2. medical examination was performed by establishment:

   a. No more than one week before initial plasmapheresis -- This period of one week is allowed to accommodate those centers which have their donors examined in physician's private office. Medical examinations are not required for infrequent donors.

   b. No more than one week before immunization if donor had not received previous medical exam by the plasma center within the past 12 months -- If a donor is being immunized before the initial plasmapheresis, the medical examination may be performed no more than one week before the first injection and need not be repeated, provided that the initial plasmapheresis occurs within 21 days of the first injection. This provision permits the donor to be examined before beginning the immunization procedure to determine if he/she may have any medical problems with the injection. The additional 21 days allowed before plasmapheresis provides for antibody production in the immunized person. If a regular donor of "normal" plasma enters the immunization program, an additional physical examination is unnecessary. See Compliance Policy Guide No. 7134.18 for additional information.

   c. At one year intervals for repeat donors

3. when the next medical examination is due;
4. that a sample of blood from each donor was tested initially and once very 4 months for:

   a. total plasma or serum protein determination,
   b. plasma or serum protein composition (electrophoretic or chemical quantitation of components),
   c. serologic test for syphilis.
   [Either the physician or approved substitute must examine the total protein, protein composition, and STS results as well as the cumulative collection records of the preceding 4 months to determine the donor's suitability for continued plasmapheresis. The review must occur within 21 days after collection of the test sample and must be signed by the reviewing physician or the physician substitute. Any abnormal findings must be evaluated by a physician. If there are delays in the review of the accumulated laboratory data due to unsuitable sample, etc., the donor may be plasmapheresed if less than six months have passed since the last sample was collected and approved by the physician.]

PLASMA SEPARATION AND POOLING

Before filling, the plasma pooling container (final container) must be labeled with the donor number and/or the bleed number that will relate it directly to the donor.

Before centrifugation each unit of whole blood must be weighed, and the weight recorded concurrently.

To minimize chances of contamination, careful aseptic techniques must be used during the transfer procedure. The connection should remain intact and the tubing properly sealed.

Red blood cells should not routinely be permitted to enter the plasma container. Most donors are plasmapheresed regularly, and over a period of time the loss of these red cells would result in anemia.

If an air vent is used for the pooling container when pooling plasma from a donor it must be sterile and capable of excluding microorganisms, e.g., a needle filled with sterile cotton, and must be inserted aseptically. It should be kept dry throughout pooling procedure. CBER has received reports of contamination of plasma during the pooling process because of bubbling or leaking of plasma through the vent ports of some pooling bottles.

Pooling bottles should be secured in a manner to remain upright during pooling operations.

The pooling of plasma from two or more donors is permitted provided pooling occurs after the plasma is removed from the red blood cells, and the red blood cell containers are sealed. The product container should be labeled for Non-Injectable Use Only.

Inadvertent cross-pooling of plasma from two or more donors is a potential problem because of the possibility of cross contamination between the two donor's red blood cells. In addition, the likelihood of a wrong red blood cell reinfusion may accompany inadvertent cross-pooling.

REINFUSION OF RED BLOOD CELLS (MANUAL COLLECTION)

Both the donor and the phlebotomist should participate in the identification of the donor's red blood cells. The regulations do not specifically require that the donor participate in self identification of red blood cells, but CBER has not issued approval unless procedures in the SOP Manual are adequate to avoid infusing the wrong red blood cells. Sufficient time and care should be taken for this important process and it should be performed as described in the approved SOP manual. The regulations do not prohibit or prevent visually or hearing impaired persons from being plasma donors as long as sufficient safeguards are used to assure that the donor receives his/her own cells back.

There will always be some red blood cells remaining in the bag; however, the maximum feasible volume of red blood cells must be returned to the donor. This is important since the donor can be bled twice a week, removing two units of blood each time. If most of her/his cells are not returned each time, s/he will suffer loss of red blood cells over a prolonged period, resulting in anemia.

Wrong red blood cell infusion is a serious error usually involving two donors that may fail to be adequately documented. Whenever such an error is found, there should be at least two DRFs documenting deferral for both donors.

HBSAG, ANTI-HIV AND ANTI-HCV TESTING

The laboratory actually performing the testing should be the laboratory approved by CBER. Notify Division of Inspections and Surveillance at (301) 594-1191 of discrepancies. Determine if the firm uses an outside laboratory to perform HBsAg, anti-HCV and/or anti-HIV testing. If so, report the name(s) and address(es) of each laboratory. Each laboratory must be approved in the license application.

HBsAg, anti-HCV and anti-HIV testing must be performed on a sample collected from the donor at the time the plasma is collected. The sample may be drawn directly from the donor, or a segment of tubing from the collection set filled with blood or plasma may be used. Plasma from more than one donor may not be pooled for testing. If the donor is known to be HBsAg reactive or anti-HIV positive, and IF THE ESTABLISHMENT HAS CBER APPROVAL TO COLLECT THIS PLASMA, testing may be eliminated for the known infectious agent after the first donation. See CBER memoranda "Revision to 26 October 1989 Guideline for Collection of Blood or Blood Products from Donors with

Positive Tests for Infectious Disease Markers ("High Risk" Donors)," dated April 17, 1991; and "Guideline for Collection of Blood or Blood Products from Donors with Positive Tests for Infectious Disease Markers ("High Risk" Donors)," dated October 26, 1989.

Results of HBsAg, anti-HCV and anti-HIV testing must be available in written form (hard copy) before the product is released or shipped for further manufacturing. Telephone reports of reactive results re encouraged so that appropriate deferral of the donor will occur promptly. No units may be shipped without all written laboratory results having been received. NOTE: With CBER approval, Source Plasma may be shipped between collection locations and fractionating plants under the same establishment license before all written test results have been received, provided the testing is performed by the licensed establishment. These plasma units may bear labeling indicating they are nonreactive for HBsAg and negative for anti-HCV and anti-HIV tests. The material is considered to be moving (sometimes interstate) under quarantine as a partially processed product. It is not a final product until all tests are complete and is not considered misbranded while moving under these conditions.

The tests must be performed according to the manufacturer's instructions for the reagent test kits and equipment used during the testing procedures. Review of control results should precede certification of any run results of donor samples. Testing data must be reviewed by personnel (usually supervisory) to assure test runs are valid, results are calculated accurately and run failures, if occurring, are resolved. See CBER memorandum "Recommendations for the Invalidation of Test Results When Using Licensed Viral Marker Assays to Screen Donors," dated January 3, 1994.

A proposed rule was published in the June 6, 1989 Federal Register to require that each establishment or laboratory responsible for performing FDA required tests for HBsAg and anti-HIV participate in an approved program to demonstrate proficiency in performing these tests. The final regulation proposed by FDA has not been published. Proficiency samples should be tested by personnel who routinely perform the tests using the laboratory's usual methods and under routine conditions and workload. Corrective action should be documented if performance is substandard.

Report other testing performed, if applicable.

HANDLING OF HBSAG, ANTI-HIV AND ANTI-HCV REPEATEDLY REACTIVE UNITS

Each HBsAg, anti-HCV or anti-HIV repeatedly reactive unit must be CLEARLY labeled as HBsAg, anti-HCV or anti-HIV Reactive or Positive and immediately removed from nonreactive/negative units and should be double wrapped to reduce exposure to personnel. The procedure used by the establishment should be adequate to assure that reactive/positive units are not shipped inadvertently. If reactive units are stored in a container, the container should be clearly labeled as "Biohazard."

See the following CBER memoranda for additional information:

1. "Revised Recommendations for Testing Whole Blood, Blood Components, Source Plasma and Source Leukocytes for Antibody to Hepatitis C Virus Encoded Antigen (Anti-HCV)," dated August 5, 1993.

2. "Revised Recommendations for Testing Whole Blood, Blood Components, Source Plasma and Source Leukocytes for Antibody to Hepatitis C Virus Encoded Antigen (Anti-HCV)," dated April 23, 1992.
3. "Revised Recommendations for the Prevention of HIV Transmission by Blood and Blood Products," dated April 23, 1992.
4. "Recommendations for the Management of Donors and Units that are Initially Reactive for Hepatitis B Surface Antigen (HBsAg)," dated December 2, 1987.

Units collected from the same donor subsequent to a reactive unit, should be removed from storage with other blood products.

Records of test results should adequately identify reactive units. There should be a system for matching reactive/positive test results with stored plasma units and making certain that such reactive/positive units are not shipped.

As soon as an HBsAg, anti-HCV or anti-HIV repeatedly reactive/positive donation is discovered, the DRF must be immediately updated. Nonreactive HBsAg and negative anti-HCV or anti-HIV testing results need not be individually entered on the DRF, but these results must be maintained on the premises. Names of reactive positive donors must be added promptly (before next possible donation) to the donor deferral records.

There should be a system for limiting access to data regarding sensitive HBsAg, anti-HCV and anti-HIV testing status to only authorized personnel.

If donor re-entry protocols are used, they should meet current guidelines. Thee are donor re-entry protocols available, but optional, for both HBsAg reactive (dated December 1987) and anti-HIV repeatedly reactive Western Blot negative (dated April 23, 1992) donors. They are in the form of algorithms and require sequences of testing with appropriate results over a period of time to determine the "true" status of the donor regarding the particular viral marker. Copies of each memoranda should be on file in each establishment and should be followed if donors are re-entered. Call Division of Inspections and Surveillance at (301) 594-1191 for additional copies or interpretation.

If reactive/positive units are not destroyed by autoclaving or incineration, the disposition of these units, i.e., if they are shipped to an establishment that manufactures reagents or plasma-based vaccine should be documented. Under Sections 610.40(d) and 610.50(d), a manufacturer who ships reactive/positive plasma must notify the Director, CBER, of such shipment. Etiologic agent labeling and packaging is required by 42 CFR, Part 72 -- Interstate Shipment of Etiologic Agents.

In-house shipments of reactive/positive units are permitted for research purposes.

There is currently no policy for look-back for Hepatitis marker positive donors but the firm should be following its SOP.

SERUM PROTEIN QUANTITATION

Serum protein electrophoresis, quantitative immunodiffusion test, or an equivalent test should be performed. The ranges for each protein component should be established by the testing laboratory based on a defined reference population. The ranges may either appear on each test report or be on file centrally in the establishment, identified by the testing laboratory name and address. The name and address on the records should be fore the approved laboratory indicated on the license or amendment approval.

SEROLOGIC TEST FOR SYPHILIS

Plasma collected before the result of a reactive initial test on a new donor is known may be used as Source Plasma without additional labeling. See Compliance Policy Guide 7134.15 and CBER memorandum, "Clarification of FDA Recommendations for Donor Deferral and Product Distribution Based on the Results of Syphilis Testing," dated December 12, 1991.

STORAGE

An automatic recording device may be used for continuous monitoring of storage temperature; however, a daily handwritten record of observe temperature must be maintained, and compared against the automatically recorded temperatures. Discrepancies between the recorder temperature and the handwritten observations should be noted and resolved. (If the firm uses an electronic central system for monitoring temperatures, see the "Guide to Inspections of Blood Banks"). Charts should be changed regularly.

If only daily handwritten records are kept, they should be made at intervals throughout the day. A device to indicate that the temperature did not rise above freezing should also be used, e.g., a tube (or any container) of water frozen on a slant and stored in an upright position.

If the records indicate that the freezer temperature was > -20°C, there should be an explanation in the records that includes the period of time the temperature was not at the proper level, a description of the action taken to correct the problem, and the condition of the plasma.

The storage area should be clean and organized. Plasma for manufacture into injectable products should be stored at -20°C immediately after filling and maintained at -20°C or colder until dispensed. "Immediately" means "without undue delay," e.g., a carton that holds 12 units of plasma might be filled and then placed in the freezer, rather than putting each unit in separately. Plasma should be frozen as soon as possible to preserve antihemophilic factor (factor VIII). Plasma may be "flash frozen" in an acetone/dry ice mixture, then stored at -20°C. Source Plasma is frequently stored in the shipping carton. Temperature records should support this.

Units which are damaged, i.e., conditions which may affect the safety or quality of the plasma, during storage should be removed from storage with undamaged units. These units should be stored separately in a manner to prevent exposure of biohazardous material to personnel and contamination of undamaged units. Any relabelling and shipment of damaged plasma must be approved by CBER.

Per 42 U.S.C. 262(a) (PHS Act), plasma collected under an unapproved licensed application or amendment reference number for shipment must be quarantined and not shipped until CBER approval. Records should show that collection did not begin prior to receipt of a reference number and shipment did not occur prior to approval.

If interim off-site storage is used, CBER approval is needed prior to transfer of Source Plasma to that site.

The firm should routinely audit off-site storage facilities to ensure compliance with applicable GMPs and records maintained of the same.

DISTRIBUTION RECORDS

Source Plasma, Liquid must be visually inspected for changes in color, transparency, particle formation, evidence of microbial contamination, etc., before shipped.

Records should indicate shipment of Source Plasma to brokers. Report the name and address of broker(s), if applicable. Brokers who store or manipulate Source Plasma in any manner must be registered with CBER.

DISPOSAL OF INFECTIOUS WASTE

The SOPs should contain specific language for bagging and disposal of contaminated waste. Needles should be disposed of in a container designed to prevent accidental puncturing of personnel. Specific instructions regarding color and size of bags used for trash, the handling of such bags inside and outside of the establishment and whether autoclaving or incineration is used for suspected contaminated waste must be written in the SOP. If contaminated waste is disposed of by a firm other than the plasma center, a contractual agreement should be on file at the plasma center.

The firm should have provisions for autoclaving or incinerating potentially infectious trash and items used in collection, processing, and testing procedures; unreturned whole blood or red blood cells; and plasma unsuitable for use. Report the firm's provisions, e.g., on-site; off-site; contract. If the firm contracts out for disposal of infectious waste, report the name and address of the contract firm.

RECORDS

When computer systems are used for records, changes to records should be traceable as to the time and person making the change so that integrity and reproducibility of data are assured. Persons authorized to make changes should be specifically identified. Periodic audits of stored data should be undertaken to assure that timely retrieval and accurate information reporting are available. Records may be kept entirely electronically provided that hard copy can be retrieved within 4 to 6 hours of request.

Source Plasma has a ten year dating period and records of its manufacture are required to be retained for 10.5 years. If there is no specified expiration date, records must be maintained indefinitely.

DISEASE STATE DONORS

All laboratory values that fall outside the expected values for a reference population should be specifically identified and approved by a physician who may authorize collection of the patient. Plasma from donors in the following categories may be collected if CBER approval has been given:

Elevated IgE
STS Positive
Multiple myeloma plasma
Hemophilus influenza antibody

Rheumatoid Factor
Anti-microsomal antibody
Anti-mitochondrial antibody
Anti-thyroglobulin antibody
Antistreptolysin-O
Anti-smooth muscle antibody
Antinuclear antibody
Anti-DNA
Heterophile antibody
Hemophilia (or other coagulation factor deficient plasma)

Thirty-day prior notification to CBER is acceptable for the following disease-associated antibodies:

C-Reactive Protein
Mononucleosis (Epstein Barr)
Cytomegalovirus (CMV)
Herpes Type I
Herpes Type II
Varicella Zoster
Coccidioidomycosis
Histoplasmosis
Pseudomonas
Rubella
Mumps
Hepatitis A (Anti-HAV)
Hepatitis B (Anti-HBs)
Hepatitis B Core (Anti-HBc)
[Anti-HBc collections are included only when the donor is known to also have Anti-HBs]
Toxoplasmosis

It may be discovered upon investigation that other disease state donors are being collected without approval. Questions regarding whether the disease state indicated could yield a product for which the establishment has approval should be directed to CBER, Division of Blood Applications (HFM-370), (301) 594-2012. The letter of approval from CBER will specifically identify the acceptable plasma. Report the disposition of plasma collected from donors with conditions for which CBER approval has not been given. Distribution records should document shipment in interstate or only that plasma for which there is specific CBER approval.

SOPs must be specifically CBER approved and will include donor selection criteria, and the procedures for the collection, testing, labeling, quarantine, and disposition of the plasma. The collection, handling, storage, and disposition of reagents, samples, and plasma should be performed in accordance with the current biosafety guidelines as established by FDA, CDC and/or OSHA.

Each disease condition for which plasma is licensed must have the label approved by CBER.

Plasma units collected from disease state donors should be segregated from other Source Plasma.

A current written agreement with the final manufacturer must be maintained on file.

Accountability for all products collected should be able to be determined.

"HIGH RISK" DONORS

"Guideline for Collection of Blood or Blood Products from Donors with Positive Tests for Infectious Disease Markers ("High Risk" Donors)," was issued October 26, 1989 and "Revision to 26 October 1989 Guideline for Collection of Blood or Blood Products from Donors with Positive Tests for Infectious Disease Markers ("High Risk" Donors)," was issued April 17, 1991. Refer to these CBER memoranda for donor qualifications, laboratory testing, collection, and labeling. SOPs must be CBER approved.

SOURCE LEUKOCYTES

Some establishments may have received approval for automated collection of Source Leukocytes in accordance with the guideline for collection of Platelets, Pheresis, revised October 7, 1988. Refer also to CBER "Guidelines for the Collection Of Human Leukocytes For Further Manufacturing," dated January 28, 1981.

Donors must meet the criteria for whole blood donation, Section 640.3, or for plasmapheresis donation, Section 640.63, as appropriate. Collection of Source Leukocytes may not occur more frequently than once in 2 days and twice in 7 days and shall not exceed collection of 32 units of source leukocytes per year from an individual donor.

Some product license applications for Source Leukocytes have been amended to include the collection of Source Leukocytes by plasmapheresis without any additional donor monitoring with a frequency of not more than once every 8 weeks for a total of 6 units per year. This product must be collected from the first unit of whole blood during a plasmapheresis procedure.

If frequency of donation is more than six donations a year, white cell counts should be done within seven days prior to collection and counts should be above $4000/mm^3$.

Source Leukocytes should be tested for disease markers as Whole Blood, i.e., HBsAg, anti-HIV, anti-HCV, syphilis, anti-HTLV-I, anti-HBc, and ALT. HBsAg/ anti-HIV-1/anti-HCV testing results that may not be available prior to shipment must be specifically identified and CBER approved for the method by which these results are communicated.

Approved labels must be used.

THERAPEUTIC EXCHANGE PLASMA (TEP)

Therapeutic plasmapheresis is a medical procedure for treatment of a disease. The plasma is incrementally removed and other fluids, usually electrolyte and/or protein solutions are substituted. The plasma derived from such procedures has been limited in use to further manufacture of SPECIAL in vitro diagnostic reagents with no alternative source, for which specific CBER approval is required. It may not be labeled as Source Plasma. This plasma, designated Therapeutic Exchange Plasma, intended for further manufacture, is a biological product subject to the licensing provisions of Section 351(a) of Public Health Service Act. Refer to CBER memorandum "Plasma Derived from Therapeutic Plasma Exchange," dated December 14, 1984.

If plasma is sold, a current written agreement between this collecting facility and the final product manufacturer should be maintained on file. Like disease state donors, there must be a specific agreement with a final manufacturer to sell these plasma products.

Products for which approval has not been obtained may not be labeled with the license number of the establishment, nor may they be shipped in interstate commerce.

Records should indicate the disposition of every container of TEP collected. In these exchange procedures, variable numbers of containers may be collected at each exchange. Documents should show that the collecting facility records the exact number of containers, that a specific designator is used for the sequence in which these products are collected, and that all units collected are accounted for in the disposition records.

Date: December 20, 1994

From: Director, Center for Biologics Evaluation and Research

Subject: Recommendations to Users of Medical Devices that Test for Infectious Disease Markers by Enzyme Immunoassay (EIA) Test Systems

To: All Registered Blood Establishments

This memorandum provides recommendations to assist users of medical devices that test for infectious disease markers by enzyme immunoassay (EIA) in complying with the Food, Drug, and Cosmetic Act, the Public Health Service Act, and all applicable regulations, including those found in 21 CFR, Parts 200 and 600. Recommendations provided address the areas of test equipment documentation, i.e., performance data, installation instructions, validation procedures, calibration methodology, verification and recalibration, precision and accuracy determinations, problem detection and correction, troubleshooting, and preventive maintenance.

The attached recommendations are directed to all establishments engaged in the collection, processing, testing, storage, and distribution of blood and blood products, specifically including blood banks, laboratories engaged in blood product related testing, plasmapheresis centers, and transfusion services.

The attachments to this memorandum provide blood establishments with a brief listing of necessary information to successfully operate the equipment (Attachment I), typical problems they may encounter, and their possible solutions (Attachment II). The development of the recommendations on validation and the attachments into a checklist format adapted to your establishment's particular needs should assist you in applying the recently issued Quality Assurance Guidelines.

This memorandum does not supersede any previous memoranda but should be used in conjunction with the References/Resources listed to obtain the optimum benefit of the test system.

Kathryn C. Zoon, Ph.D.

RECOMMENDATIONS TO USERS OF MEDICAL DEVICES THAT TEST FOR INFECTIOUS DISEASE MARKERS BY ENZYME IMMUNOASSAY (EIA) TEST SYSTEMS

I. Introduction

With the increased emphasis of blood centers and transfusion services on preventing transfusion transmitted diseases, blood establishments need additional training, quality control, preventive maintenance checks, and troubleshooting and problem solving skills to ensure successful operation of enzyme immunoassay test systems.

Manufacturers of the equipment discussed in these recommendations are required to document specifications as well a s installation, validation, calibration, calibration verification, quality control and preventive maintenance procedures as described in 21 CFR 809.10(b)(6), (8)(v) and (vi), 21 CFR 820.100(a), and 21 CFR 820.152. Frequently the equipment manufacturer's Customer Service Representative (CSR) or Technical Service Representative (TSR) will provide much first-hand assistance. However, the blood establishment should ensure that the above documentation is provided in order that it can comply with 21 CFR 211.68 and 21 CFR 606.60.

The amount of testing to be done and performance data for the above procedures depends on the equipment used and the particular test(s) associated with the equipment. The resources listed in part IX maybe helpful in determining acceptable amounts of testing for performing each of the above procedures.

II. Equipment (Device Definitions)

A. Sampler/Sample Identifier/Micropipettor/Dilutor -- a multifunctional device that may identify, dilute and/or aspirate a predetermined amount of sample and/or dispense the necessary reagents.

B. Microplate washer -- a device that adds a predetermined amount of wash solution and aspirates the solution a pre-set number of times. The manufacturer determines the amount of solution and the number of washes necessary to prepare the sample for the next stage of the testing procedure.

C. Incubators (dry or water) -- a device used to bring the sample/reagent mixture to the proper temperature and maintain it at a pre-set temperature for the amount of time determined necessary for a reaction to take place.

D. Microplate readers (Spectrophotometer) -- device used to measure the amount of color development in each well (reaction chamber) of the microplate as compared to a standard.

E. Refrigerators, Centrifuges, Rotators & Shakers, and Balances -- definition of these instruments is self-explanatory. While not integral components of the EIA test system, these instruments can have significant effects on the outcome of this testing if they are not functioning properly.

III. Installation

The blood establishment should ensure that the manufacturer installs the equipment or provides complete installation/setup instructions. The installation/setup instructions should include, when applicable, a linearity check, volume verification for any required volume measurements, comparison and confirmation of configuration/parameter settings, diagnostic communication checks for any interfaced instruments, testing of samples for each protocol, and verification of accuracy of readouts.

IV. Specifications

Specifications are the operating parameters of the instrument or test system. The device must be capable of delivering the level of performance (accuracy, precision) consistent with the requirements of the test kit package insert. The blood establishment should ensure that the manufacturer provides these performance specifications in the instrument Operator's Manual and the specifications should address random and systematic (both constant and proportional) error occurring in both intra-assay and inter-assay runs, the calibration methods used, a direct comparison of the method to an established reference method (with total error below that deemed clinically significant), the reportable range of test results, and, when applicable, results of linearity studies using, for example, National Committee for Clinical Laboratory Standards (NCCLS) listed in part IX or equivalent guidelines.

V. Validation

Validation establishes documented evidence which provides a high degree of assurance that a specific process will consistently produce a product meeting its predetermined specifications and quality attributes. The data should be collected using appropriate samples to support test performance characteristics of accuracy, precision, sensitivity, and specificity, and when applicable, reportable and reference range(s), and any other applicable performance characteristics. Process validation is performed to evaluate the performance of a system with regard to its effectiveness based on the intended purpose, such as meeting the test kit manufacturer's reagent specifications.

A. EQUIPMENT DOCUMENTATION

Infectious disease test products are medical devices. Firms that manufacture and distribute such devices must comply with regulatory requirements applicable to these devices. The blood establishment should ensure that manufacturers provide an operator/user manual that has accurate and complete instrument user instructions plus a listing of the required equipment to validate the instrument. The manual should contain these items:

1) a functional description of the microprocessor-controlled test system, i.e., its intended use, the applications and reagents that have been cleared for use with the device, and a mechanism to update the manual when new tests are available;

2) general operating instructions and system limitations with warnings/cautions;

3) a description of the hardware which includes the installation instructions (even if performed by the instrument manufacturer), all integrated functional components, configuration and linkage of integrated components, environmental and maintenance requirements, interface specifications, and all warning/error messages and their appropriate corrective action;

4) a description of the software that includes a file/record structure, field descriptions, user definable database options, warning/error messages and their appropriate corrective action, an assurance of data accuracy, and software performed calculations;

5) specimen specifications (which, although provided in each test kit package insert, should also be in the operator/user manual), such as serum or plasma (and type of anticoagulant used), maximum age at which sample can be tested, storage and preparation required, amount needed for testing, and a list of substances known to interfere with the test; and

6) change control and problem reporting procedures, document control process, and customer support services.

The blood establishment should provide adequate staff training and guidelines for the development of a test plan that addresses software control functions, calculations performed by the software, conditions that trigger alarms, error and warning messages, data entry methods, boundaries, field limits and ranges, security procedures and program overrides, and data storage, transfer, and retrieval. The software should identify the version and all default parameters used in running the test, both at the time of running the test and on the printout, and should be validated by the manufacturer for all calculations, "flags" for limits, and failures.

B. RECOMMENDATIONS FOR BLOOD ESTABLISHMENTS

The user should perform the necessary equipment validation to establish confidence that the device and its ancillary systems are capable of consistently operating within established limits and tolerances.

SAMPLER/SAMPLE IDENTIFIER/MICROPIPETTOR/DILUTOR -- can be validated with calibrated pipet tips which hold the proper volume or by a gravimetric method if a balance sensitive enough to weigh very small volumes is available. Sample identification can be validated by testing with barcode controls of known definition.

MICROPLATE WASHER -- can be validated by obtaining an exact measurement of the fluid being dispensed, either from each dispensing probe or from all probes combined into a calibrated volumetric flask.

INCUBATOR -- can be validated with the use of a total immersion verification thermometer preferably with 0.2° temperature demarcations; however, 1.0° temperature demarcations are acceptable.

MICROPLATE READER (Spectrophotometer) -- can be validated using the manufacturer's instructions, keeping in mind that an operational verification, an alignment verification, and a linearity test are necessary.

REFRIGERATORS, CENTRIFUGES, ROTATORS & SHAKERS, AND BALANCES -- can be validated by well documented procedures.

VI. Calibration and Calibration Verification

Calibration is the process of testing and adjusting an instrument, kit, or test system to provide a known relationship between the measurement response and the value of the substance that is being measured by the test procedure. Calibration verification is the process used to assure that the calibration of the instrument, kit, or test system remains stable throughout the laboratory's reportable range for test results. Calibration and recalibration of less complex instruments can be performed by following the instructions in the operator's manual. Some more complex instruments also have components that can be calibrated and recalibrated by the user, and many have troubleshooting guides which are helpful in correcting minor instrument problems. The device or test system manufacturer should provide the necessary calibrators or specify what calibrators may be used and information on the items needed to perform this process.

The blood establishment should ensure that the instrument meets the requirements described in 21 CFR 211.68 and 21 CFR 606.60 for calibration.

VII. Quality Assurance/Quality Control

Quality assurance consist of the actions, planned and taken, that provide confidence that all systems and elements that influence the quality of the product are working as expected (individually and collectively).

Quality Control is a component of quality assurance that includes the activities and controls used to determine the accuracy of the establishments' personnel, equipment, and operations.

Quality control procedures recommended by the Center for Biologics Evaluation and Research (CBER) for blood establishments are set out in CBER's Draft Guideline for Quality Assurance in Blood Establishments, dated July 2, 1993.

VIII. Preventive Maintenance/Function Checks

Preventive maintenance and function checks performed on a regular basis will help to ensure that the instrument will perform correctly, will have less downtime, and will reduce other maintenance/repair costs. The blood establishment should ensure that an acceptable maintenance and service schedule is provided, which includes cleaning, adjusting, and testing assay equipment.

There should be protocols for maintenance and function checks with documentation that calibration verification and quality control will remain stable between the intervals recommended.

IX. References/Resources

The following resources may be obtained from the Center for Biologics Evaluation and Research, Division of Congressional and Public Affairs, HFM-12, 1401 Rockville Pike, Suite 200N, Rockville, MD 20852-1448, FAX (301) 594-1938:

1. Draft Guidelines for Quality Assurance in Blood Establishments, Docket No. 91N-0450, July 2, 1993.

2. FDA Memorandum, Recommendations for Implementation of Computerization in Blood Establishments, April 6, 1988.

3. FDA Memorandum, Requirements for Computerization of Blood Establishments, September 8, 1989.

4. FDA Memorandum, Changes in Equipment for Processing Blood Donor Samples, July 21, 1992.

5. Draft Guideline for the Validation of Blood Establishment Computer Systems, Docket No. 93N-0394, October 28, 1993.

The following resource may be obtained from the Superintendent of Documents, U.S. Government Printing Office, Mail Stop: SSOP, Washington, DC 20402-9328.

6. Title 21 Code of Federal Regulations, Parts 200-299, 600-799, and 800-1299, revised annually and available April 1 of each year.

The following resource may be obtained from the Center for Devices and Radiological Health, Division of Consumer Affairs, HFZ-210, 1901 Chapman Avenue, Rockville, MD 20857, (301) 443-4150.

7. Guideline on General Principles of Process Validation, May 1987.

The following resource may be obtained from the American Association of Blood Banks, 8101 Glenbrook Road, Bethesda, MD, 20814-2749, (301) 907-6977.

8. Blood Bank/Transfusion Service Computer Systems, prepared by AABB Information System Committee and HIMA ad hoc computer committee, August 27, 1993.
   a) User Validation Guidelines.
   b) Software Manufacturing Process Guidelines.
   c) Validation Guidelines for Microprocessor Controlled Test Instruments.

OTHER USEFUL RESOURCES:

9. National Committee for Clinical Laboratory Standards, Villanova, PA.
   a) Internal quality control testing: principles and definitions. Approved Standard. 1990.
   b) User Comparison of quantitative clinical laboratory methods using patient samples. Proposed Guideline. 1992.
   c) Preliminary Evaluation of Clinical Chemistry Methods. Tentative Guideline. 1989.
   d) Evaluation of precision performance of clinical chemistry devices -- second edition. Tentative Guideline. 1991.
   e) How to define, determine and utilize reference internals in the clinical laboratory. Proposed Guideline. 1991.
   f) Specifications for Immunological Testing for Infectious Diseases. Proposed Guideline. 1991.

Technical questions, information, and comments may be directed to the:

Division of Blood Applications, HFM-380
1401 Rockville Pike, Suite 200N
Bethesda, MD 20852-1448
Phone: (301) 594-6487 FAX: (301) 594-6431

ATTACHMENT I.

LABELING APPLICABLE TO MEDICAL DEVICE MANUFACTURERS
(not to be considered all inclusive)

A. All devices, either as individual products or as components in a total system, should contain the following in the labeling section (package insert) [see 21 CFR 809.10(b)]:

   1. Intended use or function.
   2. Installation instructions and special requirements.

3. Performance characteristics and specifications.
5. Operator's Control Manual and operating instructions (which should include a Troubleshooting Guide).
6. Calibration procedures including materials and/or equipment to be used.
7. Operational precautions and limitations.
8. Hazards.
9. Preventive Maintenance Procedures and service information.
10. Interpretation of Data.
11. Quality Control Procedures and Schedules.

Information contained in the performance section should be from a statistically significant number of samples tested. Sample size depends on the particular application of the device being demonstrated.

B. <u>Sampler/Sample Identifier/Micropipettor/Dilutor</u>

1. Performance data should include test results that demonstrate:
   a) the sampler provides correct identification;
   b) the micropipettor processes the correct amount of sample;
   c) the dilutor makes the appropriate dilution;
   d) reproducibility of results; and
   e) stability and accuracy in volumes pipetted and diluted.

2. Specific labeling information should include the following:
   a) recommended pipet tips;
   b) instructions for repositioning and changing tips;
   c) instructions for verifying that pipet tips were properly changed;
   d) information on battery, shelf life, and effects of power outage;
   e) validation, calibration, and quality control procedures.

C. <u>Microplate Washer</u>

1. Performance data should include test results demonstrating the following:
   a) the washer adds the recommended amount of wash solution;
   b) the washer removes all wash solution each wash cycle;
   c) the effects of improper washing; and
   d) the washer has performed the appropriate number of soaks and the duration of each.

2. Specific labeling information should include the following:
   a) instructions for repositioning the pipet tips;
   b) instructions for replacing or correcting a seal between the pipet tip and holder;
   c) instructions for priming the washer;
   d) schedules and instructions for changing tubing and connectors;
   e) disinfection procedures;
   f) instructions for pressure problems;

g) instructions for determining proper fluid level heights and fluid volume settings;
h) recommended microplates that will be compatible with the device; and
i) validation, calibration, and quality control procedures.

D. Incubators

1. Performance data should include test results that demonstrate the following:
   a) the recommended temperature is that required by the test kit manufacturer;
   b) the effects on samples incubated outside of the recommended range; and
   c) the recommended time of incubation is that required by the test kit manufacturer.

2. Specific labeling information should include the following:
   a) procedures for making minor adjustments in temperature;
   b) procedures for detecting and correcting hot or cold spots; and
   c) validation, calibration, and quality control procedures.

E. Microplate readers (Spectrophotometers)

1. Performance data should include test results demonstrating the following:
   a) the effects of power surges;
   b) the interference effects, if any, of other equipment;
   c) reproducibility of results;
   d) stability in maintaining the manufacturer's recommended settings; and
   e) that wavelength used is optimal.

2. Specific labeling information should include the following:
   a) wavelength required for each test that can be performed;
   b) linear limits (absorbance range);
   c) linearity or accuracy;
   d) drift;
   e) repeatability;
   f) bandwidth;
   g) a method to prevent and detect the use of the wrong wavelength;
   h) a method to detect incorrect programming;
   i) information on battery -- shelf life, effects of power outage and surges;
   j) recommended placement of the device to avoid environmental interferences;
   k) recommended procedures for realigning the light source;
   l) a procedure for daily quality control of the barcode reader; and
   m) validation, calibration, and quality control procedures.

ATTACHMENT II.

SAMPLE TROUBLESHOOTING GUIDE (not to be considered all inclusive)

The purpose of this guide is to assist the blood establishment in correcting minor problems encountered any time a device is in use, thus minimizing downtime. This guide should be included in the User's or Operator's manual. It should be prepared and provided by the device manufacturer with input from users. With the continuing advances in technology, this guide should be in a format that facilitates additions, deletions, and corrections.

A list of equipment, problems associated with each, and the reasons for their occurrence follows:

A. <u>Sampler/sample Identifier/Micropipettor/Dilutor</u>

1. <u>Incorrect amount or no sample or reagent added or aspirated</u>
    a) the pipet tip is incorrectly positioned which in turn can cause air to be aspirated, thus reducing the amount of sample drawn or causing the incorrect amount of reagent to be added to the reaction chamber;
    b) the wrong pipet tips are used (not recommended by the device manufacturer or the device manufacturer has not specified which pipet tips have been found to be compatible with the device through parallel testing);
    c) the tips are not changed in accordance with the manufacturer's instructions (or the manufacturer has failed to provide adequate instructions);
    d) the tip threshold calibration has not been performed when the pipet tip nozzle, tubing syringe, or syringe plunger were changed;
    e) there are occasions when programmable dilutors do not hold the program and revert back to an unprogrammed state;
    f) the liquid level detector is not performing properly and does not signal the microprocessor to relay an error message;
    g) there are clogged lines or air bubbles in the system which have gone undetected; or
    h) the lines have been mixed and the wrong reagent has been used.

2. <u>Misalignment of pipet tip</u>
    a) the device picks up the pipet tips improperly;
    b) the wrong tips are used;
    c) the wrong micropipettor is used;
    d) there is a misalignment of either the pipet tip shaft of the device or the sample, dilution or reagent holding rack;
    e) the user has failed to verify that the device is operating properly or has failed to follow SOPs; or
    f) the manufacturer has failed to adequately validate this stage of the process or has not provided adequate instructions.

3. Incorrect sample identification
   a) there are undetected errors with the built-in barcode reader;
   b) the sample tube is incorrectly aligned;
   c) the barcode has been damaged in some way;
   d) the label supplier has been changed without verifying that the labels are compatible with the device;
   e) the wrong label is placed on tube;
   f) the user fails to quality control the barcode reader each day of use;
   g) the manufacturer fails to thoroughly test the barcode reader before distribution; or
   h) the manufacturer fails to provide adequate quality control instructions to the user.

4. Missing samples
   a) the micropipettor fails to pick up a pipet tip;
   b) the micropipettor fails to aspirate a sample; or
   c) no sample is put into the sample well of a reaction cell.

B. Microplate Washer

1. Residual wash solution in well
   a) the pipet tip is not properly positioned and does not dip far enough into the well;
   b) the seal between the pipet tip and the shaft is not complete;
   c) the wrong microplates are used;
   d) the washer not properly primed;
   e) the user fails to follow the appropriate SOPs;
   f) the manufacturer fails to properly test the device before distribution;
   g) the manufacturer fails to provide adequate instructions for the user;
   h) the device does not have an adequate vacuum source;
   i) the vacuum tubing is not of the recommended diameter and wall thickness;
   j) the lines have become clogged;
   k) the waste container is full;
   l) the line filter is wet or clogged; or
   m) the reaction tray (microwell plate) is not positioned correctly onto washer.

2. Bacterial/fungal growth in wash reservoir
   a) the system is not used on a regular basis;
   b) the tubing and connectors are not changed on a regular basis;
   c) the regularly scheduled maintenance is not performed;
   d) the wash solution is contaminated by improper handling;
   e) the recommended disinfectant is not used; or
   f) the disinfectant procedure is not followed.

3. Inadequate washing cycle/uneven filling of wells
   a) the wash solution reservoir is not checked to determine that an adequate amount of solution is in the reservoir;
   b) the washer is not properly primed before each use;
   c) salt crystals or particulate matter clog the dispensing tip;
   d) the pressure is insufficient to deliver the appropriate amount of wash solution;
   e) there is a crimp or blockage in the fill line;
   f) the volume of wash solution is inadequate due to an incorrect volume setting or the plate is not properly seated in the washer;
   g) the proper working height is not used;
   h) the correct fluid levels are not maintained;
   i) the inlet tubing is not the recommended diameter and wall thickness;
   j) the plates are used with strip wells that are not properly seated; or
   k) plates are used that are not recommended by the manufacturer.

C. Incubators

1. Temperature fluctuations
   a) the chamber door or lid is opened too often;
   b) the temperature controller is not functioning properly;
   c) the device is not located properly (i.e., placed close to sunlight, heat or air conditioning vents);
   d) the water pan (found in some incubators) dries out; or
   e) the device is moved without recalibrating.

2. Hot spots/cold spots
   a) the circulating fan is not working properly; or
   b) the device is moved and the heating elements are out of alignment.

D. Microplate Readers (Spectrophotometers)

1. Incorrect readings (falsely lowered or elevated)
   a) a particular reagent is omitted;
   b) the wrong reagent or the wrong amount of reagent is added;
   c) the microplate wells have scratches on the bottom;
   d) the wrong plates or dirty plates are used;
   e) the microplate wells have been allowed to dry out;
   f) the wrong wavelength is used;
   g) the device is programmed incorrectly;
   h) the microprocessor operating the system fails to use the correct wavelength;
   i) the plate is not seated correctly in the reader;
   j) the bottom exterior of plates contains fingerprints or moisture;
   k) the reader is exposed to excessive dust, vibration, strong magnetic fields, direct sunlight, draft, excessive moisture or large temperature fluctuations;
   l) the linearity verification and/or drift verification were not performed either at all or according to manufacturer's directions;

m) residual amounts of diluted Plate Wash Buffer remain in test wells prior to the addition of Substrate;
n) the Substrate tablets are not completely dissolved;
o) the Substrate tablet has been contaminated by moisture or metal forceps, or is not intact (< a whole tablet);
p) the temperature equilibration of the test components is incomplete;
q) the reader is not properly warmed up;
r) the position of the blank well may have been changed and the wrong value was subtracted from each reading; or
s) results may not have been read within the allowable time.

2. Data not transferred
   a) the reader and the microprocessor have different character sets;
   b) the baud rates do not match;
   c) the transmit/receive pin configuration is not set properly;
   d) the communication interface software routines fail; or
   e) the connection cable is improperly connected.

3. Light beam misalignment
   a) the device is shipped or moved without the necessary precautions;
   b) the installation or preventive maintenance is performed; or,
   c) the light source (lamp) is changed and is not seated properly or aligned correctly.

4. Incorrect sample identification
   a) the plates are not loaded properly; or,
   b) the wrong sample identification has been entered into the reader.

E. Computer

1. Failure to give error messages
   a) software section controlling conditions that trigger alarms, errors, and warning messages has not been thoroughly tested and validated by the manufacturer.

2. Failure to detect malfunctions
   a) functional components of the system (liquid level detection system, fluid handling, etc.) or the interface routines have not been thoroughly tested.

FDA COMPLIANCE PROGRAM GUIDE 7342.002 -- INSPECTION OF SOURCE PLASMA ESTABLISHMENTS

PROGRAM: 7342.002

SUBJECT: Inspection of Source Plasma Establishments (FY 91)

IMPLEMENTATION DATE: Upon Receipt

COMPLETION DATE: Continuing

DATA REPORTING:
PRODUCT CODES: For inspections
57DY-[ ] [ ]
57YY-[ ] [ ]

PROGRAM/ASSIGNMENT CODES: 42002 Source Plasma Establishments
42R825 AIDS Related Activity
Note: For all joint prelicense inspections of blood or plasma related establishments, use PAC 42832

DATE OF ISSUANCE: 09/07/94

[**BOLD TEXT** denotes current changes]
Field Reporting Requirements

EIR's and Form FDA-483's are not to be routinely sent to CBER. However, district office requests for review of an EIR regarding a particular issue that may need policy development or clarification should be sent in a memorandum to the Surveillance and Policy Branch, (HFM-660). Include copies of the EIR and exhibits for these issues, if applicable, to facilitate CBER evaluation.

PART I -- BACKGROUND

Section 640.60, Title 21, **Code of Federal Regulations** (CFR), defines the product, Source Plasma, as the fluid portion of human blood collected by plasmapheresis and intended as source material for further manufacturing use whether intended for use in injectable, noninjectable, licensed, or unlicensed products. **Source Plasma is not** intended for direct transfusion into patients.

Plasmapheresis is the procedure in which blood is removed from the donor, the plasma is separated from the formed elements, and at least the red blood cells are returned to the donor. This procedure may be performed manually or by use of automated equipment. The collection procedures, **whether manual or automated**, must be described in detail in the establishment's standard operating procedures manual.

Source Plasma is a product subject to the licensure provisions of Section 351 of the Public Health Service Act (42 U.S.C. 262). For Source Plasma is suitable for interstate (including international) shipment, **if the manufacturer has** an unsuspended and unrevoked U.S. license for the establishment (collection facility) and the product [21 CFR 601.10 and 601.20]. Before a product **may** be labeled as Source Plasma, it must comply with the standards set forth in 21 CFR, Part 640, Subpart G,and other applicable standards prescribed in the regulations contained in **21 CFR 600, Subpart B and 21 CFR 601, 606, 607, 610, subparts E & F and 21 CFR 211**.

NOTE: Some Source Plasma firms have product license applications for the collection of Source Leukocytes and/or Whole Blood for further manufacture. The collection of these products is addressed in Attachments L, R, and X of this compliance program

PART II -- IMPLEMENTATION

A. OBJECTIVE

The FDA objective in issuing this program is to assure that Source Plasma is safe, effective, adequately labeled and possesses the quality purported. To achieve this objective, the Agency needs to:

1. **Determine** if plasmapheresis establishments are operating in compliance with applicable regulations.
2. **Assure** donor protection and a safe, pure, and potent product for use in further manufacturing.
3. **Provide** guidance for the **performance of annual FDA inspections** of all licensed establishments collecting Source Plasma.
4. **Promote** voluntary compliance **and principles of quality assurance**.

The firms covered under this program include all establishments engaged in the manufacturer of Source Plasma **and auxiliary establishments (e.g., off-site storage sites, laboratories that perform testing to determine suitability of donors and product, and interstate carriers)**.

B. PROGRAM MANAGEMENT INSTRUCTIONS -- INSPECTIONAL

1. Inspections of all establishments licensed to manufacture Source Plasma will be conducted **annually** and will cover Parts 600, 601, 606, 607, 610.40, 610.41, 610.45, 610.53, and 640.60 through 640.76 **and 211** of Title 21, **Code of Federal Regulations.**
2. Each **annual** inspection under this Compliance Program will encompass the total operation of the establishment. **Inspections performed** to verify correction of previous violations need not cover all operations.

NOTE: If a licensed plasma center has a change in operations which requires a license **supplement [for example, a change in location, (21 CFR 601.12)]**, the Center for Biologics Evaluation and Research (CBER) **will** request a directed inspection.

3. CBER **provides** a checklist with accompanying instruction booklet for this program. The Plasmapheresis Checklist and Report (Form FDA-2722) is designed as an inspectional tool **and guide** for the investigator. Copies of the checklist and instruction booklet can be obtained from the Biological Product Inspections Branch **(HFM-655)**. **Completion of the checklist as part of the inspection report is no longer required. The information and guidance contained in the checklist and instruction booklet are being reformatted for issuance as an inspectional guide.**

PART III -- INSPECTIONAL

A. OPERATIONS

2. Prelicense Inspections

Staff of the Center for Biologics Evaluation and Research (CBER) and the **local district office** will conduct a joint prelicense inspection after the establishment and product license applications have been reviewed by CBER and found to be **satisfactory and acceptable**. The prelicense inspection is an in-depth review of the physical facilities, manufacturing methods, equipment, etc. **During** this inspection, the establishment must be **operating and processing Source Plasma**, have records for at least 30 days of operation **available for review**, and **must** demonstrate its ability **to** operate in compliance with applicable regulations **before** a product and/or establishment **license(s) is** issued.

A licensed firm which converts to automated collection procedures must receive approval of an ELA and PLA supplement prior to shipment of the Source Plasma. After CBER notifies a firm of the reference numbers assigned to its submissions, the firm may collect plasma and quarantine it until final approval is received. For additional information on the conversion of plasma centers to automated methods of Source Plasma collection, see Appendix S.

NOTE: **Major manufacturing changes such as changes in the collection of source plasma from manual to automated procedures usually require a preapproval inspection of the facility. Under a program commonly referred to as "me-too" conversions, plasmapheresis centers may be approved for**

automated collection procedures without an on-site inspection. (See Attachment BB.)

Arrangements for joint prelicense inspections are made as follows:

a) **CBER, Biological Product Inspections Branch (HFM-655), will notify the Director, Investigations Branch of the proposed date(s) of inspection;**
b) The district may select an Investigator to accompany the CBER Inspector(s); and
c) The District Investigator will contact the CBER Inspector(s) to make final arrangements **for the inspection**. The CBER Inspector is **the lead Inspector and is** responsible for completing the Establishment Inspection Report (EIR) and **generating any additional correspondence with the firm**. The Biological Product Inspections Branch, **(HFM-655) will send copies of the EIR, the Form FDA-483, and post-inspectional correspondence to the district office**.

2. Activities Approved Under License

CBER, Division of **Blood Applications, (HFM-370)** provides establishment licensing information to **the Office of Compliance. The Surveillance and Policy Branch (HFM-660) forwards this information to district offices. Attachment F contains a listing of typical licensing changes reported by HFM-370.**

Each firm must have CBER approval for significant changes in manufacturing methods and labeling prior to implementation [21 CFR 601.12(b)]. Observations of unapproved significant changes should be cited on the Form FDA-483. If questions concerning the significance of a change arise, the Division of Blood Applications (HFM-370) should be notified. (See Part VI for contacts.)

3. Inspections of Licensed Source Plasma Establishments

This Compliance Program contains specific detailed instructions for inspection of plasmapheresis establishments and should be used with the general guidance provided in Chapter 5 of the Investigations Operations Manual (IOM).

The districts perform annual inspections of licensed Source Plasma establishments and issue post-inspectional correspondence

Inspections should be conducted in accordance with current guidance and inspectional tools which have been provided to the Field.

Under the **Federal** Food, Drug, and Cosmetic Act and provisions of Section 351 of the Public Health Service Act, **Investigators are authorized** to examine all parts of the manufacturing facility, the manufacturing processes and controls, **computerized manufacturing** operations, and all pertinent records. Subpart C of 21 CFR, Part 600 discusses inspections of biological product establishments and outlines the duties of the **Investigator**.

If, **for any reason**, an Investigator is denied access to an establishment to conduct an inspection or is denied access to records for review or copying during an inspection, the **Investigator should** follow the applicable instructions in Chapter 5 of the IOM. Refusal to permit inspection of an establishment and refusal to permit access to or copying of required records are both bases for revocation of a U.S. license according to 21 CFR 601.5.

Objectionable conditions observed during the inspection are noted on a Form FDA-483 which, at the close of the inspection, is presented to the most responsible person on the premises. If the Responsible Head is not present to receive the original Form FDA-483, a cover letter and copy of the FDA-483 should be sent to the Responsible Head to assure that management is aware of the observations. If inspection is violative, refer to Part V for guidance.

B. <u>RECORD REVIEW</u>

Record reviews are most productive when the review is directed by inspectional observations indicating potential problems. In these instances, the number of records to review is determined on a case-by-case basis, but must be sufficient to document the existence, extent, and impact of the problem.

When inspectional findings have not suggested the existence of any specific problem areas, a statistical approach for the review of any type of record may be used as an audit procedure. When using the statistical approach, randomly select 80 records from the records to be audited. Normally, this group consists of all of those records which were produced since the previous inspection. When multiple reviews are conducted, for example as in the inspection of a laboratory, i.e., anti-HIV testing, hepatitis testing, etc., the sample size of 80 applies to each type of test record for review. Statistically, it does not matter whether or not the donor remains the same for the different reviews. If no significant errors are found, further review of records is not required. When significant errors are found, judgment must be used in determining appropriate follow-up.

This statistical approach is intended to supplement normal inspectional activities and is only used when there is no indication of problems. However, it is not to be used alone to determine compliance status of an establishment. Furthermore, this review of 80 records provides essentially the same confidence level for all quantities of records. However, this review ill not find, with any degree

of certainty, those rare events where individual, non-systematic errors have occurred. Rather, the review provides additional confidence that widespread systematic testing and processing errors do not exist.

Revised Requirements for Notification of Personnel Changes

Since October 1993, the responsibility for personnel changes, at operating locations where the Responsible Head is not on-site has been the sole responsibility of the Responsible Head. That responsibility includes ensuring that personnel standards outlined in 21 CFR 600.10 are met. Therefore, changes in center management or physicians need no longer be submitted to CBER for review and approval. Current curricula vitae, signed statements attesting to an understanding of applicable requirements, and documentation of credentialling (where appropriate, e.g., state licensure for physicians) should be on file at the operating facility. These records should be reviewed by the Investigator during inspection. CBER has recommended that each licensed location maintain on its premises, a current listing of the names of the center's on-site managers and physicians, including the dates of employment. Investigators should review the records identified above during inspections.

The above practice, at this time, does not extend to physician substitute candidates because records and documents submitted to CBER for review are often incomplete/unacceptable. For approval of a physician substitute, CBER continues to require that specific records/documents be submitted for review. If the submission is complete and acceptable, CBER will apply a dated, signed, stamped acknowledgement statement to the manufacturer's cover letter accompanying the submission. A copy of the cover letter is then returned to the Responsible Head. Investigators should review these letters for documentation of physician substitute approval.

C. SAMPLING

Generally, documentary samples rather than physical samples are collected during inspections of plasmapheresis establishments. Collection of physical samples for analysis should be coordinated through the Surveillance and Policy Branch (HFM-660) prior to collection.

Interstate jurisdiction may be established by documenting the interstate movement of either the plasma product, if applicable, or the anticoagulant in the blood collection container.

In general, documentary samples are almost always sufficient to support legal action. For samples collected at plasmapheresis centers, it is often helpful to include the standard operating procedures applicable to the violation being documented. Documents which may be worthwhile to include in a documentary sample are: donor records, quality control records, laboratory testing records, processing **records**, destruction/disposition/shipping records, and photographs of the violative conditions. **See Sub-Chapter 405, Investigations Operations Manual and Part IV, Analytical of this program.**

D. DOCUMENTATION

Investigators should contact their supervisors as soon as possible when violations which may warrant **regulatory/administrative action are encountered.**

To assist in determining whether regulatory action is warranted, two lists of significant deviations are provided as Attachments A and B.

To support regulatory action, deviations must be well documented and represent a continual pattern of operation rather than an isolated occurrence, unless an isolated deviation has jeopardized the safety of donors or products, or has violated a mandated requirement of licensure. **Refer to the Investigations Operations Manual Sub-Chapter 430, for sample documentation.**

Do not hesitate to take photographs to support verbal description of violative conditions in a plasmapheresis center. Investigators should collect sufficient evidence to document each **Form** FDA-483 observation. The following guidance is **provided:**

1. When documenting improper testing **(for a firm's performing their own testing)**, obtain copies of the **firm's** SOPs, appropriate test result records that contain the unit number, bleed date in question, date of testing, and **the manufacturer's instructions.**
2. When documenting lack of records, copy records before and after bleed date and unit number to show which records are missing, **and other records which show that unit existed. Supplement with SOP's and affidavits as appropriate.**
3. When documenting labeling deviations, make copies of the labels.
4. For deviations concerning shipment of units prior to receiving written test results, obtain copies of invoices and shipping (distribution) records **which show shipping dates** in addition to the appropriate **SOP's and** test result records.
5. **When documenting shipment of units untested or reactive for infectious disease test markers but which were labeled as nonreactive and shipped,** obtain a documentary sample, and immediately notify **district management.** (See Attachment **G and** L). **If the plasma center uses an outside testing facility which appears to have caused or contributed to the problem, an inspection of the testing facility should be requested.**
6. If misuse or malfunction of equipment is observed, the following documentation will be important to include in the report:

a) **Type of equipment**, manufacturer and identifying number.
b) Number of experiences (failures).
c) **A copy of the SOP and manufacturer's instructions (i.e., operator's manual)**, and a statement indicating whether the personnel were following prescribed procedures.
d) **A copy of the firm's notification of the problems/defects which was sent to the manufacturer, supplier, or FDA and the response received.**
e) **Records documenting that** personnel were **not** adequately trained in the use of the equipment.

NOTE: If observations noted with the use or performance of equipment or supplies represent potential design problems or possible inadequate instructions for use, alert CBER **(HFM-660)** and the **district office** where the manufacturer is located.

7. **If a danger to health is documented with respect to product or donor safety** and a recommendation for suspension is contemplated, an inventory of the plasma at the establishment on the last day of inspection should be obtained. This **theoretical** inventory may be determined by:

   a) subtracting the bleed number of the last unit shipped from the bleed number of the last unit collected; or
   b) **Collecting copies of** the packing records (with unit numbers) which are often used as shipping records and **which** list the **units** currently in storage. CBER will make a decision as to the disposition of the plasma in inventory at a suspended firm after review of all information.

8. Overbleeding

   a) When documenting overbleeding by volume **for manual plasma collection**, obtain copies of records that include the unit number, date, volume or weight of each unit of whole blood, the volume/weight of pooled plasma, and the weight of the donor on the day of donation. Also, consider collecting scale calibration and **repair**/maintenance records.
   b) When documenting overbleeding by volume for automated plasmapheresis, obtain copies of records that include the unit number, date, total volume collected, and donor hematocrit, weight, **etc.**, on the day of donation **or other criteria used to determine volume of plasma to be collected**. Also, consider collecting **donor** scale calibration **records**, repair/maintenance records, and employee training records.
   c) When documenting overbleeding by frequency, obtain copies of records that indicate which donors, unit numbers and bleeding dates are involved. Documenting the donor's name should be avoided unless

absolutely essential. **Generally, initials or other unique identifiers should be used.**

<u>NOTE</u>: Overbleeding is further discussed in Attachment B, Item 5.

D. <u>SPECIFIC INSTRUCTIONS</u>

In addition to other guidance provided to the Field, specific instructions relative to the **topics listed below** are provided in Attachments G through Z, **AA and BB**.

	Attachment G:	**Handling of Units Reactive for or Untested for Markers of Infectious Disease**
	Attachment H:	Testing Blood and Plasma for Antibodies to HIV
	Attachment I:	Off-Site Plasma Storage
	Attachment J:	Reference Laboratories and Other Outside Testing Laboratories
	Attachment K:	Imported Biological Products
	Attachment L:	Shipments of Plasma Prior To Receipt of Written Test Results
	Attachment M:	**Incidents of Wrong Red Blood Cell Infusion**
	Attachment N:	Infrequent Plasmapheresis **Programs**
	Attachment O:	Physician Substitutes
	Attachment P:	**Use of AIDS Educational Materials in Donor Screening to Elicit a History of High Risk Behavior and the Presence of Signs and Symptoms of AIDS and Other HIV Infections**
	Attachment Q:	**Shipments of Plasma Repeatedly Reactive for Anti-HIV or Anti-HCV or Positive for HBsAg**
	Attachment R:	Disposition of **Unsuitable Units**
*	Attachment S:	Automated Plasmapheresis
	Attachment T:	Misuse or Malfunction of Equipment or Supplies
	Attachment U:	Arm Preparation for Venipuncture
	Attachment V:	Damaged Bags and Bottles of Source Plasma
	Attachment W:	Computerization
	Attachment X:	**Additional Laboratory Testing: ALT, Anti-HBc, and Anti-HTLV-I**
	Attachment Y:	**Collection of Source Plasma from Special Donor Populations: Disabled Donors and Non-English Speaking Donors**
	Attachment Z:	**Reporting of Errors and Accidents**
	Attachment AA:	**FDA Investigators' Witnessing of Donor Interviewing**
	Attachment BB:	**Manufacturing Changes in Source Plasma Collection**

E. <u>Reporting</u>

Field Reporting Requirements are listed on the coversheet of this manual.

PART IV -- ANALYTICAL

No field analyses are projected under this Compliance Program.

Samples collected on assignment will be analyzed by **laboratories within CBER, e.g., the Division of Blood Applications (HFM-370) and the Division of Biological Product Quality Control (HFM-230), or CBER will otherwise coordinate testing.**

Contact the Surveillance and Policy Branch (HFM-660) for guidance and instructions prior to collection and submission of physical samples to determine: sample type and size, mode of shipment, and method of packaging for shipment since the products being sampled may contain etiological agents (See 42 CFR, Part 72).

Results of analysis will be forwarded to the District Compliance Branch. **Questions concerning analytical results should be addressed to HFM-660.**

All biological product samples for analysis submitted by field Investigators should be sent by courier as follows:

Sample Custodian (Attn: HFM-235)
Center for Biologics Evaluation and Research
Room 175
5516 Nicholson Lane
Kensington, MD 20895

Note: Notify HFM-235 prior to shipment to assure that staff will be available in the CBER laboratory on the expected date of sample arrival. Telephone: 301-594-6517 FAX: 301-594-6924

PART V -- REGULATORY/ADMINISTRATIVE STRATEGY

Violative conditions must be promptly evaluated during and following an inspection in order to assure donor protection and product safety. The strategy and actions which are outlined below are those we believe are most appropriate and expeditious when donor protection or product quality is in question. Consideration of other regulatory actions is not precluded.

When allegations are received by the District Office relating to the concealment of violative practices by means of inaccurate record keeping or other forms of cover-up, telephone contact with the **Division of Case Management (HFM-610)** early in the investigation is encouraged **(See Part VI for contact persons.)**. Prior to, or concurrent with, the initiation of the establishment inspection, visit **the informant(s)** and obtain an affidavit documenting how the true records can be distinguished from the reportedly inaccurate records. During the establishment inspection, attempt to corroborate the information supplied by the informant(s).

Lists of significant deviations which are included as Attachments A and B, will assist in determining the appropriate course of action.

A. SUSPENSION/**REVOCATION** OF OPERATIONS UNDER LICENSE

If donor health and/or product quality is jeopardized, necessitating immediate corrective action, and the deviations are well documented, Investigators should recommend suspension of operations under license to the CBER, Office of Compliance, Division of Case Management **HFM-610** as soon as possible and no later than 24 hours following the completion of the inspection. **When the district believes it has encountered such a situation, it should notify HFM-610 promptly and maintain contact with HFM-610 throughout the inspection and prior to the issuance of the Form FDA-483. Examples of violations that may warrant this level of corrective action are listed in Attachment A. Determine the effect that** the violative condition(s) **may have** on product quality **or purported quality**. The firm must make proper disposition of any product whose quality has been compromised.

Refer to the Regulatory Procedures Manual (RPM), Chapter 8-80, License Revocation or Suspension.

B. WARNING LETTERS

1. The lists of significant deviations serve as a guide for determining the recommended course of action (See Chapter 8-80 of the RPM for general guidance). Any **single** deviation **listed** on Attachment B, if well documented and representative of a continual pattern of operation, as opposed to an occasional deviation, may warrant a **Warning Letter**. Additionally, there may be some continually occurring deviations which do not affect the donor's health or the integrity **or quality** of the product, but reflect an overall inadequately operated establishment **which may warrant a Warning Letter**.
2. When a District Compliance Officer determines that a **Warning Letter** should issue, **Chapter 8-10 of the Regulatory Procedures Manual (RPM) provides guidance. The following specific areas require CBER concurrence prior to a district issuing a Warning Letter: viral marker testing violations; donor re-entry violations; violations in areas where specific guidance has not been provided; labeling violations (except areas where specific guidance has been provided, e.g., Compliance Programs, CPGs, Drug/Health Fraud Bulletins); computer application and software violations; 21 CFR 211 GMP deviations; Warning Letters to Federal agencies; and product advertising violations.** Copies of all **Warning Letters initiated by the district and follow-up correspondence should be submitted to the Division of Case Management (HFM-610).**

3. CBER regulatory policies are stated in the Compliance Policy Guide Manual, Chapter 34 -- Biologics. The **titles of the** current Compliance Policy Guides **that relate to Source Plasma** are listed in Part VI.

   NOTE: The Compliance Policy Guides are under review for updating and revision. If necessary, contact HFM-660 regarding the status and impact of the proposed revisions.

C. FEDERAL/STATE RELATIONS

Currently, the Agency has no formal cooperative program with State or local jurisdictions for the inspection or regulation of Source Plasma manufacturers; however, cooperation with these authorities is encouraged, especially if a state or local jurisdiction has a regulatory program. Whenever feasible, exchanges of information should occur between all levels of government. **Districts should routinely provide copies of Warning Letters to the appropriate state agency or agencies. If the state regulatory structure for Source Plasma products is not known, initiate appropriate inquiries.**

PART VI -- REFERENCES, PROGRAM CONTACTS AND ATTACHMENTS

REFERENCES

1. **Federal Food, Drug, and Cosmetic Act, as Amended, and Related Laws. This booklet contains the Public Health Service Act, Biological Products.**
2. Title 21, Code of Federal Regulations, Parts 211, 600, 601, 606, 607, 610, and 640.
3. Preamble to Regulations for Additional Standards for Source Plasma (Human) Used in Preparation of Blood Derivatives Intended for Injection, Federal Register, Vol. 38, No. 139, July 20, 1973.
4. Preamble to Regulations for Amendments to the Additional Standards for Source Plasma (Human), Federal Register, Vol. 41, No. 40, March 12, 1976.
5. **Preamble to Good Manufacturing Practices for Human Blood and Blood Components, FR Vol. 40, No. 223, November 18, 1975.**
6. Title 42, Public Health Service Act, Part 72, Interstate Commerce, Subpart C. (See also FR, Vol. 45, No. 141, July 21, 1980.)
7. Technical Manual, Current Edition, American Association of Blood Banks, 8101 Glenbrook Rd. Bethesda, MD 20814.
8. **Guide to Inspections of Plasma Centers.**
9. Regulatory Procedures Manual (RPM): Chapter 8-10 (**Warning Letters**); Chapter 8-80 (License Revocation and Suspension); Chapter 9-77 (Biological Products); and Chapter 9-79 (Import Alert 57-01).
10. **Investigations Operations Manual, Subchapter 560 (Biologics), Subchapter 920 (Investigations -- Injury and Adverse Reactions -- Biologics, et al).**

GUIDELINES

11. **Revision to 26 October 1989, Guideline for Collection of Blood or Blood Products from Donors with Positive Test for Infectious Disease Markers ("High Risk" Donors). Memorandum of April 17, 1991, to all registered blood establishments.**
12. Guideline for Collection of Blood and Blood Products from Donors with Positive Tests for Infectious Disease Markers ("High Risk" Donors), Memorandum of October 26, 1989 to blood establishments.
13. **Uniform Labeling of Blood and Blood Components, August 1985.**
14. Adding Heparin to Empty Containers for Collection of Heparinized Source Plasma, August 1981.
15. Guidelines for the Collection of Human Leukocytes For Further Manufacturing (Source Leukocytes), Memorandum of January, 1981, to all registered blood collecting facilities.
16. Immunization of Source Plasma (Human) Donors with Blood Substances, June 1980.
17. Reviewing Amendments to Include Plasmapheresis of Hemophiliacs, July 20, 1976.
18. Interpretive Guidelines of the Additional Standards for Source Plasma (Human), September 28, 1973.

DRAFT GUIDELINES

19. **Draft Guideline for the Validation of Blood Establishment Computer Systems, September 28, 1993.**
20. **Draft Guideline for Quality Assurance in Blood Establishments, June 17, 1993.**

MEMORANDA

21. Use of an FDA Cleared or Approved Sterile Connecting Device (STCD) in Blood Bank Practices, Memorandum of July 29, 1994 to all registered blood and plasma establishments.
22. Recommendations for Deferral of Donors for Malaria Risk, Memorandum of July 26, 1994 to all registered blood establishments.
23. **A Letter to Blood Establishment Computer Software Manufacturers, March 31, 1994.**
24. **Recommendation for the Invalidation of Test Results When Using Licensed Viral Marker Assays to Screen Donors, Memorandum of January 3, 1994, to all registered blood and plasma establishments.**
25. **Donor Suitability Related to Laboratory Testing for Viral Hepatitis and a History of Viral Hepatitis, Memorandum of December 22, 1993, to all registered blood establishments.**
26. **Guidance Regarding Post Donation Information Reports, Memorandum of December 10, 1993, to registered blood and plasma establishments.**

27. Clarification of the Use of Unlicensed Anti-HCV Supplemental Test Results in Regard to Donor Notification, Memorandum of August 19, 1993, to all registered blood and plasma establishments.
28. Revised Recommendations for Testing Whole Blood, Blood Components, Source Plasma and Source Leukocytes for Antibody to Hepatitis C Virus Encoded Antigen (Anti-HCV), Memorandum of August 5, 1993, to all registered blood establishments.
29. Deferral of Blood and Plasma Donors Based on Medications, Memorandum of July 28, 1993, to all registered blood establishments.
30. Recommendations Regarding License Amendments and Procedures for Gamma Irradiation of Blood Products, Memorandum of July 22, 1993, to all registered blood establishments.
31. Revision of October 7, 1988, Memorandum Concerning Red Blood Cell Immunization Programs, Memorandum of December 16, 1992, to all licensed establishments performing red blood cell immunization.
32. Volume Limits for Automated Collection of Source Plasma, Memorandum of November 4, 1992, to all licensed Source Plasma establishments.
33. Changes in Equipment for Processing Blood Donor Samples, Memorandum of July 21, 1992, to all licensed blood establishments.
34. Exemptions to Permit Persons with a History of Viral Hepatitis Before the Age of Eleven Years to Serve as Donors of Whole Blood and Plasma: Alternative Procedures, 21 CFR 640.120, Memorandum of April 23, 1992, to all registered blood establishments.
35. Revised Recommendations for the Prevention of Human Immunodeficiency Virus (HIV) Transmission by Blood and Blood Products, Memorandum of April 23, 1992, to all registered blood and plasma establishments.
36. Revised Recommendation for Testing Whole Blood, Blood Components, Source Plasma and source Leukocytes for Antibody to Hepatitis C Virus Encoded Antigen (Anti-HCV), Memorandum of April 23, 1992, to all registered blood establishments.
37. Use of Fluorognost HIV-1 Immunofluorescent Assay (IFA), Memorandum of April 23, 1992, to all registered blood and plasma establishments.
38. Clarification of FDA Recommendations for Donor Deferral and Product Distribution Based on the Result of Syphilis Testing, Memorandum of December 12, 1991, to all registered blood establishments.
39. Disposition of Blood Products Intended for Autologous Use that Test Repeatedly Reactive for Anti-HCV, Memorandum of September 11, 1991, to all registered blood establishments.
40. FDA Recommendations Concerning Testing for Antibody to Hepatitis B Core Antigen (Anti-HBc), Memorandum of September 10, 1991, to all registered blood establishments.
41. Revision to 26 October 1989, Guideline for Collection of Blood or Blood Products from Donors with Positive Test for Infectious Disease Markers ("High Risk" Donors), Memorandum of April 17, 1991, to all registered blood establishments.

42. **Deficiencies Relating to the Manufacture of Blood and Blood Components, Memorandum of March 20, 1991, to all registered blood establishments.**
43. **Responsibilities of Blood Establishments Related to Errors & Accidents in the Manufacture of Blood & Blood Components, Memorandum of March 20, 1991, to all registered blood establishments.**
44. Abbott Laboratories' HIVAG-1 test for HIV-1 antigen(s) not recommended for use as a donor screen, Memorandum of October 4, 1989, to all registered blood establishments.
45. Requirements for Computerization of Blood Establishments, Memorandum of September 8, 1989, to all registered blood establishments. **(To be superseded by proposed guideline upon completion.)**
46. Use of the Recombigen HIV-1 **Latex Agglutination (LA)** Test, Memorandum of August 1, 1989, to all registered blood establishments.
47. HTLV-I Antibody Testing, Memorandum of July 6, 1989, to licenses Source Plasma manufacturers approved for immunization with Red Blood Cells.
48. Blood Bank and Plasma Establishment Inspectional Guidance (to check equipment and supplies for misuse or defects), Memorandum of May 10, 1989, to Regional Directors and District Directors.
49. Guidance for Autologous Blood and Blood Components, Memorandum of March 15, 1989, to all registered blood establishments.
50. Use of the Recombigen HIV-1 LA Test, Memorandum of February 1, 1989, to all registered blood establishments.
51. HTLV-I Antibody testing, Memorandum of November 29, 1988, to all registered blood establishments.
52. Physician Substitutes, Memorandum of August 15, 1988, to all licensed manufacturers of source plasma.
53. Recommendations for Changeover from Use of Fresh Immunizing Red Blood Cells to Use of Frozen Immunizing Red Blood Cells Stored a Minimum of Six Months Prior to Use, Memorandum of October 7, 1988, to licensed Source Plasma establishments approved or applying to perform Red Blood Cell immunizations.
54. Revised Guideline for the Collection of Platelets, Pheresis, Memorandum of 7 October 1988, to Facilities Preparing Platelets, Pheresis.
55. Discontinuance of Pre-License Inspection of Immunization Using Licensed Tetanus Toxoid and Hepatitis B and Rabies Vaccine, Memorandum of July 7, 1988, to all licensed source plasma manufacturers.
56. Control of Unsuitable Blood and Blood Components, Memorandum of April 6, 1988, to all registered blood establishments. **(To be superseded by proposed guideline upon completion.)**
57. Recommendations for Implementation of Computerization in Blood Establishments, Memorandum of April 6, 1988, to all registered blood establishments.
58. Recommendations for the Management of Donors and Units that are Initially Reactive for Hepatitis B Surface Antigen (HBsAg), Memorandum of December 2, 1987, to all registered blood establishments.

59. Deferral of Donors Who Have Received Human Pituitary-Derived Growth Hormone, Memorandum of November 25, 1987, to all registered blood establishments.

60. Plasma Derived from Therapeutic Plasma Exchange, Memorandum of December 14, 1984, to all Source Plasma establishments.
61. Physician Substitutes, Memorandum of December 14, 1984, to all manufacturers of **Source Plasma**.
62. Equivalent Methods for Compatibility Testing, Memorandum of December 14, 1984 to all registered blood establishments.
63. Deferral of Blood Donors Who Have Received the Drug Accutane (isotretinoin/Roche; 13-cis-retinoic acid), Memorandum of February 28, 1984, to all establishments collecting **Source Plasma**.
64. Requirements for Infrequent Plasmapheresis Donors, Memorandum of August 27, 1982, to all licensed manufacturers of **Source Plasma**.

<u>COMPLIANCE POLICY GUIDES</u>

7134.01:	Registration of Blood Banks and Other Firms Collecting, Manufacturing, Preparing or Processing Human Blood or Blood Products
7134.02:	Human Blood and Blood Products as Drugs
7134.03:	**Standards and Minimum Requirements for Biologic Products**
7134.05:	Licensing -- Changes to be Reported per 21 CFR 601.12
7134.08:	Source Plasma (Human) -- Guidelines for Informed Consent Forms
7134.09:	**Importation of Licensed Biologic Products for Human Use**
7134.10:	Quantitative Testing for Serum Proteins in Plasmapheresis Donors
7134.12:	Blood and Blood Products -- Definitions
7134.14:	Source Plasma (Human) -- Regulatory Action Based on Overbleeding
7314.15:	Source Plasma (Human) -- Use of Units from Donors Subsequently found to be Reactive To a Serologic Test for Syphilis
7314.17:	Adequate Space for Determination of Donor Suitability
7134.18:	Schedule of Physical Examinations for Donors Receiving Immunization Injections.
7134.22:	Plasma Brokers -- Registration and Compliance with Good Manufacturing Practices
7134.23:	Plasmapheresis - 48-hour Period Between Plasmapheresis Procedures
7155e.03:	**MOU with Health Care Financing Administration Concerning Blood Banking and Transfusion Programs**

NOTE: The Biologics Compliance Policy Guides are being reviewed and updated by HFM-660.

LISTS OF SIGNIFICANT DEVIATIONS

Violations must be evaluated to determine if they warrant regulatory action. To assist in this determination, lists of significant deviations from the GMP regulations were developed:

Attachment A: Deviations Which May Warrant Suspension Of Activities Under License.

Attachment B: Deviations Which May Warrant a **Warning Letter**/Administrative Follow-up.

NOTE: Because all possible situations and variables cannot be anticipated, the lists are not all inclusive, nor are they mutually exclusive. The deviations are referenced to specific sections of the regulations.

LIST OF LICENSED TEST KITS

Attachment C: **Licensed Test Kits for the Detection of Anti-HIV**

Attachment D: Licensed Test Kits for the Detection of HBsAg **and Anti-HCV**.

Attachment E: **Licensed Test Kits for the Detection of Anti-HBsAg and Anti-HBcAg**

SPECIFIC INSTRUCTIONS

The topics of Attachments G through **BB** are listed in Part III. The information provided in these attachments is designed to answer questions which are frequently asked by field Investigators.

FDA/HEALTH CARE FINANCING ADMINISTRATION REGIONAL CONTACTS

Northeast Region	
Ellen Madigan, CSO	**617-279-1675**
Boston District	
Richard Katz, **Prg. Mgmt. Officer**	718-965-5319
New York District	
Mid-Atlantic Region	
Joseph Phillips, Asst. to RFDD	215-597-0492
Philadelphia Regional Office	
Southeast Region	
Roger Kline, Director	404-347-**3218**
Investigations Branch	
Atlanta District	
Midwest Region	
Gerald Berg, Supv. CSO	**708-688-5462**
Chicago District, Gurnee R.P.	
Jeanne Morris, Investigator	**708-323-2763**
Chicago District, Hinsdale, R.P.	
Southwest Region	
Phillip Waldron, Supv. CSO	214-655-5329
Dallas District	
James McDonald, Director	**913-752-2101**
Compliance Branch	
Kansas City District	
William Sherer, Asst. Director	**303-236-3000**
Denver District	**x357**
Pacific Region	
John Duke, CSO	
Los Angeles District	714-836-2379
Irvine R.P.	
Miriam Burbach, Supv. CSO	206-**483-4912**
Seattle District	

HCFA REGIONAL HEALTH STANDARDS AND QUALITY STAFF LABORATORY PROGRAM

Region I (CT,ME,MA,NH,RI,VT)
DHHS/HCFA/HSQB
Boston Regional Office J.F.K. **Fed.** Building, Rm. 1211
Boston, MA 02203
617-565-1299

Margaret Leoni-Lugo, Lab Consultant
Charles Reynolds, Lab Consultant

Region II (NJ,NY,PR,VI)
DHHS/HCFA/HSQB
New York Regional Office
26 Federal Plaza, **Rm. 3804**
New York, NY **10007**
212-264-**1121**

Carol Mattheiss, Lab Consultant
Ben Snyder, Lab Consultant

Region IV (AL,FL,GA,KY,MS,NC,SC,TN)
DHHS/HCFA/HSQB
Atlanta Regional Office
101 Marietta Tower
Atlanta, GA **30323**
404-331-0083/2488

Ruth McArthur, Lab Consultant

Region V (IL,IN,MI,MN,OH,WI)
DHHS/HCFA/HSQB
Chicago Regional Office
105 West Adams **Street,** 15th Floor
Chicago, IL 60603
312-**886-4392/3595**

Elizabeth Clay-Day, Lab Consultant
Bonnie Scimeca, Lab Consultant

Region IV (AR,LA,NM,OK,TX)
DHHS/HCFA/HSQB
Dallas Regional Office
1200 Mail Tower Bldg.
Dallas, TX 75202
214-767-**4414**

Sandra Pearson, **Lab Consultant**
Greg Scocio, Lab Consultant

Region VII (IA,KS,MO,NE)
DHHS/HCFA/HSQB
Kansas City Regional Office
Federal Office Building
601 East 12th Street
Kansas City, MO **64106**
816-426-2011/**3184**

Bambi Huval, Lab Consultant

Region VIII (CO,MT,ND,SD,UT,WY)
DHHS/HCFA/HSQB
Denver Regional Office
1961 Sout Street
Denver, CO 80294
303-844-4726, ext. 451 or 441

Karalou Eastmo, Lab Consultant

Region IX (AZ,CA,HI,NV,SOM., GUAM)
DHHS/HCFA/HSQB
75 Hawthorne Street, 4th Floor
San Francisco, CA 94105
415-**744-2835/3727**

Kent Feldsien, Lab Consultant
Esther-Marie Carmichael, Lab Consultant

Region X (AK,ID,OR,WA)
DHHS/HCFA/HSQB
Seattle Regional Office
2201 6th Avenue Rx-42
Seattle, WA 98121
206-**615-2321/2324**

Edwin Prichard, Lab Consultant
Gary Utter, Lab Consultant

Central Office, HCFA
HSQB/OSC/**DLSP**
Meadowwood East
6325 Security Blvd.
Baltimore, MD 21207

Judy Yost, **Division Director**
410-597-5907
Joan Simmons, Branch Chief
410-597-5882

CENTER FOR BIOLOGICS EVALUATION AND RESEARCH PROGRAM CONTACTS

General Compliance Matters, Imports, Exports, Inspections, Warning Letters, Samples, Complaints, Product Defects, Errors and Accidents, Fatalities, Viral Marker Testing, Computer Issues, GMP Guidance:

NOTE: Any questions regarding the conduct of on-going inspections, Good Manufacturing Practices, or policy should be directed to the Biological Product Inspections Branch (HFM-655) or the Surveillance and Policy Branch (HFM-660). The CBER offices referenced below may be contacted for technical or scientific questions in the areas indicated.

Office of the Director, Office of Compliance
James C. Simmons, Director
P. Michael Dubinsky, Deputy Director

American Red Cross Working Committee (FM-603)
Telephone: 301-594-2133/3132 FAX: 301-594-1944

James C. Simmons, CBER, Director Office of Compliance, Co-Chair
Kenneth C. Shelin, BLT-DO, District Director, Co-Chair

Staff:

Mary Ann Denham, CBER, Office of Blood
Larry Fenner, CBER, Office of Compliance
Jack Goodson, ORA, Division of Field Investigations
Ellen Morrison, BLT-DO Biologics Monitor
Jennifer Thomas, BLT-DO Compliance Officer

Office of Compliance
Division of Inspections and Surveillance **(HFM-650)**
Boyd Fogle, Jr., Director Telephone: 301-**594-1194** FAX: 301-**594-1944**

Biological Product Inspections Branch **(HFM-655)**
Chief (Vacant) Telephone: 301-**594-1191**

Staff:

Barbara Burke	Lois Simmons
Janet Ishimoto	Mary Ann Tourault
Sharon O'Callaghan	Rose Wilson

Contact persons:

Blood Program (CP 7342.001)	**Burke**
Errors & Accidents	O'Callaghan, **Ishimoto**
Fatalities	Tourault, **O'Callaghan**
Form FDA 483's	**Burke, Ishimoto**
Inspection Program (CBER)	**Wilson, Simmons**
Training	**Tourault, O'Callaghan**
Viral Marker Testing	O'Callaghan, **Burke**

Surveillance and Policy Branch **(HFM-660)**
Frederick W. Blumenschein, Chief Telephone: 301-**594-1070**

Staff:

Kim Cressotti	Patricia Y. Jones
Jerome Davis	Michael Marcarelli
Alice Godziemski	Fred Rangel
Linda Hamilton	

Contact persons:

Compliance Policy Guides	**Blumenschein**
Computer Policy	**Godziemski**
CLIA 88	**Hamilton**
Exports and Imports	**Cressotti**, Rangel
Field Notification of Licensing Changes	**Godziemski, Hamilton**
Health Fraud	**Hamilton, Davis**
Product Defect Reports	**Rangel**, Davis
Source Plasma Program (CP 7342.002) & Inspectional Guidance	**Jones, Davis**
Physical Sample Collections	**Davis, Jones**

Case Guidance, **Warning Letters**, License Revocations, License Suspensions, Recalls

Office of Compliance
Division of Case Management **(HFM-610)**
Steve Masiello, Director Telephone: **301-594-2006**

Recalls and Administrative Actions Branch **(HFM-615)**
Cathy Conn, Chief Telephone: 301-**594-2007**

Staff:

Tom Adams	George Dawson	Anita Richardson
Barbara Browne	Larry Fenner	Armando Zamora
Sue Cannon	Evelyn Rivera	

Contact Persons:

Recalls, Suspensions Revocations, Denials	Conn	Richardson
Inadvertent Contamination of Plasma Derivatives	Adams	Conn

Case Guidance Branch **(HFM-620)**
Emilie Sanborn, Chief Telephone: 301-**594-2009**

Staff:

Mary Taylor Burtis	Gloria Hicks
Mary Davis-Lopez	Cecily Kodis Kaufman
Anita Flynn	Jan Welch

Contact Persons:

Warning Letters, Seizures, Injunctions, Prosecutions	Sanborn, Kaufman

Consumer Affairs Representative

Office of the Center Director
Division of Congressional and Public Affairs **(HFM-11)**
Congressional and Consumer Affairs Branch (HFM-12)
Michael L. Hooton, Chief Telephone: 301-**594-2000**

Licensing and Labeling of Products, Variances: Approvals for Changes in Personnel, Location, Off-site Storage, Manufacturing Methods, Etc. SOPS, Donor Medical History Forms, AIDS Educational Materials, Consent Forms, and Technical Information Relating to Blood Products

Office of Blood Research and Review, HFM-300
Division of Blood Applications (HFM-370)
Mary Gustafson, Acting Director, 301-594-2012

Blood and Plasma Branch (HFM-375)
Ruth Solomon, M.D., Chief Telephone: 301-594-2012
Leslie Holness, M.D., Medical Officer

Source Plasma Staff:

Linda Alms	Michele Falchek
Elizabeth Callaghan	Christina Santos

Contact Persons:

Alms, Callaghan, Falchek, Santos

ORA CONTACT

Office of Regional Operations
Division of Field Investigations (HFC-130)
Robert C. Fish, Director

Field Investigations Branch, HFC-132
David K. Haggard, Chief Telephone: 301-443-3340

Division of Import Operations and Policy, HFC-170
Thomas Gardine, Chief Telephone: 301-443-6553

International and Technical Operations Branch, HFC-133
Richard R. Klug, Director Telephone: 301-443-1855

PART VII -- CENTER FOR BIOLOGICS EVALUATION AND RESEARCH RESPONSIBILITIES

A. LICENSING

Manufacturers of Source Plasma are required to submit establishment and product license applications to CBER. These applications must describe the manufacturing procedures in detail. The establishment must be operational at the time of the prelicensing inspection to demonstrate its ability to meet the specifications in the license application and to comply with the applicable regulations. CBER is responsible for reviewing the license applications and performing prelicensing inspections. **These inspections are to be performed jointly with district office staff.**

The Division of Blood Applications (HFM-370) provides the Surveillance and Policy Branch (HFM-660) with copies of correspondence relating to licensing changes. This information will be furnished to the appropriate district offices by HFM-660 (See Attachment F).

CBER requests that all manufacturers have readily available for review, copies of their CBER-approved applications/supplements so that during an inspection of the facility, Investigators can determine if a procedure has been approved for use.

B. PROGRAM REVIEW AND EVALUATION

An evaluation of this Compliance Program, with conclusions and recommendations, will be coordinated and/or prepared, if appropriate, by **CBER, Office of Compliance, Surveillance and Policy Branch. Revision of this program** will cover work completed in the preceding fiscal year. **The ORO will receive a copy of the revised program for comment before it is finalized.**

ATTACHMENT A

DEVIATIONS WHICH MAY WARRANT SUSPENSION OF ACTIVITIES UNDER LICENSE

An inspection that reveals ongoing deviations which constitute a danger to health **may warrant** suspension of activities under license as prescribed by 21 CFR 601.6 (a). **Such activities include the following:**

1. ANY PRACTICE WHICH POSES AN IMMINENT HEALTH HAZARD **plus any ground for licensee revocation exists. [21 CFR 601.6(a)]**
2. Failure to notify CBER*of any of the following which may cause a danger to health [601.12(a) and (b)]:

   a. **Changes in Responsible Head**
   b. Change in location, which may be inadequate, i.e., of unsuitable size, construction, or location to facilitate adequate cleaning, maintenance, and proper operations.
   c. Important changes in equipment, manufacturing methods, testing procedures, or labeling.

3. Personnel:

   a) Inadequately trained or supervised in functions they perform **to which they are assigned** [21 CFR 606.20(b), 640.63(a), and 640.64(a)]
   b) Unaware of or disregarding medical history questions to be asked of donors; unaware of or disregarding the **conditions** that make a donor unacceptable for donation; unfamiliar with or not performing hemoglobin, total protein,

blood pressure, or temperature determinations. [21 CFR 640.3; 606.20(b); 640.63(C)].

4. Donor Suitability: Donors continued on plasmapheresis **although the following conditions exist**:

   a. **Total** protein value is below 6 gm per 100 ml [21 CFR 640.63(c)(5)].
   b. **Hematocrit** reading is below 38% or when hemoglobin value is below 12.5 gms [21 CFR 640.63(c)(3)];
   c. **Medical** history questions were not asked;
   d. **The** refractometer is not adequately cleaned between total protein determinations, or **the** instrument is such that correct readings cannot be obtained [21 CFR 606.60];
   e. **Responses** to medical history questions indicated that the donor was unsuitable for plasmapheresis; or
   f. **Donors** are determined to be unsuitable in any other respect.

   NOTE: See Part VI, References for memoranda on current recommendations regarding donor suitability.

5. Immunizing donors by unacceptable procedure(s) and/or using unapproved antigens [21 CFR 640.66].
6. **Blood Collection**

   a. Lack of proper identification procedures including a number on the container so as to relate a unit of blood, **or** red blood cells, **or Source Plasma** directly to the donor [21 CFR 640.64(d) and 606.100(b)(4)].
   b. **Failure to collect blood, separate plasma, and return red blood cells by an aseptic method in an approved system.**
   c. **Failure to recognize and adequately respond to donor adverse reactions.**
   d. **Failure to prevent the infusion of one donor's red blood cells into another donor [21 CFR 640.65(b)(3)].**
   e. **Direct reinfusion of whole blood by squeezing or elevating the blood bags when more blood than allowed by the SOP is removed.**

7. Storage and Shipment:
   a. Failure to have an adequate quarantine system which would prevent issuance of unsuitable units [21 CFR 606.40(a)(3)(4) and (6)] and [606. 160 (e)]; and,
   b. A demonstrated pattern of shipping Source Plasma **which is reactive** for HBsAg, **anti-HCV**, and/or anti-HIV, labeled as nonreactive, [21 CFR 640.67(a) and (b)].

   NOTE: Some establishments have been approved to collect Source Plasma from donors who are **positive** for HBsAg and/or **repeatedly reactive** for anti-HIV. **Source Plasma firms usually do not re-enter persons repeatedly reactive for anti-HIV into the donor pool; therefore, the test results**

of additional more specific tests, i.e., the Western Blot or the Fluorognost immunofluorescence test, are usually not available.

For additional information on shipment of Source Plasma reactive for viral markers, see Attachments L and Q.

c. **Shipping product, except as authorized by CBER, without written test results for markers of infectious diseases,** e.g., HBsAg [21 CFR 610.40(b)(4] or anti-HIV [610.45(a)] at the collection facility.

8. Records grossly deficient or non-existent for:

   a. Determination of donor suitability [21 CFR 640.63(c) and 606.160(b)(1)]
   b. **Evaluating high risk behavior and signs and symptoms of AIDS.**
   c. Physician's review of laboratory and collection data in donor's chart [21 CFR 640.65(b)(2)].
   d) Whole blood weights [21 CFR 640.65(b)(4), (5), (6) and 606.160(a)(1)]. (Except for automated plasmapheresis.)

   NOTE: **Investigators should determine the underlying reasons for the lack of records.**

9. Changing or altering product labels which results in the release of unsuitable units.

ATTACHMENT B

DEVIATIONS WHICH MAY WARRANT A WARNING LETTER/ ADMINISTRATIVE FOLLOW-UP

Any of the listed deviations, if well-documented and representative of a continued pattern of operation, as opposed to isolated instances, could warrant regulatory/administrative follow-up.

1. Facilities: Inadequate or filthy [21 CFR 606.40].
2. Medical Supervision:

   a. Physical **examination of donor** inadequately performed or no record of exam [21 CFR 640.63(b)].
   b. **Physical examination for donor participation in immunization programs or actual immunizations, performed by other than a licensed physician unless approved as a variance under 21 CFR 640.120.**
   c. Physician not physically on premises when **red blood cell** immunizations are performed [21 CFR 640.62].
   d. Hazards or risks not explained **to donor** by physician; no signed informed consent form in donor's records [21 CFR 640.61].

e. **A Physician Substitute supervising an immunization program, prior to receiving CBER approval for participation and/or an individual performing the duties of a Physician Substitute prior to CBER approval.**

3. Donor Suitability:

a. Photographs, or some other equivalent method of positive identification of donors not available [21 CFR 640.65(b)(3)].
b. Personnel not following established procedures when determining donor suitability, including the exclusion of donors in high risk categories for AIDS [21 CFR 606.100(b)(1) and (2); 640.63(c)]:
c. Donors plasmapheresed whose plasma protein composition was outside the normal range established by the testing laboratory [21 CFR 640. 65(b)(2)(i)].
d. Donors not weighed prior to plasmapheresis [21 CFR 640.63(c)(6)].

4. **Source Plasma** Collection: **Unapproved** container used for collection of **product [21 CFR 640.68(b)].**
5. Plasmapheresis (Overbleeding):

a. Manual collection:

A substantial degree of deviation from the requirements is significant. **In determining a substantial degree of deviation, documentation must establish a pattern of conduct which illustrates a normal or customary practice of overbleeding for a period of time.**

The critical factors in determining overbleeding are: 1) the weight (converted to volume) of the whole blood collected; and 2) the volume of plasma obtained from whole blood collections. While such factors as biological variability of the donors, human errors in blood collection, variability in bag weights and scale accuracy, and the amount and kind of anticoagulant must all be considered, the following general statements may be used as guides for determining substantial deviation:

1) Whole blood volumes -- overbleeding is significant when the volume of whole blood removed routinely exceeds the maximum allowable limits [21 CFR 640.65(b)(4), (5), and (6)] **and** when there is no indication that overbleeding is recognized and/or corrected.
2) Plasma Volumes -- overbleeding is considered significant when units of Source Plasma routinely exceed the maximum allowable limit of 770 ml (790 gms) for donors weighing less than 175 pounds and 925 ml (949 gms) for those weighing 175 pounds or more. See CPG No. 7134.14 (Plasma collected in heparin solution will less due to the smaller volume of anticoagulant.)

b. Automated Collection

Overbleeding is considered significant if the firm, through neglect or intent routinely fails to perform those operations which are necessary to assure that the amount of plasma collected is in accordance with that listed in a **CBER-approved** nomogram. It **will** be necessary to document the apparent cause(s) of the overbleeding, for example: inadequate initial training and periodic retraining of device operators; failure to standardize the device internal weight monitors each day of use; failure to use or improper use of an approved and/or gender-specific nomogram; **or use of an unapproved nomogram. NOTE: See CPG 7134.14.**

6. Immunization:

   a. Inadequate records. For example, not recording, when appropriate: the antigen administered; **pre- and post-antibody** titers, **if applicable**; lot number and manufacturer; dose; date; and route administration [21 CFR 606.160(a)(1) and (b)(1)(v)].
   b. Lack of written procedures for all tests related to red blood cell immunization, including antibody detection, identification, and titration [21 CFR 640.66].
   c. **Failure to implement FDA recommended safeguards for red blood cell immunization designed to minimize the potential for transmission of unexpected infectious agents. See Compliance Policy Guide 7134.08.**

7. Tests for **Serological Markers of Infectious Disease**:

   a. Failure to perform test according to manufacturer's directions and specifications, e.g., use of outdated reagents; using/mixing reagents from different lots **except as directed by manufacturer**; failure to run the proper number of positive and negative controls concurrently with the test samples; **failure to use the required dilution of sample**; and in **radiometric procedures,** failure to perform background counts [**21 CFR 610.40** and 606.65(e)].
   b. Calculations **of controls and/or sample values** are incorrectly **performed** [21 CFR 606.65(e)].
   c. Test samples not adequately identified to relate them to donor [21 CFR 606.140(c)].

8. Storage:

   a. **Failure to store plasma for use in manufacturing injectable products at the proper temperature [21 CFR 640.69(b)].**
   b. **Failure to provide adequate storage facilities, including off-site storage, and to assure that storage conditions are adequate** [21 CFR 640.71(a)].
   c. **Daily records of storage temperature not maintained** [21 CFR 606.160(b)(3)(iii)].

d. Final containers for Source Plasma are not stored or handled in a manner to prevent them from being damaged which results in the exposure of the product to external factors that may cause deterioration or contamination [21 CFR 640. 68(b).

9. A history of similar violations about which the firm has been advised.
10. **Any product untested for infectious disease markers shipped interstate without CBER approval.**

NOTE: Approval for shipment should be product specific. If old approval letters are encountered, contact the Division of Blood Applications (HFM-370) to determine whether approval is currently valid.

11. **Use of unapproved SOPs or implementation of unapproved changes in manufacturing methods.**
12. **Computerization:**

    a. **Lack of proper validation protocols and the maintenance of complete and accurate documentation of the performance of the validation protocols and an analysis of the results [21 CFR 606.160; 21 CFR 211.68(b)].**
    b. **Failure to establish and implement adequate computer security (passwords, USERIDs, and modem access) to assure data integrity [21 CFR 211.68(b)].**
    c. **Duplicate, discrepant, or invalid records existing in the computer's donor deferral files that could lead to the acceptance of unsuitable donors and release of unsuitable products [21 CFR 606.160(b)(7)(iii), 606.100(c) and 211.68].**
    d. **Any problem with a computer system(s), such as a donor deferral system,testing equipment interface, labeling system, and/or quarantine/distribution system that could lead to the release of unsuitable products [211.68(b)].**

13. **Failure to promptly notify CBER of errors or accidents in the manufacture of products that may affect the safety, purity, or potency of any product [21 CFR 600.14], or to thoroughly investigate discrepancies in manufacture prior to release of product [21 CFR 606.100(c).**

ATTACHMENT C

LICENSED TEST KITS FOR THE DETECTION OF HIV**

Enzyme Immunodeficiency (EIA), Whole Viral Lysate Derived

LAV EIA, Genetic Systems Corporation
HIVAB HIV-1 ELISA, Abbot Laboratories
Retro-Tek HIV ELISA, Cellular Products, Inc.
Vironostika **HIV** Microelisa System, Organon Teknika Corp.

Enzyme Immunodeficiency (EIA), Recombinant Antigen

Murex SUDS HIV-1 Test, ELISA, Murex Corporation
MicroTrack, HIV-1 EIA, Cambridge **Biotech Corp.**
Recombigen (env & gag) HIV-1 EIA, Cambridge Biotech Corp

Enzyme Immunodeficiency (EIS), Synthetic Peptide

UBI Olympus HIV-1 EIA, United Biomedical, Inc.

Latex Agglutination, Recombinant Antigen

Recombigen HIV-1 LA Test, Cambridge **Biotech Corp.**

Western Blot

Epiblot HIV-1, Epitope, Inc.
Novapath HIV-1 Immunoblot Kit, **Bio-Rad Laboratories, Inc.**
Cambridge Biotech HIV-1 Western Blot Kit, Cambridge Biotech Corp.

Immunofluorescence

Fluorognost HIV-1 IFA, Waldheim Pharmazeutika

Anti-HIV 1/2 (Combination) Enzyme Immunoassay, Whole Viral Lysate Derived

Genetic Systems HIV-1/HIV-2 EIA, Genetic Systems Corporation

Anti-HIV 1/2/ (Combination) Enzyme Immunoassay, Recombinant Antigen

Abbot HIVAB HIV-1/HIV-2 (rDNA) EIA, Abbot Laboratories

Anti-HIV-2 Enzyme Immunoassay

Genetic Systems HIV-2 EIA, Genetic Systems Corporation

HIV Antigen Enzyme Immunoassay

Abbot HIVAG-1, EIA, Abbot Laboratories

**The listed licensed test kits comprise those currently being marketed for use by blood and plasma establishments.

ATTACHMENT D

LICENSED TEST KITS FOR THE DETECTION OF HBsAg**

MANUFACTURER	TRADE NAME	TEST TYPE
Abbot Laboratories	Ausria II Auszyme Monoclonal Auscell	RIA EIA RPHA
Organon Teknika B.V.	Hepanostika	EIA
Organon Teknika Corp.	NML ELISA	EIA
Ortho Diagnostic Systems, Inc.	Ortho Antibody to HBsAg ELISA Test System 2	EIA
Sorin Biomedica S.P.A.	AUK-3	RIA
Genetic Systems Corp.	Genetic Systems HBsAg EIA	EIA

LICENSED TEST KITS FOR THE DETECTION OF ANTI-HCV**

MANUFACTURER	TRADENAME	TEST TYPE
Abbot Laboratories	**Abbot HCV EIA 2.0**	**EIA**
Chiron Corporation	**Chiron RIBA HCV 2.0 Strip Immunoblot Assay**	**Strip Immunoblot**
Ortho Diagnostic	**Ortho HCV 2.0 ELISA Test System**	**EIA**

**The listed licensed test kits comprise those currently being marketed for use by blood and plasma establishments.

ATTACHMENT E

LICENSED TEST KITS FOR THE DETECTION OF ANTIBODY TO HBsAg**

MANUFACTURER	TRADENAME	TEST TYPE
Abbott Laboratories	**Ausab**	**RIA**
Abbott Laboratories	**Ausub EIA**	**EIA**
Sorin Biomedica S.P.A.	**Ab-Auk-3**	**RIA**

LICENSED TEST KITS FOR THE DETECTION OF ANTIBODY TO HBcAg**

MANUFACTURER	TRADENAME	TEST TYPE
Abbott Laboratories	**Corzyme**	**EIA**
Abbott Laboratories	**Corab**	**RIA**
Organon Teknika B.V.	**Hepanostika**	**EIA**
Ortho Diagnostic Systems, Inc.	**Ortho HBc Elisa Test System**	**EIA**
Sorin Biomedica S.P.A.	**ETI-AB-Corek**	**EIA**
Sorin Biomedica S.P.A.	**AB-Corek**	**RIA**

**The listed licensed test kits comprise those currently being marketed for use by blood and plasma establishments.

ATTACHMENT F

ACTIVITIES APPROVED UNDER LICENSE

The following establishment/product licensing information, which is provided by the Division of Blood Applications (HFM-370) to the Surveillance and Policy Branch (HFM-660), is sent to the District Offices as it is received:

1. **The name, address, U.S. license number, and registration number of each operating location for each licensed facility**
2. **The name and mailing address of the Responsible Head of each U.S. licensed firm.**
3. The name, address, registration number (if available) for each approved outside testing laboratory performing:

   a. Serum protein electrophoresis (SPE) or immunoglobulin composition (IGC);
   b. Hepatitis B surface antigen (HBsAg);
   c. Serologic test for syphilis (STS);
   d. Antibody to human immunodeficiency virus (anti-HIV);
   e. **Antibody to the hepatitis C virus and anti-HBc, if performed for donor re-entry.**

5. The approved immunizing antigens used by each establishment; **and approved PLA supplements for collection of plasma from** special donor categories, such as **those who are/have certain preexisting antibodies,** HBsAg reactive, anti-HIV positive, or STS reactive.
6. **Use of Source Plasma for** the manufacture of injectable and/or noninjectable products and the approved anticoagulant(s).
7. **Additional product license applications, such as those for** Source Leukocytes and Therapeutic Exchange Plasma.
8. **Change of Standard Operating Procedures Manual.**
9. **Manufacturing change from manual to automated plasmapheresis.**

ATTACHMENT G

HANDLING OF UNITS REACTIVE FOR OR UNTESTED FOR MARKERS OF INFECTIOUS DISEASE

The Investigator should determine the current status of units of Source Plasma which are reactive for an infectious disease marker(s), or which were improperly tested, or which were not tested, but labeled as nonreactive. If products **are** still under the firm's control, **the Investigator should** advise the responsible personnel to quarantine **the units.**

If the above-mentioned units have been shipped to a consignee, the Investigator should obtain a documentary sample consisting of:

1. The name and address of the consignee(s);
2. **Copies of any correspondence notifying the consignee;**
3. Invoice and/or packing slip which includes date of shipment;
4. The donor or bleed number(s);
5. The collection date(s) of the unit(s) in question; and,

6. **Copies of infectious disease test results which show that the unit(s) is reactive; and/or if the unit(s) was collected from an individual with a history of reactive infectious disease test results, copies of the relevant donor medical history records.**

Immediate action is imperative since units positive for infectious disease markers are potentially infectious. While the manufacturing processes for many final products manufactured from Source Plasma include procedures intended to inactivate or remove viruses transmissible by blood and blood products, it is CBER's view that the safety of such use must be assured in all cases. Therefore, shipment of positive units must be fully documented to assess the impact on the public.

Investigators should immediately notify their **supervisor(s)** who **should**, after consultation with the district management, notify the Surveillance and Policy Branch **(HFM-660) at 301-594-1070.**

After notification, CBER will determine the appropriate course of action with regard to disposition and use of intermediate and final products.

ATTACHMENT H

TESTING BLOOD AND PLASMA FOR ANTIBODIES TO HIV-1

The final rule requiring a serologic test for antibody to HIV-1 on blood and blood components was published on January 5, 1988, and became effective February 4, 1988.

Investigators should familiarize themselves with **current** CBER memoranda pertaining to HIV before conducting inspections. These memoranda are listed in Part VI, References.

A listing of the licensed anti-HIV test kits is included under Attachment C. The test kits licensed for the detection of HBsAg, anti-HCV, anti-HBsAg, and anti-HBcAg are listed as Attachments D and E.

ATTACHMENT I

OFF-SITE PLASMA STORAGE

Prior to licensure or in emergency situations, plasmapheresis establishments may temporarily store plasma at an off-site storage facility, provided that the plasma establishment monitors the storage conditions for the product and CBER has approved use of the off-site location. Investigators should determine whether an outside storage facility is being used and **whether the firm has** received written approval from the Division of Blood Applications (HFM-370). Most off-site storage facilities are located reasonably close to the plasma center.

If a plasma center is utilizing an off-site facility to temporarily store Source Plasma, at some point during the course of the inspection of the plasmapheresis center, an inspection of this storage facility should be performed. Investigators should request that a plasma center official accompany them during the inspection of the storage facility. If violative conditions exist, issue a Form FDA-483 to the owner, operator, or agent-in-charge of the storage facility. Discuss the observations with the storage facility management and also provide a copy of the Form FDA-483 to the Responsible Head of the establishment and to the center management of the plasmapheresis location.

If an off-site facility is being utilized for long-term storage of Source Plasma under specific controlled conditions, the establishment should register with CBER and be entered into the Official Establishment Inventory for routine inspection.

ATTACHMENT J

REFERENCE LABORATORIES AND OTHER OUTSIDE TESTING LABORATORIES

Establishments frequently use outside testing laboratories (off-site) to perform the **laboratory** testing which serves as the basis for determining product quality or donor suitability. These tests (e.g., serum protein electrophoresis, a serological test for syphilis, hepatitis B surface antigen (HBsAg), antibody to human immunodeficiency virus (anti-HIV), antibody to hepatitis C virus (anti-HCV, western blot or immunofluorescent assay for HIV-1, etc.) are specified in the **firm's standard operating procedures manual** and **in** product labeling. **Most testing laboratories are under the jurisdiction of the Health Care Financing Administration (HCFA); however,** any laboratory that performs testing for a Source Plasma manufacturer is an FDA **inspectional** obligation. The testing lab should be encouraged to register with FDA and should be scheduled for inspection by district **office staff**. If a plasma center is using an outside testing facility which is located in another district, that district should be notified to schedule the facility for inspection.

NOTE: Currently, clinical laboratories are specifically exempt from registration per 21 CFR 607.65(g). However, this exemption from registration does not preclude inspection.

According to 21 CFR 640.71, any testing laboratory that performs "required" testing for a Source Plasma manufacturer must be licensed **or meet the standards of the Clinical Laboratories Improvement Amendments of 1988 (CLIA 88), provided the testing facility is certified by HCFA to perform the assigned tests.** The plasma center must obtain a written agreement that the testing laboratory will permit FDA inspection of the operations performed for the plasma center **and under CLIA 88, may be required to**

participate in proficiency testing (PT) programs for specific levels of testing. The laboratory should have a quality assurance program to assess its testing operations. Proficiency testing is commonly used as one indicator that laboratory test accuracy is being maintained.

The Investigator should review PT records to determine whether testing was performed by staff normally performing routine testing and to evaluate the firm's standard operating procedures for investigating discrepancies/errors in results and implementing corrective actions, as applicable.

On June 6, 1989, FDA proposed to amend the regulations concerning blood and blood components to require that firms performing tests for anti-HIV and HBsAg participate successfully in a program to demonstrate proficiency in performing those tests (enroll in a proficiency testing program). Although PT testing represents only a snap-shot in time, general poor performance in proficiency testing may have an impact on the reliability of testing performed for other infectious disease markers and may require a more in depth review of other testing procedures.

The inspection of these laboratories should be limited to the testing the laboratory performs for **blood establishments**. The laboratory should have standard operating procedures relative to shipment, receipt, storage, documentation, testing of the plasma center samples, and the reporting of test results to the plasma center. The Memorandum of Understanding (MOU) with the Health Care Financing Administration (HCFA) concerning Blood Banking and Transfusion Programs (CPG 7155e.03) states that an FDA investigation of a clinical laboratory will be coordinated with the appropriate HCFA regional contact (See Part VI).

ATTACHMENT K

IMPORTED BIOLOGICAL PRODUCTS

Irrespective of its intended use, Source Plasma must be licensed before it can be lawfully imported.

See the **Regulatory Procedures Manual** (RPM) Chapter 9-77 and Import Alert #57-01 (Blood and Blood Components Including Human Plasma and Serum) for instructions concerning the importation of biological products.

ATTACHMENT L

SHIPMENTS OF PLASMA PRIOR TO RECEIPT OF WRITTEN TEST RESULTS

The Center for Biologics Evaluation and Research (CBER) has authorized certain manufacturers of plasma derivatives to receive units of Source Plasma from subsidiary collection facilities before all written test results are received. **This usually applies to the results of infectious disease testing.** The criteria **the firm must meet** for this approval **have** been that the collection facility:

a. is owned by the manufacturer receiving the plasma, and
b. is licensed under the same establishment license number.

In addition, manufacturers with a product license application to collect Source Leukocytes for further manufacture **may be authorized to** ship the leukocytes prior to receipt of written **infectious disease test results. This is necessary because of the short shelf life (24 hours) of Source Leukocytes. Shipment is permitted provided that:**

a. appropriate control procedures are utilized, and
b. written authorization **in the form of a license supplement from the Director, CBER has been received by the firm.**

Investigators should verify that the firm has written approval and **is following its standard operating procedures.**

ATTACHMENT M

INCIDENTS OF WRONG **RED BLOOD CELL** INFUSION

CBER has received reports of incidents where the wrong red blood cells were returned **(i.e., reinfused)** to a plasmapheresis donor. **Some** of these incidents have resulted in donor fatalities. **If such incidents occur, it represents a serious departure from the firm's standard operating procedures** [21 CFR 606.100(b)(17)] **and a lack of control over operations.**

Investigators should review the records of all **episodes of wrong** cell infusion which were not reviewed during a previous FDA inspection. Copies of the records associated with any unreported incident(s) **and a copy of the Establishment Inspection Report** should be forwarded to the **Biological Product Inspections Branch (HFM-655).**

Investigators should cite each significant objectionable condition associated with incidents of wrong cell infusion on the **Form** FDA-483. During the evaluation of such an incident, the Investigator should also review the firm's actions following the incident to determine whether:

1. **The donor who received the incorrect cells was provided** the appropriate emergency medical attention, and the error was documented in detail. [21 CFR 640.72(d)].
2. **Each donor** involved in the incident was properly deferred.

   a. Donors who did not receive **their** cells should be treated as whole blood donors and deferred for eight weeks [21 CFR 640.63(e)].
   b. Donors who received **the incorrect cells** should be deferred for one year, because, in effect, **they** received a blood transfusion.

NOTE: 21 CFR 640.63(c)(13) specifies a six month deferral period; however, the recommended deferral period for these exposures is now twelve months (see Part VI, **References, "Revised Recommendations for the Prevention of Human Immunodeficiency Virus (HIV) Transmission by Blood and Blood Products" dated April 23, 1992**). In addition, the **15th** Edition of **Standards for Blood Banks and Transfusion Sciences** published by AABB specifies a 12 month deferral period.

3. **In the event of a fatality, the Director, Office of Compliance, CBER** was notified as soon as possible [21 CFR 640.73 and 606.170(b)].

ATTACHMENT N

INFREQUENT PLASMAPHERESIS PROGRAMS

Some Source Plasma manufacturers have received **a variance** under 21 CFR 640.120, Alternative Procedures, to include an Infrequent Donor Program. (A copy of the approval letter should be available for review.) This variance eliminates only the requirements for a physical examination and plasma protein tests on healthy, non-immunized individuals who undergo plasmapheresis every eight weeks or less frequently. Often, first-time donors to a center are considered infrequent donors.

Infrequent donors must meet all other Source Plasma donor suitability requirements. If the infrequent donor returns for plasmapheresis in less than eight weeks, plasma protein tests, a serological test for syphilis, and a physical examination must be performed. These donors may not be seen by a physician or a physician substitute; therefore, CBER recommends that the process of obtaining an informed consent and explaining the hazards of plasmapheresis be performed by a trained health professional.

Because of variations in Infrequent Donor Programs, Investigators should compare the firm's procedures with the approved SOP. **Investigators should verify that infrequent donors meet all Source Plasma donor suitability requirements and that the firm has**

a procedure to determine that they are not simultaneously participating in other blood and/or plasma programs in the immediate geographic area.

NOTE: the eight-week donation interval for collection of Source Plasma from infrequent donors is being re-examined by CBER. Issues related to infrequent plasmapheresis were discussed in the June 21, 1994, meeting of the CBER Blood Products Advisory Committee. The committee recommended the following:

> **Infrequent donors donating at intervals of greater than or equal to 4 weeks or at a volume less than or equal to 8.1 to 10.4 L/year should be monitored according to Whole Blood regulations, i.e., no physical examinations, total protein, or serum protein electrophoresis, and a serological test for syphilis is to be performed at each donation. Individuals donating at intervals less than 4 weeks or at a greater volume per year should be monitored according to Source Plasma regulations.**

Recommendations concerning changes in frequency, SOP revisions, and notification to CBER will be communicated to the field and industry by memoranda.

ATTACHMENT O

PHYSICIAN SUBSTITUTES

Some plasmapheresis **manufacturers** have received written approval under 21 CFR 640.120, **Alternative** Procedures, to **utilize** physician substitutes. This written approval, **which represents a variance from 21 CFR 640.61 and 640.63 (requirements that certain duties be performed by a qualified, licensed physician),** should be maintained on the premises of the **plasma establishment. The** responsibilities of the physician substitute may include:

1. **Performing** initial history and physical examinations on new donors and annual examinations of repeat donors.
2. **Supervising or performing** immunizations of donors, **excluding** red blood cell immunizations **unless the physician is present at the facility, provided that** the physician substitute has successfully completed the firm's training program. **The training program must have been approved by CBER prior to use. The approval letter, which is specific to a center, will indicate those programs which can be supervised by the physician substitute. Currently, those programs may include tetanus, rubella, rabies, or hepatitis B. A copy of the center physician's evaluation of the physician substitute's performance during training and the CBER approval letter should be readily available for review at the center where the physician substitute is employed.** The plasma center physician is responsible for at least weekly evaluation and review of the donor immunization record and approval of scheduled injections.
3. **Evaluating** donor reactions and **initiating** appropriate therapy.

4. **Initiating** emergency medical procedures as needed at the plasmapheresis facility.
5. **Referring** donors to physicians or other medical facilities as necessary; serving as a liaison between the plasmapheresis facility and such medical facilities or physicians.
6. **Reviewing** the donor's accumulated laboratory data and collection records to determine the donor's continued suitability.

NOTE: Unless otherwise specified in a CBER approved SOP:

a. **A physician substitute may only determine the continued suitability for plasmapheresis of normal, healthy donors.**
b. **Only a qualified licensed physician may make the medical evaluation with respect to the continued suitability for plasmapheresis of a donor with abnormal medical/laboratory findings.**

A physician substitute is approved by CBER to perform some of the duties of a physician provided the physician substitute is under the direct supervision of the on-site center physician/medical director. For a more detailed description of the physician substitute's responsibilities, refer to the firm's SOP/training program maintained on-site.

Part VI, References, of this document lists an FDA Memorandum dated August 15, 1988, regarding Physician Substitutes. This memorandum updates the duties and requirements of physician substitutes previously identified in a December 14, 1984, FDA memorandum (also referenced in Part VI, References).

NOTE: FDA does not intend to **supersede** a state's licensing authority, i.e., if a state does not permit physician substitutes to conduct certain activities, FDA **will** respect the state restriction(s). **It is the responsibility of the plasma center management to be familiar with the regulations of the state in which a firm(s) is located.**

ATTACHMENT P

USE OF AIDS EDUCATIONAL MATERIAL IN DONOR SCREENING TO ELICIT A HISTORY OF HIGH RISK BEHAVIOR AND THE PRESENCE OF THE SIGNS AND SYMPTOMS OF AIDS OR OTHER HIV INFECTIONS

All plasmapheresis establishments should have implemented procedures which assure that the donor is provided, at each donation visit, with educational materials regarding the spread of AIDS and related HIV infections. CBER recommends that this information include the current risk factors for transmission of AIDS and a description of HIV associated signs and symptoms. It is also recommended that plasma center physicians ask new donors direct questions on high risk behaviors.

A suggested list of questions is found in the memorandum mentioned later in this Attachment.

Serial Source Plasma donors, who may potentially donate as often as twice in a seven day period and who because of the frequency of their visits, are familiar with the screening program/materials utilized at the plasma center, may be provided abbreviated AIDS educational materials. Even though this material may be in an abbreviated form, it should contain current FDA recommendations. It is not required that donors be asked individual questions about specific behaviors or sign and symptoms of AIDS, but the donors must be given information about these factors so that they can make an informed decision regarding self deferral or continued participation in a plasmapheresis program.

Although there is not specific format for the abbreviated materials, some materials have incorporated pictures to help convey information on risk behavior, or have included use of posters, which may have specific areas highlighted or information strategically placed on the poster so that information can be more easily read. It is possible that materials may include risk behavior and signs/symptoms in a "bullet" format.

Various means are used to present this information to donors, which as use of posters placed at various locations within a center and in the screening area, or videos. Investigators should note whether donors actually read the AIDS materials before making a decision regarding self-deferral. It has been reported to CBER that some centers do not specifically ask donors to read the AIDS material. Such an observation should be cited on the Form FDA-483. Investigators should also observe the firm's procedure for providing new donors with an opportunity to voluntarily exclude themselves from donation. This procedure also requires donors to read and consider the AIDS material before making a decision about self deferral.

Investigators should review the AIDS educational materials for consistency with current recommendations. If the materials do not comply, the deficiencies should be documented, cited on the Form FDA-483, and fully explained in the establishment inspection report.

The current high risk behavior and sings and symptoms for AIDS are outlined in the April 23, 1992, FDA Memorandum, "Revised Recommendations for the Prevention of Human Immunodeficiency Virus (HIV) Transmission by Blood and Blood Products.

Questions concerning educational materials and self exclusion practices should be directed to the Surveillance and Policy Branch (HFM-660).

NOTE: See Attachment Y for communication of AIDS educational material to special donor populations.

ATTACHMENT Q

SHIPMENTS OF PLASMA REPEATEDLY REACTIVE FOR ANTI-HIV OR ANTI-HCV, OR POSITIVE FOR HBsAg

Several Source Plasma manufacturers have product license application supplements to collect plasma from individuals who test repeatedly reactive for viral markers, i.e., anti-HIV or anti-HCV, or who test positive for HBsAg. Each location which collects Source Plasma from these "high risk" donors should have a copy of the CBER approval letter at the collection facility for review during an FDA inspection.

Source Plasma units, which after testing are determined to be positive or repeatedly reactive for HBsAg, anti-HCV, and/or anti-HIV (inadvertently collected), may be used for specific purposes and shipped without prior CBER approval. Criteria for use of this product, shipment, and reporting must be consistent with 610.40(d) and 610.45(c). If inadvertently collected, repeatedly reactive/positive units of Source Plasma are shipped for any purpose, verify that the plasma was labeled appropriately and was shipped consistent with all applicable regulations for the interstate shipment of etiologic agents. In addition, verify that CBER was notified of the shipments as set forth in 21 CFR 610.40(d)(1)(v) and (2)(iv) and 610.45(c).

NOTE: Additional information concerning the shipment of biological products is found in 39 CFR Part 111, U.S. Postal Services, 49 CFR Part 173, U.S. Department of Transportation, and 42 CFR Part 72, The Centers for Disease Control U.S. Department of Health and Human Services.

Also, review the following pertinent FDA Memoranda:

1. December 9, 1985: Collection and Shipment of HTLV-III Antibody-Positive Blood Products.
2. April 23, 1992: Revised Recommendations for the Prevention of Human Immunodeficiency Virus (HIV) Transmission by Blood and Blood Products.

ATTACHMENT R

DISPOSITION OF UNSUITABLE UNITS

Disposition

Units of Source Plasma and Source Leukocytes positive for a serological marker of an infectious disease, e.g., anti-HIV or HBsAg, are unsuitable for further manufacture unless otherwise approved by CBER. The disposition of all such units, including those distributed for research, should be verified. Investigators should include in the establishment inspection report the name and address of each consignee who received units for research purposes.

CBER issued a memorandum to all registered blood establishments (April 23, 1992) which recommended that all units of Source Plasma and Source Leukocytes be screened by an FDA-licensed test for anti-HCV and that repeatedly reactive units not be used for further manufacture (see Part VI References, Item 27). Investigators should verify the appropriate disposition of these units. Appropriate labeling statements are found in the memorandum mentioned above.

If plasma collected from a donor known to be positive for a marker of an infectious disease was shipped for further manufacture without written approval from CBER, refer to Attachment G.

Acceptable methods for destruction of unsuitable units include total incineration or steam under pressure at 121.5°C for at least 60 minutes (see FDA memorandum "Control of Unsuitable Blood and Blood Components" dated April 6, 1988).

Unsuitable units may also be disposed of by a hazardous waste carrier provided there is a contract between the plasma center and the waste carrier which assures that unsuitable units are disposed of in a safe and sanitary manner [21 CFR 606.40(d)(2)]. A copy of the contract should be on file at the plasma center. The disposition/destruction records should provide sufficient information to relate the unit destroyed to the donor [21 CFR 606.160(c)].

Lookback

In a February 5, 1990 memorandum, "Recommendations for the Prevention of Human Immunodeficiency Virus (HIV) Transmission by Blood and Blood Products," CBER first recommended that units of blood or blood products collected from donors who later test positive for anti-HIV-1 or who later exhibit signs or symptoms of AIDS not be used for further manufacture.

This memorandum was superseded by an April 23, 1992, FDA memorandum which includes the current recommendations for excluding from use previously collected units of Source Plasma from any person who later exhibits signs or symptoms of AIDS, or later tests repeatedly reactive by a screening test for anti-HIV and does not have a negative licensed Western blot or IFA for antibodies to HIV-1 and a negative licensed EIA test for antibodies to HIV-2.

Instructions regarding exclusion and retrieval of potentially contaminated units from prior collections and notification of consignees can be found in the April 23, 1992, memorandum "Revised Recommendations for the Prevention of Human Immunodeficiency Virus (HIV) Transmission by Blood and Blood Products", issued by CBER to all registered blood establishments.

On June 30, 1993, a proposed rule entitled "Current Good Manufacturing Practices for Blood and Blood Components: Blood and Blood Components at Increased Risk of Transmitting HIV Infection" was published in the Federal Register. This proposed rule includes requirements for written procedures, quarantine, notification to consignees, and records in relation to the lookback of units for which subsequent units from the same donor are repeatedly reactive for antibody to HIV.

The final rule will require an establishment to take appropriate action when a donor of blood or blood components tests repeatedly reactive for antibody to HIV. For Source Plasma collected within 6 months prior to the repeatedly reactive test result, blood establishments would be required to promptly: (i) quarantine all such blood and blood components being held at the establishment; (ii) notify consignees so that such products held by them may be quarantined. Pooled blood products intended for further manufacture will not be required to be quarantined.

Blood establishments that have distributed such units will be required to perform a licensed, more specific test for antibody to HIV and to notify the consignees of the results. If the results of the licensed more specific test is negative, products which have been quarantined may be released. If the result of the licensed, more specific test is positive, indeterminate, or unavailable, the units must not be released from quarantine.

The final rule will also require written SOP's for all steps involved with the lookback process, and records of all steps performed with each lookback process.

For further guidance on lookback, contact Biological Product Inspections Branch (HFM-655).

ATTACHMENT S

AUTOMATED PLASMAPHERESIS

Many plasmapheresis facilities have received approval to collect Source Plasma using an automated device. The two plasma collection systems which are approved for automated collection of Source Plasma are the Haemonetics PCS system and the Fenwal Autopheresis C system. The license approval letter will indicate the name and model number of the device(s) approved for use.

Investigators should direct specific attention to the following areas: personnel training (initial and ongoing); the number of automated devices each employee is assigned to monitor, especially during periods of reduced staff, for example, at lunchtime or assigned breaks; device logs, device alarms/messages, and testing and maintenance of each device. Device logs should be a record of incidents/follow-up actions which occurred during the collection of Source Plasma, and should be reviewed periodically by management to determine if additional operator training or instrument maintenance are needed

The approved automated devices use various set of criteria, i.e., a nomogram, to determine the maximum acceptable plasma volume. The plasma center's nomogram must be approved by CBER before use. (See exception noted in the CBER memorandum dated November 4, 1992, Volume Limits for Automated Collection of Source Plasma.) Both approved devices include body weight as one of the variables in deriving the maximum plasma collection volume. It is, therefore, very important that the scale(s) used to weigh donors be accurate and routinely monitored. The investigator should observe the scales during use. The scales shall be standardized and calibrated on a regularly scheduled basis as prescribed in the firm's standard operating procedure manual, and shall perform in the manner for which they were designed [21 CFR 606.60(a)]. It is also important that the collection device's built-in plasma weigher (container scale) be checked on each day of use [21 CFR 606.60].

In a November 4, 1992, memorandum, CBER recommended volume limits for automated collection of Source Plasma using a simplified nomogram. This nomogram can be used with either of the devices mentioned above, without advance CBER approval. If a manufacturer elects to use the CBER-recommended nomogram, however, it must be the only nomogram in use at the firm and must be adopted as a complete set of limits. Portions of the tables should not be selectively applied in combination with some other system of limits, even previously approved nomograms, without CBER approval.

The facility's approved SOP Manual, the device manual and instructions, and the license approval letter from CBER should be used as guides for inspecting these automated devices.

NOTE: Some firms' approved SOPs will only reference use of the collection device operator's manual, which may not include all of the procedures required for automated collection. For example, the firm's method for determining the amount of red blood cell loss is necessary to supplement the operator's manual if the device or its disposable set malfunction.

ATTACHMENT T

MISUSE OR MALFUNCTION OF EQUIPMENT OR SUPPLIES

During the course of an inspection, **an** Investigator may observe instances where equipment or supplies are either being misused or are not functioning as designed. Such instances may adversely affect the product, the operator/user, or the donor. It is, therefore, important to examine the firm's overall use of equipment and supplies to be sure that they are used in accordance with the manufacturer's instructions, are satisfactorily inspected and calibrated, and are in good repair.

On May 10, 1989, the Office of Regulatory Operations (HFC-100) issued a memorandum to field investigators discussing the need to review the use of key equipment and supplies for misuse and defect related problems. Investigators should refer to that memorandum (listed in Part VI, References) for detailed inspectional guidance relating to:

-- Key Equipment
-- Examples Of Problems Encountered/Observed
-- Reasons Why Such Problems May Occur
-- Items To Be Alert For During Inspections
-- Steps To Take (appropriate follow-up)

ATTACHMENT U

ARM PREPARATION FOR VENIPUNCTURE

For each phlebotomist present, observe at least one arm preparation procedure. The arm prep should be performed in accordance with the firm's written standard operating procedures. The key factors to consider in achieving the satisfactory removal of superficial microbes from the venipuncture site involve both the duration and the vigor with which the scrubbing is performed. The final step is to apply a bacteriostatic agent in a non-overlapping outward concentric spiral beginning at the intended needle puncture site.

Following the preparation of the venipuncture site, the vein may be palpated above or below the prepared area, **but** the site of the needle insertion should <u>not be touched in any manner</u> prior to venipuncture. **If the site is compromised prior to venipuncture, the**

<u>entire</u> arm prep procedure must be repeated. It is <u>not</u> permissible to put iodine or sterile gauze on the finger to palpate the **prepped** venipuncture site.

ATTACHMENT V

<u>DAMAGED BAGS AND BOTTLES OF SOURCE PLASMA</u>

During past inspections of plasmapheresis centers, a significant number of damaged/defective containers (bags or bottles) of frozen Source Plasma were found. **Defects in bags** included tears, holes, cracks, split seams, and/or broken entry tubes, and instances where a corner of the bag was completely missing. **Damage to** plasma pooling bottles included broken tubes and/or elbow ports.

During each inspection of a plasma center, examine containers of Source Plasma in storage for evidence of defects or damage.

Although CBER believes that **it is unlikely** that final injectable products will be adversely affected by the use of plasma exposed to environmental conditions during storage, including **plasma that has been stored** among units found to be unsuitable based on viral marker testing results or elevated ALT values, there **are** no data available upon which adequate assurance of the safety, purity, and potency of final products manufactured from such plasma can be judged. **Furthermore, 21 CFR 640.68(b),** specifically **requires** that final containers used for Source Plasma provide adequate protection against external factors that may cause deterioration or contamination. For these reasons, it is viewed as appropriate that units of Source Plasma with visible damage, resulting in exposure of the product to the environment, should be restricted to noninjectable use only.

The firm should have a standard operating procedure for the handling and disposition of broken or damaged units of Source Plasma. This procedure must **state how the damaged product will be labeled,** as well as its intended use and/or disposition. All labels must be preapproved by **the Division of Blood Applications (HFM-370)**.

Damaged units should be segregated from undamaged units and relabeled for use in manufacturing noninjectable products, **if approved by CBER**. Boxes containing broken or damaged units in which leakage from broken units may have occurred should be overwrapped with plastic or the contents should be placed in new boxes and the old boxes appropriately discarded. Units that are undamaged and suitable for use as labeled should be repackaged as described in the firm's SOP.

If any significant problems with damaged bags are encountered, please notify CBER, Surveillance and Policy Branch **(HFM-660) at 301-594-1070** as soon as possible for further guidance.

ATTACHMENT W

COMPUTERIZATION

Blood establishments have rapidly expanded the use of computers as tools in the manufacture of blood and blood components to assist in decision making and in the management of critical data (e.g., donor registry, laboratory testing, storage of required records and data). As manufacturers of biological products, Source Plasma establishments are responsible for ensuring the proper operation and validation for all procedures performed and supported by the computerized systems relating to the manufacture and distribution of blood and blood components (21 CFR 211.68 and 21 CFR 606.60 and 606.100).

The computerized system consists of computer hardware, application software, peripheral devices or instrumentation, personnel, and manuals. Validation of these systems must be performed. For new systems, validation should be accomplished prior to integrating the computer system into manufacturing, but for existing systems retrospective validation is recognized. In addition, significant changes in the system require revalidation of the affected portions of the system. It is important that the plasma center have a concise validation protocol, change control procedures, and enforce access security procedures.

The September 8, 1989, memorandum "Requirements for Computerization of Blood Establishments" from CBER to all registered blood establishments **discusses validation of computerized systems and related operations**. Another CBER memorandum on this topic "Recommendations for Implementation of Computerization in Blood Establishments" was issued on April 6, 1988. Both of these memoranda are included in Part VI, **References**.

In addition, by means of a <u>Federal Register</u> announcement dated November 9, 1993, the Agency announced the availability of a "Draft Guideline for the Validation of Blood Establishments Computer Systems," and a public workshop to discuss the document was held on December 6-7, 1993. When finalized, the guideline will supersede the FDA memoranda of September 8, 1989,and April 6, 1988.

On March 31, 1994, CBER issued a letter to blood establishment software manufacturers advising them that if they manufacture and distribute software products intended for use in the manufacture of blood and blood components, they are subject to the device provisions of the FD&C Act and FDA's device regulations, including establishment registration, product listing, premarket notification or approval, Current Good Manufacturing Practices (CGMP), and adverse event reporting (MDR).

If the software is vendor supplied, the Investigator should identify the vendor and the version number of the software. The district in which the software vendor is located should be noticed.

ATTACHMENT X

ADDITIONAL LABORATORY TESTING: ALT, ANTI-HBc, AND ANTI-HTLV-I

Currently, Source Plasma establishments are not required by regulation to test donor blood for alanine aminotransferase (ALT) or the antibody to hepatitis B core antigen (anti-HBc). Nevertheless, many plasma centers have donor blood tested for ALT **and/or anti-HBc**.

Firms who have a product license application or a supplement for Source Leukocytes or a red blood cell immunization program, must test cells for anti-HTLV-I. These firms, according to 21 CFR 606.100(b)(7), are required to have written standard operating procedures (SOPs) for this testing.

If ALT, anti-HBc and/or **anti-HTLV-I** tests re performed and test results are a factor in determining donor suitability or product quality, the testing procedures and records should be evaluated to determine if the testing is being conducted in a manner consistent with the instructions provided by the manufacturer [21 CFR 606.65(e)], and that test records include the results and interpretation of all tests and retests [21 CFR 606.160(b)(2)(i)].

Investigators should review the firm's SOP manual to determine, among other things, that it includes all of the steps for the test procedures performed, and includes instructions for evaluations of donor suitability and product quality based on test results. Investigators should also determine proper disposition of any units judged unsuitable based on ALT, anti-HBc, or anti-HTLV-I test results. See Attachments G, L, and Q.

ATTACHMENT Y

COLLECTION OF SOURCE PLASMA FROM SPECIAL DONOR POPULATIONS: DISABLED DONORS AND NON-ENGLISH SPEAKING DONORS

As required in 21 CFR 606.100, a manufacturer must have standard operating procedures which identify criteria used to determine donor suitability, including acceptable medical history criteria. Manufacturers who collect Source Plasma from donors who are vision and/or hearing impaired must have standard operating procedures (SOP) which permit each donor's suitability for plasmapheresis to be determined. The donor history, informed consent, communication of AIDS information, and the opportunity for self exclusion from donation must be presented in a manner appropriate to each donor and that ensures comprehension and confidentiality.

Manufacturers who collect Source Plasma from donors who are predominantly non-English speaking must also have written procedures to communicate appropriate information to donors and must employ measures which ensure comprehension and confidentiality as necessary. These procedures may extend to the availability of personnel who can communicate in a language understood by these donors, or the use of donor records written in a language understood by the donors, etc.

Investigators should review the SOP to determine if procedures applicable to a specific donor population exist at the firm. If the FDA Investigator observes a manufacturer collecting Source Plasma from a donor who requires special considerations, but such collection is not covered in the firm's SOPs, it should be cited on the Form FDA-483.

ATTACHMENT Z

REPORTING OF ERRORS AND ACCIDENTS

Title 21 CFR 600.14 requires that licensed manufacturers notify CBER of errors and accidents in the manufacture of biological products that may affect the safety, purity, or potency of the product. In a March 20, 1991 memoranda, "Responsibilities of Blood Establishments Related to Errors & Accidents in the Manufacture of Blood and Blood Components," CBER reminded all licensed blood establishments of their responsibility to report errors and accidents which occur during the manufacture of blood and blood components.

Some examples of reportable errors and accidents in manufacturing which may affect product quality include, but are not limited to, the release of units (1) that tested repeatedly reactive; (2) were improperly tested; (3) were collected from donors who are or should have been deferred; (4) prior to completion of all testing; (5) incorrectly labeled (e.g., ABO, expiration date); and (6) which are contaminated when the contamination is caused by an error in manufacturing.

Source Plasma manufacturers reported the following types of errors and accidents for the reporting period October 1, 1992 - September 30, 1993:

Type of Error/Accident	Percentage of Reports Received
Donor Screening	34
Product Quarantine	15
Donor Deferral	14
Collection	11
Viral Testing	11
Distribution	7
Labeling	4
Post Donation Information	3

In addition, manufacturers are required to investigate and document the circumstances associated with each occurrence. There must be written procedures which include provisions for recording errors and accidents [21 CFR 606.160(b)(7)(iii)].

Investigators should review files and pertinent records concerning errors and accidents. This review may be used to target certain areas of an establishment's operations for further investigation. Investigators should also assess whether the systems for handling errors and accidents are adequate. SOPs should be sufficiently detailed to provide for thorough investigation, documentation, and review by appropriate responsible personnel.

Post Donation Information

Post donation information refers to information provided by a donor to a blood establishment that would have prompted deferral of the individual if the information had been known at the time of donation. This information may also be provided by sources other than the donor. Post donation information is provided either via telephone (a.k.a., "callbacks") or at the time of a subsequent donation.

Situations related to post donation information include donors who inform the establishment: (1) that their blood should not be used, and do not provide additional information, (2) that they become ill after donating, and (3) of information that would have deferred them if it had been known at the time of donation, e.g., participated in high risk behavior.

Establishments should have control procedures that provide for: (1) the receipt and documentation of post donation information reports, (2) prompt medical evaluation, (3) the timely investigation of the reports to determine if an error or accident in manufacturing occurred and if the quality of distributed blood and blood components was affected, and (4) appropriate consignee notification and determination regarding the disposition of distributed products. See 21 CFR 606.100 and 211.180, et seq.

Reports of post donation information must be submitted to CBER if the product was made available for distribution by the blood establishment and if: (1) the donor should have been deferred had the information been known at the time of donation (e.g., per regulation or CBER recommendation) and the product quality may have been affected, or (2) the medical evaluation otherwise suggests that product quality may be affected, or (3) the information is insufficient to conclude that product quality is not compromised.

Establishments should also have and follow procedures that include provisions for consignee notification concerning in-date and expired products and outline appropriate actions for all affected products, including those intended for transfusion and those intended for further manufacturing use where the quality of the final manufactured product may be compromised. The agency does not ordinarily intend to evaluate these actions as recalls, provided that the blood establishment adhered to applicable regulations and SOPs. See the memorandum of December 10, 1993, titled "Guidance Regarding Post Donation Information Reports."

ATTACHMENT AA

FDA INVESTIGATORS' WITNESSING OF DONOR INTERVIEWS

The donor interview is a critical step in manufacturing blood and blood products, so it is appropriate and necessary for the Investigator to observe this procedure during an inspection of a plasma center. Our inspectional authority is found in the FD&C Act and in the PHS Act. Donor interviews should be witnessed as part of an inspection to assure that all required and recommended donor screening steps are being taken to protect both the donor and the blood supply. A mock interview using firm employees or the investigator is not an acceptable substitution unless the firm is not screening donors at all during the inspection. Refer to IOM section 514, if the firm refuses to allow the witnessing of donor interviews.

The Investigator should first obtain the donor's consent by tactfully explaining to the donor that, as part of the inspection, FDA staff assesses the completeness and effectiveness of donor screening. In order to do so, Investigators must witness a number of donor interviews. If a donor requests that his or her interview not be witnessed, the Investigator should honor this request.

The Investigator must be able to see and hear both the donor and the screener, and must be able to determine whether or not the screener looks at the donor.

ATTACHMENT BB

MANUFACTURING CHANGES IN THE COLLECTION OF SOURCE PLASMA

Licensed Source Plasma manufacturers who intend to change manufacturing methods and equipment shall report those changes to CBER in accordance with 21 CFR 601.12. Manufacturers that convert from manual to automated collection procedures, or from one type of automated collection device to another, or that convert from automated to manual procedures, must submit a product license and establishment license supplement. These major manufacturing changes usually require a preapproval inspection of the facility. Under a policy commonly referred to by CBER as "me-too" conversion, the Division of Blood Applications (HFM-370) may approve a firm's conversion to automated collection procedures without an on-site inspection.

The conversion of a center under a "me-too" status is applicable only to locations under the same license number and to those manufacturers who have been previously approved by CBER for automated collection. A manufacturer who has previously demonstrated compliance with good manufacturing practices in the conversion of two or more centers to automated collection procedures (verified by CBER inspection) may be considered for "me-too" status for the conversion of additional locations.

For conversion of additional locations, the Division of Blood Applications (DBA) will determine whether to approve the conversion of other locations under a "me-too" status, based on the following considerations:

a. **Review of the PLA/ELA supplements for pertinent information/appropriate documentation, including verification of staff training and presence of documentation to support the consistency of SOPs, collection/processing procedures, quality assurance parameters, etc., with procedures previously approved by CBER.**
b. **An evaluation of the inspectional history of the specific location being converted, and**
c. **The overall compliance status of the corporate operations.**

The CBER, Office of Compliance (OC) will evaluate the inspectional history and compliance status of the firm through information provided by the Division of Case Management (HFM-610), the Surveillance and Policy Branch (HFM-660), and the local district office. HFM-660 contacts the district office, provides notice that approval of a "me-too" request is being considered by CBER, and requests

information and the classification of the most recent inspection of the firm. If there is not reason to withhold approval due to a pending or ongoing regulatory action, OC will recommend approval. If regulatory action(s) is pending, OC will recommend non-approval. An establishment is removed from "me-too" consideration if there are outstanding regulatory actions.

POINTS TO CONSIDER

Computer Assisted Submissions
for License Applications

Center for Biologics Evaluation and Research
Food and Drug Administration

July 17, 1990

I. Introduction

A. Background

The Center for Biologics Evaluation and Research (CBER), Food and Drug Administration (FDA) believes that the use of computers may improve the efficiency and quality of the review process for biologic product license applications. These points are intended to facilitate the use of computers in the licensing process by giving manufacturers some points to consider on the factors CBER believes may be desirable for a Computer Assisted Product License Application Review (CAPLAR).[2]

Any new biological product must be approved by FDA before it may be marketed in the United States. As part of the approval process, the manufacturer of a biological product must submit to CBER adequate data concerning the safety and efficacy of its product. Such submissions may include many pages of reports, analyses, tabulations, and case reports.

A continuing goal of FDA is to improve the speed and quality of the review process for all products. The agency views computer technology as a promising means of making the review process more efficient. In order to evaluate the utility of computer assisted submissions, CBER has instituted a program to provide the opportunity to gain practical experience and to develop procedures leading to generalized acceptance of computer assisted submissions.

CBER plans to explore the use of computer technology to enhance the timeliness, effectiveness, and efficiency of the license application review process and to reduce paper handling. Nevertheless, at this time a paper copy of the

---

[2]These "Points to Consider" have been developed through t he CAPLAR Steering Committee, a CBER committee.

application will still be required to accompany any computer assisted submission. At some future date, a separate document or procedure for computer assisted review of investigational new drug (IND) applications may be developed.

These points are intended as reference information for the manufacturer who wishes to submit an application by computer assisted means to CBER.

B. Purpose

The purpose of these "Points to Consider" is to provide a manufacturer of a biologic product with information regarding a computer assisted submission of a product license application (PLA) or establishment license application (ELA). These points provide general information about CBER's CAPLA program as well as identify topics that a manufacturer should be prepared to discuss with CBER prior to requesting to file a computer assisted submission.

C. Capabilities of CBER

1. *Present Limitations*. CBER has limited experience with computer assisted submissions at the present time. This experience has not been sufficiently extensive to permit CBER to draw conclusions about the degree to which a computer assisted submission may ultimately improve the efficiency of the biologic licensing process, nor to allow conclusions about the superiority of one technical approach over another. However, there are certain approaches to implementation of a computer assisted submissions program that have been identified by CBER's CAPLAR Steering Committee as reasonable for the program. CBER anticipates handling various document types. For the current program, the format used should be compatible with equipment and software currently available within the Center. As Center experience grows and computer technology improves, suggested formats and modes of submission will be modified. These modifications will be reflected in future, updated versions of these "Points to Consider."

2. *Future Goals*. Also, it is anticipated that in the future the various CBER units that will be involved in the review process will be connected by wide area networks (WANS) and local area networks (LANS). Through networking, computer assisted submissions of license applications, or portions thereof, may be accessed from a central computer. For example, with the implementation of the appropriate software on a centralized electronic file system, full text searching, image display, and other relevant capabilities will become available to CBER reviewers.

II. General Information

A. Program

CBER is limiting computer assisted submissions to a program for PLA and ELA applications. It is CBER's intent to accept into this program, selected computer assisted submissions that will provide experience with a variety of types of submissions and provide a means for evaluating utility to both CBER staff and the manufacturing industry.

The computer assisted submission can be defined as a paper copy supplemented by portions or all of a license application on computer readable media. Thus, it can be viewed as a method of using computer technology to improve the transmission, storage, retrieval, and analysis of data submitted to CBER as part of the development, licensing, and marketing of a biological product.

1. *Early Notification.* The manufacturer considering a computer assisted submission should contact the Division of Product Certification, CBER prior to the time the application would be submitted. The earlier the contact with CBER, the more likely an application can be developed that will be acceptable to CBER. At least three months advanced notice would be desirable to allow for adequate discussion.

   This notification should include a description of the product and its use; identification of the sections of the PLA or ELA to be submitted in computer assisted form; and proposed formats and modes of data submission. During this time period the manufacturer should be prepared to discuss the methods and anticipated benefits of this effort. Prior notification will allow the CAPLAR Steering Committee, or a subcommittee thereof, to review and evaluate the request and to make the necessary arrangements for proper clearances or concurrences with the Office of the Director, and/or Division Directors; and to meet other requirements as appropriate.

2. *FDA/CBER Contact.* For a computer assisted submission, CBER will identify a contact within CBER's Division of Product Certification who will be responsible for the initial and overall coordinating activities with the scientific divisions/offices involved in the licensing review process and the Chairperson of the CAPLAR Steering Committee. Once a computer assisted application is submitted, the Chairperson and members of the designated internal licensing review committee will be responsible for conducting the review and interacting with the manufacturer and the Division of Product Certification.

   The Chairperson, CAPLAR Committee should be contacted for general information unrelated to specific computer assisted submissions.

3. *Manufacturer Contact.* The applicant should identify an appropriate contact such as the Responsible Head or other designee.

4. *Shipping.* The computer assisted PLA/ELA document should be sent to CBER using the same procedures as a paper submission through the Document Control Center to the Division of Product Certification. Shipping can be by mail or express services. Any computer assisted submission must be accompanied or preceded by an official paper copy of the application(s).

5. *Authenticity.* All computer assisted submissions should be an exact copy of the paper submission. However, in some instances there may be differences, such as in presentation of the data. The manufacturer would be expected to submit a letter of certification stating either that the computer assisted submission is a true and exact copy of the paper submission or itemizing any differences between the two and the reasons for these differences. If there are any variations to presentation of the data, such variations should be identified and described.

6. *Identification of Data.* The paper copy should identify which portions of the application are being submitted in computer assisted format. The computer assisted copy should identify what portions (or figures, etc.) are not present.

7. *Review Policy.* The fact that a PLA/ELA submission is made in computer assisted form would not affect its priority for review.

B. Required Paper Submission

A paper submission, in addition to the computer assisted submission, would still be required under the applicable laws and regulations. The paper copy would be the official and archival copy for the PLA/ELA submission. The computer assisted submission would also be considered a submission for review purposes and subject to the same penalties for fraudulent submission as a paper copy (21 U.S.C. 1001).

III. Submission of Application Document

The format, media, and software mentioned below are suggestions based upon current CBER capabilities. Alternative approaches should be discussed in advance with CBER staff.

A. Document Submission

Documents may be submitted on several different media types. Depending on the capability of the manufacturer, some media may offer advantages over others in terms of ease of data handling and maintenance of data integrity. Applications may be placed on 3 1/2 inch diskettes (double or high density) for use with IBM or Apple Macintosh or compatible computers; or compact disk read only memory (CD-ROM); or Bernoulli disks. This should be discussed with CBER during the early notification period. CBER will work with the manufacturer to select the technology that is compatible with the needs of both that maximizes security, accessibility, and other concerns.

B. Document Format

1. *Text Format.* The PLA or ELA submissions can be accepted by CBER in either PC-DOS/MS-DOS or Apple Macintosh OS using WordPerfect as a word processing software package. Submission formats should follow official license application formats.

2. *Directory Structure.* The organization of the computer assisted submission should parallel that of the paper submission as closely as possible. Individual files should be limited to about 50 to 100 text pages. There should be a brief note outlining any peculiarities of the submission or any special instructions in an ASCII file, with an obvious name such as INTRO.DOC or READ.ME; this should be at the top directory level on the first disk of the application. Additionally, written instructions are needed on how to use INTRO.DOC or READ.ME.

   As appropriate, the file should include information regarding the program(s) needed to review the application and other technical notes that are to be included. The table of contents for the paper submission should indicate which portions of the application are being submitted by computer, and should include full pathnames of the files, and the disk number for a multiple disk submission. This should also be provided as an ASCII file at the top directory level on each disk submitted with the application.

3. *Graphic Data.* At present, no clear consensus exists within CBER for handling graphic data in computer assisted format. Data traditionally submitted in photographic form (e.g., electrophoretic gels, electron micrographs) may be particularly difficult to handle in computer assisted format. As with any other data present on paper but not in the computer assisted submission, the manufacturer should specifically note what graphic data is absent from the computer assisted submission. In addition, it may be of value to submit an additional paper copy or copies of such data in a separate volume as a supplement to the computer assisted submission.

   CBER is interested in initiating trials of CAPLAR including graphic data. We are not yet prepared to suggest a standard format; however, graphic data imbedded in a WordPerfect document or in standard formats such as Postscript, tagged image format, (TIF), and Hewlett Packard Graphics Language (HPGL) may be useful, depending upon reviewer(s) involved. Any proposals to submit graphic data in computer assisted form should be discussed with the CBER contact and the reviewing division.

4. *Numerical Data.* Any numerical and/or tabular data submitted in the paper copy should be submitted in the corresponding computer assisted file in the format of the word processor being used.Additionally, CBER often wishes to receive raw data (usually clinical) for in-house statistical analysis. For this purpose, Statistical Analysis System (SAS) file format, Data Interchange File (DIF) format, or spreadsheet or database formats may be useful; and the ASCII files offer considerable flexibility for use with various software packages.

C. Communication Between FDA and Applicants

During the course of reviewing an application, FDA communicates with applicants about scientific, medical, and procedural issues that arise during the review process. Such communication may take the form of telephone conversations, letters, or meetings, whichever is most appropriate to discuss the particular issue at hand. Communications are to be appropriately documented in accordance with 21 CFR 10.65. Further details on the procedures for communication between FDA and applicants are contained in a staff manual guide that is publicly available. (Reference: "External Relations," FDA Staff Manual Guide No. 2126.2;available from Freedom of Information Staff (HFI-35), Food and Drug Administration, 5600 Fishers Lane, Rockville, MD 20857).

All official revisions to a pending application are to include both a paper copy and computer format submitted by the Responsible Head of the company through CBER's Division of Product Certification. All official correspondence (paper copy) from CBER concerning the application would be transmitted to the applicant through the normal clearance channels within CBER.

The following are various modes of communication that may be used between CBER and the applicant:

1. Official.

   Mail. Mail provides for official correspondence and documentation. All other modes of communication should be followed-up with submission of official paper copy.

2. Unofficial. Information transmitted by the following means is unofficial until documented as official correspondence by the submission of paper copy.

a. *E-Mail.* Electronic (E) mail unofficially provides the exchange of some information between FDA and applicants. Information or instructions for use of E-Mail maybe obtained by contacting the appropriate reviewing division. E-mail may be used with receipt terminals located within the scientific review division and the Division of Product Certification.

b. *FAX.* Correspondence by facsimile (FAX) provides an unofficial draft copy. FAX copies should be sent to CBER only upon request, and should be followed-up with an official paper submission.

c. *Telephone.* Communication by telephone provides open lines of discussion, and is to be documented in accordance with FDA policy (see SMG 2126.2 referenced above).

D. Amendments

Manufacturers should amend their submissions with updates as they occur. This should be done by submitting updated 3 1/2 inch diskettes or other appropriate media which incorporate and highlight the revisions. All submissions of updated diskettes are to be accompanied by appropriate updated paper copy for incorporating into the official archive paper copy of the application. Security procedures for amendments would be the same as with the original submission.

Chapter 11

MEDICAL DEVICE INSPECTIONS

OVERVIEW

In 1976, the Medical Device Amendments to the Federal Food, Drug and Cosmetic Act gave the FDA specific authority to regulate medical devices. The agency received additional authority under the Safe Medical Devices Act (SMDA) of 1990.

The level of regulation depends on the class in which a medical device is placed -- Class I, II, or III -- with Class I devices receiving the lowest level of regulation and Class III receiving the greatest amount of regulation.

Validation concerns are generally associated with FDA investigations under the Good Manufacturing Practice (GMP) regulations, which encompass the methods, facilities and controls used in pre-production design, manufacturing, storing, packaging, and installing medical devices.

Perhaps more is happening in the area of medical device regulation than in any other aspect of FDA's regulatory program. More and more devices are being improved and brought on the market, while pressure from consumer groups and Congress has forced the FDA to reinvent its regulatory program in terms of device classification, Premarket Approval Applications, and Premarket Notifications (510(k)s). It has been apparent over the last year, since the publication of the *Validation Compliance Annual: 1995*, that what transpires within the Center for Devices and Radiological Health is usually a good indication of what transpires throughout the agency (see, for example, document No. 1, below).

The documents which follow all relate to FDA's regulation and inspection of medical devices. These documents are:

1. A letter of October 30, 1995, from FDA regarding a Citizen's Petition presented to the agency relating to FDA regulatory guidance documents.
2. CDRH's latest (August 1995) draft of its guidance document, "Deciding When to Submit a 510(k) for a Change to an Existing Device"
3. FDA's newest revision of its Compliance Program Guide 7382.830, "Inspection of Medical Device Manufacturers."

October 30, 1995

Bradley Merrill Thompson
BAKER & DANIELS
300 North Meridian Street
Suite 2700
Indianapolis, Indiana 46204-1782

Re: Citizen's Petition, Docket No. 95P-0110

Dear Mr. Thompson:

This letter responds to your May 2, 1995 petition and your October 3, 1995 letter on behalf of the Indiana Medical Device Manufacturers Council (IMDMC). In your petition, you ask the Food and Drug Administration (FDA) to control the initiation, development, and issuance of guidance documents by written procedures that assure the appropriate level of meaningful public participation and to adopt stricter, written internal controls over communications by Agency officials to broad, public audiences. You also ask the Agency to initiate rulemaking proceedings to amend 21 C.F.R. 10.40 to require notice-and-comment rulemaking not only for substantive rules, but also for interpretive rules and rules of agency practice and procedure. In your letter, you ask FDA to hold a conference and to solicit written comments on the issues raised in your petition.

As discussed in more detail below, FDA believes that there is merit to your concern about the initiation, development, and issuance of guidance documents. FDA agrees that public participation benefits the guidance document development process. Moreover, FDA believes that the Agency can do a better of job of communicating to its employees and to the public the non-binding nature of guidance documents. Accordingly, FDA is granting your request that FDA improve its guidance document procedures.

However, FDA is denying your request to amend section 10.40. Because FDA does not treat guidance documents that set forth the Agency's general policies and interpretations of the laws or regulations as binding, the Administrative Procedure Act (APA) does not require notice-and-comment rulemaking. In addition, notice-and-comment rulemaking would significantly delay the issuance of guidance documents, or more likely, make it impracticable to issue them at all. The Agency believes that the proper balance between the need for public input and the need for timely guidance can be struck if FDA modifies its guidance document procedures. This approach addresses your concerns regarding adequate public participation but does not make it impossible for FDA to continue making guidance available.

We plan to publish a Federal Register notice and schedule a public meeting to solicit comments on the issues raised in your petition and discussed herein.

I. FDA Guidance Documents

For purposes of responding to this petition, when FDA uses the term "guidance documents," it means: (1) documents prepared for review staff and applicants/sponsors relating to the processing, content, and approval of applications and relating to the design, production, manufacturing, and testing of regulated products; and (2) documents prepared for FDA personnel and/or the public that establish policies intended to achieve consistency in the Agency's regulatory approach and establish inspection and enforcement procedures. Guidance documents do not include Agency reports, general information provided to consumers, documents relating to solely internal FDA procedures, speeches, journal articles and editorials, media interviews, warning letters, or other communications or actions taken by individuals at FDA or directed to individual persons or firms.

One major function of FDA's guidance documents is to provide assistance to the regulated industry in their efforts to comply with the requirements of the applicable laws and implementing regulations. Guidance documents help industry by explaining how they may comply with statutory and regulatory requirements. An equally important purpose for guidance is to help ensure that FDA's employees implement the Agency's mandate in a fair and consistent manner.

Although guidance documents serve important and useful purposes, FDA recognizes that the Agency's development, dissemination, and use of guidance documents can be improved. Accordingly, over the past several months we have been reviewing the Agency's development, dissemination, and use of guidance documents to determine what steps FDA can take to make the processes more transparent and consistent throughout the Agency. We are still conducting this review, but our current thoughts on ways to improve the procedures are set forth below. We expect to complete our review and develop more definite ideas on changing the procedures within the next three months. At that time, we will issue a Federal Register notice that seeks comment on the issues raised in your petition and the Agency's proposals for improving FDA's guidance document procedures. During the comment period, we will hold a public meeting to discuss those issues. Shortly after the comment period closes, we will publish our final plan.

A. Nomenclature

The first step the Agency intends to take in revising its guidance document procedures is to standardize what we call these documents. Guidance documents currently are issued under a number of different names (e.g., guidelines, guidance memoranda, points to consider, blue book memos, compliance policy guides, etc.). While a distinction can be drawn between certain types of guidance (e.g., compliance policy guides versus points to consider), there is overlap in the types of information contained in many such documents (e.g., guidance memoranda and points to consider). A standardized nomenclature would help to eliminate any confusion regarding which documents are guidance documents and their legal effect. We intend to publish a proposed nomenclature in the Federal Register notice discussed herein.

B. Effect of Guidance Documents

FDA guidance documents are not binding. The Agency states this principle explicitly in many of its guidance documents. Moreover, when we train our employees, we instruct them that guidance documents are not binding. Nevertheless, some industry representatives say that they feel bound by guidance documents and the Agency knows that FDA employees have not always been clear about the effect of such documents. Therefore, we must better communicate the message that guidance documents are not binding.

First, the Agency intends to state this principle explicitly on each guidance document. Second, the Agency intends to take steps to ensure that guidance documents do not use compulsory language such as "shall" and "will" but instead use language that clearly conveys their non-binding nature. Third, the Agency intends to ensure that training materials include a section that accurately describes the legal effect of guidance documents. The Agency also intends to take steps to ensure that FDA personnel receive adequate training and to explore the development of an internal policy and procedure document that explicitly defines how the Agency will use guidance. Finally, we will set forth our position that guidance documents are not binding in the Federal Register notice discussed herein and in connection with the completion of rulemaking on our proposal to amend 21 C.F.R. 10.85(e) and 10.90(b)(2), which we will do in conjunction with the timetables set forth herein.

C. Development/Public Input

In your petition, you argue that FDA should institute greater controls over the initiation, development, and issuance of guidance documents to assure the appropriate level of meaningful public participation. Although FDA currently has certain review and clearance procedures for its guidance documents and increasingly solicits public input during guidance document development, we have not always been consistent in these respects. Therefore, the Agency plans to implement consistent procedures for guidance document review and clearance and for the solicitation of public comment on such documents.

With respect to review and clearance, we want to ensure that the same types of documents are receiving the same type of review in all Centers/Offices at FDA. We also want to ensure that guidance is being approved at the appropriate levels. For example, where warranted by such factors as the broad scope or potential implications, we will consider whether to involve the Office of the Chief Counsel and/or the Office of Policy.

With respect to public comment, we will develop an Agency-wide practice whereby all of the Centers and Offices are soliciting some type of public input in connection with the development of their guidance documents. We want the level of public input to be such that the public has fair and sufficient opportunity to comment, but not such that the burden and inherent delay would make it too difficult for the Agency to issue guidance.

One approach that the Agency is considering is to adopt a procedure whereby FDA notifies the public of recently developed guidance and solicits comment on those documents before they are disseminated in final form. When we are developing a more significant, more controversial guidance or one that raises complex issues about which we would like to have significant input, we might also hold a public meeting or workshop to discuss the document. In addition, where appropriate (e.g., when complex scientific issues are raised), advisory committees could be involved. When we issue the Federal Register notice discussed herein, we will seek comment on these and additional ideas for soliciting public input during the development of guidance documents.

An important aspect of whatever approach the Agency decides to adopt will be the principle that FDA is receptive to comments on all of its guidance documents -- old and new -- at any time. FDA also is receptive at any time to comments from the public that propose areas for new guidance. These principles will be set forth in the Federal Register notice.

D. Dissemination/Availability to Public

Another important step in revising the Agency's use of guidance documents and assuring adequate public participation is to make sure that the public knows which guidance documents exist and can easily obtain those documents. The Centers and Offices have lists of their guidance documents that are routinely made available to the public.

In addition, the Centers for drugs, biologics, and devices (CDER, CBER, and CDRH) currently have information systems through which the public can request copies of guidance documents to be sent by telecopy. CDRH and CDER also maintain electronic dockets through which subscribers can access their guidance documents. CBER is in the process of implementing a similar program. The Center for Food Safety and Applied Nutrition (CFSAN) and the Center for Veterinary Medicine (CVM) guidance documents are available directly from those Centers. Both CFSAN and CVM are in the process of making their guidance available on the World Wide Web. The Office of Regulatory Affairs (ORA) makes its "Guide to the inspection of . . . " series available via a dial-in PC. A large number of FDA guidance documents are available through the National Technical Information Service (NTIS). Finally, when new guidance is issued, the Centers and Offices often publish notices in the Federal Register and/or mail copies of the documents to the regulated industry and/or trade associations.

FDA's plan is to ensure that the lists maintained by the Centers and Offices are current, the current access systems are adequate, and the public is aware that the access systems exist. We also will explore ways of informing the public when new guidance becomes available and when existing guidance becomes obsolete.

E. Appeals

One of the most important aspects of a fair guidance document program is the existence of an effective appeals process. Such a process assures the public that there will be full and fair reconsideration and review of how guidance is being applied. The process would further protect against guidance documents being applied as binding requirements.

Under the general provisions set forth in Part 10 of Title 21, FDA provides a number of vehicles that any person or firm may use to seek an appeal of an Agency employee's decision. Pursuant to section 10.75, an interested person may request internal Agency review of an Agency decision made by anyone other than the Commissioner. Such review ordinarily would be by the employee's supervisor, but may move up to the Center Director or Commissioner's Office if the issue cannot be resolved, important policy matters are present, or it would be in the public interest. Sections 10.25 and 10.33 permit an interested person to petition the Commissioner to review any administrative action. This would permit a person or firm to petition the Agency regarding guidance documents. The regulations also include less formal methods of appeal. For example, pursuant to section 10.65, an interested person may correspond or meet with FDA about any matter under FDA's jurisdiction.

In addition, there are specific provisions and procedures that apply to or are used by the Centers. For example, FDA's new drug regulations provide procedures for dispute resolution regarding new drug applications. These procedures include informal meetings with the division reviewing the application, meetings with an ombudsman, and referrals to advisory committees. The new drug regulations also provide the sponsor an opportunity for a hearing on the question of whether there are grounds for denying approval of the application. CBER's review letters ("approvable" and "not approvable") state the sponsor's options for appeal. Specifically, CBER's "not approvable" letter informs the sponsor that it may request a meeting with CBER to discuss the steps needed for approval or may request an opportunity for a hearing.

Finally, persons with concerns about the application of guidance documents may contact the Office of the Chief Mediator and Ombudsman (the Ombudsman's Office). The Ombudsman's Office, which reports directly to the Commissioner, works on resolving issues and conflicts that arise in any FDA component. The Ombudsman's staff is available to discuss options, explain FDA's practices and procedures, and suggest approaches for resolution. When appropriate, the staff of the Ombudsman's Office may contact the FDA staff involved in the issue and mediate a dispute.

As the above discussion indicates, FDA already has a significant number of appeals mechanisms. It may be, however,that FDA needs to take steps to make the public aware of these vehicles and to better ensure that the Agency provides the opportunity for a meaningful appeal. We are exploring ways to meet these goals and when we publish our Federal Register notice,we will solicit comment on ways to improve our appeals processes.

F. Federal Register Notice/Public Meeting

Our plan is to publish a Federal Register notice by January 31, 1996. The notice will set forth the Agency's proposed ideas for revising FDA's guidance document procedures and will solicit comment on the issues raised in your petition and the Agency's proposals. Specifically, the notice will:

1. set forth a proposed nomenclature for guidance documents and seek comments on the proposal;

2. discuss the non-binding nature of guidance documents;

3. notify the public of the continuing opportunity to comment on existing guidance documents or to propose new guidance when a need is identified;

4. explain the Agency's ideas for developing future guidance documents, including how we plan to solicit public comment, and seek comment on additional or alternative ways to assure adequate public participation; and

5. explain available appeals processes and seek comment on ways to improve those processes.

We will have a 90-day comment period, during which we will conduct a public meeting to discuss the issues raised in your petition and the proposal. Shortly after the close of the comment period, we will publish, in the Federal Register, a summary of the actions that FDA will take to revise its guidance document procedures. At that time, we will provide information regarding how to obtain current lists of guidance documents and how to obtain the actual documents.

II. Other Types of Agency Statements or Documents

The procedures and changes discussed in section I relate to FDA's guidance documents. Your petition identifies other Agency statements or documents (specifically warning letters, speeches, journal articles and editorials, press releases, and media interviews) for which you ask FDA to use notice-and-comment rulemaking or adopt stricter internal written controls.

Warning letters are neither guidance documents nor general agency statements. They notify an individual or a firm that the Agency considers a product or activity to violate the law. Warning letters do not create law; rather they inform the recipient about existing law and provide an opportunity for voluntary compliance. Even though FDA does not generally have a legal obligation to warn individuals or firms prior to taking enforcement action, FDA usually does so in the form of a warning letter. FDA has adequate procedures for clearing and issuing warning letters.

The remaining Agency statements and documents discussed in your petition (speeches, journal articles and editorials, press releases, and media interviews) are neither binding rules nor guidance documents. Their purpose is to explain and provide information about already existing policies or to discuss current relevant issues. FDA already has procedures for issuing these other types of statements. Because none of these statements set out binding rules, FDA is not required to utilize notice-and-comment rulemaking. In addition, because these statements are not guidance documents, they will not be a part of FDA's revised guidance document program. Nevertheless, we will review our current procedures for clearing and issuing these types of statements to ensure that they are sufficient and consistent. We also will determine whether it is necessary to clarify their status. However, we do not anticipate making significant changes to the processes.

III. Section 10.40

Section 10.40 defines the circumstances under which FDA will use notice-and-comment rulemaking. Under current section 10.40, the provisions for notice-and-comment rulemaking apply only to the extent required by the APA. The APA does not require notice-and-comment rulemaking for interpretive rules, general statements of policy, and rules of Agency organization, procedure, and practice. In your petition, you ask FDA to adopt notice-and-comment rulemaking for interpretive rules and rules of agency practice and procedure.

You first argue that FDA should adopt notice-and-comment rulemaking for interpretive rules and rules of agency practice and procedure because FDA, in violation of the APA, uses guidance documents to develop and communicate new rules. FDA disagrees with your assertion that certain guidance documents (cited in your petition) were issued in violation of the APA. The APA does not require notice-and-comment rulemaking for non-binding documents. None of the guidance documents cited in the petition are binding. Therefore, none were issued in violation of the APA.

You next argue that FDA should amend section 10.40 to adopt notice-and-comment rulemaking for interpretive rules and rules of agency practice and procedure because prior to 1991, section 10.40 included that requirement. Some background information about the development of section 10.40 explains why the Agency is not persuaded by that argument. When section 10.40 was promulgated in 1977 and FDA stated that it ordinarily also would use notice-and-comment rulemaking for interpretive rules or rules related to agency practice and procedure, it was codifying its past practice.

After FDA promulgated section 10.40, the courts began to embrace an expansive interpretation of what constitutes a binding rule. Under this interpretation, FDA could have been required to undertake notice-and-comment rulemaking even for actions that were not binding rules. This was especially problematic in that the courts suggested that section 10.40 might independently limit FDA's ability regarding APA rulemaking exemptions and thus, require notice-and-comment rulemaking for an even greater number of statements that FDA never intended to be binding rules. See Caribbean Produce

Exchange, Inc. v. Secretary of HHS, 893 F.2d 3 (1st Cir. 1989). It was not feasible for FDA to follow notice-and-comment rulemaking for all Agency statements that a court might find to be a rule.

FDA amended section 10.40 in 1991, therefore, to conform its requirements to those set forth in the APA and to reduce the possibility that it would be required to use notice-and-comment rulemaking for statements that are not binding rules.

Finally, you argue that even if the APA does not require notice-and-comment rulemaking for guidance documents, FDA should adopt it because the benefits of using the process outweigh the costs. Again, the Agency disagrees. Notice-and-comment rulemaking is relatively long and burdensome. Because technology in the medical products area in particular advances so rapidly, many guidance documents would be obsolete by the time notice-and-comment rulemaking was completed. Because the information contained in guidance frequently must be updated or changed to meet emerging situations, flexibility is essential. If all guidance documents were developed through notice-and-comment rulemaking, that flexibility would be lost. FDA could not revise a regulation that, in light of new scientific or technical knowledge, was overly restrictive or burdensome without again going through notice-and-comment rulemaking.

Another problem with notice-and-comment rulemaking is that FDA simply does not have the resources to undertake the cumbersome process for each guidance document it issues. If notice-and-comment rulemaking were instituted for all FDA guidance documents, FDA either would have to shift resources away from other essential activities (such as application review) or (more likely) would be unable to offer much guidance. The statutory and regulatory requirements would not disappear, but companies would have less information about how to comply with those requirements and regulations and FDA personnel would have fewer tools to ensure their consistent application. Such a result would not benefit the public or FDA.

* * * * * * * * * * *

Your petition raises a number of legitimate concerns about the Agency's guidance document procedures. Although FDA is rejecting your request to require notice-and-comment rulemaking for guidance documents, FDA plans to take steps to improve the Agency's guidance document procedures. FDA wants to ensure that the public has a sufficient opportunity to participate in the development of guidance documents, that guidance documents are easily obtainable, that the Agency applies guidance documents in a non-binding manner, and that the public will have a meaningful opportunity to challenge the application of principles set forth in guidance. It is the Agency's hope that in clarifying its policy regarding guidance documents, the public and Agency employees will better understand the appropriate legal force of such documents. If you have additional questions or concerns regarding the petition, please feel free to contact me.

William B. Schultz
Deputy Commissioner for Policy

Deciding When to Submit a 510(k)
for a Change to an Existing Device

Draft #2
August 1, 1995

Preface

On April 8, 1994, FDA circulated for comment the first draft of a document intended to provide guidance to manufacturers on deciding when to submit a new 510(k) for changes to an existing device. That draft was the subject of an FDA/FDLI teleconference on May 12, 1994, and also was the subject of discussion at several trade and industry association meetings. The invitation to comment on the draft guidance elicited over 60 distinct comment letters.

Six of the letters FDA received were from organizations representing multiple manufacturers, while the remainder were from individual manufacturers, consultants and other interested parties. Almost all of these letters contained detailed and thoughtful comments on the guidance and its role in a manufacturer's 510(k) decision making. No attempt will be made here to detail the hundreds of specific comments and changes that were suggested and how they have been incorporated in the second draft of the guidance document, which follows. Nevertheless, FDA staff have attempted to include all pertinent suggestions in this draft of the guidance, and individual submitters of comments should easily recognize where their specific suggestions have lead to revisions.

The comment letters, however, reflected several general themes and concerns regarding the philosophy of 510(k) decision making (particularly for changes to existing products, as well as the respective roles of the manufacturer and FDA in this process. These themes and concerns warrant a discussion since they are a large part of the underpinning of this new draft of the guidance document.

Several writers believed that the April 8 draft ignored the language of the current regulations on premarket notification, Section 807.81(a)(3), where it is stated that manufacturers must submit a premarket notification whenever their product is "about to be significantly changed or modified in design, components, method of manufacture, or intended use." The text of this part of the regulations is provided on the first page of this document and was also provided on the first page of the April 8 draft. The previous draft did not ignore the regulations, but rather, it was an attempt to expand on the phrases "could significantly affect," "major change," "significant change," etc. The April 8 draft as a whole was intended to better define these terms, at least from the FDA reviewer's point of view. Thus, the document did not ignore the regulations, but tried to demonstrate how compliance with the regulation could be achieved. The current draft tries to correct those areas where the April 8 draft fell short of this goal and should provide a reasonable basis for a manufacturer's decision making.

Some of the authors believed that the decision-making process had been inappropriately shifted from the manufacturer to FDA. Although this was not the intent of the April 8 draft, some of the statements can be interpreted in this way, especially when taken out of context. FDA still believes that it is important to define as clearly and carefully as possible when manufacturers ought to submit a 510(k) for a change to an existing device. In some instances, FDA's advice to manufacturers is clear cut, and when asked whether a 510(k) is necessary, FDA's response will always be "yes" or "no." In other instances, the response will depend on a variety of factors. <u>This document attempts to provide for manufacturers FDA's response, were the manufacturer to ask FDA the question of whether a 510(k) would be appropriate for a particular type of change.</u> It is still the manufacturer's responsibility to make that decision for, indeed, the manufacturer knows more about its product and that product's use than anyone else. The guidance contained in this document is only intended to be FDA's best advice. FDA believes that the changes made to the April 8 draft will demonstrate this fact. FDA notes, however, that confusion exists in some of the comments regarding what is the manufacturer's and what is FDA's decision-making responsibility. Manufacturers have the responsibility for determining when a change to their product is significant enough to merit submission of a premarket notification. (If FDA finds, after the fact, that a change was significant and the manufacturer did not submit a 510(k), FDA retains the right to notify the manufacturer of this fact and to require the submission of the 510(k).) FDA's responsibility is the determination of substantial equivalence, based upon the evidence submitted by the manufacturer in a premarket notification.

Several letters noted that FDA's GMP regulations provide adequate control over most design and process changes. They believe that it is unnecessarily duplicative to submit 510(k)s for most changes and that the 510(k) process and the GMP regulations should perform complementary functions. FDA completely agrees with this philosophy. In the April 8 draft, an attempt was made to make explicit the complementary nature of the 510(k) and GMP regulations. Obviously, that attempt was made clear to some reviewers and not to others, since some of the comments complimented FDA for trying to rely more on GMP than upon premarket notification. This difference in viewpoint may derive from individual biases regarding the limits of design validation for changes. In this draft guidance, the reliance on design validation of changes under GMP is much more explicit, and, in many instances, the results of design validation activities are identified as indicators of when a change is significant enough to merit a premarket notification.

The comments also discussed what should be included in a 510(k), once the decision to submit a new 510(k) has been made. The statement in the April 8 draft that "when any one change requires the submission of a 510(k), then a 510(k) incorporating all the changes, including any prior incremental changes, must be submitted ... the new 510(k) should incorporate the old 510(k) and identify and discuss <u>all</u> the changes made to date," caused some confusion. In some instances, the statement was interpreted to mean a description would be necessary of each and every engineering change made since the most recent 510(k) submission. This was definitely not intended. The 510(k) for the changed product, however, must discuss all of the differences between the old and the new product, in order for a determination of substantial equivalence to be possible. In most instances, this is currently accomplished by the provision of a table describing major differences in performance specifications and indications for use or other labeling issues. Further, it is not necessary to "physically" include the old 510(k) with the new 510(k), but the original 510(k) should always be identified and referenced.

The question of consultation with FDA was discussed at great length. In order for this or any other guidance to work well, it is important that feedback be provided to FDA of where it works well and where it does not work so well. The April 8 draft was intended to stimulate manufacturers to contact FDA with those decisions where the guidance was not useful (called decisions in the gray area) in order to refine the guidance at some future date. It was not intended that manufacturers refrain from their decision-making pending an FDA determination; the choice of words in the April 8 draft was unfortunate in this regard. Manufacturers certainly make their own decisions on whether to submit a premarket notification, however, in those instances where this guidance does not prove useful, FDA would like to be notified so that these issues can be clarified in future revisions.

The model that was discussed in the April 8, 1994, document was described as a matrix model. In reality, the model combined the use of a flowchart with a matrix-like array of advice for each type of device and device change to guide decision making. The advice contained in the matrix was whether to submit a 510(k) or merely to document the decision-making under the current Good Manufacturing Practices regulation (21 CFR 820) for changes in device design, components, packaging or labeling.

The current model no longer contains the matrix. This was done for two reasons. First, the changes made as a result of the comments diminished the need to rely on a matrix-like illustration of the advice being given. A more generic approach was taken in the model, and there was less need to differentiate among specific device types. In the few instances where the type of device would have an impact on the guidance, this was taken into account in the flow diagram. Thus, the flow diagram in this model is more complex than the April 8 version, but the matrix has been eliminated.

Second, the use of a matrix appeared to confuse many more people than it helped. If the purpose of a guidance document is to clarify regulatory responsibilities, elimination of any source of confusion is beneficial, thus the demise of the matrix.

Many of the writers believed that this guidance is of sufficient importance that it is necessary to provide notice and comment through the Federal Register. FDA does not believe that this document constitutes either a substantive or procedural rule, and, consequently, notice and comment through the Federal Register is not necessary. FDA is cognizant, however, of the potential widespread use of this document, and, therefore, a Notice of Availability has been prepared for publication in the Federal Register to announce the second draft of this guidance for further public comment.

Introduction

Almost from the enactment of the Medical Device Amendments to the Federal Food, Drug and Cosmetic Act in 1976,FDA staff have attempted to define with greater accuracy when a change in a medical device would trigger a need for a manufacturer to submit a premarket notification (510(k) to the Agency. The regulatory criteria contained in §807.81(a)(3) state that a premarket notification must be submitted when:

(3) The device ... is about to be significantly changed or modified in design, components, method of manufacture, or intended use. The following constitute significant changes or modifications that require a premarket notification:

> *(i) A change or modification in the device that could significantly affect the safety or effectiveness of the device, e.g., a significant change or modification in design, material, chemical composition, energy source, or manufacturing process.*
>
> *(ii) A major change or modification in the intended use of the device.*

The key issue here is that the phrase "could significantly affect the safety or effectiveness of the device" and the use of the adjectives "major" and "significant" sometimes lead to subjective interpretations. Because of this, manufacturers have frequently expressed the need for more specific guidance in applying the regulatory standard in their decision making.

Previous internal attempts to develop such guidance on interpreting the regulation have focused generally on defining broad issues or principles that should be used in deciding when to submit a 510(k). These attempts have been only partially successful in clarifying the situation. The primary reason for partial success is that the variety of device types currently marketed, as well as the myriad changes that occur as technology evolves, are so diverse that one or two unifying principles cannot possibly account for all decision-making opportunities.

To be certain that a decision on when to submit a 510(k) is correct, one would probably need to enumerate all device types and all potential types of changes and then match each combination of device and change with a decision. Given that there are thousands of individual device types and possibly tens or hundreds of enumerable changes, this would be an extremely difficult task. Furthermore, the resultant guidance would fill volumes, would probably be difficult to use and would be unlikely to keep pace with an ever-changing technology.

Between the two extremes of broad principles and detailed enumeration is the area where models can be developed to assist in the decision making. If created and used properly, such a model could provide guidance leading toward a rational answer as to whether a 510(k) is necessary in the large majority of circumstances. This document proposes a flowchart model that can be used by manufacturers in their decision making to analyze how changes in devices may affect safety or effectiveness. In the model,we attempt to address changes to devices at a level detailed enough so that application of the broad principles espoused by the agency would assure a very high percentage of "correct" decisions. The goal of the model is to provide guidance in answering a manufacturer's (or other 510(k) holder's) questions on whether a 510(k) is necessary for a particular type of change and to minimize the number of situations where the answer would be uncertain.

The 510(k) Process and Good Manufacturing Practices

Any guidance on 510(k)s for changes to a marketed device must consider the role of the Good Manufacturing Practices (GMP) regulation in changes in device design. For some types of changes to a device, the Agency continues to find that a 510(k) is not necessary and that reliance on existing GMP requirements may reasonably assure safety and effectiveness of the changed device.

It is important to note that the current GMP regulation does not directly address the original design of a device. In fact, it was the recognition of the need for this type of control for many types of devices that led to the inclusion of pre-production design controls in the Safe Medical Devices Act of 1990.[1] The recently proposed new GMP regulation will implement the new authority and directly address design controls for new devices.

The current GMP regulation, however, is not silent on device design. It requires manufacturers to document in the device master record (§820.181) any changes (and internal approval to changes) to device design and any associated testing (§820.100). It also requires process validation to assure that products meeting the designed quality characteristics will consistently be produced (§820.5 and (§820.100). Finally, manufacturers must have a formal approval procedure for any change in the manufacturing process of a device (including those dictated by design changes) ((§820.100(b)(3)).

The net effect of the GMP regulation is to require that, when manufacturers make a change in the design of a device, they must have a process in place to demonstrate that the manufactured product meets the change in design specifications (or the original specifications, if no change was intended). They must keep records, and these records must be made available to an FDA inspector. Thus, although the current GMP regulation does not address the initial design of a device (the proposed revision to the GMP regulation does address the need for design controls for many products), it does impose certain requirements when changes to a device are implemented. For many types of changes to a device, it may be found that a 510(k) is not necessary, and the Agency may reasonably rely on the existing GMP requirements to assure safety and effectiveness of the changed device. This reliance is enhanced when manufacturers document their decision making based on their testing results or other design validation criteria.

---

[1]Section 520(f) of the Food, Drug and Cosmetic Act.

Scope of the Proposed Guidance

The guidance outlined in this document has been developed to aid manufacturers, specification developers or distributors of class I, class II or pre-enactment class III devices (for which premarket approval has not yet been required under section 515(b)) who intend to modify their device and are in the process of deciding whether the modification meets the statutory threshold for requiring a new 510(k) submission. This guidance for changes to an existing device is intended to supplement the general guidance on submitting 510(k)s contained in the ODE bluebook memorandum on the premarket notification program.[2]

The types of modifications addressed in this document include modifications to design as well as modifications to device labeling, such as modified indications for use, claims for safety and effectiveness, etc. Furthermore, this guidance can be applied to those situations when a legally-marketed device is recalled and a change in device design or labeling is indicated and has not been previously evaluated by the Agency.

This guidance is not intended to apply to device kits (as defined by the 510(k) certification policy memo of April 23, 1994)[3], nor is it intended to apply to combination products, such as drug/device or biologic/device combinations. Furthermore, this guidance is not intended to address the need for submitting 510(k)s by refurbishers or remanufacturers of devices. FDA intends to develop additional guidance specific to these situations.

This document attempts wherever possible to incorporate existing guidance and policy regarding when 510(k)s are necessary for modifications to a currently legally-marketed device.[4] In some cases, the existing guidance derives from advice given to only a few manufacturers for a limited number of devices. In such instances, we have attempted to generalize the concepts to apply to a broader range of devices. However, special cases exist where both manufacturers and FDA have worked to establish definitive guidance for modifications to specific devices, e.g., for daily wear contact lenses.[5] <u>This guidance is not intended to supplant such existing device-specific guidance</u>.

---

[2]ODE Bluebook Memorandum No. K86-3 (June 30, 1986), "Premarket Review Notification Program."

[3]"Kit Certification and Information for Kit 510(k)s," memorandum to manufacturers(?), June 7, 1993.

[4]See, for example, ODE Bluebook Memoranda K86-3, K90-1, etc., as well as device-specific guidance documents.

[5]"Premarket Notification Document (510(k)) for Daily Wear Contact Lenses," May 12, 1994.

Assumptions/Axioms

In developing this guidance for aiding in the decision on when to submit a 510(k), a number of assumptions had to be made. Some derive from existing 510(k) policy and are widely known, others are necessary for using the logic scheme contained in this guidance. Thus, anyone using this guidance needs to bear in mind the following assumptions:

- Any person required to register under 21 CFR 807.20, who plans to market a device requiring premarket notification, will always have to submit a new 510(k). (Note that private label distributors and repackagers are exempt from submitting a 510(k) if they satisfy the requirements of 21 CFR 807.85(b).)

- The guidance applies to the change intended by the manufacturer and not any unforeseen results of implementing that change (although unforeseen results may impact safety and effectiveness and, thus, may be key in deciding to submit a 510(k)).

- The basis for comparison of any changed device is the device described by a cleared 510(k) (or a legally marketed pre-1976 device). That is, manufacturers may make a number of changes without having to submit a 510(k), but each time they make a change, the device they should compare to is their most recently cleared device (or their pre-1976 device), not the current legally marketed device. Although a single incremental change to a device may not substantially affect safety or effectiveness, the sum of many incremental changes can have a substantial effect on safety or effectiveness. Thus, manufacturers can only assure themselves that the last change has not had a significant effect by comparing to the most recently cleared device (or the device marketed prior to May 28, 1976). In effect, manufacturers need to submit a new 510(k) only when the sum of the incremental changes (taken together as though they were in fact one change) exceeds the regulatory threshold, described by the phrase, "could significantly affect the safety or effectiveness," for submitting a new 510(k).

- Because many simultaneous changes may abe considered in the evolution of device design, each type of change must be assessed separately, and, when any one change leads a manufacturer to decide to submit a 510(k), then a 510(k) incorporating all the current changes and comparing the new device to the originally cleared device (or one marketed prior to May 28, 1976) must be submitted. Note that this comparison may be done via a table or other means. The new 510(k), once cleared, would form the basis of comparison for the next sequence of changes.

- Whenever manufacturers change their device, they must comply with the Good Manufacturing Practice regulation.[6] This regulation requires that specification changes be subject to controls as stringent as those applied to the original design specifications of the device, and that such changes be approved and documented by a designated individual(s), which includes the approval date and the date the change becomes effective (21 CFR 820.100(a)(2)). This means that when a change is made to the device, there is verification through testing that the change does not adversely affect the device's safety or effectiveness. Only then can manufacturers assure an accurate assessment of the change(s) in the device in applying this guidance. They must maintain records of the validation under the current GMP. it is this validation of the changes and the documentation maintained by manufacturers that can support the decision on whether to submit a 510(k).

- To derive maximum benefit from this guidance, manufacturers should have in place a mechanism for evaluating whether a proposed change meets the regulatory threshold for a new 510(k). This mechanism could also document use of this guidance, if applicable, or other decision-making aids or bases for deciding whether a 510(k) is necessary.

- This guidance can not address <u>every</u> type of change to <u>every</u> type of device. No matter how ingeniously this guidance is applied, there will still be decisions in the "gray areas" that manufacturers will have to make. If manufacturers notify the Office of Device Evaluation of these instances, the gray areas can be better defined and understood, and, ultimately, this guidance can be refined to encompass the new areas.

- Manufacturers should understand that, even though they use this guidance and submit a 510(k), <u>a substantial equivalence determination is not assured</u>. Some changes to a device may be sufficiently significant that the changed device would be determined to be not substantially equivalent and a premarket approval application would be required before the device could be marketed.

Proposed Model

The model proposed here uses a flowchart to help manufacturers through the logic scheme necessary to arrive at a decision on when to submit a 510(k) for a change to an existing device. A single flowchart containing all the logical steps necessary is large and cumbersome and could be quite daunting. Therefore, it is not included in this document. Rather, for ease of use, Figure 1 depicts the three logical breakouts of changes that might be made to a device:

- labeling changes,
- technology or performance specifications changes, and

---

[6]21 CFR Part 820.

- materials changes.

The reader is referred to Figure 2 for labeling changes and to Figure 3 for technology or performance specifications changes. Materials changes are addressed by the flowchart in Figure 4, with the exception of materials changes for in vitro devices (IVDs), which are addressed by the flowchart in Figure 5.

To use the model, one must answer the questions posed in the flowchart until either directed to consider submitting a 510(k), document the decision-making, or notify the Agency of the change being effected. This last option occurs for the addition of a contraindication and the necessary documentation would constitute an administrative addition to the 510(k) currently on file.

When contemplating changes to a device, manufacturers should use the flowchart for each individual type of proposed change, e.g., performance specification change, material change, etc. If any one of the changes results in a manufacturer's decision to submit a 510(k), then the 510(k) should be submitted and should incorporate all of the intended changes, as well as a comparison of the changed device to the originally cleared device described by the 510(k) currently on file with the Agency. If a manufacturer's consideration of all proposed changes results in a decision merely to document the decision-making, it should document the application of the model along with the necessary records of the validation of changes to the device.

For those circumstances where the proposed change is not addressed in the flowchart or in a device-specific guidance document, manufacturers are encouraged to contact the Office of Device Evaluation in CDRH to find out whether other, specific, guidance exists or if additional help is available.

Before using the flowchart, the reader is reminded that specific guidance has been developed for changes to a legally-marketed device that result from a recall. That guidance has been developed separately, but its philosophy is similar to this document in that changes to a product that are intended to bring a product back to its original specifications and that can be validated under GMPs do not require a 510(k). On the other hand, changes in specifications that raise issues of safety or effectiveness would require a 510(k).

Note that the flowchart entries, "new 510(k)" and "documentation," are written in this way only for conciseness. The reader should interpret "new 510(k)" as **strongly consider submitting a 510(k)** and "documentation" as **document your analysis and file it for future reference.** This is, after all, a guidance document, and it is not intended to be prescriptive. It is intended only to provide the outline of a logic scheme for enhancing the likelihood of good decisions.

Each of the questions listed on the detailed flowcharts (Figures 2-5) are identified by the flowchart letter (A through D) and a sequential number. Those questions on the main spine of the flowcharts relate to major questions to be asked and are identified by a letter and an integer, such as **A1, A2,** etc. Subsidiary questions that are asked in response to a "yes" answer are identified by the integer for the question, a decimal point and a sequential integer, e.g., **C2.1** in Figure 3 labels a decision point containing the question "(Is there a) Change in device performance specifications?" which follows the determination made in decision point **C2** that a change in material type is contemplated.

Labeling changes

As noted above, the types of changes are divided into labeling changes, technology or performance specifications changes, and materials changes. All labeling changes are handled with a separate logic scheme that concentrates on changes in indications for use as the threshold for contemplating the submission of a 510(k). Other labeling changes are more frequently recommended for documentation only.

Chart A (Figure 3) describes the logic scheme to be used when contemplating the need for a 510(k) for a labeling change. Changes in device labeling often pose the most difficult questions to be addressed by device manufacturers when deciding whether a new 510(k) submission is necessary. Frequently, a rather subtle change in a device labeling can have a significant impact on the safe and effective use of the device.

A1 **Is the labeling being revised for clarity, safer use or effectiveness?** Device labeling may be changed for a multitude of reasons. Probably, most labeling changes result from attempts to clarify instructions or to make the device easier, safer or more effective to use. In most instances, such labeling changes would not result in the need to submit a 510(k). For example, the instructions for use of an automated clinical chemistry analyzer may be modified to clarify how routine batch testing operation may be temporarily interrupted to allow efficient processing of high priority samples. No 510(k) would be necessary in this instance. If the question of a 510(k) submission did arise, however, manufacturers should document the rationale for their decision.

A2 **Does the change affect the indications for use?** Changes in the indications for use section of labeling raise more agency concern than any other aspect of labeling. In fact, changes in this part of the labeling will in most instances require the submission of a 510(k). For all practical purposes, the only substantial change in the indications for use that could occur without the submission of a 510(k) is when the change limits use to within the currently cleared indication. For example, the device was cleared for use with three specific indications and the firm decides to market the device for only two of those indications. Another example would be further limiting the patient population by age or weight. In determining whether a change to the indications for use raises issues of safety and effectiveness, the manufacturer should ask whether the specific uses pose any additional risks or expand the use to a new patient population.

Confusion often results when discussing the distinction between "indications for use" and the "intended use" of the device. The regulatory term, "intended use," refers to objective intent of the persons legally responsible for the labeling of the device. Intent may be determined by written expressions or may be shown by the circumstances surrounding the distribution of the device. The concept of intended use has particular relevance in determining whether a device can be cleared for marketing through the premarket notification (510(k)) process or must be evaluated in a premarket approval application (PMA). Manufacturers should recognize that if a particular labeling change results in a "new intended use for the device, the Agency will find the device to be not substantially equivalent and require premarket approval.

For the purpose of this guidance, we have elected to avoid referring to "intended use" as a determinant in deciding when to submit a 510(k). We have chosen, rather, to identify the specific changes or modifications that we believe have a major impact on intended use and thus would require the submission of a 510(k) (§807.81(a)(3)). Two common labeling changes that impact intended use and would usually require submission of a 510(k) are:

(1) reuse of devices previously labeled "single use only;" and
(2) changes from prescription to over the counter or OTC.

One exception to (2) above is providing home-use instructions for devices that remain prescription and whose movement to use in the home is accepted medical practice in the United States. Many prescription devices are used in the home with increasing frequency and the Agency believes that 510(k)s are not necessary for the labeling changes. The reader is referred, however, to the FDA publication, "Write It Right,"[7] for techniques to provide clear and understandable home use instructions.

A3 **Is it a change in warnings or precautions?** In order to facilitate a continuous upgrading in device labeling, manufacturers should monitor device usage and promptly revise the warnings and precautions section based on use experience. 510(k)s for such changes are generally unnecessary unless new issues of safety or effectiveness have arisen to cause the change in labeling, however, manufacturers should document these changes in their files.

A4 **Does the change add a <u>contraindication</u>?** While all changes in the labeled contraindications for device use should be reviewed by the Agency, CDRH recognizes that the addition of a contraindication based on new information is important to public health and should be implemented immediately. Because of

---

[7]"Write It Right, Recommendations for Developing User Instruction Manuals for medical Devices Used in Home Health Care," HHS Publication (FDA) 93-4258, August 1993.

this, manufacturers are encouraged to add new contraindications to their labeling and to notify existing users of their device as expeditiously as possible whenever a pressing public health need arises. The new labeling can be submitted to FDA after the fact as an addition to the existing 510(k) (indicating that the change is already effected). Manufacturers may continue to market their product with the modified labeling, unless otherwise notified by FDA. Manufacturers should be thoroughly familiar with what constitutes a true contraindication to do this.

A5 **Does the change delete a contraindication?** On the other hand, deletion of a contraindication usually requires the submission of a 510(k) prior to effecting the change. In effect, deletion of a contraindication expands the Indications for use and, thus, would normally require submission of a new 510(k).

FDA believes that, if manufacturers follow this approach to changes in device labeling, the maximum number of necessary 510(k)s (those changes that pose the potential to significantly impact safety and effectiveness) will be submitted while minimizing the submission of unnecessary 510(k)s (those where safety and effectiveness are unlikely to be affected). At the same time, manufacturers should be able to retain the flexibility to update and clarify their labeling to maximize the safe and effective use of their devices.

Technology or Performance Changes

Chart B illustrates the decision making logic scheme for technology or performance specifications changes to a device.

B1 **Is it a control mechanism change?** Almost all changes in the control mechanism for a device raise questions of safety and effectiveness. Therefore, such changes will normally require the submission of a new 510(k). This is also true for changes in operating principle (decision point **B2**) as well as for changes in energy type (decision point **B3**). (This last was recognized as a significant change both in the statute[8] and the implementing regulations.[9]) Changes of these types tend to be more revolutionary than evolutionary.

One obvious example of a control mechanism change that would raise new questions of safety and effectiveness would be the change from analog to digital control of a medical device. While the change to digital control can markedly improve device performance specifications and effectiveness, the integration of a digital control into a previously all analog system is complex and usually undertaken only as part of a major redesign of a product. Thus, it would be rare that a new 510(k) would not be necessary. Most often, such changes in control mechanism represent the introduction of a new product line.

---

[8]Section 513(i)(1) of the Food, Drug and Cosmetics Act.

[9]21 CFR 807.81(a)(3).

Other changes in control mechanism of a similar nature would also lead to submission of a new 510(k). Examples of such changes would be the change from pneumatic to electronic control of a respiratory care device or a change from manual power to compressed gas power for a mechanical surgical stapler.

B2 **Is it an operating principle change?** Similar to a control mechanism change, a change in operating principle would also normally lead to the submission of a 510(k). A typical example of a new operating principle for a device would be changing the image reconstruction algorithm used in a computed tomography x-ray system from simple back projection to a new, more radiation-efficient method. In this case, testing both at the bench and in the clinic would be necessary to support a finding of substantial equivalence for the new device. Such changes may also be accompanied by significant labeling changes and, sometimes, by a need for operator retraining to assure continued safe and effective operation. Note, however, that some minor changes to the algorithm that can easily be validated by the manufacturer may not require the submission of a 510(k). Such incremental software changes are discussed under decision point **B8** below.

B3 **Is it a change in energy type?** Here, too, the submission of a 510(k) will be the likely result of a manufacturer's decision making. For example, changing from AC to battery power is usually part of a redesign to provide a portable device that can be used under different environmental conditions than the original device. Such a change would normally be accompanied by significant labeling changes, including a new or expanded indication for use. Note that this type of change does not include a change from 3V to 9V operation or a change from NICad to lead acid storage batteries. Such changes would be considered changes in performance specifications or technical specifications and are discussed at decision point **B5** below.

B4 **Is it a change in environmental specifications?** See **B8** below.

B5 **Is it a change in performance specifications?** See **B8** below.

B6 **Is it a change in ergonomics of the patient/user interface?** See **B8** below.

B7 **Is it a change in dimensional specifications?** See **B8** below.

B8 **Is it a change in software or firmware?**

The types of changes identified at decision points **B4** through **B8** have frequently been called design changes or engineering changes. They encompass everything from the routine specification changes necessary to maintain or improve device performance as a result of feedback from users, field or plant personnel, etc., up to an including significant product redesign. The major difficulty lies in sorting out which of these changes is significant enough to trigger the need for a 510(k). The logic scheme that follows is intended to lead a manufacturer through a thought process that will allow routine engineering change orders to occur, while identifying those changes for which a 510(k) would be indicated.

B8.1 Does the change affect the indications for use? As with a labeling change, if the indications for use are affected, a 510(k) is probably necessary. An example of this would be changing the length of a surgical scissor from 4 inches to 12 inches so that the device could be used in laparoscopic procedures. The original indication was for open surgical procedures, while the new indication is for closed, endoscopically-controlled, procedures. Note that even though a surgical scissor is exempt from 510(k) by classification rule, one must still evaluate the change to assure that the change does not affect the device's classification.

B8.2 Are clinical data necessary to establish safety and effectiveness for purposes of substantial equivalence? Whenever a manufacturer recognizes that clinical data are needed and that bench testing or simulations are not sufficient to establish the substantial equivalence of a new design, it is clear that a 510(k) is necessary.

B8.3 Are clinical data necessary to show continuing conformance to a recognized standard? Although the need to collect clinical data for establishing substantial equivalence usually means a 510(k) is necessary, the mere fact that clinical data are collected does not imply the same thing. Frequently, clinical data of some sort may be necessary to demonstrate that device performance continues to conform to a voluntary standard. This frequently occurs for relatively minor changes to *in vitro* diagnostics that are labeled as meeting particular NCCLS standards. Such changes do not normally require a new 510(k).

B8.4 Do results of design validation raise new issues of safety and effectiveness? All changes to product design will require some level of design validation to assure that the product continues to perform as intended. The successful application of routine design validation activities will logically result in manufacturers documenting their efforts and proceeding with the design change, i.e., no issues of safety or effectiveness are raised. Occasionally, however, either routine design validation activities produce unexpected results or otherwise prove to be inadequate to validate the design change. In such instances, questions of safety and effectiveness may be associated with the design change, and the manufacturer may need to submit a new 510(k).

For example, a manufacturer sees the need to add a higher kilovoltage position on the control of a conventional diagnostic x-ray system. The results of the design change are predicted based on models, calculations, etc. The new system is used to image a phantom and all results are as predicted. The manufacturer documents the efforts and proceeds to production. On the other hand, a manufacturer of monitoring devices wants to use a more sensitive comparator circuit and makes other design changes to accommodate the more sensitive component. Tests with a simulator produce unexpected results, and additional work is necessary to rationalize what has occurred. The manufacturer should carefully assess what has been done and whether new questions of safety or effectiveness have been uncovered. One key to the answer (but not the only one) is whether a significantly different scheme for design validation was necessary.

B9 Is there a change in packaging or expiration dating? Generally, changes in device packaging or changes in the expiration date for device use usually do not result in the need for a new 510(k). Such changes are properly within the function of GMPs. This is true whether the manufacturer applies an expiration date because of package integrity considerations or because of finite shelf life of the device. The reader should note, however, that changes in packaging wherein two or more separately packaged products are combined as a kit will be addressed in a separate guidance document on kits.[10]

B10 Has there been a change in sterilization? Changes in sterilization have the potential for affecting the safety or effectiveness of the device and, thus, must be carefully assessed.

B10.1 Has there been a change in performance specification of the device or in the sterility assurance level attained as a result of the change in sterilization? Changes in the method of sterilization have the potential for changing performance characteristics of a device. This is particularly true of materials properties of plastics. When manufacturers make changes in sterilization methods, they must document that important properties of the device do not change as a result. In addition, if the sterility assurance level is changed, manufacturers must assess whether device safety or effectiveness may have been compromised by the new (lower) level. Manufacturers need, in these instances, to assess critically the need for a new 510(k) for their product.

Materials Changes

Firms making changes to the materials from which their device is manufactured first need to consider the other types of changes discussed above and their impact on the decision regarding the need for a new 510(k). For example, a change of a material type, as discussed below, might also engender a change in the labeling of the device, e.g., the removal of a contraindication or the addition of a new warning. These changes should be considered first, before applying the logic scheme describe in this section. See Chart C.

C1 Is the device an in vitro diagnostic product (IVD)? If the device is an IVD, refer to the later section of this Guidance which is specific to materials changes in IVD's (Chart D).

---

[10](Reference to draft guidance on Kits).

C2 Is this a change in type of material from which the device is manufactured? Is the generic type of material being changed? There is considerable discussion available regarding what is meant by generic materials types. FDA is developing a Biomaterials Compendium for implant devices which will give form and structure to this discussion. The goal of this Compendium is to relate the type of device to the materials of manufacture. Appendix A to this Guidance is a draft of the current tables of generic materials from that Compendium. (It is anticipated that the tables from the Biomaterials Compendium will become final by the time this guidance on when to submit a 510(k) becomes final).

C2.1 Is there a change in performance specifications? Frequently, a change in materials is made to purposefully alter the performance specifications of a device. In this case, decision point **B5** should be used (in addition to this one) to help decide whether a 510(k) is necessary. Sometimes, however, changes in materials can inadvertently affect performance of a device. Under GMPs, manufacturer are responsible for assessing this possibility. If performance specifications are inadvertently changed, it is possible that a new 510(k) is necessary. Manufacturers should still use the logic scheme beginning at decision point **B5** to help decide whether a 510(k) is necessary.

C2.2 Is the device an implant? Implant devices are those described in the "permanent contact" category of ISO 10993-1.[11]

C2.2.1 (Since the device is an implant) Will the material of the affected part of the implant be likely to contact body tissues or fluids? Changes in materials that contact body tissues or fluids may critically affect the device's safety or effectiveness, either because of potentially new interactions of the device material on the body or because of the body's environmental effects on the new material in the device. Except in very restricted circumstances, manufacturers are likely to opt for submitting a new 510(k) in this instance. Examples of devices for which changes in material type would normally require a new 510(k) are total joints or their components. On the other hand, changes in materials of an implant that are not intended to contact body tissues or fluids are not likely to require a 510(k). Examples of such changes in material type are changes in the interior materials of an implantable electric stimulator (e.g., a single chamber cardiac pacemaker) which are sealed from ingress of body fluids or tissues.

C2.3 Will the material of the affected part of the device be likely to contact body tissues or fluids in vivo? Since the decision in box **C2.2** above sorted from consideration the implant devices, the remaining devices, i.e. those for which a "yes" answer is appropriate, include both "limited exposure" and "prolonged exposure" devices, as described in ISO 10993-1: Biological Evaluation of Medical

---

[11]ISO 10993-1, Biological Evaluation of Medical Devices, Part 1: Evaluation and Testing.

Devices, Part 1: Evaluation and Testing. Examples of prolonged exposure devices which might contact in vivo fluids or tissues are: a parenteral feeding catheter; a wound drain; an infusion catheter; sutures; etc.

C2.3.1 Considering that the material is likely to contact in vivo body tissues or fluids and the requirements of ISO 10993-1, is additional testing required? ISO 10993-1 outlines a rational process by which a manufacturer can determine which types of biological evaluation should be performed on a device prior to its use with patients. proper consideration of the various aspects of this evaluation should lead the manufacturer to consistent decisions regarding the changes to the material and the necessity of additional testing. if such additional testing is required, a 510(k) is usually necessary.

C3 **Is this a change in the formulation of the material, but not a change in material type?** These are changes within a single generic material type that can affect the chemistry, metallurgy or other property or stability of the material. These do not include changes in processing aids, catalysts, residual contaminants or manufacturing aids that are not intended to be part of the material. An example of a change in material formulation is a change from one type of polyurethane to another or a change from a AISI Type 316 stainless steel to a AISI type 400 stainless steel. To determine the need for a 510(k) for a change in material formulation, the same logic used for a change in material type should be followed. (See **C2.1** above.)

C4 **Is there a change in the vendor of the raw material from which the device is manufactured?** Changes in the suppliers of raw materials to the manufacturers of medical devices are described in both the existing GMP Regulations and 510(k) Regulations. These regulations establish the responsibility of the device manufacturer to purchase those materials against a materials specification. Such a specification requires performance characteristics of the raw materials related to the desired performance of the finished device. The controlling aspect of the logic scheme for this change is the existence of such a material specification.

C4.1 **Is the new material being supplied to a specification?** If the material is being supplied to a specification, a 510(k) is probably not required. For example, a device manufacturer might include a transparency requirement in the purchase specification for tubing for an implantable catheter. Such a requirement might be related to the later processing of the tubing into the finished device. To change the supplier of that material, the specification must include a transparency requirement, and the device manufacturer's design validation, as required by the GMP regulation, must describe the rationale for that transparency requirement. Further, the manufacturer should document that component specifications are still met and that the performance specifications (characteristics) of the device are not significantly affected.

Materials Changes for In Vitro Diagnostic Products

D1 Is there a change in performance specifications? Changes in the material used in an IVD can affect the performance of the product and should be assessed as to their impact on safety and effectiveness.

D1.1 Does the change in performance specifications of the IVD affect its indications for use? A change in the indications for use will likely include a change in the labeling for the IVD. Examples of changes in indications for use that require a new 510(k) are: the expansion of the use of the test to a new population of clinical applications; and the addition of a new matrix for analysis, e.g. for an IVD approved for serum, the addition of use with other matrices. An example of a change in an indication for use not requiring a new 510(k) is the expansion of the use of the test to other equivalent instrumentation beyond that noted in the original submission. (For details on the procedure to be followed for this change, see "FDA Guidance for Review of Primary and Secondary Reagents.")[12] *It should be noted that a reduction in the indications for use for an IVD will not always trigger the need for a 510(k).*

D1.2 Does the change in the performance specifications of the IVD mean that new clinical data will be necessary to establish the safety and effectiveness of the device for the purpose of demonstrating substantial equivalence? An example of a change where a new 510(k) would be required is if the material change results in a change in cut-off. In that case, clinical testing would be required, and the results should be part of the new 510(k). (Note that clinical testing refers to testing of clinical samples either at the manufacturing site or at sites of intended use.) An example where new clinical data are necessary, but a new 510(k) is not necessary, is when the testing of the new material, using splits of patient specimens, is necessary to compare the new material to a previously used material.

D1.3 Does the change in the performance specifications of the IVD mean that new clinical data will be necessary to show continuing conformance of the device to a recognized standard? Voluntary standards such as those developed by the National Committee of Clinical Laboratory Standards (NCCLS), the National Cholesterol Educational Program and other professional groups may be the basis of a substantial equivalence determination for an IVD. Deviation of IVD performance specifications from the performance values of the standard may indicate that substantial equivalence of the device is in question.

D1.4 Do the results of the design validation performed as a result of this change in materials raise new questions of safety and effectiveness? As noted above, design validation is required when changes are made to any device, including an IVD. The results of such a validation showing that questions of safety and effectiveness have been raised may indicate that the performance of the device is not known or well established, and a 510(k) may be necessary.

---

[12]"FDA Guidance for Review of Primary and Secondary Reagents."

D2 Does the change in material alter the operating principle of the IVD? Examples of changes in materials that alter the operating principle of the device and would routinely require a 510(k) are: changes from liquid to solid reagent; change from an RIA to a non-RIA; changes in the source and type of an antibody, e.g. from one animal species to another or from polyclonal to monoclonal or vice-versa; change from immunofluorescence to ELISA; and a change in conjugates. Examples of changes that might affect the operating principle of the IVD are changes in reaction components or materials such as calibration materials and quality control materials or changes in methods such as specimen pretreatment, incubation times and temperatures. If these result in changes in the reporting of performance in labeling, they would require a new 510(k). Examples of changes in materials which do not affect the operating principles are changes in preservatives and changes in formulations of the existing materials.

Flowchart Definitions

The following definitions are provided to clarify the meaning of all terms used in the flowchart and to assist in determining when a 510(k) is necessary for a change to an existing device. Wherever possible, existing definitions from the Food, Drug and Cosmetic Act, the medical device regulations or ODE Bluebook memoranda have been used. In some cases, where regulatory definitions are unavailable, we have relied on strict dictionary definitions of terms.

1. Indications for use: An indication for use is "a general description of the disease or condition the device will diagnose, treat, prevent, cure, or mitigate, including a description of the patient population for which the device is intended."[13] The indications include all the labeled patient uses of the device, as for example:

- the condition(s) or disease(s) to be screened, monitored, treated or diagnosed,
- prescription versus over-the-counter use,
- part of the body or type of tissue applied to or interacted with,
- frequency of use, or
- physiological purpose (e.g., removes water from blood, transports blood, etc.).

The indications for use are normally found in the indications section of the labeling, but indications may also be inferred from other parts of the labeling such as the precautions, warnings or the bibliography sections. In some instances, a change in the indications for use may imply a change in the intended use for the device, in which case, the 510(k) for

[13]21 CFR 814.20(b)(3)(i).

the change would be found not substantially equivalent and a premarket approval application or a reclassification petition would be necessary.[14]

2. Labeling: *The term "labeling" means all labels and other written, printed, or graphic matter (1) upon any article or its containers or wrappers, or (2) accompanying such article.*[15] This can include, among other things, any user or maintenance manuals, as well as advertising and promotional literature.

3. Warnings, precautions and contraindications:

- Warnings describe serous adverse reactions and potential safety hazards that can occur in the proper use or misuse of a device, along with consequent limitations in use and mitigating steps to take if they occur.

- Precautions describe any special care to be exercised by a practitioner or patient for the safe and effective use of a device.

- Contraindications describe situations in which the device should not be used because the risk of use clearly outweighs any reasonably foreseeable benefits.[16]

4. Documentation: Recording the results of applying the model to proposed changes in a device. Consideration of each decision point should be recorded, as well as the final conclusions reached. A copy of this documentation should be maintained for future reference.

5. Control Mechanism: The manner by which the actions of a device are directed. An example of a change in control mechanism would be the replacement of an electro-mechanical control with a microprocessor control.

6. Operating Principle: The mode of operation or mechanism of action through which a device fulfills (or achieves) its intended use. An example of a change in operating principle would be using a new algorithm to compress images in a picture archiving and communications system.

7. Energy Type or Character: The type of power input to or output from the device. Examples of a change in energy type or character would be a change from AC to batter power (input) or a change from ionizing radiation to ultrasound to measure a property of the body (output).

---

[14]ODE Bluebook Memorandum K86-3, "Guidance on the CDRH Premarket Notification Review Program."

[15]Section 201(m) of the Food, Drug and Cosmetic Act.

[16]For further information see ODE Bluebook Memorandum G91-1, "Device Labeling Guidance."

8. Environmental Specifications: The (range of) acceptable levels of environmental parameters or operating conditions under which the device will perform safely and effectively. Examples of changes in environmental specifications would be expanding the acceptable temperature range in which the device will operate properly or hardening the device to significantly higher levels of electromagnetic interference.

9. Performance Specifications: The performance characteristics of a device as listed in device labeling or in finished product release specifications. Some examples of performance specifications are measurement accuracy, output accuracy, energy output level, stability criteria, etc.

10. Ergonomics or Patient/User Interface: The way in which the device and the patient /user are intended to interact. Examples of this would be the various audible or visible alarms intended to alert the user to a hazardous condition, the layout of a control panel, or the mode of presentation of information to the user.

11. Packaging: Any wrapping, containers, etc., used to protect, to preserve the sterility of, or to group medical devices.

12. Expiration date: The date beyond which there are no data to assume that the product may perform safely or effectively.

13. Method of Sterilization: The physical or chemical mechanism used to achieve sterility or to achieve a specific sterility assurance level (SAL).

14. Software: The set of instructions used to control the actions or output of a medical device, to provide input to or output from a medical device, or to provide the actions of a medical device. This definition includes software that is imbedded or permanently a part of a medical device, software that is an accessory to a medical device, o software that is itself a medical device.

15. Material Supplier: The firm supplying the raw material to a finished device manufacturer.

16. Reuse: Use of a device more than once on a single patient or on more than one patient. Actions necessary for reuse of a device may include instructions for assembly/disassembly, on-site sterilization or disinfection, etc. This definition does not include the refurbishing or repair of a device for redistribution or resale.

17. Material Type: The generic name of the material from which the device is manufactured. (Use the generic name in the biomaterials compendium.) An example of a material type change would be the change from natural latex rubber to synthetic rubber.

18. Material Formulation: The base polymer formulation or the alloy, additives, colors, etc., used to establish a property or the stability of the material. This does not include processing aids, mold release agents, residual contaminants, or other manufacturing aids

that are not intended to be a part of the material. An example of a change in material formulation would be a change from a series 300 stainless steel to a series 400 stainless steel.

19. Dimensional Specifications: The physical size and shape of the device. Such specifications may include the length, width, thickness or diameter of a device, as well as the location of a part or component of the device.

20. Implant: A device which is intended to reside within a surgically or naturally formed channel or cavity of the human body for a period of 30 days or more (excluding dental restoration materials).

21. In Vitro Device: *Those reagents, instruments, and systems intended for use in the diagnosis of disease or other conditions, including a determination of the state of health, in order to cure, mitigate, treat, or prevent disease or its sequelae. Such products are intended for use in the collection, preparation, and examination of specimens taken from the human body.*[17]

22. Change: As used in the model, this means a proposed change and not the impact of a proposed change. Important impacts of a proposed change are identified on the flowchart. For example,, a manufacturer may propose a change in method of sterilization. This change could impact on performance specifications because of potential chemical or physical damage to the device. The proposed change (in method of sterilization) is the change that must be used in the model.

[17]21 CFR 809.3(a).

APPENDIX: DRAFT CONTROLLED VOCABULARY FROM THE FDA BIOMATERIALS COMPENDIUM

Material Class

metal
polymer
ceramic
composite
biological origin

Material Subclasses

Metals
stainless steels
co & ni alloys
titanium alloys
zirconium alloys
tantalum alloys
precious metals
amalgams

Polymers
thermoplastics
thermoset/elastomers
absorbable
adhesives
fluids

Ceramics
Al compounds
Ti compounds
Zr compounds
Ca compounds
carbons
glasses

Composites
polymer matrix
metal matrix
ceramic matrix

Biologic Origin
allograft
xenograft
engineered tissues
biomolecules

METALS

Generic Material Names

Stainless Steels

316L FeCrNiMo
nitrogen strengthened
ferritic ppt hardened
?martensitic, ppt hardened

Co & Ni Alloys

CoCrMo period

CoCrWNi
CoNiCrMo
CoNiCrMoWFe
CoCrNiMoFe

Ti Alloys

CpTi (grade 1-4)
Ti 6Al 4V
Ti 6Al 7Nb
Ti 5 Al 2.5 Fe
Ti 3.8Al15Mo5Zr
Ti 13Nb13Zr
Ti 12Mo6Zr2Fe
Ti 15Mo2.8Nb.2Si
NiTi alloy

Zr Alloys
Zr2.5Nb

Amalgams

*Ag-Hg amalgams

Ta Alloys
unalloyed Ta

Precious/Noble
gold
silver
platinum
palladium
iridium
Pt/Ir

POLYMERIC MATERIALS

Thermoplastic

acetals
*acrylic, powder/liquid
acrylic, PMMA
acrylic, hydrogels
fluorocarbons
polyamides (nylons)
poly(aryl)ether ketones
poly(aryl)sulfones
polycarbonate
parylene
polyesters (PET, PBT)
polyethylene
polyethylene, UHMWPE
polyimides
polypropylene
polystyrene
polyurethane
polyvinyl alcohol
polyvinyl chloride PVC
polyvinylidine chloride

PEO hydrogel

Thermoset/Elasto

butyl
epoxy
hydrogel based
natural hevea
polyesterurethanes
polyetherurethanes
polyurethanes (other
polyisoprene
rubbermodified acrylic
silicone gel
silicone elastomer
bis/GMA

Absorbable

polyesters
polyesters
polyanhydrides
polyorthoesters
polyetheramide

copolymers

Adhesives

cyanocrylates
acrylic based
epoxy
silicone

Fluids

polyvinylpyrrolidone
silicone (PDMS)

Carbons

fibers
graphite
LTI pyrolytic
LTI-Si alloy
ULTI pyrolytic
vapor deposited
vitreous

Glasses

bioactive glass
silica based

COMPOSITES

Polymer Matrix

CFR epoxy
CFR poly(etherketones)
CFR poly(imide)
CFR Poly(sulfone)
glass reinforced
ceramic particle reinforced
metal fiber reinforced
PUIPC
*PTFE composites
CFR UHMWPE
bis/GMA composites
urethanedimethacrlate

Metal Matrix

*Ta-Eigiloy wire
Ag-MP35 I?

Ceramic Matrix

carbon-carbon
zinc phosphate cement
porcelains
*zinc oxide eugenol
glass ionomer cements

zinc polycarboxylate cement

Users of Compliance Program 7382.830 (Inspection of Medical Device Manufacturers)

The Food and Drug Administration (FDA) has refined and refocused its compliance program linking Class I and II 510(k) submissions and PMA Supplements with Good Manufacturing Practice (GMP) requirements. These revisions have been incorporated into the Compliance Program 7382.830, Inspection of Medical Device Manufacturers, new Attachments A-1 and G, and in Parts II, III, and V, (Attachment 1).

1. The new attachments A-1 and G describe:

   a. The criteria for linking GMPs with marketing clearance for Class I and II 510(k) devices.

   b. How FDA will notify firms that clearance of their Class I or II 510(k) submission may be deferred due to serious GMP violations;

   c. What actions FDA will take to reply promptly to a firm's response to a Form FDA 483 and/or GMP Warning Letter.

   d. Agency time frames so that medical device firms may conduct their business planning with reasonable predictability.

   The same procedural changes noted above in paragraphs 1a through 1d also apply to PMA Supplements that are not subject to the PMA Pre-Approval Inspection Program.

2. The revisions contained in Attachment A, Exhibits I through IV contain additional guidance for determining when a GMP inspection has identified violations that meet the criteria for Situation I in Part V, A.1. of this program.

3. Changes made to this compliance program which are related to Attachments A-1 and G will be found on the pages that are annotated with the comment, Advance Copy May 1, 1995.

If you have any questions about information contained in Attachments A-l and G please contact me (301)-594-4695.

Please refer to Part VI of the program, References, Attachments, and Program Contacts, for questions or comment about any other part of this compliance program.

Marje Hoban
Acting Chief
Field Programs Branch, HFZ-306
Division of Program Operations
Office of Compliance
Center for Devices and Radiological Health

SUBJECT:

INSPECTION OF MEDICAL DEVICE MANUFACTURERS

IMPLEMENTATION DATE

Upon Receipt

COMPLETION DATE
September 30, 1998

DATA REPORTING

PRODUCT CODES

PRODUCT/ASSIGNMENT CODES

73-91

82830L 42830L -- All Routine Inspections
82830C 42830C -- All F/U Inspections
82011 -- Report Time Spent on Assessment of Firm's MDR Practices

Field Reporting Requirements

Send an EIR to CDRH only if the inspection resulted in the issuance of a direct reference Warning Letter to the firm, or is part of a recommendation for other administrative/regulatory action.

A copy of all Warning Letters should be sent to HFZ-306.

If the district wishes to obtain comment from CDRH for other EIRs you must attach a cover memo outlining the issues to be considered by OC. This policy does not relieve the district from COMSTAT reporting requirements.

All EIRs and administrative/regulatory action recommendations shall be sent to HFZ-306.

As soon as the district becomes aware of any information which might effect the agency's new product approval decision with respect to a firm, the district should immediately notify HFC-240, Division of Medical Products Quality Assurance, via EMS or FAX. They will convey the information by FAX or equivalent expeditious means to the appropriate Center regulatory units.

PART I BACKGROUND

This compliance program provides guidance to the FDA field staff for the enforcement of the requirements of the GMP regulation (21 CFR 820), the MDR regulation (21 CFR Part 803), and the Medical Device Tracking regulation (21 CFR Part 821).

A. THE GMP REGULATION

This compliance program, introduced in FY'94, provides guidance for a new GMP inspectional strategy designed to compensate, in part, for the increasing strain on field resources while, at the same time, providing for rapid regulatory action when required. The strategy also places additional emphasis on manufacturers' responsibility to monitor their compliance with the requirements of the GMP regulation, and to make appropriate and timely corrections of problems in their manufacturing and quality assurance systems.

B. THE MDR REGULATION

The Medical Device Reporting (MDR) regulation became effective on December 13, 1984. This rule requires manufacturers and importers (initial distributors) of medical devices, including in-vitro diagnostic devices, to report to FDA whenever the manufacturer or importer receives or otherwise becomes aware of information that reasonably suggests that one of its marketed devices:

1) may have caused or contributed to a death or serious injury or,

2) has malfunctioned and that the device or any other device marketed by the manufacturer or importer would be likely to cause or contribute to a death or serious injury if the malfunction were to recur.

Every time a GMP inspection is conducted under this Compliance Program, an MDR inspection will also be conducted.

C. THE MEDICAL DEVICE TRACKING REGULATION

Congress recently amended the Federal Food, Drug, and Cosmetic (FD&C) Act to add new requirements and provisions concerning the regulation of medical devices. These requirements are described in the Safe Medical Devices Act (SMDA) of 1990, signed into law by President Bush on November 28, 1990. One of those provisions was the Medical Device Tracking Requirements promulgated as 21 CFR 821.

The Tracking regulation requires manufacturers to implement a method of tracking permanently implanted or life sustaining/supporting devices used outside a device user facility, the failure of which would be reasonably likely to have serious adverse health consequences.

This regulation is intended to ensure that tracked devices can be traced from the device manufacturing facility to the patient.

PART II - IMPLEMENTATION

A. OBJECTIVE

GMP's

1. To identify domestic and foreign manufacturers who are not operating in a state-of-control. *To assure that firms that have submitted premarket 510(K) notifications for devices are in substantial compliance with the GMP requirements that reasonably relate to the submission.* To bring such manufacturers into a state-of-control through voluntary, administrative or regulatory means, as appropriate.

MDR

2. To identify manufacturers and importers who are not in compliance with the Medical Device Reporting regulation. To bring such firms into compliance through voluntary, administrative or regulatory means, as appropriate.

Medical Device Tracking

3. To determine if manufacturers are making good faith efforts to comply with the Medical Device Tracking regulation. To assure industry awareness of this requirement and encourage compliance as a prerequisite to a phased enforcement policy assuring compliance.

B. PROGRAM MANAGEMENT INSTRUCTIONS

1. The following guidelines are suggested for implementing this compliance program:

   a. This compliance program is to be used to conduct Compliance Status Information System (COMSTAT) inspections (formerly GWQAP inspections) of devices when directed by HFC-240. This is in accordance with the current COMSTAT Manual and to obtain data for COMSTAT profiles and/or updates during regularly scheduled GMP inspections.

   b. *As part of Reinventing Government, certain modifications have been made to the process by which FDA assures that manufacturers of Class I and II devices that are the subjects of premarket 510(K) submissions or certain types of PMA supplements, are in compliance with the GMP requirements. The agency is required to make these modifications more transparent and available to the public. As a result, this program now includes detailed instructions for the process of assuring that devices that are the subject of premarket submissions are manufactured in substantial compliance with the GMP requirements. The procedures are mandatory and must be followed closely. Please refer to Part V, A, 3. and Attachments A-1 and G.*

c. If the device is labeled as sterile, also use Compliance Program Circular 7382.830A to inspect the sterilization process.

d. If the establishment is a contract sterilizer, (see 7382.830B for the definition of a contract sterilizer), it is subject to applicable requirements of the GMP regulation and should be covered using this compliance program as well as 7382.830B.

e. Some manufacturers produce their own devices labeled as sterile and act as a contract sterilizer for other manufacturers. Such manufacturers should be covered under 7382.830A and 7382.830B as well as this compliance program.

NOTE: A device which is subjected to a process designed to reduce its microbial load, but which is NOT labeled as sterile, is to be covered under 7382.830, not under 7382.830A/B.

f. Medical Devices related to blood banking and/or human blood processing will be inspected under this Compliance Program. For guidance, see: Working Relationships Agreement Among the Bureaus of Medical Devices (BMD), Radiological Health (BRH), and Biologics (BOB), April 1, 1982.

2. Intensified Review of the Complaint File

Part III and Attachments B and E of this Compliance Program contain special guidance for reviewing manufacturers' product experience reports to determine compliance with the GMP/MDR requirements for handling complaints.

When a location other than the manufacturing facility or importer is responsible for investigating complaints and submitting MDR reports, the home district should forward a copy of the EIR to the district office where the complaints are handled, with a request for additional follow-up. That district should send an "FYI" copy of its complaint file review to the home district. (Both districts should also follow the reporting requirements shown on the cover page of this program).

3. Scheduling Biennial Inspections of Medical Device Manufacturers

a. Priorities for GMP Inspections

In order to assure the best use of resources, and to assure that manufacturers of devices which present a greater risk to the public are inspected before those which pose a lesser risk, the following manufacturers should, for scheduling purposes, be given top priority:

(1) Manufacturers whose last GMP inspection was violative and there is no compelling evidence of correction.

(2) Manufacturers who received a 510(K) approval decision for a critical/significant risk device within the past year, and have not been inspected within the last two years for processes similar to those used to manufacture the 510(K) device.

(3) Manufacturers of all other Class II or III devices that have never received a GMP inspection.

(4) Manufacturers of all other critical/significant risk devices. Within this group assign the highest inspectional priority to those establishments which have gone the longest without a GMP inspection.

Note: Inspections of manufacturers of devices with a pending PMA approval will be assigned under the PMA Compliance Program (7383. 001). Inspections of manufacturers that have submitted 510(K)s for preamendment Class III devices will be assigned under Compliance Program 7383.003. All other manufacturers should be inspected as resources permit. The primary goal of emphasizing inspection of the above device manufacturers is to change the scheduling of inspections from one that is purely based on the interval since the last inspection to one that also considers the health-hazard significance of the device. Conducting the inspection shortly after a 510(k) has received approval will also allow an evaluation of manufacturers of significant devices at the most critical stage of production. Because most manufacturing and design problems develop or become apparent within the first year of the device's life cycle, inspecting at this time should provide a better opportunity for identifying manufacturing and design problems. GMP inspectional coverage will be focused on that segment of the industry that is actively bringing devices to market and thus presenting the most risk to the public. Those firms that may not receive a biennial inspection should be those producing lower risk products.

b. GMP Pre-Clearance Inspection Program for Class III 510(k)Pre-amendments Devices (CP 7383.003)

When top-priority inspectional assignments that support this program cover all profile classes (except those associated exclusively with certain Class I devices) the district may count the inspection as a qualifying GMP inspection.

c. Initial Inspections
Newly registered and listed firms should receive a directed inspection per Section I of Attachment A as soon as possible after manufacturing operations commence. Generally, firms that manufacture Class III devices and critical/significant risk devices should be inspected within 6 months and firms that manufacture all other Class II devices within 12 months. If the device(s) classification is not known in advance and cannot be determined otherwise, i.e., phone contact, catalog review, etc., schedule the inspection and determine the appropriate inspectional approach after identifying the device(s). For guidance in determining if an establishment should be subject to the GMP regulation refer to page 50 of Medical Device GMP Guidance for FDA Investigators.

If it cannot be determined that at least one device is Class II, III, or Non-GMP exempt Class I, as discussed in section B.3.f. below, review the firm's complaint handling practices; then terminate the inspection. Report the time against PAC 82R800 (District Initiated Assignment).

d. Routine Reinspections

All manufacturers of Class II or III devices should receive a directed GMP/MDR inspection within 24 months of the previous "qualifying inspection". See part II, B.3.a [Priorities for GMP Inspections.]

e. Statutory Coverage List (formerly the Alert List)

Any registered firm that manufactures Class II or III devices and has not had a "qualifying inspection" during the 24 months since they registered will appear on the district's Statutory Coverage List (formerly the alert list). The Statutory Coverage List (formerly the Alert List) will be based on the date of the last "qualifying inspection" (i.e., the last GMP/MDR inspection under PAC's 82830 C, L, or F, 83001, 83003, or 42830 C, L, or F).

f. Class I Device Manufacturers

NOTE: All Class I devices, including those exempted from most of the GMP requirements, must comply with the complaint file requirements of the GMP regulation as well as the reporting requirements of the MDR regulation. Class I manufacturers should receive lowest inspectional priority unless addressed by a special assignment or a health hazard is apparent. See Attachment C for those Class I devices which are exempt from most GMP requirements.

g. Follow-up Inspections

Part III of this program instructs investigators to discontinue the inspection when they encounter conditions that meet the criteria for Situation I in Part V.A.1.. The Warning Letter to the manufacturer warns the manufacturer of its responsibility for reviewing all manufacturing and quality assurance systems. Because other problems may have existed which the manufacturer should have identified and corrected, the follow-up inspection should, *with the exception noted below, be a comprehensive inspection.

Some firms may have clearance of pending 510(K) premarket submissions or certain types of PMA supplements withheld because of significant GMP violations reasonably related to the submission. Such firms may indicate in their Warning Letter or FDA 483 responses that they have only corrected the violations related to the pending submissions. Even though the firms have not claimed to correct all violations, the District is obligated, provided that the corrections appear to be adequate, to conduct a follow-up inspection within 60 days after the firm notifies the district that it is ready for inspection (see Part V, A, 3). The follow-up inspection should be limited to verification that the commitments made in the firm's response were actually made. A comprehensive follow-up inspection may still be necessary at a later date to verify that all other violations, including systems problems, have been corrected. See Attachments A-1 and G.*

Follow-up inspections conducted to determine if violations have been corrected may be counted as qualifying inspections and should be reported against PAC 82830C. All other follow-up inspections, including washouts, are to be reported against PAC 82R800.

4. Scheduling Inspections of Importers (Initial Distributors)

MDR Compliance inspections of importers (initial distributors) will be under separate assignment.

5. Resource Instructions

When possible, Electro-Optical Specialists should be used for inspections of laser devices. Experienced and knowledgeable investigators should conduct inspections of establishments that are manufacturing critical devices. Contact DFI/International and Technical Operations Branch (HFC-133) should the need for expertise, not available in the Region, become apparent (Refer to FMD No. 142).

PART III - INSPECTIONAL

BACKGROUND

This program includes guidance for determining compliance with the Good Manufacturing Practices (GMP), Medical Device Reporting (MDR) and Medical Device Tracking regulations.

A. OPERATIONS

Inspectional Strategy

A "qualifying inspection" is a GMP inspection conducted under this program as per the inspectional strategy presented in Attachment A.

This compliance program, introduced in FY'94, initiated a major change in inspectional strategy. Investigators will conduct a comprehensive inspection only when conducting a follow-up inspection following enforcement action. All other inspections will be directed inspections as per Section I in Attachment A.

When conducting all routine GMP inspections you are required to start the inspection with a review of: (1) complaints and MDR reports (see Attachment A, Section I (B), (2) changes which the manufacturer has made in the design or manufacturing process, and (3) records of production lots which failed in-process or finished device testing. Any indications of problems that your review identifies will provide a focus for your inspection. If you do not find indications of problems after reviewing the above records, complete the inspection as directed in Section I in Attachment A and issue a FDA 483, listing any objectionable conditions that you have observed.

WHEN THE INSPECTION IDENTIFIES SYSTEM WIDE DEFICIENCIES WHICH, IN TOTAL, MEET THE CRITERIA FOR SITUATION I IN PART V,A,1. OF THIS PROGRAM, DOCUMENT THE CONDITIONS THAT CONTRIBUTED TO THE PROBLEM(S), AND CLOSEOUT THE INSPECTION.

SEE ATTACHMENT A, PAGE 3 FOR ADDITIONAL GUIDANCE ON SITUATION I CONDITIONS.

The FDA 483 should contain the following statement:

THE OBSERVATIONS NOTED IN THIS FDA 483 ARE NOT AN EXHAUSTIVE LISTING OF OBJECTIONABLE CONDITIONS. UNDER THE LAW, YOUR FIRM IS RESPONSIBLE FOR CONDUCTING INTERNAL SELF-AUDITS TO

IDENTIFY AND CORRECT ANY AND ALL VIOLATIONS OF THE GMP REGULATION.

1. Initial Inspections

   Follow the instructions in Section I of Attachment A only. Consider the guidance in A.3 through A.8 below.

2. Routine Inspections

   You will be required to tailor your inspection to the particular situation presented by each manufacturer. For all routine inspections follow only the instructions in Section I of Attachment A. It will also be necessary to consider the guidance in A.3 through A.8. below.

3. Important Points to Remember for an Inspection

   a. Cover all of the profile classes (except those associated exclusively with certain GMP exempt Class I devices - see Part II, B.3.f.). If your district has not determined all ofthe major profile classes the establishment is involved in, consult with the COMSTAT coordinator to obtain this information.

   b. In determining whether a firm's systems are adequate and appropriate, consider applicable factors such as:

      - size of manufacturer;
      - the devices and their intended use; and
      - whether the total manufacturing and quality assurance program fulfills the intent of the GMP regulation.

   c. Device manufacturers subject to existing FDA performance standards such as 21 CFR 1020, 21 CFR 1040, 21 CFR 1050 should include in their device master and history records those procedures and records demonstrating compliance with the applicable standard.

   d. Special Instructions for Sterilization Process

      A device subjected to a process designed to reduce its microbial load, but which is NOT labeled as sterile, is to be covered only under this program.

      If the device is labeled as sterile, inspectional coverage indicated in 7382.830A and B, where appropriate, is to occur.

e. Special Instructions for Inspecting Small Manufacturers

Refer to Section I (I) of Attachment A.

f. Inspection of Radiation Emitting Devices

When conducting GMP inspections of radiation emitting devices, also cover the requirements of the applicable standard promulgated under PL90-602.

g. Recalls

Under the provisions of the Safe Medical Device Act manufacturers must now report all recalls to FDA. Confirm that all recalls conducted by the firm since the last inspection have, in fact, been reported to the district office.

4. Selection of Device(s) for Inspection

See Part II, B.3.a. for information on GMP Inspection Priorities.

The selection of the device or devices to be covered should be based first on evidence of defective and/or nonconforming devices identified by your review of the complaint files, change control records, and finished device testing records. The selection also will depend on the total number of appropriate profile classes used in the establishment. This can best be accomplished at the beginning of your inspection by reviewing current product catalogs and/or product listings. Cover all profile classes (except those associated exclusively with GMP exempt Class I devices (see Part II, B.3.f.) by examining the manufacturing of as few device lines representative of those classes as possible.

5. Critical Devices

Investigators should refer to Attachment A regarding questions on the proper application of the critical device requirements. Questions relevant to the critical status of a device should be directed to the contact person designated in Part VI of this program.

6. FDA Compliance Status Information System (COMSTAT) (formerly known as GWQAP)

a. When selecting specific devices to represent profile classes, give preference to critical devices and devices which have had problems. Where possible, select those devices which represent multiple, mutually exclusive profile classes. A list of the device related profile classes appears in the current FDA COMSTAT Manual(formerly the GWQAP Manual).

b. Inspections conducted under a COMSTAT assignment should include:

(1) coverage of the device(s) specified in the assignment, or devices representing all the same profile classes as the assigned device; and,

(2) other devices as required to provide coverage of any remaining profile classes, except GMP exempt Class I devices (See Part II, B.3.f.).

7. Complaint File Review

a. GMP Complaint Handling Requirements

FDA surveys of firms' complaint files have shown that some firms were deficient in their complaint handling practices. These deficiencies were caused by a firm's failure to:

(1) follow its own procedures for processing product experience reports; or,

(2) develop procedures which meet the requirements of 21 CFR 820.198, .162, .20 and .5, and

(3) firms identify problems but do not take timely corrective actions.

The GMP regulation requires that all complaints be reviewed, evaluated and maintained by a formally designated unit. This unit could be one appropriately trained individual, or a department which is staffed with appropriately trained individuals. This unit must decide whether an investigation of the complaint needs to be performed. There is NO requirement in the GMP regulation that all complaints must be maintained in one file, or that there be a written procedure for processing complaints. Typically, manufacturers will keep complaints in a customer sales file, product returns/credits file, warranty file, medical file, or legal file. The inspection should ascertain what files are maintained that meet the definition of a complaint, as found in 21 CFR 820.198(a).

[Note: If GMP defined complaints are not maintained by the formally designated unit, it should be noted on the FDA 483.]

By placing these complaints in different files, manufacturers have not noted instances of repeated component/device failure with a common cause. Ask the firm if it trends complaints. If no trending or problem identification is done, then the inspection should begin with the investigator conducting a trend of the complaints.

Evaluate the firm's complaint handling practices for the devices you are inspecting since the last GMP/MDR inspection. Refer to Attachments A, B, and E for guidance.

Note: The actual complaints or deficiencies in complaint handling practices may provide leads in identifying product defects which have not been adequately corrected by the firm, either by recall, and/or changes in the device and/or manufacturing process.

Reference #4 in Part VI explains how the GMP complaint files relate to reports required under MDR.

b. Operational Procedure for Determining Compliance with the MDR Regulation.

The guidance formerly included in C.P. 7382.011, Enforcement of the Medical Device Reporting (MDR) Regulation, is now part of this compliance program. See Attachments A and E for detailed directions.

8. Medical Device Tracking

Determine if the firm makes any medical device(s) subject to the Medical Device Tracking regulation. See Attachment F for a list of subject devices. If so, follow the guidance in Attachment A, Section I G.3 to determine if the firm has device tracking procedures in place and, if it does, to assess their adequacy.

If there is a complete absence of a tracking system, you should cite this violation on the FDA 483. Other violations should be addressed verbally and reported in the EIR.

9. Sample Collection

For GMP or MDR violations, documentary samples will be collected as necessary.

Physical samples are not required to support GMP violations, and should not be routinely collected for GMP cases. If the district should reference violative documentary or physical samples as evidence to support GMP deviations, the condition of the samples should be tied to the GMP deviation to show a cause/effect relationship.

If you are uncertain as to whether a sample should be collected, the district should consult with the CDRH Headquarters Laboratory Liaisons or the division of field science in ORA on the laboratory capability to conduct the analysis (See Part VI, C. Program Contacts).

10. Imports

No import wharf examinations or sample collections are scheduled under this program.

11. Exports

Devices that are manufactured in the U.S., but not marketed in the U.S., are not subject to the GMP requirements, provided that the manufacturer has documented proof that its devices are offered for sale only in foreign countries.

Devices that are manufactured in the U.S., but not marketed in the U.S. are subject to the MDR requirements including those exported under the provisions of Section 801(e) of the Act. This includes devices exported for investigational use.

12. Follow-up Inspections

Unless the previous violative inspection was a comprehensive inspection, the Situation I violations that were identified may have been part of more widespread systems problems that the investigator did not have an opportunity to evaluate. After receiving the Warning Letter the manufacturer should have investigated its manufacturing and quality assurance systems and initiated appropriate corrections. To assure that the manufacturer has fulfilled its responsibility, the follow-up inspection should be conducted as a comprehensive inspection using both sections I and II of Attachment A. If problems similar to those originally identified, or new problems which meet the criteria for Situation I are identified, complete the comprehensive inspection and document all observations.

*The only exception to the above policy concerns those firms that have had clearance of pending premarket 510(k) submissions or certain types of PMA Supplements withheld due to GMP violations.

Such firms may indicate in their Warning Letter or FDA 483 response that they have corrected the violations related only to these pending submissions. Even though the firms have indicated that they have not corrected all violations, the district is obligated, provided that the corrections appear to be adequate, to conduct a follow-up inspection within 60 days after the firm notifies the district that it is ready for inspection (see Part V, A3). The follow-up inspection may be limited to verification that the corrections made in the firm's response were actually made. A comprehensive follow-up inspection may still be necessary at a later date to verify that all other violations, including systems problems, have been corrected. See Attachments A-1 and G.*

13. Foreign Inspections

    Use FDA/CDRH document entitled, "Inspections of Foreign Device Manufacturers" for guidance in conducting foreign device inspections.

B. REPORTING

1. General Reporting requirements are listed on the cover page. As a general rule the time used for preparing the EIR should not exceed the time spent conducting the inspection.

2. MDR Observations - If the establishment failed to comply with any of the MDR requirements, include those observations on the FDA 483.

3. GMP Observations - If you observe any violations of the GMP requirements, you should place them on the FDA 483. The most serious violations (e.g., those that could potentially result in production of defective devices, or identification of production/design problems) should be noted on the FDA 483 first. Your FDA 483 comment, however, should differentiate between problems which are indicative of a system failure and rare isolated situations.

4. Critical Devices - The EIR must indicate whether the device being covered is a critical device or non-critical device. Please place this information in the Summary of Findings.

5. 510(k) Observations - If the firm failed to have a 510(k), or made significant changes which requires a new 510(k), do not place them on the FDA 483 unless you obtain concurrence from CDRH/OC.

6. Medical Device Tracking Requirements - The EIR must indicate whether or not the establishment inspected:

   (a) makes any devices subject to the "Medical Device Tracking Requirements" and, if so,whether the firm is meeting its tracking obligations. These new requirements have an implementation date of 8/29/93. See Attachment F for questions and answers about the Medical Device Tracking regulation. See the April 1994 Edition of 21 CFR, Part 821 for a corrected copy of the medical device tracking regulation.

If there is a complete absence of a tracking system, you should cite this violation on the FDA 483. Other violations should be addressed verbally and reported in the EIR.

PART IV - ANALYTICAL

A. ANALYZING LABORATORIES

The district will make all the necessary arrangements for proper handling of samples with the following designated testing facilities:

TYPES OF DEVICES	ANALYZING LABORATORIES
All General Medical Devices	WEAC
Radioimmunoassay Devices	PHI
All other In Vitro Diagnostic Devices	BLT
Sterility Analysis, Bioburden, Bioindicators, and Endotoxins	Refer to CP 7382.830A

See PART VI regarding those persons designated as contacts for designated laboratories and specific products.

B. ANALYSES TO BE CONDUCTED

Sample collection and analysis will be determined on a case-by-case basis through consideration of inspectional findings, compliance and scientific capabilities and expertise. Full collaboration between investigations and analytical personnel is essential at this phase. See Part III A.9 for additional information.

PART V - REGULATORY/ADMINISTRATIVE FOLLOW-UP

A. REGULATORY/ADMINISTRATIVE FOLLOW-UP.

1. Situation I

The district has evidence indicating that the manufacturing process is producing nonconforming and/or defective finished devices. Such evidence would include information from sample analyses, complaint files, the firm's failure analyses, MDR/PRP reports, or GMP observations made during the inspection.

OR

The inspection documents GMP deviations of a significant type or quantity to conclude that there is a reasonable probability -- in light of the relationship between GMP deviations observed, and the particular product and manufacturing process involved -- that the firm will likely produce nonconforming and/or defective finished devices. Such deviations include one or more of the following:

- failure to establish and document a formal quality assurance program;

- failure to validate significant manufacturing processes and quality assurance tests (refer to document titled "Guidelines on the General Principles of Process Validation" for the validation requirements)

- failure to document, review, approve, implement and validate changes to components, finished devices, labeling, packaging or manufacturing process specifications;

- failure to establish and implement adequate recordkeeping procedures (e.g., device history record, device master record);

- failure to establish and implement an adequate complaint handling program;

- failure to establish and implement an adequate failure investigation program; and,

- failure to ensure that finished devices meet all specifications prior to distribution.

If any of these deviations exist, and the significance of the deviation and the device warrants it, the district should consider administrative and/or regulatory action, e.g., Warning Letter, injunction, administrative detention, seizure, civil penalty, and/or prosecution. The district is expected to classify the EIR as OAI.

SEE ATTACHMENT A, PAGE 3, FOR ADDITIONAL GUIDANCE ON SITUATION I CONDITIONS.

If any of these deviations exist for foreign manufacturers, and the device warrants it, regulatory action should be considered, i.e., Warning Letter and/or Warning Letter with automatic detention.

If a serious health hazard is identified, FDA initiated recall and/or administrative detention/seizure should be considered as the initial action to bring the situation under prompt control.

2. Situation II

The inspection documents GMP deviations of a quantity or type to conclude that there is minimal probability -- in light of the relationship between GMP deviations observed and the particular product and manufacturing processes involved -- that the firm will produce nonconforming and/or defective finished devices. The Form FDA 483, Inspectional Observations, will serve to inform the firm of any objectional findings.

Firms receiving an FDA 483 under Situation II should receive follow-up coverage as district resources permit. However, the continued presence of GMP deviations which have a low probability of leading to an unsafe or ineffective device will not usually warrant recommendation of an administrative and/or regulatory action if no new or more serious deviations are found.

3. Policy

Agency policy requires that products sold to the federal government should be treated in the same manner as products sold to commercial accounts. Consequently, when FDA recommends against acceptance of a device because that device, or its manufacturer, is in violation of the FD&C act, FDA must also include appropriate regulatory/administrative action against thesame or similar device sold to commercial accounts.

Compliance Policy Guide 7132c.02, and 7153.04 should be followed in making an evaluation of the violations and the action of choice.

If a firm has shipped a violative product to a Government agency, regulatory action consistent with the nature of the violation(s) may be taken even though there have been no shipments to commercial customers. Formal regulatory action in connection with a violative shipment may not be necessary in some cases. For example, the firm promptly corrects the violative condition, and existing Agency policy would not require further action if the matter involved a product shipped to a non-government customer. However, where corrections are not or can not be made promptly, the main concern is preventing the subsequent shipment of the product to another customer. When the product has been shipped solely to a Government agency and is under the control of that agency and there is no threat to the public, DMPQA should ascertain the intention of the agency holding the goods (e.g., will they return or destroy the goods; will they request FDA to initiate seizure, etc). If the procuring agency requests FDA action, DMPQA will refer the matter to the home district for their consideration of an appropriate recommendation.

Action which may be considered are FDA-requested recall, Warning Letter, seizure, injunction, and prosecution.

Corrective action proposals should be submitted by a responsible official of the firm in writing, detailing the action(s) to be taken to bring the violative process or product into compliance within a specified time frame. Voluntary correction does not preclude the initiation of administrative and/or regulatory action.

In determining whether GMP deviations are sufficient to support legal action, consideration should be given to the significance of the device, the firm's quality history, and whether the problem is widespread or continuing.

When CDRH does not agree with a district's recommendation for a regulatory action, the district will be notified of the reasons for disapproval in writing.

*As part of Reinventing Government, certain modifications have been made to the process by which FDA assures that manufacturers of Class I and II devices that are the subject of premarket 510(k) submissions or certain types of PMA Supplements are in compliance with the GMP requirements.

See Attachment A-1. It is now a mandatory requirement for the District to promptly acknowledge receipt of a firm's response to either a FDA 483 or Warning Letter and to indicate in the acknowledgement letter the extent to which the response appears to be satisfactory. When a firm has clearance of a pending premarket submission withheld because of GMP violations the District must schedule a prompt follow-up inspection after the firm has submitted what appears to be an adequate response and indicates it is ready for inspection. In some situations the firm's response may be sufficient and a follow-up inspection may not be necessary (e.g., when the firm submits complete validation data). When this occurs, the District must promptly notify the firm, and CDRH, that clearance of its premarket submissions will no longer be withheld because of GMP problems.

See Attachment G for detailed mandatory procedures for dealing with firms that have clearance of a premarket submission being withheld.*

a. Warning Letter

Issuance of a Warning Letter should be in accordance with Chapters 8-10 of the Regulatory Procedures Manual. See also Attachment G for model Warning Letters.

If the District determines that issuance of the Warning Letter has resulted in corrective action by the firm, the district shall, within five (5) days after confirmation, update the firm's Profile Data Sheet and inform CDRH, Field Programs Branch of the firm's return to compliance.

b. Violative Follow-Up Inspections

With the issuance of the FY'94 compliance program, investigators were instructed to close out an inspection as soon as they documented conditions which meet the criteria of Situation I. The new model Warning Letter (Attachment G) advises manufacturers that the conditions identified by the investigator may be symptomatic of systems problems, and that the manufacturer is responsible for investigating, identifying, and correcting systems problems. The model Warning Letter further directs the firm to discuss in its response how it will address the systems problems related to the conditions identified by the investigator.

To assure that the manufacturer did, in fact, review all manufacturing and quality assurance systems, investigators are instructed to conduct a comprehensive follow-up inspection. When investigators identify the same or additional conditions that meet the criteria for Situation I., the District should consider seizure or injunction as the regulatory action of choice.

c. Enforcement Strategy For Firms With Repeated Violative Inspections

(1) Some firms have a high rate of recidivism. They have developed a pattern of correcting violative conditions in response to Warning Letters or other regulatory actions, and usually maintain those corrections long enough to pass the follow-up inspection. When FDA inspects the firm several years later, however, the investigator identifies similar conditions that again meet the criteria for Situation I. This tendency toward recidivism is often due to the failure of the firm to have in place basic manufacturing and quality assurance systems which meet the requirements of the GMP regulation.

(2) When dealing with another violative inspection for such a firm the District should consider using the following strategy, which may be used either alone or in conjunction with a seizure when the severity of the conditions do not justify mass seizure:

(a) Issue a Warning Letter that follows the model Warning Letter in Attachment G-8. This Warning Letter requests the manufacturer to submit to the District (for up to 2-years) an annual certification by an outside expert consultant stating that it has conducted a complete audit of the firm's manufacturing and quality assurance systems relative to the requirements of the device GMP regulation. The firm must submit a copy of the consultant's report, and certification by the firm's CEO that he or she personally has received and reviewed the consultant's report and that the firm has made all corrections called for in the report.

(b) You have the option of limiting your review of the certifications only to the extent necessary to confirm that the consultant and the firm have met the requirements set forth in the Warning Letter. You may also request a technical evaluation of the consultant's report by the appropriate branch within the Office of Compliance. You have no obligations, however, to send to the firm comments regarding the adequacy of the consultant's report or the firm's corrections.

(c) It will not be necessary to schedule a follow-up inspection for at least 6-months after the firm certifies that it has completed all corrections.

The District may remove the firm from the COMSTAT list as soon as the firm has certified that it has completed all corrections recommended by the consultant.

(d) If the follow-up inspection indicates that the corrections are satisfactory, notify the firm that you have no objections to the corrections. Remind the firm, however, that it must continue to submit to the District, on the schedule specified in the Warning Letter, certification by an outside expert consultant that it has conducted an update audit and certification by the firm's CEO that any corrections noted to be needed by the consultant have been made, and that it remains in compliance with the requirements of the GMP regulation. The firm should also continue to submit copies of the audit results.

(3) If conditions identified by the follow-up inspection meet the criteria for Situation I, initiate action per 3b above.

(4) If the evidence indicates that the consultant's or firm's certifications are fraudulent, the District may wish to request participation by the Office of Criminal Investigations. When there is clear evidence that the firm falsified its status reports to the District, initiate appropriate charges under USC 18, Section 1001.

d. Recall

If the district believes that prompt removal of a violative product from channels of commerce is necessary, it shall proceed in accordance with established recall procedures in the RPM and 21 CFR, Part 7 (Enforcement Policy), Subpart C (Recalls). In the event there exist serious adverse health consequences or a death, CDRH may request discontinuation of distribution and recall a device to the user level in accordance with Section 518(e) of the Act.

e. Administrative Detention/Seizure

Prior to approving an administrative detention, the District Director must have reason to believe the device is misbranded or adulterated and the firm holding the device is likely to quickly distribute or otherwise dispose of the device, or detention is necessary to prevent use of the device by the public until appropriate regulatory action may be taken by the Agency. District Directors must consult with the Center for Devices and Radiological Health by telephone. Contact the appropriate Div./Branch in OC for the subject device by consulting the CDRH/OC organization chart in PART VI, C. Program Contacts. Concurrence must be given by the Director, OC, CDRH, based on a recommendation by his/her staff.

The district must immediately recommend a seizure.

Of course, a seizure action can be recommended without administrative detention to remove violative devices from commercial distribution, either at the manufacturer, distributor, repacker or at a device user location.

f. Injunction

If a firm has a continuing pattern of significant deviations in spite of past warning, or its operations constitute a serious health hazard, injunction shall be the recommended action of choice. If a serious health hazard exists, the recommendation should include a request for a temporary restraining order (TRO) to prevent the distribution of devices which have been manufactured under the violative conditions documented in the inspection report. (See RPM 8-60-30C). The recommendation shall be accompanied by copies of all necessary documentation, e.g., complete inspection reports, Warning Letters issued, sample analysis reports, firm's response to Warning Letters and/or FDA 483s. In the absence of samples, the inspectional evidence must clearly show that the establishment has substantially deviated from GMPs. These deviations must be well documented.

g. Citation

A citation shall be recommended if appropriate as stated in RPM 8-30.

h. Prosecution

The criteria stated in RPM 8-50 shall be the criteria for consideration of prosecution of individuals in violative GMP cases.

i. PMA Disapproval/Withdrawal

Refer to Compliance Program 7383.001, Part V.

j. Automatic Detention

Automatic detention should be recommended whenever there is clear documented evidence to suggest that the foreign manufacturer is producing nonconforming and/or defective devices.

k. Civil Penalties

Section 303(f)(1)(B)(i) of the Act states that civil penalties shall not apply to GMP violations "unless such violation constitutes (I) a significant or knowing departure from such requirements, or (II) a risk to public health." Section 303(f)(1)(B)(iii) further stipulates that civil penalties shall not apply to "section 501(a)(2)(A) which involve one or more devices which are not defective." However, if it can be shown that the firm has received appropriate warning (e.g., FDA 483, Warning Letter, etc.) and that the same GMP violations are subsequently encountered, civil penalties should be considered. Recommendations for civil penalties should be made to address past violations as opposed to ongoing violations.

l. Facilitating Review of Regulatory Recommendations

(1) The district is expected to consult with OC both prior to, and especially during the inspection, once it is determined that regulatory action is being considered. Contact the appropriate Div./Branch in OC for the subject device by consulting the CDRH/OC organization chart in PART VI, C. Program Contacts.

(2) When the district knows a regulatory action will be forthcoming as a result of the inspection, FAX a copy of the issued FDA 483 to the appropriate division in the Office of Compliance. A copy of the FDA 483 annotated with exhibit numbers, and EIR page numbers, helps the reviewers. Therefore, the review process can begin within CDRH while the EIR and recommendation are being written by the district.

(3) It is the responsibility of district management to ensure that the documentation and evidence presented with each legal action

recommendation is sufficient to justify each charge. The volume of material submitted must be minimized, and should include only the basic documentation needed to support each GMP charge/example.

(4) All necessary samples and other supporting documentation must be tabbed and their location cross referenced in the recommendation in order to assist in a timely review. It is highly recommended that you provide a table which cross references the violation with the FDA 483 item number, the inspection report page number and the exhibit.

Photographs submitted with recommendation for 502(a) and/or 502(f) charges must be prints rather than reproductions, such as copies. Photos will be returned upon request.

(5) It is essential that all significant questions, problems, or other weaknesses in the evidence regarding the recommended action be stated, along with pertinent district comments. Otherwise, reviewers may miss a problem entirely until litigation is commenced.

(6) The recommendation must begin with the most serious violation of the regulations with reference to the EIR pages, exhibits and sample results which document the violation. Each charge must be parenthetically referenced in the Recommendation Memorandum and the page location of the supporting evidence given. Violations must be listed in decreasing order of importance. Each violation should be related to its effect on device quality in light of overall controls, and should be separated according to the type of manufacturing activity.

(7) Physical samples are not required to support GMP violations, and should not be routinely collected for GMP cases. If the district should reference violative documentary or physical samples as evidence to support GMP deviations, the condition of the samples should be tied to the GMP deviation to show a cause/effect relationship.

(8) Evidence of previous warning and other regulatory actions should be referenced along with a description of corrective actions. If the recommendation or current EIR references a previous report, either copy the cited EIR pages, or summarize the information.

B. MDR REGULATORY/ADMINISTRATIVE FOLLOW-UP

1. General Information

MDR violations must be fully documented in any of the situations described below.

If there are no complaint or MDR records due to the firm's failure to maintain such records, the EIR narrative should document those findings. When review of the firm's complaint files and service/repair/warranty records has identified MDR reportable events involving deaths, serious injuries, or malfunctions that were not reported to the FDA, failure to report should be listed on the FDA 483.

CDRH requests that copies of Warning Letters and information regarding Direct Reference Seizures be sent to HFZ-306 and HFZ-530. If an alternate means is selected to advise the firm that they are in violation of the MDR regulation, please inform HFZ-306 and HFZ-530 accordingly, by forwarding copies of the FDA 483, memorandum of meeting, or any other pertinent information as appropriate.

2. Situation I

NATURE OF VIOLATION	ACTION
o The firm has received prior notice, but continues to not submit MDR's for deaths or seriousand/or Civil Penalties injuries (other than those which are "unanticipated temporary injuries"), and has not established procedures (GMP) for complaint handling and investigation.	Direct Reference Seizure
o Reinspection determines that a firm is a chronic or flagrant violator which continues to fail to report MDR's, particularly those involving deaths or serious injuries. Consideration should be given to the significance of the unreported incident(s) (the apparent risk to health or level of injury involved), the scope of the problem (whether the problem is widespread or continuing), and the firm's history.	Injunction

o Inspection reveals the firm has failed to obtain the data needed to file a complete MDRreport as described in Part 803.24(c). This failure is the result of an incomplete investigation or the failure to conduct any investigation into the event. The failure to conduct afollow-up investigation may be a violation of the GMP requirements for failure investigation and Complaint Files (Parts 820.162 and 820.198) and a GMP charge should be considered. The MDR regulation itself does not require that an investigation be performed.	Direct Reference Warning Letter
o Failure to submit MDR's for deaths or serious injuries (other than those which are "unanticipated temporary injuries") of which the firm is aware.	Direct Reference Warning Letter

Authority does not extend to citing the failure to report medical intervention to relieve unanticipated temporary injuries as defined in 21 CFR 803.3(h)(3)(ii), or the failure to report malfunctions as defined in 803.3(c). CDRH must be consulted prior to citing failures to report serious injuries that necessitate medical or surgical intervention, by a health care professional, to relieve unanticipated temporary impairment of a body function or unanticipated temporary damage to a body structure, or malfunctions that would be likely to cause or contribute to a death or serious injury.

o Failure to submit MDR's for unanticipated temporary injuries or malfunctions. (Requires CDRH Concurrence)	Warning Letter

3. Situation II

NATURE OF VIOLATION	ACTION
Review of the firm's complaint files finds they have failed to maintain the records required by Part 803.31 (Complaint files) of the MDR regulation.	FDA-483 Only

The failures should be listed on the FDA 483. The district should follow the guidance in the Regulatory Procedures Manual (particularly where continuing violations have been documented) when determining whether to utilize additional means of notification beyond the FDA 483, such as issuing a Warning Letter or setting up a conference between the firm and the District.

4. Guidance on Citations Used in and Content of Warning Letters

 If reinspection reveals continuing problems with compliance with FDA regulations, particularly where a potential risk to health exists, the district should consider taking additional regulatory action (seizure, injunction, 518(e), etc.).

 Cite Formats for Warning Letters

 a. When cite involves failure to report use one of the following paragraphs:

 502(t)(2) The devices are misbranded in that information required to be submitted to the Food and Drug Administration (FDA) by the Medical Device Reporting (MDR) regulation specified in 21 CFR Part 803 was not submitted as follows:

 b. In addition to the cite, include all of the following paragraphs that apply to your situation:

 USE WHEN FIRM HAS NOT FILED ANY REPORTS

 Failure to submit reports to FDA after receiving information which reasonably suggested that one of your marketed devices [CHOOSE ALL OF THE FOLLOWING THAT APPLY TO YOUR CHARGE] has malfunctioned or caused or contributed to a death or serious injury, as required by 21 CFR 803.24. Failure to make telephone reports to FDA within five (5) calendar days after receiving information which reasonably suggests one of your marketed devices may have caused or contributed to [CHOOSE ONE OR BOTHOF THE FOLLOWING DEPENDING ON YOUR CHARGE] a death or serious injury, as required by 21 CFR 803.24.

 NOTE: Days are counted starting the day AFTER a firm learns of an incident.

C. MEDICAL DEVICE TRACKING

In the initial phase of surveillance for compliance with requirements of the Device Tracking regulation districts should send Warning Letters only when firms have failed to implement any form of a tracking system. If districts wish to include in the Warning Letter other charges relating to the Device Tracking regulation, the draft Warning Letter must first be approved by CDRH.

PART VI REFERENCES, ATTACHMENTS AND PROGRAM CONTACTS

A. APPLICABLE REFERENCES OR AIDS

1. Code of Federal Regulations, Title 21, Part 820 Good Manufacturing Practice for Medical Devices: Regulations Establishing Good Manufacturing Practices for the Manufacturer, Packing, Storage, and Installation of Medical Devices.

2. Federal Food, Drug, and Cosmetic Act, As Amended.

3. Investigations Operations Manual - Chapter 5, Subchapter 550.

4. GMP Complaint Files: How They Relate to Reports Required Under MDR, originally published in Medical Device & Diagnostic Industry, Volume 7, Number 5, May 1985, Revised version (4/10/85) distributed to all district offices.

5. Medical Device Reporting - Questions and Answers, HHS Publication (FDA) 88-4226, February 1988.

6. An Overview of the MEDICAL DEVICE REPORTING Regulation, DHHS publication No. (FDA) 85/4194, December 1984.

7. MDR - Federal Register, Vol. 49, No. 180, Friday September 14, 1984, Rules and Regulations, pages 36326 through 36351, Medical Device Reporting; Final Rule. 21 CFR PARTS 600, 803, 1002, and 1003.

8. STERILIZATION - QUESTIONS AND ANSWERS, MARCH 1985

9. DEVICE GOOD MANUFACTURING PRACTICES MANUAL. HHS Pub. No. (FDA) 91-4179, Fifth Edition August 1991.

10. NBS special Publication 250 - May 1984 (or update) Calibration and Related Measurement Services U.S. Dept. of Commerce NBS Washington, D.C. 20234.

11. Medical Device GMP Guidance for FDA Investigators, DHHS Publication No. (FDA) 84-4191, April 1984.

12. Sterile Medical Devices. A GMP Workshop Manual. fourth Edition November 1984. Prepared by Division of Small Manufacturers Assistance, Office of Training and Assistance, HHS Publication FDA 84-4174.

13. Guideline on General Principles of Process Validation: Notice of Availability published in the Federal Register on May 1987.

14. Working Relationships Agreement Among The Bureau of Medical Devices (BMD), Radiological Health (BRH), Biologics (BOB), Published April 1982, Refer to District Reference File for copies.

15. Device GMP - Applications and Interpretations for the Field Investigator, November 1980.

16. Plastic Medical Devices: A Study of Quality in The Making. September 1980. A copy of this film has been supplied to each FDA District office. This film is intended for use with the Medical Device Reference Files on plastics (see reference #17 below).

17. Medical Device Reference Files on syringes, catheters, tubes and airways, IOL and contact lenses, IUDs and filters. September 1980. One hard copy and one microfiche copy of each of these reference files has been supplied to each FDA District office.

18. GMP Questions and Answers, November, 1979.

19. Device Good Manufacturing Practices. A Quality Audit Program for Industry. September, 1979.

20. Quality Control Handbook, Juran, J.M., 3rd edition, McGraw-Hill, 1974.

21. ANSI Z80.1-1972 Revision of X80.1-1964 American National Standard Requirements of First-Quality Prescription Ophthalmic Lenses.

22. MIL-STD-105E, Sampling Procedures and Tables for Inspection by Attribute.

23. GWQAP Manual

24. Classification Names for Medical Devices and In Vitro Diagnostic Products, HHS Publication No. (FDA) 91-4246, August 1991. This directory is organized by "keywords" in alphabetical order. The classification number (5 digit product code), class, and CFR regulation number is given for each entry listed.

25. Code of Federal Regulations, Title 21, Part 809.10, Labeling for In Vitro Diagnostic Products.

26. Advisory List of Critical Devices - 1988; Notice Published in the Federal Register on March 17, 1988.

27. Overview of Metallic Orthopedic Implants; Technical report, reference material and training aid for investigators prepared by the Division of Field Investiga-

tions (HFC-130), Office of Regional Operations, Office of Regulatory Affairs, HHS, Public Health Service, FDA, June, 1988.

28. AQL Inspector's Rule and Manual. This special purpose plastic slide rule that rigidly adheres to MIL-STD-105E can be obtained from Infor. Inc., P.O. Box 606, Ayer, MA. 01432. Phone (508) 772-0713. Cost is approximately $20 each excluding shipping and packaging.

29. Inspections of Foreign Device Manufacturers, February, 1993. This eight page document was prepared by CDRH, Office of Compliance.

30. Code of Federal Regulations, Title 21, Part 821, Medical Device Tracking Requirements.

31. Medical Device Tracking - Questions and Answers Based on the Final Rule, HHS Publication No. (FDA) 93-4259, August 26, 1993.

Copies of CDRH GMP publications are available from the Division of Small Manufacturers Assistance (DSMA), Telephone: 800-638-2041 or FAX 301-443-8818.

B. ATTACHMENTS

ATTACHMENT A - DEVICE GMP INSPECTIONS

ATTACHMENT A-1 - PROCESS FOR COMPLIANCE REVIEW OF CLASS I AND II 510(K) SUBMISSIONS/PMA SUPPLEMENTS

ATTACHMENT B - INSTRUCTIONS FOR REVIEWING RECORDS KEPT IN A MANUFACTURER'S DEVICE COMPLAINT FILE(S)

ATTACHMENT C - CLASS I DEVICES EXEMPT FROM MOST OF THE GMP REQUIREMENTS BY CLASSIFICATION REGULATIONS

ATTACHMENT D - ADVISORY LIST OF CRITICAL/SIGNIFICANT RISK DEVICES

ATTACHMENT E - INSPECTIONAL GUIDANCE FOR PERFORMING A REVIEW OF THE FIRM'S COMPLIANCE WITH THE MDR REGULATION.

ATTACHMENT F - MEDICAL DEVICE TRACKING, QUESTIONS AND ANSWERS BASED ON THE FINAL RULE, AUGUST 26, 1993

ATTACHMENT G - MODEL WARNING LETTERS

C. PROGRAM CONTACTS

1. ORA Contacts

a. Questions regarding inspectional requirements:

Investigations Branch (HFC-132)
Telephone: (301) 443-3340, 3341, 3342

b. For technical assistance or advice while conducting an inspection:

International and Technical Operations Branch (HFC-135)
Telephone: (301) 443-9894

c. Questions about accessing or connecting to the Parklawn Computer Center and Model 204

Leo Bauman
Division of Information Systems, ORA
(301) 443-1314

An easy method for Field Users to access the system is to log on to the regional VAX, then type:

TELNET PCCSNA.FDA.GOV <return>

Field Users should set up their communication program to emulate a VT100 or other of the options before logging on to the Regional VAX.

d. Questions regarding sampling of devices and laboratory capabilities:

Division of Field Science (DFS), HFC-140
Telephone: (301) 443-3320

e. The WEAC contact point for testing medical devices is:

Director
WEAC Engineering Branch, HFR-NE480
Telephone: (617) 729-5700

f. Questions regarding COMSTAT

Gillie Kovalsky
Division of Medical Products Quality Assurance (DMPQA), HFC-240
Telephone: (301) 827-0385

2. District Office Contacts For Industry Management Concerns About Their GMP Compliance Status.

Atlanta	Ballard Graham
Baltimore	Gary Pierce
Boston	James Rado
Buffalo	E. Pitt Smith
Chicago	Ray Mlecko
Cincinnati	Debbie Grelle
Dallas	Austin Templer
Denver	William Sherer
Detroit	John Dempster
Kansas City	Robert Wilson
Los Angeles	Elaine Messa
Minneapolis	John Feldman
Nashville	Ray Hedblad
New Orleans	James Green
Newark	Marguerite Eagan
New York	Rick Trainor
Orlando	Timothy Couzins
Philadelphia	Diana Kolaitis
San Francisco	Andrea Scott
San Juan	Sam Jones
Seattle	David Pettenski
St. Louis	Charles Bringman
Foreign Firms	Joy Lazaroff/Marje Hoban

3. CDRH Contacts

See the New CDRH/OC Organizational Chart at the end of of Part VI to identify which unit within OC is responsible for answering your question or giving you guidance, depending on the type of device.

a. Questions regarding Compliance Review of Class I and II 510(K) Submissions/PMA Supplements:

Joy Lazaroff
Office of Compliance
Division of Program Operations
Field Programs Branch, HFZ-306
Telephone: (301) 594-4695, ext. 109
FAX: (301) 594-4715

b. Questions regarding the interpretation and applicability of the MDR regulation:

Wayne Robinson
Division of Surveillance Systems, OSB
Reporting Systems Monitoring Branch, HFZ-533
Telephone: (301) 594-2735
Electronic Mail (EWR)

c. MDR Report and Data Summaries:

Arlene Underdonk
Division of Surveillance Systems, OSB
Information and Analysis Branch, HFZ-531
Telephone: (301) 594-2731

d. Industry MDR Report: (301) 427-7500. Do not call this phone number to make inquiries.

e. Questions about using MDRAPSY (MDR/PRP database).

Del Futrell
Division of Surveillance Systems, OSB
Information and Analysis Branch, HFZ-531
Telephone: (301) 594-2731

f. Questions regarding sampling and/or testing of general medical devices.

Edward Mueller or Donald Marlowe
Division of Mechanics and Material Sciences, HFZ-150
Telephone: (301) 443-7003

g. Express Mail Address for All Regulatory Action Recommendations:

Field Programs Branch, HFZ-306
Office of Compliance
Center for Devices and Radiological Health
2094 Gaither Road
Rockville, Maryland 20850

h. Questions regarding the interpretation and applicability of the device GMP regulation and GMP exemptions:

Contact the appropriate Division/Branch in OC for the subject device by consulting the CDRH/OC organization chart at the end of Part VI.

i. Questions regarding GMP critical device status:

Sharon M. Kalokerinos
Division of Enforcement II, HFZ-331
Telephone: (301) 594-4613 ext. 139

j. Questions regarding this Compliance Program:

Bill Keene
Field Programs Branch, HFZ-306
Telephone: (301) 594-4695 ext. 111

k. Questions regarding New Attachments A-1 and G:

Marje Hoban
Field Programs Branch, HFZ-306
Telephone: (301) 594-4695 ext. 103
FAX: (301) 594-4715

PART VII - CENTER RESPONSIBILITIES

A. Program Evaluation for GMP Inspections

1. Unit Assigned: Field Programs Branch
Associate Director for Program Operations
Office of Compliance

2. An evaluation of this program will be conducted upon request of the Director, Office of Compliance, CDRH.

B. MDR Report Summaries

1. Unit Assigned

Office of Surveillance and Biometrics
Information Management and Analysis Branch, HFZ-531

2. Summary of Work

Upon request, OSB will forward to an investigator, reports containing summaries of MDR reports submitted to CDRH by a particular firm.

C. Selection and Assignment of Importers and Manufacturers Requiring Special Inspection.

a. Unit Assigned

Office of Surveillance and Biometrics
Device Administration Experience and Monitoring Branch

b. Summary of Work

On a fiscal year as well as an ad hoc basis, firms will be selected and assigned for inspection.

ATTACHMENT A

DEVICE GMP INSPECTIONS
Medical Device Manufacturers
and
GMP Regulation Requirements

GMP INSPECTIONAL STRATEGY

This compliance program, introduced in FY'94, initiated a new strategy for conducting GMP inspections. The traditional "Two-Track GMP Strategy" required, at a minimum, that every manufacturer receive a comprehensive GMP quality audit once every four years, and a limited GMP quality audit at the intervening two-year interval. If manufacturers experienced problems, however, they were subject to comprehensive quality audits more frequently.

The new strategy requires that only a follow-up inspection must be a comprehensive quality audit. All other inspections, including initial inspections, will be directed inspections per Section I of this Attachment.

All inspections must be started with a review of the complaint and MDR files, recalls, manufacturing change control records, and finished device testing records. When these records indicate that the manufacturer has experienced device failures and/or produced nonconforming devices you should focus on those particular devices and manufacturing processes. WITH THE EXCEPTION OF COMPREHENSIVE FOLLOW-UP INSPECTIONS, AS SOON AS YOU HAVE IDENTIFIED AND DOCUMENTED SYSTEM WIDE DEFICIENCIES WHICH, IN TOTAL, MEET THE CRITERIA FOR SITUATION I IN PART V.A.1., DISCONTINUE THE INSPECTION AND ISSUE THE FDA 483.

The FDA 483 should contain the following statement:

THE OBSERVATIONS NOTED IN THIS FDA 483 ARE NOT AN EXHAUSTIVE LISTING OF OBJECTIONABLE CONDITIONS. UNDER THE LAW, YOUR FIRM IS RESPONSIBLE FOR CONDUCTING INTERNAL SELF-AUDITS TO IDENTIFY AND CORRECT ANY AND ALL VIOLATIONS OF THE GMP REGULATION.

When you identify objectionable conditions that, in total, do not meet the criteria for Situation I in Part V.A.1., complete the inspection as directed per Section I of this Attachment and issue the FDA 483.

This Attachment includes the following information:

-- *Section I of this Attachment is used to conduct directed inspections.* It is designed to help you identify and obtain evidence of nonconforming product and also contains general inspectional guidance. This information should be used for both initial inspections and reinspections.

-- Section II, which should be used in addition to Section I for comprehensive follow-up inspections, contains the GMP Regulation requirements presented as questions, with important references cited where needed. The sequence in which the GMP requirements are listed follows the sequence of the narrative guidance in Section I.

The inspectional guidance that follows requires not only a review of procedures and records, but also familiarity with the actual manufacturing process and the conditions in the manufacturing area. Problems in the manufacturing process or quality assurance system, which you identify, should be documented. When these problems are related to any nonconforming devices uncovered during the inspection, refer to Part V of C.P. 7382.830 for guidance on documenting potential regulatory actions.

*SEE EXHIBITS I THROUGH IV AT THE END OF ATTACHMENT A FOR FLOW CHARTS AND NARRATIVE THAT GIVE ADDITIONAL GUIDANCE WHEN CONDUCTING DIRECTED INSPECTIONS PER SECTION I.

Exhibit I shows the relationship between problem indicators and possible situation I conditions for both manufacturing and sterilization operations. Exhibits II and III are working examples of an inspection of a manufacturing operation of heated wire breathing circuits. These exhibits demonstrate how the problem indicators shown in Exhibit I may lead to specific GMP charges. Exhibit IV is a narrative explaining directed inspections.*

SECTION I

Inspections of Medical Device Manufacturers

A. General GMP Application Notes

B. Pre-Inspectional Activity

C. 510(k) Device Inspections

D. PMA Device Inspections

E. Device Selection for Inspection

F. Critical Devices

G. All Device Inspections
   1. Complaint Handling System
   2. MDR Compliance
   3. Medical Device Tracking
   4. Failure Investigation
   5. In-Process & Finished Device Rejects & Rework
   6. Evaluation of Procedures for Change Control
   7. Validation
   8. Components
   9. Audits
   10. PMA Devices

H. Sterile Devices

I. The Small Manufacturer

J. Written Procedures

See Exhibits I, II, III, and IV at the end of Attachment A for Flow Charts and Narrative that give additional guidance when conducting Directed Inspections as per Section I.

SECTION II

820.198 Complaint Files

820.162 Failure Investigation
820.100 Manufacturing Specifications and Processes
820.101 Critical Devices, Manufacturing
Specifications and Processes
820.115 Reprocessing of Devices or Components
820.116 Critical Devices, Reprocessing of
Devices or Components
820.160 Finished Device Inspection
820.161 Critical Devices, Finished Device Inspection
820.180 General Requirements
820.181 Device Master Record
820.182 Critical Devices, DMR
820.80 Components
820.81 Critical Devices, Components
820.20 Organization
820.25 Personnel
820.40 Buildings
820.46 Environmental Controls
820.56 Cleaning and Sanitation
820.60 Equipment
820.61 Measurement Equipment
820.184 Device History Record
820.185 Critical Device, DHR
820.195 Critical Devices, Automated Data Processing
820.120 Device Labeling
820.121 Critical Devices, Labeling
820.130 Device Packaging
820.150 Distribution
820.151 Critical Devices, Distribution Records
820.152 Installation

SECTION I

INSPECTIONS OF MEDICAL DEVICE MANUFACTURERS

By following this approach, investigators will be able to focus on actual/potential problem areas during the Establishment Inspection (EI). Briefly summarized below are some of the key points that should be kept in mind during GMP inspections.

A. GENERAL GMP APPLICATION NOTES

The Center for Devices and Radiological Health has determined that certain types of establishments are either wholly or partially exempt from the GMP. Refer to pages 50-53 of the guideline booklet entitled "Medical Device GMP Guidance for FDA Investigators" for information regarding the application of device GMPs in the following subject areas:

1. Determining when a product is exempt from the regulation
2. Applicability of the GMP to the component manufacturers
3. Applicability of the GMP to service firms
4. The GMP and custom device manufacturers
5. Responsibilities of contract manufacturers and specifications developers under the GMP
6. Application to Contract Test Laboratories

B. PRE-INSPECTIONAL ACTIVITY

A review of the MDR/PRP data base should be the starting point of any inspection.

A summary of the data can be obtained from DSS, Information and Analysis Branch, HFZ-531, by calling Arlene Underdonk at 301-594-2731. Please make requests at least five days in advance of the date the inspection will begin.

C. 510(k) DEVICE INSPECTIONS

A majority of inspections will cover devices that were marketed after May 28, 1976, via a 510(k) premarket notification. The inspection should determine whether all the devices being covered have a 510(k) substantial equivalence (SE) ruling. For those devices selected for inspection, it should be determined if the company is complying with any product and/or process specifications listed in the 510(k). Also, if significant product and/or process changes have occurred, it should be determined if a new 510(k) has been submitted and found to be substantially equivalent. *See Part III, page 7, about restrictions on placing observations about 510(k) matters on the FDA 483.*

D. PMA DEVICE INSPECTIONS

Refer to C.P.7383.001 for guidance regarding PMA inspections.

E. DEVICE SELECTION FOR INSPECTION

All profile classes (manufacturing technologies) used by the manufacturer must be covered by selecting as few device lines as possible. If a device is selected because it appears to be a nonconforming product, and that device does not involve all of the profile classes used by the manufacturer, select an additional device(s) which includes the missing profile class(es).

F. CRITICAL DEVICES

If the firm manufactures any critical devices, usually one or more of these should be covered. Determine whether the critical device sections of the GMPs have been applied to the critical device. In particular: (a) Which components are critical? What mechanism was used to define critical components, and what types of tests are performed on incoming critical components prior to acceptance? (b) Which operations are considered critical? What mechanism was used to define critical operations, and how are critical operations documented?

G. ALL DEVICE INSPECTIONS

1. Complaint Handling System - 820.198

This should be the beginning point of every inspection to determine whether the firm has received complaints of possible (or potentially) defective devices.

If the firm states that because it has never received complaints it does not need a complaint file system, it should be cited on the 483 for failure to have a complaint file system.

-- Does the firm repair and service devices it produces? If so, the firm must have an adequate system in place for screening repair and service requests to assure whether any of these meet the definition of a complaint. Report in the EIR any refusals to provide access to service or repair records.

-- Also, does the firm screen repair and service requests and conduct trend analyses to determine whether there are premature failures within the warranty period, and to detect problems with particular components, subassemblies, or design?

-- Does the firm review complaints, service records, or repair records for MDR reportability?

-- Who has responsibility for MDR related event review and reporting?

Remember when reviewing complaints to check for MDR reportable events.

-- Does the firm perform trending of components that are replaced? If so, attempt to review the data. Report refusals.

2. MDR Compliance-803.24

-- Has the firm submitted MDR reports to FDA for the types of events considered to be reportable?

Prior to an inspection, obtain copies of MDR reports submitted to CDRH in the last 12 months for the firm to be inspected. When writing the EIR, be sure to state which 12 month time period was covered. If a firm has submitted no MDR report, you have the option to ask the D.P.U. to search through the MDR data base for another firm that manufacturers or imports similar types of devices and familiarize yourself with those incidents. Look at examples of each type of event reported in each MDR category (death, serious injury, malfunction). These can be discussed with the firm to illustrate the types of problems that should be reported or used by the investigator as a guide in determining if the firm has information that should have been reported under the MDR regulation.

Examine records at the firm according to instructions in Attachment E. THIS ACTIVITY IS NOT OPTIONAL.

3. Medical Device Tracking

-- Does the firm manufacture any of the devices listed in Attachment F, (page 6)?

If so, determine if the firm has device tracking procedures in place and, if it does, assess their adequacy. If there is a complete absence of a tracking system, you should cite this violation on the FDA 483. Other violations should be addressed verbally and reported in the EIR.

4. Failure Investigation - 820.162

-- Does the firm have procedures in place for a formally documented failure analysis program?

Written procedures are not required but must be followed if established by the firm. Review the firm's procedures to determine if they are adequate.

It should be remembered that the complaint section of the GMP (21 CFR 820.198) refers to possible complaints, and performing investigations to determine whether the complaint can be confirmed. Once the complaint is confirmed as an actual failure of the device, the failure investigation section of the GMP (21 CFR 820.162) takes effect.

-- Review records for indications of defects.

-- Determine whether investigations have been performed and ascertain if there is a common failure trend (e.g., a particular component(s), subassembly(ies), manufacturing error(s), or employee training).

These common failure trends would provide clues on areas or products to focus on during the inspection.

5. In-Process and Finished Device Rejects and Rework - 820.20(a)(2),.100 & .101; .115 & .116; 820.160 & .161

-- Does an examination of the device history records show that any lots, or portions of lots, have been rejected during either in-process or finished device inspection for failing to meet any or all of the product's specifications?

Instances of release and distribution of lots that failed to meet any specifications should be reported and documented. Records should be examined to determine whether any lots that have failed specifications were reworked/reprocessed, and whether this reprocessing is adequate to assure that specifications will be met without affecting the safety or performance of the device.

-- Are all sampling plans for inspection and rework based on an acceptable statistical rationale (i.e., MIL STD 105E)?

The sampling plan used should be examined to determine whether adequate samples are obtained for inspection, and should be described completely in the EIR.

6. Evaluation of Procedures for Change Control - 820.100(a)(2), 820.100(b)(3), and 820.181 (DMR)

-- Have any changes been made to the device and/or process?

These may have occurred as a result of complaints, services/repairs, failure investigations, in-process/finished device inspection, or rework/reinspection, and can be determined by examining the change control documents. If changes have been made to either the device design or the device manufacturing process, determine whether they have been adequately validated to assure that:

- The change corrected the problem

- The change did not adversely affect any other component/subassembly in the device

Avoid getting bogged down in cosmetic changes. In addition, changes for economic reasons (e.g., to cut costs) may sometimes lead to performance problems with the device.

-- Can you find a series of change orders (found by reviewing the DMR and various engineering drawings) for the same device which are intended to sequentially correct the same or similar problems?

If yes, this may indicate that the original change was not effective and that subsequent changes have been or are being made to correct the problem.

-- Did the firm receive complaints on lots manufactured after the design change?

All these are potential indications that the change was not effective and was not adequately validated.

-- Review the change control system.

-- Who must authorize changes?

-- What are the provisions for validating the changes?

-- When is a change actually effective?

-- Are the changes adequately documented?

-- Were personnel affected by any changes adequately notified?

7. Validation - 820.100(a)(1)

Process validation is the act of establishing documented evidence that provides a high degree of assurance that a specific process will consistently produce a product meeting its pre-determined specifications and quality attributes.

Prospective process validation is validation conducted prior to the distribution of either a new product, or a product made under a revised manufacturing process, where the revisions may affect the product's characteristics.

Retrospective process validation is validation of a process for a product already in distribution based upon accumulated production, testing and control data.

The inspection should determine whether adequate prospective or retrospective validation of the manufacturing process has been performed. Validation should ensure that the quality of the product will be maintained if the process is controlled within established parameters and that the validation, either prospective or retrospective, has addressed the limits of these parameters. Validation, depending on the scope of the operation, can cover all aspects from the selection of components to various manufacturing processes to end-product testing.

-- What are the process controls and were they validated?

-- The firm should be able to document that the product will meet specifications when the process is operated at the optimal process parameters.

-- The firm should be able to document that it can control the process within limits.

-- Pay attention to the process parameters; i.e., temperature, humidity, tensile strength, viscosity. Did the manufacturer include all the necessary parameters in the processing procedures?

8. Components - 820.20(a)(2), 820.80 and .81

The selection and acceptance of components is critical to the confidence that will be built into a finished device.

-- Does the manufacturer have component specifications?

-- If yes, do the methods used to evaluate components assure that received components meet these specifications?

Particular attention should be paid to critical components and those components that would require special handling or storage to maintain their integrity. Be wary of numerous changes of suppliers without having adequate change control. Also, be wary of situations where a manufacturer has relaxed specifications because the supplier could not meet the original specifications. Check device history records for affected lots and preceding lots to determine if finished device reject rates have changed. Also check complaint files to determine if the changes have resulted in increased complaints.

Vendor audits, at this time, are not required and failure to perform them should not be placed on the FDA 483, unless it is required by the firm's written procedures.

Where a manufacturer uses Just In Time (JIT) component acceptance, and relies on component vendor audits in place of incoming component inspection, review the manufacturer's audit schedule and audits. Pay special attention to the vendor's rejection rate.

9. Audits - 820.20(b)

-- Does this finished device manufacturer conduct planned and periodic audits of its quality assurance program as required?

The quality audit is the foundation of the quality assurance program. Determine if the manufacturer has a written procedure for conducting quality audits and how often these audits are conducted. It is recommended that the time between audits not exceed a 12-month period. More frequent audits may be recommended if the firm has a serious GMP problem.

Does the written audit procedure include the following?

- Objectives
- Responsibilities
- Scope
- Evaluation criteria
- Scheduling

Consult Subchapter 550 of the IOM for updated coverage of Audits (Revised 2-9-90).

10. PMA Devices - 814.39(a)(4)

-- If any PMA devices are manufactured at this facility, is this manufacturing site approved in either the original PMA application or a PMA Supplement, even if only partially manufactured here?

H. STERILE DEVICES

1. If the firm is manufacturing sterile devices, the criteria for Track I and II inspections in C.P. 7382.830 A, Attachment A, must be followed.

2. There should be records to show how the sterilization process was validated. Obtain appropriate records to document any deficiencies.

3. Determine whether the firm may be producing nonsterile devices. Some of the ways of determining this are:

   -- Have any lots had positive sterility test results and/or positive BI results? Were these lots adequately reprocessed, or were they just released? Was an adequate investigation performed prior to discrediting the initial test results?

   -- Are reprocessing (resterilization) procedures adequate, and do they assure that device performance or package seals will not be adversely affected?

   -- Are test methods based on accepted test methodologies?

- Are they performed properly so as to be able to detect positive sterility/BI test results?

- If an in-house test method is used, has it been validated to show it is capable of detecting nonsterile units?

Note: In evaluating in-house sterility test methods, the sample size is an important factor. (For example, the USP test of 40 samples is unable to detect 15% of nonsterile units with 95% confidence.)

-- Have any lots been released even though cycle specifications have not been met?

-- Does the cycle have an adequate Sterility Assurance Level (SAL) for the device being sterilized?

The SAL is the probability of a unit being nonsterile after exposure to a valid sterilization process. The SAL varies according to the intended use of the device. Sterilized articles not intended to contact compromised tissues are generally thought to be safe for use with an SAL of 10-3; that is, a probability of one nonsterile unit in a thousand. Invasive and implantable devices should have an SAL of 10-6; that is, no more than one nonsterile unit in a million. In practice, many firms use overkill cycles which assure an even lower probability that a device will be nonsterile.

-- What is the bioburden on the device, and is the sterilization cycle able to destroy the bioburden to an appropriate SAL?

I. THE SMALL MANUFACTURER

At the present time there is no official definition of a small manufacturer, although 10 or fewer employees is often used as the definition. An investigator should realize, however, that a small firm usually does not need the same degree of documentation necessary for a large firm to achieve a state of control and that many of the GMP written procedure requirements are not necessary for a small manufacturer, unless the firm is producing
nonconforming devices.

At a minimum, the small manufacturer must have the following:

- Space for components, finished devices, and manufacturing processes

- A DMR containing these specifications

- Testing/release according to specifications

- A DHR containing the test results

- Training/qualifications of employees

- Audit procedures

The flexibility of the GMP requires greater effort from the investigator in determining compliance with the GMP than a GMP where all requirements are more specific. An investigator who understands the purpose and meaning of the various GMP requirements will not merely follow the words verbatim, but will apply the GMP properly to any operation to which the GMPs apply. If the guidelines in the booklet entitled "Medical Device GMP Guidance for FDA Investigators" are followed, an investigator can apply the GMP to any situation, irrespective of the size of the firm or the complexity of its operations.

J. WRITTEN PROCEDURES

The purpose of written procedures is to provide guidance, to assure uniformity and completeness, and for communicating and managing operations.

In large manufacturing operations involving many unskilled people and operations, written procedures are usually necessary. In a small firm, communication lines are usually short, few people are involved, and management is readily available to provide guidance so that the need for written procedures is usually considerably less.

Often, training and work experience are valid substitutes for written procedures. For example, machinists are typically skilled personnel who fabricate components and finished devices using dimensioned drawings for guidance instead of written procedures. The investigator must evaluate each situation and determine the need for a written procedure on the basis of training and the knowledge possessed by the operator and the control needed. Typically, a written procedure may not be necessary when:

- The activity is very simple

- Straightforward quantitative rather than qualitative standards determine acceptability

- The operation is performed by skilled personnel

In a small firm, the investigator may conclude that a GMP requirement for a written procedure is not needed. Such a decision, however, must be supported by observation that sufficient control is present to meet the intent of the written procedure, and the fact that nonconforming devices are not being produced.

When the GMP requires a procedure, but does not specify a "written" procedure (e.g., 820.115, 820.100(a)) and the manufacturer does not have the applicable procedure, the investigator must evaluate the controls in place and determine if they are adequate without written procedures. Such decisions can be made taking into consideration factors such as the number of mistakes, rework, rejects, complaints, etc., that can be related to the operation for which the procedures are intended.

An investigator should not insist that a manufacturer meet GMP requirements that do not contribute to assuring conformance to specifications simply because they are part of the GMP. Section 519(a)(1) of the Act prohibits recordkeeping requirements that are unduly burdensome to a device manufacturer.

SECTION II

820.198 Complaint Files
820.162 Failure Investigation
820.100 Manufacturing Specifications and Processes
820.101 Critical Devices, Manufacturing Specifications and Processes
820.115 Reprocessing of Devices or Components
820.116 Critical Devices, Reprocessing of Devices or Components
820.160 Finished Device Inspection
820.161 Critical Devices, Finished Device Inspection
820.180 General Requirements
820.181 Device Master Record
820.182 Critical Devices, DMR
820.80 Components
820.81 Critical Devices, Components
820.20 Organization
820.25 Personnel
820.40 Buildings
820.46 Environmental Controls
820.56 Cleaning and Sanitation
820.60 Equipment
820.61 Measurement Equipment
820.184 Device History Record
820.185 Critical Device, DHR
820.195 Critical Devices, Automated Data Processing
820.120 Device Labeling
820.121 Critical Devices, Labeling
820.130 Device Packaging
820.150 Distribution
820.151 Critical Devices, Distribution Records
820.152 Installation

SECTION II

820.198 COMPLAINT FILES

-- Are written and oral complaints related to the identity, quality, durability, reliability, safety, effectiveness, or performance of a device reviewed, evaluated, and maintained by a formerly designated unit?

-- Does this unit determine whether or not an investigation is necessary?

-- When no investigation is made, does the unit maintain a record that includes the reason and the name of the individual responsible for the decision not to investigate?

-- Are all complaints involving the possible failure of a device to meet any of its performance specifications reviewed, evaluated, and investigated?

-- Are all complaints pertaining to injury, death, or any hazard to safety immediately reviewed, evaluated, and investigated by a designated individual(s), and are they maintained in a separate portion of the complaint file?

-- When an investigation is made, is a written record of each investigation maintained by the unit that is designated to process complaints?

-- Does the record of investigation include the following?

- The name of the device
- Any control number used
- Name of complainant
- Nature of complaint (i.e., specific details of the complaint, as obtained from the complainant)
- Reply to complainant

-- Where the formally designated unit is located at a site separate from the actual manufacturing establishment, is there a duplicate copy of the record of investigation of any complaint transmitted to and maintained at the manufacturing establishment where the device is produced in a file designated for device complaints?

820.162 FAILURE INVESTIGATION

-- After a device has been released for distribution, is any failure of that device or any of its components to meet performance specifications investigated?

-- Is a written record of such investigation, including conclusions and followup, made?

820.100 MANUFACTURING SPECIFICATIONS AND PROCESSES

(a) Specification controls

-- Are written manufacturing specifications and processing procedures (including sterilization) established, implemented, and controlled to assure the device conforms to its original design or any approved changes in that design?

-- Are there procedures for the development of manufacturing specifications for the device?

-- Are there procedures for making changes in the manufacturing specifications of the device? Are such changes approved and documented by a designated individual(s) and do they include the approval date and the date the change becomes effective?

(b) Processing controls

-- Are there written procedures describing any processing controls necessary to assure conformance to specifications?

-- Is any approved change in the manufacturing process communicated to appropriate personnel in a timely manner?

820.101 CRITICAL DEVICES, MANUFACTURING SPECIFICATIONS AND PROCESSES

(a) Critical operation performance

-- Are all critical operations performed by a designated individual(s) on suitable equipment and are they verified?

(b) Record of critical operation

-- Does the device history record contain or refer to the location of each critical operation identifying the date performed, the designated individual(s) performing the operation and, when appropriate, the major equipment used?

820.115 REPROCESSING OF DEVICES OR COMPONENTS

-- Are reprocessing procedures established, implemented, and controlled to assure that the reprocessed device or component meets the original, or subsequently modified and approved, specifications?

820.116 CRITICAL DEVICES, REPROCESSING OF DEVICES OR COMPONENTS

(a) Reprocessing procedures

-- Are there written procedures for any reprocessing associated with the production of a critical device or component?

-- Do these procedures describe the reprocessing equipment and include any special quality assurance methods or tests?

-- Is the device or component to be reprocessed clearly identified and separated from like devices or components not to be reprocessed?

-- When there is a constant reprocessing of a device or component, is a determination of the effect of the reprocessing upon the device or component made and documented?

(b) Reprocessing control

-- Is any prior quality assurance check repeated on the reprocessed device or components if reprocessing could adversely affect any performance characteristic previously inspected?

820.160 FINISHED DEVICE INSPECTION

-- Are there written procedures and acceptance criteria for the finished device inspection to assure that device specifications are met?

-- Prior to release for distribution, is each production run, lot, or batch CHECKED for conformance with device specifications?

-- Where necessary, is each production run, lot, or batch TESTED for conformance with device specifications?

-- Where practical, is a device selected from a production run, lot, or batch, and tested under simulated use conditions?

-- In your narrative report, describe the sampling plan for checking, testing, and release of the device. CDRH has applied this requirement for noncritical devices using 820.20 as justification.

-- Are finished devices held in quarantine or otherwise adequately controlled until released?

-- Are devices that fail finished device inspection, and are not to be reprocessed, disposed of in a manner which precludes their release as finished goods?

820.161 CRITICAL DEVICES, FINISHED DEVICE INSPECTION

-- Is a critical device or component that does not meet its performance specifications investigated?

-- Is a written record of the investigation, including conclusions and followup, made?

-- Does a critical device leave the control of the manufacturer for distribution only after all acceptance records and test results have been checked by a designated individual(s)?

-- Does a designated individual(s) assure that all records and documentation required for the device history record are present, complete, and show that release of the device was consistent with the release criteria?

-- Does a designated individual(s) authorize, by signature, the release of the device for distribution?

820.180 GENERAL REQUIREMENTS

-- Are the following records maintained at the manufacturing establishment or other reasonably accessible location and are they available for review and copying?

  - Device master record

  - Device history record

  - Complaint files

-- Are all records required for this device retained for a period of at least 2 years from the date of release for commercial distribution by the manufacturer?Continued on Following Page

820.181 DEVICE MASTER RECORD

-- Is the device master record prepared, dated, and signed by a designated individual(s)?

Records and Change Control Procedures

  -- Are any CHANGES in the device master record authorized in writing by the signature of a designated individual(s)?

  -- Do such CHANGES include the approval date and the date the changes become effective?

-- Are any forms used to approve CHANGES in the device master record part of the device master record?

Does the device master record for this device include, or refer to the location of, the following information?

-- Device specifications including appropriate drawings, composition, formulation, and component specifications

-- Production process specifications including appropriate equipment specifications, production methods, production procedures, and production environment specifications

-- Quality assurance procedures and specifications including quality assurance checks used and the quality assurance apparatus used

-- Packaging and labeling specifications including appropriate methods and processes used

820.182 CRITICAL DEVICES, DEVICE MASTER RECORD

In addition to the requirements of 820.181, does the device master record for a critical device include or refer to the location of the following?

-- Full information concerning critical components and critical component suppliers, including the complete specifications of all critical components, the sources where they may be obtained, and written copies of any agreements made with suppliers

-- Complete labeling procedures for the individual device and copies of all approved labels and other labeling

820.80 COMPONENTS

-- Are components used in manufacturing received, stored, and handled in a manner designed to prevent damage, mixup, contamination, and other adverse effects?

-- Are components quarantined prior to acceptance or clearly identified as not yet accepted?

(a) Acceptance of components

-- Are there written acceptance criteria and procedures for the acceptance of components?

-- Does a designated individual(s) accept or reject components?

-- Is a record maintained of component acceptance and rejection?

-- Where deviations from component specifications could result in the device being unfit for its intended use, are components inspected, sampled, and tested for conformance to specifications?

(b) Storage and handling of components

-- Are components whose fitness for use deteriorates over time stored in a manner to facilitate proper stock rotation?

-- Are component control numbers or other identifications easily viewable?

-- Are all obsolete, rejected, or deteriorated components clearly identified and segregated from accepted components?

-- Are records maintained of the disposition of all obsolete, rejected, or deteriorated components?

820.81 CRITICAL DEVICES, COMPONENTS

(a) Acceptance of critical components

-- Are there written procedures for sampling, testing, and inspecting all lots of critical components to assure that critical components conform to specifications?

-- You may want to collect or record the firm's procedures for your EIR regarding the statistical rationale for sampling critical components.

-- Is each lot of critical components identified with a control number(s) upon receipt?

-- Are the percentage of defective critical components in each sample and the percentage of lots rejected recorded and identified by supplier name?

(b) Critical component supplier agreement

-- Where possible, has the manufacturer secured from the critical component supplier a written agreement whereby the supplier agrees to notify the manufacturer of any proposed change in a critical component?

820.20 ORGANIZATION

-- Where possible, does a designated individual(s), not having direct responsibility for the performance of a manufacturing operation, take responsibility for the quality assurance program?

(a) Quality assurance program requirements

-- Does the QA program assure that production records are reviewed?

-- Are there acceptance criteria for the receipt of all components, manufacturing materials, in-process materials, packaging materials, labeling, and finished devices?

-- Does the quality assurance program consist of procedures that assure the approval or rejection of devices manufactured, processed, packaged, or held under contract by another company?

-- Does the QA program assure the function of identifying, recommending, or providing solutions for quality assurance problems and verifying that the implementation of such solutions is performed?

-- Is the QA program assuring that all quality assurance checks are appropriate and adequate for their purpose and are performed correctly?

(b) Audit procedures

-- Are there written audit procedures? (Note: The investigator should not seek access to the firm's internal audit reports.)

-- Are planned and periodic audits of the quality assurance program implemented to verify compliance with the quality assurance program?

-- Are audits performed in accordance with written procedures by appropriately trained individuals not having direct responsibilities for the matters being audited?

-- Are audit results documented in written audit reports that are reviewed by management having responsibility for the matters audited?

-- Is followup corrective action, including re-audit of deficient matters, taken when indicated?

The following additional requirements should be covered for Track II inspections.

820.25 PERSONNEL

(a) Personnel training

-- Where training programs are necessary to assure that personnel have a thorough understanding of their jobs, are such programs conducted and documented?

-- Are employees made aware of device defects which may occur from the improper performance of their specific jobs?

-- Are quality assurance personnel made aware of defects and errors likely to be encountered as part of their quality assurance functions?

-- Are personnel in contact with the subject device or its environment: clean, healthy, and suitably attired where lack of cleanliness, good health, or suitable attire could adversely affect the device?

-- Are any personnel who, by medical examination or supervisory observation, appear to have a condition which could adversely affect the device excluded from affected operations until the condition is corrected?

-- Are personnel with such conditions instructed to report them to their supervisors?

-- Look for examples where personnel failed to perform, or inadequately performed, a task because of training; e.g., bonding, molding, assembly.

820.40 BUILDINGS

-- Do the facilities provide ADEQUATE SPACE DESIGNED to PREVENT mixups and to assure orderly handling of the following?

- Incoming components
- Rejected or obsolete components
- In-process components
- Finished devices
- Labeling
- Devices that have been reprocessed, reworked, or repaired
- Equipment
- Molds, patterns, tools, records, drawings, blueprints
- Testing and laboratory operations
- Quarantined products

840.46 ENVIRONMENTAL CONTROLS

-- Where environmental conditions at the manufacturing site could have an adverse effect on a device's fitness for use, are these environmental conditions controlled to prevent contamination of the device? See C2.0 of Medical Device GMP Guidance for FDA Investigators.

-- Are there periodic documented inspections of these environmental controls to verify that the environmental control system is properly functioning?

820.56 CLEANING AND SANITATION

-- Are written cleaning procedures and schedules adequate to meet manufacturing process specifications?

(a) Personnel sanitation

-- Are washing and toilet facilities clean and adequate?

-- Where special clothing requirements are necessary to ensure that a device is fit for its intended use, are clean dressing rooms provided for personnel?

(b) Contamination control

-- Are there procedures designed to prevent contamination of equipment, components, or finished devices by rodenticides, insecticides, fungicides, fumigants, hazardous substances, and other cleaning and sanitizing substances?

-- Is adherence to such procedures documented?

(c) Personnel practices

-- Where eating, drinking, and smoking by personnel could have an adverse effect on a device's fitness for use, are such practices limited to designated areas?

(d) Sewage and refuse disposal

-- Are sewage, trash, byproducts, chemical effluents, and other refuse disposed of in a timely, safe, and sanitary manner?

820.60 EQUIPMENT

(a) Maintenance schedule

-- Is equipment used in the manufacturing process appropriately designed, constructed, placed, and installed to facilitate maintenance, adjustment, and cleaning?

-- Where maintenance of equipment is necessary to assure that manufacturing specifications are met, is a written record maintained documenting when scheduled maintenance activities are performed?

-- Is such schedule visibly posted on or near each piece of equipment, or is the schedule readily available to personnel performing maintenance activities?

(b) Inspection

-- Are periodic documented inspections made to assure adherence to applicable equipment maintenance schedules?

(c) Adjustment

-- Are any inherent limitations or allowable tolerances visibly posted on or near equipment requiring periodic adjustments, or are they readily available to personnel performing these adjustments?

(d) Manufacturing material

-- Is manufacturing material subsequently removed from the device or limited to a specified amount that does not adversely affect the device's fitness for use?

-- Are there written procedures for the use and removal of manufacturing material?

-- Is the removal of manufacturing material documented?

820.61 MEASUREMENT EQUIPMENT

-- Is all measurement equipment suitable, and capable of making the required measurement?

-- Is all QA and production measurement equipment routinely checked, calibrated, and inspected according to written procedures?

-- Are records documenting these activities maintained?

-- When computers are used as part of an automated production or quality assurance system, are all program changes made by a designated individual(s) through a formal approval procedure?

(a) Calibration

-- Do calibration procedures include specific directions and limits for accuracy and precision?

-- Are there provisions for remedial action when accuracy and precision limits are not met?

(b) Calibration standards

-- Does all production and quality assurance measurement equipment have calibration standards?

(c) Calibration records

-- Are the calibration date, the calibrator, and the next calibration date recorded and displayed, or are records containing such information readily available for each piece of equipment requiring calibration?

-- Does a designated individual(s) maintain a record of calibration dates and also of the individual performing the calibration?

820.184 DEVICE HISTORY RECORD

-- Is a DHR maintained to demonstrate that the device is manufactured in accordance with the device master record?

-- In addition, does the DHR include or refer to the location of: (1) the dates of manufacture, (2) the quantity manufactured, (3) the quantity released for distribution, and (4) any control number used?

820.185 CRITICAL DEVICES, DEVICE HISTORY RECORD

-- Is there a critical device history record for each control number which identifies the specific label, labeling, and control number used for each production unit, and is this record readily accessible and maintained by a designated individual(s)?

(a) Component documentation

-- Does the DHR contain the control number designating each critical component or lot of critical components used in the manufacture of the device?

-- Does the DHR contain the acceptance record of the critical component, including acceptance date and signature of the recipient?

(b) Inspection checks

-- Does the DHR contain or refer to the location of records containing: the inspection checks performed, the methods and equipment used, results, the date, and signature of the inspecting individual(s)? (See 820.101 Record of critical operation.)

820.195 CRITICAL DEVICES, AUTOMATED DATA PROCESSING

-- When automated data processing is used for manufacturing or quality assurance purposes, are there documented procedures designed to prevent inaccurate data input, output, and programming errors?

820.120 DEVICE LABELING

(a) Label integrity

-- Are labels designed, printed, and applied so as to remain legible during the customary conditions of processing, storage, handling, distribution, and use?

-- Are labels and other labeling not released to inventory until a designated individual has proofread samples of the labeling for accuracy?

(b) Separation of operations

-- Is each labeling or packaging operation separated physically or spatially in a manner designed to prevent mixups?

(c) Area inspection

-- Prior to the implementation of any labeling and packaging operation, is there an inspection of the area where the operation is to occur by a designated individual to assure that devices and labeling materials from prior operations do not remain in the labeling or packaging area?

-- Are any such items found destroyed, disposed of, or returned to storage prior to the onset of a new or different labeling or packaging operation?

(d) Storage

-- Are labels and labeling stored and maintained in a manner that provides proper identification and that is designed to prevent mixups?

(e) Labeling materials

-- Are labeling materials issued for devices examined for identity and accuracy?

-- Is a record of such examination, including the date and person performing the examination, maintained in the device history record?

820.121 CRITICAL DEVICES, DEVICE LABELING

(a) Control number

-- Do labels issued for critical devices contain a control number?

(b) Labeling check

-- Is the signature of the individual who proofreads the label and other labeling recorded, and is the date of the proofreading recorded?

(c) Access restriction

-- Is access to labels and labeling restricted to authorized personnel?

820.130 DEVICE PACKAGING

-- Are the device package and any shipping container for a device designed and constructed to protect the device from alteration or damage during the customary conditions of processing, storage, handling, and distribution?

820.150 DISTRIBUTION

-- Are there written procedures and release criteria to assure that only those finished devices approved for release are distributed?

-- Where a device's fitness for use or quality deteriorates over time, is there a system to assure that the oldest approved devices are distributed first?

820.151 CRITICAL DEVICES, DISTRIBUTION RECORDS

-- Do the distribution records for critical devices include or make reference to the location of: the name and address of the consignee, the name and quantity of devices, the date shipped, and the control number used?

820.152 INSTALLATION

-- Where a device can be installed by someone other than the manufacturer or his authorized representatives, does the manufacturer provide instructions and procedures for proper installation?

GMP INSPECTION

The current inspectional approach is designed to look for actual problems with devices that the firm is manufacturing, rather than at paper work and procedural types of GMP violations. The inspection begins with coverage in the following areas:

1. a review of complaints, service records and MDR reports. (Refer to Part II, Page 2; Part III, Page 4; Attachment A, Page 8;and Attachment B for instructions on reviewing complaint and service records.) The review is two-fold: to determine whether the firm is complying with 820.198 and 820.162; and, more importantly, to determine whether there are nonconforming and/or defective devices (lot, models, etc.).

2. changes which the manufacturer has made to any devices and manufacturing processes, including software. Determine whether the changes: were made to correct past failures (i.e., had to or should have recalled a device); were adequately validated (i.e., tested to assure that the change corrects any previous problems with the device or manufacturing process, and does not introduce any new problems); and, did not create any new problems with the device or manufacturing process (i.e., result in an increase of nonconforming and/or defective devices).

3. device history records to determine whether the firm has produced any in-process or finished devices which were nonconforming and/or defective, whether or not they were released for distribution. The review is two-fold: to determine whether nonconforming and/or defective devices have been released into commercial distribution; and, even if the devices were not released into distribution, it provides an assessment of how high the reject rate is for the device being manufactured.

If problems are not found after reviewing the above records, complete the inspection as directed in Attachment A, Section I, and issue a FDA 483 listing any objectionable GMP conditions.

If the review of the above records does find problems, the problems will help guide the remainder of the inspection. For example:

- A review of service records for the past year for the Model 200 ventilator finds numerous "bugs" with the current software version 3.2. The inspection determines that this version was installed in all units in the field 18 months ago with new enhancements. However, the inspection determines that the software change was not adequately validated, the software was not adequately tested prior to its release to the field, and no failure investigation has been performed to determine the probable impact of the bug on the safety and effectiveness of the device, and why the software had numerous bugs.

- A review of device change orders for the past year finds that the Model 100 intracranial syringe had a needle length change from 18 mm to 15 mm. The inspection determines that numerous complaints were received by the firm for the 18 mm length, in that the needle was too long and penetrating too deeply into the brain. The inspection determines that the complaints were not adequately investigated, no failure investigation was performed, and the change of needle length was not adequately validated in that additional complaints were received concerning inability to reach the desired site in the brain with the 15 mm length needle.

- A review of device history records for the past year finds that the 3 french catheter has a widely fluctuating in-process reject rate (from 0% up to 57%), and when a lot does get to the finished inspection stage it fails sampling testing and the lot must be 100% inspected. The inspection determines that numerous complaints have been received for breakage which were inadequately investigated, no failure investigation was performed, and none of the manufacturing processes (i.e., injection molding, extrusion molding, automated gluing process) had been validated.

The above 3 examples all meet the criteria for Situation I in Part V., Section A.1. As such, document all violations encountered to this point. The inspection is to be discontinued when conditions are encountered which meet the criteria for Situation I. The FDA 483, Inspectional Observations, must then contain the following statement: "The observations noted in this FDA 483 are not an exhaustive listing of objectionable conditions. Under the law, your firm is responsible for conducting internal self-audits to identify and correct any and all violations of the GMP regulation." The Warning Letter to the manufacturer warns them of their responsibility for reviewing all manufacturing and QA systems.

Because other problems may have existed which the manufacturer should have identified and corrected, the follow-up inspection should be a comprehensive inspection, especially if regulatory and/or administrative action is being contemplated.

MDR INSPECTION

Each GMP inspection must also include coverage of the MDR regulation. (Refer to Attachment A, Page 9, and Attachment E for MDR guidance.)

Medical Device Tracking (MDT) INSPECTION

Perform an inspection if any of the 26 devices subject to the MDT regulation are being manufactured and/or distributed. Refer to Attachment F for MDT guidance.

PROCESS FOR COMPLIANCE REVIEW OF CLASS I AND II 510(K) SUBMISSIONS/PMA SUPPLEMENTS

INTRODUCTION

These revised procedures are designed to provide reasonable assurance that FDA will not clear a pending Class I or II 510(k) Submission or PMA Supplement when existing inspectional information reveals findings of GMP violations which result in an inspection classification of Official Action Indicated (OAI) and the GMP violations are reasonably related to the application. A reasonable relationship will be found when: (a) there are specific GMP violations that are directly applicable to the pending application; or (b) there are systemic violations that would be generally applicable to the pending application. Clearance of 510(k) applications will not be affected if no such reasonable relationship is found. Within FDA, the applicable District Office will have initial responsibility for determining whether such a reasonable relationship exist.

PROCESS FOR COMPLIANCE REVIEW OF CLASS I AND II 510(K) SUBMISSIONS AND PMA SUPPLEMENTS

NOTE: The Center is modifying one of its databases to establish electronic transmission of information between ODE and OC for this program. The following procedures are written in anticipation of the early completion of this modification. In the meantime, all communications between ODE and OC will be transmitted by fax or special messenger.

1. ODE receives a Class I or II 510(k) Submission or PMA Supplement (Application) in the Document Control Center.

2. ODE Document Control:

   a. Logs in the application.

   b. Checks the application against the COMSTAT database for the GMP compliance status of the applicant, and also that of manufacturer, and/or sterilizer per information provided on the 510(k) cover sheet.

      (1) If there is a match, i.e., any of the firms, applicant, manufacturer, or sterilizer are in the COMSTAT database, Document Control will refer the application to ODE/POS.

      (2) If there is no match in the COMSTAT database, i.e., neither applicant, manufacturer, nor sterilizer are in the COMSTAT database, ODE does not refer the application to OC/FPB at this point in the process. Go to step 4.

(3) If application is received without a manufacturer or sterilizer identified, Document Control will refer application to ODE/POS for transmission to OC/FPB for compliance check.

3. OC/FPB receives the application and initiates the compliance review.

a. If the application was referred without the manufacturer and/or sterilizer site information, OC/FPB contacts the applicant by fax to determine the manufacturing and/or sterilizing sites.

(1) Applicant faxes the site information back to OC/FPB who then checks the COMSTAT database to determine the GMP compliance status of the manufacturer and/or sterilizer.

(2) If the applicant cannot identify the intended manufacturer and/or sterilizer site(s), OC/FPB will send a "recommend defer" notice to ODE/POS. (See step 4 c)

(3) If all sites are currently in GMP compliance, OC/FPB transmits a "recommend proceed" notice to ODE/POS. (Go to step 4)

c. If any of the sites are not in GMP compliance or the information available needs clarification, OC/FPB will contact the district office(s) - by fax -, or the appropriate OC division of enforcement, to determine the following:

(1) the firm's current GMP compliance status and,

(2) if the firm's most recent inspection was classified as OAI because of GMP violations, OC/FPB will ask the district office/OC division for a determination as to whether or not the GMP violations are reasonably related to the pending 510(k)/PMA Supplement device, i.e., are the GMP violations system-wide, product related, or product line related.

d. If firm is in GMP compliance, OC/FPB issues a "recommend proceed" notice to ODE/POS.

e. If the firm is out of GMP compliance but the GMP violations are not reasonably related to the application, OC/FPB issues a "recommend proceed" notice to ODE/POS.

f. If the firm is out of GMP compliance and the GMP violations are reasonably related to the application:

(1) OC/FPB issues a "recommend postpone" notice to ODE/POS, giving the name and address of the relevant district office or OC division so that ODE/POS may include this information in their letter to the applicant.

(2) ODE/POS issues a letter to the applicant advising that, although the technical review of the application will continue, the 510(k) Submission or PMA Supplement cannot be cleared/approved until the manufacturer's and/or sterilizer's GMP violations which are reasonably related to the 510(k) or PMA Supplement, have been adequately corrected. This letter will advise the firm to contact their district office or the OC division of enforcement about the GMP issues.

4. ODE completes the technical review and is ready to issue a determination of Substantial Equivalence (SE) or PMA approval letter:

a. ODE/POS will re-check the COMSTAT database before a determination of SE or PMA approval letter is sent to the applicant. This re-check is necessary because a recent inspection may have found the firm out of compliance with GMPs.

(1) If none of the sites - applicant, manufacturer and/or sterilizer - are in the COMSTAT database, ODE/POS issues a determination of SE or a PMA approval letter to the applicant.

(2) If any of the sites are currently in the COMSTAT database, ODE/POS refers application to OC/FPB following step 2b(1) above.

b. If a "recommend defer" notice has been received from OC/FPB without further compliance update, ODE/POS electronically queries OC/FPB for a status update.

c. Reference step 3a(2), this document, if a "recommend postpone" notice has been received from OC/FPB because the applicant could not identify the manufacturer and/or sterilizer, e.g., applicant has not yet chosen an manufacturer/sterilizer, ODE/POS will issue a determination of SE letter to the applicant and will advise that production cannot commence until GMP compliance has been confirmed at those sites.

CDRH/ORA CLASS I & II 510(K) SUBMISSION & PMA SUPPLEMENT PROGRAM

FIELD FOLLOW-UP TO VIOLATIVE DOMESTIC DEVICE GMP INSPECTIONS

I. VIOLATIVE GMP INSPECTIONS (EIR is classified OAI)

The starting point in determining a firm is not in GMP compliance is when an inspection reveals significant GMP violations which meet the Situation I criterion and the inspection is classified as Official Action Indicated (OAI). (CP 7382.830, Part V, A 1, and Attachment A, Exhibits 1 - 4 provide guidance to be used in determining whether the GMP violations warrant an inspection classification of

OAI). If the criteria for classifying the inspection as OAI is met, the firm will be issued a Warning Letter or other enforcement action taken, i.e., a Warning Letter issued directly to the firm by the district office, or a Warning Letter or other enforcement action recommended to the Center for Devices and Radiological Health (CDRH). However, if the firm responds to the FDA-483 (483) before issuance of the Warning Letter, the Warning Letter will issue but the language of the Warning Letter will be modified accordingly.

A. ADEQUATE RESPONSE TO FORM FDA 483 (483) - NOT REQUIRING A PROMPT REINSPECTION

Situation: The firm responds to the 483 within 5 working days of its issuance. Using the criteria described in section III of this document, the district determines the response to be adequate and that no prompt reinspection is necessary. To establish prior warning a Warning Letter will issue.

Action: (See G-1 Warning Letter)

1. Within 15 working days following end of the Establishment Inspection the district issues a Warning Letter to the firm based on the GMP violations found during the inspection. However, in the Warning Letter, the district will:

   a. Acknowledge the firm's adequate response to the 483.

   b. Inform the firm that the clearance/approval of any pending 510(k)s, PMA Supplements, or Export Certificates for products manufactured or sterilized at the facility in which the GMP violations were found, will not be delayed because of GMP issues.

   c. Inform the firm that the Warning Letter does not require further response.

2. District will update firm's compliance status in ORA/COMSTAT within 5 days.

3. District will fax copy of Warning Letter to Joy Lazaroff, CDRH/OC/FPB, HFZ-306, Fax (301)-594-4715.

4. District conducts reinspection "(See Section III, General Guidelines)" within 6 to 18 months.

B. ADEQUATE RESPONSE TO 483 - REQUIRES PROMPT REINSPECTION TO VERIFY CORRECTIONS

Situation: The firm responds to the 483 within 5 working days of its issuance. Using the criteria described in section III, the district determines that although the response appears to be adequate a reinspection is necessary to verify that the corrections have been made. To establish adequate warning, a Warning Letter will issue.

Action: (See G-2 Warning Letter)

1. Within 15 working days following end of the Establishment Inspection the district issues a Warning Letter to the firm based on the GMP violations found during the inspection. However, in the Warning Letter, the district will:

   a. Acknowledge receipt of the adequacy of the firm's response to the 483 and inform firm that a reinspection is necessary to verify corrections. If firm has not indicated its readiness for reinspection in the response to the 483, it should be advised in the Warning Letter to indicate when it will be ready for reinspection in its response to the Warning Letter.

   b. Inform firm that the clearance/approval of any pending 510(k)s, PMA Supplements, or Export Certificates for products manufactured or sterilized at the facility in which the GMP violations were found, will be deferred until the adequacy of its corrections can be verified by reinspection, and that Federal agencies may take this into account when considering award of contracts.

2. District will update firm's compliance status in ORA/COMSTAT within 5 days.

3. District will fax copy of Warning Letter to Joy Lazaroff, CDRH/OC/FPB, HFZ-306, Fax (301)-594-4715.

4. District conducts reinspection within 60 days of receipt of firm's response in which it indicated it was ready for reinspection.

C. INADEQUATE RESPONSE TO 483

Situation: The firm responds to the 483 within 5 working days of its issuance and, using the criteria described in section III, the district determines that the response is inadequate.

Action: (See G-3 Warning Letter)

1. The district will issue a Warning Letter to the firm within 15 working days from end of the Establishment Inspection. The Warning Letter will:

    a. Acknowledge the firm's response to the 483 but advise firm the response was inadequate and explain why it was inadequate, and

    b. Inform firm that clearance/approval of any pending 510(k)s, PMA Supplements, or Export Certificates for products manufactured or sterilized at the facility in which the GMP violations were found, will be deferred because of GMP issues, and that Federal agencies may take this into account when considering award of contracts.

2. District will update firm's compliance status in ORA/COMSTAT within 5 days.
3. District will fax a copy of the Warning Letter issued to firm to Joy Lazaroff, CDRH/OC/FPB, HFZ 306, (301)-594-4715.

    NOTE: (See G-4 or G-5 Warning Letter)

    A district may determine that a firm's response is partially adequate because the firm has only addressed those GMP violations related to a pending 510(k)/PMA Supplement. Please refer to paragraph IV, Special Circumstances, for the procedures which may be followed in this situation.

D. RESPONSES TO 483 RECEIVED AFTER THE 5 WORKING DAY TIME FRAME:

Situation: The firm responds to the 483 after the 5 working day time frame but the response is received prior to issuance of the Warning Letter:

Action: (See G-6 Warning Letter)

1. The district will issue a Warning Letter to the firm within 15 working from the end of the Establishment Inspection. The Warning Letter will:

    a. Acknowledge receipt of the response to the 483, and tell the firm the response has not yet been evaluated.

    b. Indicate that the response to the 483 will be evaluated and will be communicated to them, but the firm should not delay its response to the Warning Letter within the time frame given.

    c. Inform firm that clearance/approval of any pending 510(k)s, PMA Supplements, or Export Certificates for products manufactured or sterilized at the facility in which the GMP violations were found, will be deferred because of GMP issues, and that Federal agencies may take this into account when considering award of contracts.

2. The district will update firm's compliance status in ORA/COMSTAT within 5 days.

3. District will fax a copy of the Warning Letter issued to firm to Joy Lazaroff, CDRH/OC/FPB, HFZ 306, (301)-594-4715.

E. NO RESPONSE TO 483

Situation: No response to the 483 is received before the district issues a Warning Letter.

Action: (See G-7 or G-8, as appropriate Warning Letter)

1. The district will issue a Warning Letter to the firm within 15 working days of the last day of the Establishment Inspection. The Warning Letter will: Inform firm that the clearance/approval of any pending 510(k)s, PMA Supplements, or Export Certificates for products manufactured or sterilized at the facility in which the GMP violations were found, will be deferred because of GMP issues, and that Federal agencies may take this into account when considering award of contracts.

2. District will update firm's compliance status in ORA/COMSTAT within 5 days.

3. District will fax a copy of the Warning Letter issued to firm to Joy Lazaroff, CDRH/OC/FPB, HFZ 306, (301)-594-4715.

F. GMP INSPECTIONAL FINDINGS WHICH NEED FURTHER TECHNICAL REVIEW:

Situation: A district determines that the response to the 483 is very voluminous and/or the inspectional findings need the technical review of the CDRH.

Action: (See G-9 Model Letter)

The district will:

1. Within 15 working days of the end of the inspection, the district will inform the firm in writing that the response to the 483 is very voluminous and/or the GMP inspectional findings need further technical review by the CDRH Office of Compliance, and this letter will also inform the firm that if the findings have been referred to CDRH for review, any pending 510(k), PMA Supplement, or Export Certificate for products manufac-

tured or sterilized at the facility, may be deferred if the district determines that the GMP violations are reasonably related to the product. The pending applications may not be cleared until the district and/or Office of Compliance determines the related GMP violations are adequately corrected. Federal agencies may take this information into account when considering award of contracts.

2. Forward the EIR and related documentation to CDRH Office of Compliance in accordance with prescribed procedures.

3. District will update firm's compliance status in ORA/COMSTAT within 5 days.

4. District will fax copy of letter sent to firm to Joy Lazaroff, CDRH/OC/FPB, HFZ-306, Fax (301)-594-4715.

II. ACTION FOLLOWING ISSUANCE OF WARNING LETTERS

A response to a Warning Letter falls into one of the following categories:

- Adequate Response Not Requiring a Prompt Reinspection
- Adequate Response Requiring Prompt Reinspection to Verify Corrections
- Inadequate response
- No response

A. ADEQUATE RESPONSE TO WARNING LETTER NOT REQUIRING A PROMPT REINSPECTION

Situation: The district determines that the firm's response to the Warning Letter is adequate and a prompt reinspection is not necessary, based on consideration of the factors contained in section III.

Action: (See G-10 Model Letter)

1. Within 15 working days of receipt of the firm's response to the Warning Letter, the district will respond in writing to firm to:

   a. Acknowledge the firm's adequate response to the Warning Letter, and

   b. Inform the firm that the clearance/approval of any pending 510(k)s, PMA Supplements, or Export Certificates for products manufactured or sterilized at the facility in which the GMP violations were found, will not be deferred due to GMP issues, and that Federal agencies may take this into account when considering award of contracts.

2. District updates firm's compliance status in ORA/COMSTAT within 5 days.

3. District will fax copy of its acknowledgement of the firm's adequate response to Joy Lazaroff, CDRH/OC/FPB, HFZ-306, Fax (301)-594-4715.

4. District schedules firm for reinspection within 6 to 18 months.

B. ADEQUATE RESPONSE TO WARNING LETTER REQUIRING PROMPT REINSPECTION TO VERIFY CORRECTIONS

Situation: The district determines that the firm's response to the Warning Letter is adequate and a prompt reinspection is necessary to verify that the corrections are adequate.

Action: (See G-11 Model Letter)

1. Within 15 working days of receipt of the firm's response to the Warning Letter, the district will respond in writing to firm to:

   a. Acknowledge the adequacy of the firm's response and inform firm that a reinspection is necessary to verify corrections (See Section III, General Guidelines).

   b. Inform firm that the clearance/approval of any pending 510(k)s, PMA Supplements, Export Certificates for products manufactured or sterilized at the facility in which the GMP violations were found, will continue to be deferred until a reinspection is conducted to verify the corrective actions are adequate. This letter should also advise firm that Federal agencies may take this into account when considering award of contracts.

   c. District will fax a copy of its acknowledgement of the firm's adequate response to Joy Lazaroff, CDRH/OC/FPB, HFZ-306, Fax (301)-594-4715.

2. When FDA reinspection is required:

   a. When a firm notifies the district, in writing, of the actions it has taken to bring about correction and indicates it is ready for reinspection, the district must reinspect the firm within 60 days of the date of readiness. That is, 60 days from the date given by the firm as being ready for a reinspection.

b. If the reinspection is not initiated within 60 days of receipt of the firm's response and time table, the agency has committed, as part of Reinventing Government, to remove from a deferred status any Class I or II 510(k), PMA Supplement, or Export Certificate which was previously postponed because the firm's GMP violations are reasonably related to the device.

(1) Districts will still reinspect as soon as practicable but will deal with the device as a marketed product.

3. When reinspection reveals that the corrections are adequate:

a. Within 15 working days of completion of the inspection inform the firm, in writing that the corrections are adequate and that the approval of any pending 510(k)s, PMA Supplements, or Export Certificates for products manufactured or sterilized at the facility in which the GMP violations were found, will not be delayed because of GMP issues, and that Federal agencies may take this into account when considering award of contracts. (See G-12 Model Letter)

b. District will update firm's compliance status in COMSTAT.

c. District will fax a copy of its acknowledgement of the firm's adequate response to Joy Lazaroff, CDRH/OC/FPB, HFZ-306, Fax (301)-594-4715.

4. If the reinspection reveals that the corrections are not adequate, that serious GMP violations continue, and are reasonably related to any device on hold, the district will:

a. Within 15 working days of completion of the reinspection:

(1) inform firm (in writing) that the corrections are not adequate, and (See G-13 Model Letter)

(2) that the approval of any pending 510(k)s, PMA Supplements, or Export Certificates for products manufactured and/or sterilized at the facility to which the GMP violations are related, will not be cleared because of continued GMP issues. This letter will advise firm that Federal agencies may take this into account when considering award of contracts.

b. Within 5 days the district will update firm's compliance status in COMSTAT.

c. Fax a copy of its acknowledgement of the firm's adequate response to Joy Lazaroff, CDRH/OC/FPB, HFZ- 306, Fax (301)-594-4715.

C. INADEQUATE RESPONSE TO WARNING LETTER

Situation: The firm's response to the Warning Letter is inadequate based upon the criteria described in section III.

Action: (See G-14 Model Letter)

1. The district responds to the firm (in writing) within 15 working days of receipt of the firm's response to the Warning Letter and:

   a. Informs the firm that the response is inadequate and explains why it is inadequate.

   b. Informs the firm that the approval of any pending 510(k)s, PMA Supplements, or Export Certificates for products manufactured or sterilized at the facility in which the GMP violations were found, will be deferred because of GMP issues, and that Federal agencies may take this into account when considering award of contracts.

2. District updates firm's compliance status in ORA/COMSTAT within 5 days.

3. District will fax a copy of its acknowledgement of the firm's adequate response to Joy Lazaroff, CDRH/OC/FPB, HFZ-306, Fax (301)-594-4715.

4. The district conducts an in-depth reinspection within 6 months based upon the public health risk.

NOTE: A district may determine that a firm's response is partially adequate because the firm has addressed only those GMP violations related to a pending 510(k)/PMA Supplement. Please refer to section IV, Special Circumstances, for the procedures which may be followed in this situation.

D. NO RESPONSE TO WARNING LETTER

Situation: The firm does not respond to the Warning Letter.

Action:

1. The firm remains in COMSTAT as OAI with Warning Letter issued.

2. The district conducts an in-depth reinspection within 6 months based upon the public health risk.

III. GENERAL GUIDELINES

A. Criteria for determining the adequacy of firm response:

It addresses each observation/violation in a specific, clear way.

It contains sufficient documentation to demonstrate that the observations/violations have been corrected or it has realistic time frames for the accomplishment of each and every action item needed to come into compliance.

It takes into account the impact that new corrections may have on the system(s) as a whole, and indicates positively that such impact has been seriously considered.

B. Criteria for determining whether an immediate reinspection is necessary:

Affirmative responses to the following questions suggest that a prompt reinspection is necessary:

Does the GMP problem(s) involved present a potential public health hazard?

Does the recall history of the firm suggest a pattern of repeat recalls for similar kinds of systemic problems? Does this recall history raise questions about the permanency of corrections made in the past?

Does the firm have a history of making promises to correct, and then backsliding on those promises?

Another consideration is the responsibility that has been demonstrated by management, particularly if there have been management changes, e.g., has the firm established a quality assurance program?

Does the firm have any 510(k) applications, PMA Supplements, or Export Certificates on hold in CDRH because of GMP violations which are reasonably related to the device? That is are the GMP violations systemic, limited to the specific product line, or limited to a specific device?

If the response to the above questions is "yes", and the corrections can only be verified by reinspection, then an immediate inspection must be conducted within 30 to 60 days, but not later than 60 days following written receipt of the firm's indication it is ready for reinspection.

NOTE: If the district office determines it cannot reinspect within 60 day time frame, or the firm has a high rate of recidivism, the district may suggest that the firm submit:

- certification by an outside expert consultant that it has conducted an update audit, and

- certification by the firm's CEO that any corrections noted to be needed have been made, and

- certification that the firm is in compliance with the requirements of the GMP regulation. (CP 7382.830. Part V, A, 3, c. (a) and (d))

IV. SPECIAL CIRCUMSTANCES

A firm's response may clearly indicate that only the GMP violations which reasonably relate to a postponed 510(k), PMA Supplement, or Export Certificate have been corrected, however, other significant GMP violations have not been corrected. In this instance, the district will advise CDRH/FPB by fax that the postponed 510(k), PMA Supplement, or Export Certificate should be released from its deferred status because of the firm's corrective action. (See G-4 or G-5)

Within 15 working days of receipt of the firm's response the district will then advise the firm (in writing) that:

- the district has advised CDRH/FPB that the firm has made adequate corrections of GMP violations related to (specify application). But, the firm will be reminded that the other GMP violations must be corrected and that approval of future applications related to those inadequately addressed GMP violations may be delayed.

The district will fax a copy of its acknowledgement of the firm's response to Joy Lazaroff, CDRH/OC/FPB, HFZ-306, Fax (301)-594-4715.

V. AUTHORITY

Those authorized to evaluate inspectional findings, classify inspection reports, and recommend enforcement actions are set forth in various agency documents such as the Investigations Operations Manual, Regulatory Procedures Manual, and Field Management Directives. These individuals are also authorized to determine whether a firm's GMP violations are system wide, limited to a specific product line, or limited to a specific product, and whether the GMP violations reasonably relate to 510(k)s, PMA Supplements, or Export Certificates.

CDRH/ORA CLASS I & II 510(K) APPLICATION & PMA SUPPLEMENT PROGRAM

CDRH/OC FOLLOW-UP TO VIOLATIVE FOREIGN GMP INSPECTIONS

I. VIOLATIVE GMP FOREIGN INSPECTIONS (EIR is classified OAI)

The starting point in determining a firm is not in GMP compliance is when an inspection reveals significant GMP violations which meet the Situation I criterion and the inspection is classified as Official Action Indicated (OAI). (CP 7382.830, Part V, A 1, and Attachment A, Exhibits 1 - 4 provide guidance to be used in determining whether the GMP violations warrant an inspection classification of OAI). If the criteria for classifying the inspection as OAI is met, the foreign firm will be issued a Warning Letter by the CDRH Office of Compliance (OC) or other enforcement action taken, i.e., a Warning Letter and Automatic Detention action. However, if the firm responds to the FDA-483 (483) before issuance of the Warning Letter, the Warning Letter will issue but the language of the Warning Letter will be modified accordingly.

All foreign Establishment Inspection Reports (EIR) are initially received by the CDRH/OC Field Programs Branch (FPB) for tracking action. Foreign EIRs which are classified as OAI by the district are handcarried to the appropriate OC Division of Enforcement (OC/DOE) for review and action.

A. ADEQUATE RESPONSE TO FDA 483 (483) - NOT REQUIRING A PROMPT REINSPECTION

Situation: When the OC/DOE receives the EIR a response to the 483 from the firm is attached. Using the criteria set forth in section III of this document, the OC/DOE determines that the response appears to be adequate but no immediate reinspection is necessary. To establish documentation of adequate warning a Warning Letter will issue.

Action: (See G-1 Warning Letter)

1. Within 15 working days following receipt of the foreign Establishment Inspection Report in the OC/DOE, OC/DOE issues a Warning Letter to the firm based on the GMP violations found during the inspection. However, in the Warning Letter, the OC/DOE will:

a. Acknowledge the firm's adequate response to the 483,

b. Inform the firm that the clearance/approval of any pending 510(k)s or PMA Supplements for products manufactured or sterilized at the facility in which the GMP violations were found will not be deferred because of GMP issues.

c. Inform the firm that the Warning Letter does not require further response.

2. OC/DOE will complete a firm profile data sheet (PDS) within 5 days.

3. OC/DOE will handcarry or fax completed PDS and copy of Warning Letter to Wayne Miller/Jenny Garvey, OC/FPB, HFZ-306, Fax 4-4715. OC/FPB will update ORA/COMSTAT database.

4. OC/DOE will recommend reinspection within 6 to 18 months.

B. ADEQUATE RESPONSE TO 483 - REQUIRES PROMPT REINSPECTION TO VERIFY CORRECTIONS

Situation: When the OC/DOE receives the EIR a response to the 483 from the firm is attached. Using the criteria described in section III, the OC/DOE determines that the response appears to be adequate and a prompt reinspection is necessary to verify that the firm's corrections are adequate. To establish documentation of adequate warning, a Warning Letter will issue.

Action: (See G-2 Warning Letter)

1. Within 15 working days following receipt of the foreign EIR in the OC/DOE, the OC/DOE issues a Warning Letter to the firm based on the GMP violations found during the inspection. However, in the Warning Letter, the OC/DOE will:

a. Acknowledge receipt of the adequacy of the firm's response to the 483 and inform the firm that a reinspection is necessary to verify that the corrections are adequate. If firm has not indicated its readiness for reinspection in their response to the 483, it should be advised in the Warning Letter to indicate its readiness in its response to the Warning Letter.

b. Inform the firm that the clearance/approval of any pending 510(k)s or PMA Supplements for products manufactured or sterilized at the facility in which the GMP violations were found, will be deferred until the adequacy of its corrections can be verified by reinspection, and that U.S. Federal agencies may take this into account when considering award of contracts.

2. OC/DOE will complete a firm profile data sheet (PDS) within 5 days.

3. OC/DOE will handcarry or fax completed PDS and copy of Warning Letter to Wayne Miller/Jenny Garvey, OC/FPB, HFZ-306, Fax 4-4715. OC/FPB will update ORA/COMSTAT database.

4. If firm's response to the 483 indicates its readiness for reinspection OC/DOE will request ITOB to arrange a reinspection within 60 days of the date indicated by the firm.

C. INADEQUATE RESPONSE TO 483

Situation: When the OC/DOE receives the EIR a response to the 483 from the firm is attached. Using the criteria set forth in section III, the OC/DOE determines that the response is inadequate.

Action: (See G-3 Warning Letter)

1. Within 15 working days following receipt of the foreign EIR report in OC/DOE, the OC/DOE issues foreign firm a Warning Letter. The Warning Letter will:

   a. Acknowledge the firm's response to the 483 but advise firm the response was inadequate and explain why it was inadequate, and

   b. Inform firm that clearance/approval of any pending 510(k)s or PMA Supplements for products manufactured or sterilized at the facility in which the GMP violations were found, will be deferred because of GMP issues, and that U.S. Federal agencies may take this into account when considering award of contracts.

2. OC/DOE will complete a firm profile data sheet (PDS) within 5 days.

3. OC/DOE will handcarry or fax completed PDS and copy of Warning Letter to Wayne Miller/Jenny Garvey, OC/FPB, HFZ-306, Fax 4-4715. OC/FPB will update ORA/COMSTAT database.

NOTE: The OC/DOE may determine that a firm's response is partially adequate because the firm has only addressed those GMP violations related to a deferred 510(k) application or PMA Supplement. Please refer to section IV, Special Circumstances, for the procedures which may be followed in this situation.

D. RESPONSES TO 483 RECEIVED AFTER RECEIPT OF FOREIGN EIR IN OC/DOE:

Situation: The EIR is received by OC/DOE and is followed by receipt of the firm's response to the 483 after the report has been in OC for more than 5 days.

Action: (See G-6 Warning Letter)

1. OC/DOE will issue a Warning Letter within 15 working days of receipt of the EIR in OC. The Warning Letter will:

   a. Acknowledge receipt of the response to the 483, and tell the firm the response has not yet been evaluated.

   b. Indicate that the response will be evaluated and communicated to them, but that the firm should not delay its response to the Warning Letter within the time frame given.

   c. Inform firm that clearance/approval of any pending 510(k) applications or PMA Supplements for products manufactured or sterilized at the facility in which the GMP violations were found, will be deferred because of GMP issues, and that U.S. Federal agencies may take this into account when considering award of contracts.

2. OC/DOE will complete a firm profile data sheet (PDS) within 5 days.

3. OC/DOE will handcarry or fax completed PDS and copy of Warning Letter to Wayne Miller/Jenny Garvey, OC/FPB, HFZ-306, Fax 4-4715. OC/FPB will update ORA/COMSTAT database.

E. NO RESPONSE TO 483

Situation: No response to the 483 is received in OC before the OC/DOE issues a Warning Letter.

Action:

1. The OC/DOE will issue a Warning Letter within 15 working days of receipt of the EIR by OC/DOE. The Warning Letter will:

   a. Inform firm that the clearance/approval of any pending 510(k) applications or PMA Supplements for products manufactured or sterilized at the facility in which the GMP violations were found, will be deferred because of GMP issues, and that U.S. Federal agencies may take this into account when considering award of contracts.

2. OC/DOE will complete a firm profile data sheet (PDS) within 5 days.

3. OC/DOE will handcarry or fax completed PDS and copy of Warning Letter to Wayne Miller/Jenny Garvey, OC/FPB, HFZ-306, Fax 4-4715. OC/FPB will update ORA/COMSTAT database.

F. GMP INSPECTIONAL FINDINGS WHICH INDICATE OTHER ENFORCEMENT ACTION BEYOND A WARNING LETTER

Situation: The OC/DOE determines that the inspectional findings indicate risk to the public so a Warning Letter and Automatic Detention action should issue. (Normally a detention action will not be initiated when the response to the 483 is adequate)

Action: The OC/DOE will:

1. Within 15 working days following receipt of the EIR in OC/DOE issue a Warning Letter and Automatic Detention action to the firm.

   This letter will also inform firm that the clearance/approval of any pending 510(k) applications or PMA Supplements for products manufactured or sterilized at the facility in which the GMP violations were found, will be deferred because of GMP issues, and that U.S. Federal agencies may take this into account when considering award of contracts.

2. OC/DOE will complete a firm profile data sheet (PDS) within 5 days.

3. OC/DOE will handcarry or fax completed PDS and copy of Warning Letter to Wayne Miller/Jenny Garvey, OC/FPB, HFZ-306, Fax 4-4715. OC/FPB will update ORA/COMSTAT database.

II. ACTION FOLLOWING ISSUANCE OF WARNING LETTERS

A response to a Warning Letter falls into one of the following categories:

- Adequate Response Not Requiring a Prompt Reinspection

- Adequate Response Requiring Prompt Reinspection to Verify Corrections

- Inadequate response

- No response

A. ADEQUATE RESPONSE TO WARNING LETTER NOT REQUIRING A PROMPT REINSPECTION

Situation: The OC/DOE determines that the firm's response to the Warning Letter is adequate and a prompt reinspection is not necessary, based on consideration of the factors contained in section III.

Action: (See G-10 Model Letter)

1. Within 15 days of receipt of the firm's response to the Warning Letter, the OC/DOE will respond in writing to firm to:

    a. Acknowledge the firm's adequate response to the Warning Letter, and

    b. Inform the firm that the clearance/approval of any pending 510(k) applications and PMA Supplements for products manufactured or sterilized at the facility in which the GMP violations were found, will not be deferred due to GMP issues, and that U.S. Federal agencies may take this into account when considering award of contracts.

2. OC/DOE will complete a firm profile data sheet (PDS) within 5 days.

3. OC/DOE will handcarry or fax completed PDS and copy of Warning Letter to Wayne Miller/Jenny Garvey, OC/FPB, HFZ-306, Fax 4-4715. OC/FPB will update ORA/COMSTAT database.

4. OC/DOE will handcarry or fax the completed PDS and a copy of the Warning Letter to Wayne Miller/Jenny Garvey, OC/FPB, HFZ-306, Fax 4-4715.

5. OC/DOE will request ITOB to schedule firm for reinspection within 6 to 18 months.

B. ADEQUATE RESPONSE TO WARNING LETTER REQUIRING PROMPT REINSPECTION TO VERIFY CORRECTIONS

Situation: The OC/DOE determines that the firm's response to the Warning Letter is adequate and a prompt reinspection is necessary to verify that the firm's corrections are adequate, based on consideration of the factors contained in section III.

Action: (See G-11 Model Letter)

1. Within 15 working days of receipt of the firm's response to the Warning Letter, the OC/DOE will respond in writing to firm to:

   a. Acknowledge the adequacy of the firm's response and inform the firm that the clearance/approval of any pending 510(k) applications or PMA Supplements for products manufactured or sterilized at the facility in which the GMP violations were found, will continue to be deferred until a reinspection to verify the corrective actions are adequate. This letter should also advise firm that U.S. Federal agencies may take this into account when considering award of contracts.

   b. OC/DOE will handcarry or fax a copy of its acknowledgement of the firm's adequate response to Wayne Miller/Jenny Garvey, OC/FPB, HFZ-306, Fax 4- 4715.

2. When an FDA reinspection is required:

   a. When a firm notifies the OC/DOE, in writing of the actions it has taken to bring about correction and indicates it is ready for reinspection, ITOB will make arrangements for a reinspection of the firm within 60 days of the date of readiness given by the firm. That is, 60 days from the date given by the firm as being ready for a reinspection.

   b. If the reinspection is not initiated within 60 days of receipt of the firm's response and time table, the agency has committed, as part of Reinventing Government, to remove from a deferred status any 510(k) application or PMA Supplement which was previously deferred because the firm's GMP violations are reasonably related to the device.

      (1) FDA will still reinspect as soon as practicable but will deal with the cleared/approved application as a marketed product.

3. When reinspection reveals that the firm's corrections are adequate the OC/DOE will:

   a. Within 15 working days of completion of the reinspection inform the firm, in writing, that the corrections are adequate that any deferred 510(k)s or PMA Supplements for products manufactured or sterilized at the facility in which the GMP violations were found, will cleared/approved that U.S. Federal agencies may take this into account when considering award of contracts; and that if firm is on Automatic Detention it will be removed. (See G-12 Model Letter)

b. If firm is on Automatic Detention, OC/DOE will initiate action to remove it from AD within 5 days.

c. OC/DOE will complete a firm profile data sheet (PDS) within 5 days.

d. OC/DOE will handcarry or fax completed PDS and copy of Warning Letter to Wayne Miller/Jenny Garvey, OC/FPB, HFZ-306, Fax 4-4715. OC/FPB will update ORA/COMSTAT database.

4. If the reinspection reveals that corrections are not adequate that serious GMP violations continue, and are reasonably related to the device which has been deferred, the OC/DOE will:

a. Within 15 working days of receipt of the reinspection report:

(1) Inform the firm, in writing, that the corrections are not adequate, and

(2) That the clearance/approval of any deferred 510(k) or PMA Supplements for products manufactured and/or sterilized at the facility to which the GMP violations are related, will not be cleared/approved because of continued GMP issues. This letter will advise firm that U.S. Federal agencies may take this into account when considering award of contracts.

(3) If firm is on Automatic Detention, it will be continued in that status.

b. OC/DOE will complete a firm profile data sheet (PDS) within 5 days.

c. OC/DOE will handcarry or fax completed PDS and copy of Warning Letter to Wayne Miller/Jenny Garvey, OC/FPB, HFZ-306, Fax 4-4715. OC/FPB will update ORA/COMSTAT database.

C. INADEQUATE RESPONSE TO WARNING LETTER

Situation: The firm's response to the Warning Letter is inadequate based upon the criteria identified in section III.

Action: (See G-14 Model Letter)

1. Within 15 working days of receipt of the firm's response to the Warning Letter, the OC/DOE responds to the firm (in writing) and:

a. Informs firm that the response is inadequate and explains why it is inadequate.

b. Informs firm that the clearance/approval of any pending 510(k) applications or PMA Supplements for products manufactured or sterilized at the facility in which the GMP violations were found, will continue to be deferred because of GMP issues, and that U.S, Federal agencies may take this into account when considering award of contracts.

c. If firm is on Automatic Detention, it will continue in that status.

2. OC/DOE will complete a firm profile data sheet (PDS) within 5 days.

3. OC/DOE will handcarry or fax completed PDS and copy of Warning Letter to Wayne Miller/Jenny Garvey, OC/FPB, HFZ-306, Fax 4-4715. OC/FPB will update ORA/COMSTAT database.

4. OC/DOE requests ITOB to schedule an in-depth reinspection within 6 months based upon the public health risk.

NOTE: The OC/DOE may determine that a firm's response is partially adequate because the firm has only addressed those GMP violations related to a pending 510(k)/PMA Supplement. Please refer to paragraph IV, Special Circumstances, for the procedures which may be followed in this situation.

D. NO RESPONSE TO WARNING LETTER

Situation: The firm does not respond to the Warning Letter.

Action:

1. The firm remains in COMSTAT as OAI with Warning Letter issued.

2. The OC/DOE determines whether Automatic Detention (if firm is not already on Automatic Detention) or a request to schedule reinspection is indicated based upon the public health risk.

III. GENERAL GUIDELINES

A. Criteria for determining the adequacy of firm response:

It addresses each observation/violation in a specific, clear way.

It contains sufficient documentation to demonstrate that the observations/violations have been corrected or it has realistic time frames for the accomplishment of each and every action item needed to come into compliance.

It takes into account the impact that new corrections may have on the system(s) as a whole, and indicates positively that such impact has been seriously considered.

B. Criteria for determining whether an immediate reinspection is necessary:

Affirmative responses to the following questions suggest that a prompt reinspection is necessary:

Does the GMP problem(s) involved present a potential public health hazard?

Does the recall history of the firm suggest a pattern of repeat recalls for similar kinds of systemic problems? Does this recall history raise questions about the permanency of corrections made in the past?

Does the firm have a history of making promises to correct, and then backsliding on those promises?

Does the firm have a quality assurance program? This is particularly important if there have been management changes.

Does the firm have any 510(k) submissions or PMA Supplements on deferred in CDRH because of GMP violations which are reasonably related to the device? That is, are the GMP violations systemic, limited to the specific product line, or limited to a specific device?

If the response to any of the above questions is "yes", and the corrections can only be verified by reinspection, then an immediate inspection must be conducted within 60 days following written receipt of the firm's indication it is ready for reinspection.

NOTE: If the ITOB determines that a reinspection cannot be conducted within the 60 day time frame, or the firm has a high rate of recidivism, the OC/DOE may suggest that the firm submit:

- certification by an outside expert consultant that it has conducted an update audit, and
- certification by the firm's CEO that any corrections noted to be needed have been made, and

- certification that the firm is in compliance with the requirements of the GMP regulation. (CP 7382.830. Part V, A, 3, c. (a) and (d))

IV. SPECIAL CIRCUMSTANCES

A firm's response may clearly indicate that only those GMP violations which reasonably relate to a deferred 510(k) submission or PMA Supplement have been corrected, however, other significant GMP violations have not been corrected. If a reinspection is not needed to verify the adequacy of these corrections, the OC/DOE will immediately advise FPB, Joy Lazaroff, in writing, that the postponed 510(k) submission or PMA Supplement should be released from its postponed compliance status because of the firm's corrective action.

a. Within 15 working days of receipt of the firm's response the OC/DOE advise the firm (in writing) that:

   it has made adequate corrections to the GMP violations related to (specify application). However, the firm will be reminded that the other GMP violations must be corrected and that approval of future applications related to those inadequately addressed GMP violations may be delayed.

b. OC/DOE will complete a firm profile data sheet (PDS) within 5 days.

c. OC/DOE will handcarry or fax completed PDS and copy of Warning Letter to Wayne Miller/Jenny Garvey, OC/FPB, HFZ-306, Fax 4-4715. OC/FPB will update ORA/COMSTAT database.

V. AUTHORITY

Those authorized to evaluate inspectional findings, classify inspection reports, and recommend enforcement actions are set forth in various agency documents such as the Investigations Operations Manual, Regulatory Procedures Manual, and Field Management Directives. These individuals are also authorized to determine whether a firm's GMP violations are system wide, limited to a specific product line, or limited to a specific product, and whether the GMP violations reasonably relate to 510(k) applications or PMA Supplements.

DISTRICT OFFICE CONTACTS FOR INDUSTRY MANAGEMENT CONCERNS ABOUT THEIR GMP COMPLIANCE STATUS

Atlanta	Ballard Graham
Baltimore	Gary Pierce
Boston	James Rado
Buffalo	E. Pitt Smith
Chicago	Ray Mlecko
Cincinnati	Debbie Grelle
Dallas	Austin Templer
Denver	William Sherer
Detroit	John Dempster
Kansas City	Robert Wilson
Los Angeles	Elaine Messa
Minneapolis	John Feldman
Nashville	Ray Hedblad
New Orleans	James Green
Newark	Marguerite Eagan
New York	Rick Trainor
Orlando	Timothy Couzins
Philadelphia	Diana Kolaitis
San Francisco	Andrea Scott
San Juan	Sam Jones
Seattle	David Pettenski
St. Louis	Charles Bringman
Foreign Firms	Joy Lazaroff/Marje Hoban

Instructions for Reviewing Records Kept in a Manufacturer's Device Complaint Files

These instructions provide guidance for auditing records kept in the manufacturer's device complaint file. The purpose of this review is to determine if the complaint records meet specific GMP complaint requirements.

During the normal course of a GMP inspection other determinations are made regarding the adequacy of the firm's complaint handling practices. Deficiencies which are identified in the complaint records may provide leads to other significant deficiencies. This data is used for early detection of defects, defect trends, and determining the effectiveness of the quality assurance program to prevent recurrences.

Reference #4 in Part VI of this program explains how the GMP complaint files relate to reports required under MDR. If, during your review of complaints, you discover an unreported complaint involving a death or serious injury that is reportable under MDR, inform management that (1) all MDR reportable events, as defined in the CFR, are to be reported to CDRH, and (2) failure to report is a violation of the FD&C Act. If you identify what you believe to be an MDR reportable event involving device malfunction that has not been reported to FDA (HFZ-533), document your findings, and flag it "MDR malfunction."

In most cases, complaint files will consist of GMP defined complaints only, i.e., oral or written complaints relative to the identity, quality, durability, reliability, safety, effectiveness or performance of a medical device. In some cases, however, other types of records may be commingled with these complaints. If these are not complaints as defined by the GMP regulation, do not include them in your review.

Each complaint record selected for review should be considered deficient if it fails to meet one or more of the following requirements:

1. The complaint was adequately reviewed, evaluated and, if necessary, investigated.

* 2. If the complaint was not investigated the record contains the reasons why and name of the individual responsible for the decision.*

3. If the complaint pertained to injury, death or any hazard to safety it was immediately reviewed, evaluated and investigated by a designated individual. Under the MDR regulation, a written report with some of this information must be sent to CDRH within 15 working days and if a death or serious injury is involved a telephone report to CDRH within 5 calendar days may also be required.

4. All complaints pertaining to injury, death or any hazard to safety are maintained by the unit designated to process complaints.

5. For each complaint where an investigation was made, a written record was maintained by the unit designated to process complaints.

6. Where an investigation was made, the record of investigation includes:

   A. the name of the device.
   B. any control number used.
   C. name of complainant.
   D. nature of complaint.
   E. results of the firm's investigation into the cause of alleged problem.
   F. if applicable, recommended corrective action.
   G. reply to complainant.

7. Investigations are conducted in a timely manner.

8. Investigations are carried out to the degree necessary to draw conclusions as to the validity and cause of the complaint to include review of the Device History Record as appropriate.

9. If the manufacturing facility is at another location, a copy of the investigation was forwarded to that location.

When trends in complaint file deficiencies are identified, list on the FDA-483 that a system defect exists with respect to the complaint handling procedures. This observation should be expanded and supported by summarizing all deviations noted during the review of all complaint records reviewed. Each deviation should not be listed as an FDA-483 item. All deviations summarized support the observation of a system defect.

INSPECTIONAL GUIDANCE FOR PERFORMING A REVIEW OF THE FIRM'S COMPLIANCE WITH THE MDR REGULATION

BACKGROUND

This attachment provides inspectional guidance for performing a review of the firm's compliance with the MDR regulation. It must be accomplished for each inspection conducted under this
compliance program.

a. Inspections of Manufacturers and Importers

   When review of the firm's MDR procedures identifies a problem in their interpretation of what constitutes a reportable death, serious injury, or malfunction under MDR, this should be handled through the Discussion with Management at the conclusion of the inspection. Attachment F provides guidance for when an MDR report is not required. If the firm still disagrees on the matter, the firm should be advised to express their position IN WRITING and address the correspondence to CDRH at the address used for filing written MDR reports.

   Some firms may believe that one or more of their devices and/or certain events are "exempt" from the MDR reporting requirements. Ask the firm's management if they believe that any of their devices are exempt from MDR reporting and the reason(s) for their position. CDRH may have determined that certain incidents are no longer "reportable" and authorized "alternative reporting" of certain types of incidents on a periodic basis, and notified the firm by letter. Verify that the firm has a copy of any such letter on file. A list of the firms which have been granted Alternative Reporting is available from HFZ-533.

The types of records to be examined will relate to the nature of the events reported to the firm. Most firms will have only two or three types of records. Firms are required to maintain copies of all MDR reports and the copies must be identified in such a way that they may be easily examined during an FDA inspection. Complaints concerning death and serious injury which are not reported under MDR may be located in the same file as those death and serious injury complaints reported under MDR.

1) MDR Reported Events: Death, Serious Injury, and Malfunction.

   Ask the firm to identify those reports it has submitted to CDRH pursuant to MDR. If there are no such reports go to (a)(2). Audit a minimum of 20 randomly selected reports. If the total number of reports is less than 20, review all of them. Determine if these reports are listed in the MDR information obtained from the D.P.U. Also verify their accuracy. If random audits identify significant deficiencies, examine all MDR reports in this category for the last two (2) years or since the last qualifying inspection.

   The Center believes that the data requested by MDR constitutes the MINIMUM information that a firm needs in order to determine whether an event is reportable under MDR and that a complete complaint investigation by the firm (per GMPs) would yield the data required by the MDR regulation. If a data element is missing because the reporter did not submit the information, the firm should have some indication in its files that it tried to obtain the data from the reporter during the receipt and/or follow-up of the report. If data is missing because the firm failed to perform an adequate investigation as required by the GMP regulation (820.198), be sure to cite this on the FDA-483.

   In verifying the accuracy of reports, review the entire report paying specific attention to the following fields in the printout:

   o Date the event occurred:

     This information is usually submitted by the reporter. If the date has not been provided with the firm's initial report the firm should attempt to determine the date in any subsequent investigation.

   o Date the firm became aware of event:

     When submitting MDR reports, firms are not required to provide the "aware" date unless the information is requested in writing by FDA. The "aware" date is used in calculating the timeframes for submission of the MDR report. Therefore, it is important that this date be verified. If the "aware" date is not on the MDR print-out, check to determine that the firm has met its reporting timeframes. This date should be part of the firm's records maintained pursuant to section 519 of the FD&C Act, thus, you don't need to request it in writing during the EI. The "aware" date appears on the MDR print-out under the field name "Date report."

NOTE: FDA considers the aware date to be the date any component of the firm is advised of the problem. Hence, if the firm's sales representatives are learning of events and there is a delay in reporting to the QA or MDR handling staff at the firm's headquarters, the firm may experience a problem meeting their timeframes for reporting. Furthermore, if the firm's representatives overseas are learning of problems and not forwarding the information to the USA in a timely fashion, the firm may not be be able to meet the MDR requirements.

o Text of the report:

Firms are requested to describe, to the extent known, the event that requires reporting under MDR. You should compare the text submitted by the reporter to the information reported to FDA by the manufacturer.

o Frequency/severity information:

Firms are required to state whether the reported event has occurred or is occurring more frequently or with greater severity than stated in the labeling, or, if not in the labeling, than is usual for the device. You should determine if the firm maintains severity and frequency data about problems with its products and note this in the EIR. Statements in MDR complaints concerning severity or frequency should be verified.

The questions that the firm needs to answer are simplified as follows:

(a) Is Frequency of Occurrence of the Event Addressed in the Device Labeling? Yes, No, Unknown

If YES, then is Frequency of the event Greater Than Labeled? Yes, No, Unknown

If NO, then is Frequency Greater for this event Than Is Usual? Yes, No, Unknown

(b) Is Severity of Occurrence of the Event Addressed in the Device Labeling? Yes, No, Unknown

- o Lot, serial, model, and other identification numbers.

The firm's MDR file should contain, at a minimum, (1) the original report received by the firm concerning a death, serious injury, or malfunction that requires a report under MDR, and (2) a copy of any MDR report submitted to the FDA. This includes:

- o the initial death or serious injury report submitted by telephone to FDA within 5 calendar days after receipt of information.
- o the written death or serious injury report submitted within 15 working days after receipt of information.
- o the written malfunction report submitted within 15 working days after receipt of information.
- o all subsequent follow-up data submitted to FDA voluntarily or as requested by CDRH. Keep in mind the importer is not required to investigate complaints.

Where review of the firm's MDR records finds they have failed to maintain the records required by the MDR regulation, each incomplete file may represent an item of noncompliance and should be listed on the FDA-483.

Although the MDR regulation does not require an investigation of the incidents reported under MDR, the failure to conduct a follow-up investigation may be a violation of the GMP requirements for Complaint Files. If the inspection reveals that the firm has failed to obtain the data required by MDR and this failure is the result of an incomplete investigation or the failure to conduct any investigation into the event, this problem should be addressed under the GMPs and according to district policy.

Also review the failure investigation report which should be filed with the MDR report. This evaluation should follow all of the requirements of section 820.162 and 820.198 of the GMP regulation.

Emphasize the Agency's interest in increasing the dialogue between firms and complainants, especially health care professionals, concerning the clinical implications of the event subject to MDR. Also document any actions the firm took as a result of an MDR report.

Please note that the MDR regulation requires that firms establish procedures, criteria, protocols, etc. The firm, however, is not required to and should not be cited on the FDA-483 for failure to establish written protocols. Firms should be encouraged to establish written procedures or protocols for determining if an event is MDR reportable.

2) Complaints Pertaining to Deaths, Serious Injury, and Malfunction (That Have Not Been Reported) Examine a maximum of 50 of these complaints (received after December 13, 1984 or since the last MDR inspection, whichever is more recent,) to determine if any are reportable. If significant deficiencies are found, examine all complaints in this category.

Under 820.198 of the GMP regulation, complaints pertaining to death and serious injury are to be kept in a separate part of the complaint file. If reports are not kept in this manner, document your findings.

When review of the firm's complaint files and service/repair records identifies MDR reportable events which were not reported to the FDA, each unreported incident should be listed on the FDA-483.

3) All Other Complaints (That Have Not Been Reported)

Select a random sample of 20 reports of those complaints that the firm claims do not pertain to death, serious injury, or malfunction and determine whether any of those records appear to be reportable. Request that manufacturers/importers explain why they did not report complaints to CDRH in those cases where they should have. If significant reporting deficiencies are found, examine all reports in this category.

4) Service, Repair and Warranty Records

Selecting a random sample of 30 service, repair and warranty records, determine whether any of these appear to be reportable under MDR or should be in the complaint file. Request that the firm explain why these reports were considered service or repair or warranty actions.

5) Legal Files and Records

Selecting a random sample of 20 legal files and records, determine whether any of these appear to be reportable under MDR or should be in the complaint file.

NON-REPORTABLE EVENTS

DEATH/SERIOUS INJURIES

Reasons for NOT submitting an MDR Death or Serious Injury Report.

An MDR report is not required for an incident if, within 5 calendar days, the firm is able to determine that the information was erroneous in that a death or serious injury did not occur or the device involved was manufactured or imported by another manufacturer or importer (21 CFR 803.24(d)(3)).

In addition, a report would not be required for an incident if, during the 5 calendar days allowed before the MDR telephone report is due, the firm is able to obtain information from the health care professional to confirm that their device did not cause or contribute to the reported death or serious injury.

Information such as this from the health care professional would mean that the firm no longer had information to reasonably suggest that their device caused or contributed to the death or serious injury and therefore would not have a reportable incident. PLEASE NOTE THAT THE LACK OF SUCH A STATEMENT FROM THE HEALTH CARE PROFESSIONAL DOES NOT EXCUSE THE FIRM FROM FILING AN MDR REPORT. IN THE ABSENCE OF INFORMATION TO THE CONTRARY, THE FIRM MUST APPLY THE PER SE RULE (21 CFR 803.3(f)).

MALFUNCTIONS:

Reasons for NOT submitting an MDR Malfunction report.

1. Even though a firm has received information that reasonably suggests that a malfunction has occurred, each of the following five (5) caveats have been met. If a single one is not met then the firm cannot use this criteria to avoid reporting a malfunction.

    a. A death or serious injury has not occurred.

    b. The device's labeling sets forth information concerning the potential for death or the type of serious injury that may occur.

    c. The device's labeling describes the malfunction and the routine service, repair, or maintenance instruction to correct the malfunction.

    d. The malfunction has occurred or is occurring at or below the frequency and severity stated in the labeling for the device, or if no such statement is in the labeling, at or below the frequency and severity that is usual for the device.

e. As a result of the malfunction, the manufacturer or importer does not undertake a remedial action involving any device other than the particular device in which the malfunction occurred.

2. The event in question is considered by the firm to be a (a) routine service and/or repair, therefore, not likely to cause or contribute to a death or serious injury. In order to qualify as a routine service and/or repair event, the malfunction must occur and/or be corrected during a scheduled service visit and the failure in question must be anticipated and within the expected life of the component in question. Obviously, no death or serious injury may be associated with the malfunction. For example, if a firm expects a valve to wear out at X hours and schedules maintenance to replace the part at that time, there is no malfunction since the part is "failing" according to its specification. Conversely, if the part is failing or being replaced prior to its scheduled (anticipated) life, it is malfunctioning and subject to consideration for reporting under MDR.

   Malfunctions that have a remote likelihood of causing or contributing to a death or serious injury are not reportable. The MDR regulation requires that a malfunction be reported when it is "likely to cause/contribute." "Likely" has been defined as more probable than not, and allows a firm to factor in subjective assessments such as fail safe mechanisms, monitoring devices, and the basic function of the device. All explanations of why a malfunction is not likely to cause/contribute and, therefore, is not reportable, must be reviewed on an ad hoc basis because of the very unique nature of medical devices. This can be established by the firm's device history. For example, if a firm can provide documented proof that a specific type of malfunction is unlikely to cause or contribute to a death or serious injury with the device, no report is required. In determining the rate of failure, firms must be able to provide adequate documentation of failure rate history and the rate must be computed by production run or equivalent. Failure to maintain this type of data will void the firm's contention that the event is occurring within a rate usual for the device.

   Note that "probable" does not equate with the mathematical concept of probability, i.e., 50 percent or more. MDR "probable" means likely to happen and the "likely" includes all environmental factors, design, use, and clinical sequelae associated with the device and its use.

In summary, the MDR rule does not require the reporting of routine scheduled maintenance, service and repair, acceptable rates of anticipated malfunctions, malfunctions where the device, clinical circumstance or environment make the probability of death or serious injury remote.

Regulation (803.24(d)(4)) authorized the Center to determine that incidents do not have to be reported. However these "determinations" must be in the form of a written letter from CDRH to the manufacturer. If an "exemption" from MDR is claimed, ask for documentation.

MODEL WARNING LETTERS

ATTACHMENT G-1

MODEL WARNING LETTER (GMPs and MDR)

Adequate Response to FDA 483 - Not Requiring a Prompt Reinspection (Reference: Guidelines, Paragraph I A 1)

CERTIFIED MAIL
RETURN RECEIPT REQUESTED

RESPONSIBLE INDIVIDUAL, TITLE
FIRM NAME
FIRM'S COMPLETE ADDRESS

Dear (Addressee);

During an inspection of your firm located in (City, State) , on (dates), our Investigator(s) determined that your firm manufactures (type of device) . (Name of device) are devices as defined by Section 201(h) of the Federal Food, Drug, and Cosmetic Act (the Act).

The above-stated inspection revealed that these devices are adulterated within the meaning of section 501(h) of the Act, in that the methods used in, or the facilities or controls used for manufacturing, packing, storage, or installation are not in conformance with the Good Manufacturing Practice (GMP) for Medical Devices Regulation, as specified in Title 21, Code of Federal Regulations (CFR), Part 820, as follows:

1. Failure to conduct planned and periodic audits of the quality assurance program in accordance with written procedures. For example, no audits of the quality assurance program have been performed for at least 3 years.

2. Failure to investigate the failure of a device to meet performance specifications after a device has been released for distribution, and to make a written record of the investigation including conclusions and follow-up. For example, there are no records of failure investigations for Model , S/N , and Model , S/N , which were returned because they did not operate properly.

3. Failure to maintain device history records for Model to demonstrate that the devices are manufactured in accordance with the device master record.

4. Failure to immediately review, evaluate and investigate any complaint pertaining to injury, death, or any hazard to safety. For example, there is no record of the

investigation of a report that a child's death associated with the use of Model at the Community Medical Ctr. on/or about Feb. 8, 1991.

Additionally, the above stated inspection revealed that your devices are misbranded within the meaning of section 502(t)(2) of the Act, in that your firm failed to submit information to the Food and Drug Administration as required by the Medical Device Reporting (MDR) Regulation, as specified in 21 CFR Part 803. Specifically, you failed to submit an MDR report to FDA after receiving information which reasonably suggested that one of your commercially distributed devices may have caused or contributed to a death. The February 8, 1991, incident report from the Community Medical Center in which a child standing in a crib fell over, caught his head in a "Y" formed by the crib rail and end post, and died, should have been reported as a death.

This letter is not intended to be an all-inclusive list of deficiencies at your facility. It is your responsibility to ensure adherence to each requirement of the Act and regulations. The specific violations noted in this letter and in the FDA 483 issued at the closeout of the inspection may be symptomatic of serious underlying problems in your firm's manufacturing and quality assurance systems. You are responsible for investigating and determining the causes of the violations identified by the FDA. If the causes are determined to be systems problems, you must promptly initiate permanent corrective actions.

We acknowledge that you have submitted to this office a response concerning our investigator's observations noted on the form FDA 483. It appears that the response is adequate, therefore no submissions for premarket clearance will be withheld for GMP reasons.

You should take prompt action to correct any manufacturing or quality systems deviations identified by your internal audits. Failure to promptly correct these deviations may be identified in a later comprehensive follow-up inspection, and may result in regulatory action being initiated by the Food and Drug Administration without further notice. These actions include, but are not limited to, seizure, injunction, and/or civil penalties.

Any correspondence concerning this matter should be sent to (Name) , Compliance Officer, Food and Drug Administration, (street address) , (City, State & Zip Code) .

Sincerely yours,

ATTACHMENT G-2

MODEL WARNING LETTER (GMPs and MDR)

Adequate Response to FDA 483 - Requires Prompt Reinspection To Verify Corrections - (Reference: Guidelines, Paragraph I B 1)

CERTIFIED MAIL
RETURN RECEIPT REQUESTED

RESPONSIBLE INDIVIDUAL, TITLE
FIRM NAME
FIRM'S COMPLETE ADDRESS

Dear (Addressee);

During an inspection of your firm located in (City, State) , on (dates), our Investigator(s) determined that your firm manufactures (type of device) . (Name of device) are devices as defined by Section 201(h) of the Federal Food, Drug, and Cosmetic Act (the Act).

The above-stated inspection revealed that these devices are adulterated within the meaning of section 501(h) of the Act, in that the methods used in, or the facilities or controls used for manufacturing, packing, storage, or installation are not in conformance with the Good Manufacturing Practice (GMP) for Medical Devices Regulation, as specified in Title 21, Code of Federal Regulations (CFR), Part 820, as follows:

1. Failure to conduct planned and periodic audits of the quality assurance program in accordance with written procedures. For example, no audits of the quality assurance program have been performed for at least 3 years.

2. Failure to investigate the failure of a device to meet performance specifications after a device has been released for distribution, and to make a written record of the investigation including conclusions and follow-up. For example, there are no records of failure investigations for Model , S/N , and Model , S/N , which were returned because they did not operate properly.

3. Failure to maintain device history records for Model to demonstrate that the devices are manufactured in accordance with the device master record.

4. Failure to immediately review, evaluate and investigate any complaint pertaining to injury, death, or any hazard to safety. For example, there is no record of the investigation of a report that a child's death associated with the use of Model at the Community Medical Ctr. on/or about Feb. 8, 1991.

Additionally, the above stated inspection revealed that your devices are misbranded within the meaning of section 502(t)(2) of the Act, in that your firm failed to submit information to the Food and Drug Administration as required by the Medical Device Reporting (MDR) Regulation, as specified in 21 CFR Part 803. Specifically, you failed to submit an MDR report to FDA after receiving information which reasonably suggested that one of your commercially distributed devices may have caused or contributed to a death. The February 8, 1991, incident report from the Community Medical Center in which a child standing in a crib fell over, caught his head in a "Y" formed by the crib rail and end post, and died, should have been reported as a death.

This letter is not intended to be an all-inclusive list of deficiencies at your facility. It is your responsibility to ensure adherence to each requirement of the Act and regulations. The specific violations noted in this letter and in the FDA 483 issued at the closeout of the inspection may be symptomatic of serious underlying problems in your firm's manufacturing and quality assurance systems. You are responsible for investigating and determining the causes of the violations identified by the FDA. If the causes are determined to be systems problems, you must promptly initiate permanent corrective actions.

We acknowledge that you have submitted to this office a response concerning our investigator's observations noted on the form FDA 483. It appears that the response is adequate. A follow-up inspection will be required, however, to assure that corrections are adequate.

Until it has been determined that corrections are adequate, federal agencies are advised of the issuance of all Warning Letters about devices so that they may take this information into account when considering the award of contracts. Additionally, no pending submissions for premarket clearance for devices to which the GMP violations are reasonably related will be cleared. Also, no requests for Certificates For Products For Export will be approved.

You should take prompt action to correct deviations whether identified by our investigator or your internal systems audit. Failure to promptly correct these deviations may result in regulatory action being initiated by the Food and Drug Administration without further notice. These actions include, but are not limited to, seizure, injunction, and/or civil penalties.

Please notify this office in writing within 15 working days of receipt of this letter, of the anticipated date that your facility will be ready for reinspection.

Your response should be sent to (Name) , Compliance Officer, Food and Drug Administration, (street address) , (City, State & Zip Code) .

Sincerely yours,

ATTACHMENT G-3

MODEL WARNING LETTER (GMPs and MDR)

Inadequate Response to FDA 483 - For use when firm responds to the FDA 483 within five days but response is inadequate. (Reference: Guidelines, Para I C 1)

CERTIFIED MAIL
RETURN RECEIPT REQUESTED

RESPONSIBLE INDIVIDUAL, TITLE
FIRM NAME
FIRM'S COMPLETE ADDRESS

Dear (Addressee);

During an inspection of your firm located in (City, State) , on (dates), our Investigator(s) determined that your firm manufactures (type of device) . (Name of device) are devices as defined by Section 201(h) of the Federal Food, Drug, and Cosmetic Act (the Act).

The above-stated inspection revealed that these devices are adulterated within the meaning of section 501(h) of the Act, in that the methods used in, or the facilities or controls used for manufacturing, packing, storage, or installation are not in conformance with the Good Manufacturing Practice (GMP) for Medical Devices Regulation, as specified in Title 21, Code of Federal Regulations (CFR), Part 820, as follows:

1. Failure to conduct planned and periodic audits of the quality assurance program in accordance with written procedures. For example, no audits of the quality assurance program have been performed for at least 3 years.

2. Failure to investigate the failure of a device to meet performance specifications after a device has been released for distribution, and to make a written record of the investigation including conclusions and follow-up. For example, there are no records of failure investigations for Model , S/N , and Model , S/N , which were returned because they did not operate properly.

3. Failure to maintain device history records for Model to demonstrate that the devices are manufactured in accordance with the device master record.

4. Failure to immediately review, evaluate and investigate any complaint pertaining to injury, death, or any hazard to safety. For example, there is no record of the investigation of a report that a child's death associated with the use of Model at the Community Medical Ctr. on/or about Feb. 8, 1991.

Additionally, the above stated inspection revealed that your devices are misbranded within the meaning of section 502(t)(2) of the Act, in that your firm failed to submit information to the Food and Drug Administration as required by the Medical Device Reporting (MDR) Regulation, as specified in 21 CFR Part 803. Specifically, you failed to submit an MDR report to FDA after receiving information which reasonably suggested that one of your commercially distributed devices may have caused or contributed to a death. The February 8, 1991, incident report from the Community Medical Center in which a child standing in a crib fell over, caught his head in a "Y" formed by the crib rail and end post, and died, should have been reported as a death.

This letter is not intended to be an all-inclusive list of deficiencies at your facility. It is your responsibility to ensure adherence to each requirement of the Act and regulations. The specific violations noted in this letter and in the FDA 483 issued at the closeout of the inspection may be symptomatic of serious underlying problems in your firm's manufacturing and quality assurance systems. You are responsible for investigating and determining the causes of the violations identified by the FDA. If the causes are determined to be systems problems, you must promptly initiate permanent corrective actions. We acknowledge that you have submitted to this office a response concerning our investigator's observations noted on the form FDA 483. We have reviewed your response and have concluded that it is inadequate. Detailed comments on your response are attached.

Federal agencies are advised of the issuance of all Warning Letters about devices so that they may take this information into account when considering the award of contracts. Additionally, no premarket submissions for devices to which the GMP deficiencies are reasonably related will be cleared until the violations have been corrected. Also, no requests for Certificates For Products For Export will be approved until the violations related to the subject devices have been corrected.

You should take prompt action to correct these deviations. Failure to promptly correct these deviations may result in regulatory action being initiated by the Food and Drug Administration without further notice. These actions include, but are not limited to, seizure, injunction, and/or civil penalties.

Please notify this office in writing within 15 working days of receipt of this letter, of the specific steps you have taken to correct the noted violations, including an explanation of each step being taken to identify and make corrections to any underlying systems problems necessary to assure that similar violations will not recur. If corrective action cannot be completed within 15 working days, state the reason for the delay and the time within which the corrections will be completed.

Your response should be sent to (Name) , Compliance Officer, Food and Drug Administration, (street address) , (City, State & Zip Code) .

Sincerely yours,

ATTACHMENT G-4

MODEL WARNING LETTER (GMPs and MDR)

Partially adequate response, i.e., firm responds to the FDA 483 within 5 days - response appears to be adequate for observations specifically related to a pending premarket submission, but a f/u inspection is required to assure the adequacy of corrections. (Reference: Guidelines, Para I C NOTE, and Para IV)

CERTIFIED MAIL
RETURN RECEIPT REQUESTED

RESPONSIBLE INDIVIDUAL, TITLE
FIRM NAME
FIRM'S COMPLETE ADDRESS

Dear (Addressee);

During an inspection of your firm located in (City, State) , on (dates), our Investigator(s) determined that your firm manufactures (type of device) . (Name of device) are devices as defined by Section 201(h) of the Federal Food, Drug, and Cosmetic Act (the Act).

The above-stated inspection revealed that these devices are adulterated within the meaning of section 501(h) of the Act, in that the methods used in, or the facilities or controls used for manufacturing, packing, storage, or installation are not in conformance with the Good Manufacturing Practice (GMP) for Medical Devices Regulation, as specified in Title 21, Code of Federal Regulations (CFR), Part 820, as follows:

1. Failure to conduct planned and periodic audits of the quality assurance program in accordance with written procedures. For example, no audits of the quality assurance program have been performed for at least 3 years.

2. Failure to investigate the failure of a device to meet performance specifications after a device has been released for distribution, and to make a written record of the investigation including conclusions and follow-up. For example, there are no records of failure investigations for Model , S/N , and Model , S/N , which were returned because they did not operate properly.

3. Failure to maintain device history records for Model to demonstrate that the devices are manufactured in accordance with the device master record.

4. Failure to immediately review, evaluate and investigate any complaint pertaining to injury, death, or any hazard to safety. For example, there is no record of the investigation of a report that a child's death associated with the use of Model at the Community Medical Ctr. on/or about Feb. 8, 1991.

Additionally, the above stated inspection revealed that your devices are misbranded within the meaning of section 502(t)(2) of the Act, in that your firm failed to submit information to the Food and Drug Administration as required by the Medical Device Reporting (MDR) Regulation, as specified in 21 CFR Part 803. Specifically, you failed to submit an MDR report to FDA after receiving information which reasonably suggested that one of your commercially distributed devices may have caused or contributed to a death. The February 8, 1991, incident report from the Community Medical Center in which a child standing in a crib fell over, caught his head in a "Y" formed by the crib rail and end post, and died, should have been reported as a death.

This letter is not intended to be an all-inclusive list of deficiencies at your facility. It is your responsibility to ensure adherence to each requirement of the Act and regulations. The specific violations noted in this letter and in the FDA 483 issued at the closeout of the inspection may be symptomatic of serious underlying problems in your firm's manufacturing and quality assurance systems. You are responsible for investigating and determining the causes of the violations identified by the FDA. If the causes are determined to be systems problems, you must promptly initiate permanent corrective actions.

We acknowledge that you have submitted to this office a response concerning our investigator's observations noted on the form FDA 483. It appears that the response is adequate where it addresses those observations relating to the manufacturing of the device covered by 510(k) #XXXXXXX/PMA Supplement #XXXXXXX. A follow-up inspection will be required, however, to assure that corrections are adequate. Please notify this office when the facility will be ready for inspection. Until the adequacy of the corrections can be confirmed clearance of the premarket submissions noted above will be withheld.

As discussed in the attached review, your response does not adequately address those violations relating to (other manufacturing processes or types of devices). The remainder of this letter applies to those violations and the devices to which they are related.

Federal agencies are advised of the issuance of all Warning Letters about devices so that they may take this information into account when considering the award of contracts. Additionally, no premarket submissions for devices to which the remaining GMP violations are reasonably related will be cleared until the violations have been corrected. Also, no requests for Certificates For Products For Export will be approved until the violations related to the subject device have been corrected.

You should take prompt action to correct these deviations. Failure to promptly correct these deviations may result in regulatory action being initiated by the Food and Drug

Administration without further notice. These actions include, but are not limited to, seizure, injunction, and/or civil penalties.

Please notify this office in writing within 15 working days of receipt of this letter, of the specific steps you have taken to correct the noted violations, including an explanation of each step being taken to identify and make corrections to any underlying systems problems necessary to assure that similar violations will not recur. If corrective action cannot be completed within 15 working days, state the reason for the delay and the time within which the corrections will be completed.

Your response should be sent to (Name) , Compliance Officer, Food and Drug Administration, (street address) , (City, State & Zip Code) .

Sincerely yours,

ATTACHMENT G-5

MODEL WARNING LETTER (GMPs and MDR)

FOR USE WHEN THE FIRM RESPONDS TO THE FDA 483 WITHIN FIVE DAYS--THE RESPONSE APPEARS TO BE ADEQUATE FOR OBSERVATIONS DIRECTLY RELATED TO THE PENDING PREMARKET SUBMISSION AND NO IMMEDIATE F/U INSPECTION WILL BE REQUIRED. (Reference: Guidelines, Para I C NOTE, and Para IV)

CERTIFIED MAIL
RETURN RECEIPT REQUESTED

RESPONSIBLE INDIVIDUAL, TITLE
FIRM NAME
FIRM'S COMPLETE ADDRESS

Dear (Addressee);

During an inspection of your firm located in (City, State) , on (dates), our Investigator(s) determined that your firm manufactures (type of device) . (Name of device) are devices as defined by Section 201(h) of the Federal Food, Drug, and Cosmetic Act (the Act).

The above-stated inspection revealed that these devices are adulterated within the meaning of section 501(h) of the Act, in that the methods used in, or the facilities or controls used for manufacturing, packing, storage, or installation are not in conformance with the Good Manufacturing Practice (GMP) for Medical Devices Regulation, as specified in Title 21, Code of Federal Regulations (CFR), Part 820, as follows:

1. Failure to conduct planned and periodic audits of the quality assurance program in accordance with written procedures. For example, no audits of the quality assurance program have been performed for at least 3 years.

2. Failure to investigate the failure of a device to meet performance specifications after a device has been released for distribution, and to make a written record of the investigation including conclusions and follow-up. For example, there are no records of failure investigations for Model , S/N , and Model , S/N , which were returned because they did not operate properly.

3. Failure to maintain device history records for Model to demonstrate that the devices are manufactured in accordance with the device master record.

4. Failure to immediately review, evaluate and investigate any complaint pertaining to injury, death, or any hazard to safety. For example, there is no record of the investigation of a report that a child's death associated with the use of Model at the Community Medical Ctr. on/or about Feb. 8, 1991.

Additionally, the above stated inspection revealed that your devices are misbranded within the meaning of section 502(t)(2) of the Act, in that your firm failed to submit information to the Food and Drug Administration as required by the Medical Device Reporting (MDR) Regulation, as specified in 21 CFR Part 803. Specifically, you failed to submit an MDR report to FDA after receiving information which reasonably suggested that one of your commercially distributed devices may have caused or contributed to a death. The February 8, 1991, incident report from the Community Medical Center in which a child standing in a crib fell over, caught his head in a "Y" formed by the crib rail and end post, and died, should have been reported as a death.

This letter is not intended to be an all-inclusive list of deficiencies at your facility. It is your responsibility to ensure adherence to each requirement of the Act and regulations. The specific violations noted in this letter and in the FDA 483 issued at the closeout of the inspection may be symptomatic of serious underlying problems in your firm's manufacturing and quality assurance systems. You are responsible for investigating and determining the causes of the violations identified by the FDA. If the causes are determined to be systems problems, you must promptly initiate permanent corrective actions.

We acknowledge that you have submitted to this office a response concerning our investigator's observations noted on the form FDA 483. It appears that the response is adequate where it addresses those observations relating to the manufacturing of the device covered by 510(k)#XXXXXXX/PMA Supplement #XXXXXXX. Consequently, clearance of preapproval submissions for those devices will not be withheld for GMP reasons.

As discussed in the attached detailed review, your response does not adequately address those violations relating to (other manufacturing processes or types of devices). The remainder of this letter applies to those violations and the devices to which they are related.

Federal agencies are advised of the issuance of all Warning Letters about devices so that they may take this information into account when considering the award of contracts. Additionally, no premarket submissions for devices to which the remaining GMP violations are reasonably related will be cleared until the violations have been corrected. Also, no requests for Certificates For Products For Export will be approved until the violations related to the subject devices have been corrected.

You should take prompt action to correct these deviations. Failure to promptly correct these deviations may result in regulatory action being initiated by the Food and Drug

Administration without further notice. These actions include, but are not limited to, seizure, injunction, and/or civil penalties.

Please notify this office in writing within 15 working days of receipt of this letter, of the specific steps you have taken to correct the noted violations, including an explanation of each step being taken to identify and make corrections to any underlying systems problems necessary to assure that similar violations will not recur. If corrective action cannot be completed within 15 working days, state the reason for the delay and the time within which the corrections will be completed.

Your response should be sent to (Name) , Compliance Officer, Food and Drug Administration, (street address) , (City, State & Zip Code) .

Sincerely yours,

ATTACHMENT G-6

MODEL WARNING LETTER (GMPs and MDR)

For use when response to FDA 483 is received after the 5 working day frame
(Reference: Guidelines, Para I D 1)

CERTIFIED MAIL
RETURN RECEIPT REQUESTED

RESPONSIBLE INDIVIDUAL, TITLE
FIRM NAME
FIRM'S COMPLETE ADDRESS

Dear (Addressee);

During an inspection of your firm located in (City, State) , on (dates), our Investigator(s) determined that your firm manufactures (type of device) . (Name of device) are devices as defined by Section 201(h) of the Federal Food, Drug, and Cosmetic Act (the Act).

The above-stated inspection revealed that these devices are adulterated within the meaning of section 501(h) of the Act, in that the methods used in, or the facilities or controls used for manufacturing, packing, storage, or installation are not in conformance with the Good Manufacturing Practice (GMP) for Medical Devices Regulation, as specified in Title 21, Code of Federal Regulations (CFR), Part 820, as follows:

1. Failure to conduct planned and periodic audits of the quality assurance program in accordance with written procedures. For example, no audits of the quality assurance program have been performed for at least 3 years.

2. Failure to investigate the failure of a device to meet performance specifications after a device has been released for distribution, and to make a written record of the investigation including conclusions and follow-up. For example, there are no records of failure investigations for Model , S/N , and Model , S/N , which were returned because they did not operate properly.

3. Failure to maintain device history records for Model to demonstrate that the devices are manufactured in accordance with the device master record.

4. Failure to immediately review, evaluate and investigate any complaint pertaining to injury, death, or any hazard to safety. For example, there is no record of the

investigation of a report that a child's death associated with the use of Model at the Community Medical Ctr. on/or about Feb. 8, 1991.

Additionally, the above stated inspection revealed that your devices are misbranded within the meaning of section 502(t)(2) of the Act, in that your firm failed to submit information to the Food and Drug Administration as required by the Medical Device Reporting (MDR) Regulation, as specified in 21 CFR Part 803. Specifically, you failed to submit an MDR report to FDA after receiving information which reasonably suggested that one of your commercially distributed devices may have caused or contributed to a death. The February 8, 1991, incident report from the Community Medical Center in which a child standing in a crib fell over, caught his head in a "Y" formed by the crib rail and end post, and died, should have been reported as a death.

We acknowledge that you have submitted to this office a response concerning our investigator's observations noted on the form FDA 483. Your response was not received within 5 days after the inspection and has not yet been reviewed. This response will be evaluated and communicated to you, however, you should not delay your response to this Warning Letter in the meantime.

This letter is not intended to be an all-inclusive list of deficiencies at your facility. It is your responsibility to ensure adherence to each requirement of the Act and regulations. The specific violations noted in this letter and in the FDA 483 issued at the closeout of the inspection may be symptomatic of serious underlying problems in your firm's manufacturing and quality assurance systems. You are responsible for investigating and determining the causes of the violations identified by the FDA. If the causes are determined to be systems problems, you must promptly initiate permanent corrective actions.

Federal agencies are advised of the issuance of all Warning Letters about devices so that they may take this information into account when considering the award of contracts. Additionally, no premarket submissions for devices to which the GMP deficiencies are reasonably related will be cleared until the violations have been corrected. Also, no requests for Certificates For Products For Export will be approved until the violations related to the subject devices have been corrected.

You should take prompt action to correct these deviations. Failure to promptly correct these deviations may result in regulatory action being initiated by the Food and Drug Administration without further notice. These actions include, but are not limited to, seizure, injunction, and/or civil penalties.

Please notify this office in writing within 15 working days of receipt of this letter, of the specific steps you have taken to correct the noted violations, including an explanation of each step being taken to identify and make corrections to any underlying systems problems necessary to assure that similar violations will not recur. If corrective action cannot be completed within 15 working days, state the reason for the delay and the time within which the corrections will be completed.

Your response should be sent to (Name) , Compliance Officer, Food and Drug Administration, (street address) , (City, State & Zip Code) .

Sincerely yours,

ATTACHMENT G-7

MODEL WARNING LETTER (GMPs and MDR)

For use when there's no response to the FDA 483 received before the Warning Letter is issued
(Reference: Guidelines, Para I E 1)

CERTIFIED MAIL
RETURN RECEIPT REQUESTED

RESPONSIBLE INDIVIDUAL, TITLE
FIRM NAME
FIRM'S COMPLETE ADDRESS

Dear (Addressee);

During an inspection of your firm located in (City, State) , on (dates), our Investigator(s) determined that your firm manufactures (type of device) . (Name of device) are devices as defined by Section 201(h) of the Federal Food, Drug, and Cosmetic Act (the Act).

The above-stated inspection revealed that these devices are adulterated within the meaning of section 501(h) of the Act, in that the methods used in, or the facilities or controls used for manufacturing, packing, storage, or installation are not in conformance with the Good Manufacturing Practice (GMP) for Medical Devices Regulation, as specified in Title 21, Code of Federal Regulations (CFR), Part 820, as follows:

1. Failure to conduct planned and periodic audits of the quality assurance program in accordance with written procedures. For example, no audits of the quality assurance program have been performed for at least 3 years.

2. Failure to investigate the failure of a device to meet performance specifications after a device has been released for distribution, and to make a written record of the investigation including conclusions and follow-up. For example, there are no records of failure investigations for Model , S/N , and Model , S/N , which were returned because they did not operate properly.

3. Failure to maintain device history records for Model to demonstrate that the devices are manufactured in accordance with the device master record.

4. Failure to immediately review, evaluate and investigate any complaint pertaining to injury, death, or any hazard to safety. For example, there is no record of the

investigation of a report that a child's death associated with the use of Model at the Community Medical Ctr. on/or about Feb. 8, 1991.

Additionally, the above stated inspection revealed that your devices are misbranded within the meaning of section 502(t)(2) of the Act, in that your firm failed to submit information to the Food and Drug Administration as required by the Medical Device Reporting (MDR) Regulation, as specified in 21 CFR Part 803. Specifically, you failed to submit an MDR report to FDA after receiving information which reasonably suggested that one of your commercially distributed devices may have caused or contributed to a death. The February 8, 1991, incident report from the Community Medical Center in which a child standing in a crib fell over, caught his head in a "Y" formed by the crib rail and end post, and died, should have been reported as a death.

This letter is not intended to be an all-inclusive list of deficiencies at your facility. It is your responsibility to ensure adherence to each requirement of the Act and regulations. The specific violations noted in this letter and in the FDA 483 issued at the closeout of the inspection may be symptomatic of serious underlying problems in your firm's manufacturing and quality assurance systems. You are responsible for investigating and determining the causes of the violations identified by the FDA. If the causes are determined to be systems problems, you must promptly initiate permanent corrective actions.

Federal agencies are advised of the issuance of all Warning Letters about devices so that they may take this information into account when considering the award of contracts. Additionally, no premarket submissions for devices to which the GMP deficiencies are reasonably related will be cleared until the violations have been corrected. Also, no requests for Certificates For Products For Export will be approved until the violations related to the subject devices have been corrected.

You should take prompt action to correct these deviations. Failure to promptly correct these deviations may result in regulatory action being initiated by the Food and Drug Administration without further notice. These actions include, but are not limited to, seizure, injunction, and/or civil penalties.

Please notify this office in writing within 15 working days of receipt of this letter, of the specific steps you have taken to correct the noted violations, including an explanation of each step being taken to identify and make corrections to any underlying systems problems necessary to assure that similar violations will not recur. If corrective action cannot be completed within 15 working days, state the reason for the delay and the time within which the corrections will be completed.

Your response should be sent to (Name) , Compliance Officer, Food and Drug Administration, (street address) , (City, State & Zip Code) .

Sincerely yours,

ATTACHMENT G-8

MODEL WARNING LETTER (GMP's and MDR)

FOR USE WHEN FOLLOWING THE ENFORCEMENT STRATEGY FOR FIRMS WITH REPEATED VIOLATIVE INSPECTIONS (Part V, A.3.c.).

CERTIFIED MAIL
RETURN RECEIPT REQUESTED

RESPONSIBLE INDIVIDUAL, TITLE
FIRM NAME
FIRM'S COMPLETE ADDRESS

Dear (Addressee):

During an inspection of your firm located in (City, State) , on (dates), our Investigator(s) determined that your firm manufactures (type of device) . (Name of device) are devices as defined by Section 201(h) of the Federal Food, Drug, and Cosmetic Act (the Act).

The above-stated inspection revealed that these devices are adulterated within the meaning of section 501(h) of the Act, in that the methods used in, or the facilities or controls used for manufacturing, packing, storage, or installation are not in conformance with the Good Manufacturing Practice (GMP) for Medical Devices Regulation, as specified in Title 21, Code of Federal Regulations (CFR), Part 820, as follows:

1. Failure to conduct planned and periodic audits of the quality assurance program in accordance with written procedures. For example, no audits of the quality assurance program have been performed for at least 3 years.

2. Failure to investigate the failure of a device to meet performance specifications after a device has been released for distribution, and to make a written record of the investigation including conclusions and follow-up. For example, there are no records of failure investigations for Model , S/N , and Model , S/N , which were returned because they did not operate properly.

3. Failure to maintain device history records for Model to demonstrate that the devices are manufactured in accordance with the device master record.

4. Failure to immediately review, evaluate and investigate any complaint pertaining to injury, death, or any hazard to safety. For example, there is no record of the investigation of a report that a child's death associated with the use of Model at the Community Medical Ctr. on/or about Feb. 8, 1991.

Additionally, the above stated inspection revealed that your devices are misbranded within the meaning of section 502(t)(2) of the Act, in that your firm failed to submit information to the Food and Drug Administration as required by the Medical Device Reporting (MDR) Regulation, as specified in 21 CFR Part 803. Specifically, you failed to submit an MDR report to FDA after receiving information which reasonably suggested that one of your commercially distributed devices may have caused or contributed to a death. The February 8, 1991, incident report from the Community Medical Center in which a child standing in a crib fell over, caught his head in a "Y" formed by the crib rail and end post, and died, should have been reported as a death.

This letter is not intended to be an all-inclusive list of deficiencies at your facility. It is your responsibility to ensure adherence to each requirement of the Act and regulations. The specific violations noted in this letter and in the FDA 483 issued at the closeout of the inspection may be symptomatic of serious underlying problems in your firm's manufacturing and quality assurance systems. You are responsible for investigating and determining the causes of the violations identified by the FDA. If the causes are determined to be systems problems, you must promptly initiate permanent corrective actions.

In order to facilitate FDA in making the determination that such corrections have been made and thereby enabling FDA to withdraw its advisory to other federal agencies concerning the award of government contracts, and to resume marketing clearance, and export clearance for products manufactured at [x] facility, we are requesting that you submit to this office on the schedule below, certification by an outside expert consultant that it has conducted an audit of your firm's manufacturing and quality assurance systems relative to the requirements of the device GMP regulation (21CFR, Part 820). You should also submit a copy of the consultant's report, and certification by your firm's CEO (if other than yourself) that he or she has reviewed the consultant's report and that your firm has initiated or completed all corrections called for in the report. The attached guidance may be helpful in selecting an appropriate consultant.

The initial certifications of audit and corrections and subsequent certifications of updated audits and corrections (if required) should be submitted to this office by the following dates:

- o Initial certifications by consultant and firm -Show actual date (allow approximately six months from issuance of Warning Letter).

- o Subsequent certifications-Show actual date(s). You may ask for annual reports for two years after the follow-up inspection.

Federal agencies are advised of the issuance of all Warning Letters about devices so that they may take this information into account when considering the award of contracts. Additionally, no premarket submissions for devices to which the GMP deficiencies are reasonably related will be cleared until the violations have been corrected. Also, no requests for Certificates For Products For Export will be approved until the violations related to the subject devices have been corrected.

You should take prompt action to correct these deviations. Failure to promptly correct these deviations may result in regulatory action being initiated by the Food and Drug Administration without further notice. These actions include, but are not limited to, seizure, injunction, and/or civil penalties.

Please notify this office within 15 days of receipt of this letter, of the specific steps you will be taking to comply with our request.

Your response should be sent to (Name) , Compliance Officer, Food and Drug Administration, (street address) , (City, State & Zip Code) .

Sincerely yours,

ATTACHMENT G-9

MODEL LETTER TO INDUSTRY (GMP's and MDR)

CERTIFIED MAIL
RETURN RECEIPT REQUESTED

RESPONSIBLE INDIVIDUAL, TITLE
FIRM NAME
FIRM'S COMPLETE ADDRESS

Dear (Addressee):

We have received your letter dated _____ to the FDA 483 which was issued to your firm on ______ . Our preliminary determination is that your response is one which requires further technical review by FDA headquarters. Accordingly, this is to advise you that your complete response has been forwarded to the Office of Compliance in the Center for Devices and Radiological Health.

Any pending premarket submissions or Export Certificates for products manufactured at your facility may be deferred if a determination is made that the GMP violations reflected on the FDA 483 and further detailed in the establishment inspection report are reasonably related to the product[s] involved in the pending applications. If such is the case, pending applications may not be cleared until the District and/or CDRH Office of Compliance determines that the related GMP violations are adequately corrected. Federal agencies may take this information into account when considering award of contracts.

Sincerely yours,

ATTACHMENT G-10

MODEL LETTER TO INDUSTRY (GMP's and MDR)

CERTIFIED MAIL
RETURN RECEIPT REQUESTED

RESPONSIBLE INDIVIDUAL, TITLE
FIRM NAME
FIRM'S COMPLETE ADDRESS

Dear (Addressee):

We have completed our review of your letter dated _____ which you provided in response to the Warning Letter which was issued to you on _______ . We find that your response adequately addresses our concerns which were stated in the Warning Letter, and therefore the approval of any pending premarket submissions, or Export Certificates for products manufactured at your facility will not be deferred due to GMP issues. This information will be made available for reference by Federal agencies when considering award of contracts.

Sincerely yours,

ATTACHMENT G-11

MODEL LETTER TO INDUSTRY (GMP's and MDR)

CERTIFIED MAIL
RETURN RECEIPT REQUESTED

RESPONSIBLE INDIVIDUAL, TITLE
FIRM NAME
FIRM'S COMPLETE ADDRESS

Dear (Addressee):

We have completed our review of your letter dated _____ which you provided in response to the Warning Letter which was issued to you on ______ . We find that while your response appears adequate, further verification is needed before we will be in a position to support the approval of any pending premarket submissions or Export Certificates for products manufactured at your facility in which the GMP violations were found to be reasonably related to those pending applications. This further verification will be in the form of a reinspection by the District Office unless, as an alternative, you arrange for an independent consultant to document corrections and certify to us that your firm is in compliance with GMP's. Please advise us if you choose the alternative course.

Sincerely Yours,

ATTACHMENT G-12

MODEL LETTER TO INDUSTRY (GMP's and MDR)

CERTIFIED MAIL
RETURN RECEIPT REQUESTED

RESPONSIBLE INDIVIDUAL, TITLE
FIRM NAME
FIRM'S COMPLETE ADDRESS

Dear (Addressee):

We have completed our review of the results of [our reinspection which ended on. _____] [the independent consultant certification which you provided to us on ______], and wish to advise you that approval of any pending premarket submissions or Export Certificates for products manufactured or sterilized at your facility in which GMP violations were found, will not be deferred because of GMP issues, and that Federal agencies may take this into account when considering award of contracts.

Sincerely yours,

ATTACHMENT G-13

MODEL WARNING LETTER (GMP's and MDR)

CERTIFIED MAIL
RETURN RECEIPT REQUESTED

RESPONSIBLE INDIVIDUAL, TITLE
FIRM NAME
FIRM'S COMPLETE ADDRESS

Dear (Addressee):

As a result of [our reinspection performed in response to your reply to our Warning Letter of ____] [our review of the outside consultant's certification done in response to our Warning Letter of ____] we have determined that serious GMP violations continue. [These violations appear to be reasonably related to pending applications that are pending in CDRH].

====END WITH STANDARD LANGUAGE ABOUT PENDING PREMARKET SUBMISSIONS--- EXPORT CERTS

Sincerely yours,

ATTACHMENT G-14

MODEL LETTER TO INDUSTRY (GMP's and MDR)

CERTIFIED MAIL
RETURN RECEIPT REQUESTED

RESPONSIBLE INDIVIDUAL, TITLE
FIRM NAME
FIRM'S COMPLETE ADDRESS

Dear (Addressee):

We acknowledge your letter of _____ responding to the Warning Letter we issued your firm on _____ . We must advise you that your response does not adequately address the problems identified in the Warning Letter for the following reasons:

(LIST THE INADEQUACIES AND SAY WHY THE RESPONSE ISN'T MET)

Until these inadequacies are corrected, the approval of any pending premarket submissions or Export Certificates for products manufactured at your facility will continue to be deferred because of GMP problems which are reasonably related to those applications, and Federal agencies may take this into account when considering award of contents.

Partially adequate:

We acknowledge your letter of ____ responding to the Warning Letter we issued your firm on ______ . Your response clearly indicates that you have corrected only those GMP violations which reasonably relate to particular pending [specify...]. This is to advise you that those [specify which types of applications] will not be deferred because of GMP problems, and that Federal agencies may take this in account when considering award of contents.

However, other GMP problems remain and must be corrected before approval of any future applications will be considered. In addition, you should be aware that enforcement action, as described in the Warning Letter, may be considered if these problems are not corrected.

Sincerely yours,

The following guidance was originally published in the CDRH, Office of Compliance Industry Letter No. 2, dated July 6, 1993

SELECTING A CONSULTANT ?

As the number of consultants has increased in the past few years, so too has our concern about their qualifications and the quality of their work. While most consultants accurately and honestly promote their capabilities, we believe the device industry should exercise diligence in the selection of a consultant. It is very disappointing to see a company which is experiencing serious problems go to the expense of hiring a consultant who fails to constructively contribute to the restoration of the company's regulatory health.

Of course, FDA cannot recommend or endorse a particular consultant, but we can offer some criteria that should be considered when selecting one. You should first determine what type of consultant you need. There are basically three types of consultants: Regulatory, Quality, and Technical. A Regulatory consultant is one that will specialize in 510(k) and PMA issues, GMP's and/or device labeling. A Quality consultant is adept at GMP auditing, writing and revising procedures. The Technical consultant basically knows how to find problems and fix them. In some cases a company may need the services of one or more of these consultants. The ideal consultant would be highly qualified in all three of these areas. Since we in compliance deal most with GMP issues, we have identified some factors that we recommend you consider when selecting a Quality consultant, but these factors may have applicability for the other types of consultants also:

- How long has the consultant worked with the device (not drug) GMP regulation?

- Is his/her knowledge current?

- How does he/she know what CDRH's "current" policies and interpretations are for device GMP's?

- Does the consultant sponsor/participate in training courses?

- Is he/she frequently asked to give presentations at FDA/Industry sponsored seminars? What have been the reactions to these presentations?

- One of the primary attributes of a good consultant is to be a "good communicator". He/she must be able to communicate problems and provide solutions in a clear, concise manner, and in such a way that the company knows how to perform corrections the "right" way, the first time.

- Has he/she been deposed and/or testified as an expert witness, either for the FDA or for industry?

- Obtain a listing of the consultant's clients over the last several years. Check these references!

- What types of certifications does the consultant have, i.e.? Is the certification recognized by professional societies, etc?

We believe that a little homework in identifying and selecting a consultant will have long term payoffs for any company.

Chapter 12

FDA ADVERSE FINDINGS

OVERVIEW

FDA-regulated industries are inspected by investigators from the agency's district offices. Following the inspection of a firm, FDA investigators submit to the firm's management a Form FDA 483 ("Inspectional Observations"), which lists violative findings.

FDA can take further regulatory actions if violations noted on the FDA 483 are not corrected, including disapproval or withdrawal of applications, the issuance of Warning Letters, FDA-requested recall of a product, product detention and seizures, injunctions, and prosecutions.

WARNING LETTERS

The most common regulatory action following the issuance of an FDA 483 is the Warning Letter. Warning Letters regarding violations of GMP regulations have been emphasizing validation-related problems for the last three or four years. A 1990 FDA memorandum relating to computerized systems in nonclinical laboratories noted that there were 84 computer-related observations cited on FDA-483 between 1979 and 1990. This was only 3% of the total reported FDA-483 observations. Most of the computer-related deficiencies were in the areas of general validation, inadequate SOPs, and inaccurate data.[1]

Although the problems have been recognized, not all of them have been solved. As we noted in *Validation Compliance Annual: 1995*, off the 167 Warning Letter citations we reviewed for the years 1992-1994, the majority (21%) dealt with general validation violations. The next largest amounts (15% and 14%, respectively) related to lack of adequate systems documentation or lack of specifications. Eleven percent of citations related to inadequate testing. Smaller numbers of violations were found relating to change control procedures (9%), SOPs and protocols (6%), systems security problems

---

[1]Memorandum, Ty Fugiwara, Nonclinical Laboratory Studies Branch (HFD-345) to Dr. Paul Lepore, HFC-230, March 29, 1990.

(5%), inadequate personnel training (4%), quality assurance/quality control (4%), maintenance (4%), 510(k) submissions (3%), and vendors (2%).

RECALLS

Another regulatory action is product recall initiated by the FDA. Often, when problems are discovered the product-manufacturing firm will initiate a recall on its own accord.

In 1992, the FDA prepared a report on software-related recalls of medical devices for the period between 1983 and 1992 to provide manufacturers with information on where to direct their quality assurance efforts. That report was attached to Chapter 18 of the *Validation Compliance Annual: 1995*. The FDA has not prepared a similar report on trends since 1992.

The following listing of recent Warning Letters, however, indicates how widespread validation problems still seem to be in the FDA-regulated industries.

Blood Bank, Kansas, 8/9/94 -- "Failure to have adequate written records of validation of your computer system to assure proper performance during all possible operations."

Blood Bank, New Jersey, 8/10/94 -- "3. Validation of the *** computer system is deficient in that the following data is not included: (a) Testing of all the system functions are not routinely conducted (e.g. - expiration dates assigned to all products, donor name, etc.). (b) Documentation of stress testing and system challenges. (c) Testing to confirm or detect illegal values, abnormal cases or empty sets: (d) A schedule for periodic system validation. (e) A lack of validation of changes to the software (e.g. - untested autologous blood units were released by the system). (f) The validation protocol does not indicate acceptable specifications for each computer function or evaluation criteria for testing results.

"4. There is not documentation of a quality assurance program to audit the input/output of computer information to assure data integrity.

"5. Blood bank personnel responsible for establishing a validation protocol for computer operations have not been properly trained and lack the experience with respect to fully understanding computer validation principles and techniques....

"8. Written standard operating procedures (SOP) have not been established for: ... (b) Donor name entry into the computer to preclude confusion (e.g. - some are entered differently...)."

Medical Device Manufacturer, Kansas, 9/1/94 -- "Failure to document validation of the *** data acquisition software (version 1 & 2) which is used with your *** system."

Medical Device Manufacturer, New York, 9/8/94 -- "Failure to have written manufacturing specifications and processing procedures to assure the devices conform to their original design or approved changes; and failure to have documentation of the approval or implementation of device changes. For example, there was no documentation demonstrating validation/approval of the following: Updating the *** in the *** devices to revisions *** and use of the software for programming in the *** devices."

Medical Device Manufacturer, Rhode Island, 9/15/94 -- "Our investigator documented significant deviations from the GMP regulations in conjunction with these devices that compromise the assurance of safety and effectiveness. For example, component testing on sub-assemblies is incomplete, automated processes have not been validated, inadequate procedures for software development and validation, ... failure to follow written procedures for burn in, and incomplete or missing testing records."

Medical Device Manufacturer, Illinois, 9/19/94 -- "Failure to maintain an adequate complaint handling system. During the inspection, your software engineer claimed that a procedure had been written; however, the procedure could not be located. Also during the inspection, you stated that your firm has not received any complaints for ***. However, our investigator found in your customer files, complaints involving a mechanical failure and a software problem."

Blood Bank, Minnesota, 9/20/94 -- "Failure to accurately enter into the computer tracking system unit *** ..., initially reactive for Anti-HCV. The unit was incorrectly entered as non-reactive."

Substance Abuse Program, New Jersey, 9/27/94 -- "Your program failed to maintain an adequate record system to document and monitor patient care. For example, validation of the computerized dispensing and record keeping system was lacking...."

"Your program failed to maintain adequate security over the manner in which drug stocks are administered and dispensed. For 1 out of 10 patients ... computer and manual dosage records documented different amounts of narcotic drug dispensed. All dispensing employees used the same computer password. The office manager and nursing supervisor are knowledgeable of the password for the 'Doctor's Module' allowing access to medication orders."

Medical Device Manufacturer, France, 9/28/94 -- "Failure to maintain an accurate written record of each complaint investigation.... For example, the type of failure that was initially reported and entered into the computer may be found incorrect at a later date, however, corrections are not possible in the database."

Medical Device Manufacturer, Colorado, 10/13/94 -- "1). Failure to review, evaluate, and maintain by a formally designated unit all records of written and oral complaints relative to the identity, quality, reliability, safety, effectiveness, or performance of a device; and failure of the formally designated unit to determine whether or not an investigation is necessary, and to maintain a record, when no investigation of a complaint is made, that includes the reason and the name of the individual responsible for the decision not to investigate.... For example: ... (c). *** facility has records that indicate software and hardware failures of the ***. Many of these events involving the *** were not found in ***'s PPR system.

"4). Failure of the quality assurance program to identify, recommend, or provide solutions for quality assurance problems and verify the implementation of such solutions. For example: ... (b). ***'s anomalies summary, and related anomaly reports ... indicate that numerous *** pacer, and software anomaly reports discovered in 1992 and 1993 still remain unresolved and open. The related records do not indicate or give any justification for this status.

"5). Failure to assure that specification changes shall bee subject to controls as stringent as those applied to the original device.... For example, anomaly report b indicates an error in a released software version that was discovered after distribution to the field. Several other anomaly reports indicate software problems being discovered in the field.

"6). Failure to have a critical device history record which documents each critical component used in the manufacture of a device and the acceptance record of the critical component used in the manufacture of a device and the acceptance record of the critical component, including acceptance date and signature of the recipient.... For example, there is no documentation maintained to show that the performance of the copying of released software to ***, or for subsequent copies made for distribution was successful (i.e., testing documentation that verifies that the copy process was performed correctly for each unit made). Service records from the *** facility appear to demonstrate that failures have been confirmed for the ***software cassettes following release for distribution."

Blood Bank, Iowa, 10/24/94 -- "Failure to maintain an adequate donor deferral system which is designed to protect the blood supply by not allowing deferred donors to re-enter your donor population.... Examples include: Between August 26 and September 20, 1994, all donor deferral information for approximately 176 donations had not been entered into your card an/or computer donor deferral systems. There are no written SOPs concerning your computerized donor deferral system, which you use for back-up to your card system."

Blood Bank, Pennsylvania, 11/14/94 -- "Failure to maintain complete and adequate written standard operating procedures (SOPs) that reflect current blood bank operations ... in that ... there are no written validation procedures to follow for changes or updates of computer hardware or software."

Medical Device Manufacturer, California, 11/21/94 -- "Failure to control all changes to written manufacturing specifications and processing controls to assure that the device conforms to its original design or any approved changes in that design.... For example, unrestricted read/write access to computer device master files allows unauthorized changes."

Drug Manufacturer, New Jersey, 11/23/94 -- "There are no SOPs which address validation of the computer programs/systems at your firm." Also: "There is insufficient validation data to demonstrate that the computer spreadsheet programs used to analyze products for release and stability, were tested and challenged, and are capable of producing accurate results."

Medical Device Manufacturer, Maryland, 12/2/94 -- "Failure to adequately validate all elements of the *** software version *** and its functions associated with the *** System....

"Failure to use real test specimens or data received from the autoread instrument in real time. Duplicate tests from at least 40 specimens for each software version were not tested side by side. Specimens should represent concentrations spread over the entire clinically significant range of values.

"Your protocol apparently failed to follow the suggested distribution of data listed in Table 1 of the NCCLS Guideline. Our review indicates that your concentrations do not follow the NCCLS Guideline and your results are clustered in or near the normal ranges for the analyses. This could result in the data being improperly weighted and not providing a complete comparison of the two software packages.

"There was no data representing NCCLS statistical calculations comparing the *** test results with the results from the previous software package."

Medical Device Manufacturer, U.K. 12/9/94 -- "Failure to have written procedures describing any processing controls necessary to assure conformance to specifications, where deviations from device specifications could occur as a result of the manufacturing process itself.... For example: Device history records do not document that operations were performed in sequence...." After review of manufacturer's response: "In your response of June 9, 1994, your firm state that this was a human error. They stated that the operator had neglected to manually fill in and sign off his operation, although subsequent examination of the relevant computer printout showed that the operation had in fact been ***. They noted that this omission would have been detected during the final review of the manufacturing records for the batch. They state that the importance of signing off each operation has been re-emphasized to all employees by means of a company notice backed up by meetings between Supervisors and the Operators under their control.... This response is not completely adequate because ongoing training documents must be revised to assure this requirement and other requirements are periodically brought to the attention of all employees affected, as well as to the attention of new employees as appropriate."

Drug Manufacturer, New Jersey, 12/14/94 -- "Failure to document complete validation of the computer system used to monitor and control the production process."

Medical Device Manufacturer, Ohio, 12/21/94 -- "Failure to validate software programs by adequate and documented testing when computers are used as part of an automatic quality assurance system. The software program for the Factory Generator Tester used during the functional test of the *** has not been validated"; "There is a malfunction (bug) in the software program for the Factory Generator Tester which causes an error to appear in the test results. When this occurs a manual *** Test Signal Generator is used to complete the test."

Medical Device Manufacturer, Illinois, 1/13/95 -- "Failure to establish procedures for specification control measures to assure that the design basis for the device, components and packaging is correctly translated into approved specifications.... For example, manufacturing processes for *** series controllers and software do not include descriptions of the software revisions, descriptions of the qualification of software revisions which assure that software revisions are properly validated and are not contributing to the failures of the *** series controllers.

"Failure to implement an adequate and complete quality assurance program to assure that the devices produced are manufactured in accordance with the requirements of the GMP regulations and with documentation showing the quality assurance audit is performed.... For example, the audits of the QA program do not verify compliance of software design, development, and process control.

"Failure to establish and maintain adequate written procedures for finished device inspection to assure that device specifications are met, and to test devices under simulated use conditions, and assure that release of devices are based on acceptable statistical rationale.... For example, not all *** series controllers receive a function test for burn-in, and established procedures do not specify a burn-in time period for the controllers....

"Failure to receive, store and handle components used in manufacturing in a manner designed to prevent damage, mix-up, contamination and other adverse effects.... For example, (a) no established written procedures exist for the acceptance of software components such as the *** series controllers...."

Medical Device Manufacturer, Texas, 1/17/95 -- "Failure to review, evaluate, and maintain by a formally designated unit all records of written and oral complaints relative to the identity, quality, reliability, safety, effectiveness, or performance of a device and to determine whether or not an investigation is necessary.... For example, repair personnel enter information about device/software malfunctions (e.g. data corruption, computer host/network problems, display malfunctions, image artifact, and programming problems) into a data base, but the data is not evaluated for complaints and recorded as required in your firm's procedures. Complaints are instead classified as repairs/maintenance.

"Failure to adequately investigate any failure of a device to meet performance specifications after the device has been released for distribution.... For example, the above referenced device failures/complaints (e.g. data corruption, computer host/network problems, display malfunctions, image artifacts, and programming problems) have not been investigated to determine their cause.

"Failure to assure that specification changes shall be subject to controls as stringent as those applied to the original device.... For example, your firm failed to adequately validate *** software version 2.5 release 1 for correct subroutine version prior to release. This failure resulted in the patient orientation indicators not being displayed. Your firm was aware of this during the validation process and classified the software error as a *** defect causing severely impaired function. Your firm also failed to document testing of *** software version 2.5.4 to show it met specifications. Version *** had been issued to repair programming errors previously identified in version ***.

"Failure to have adequate checks designed and implemented for automated data processing used for manufacturing or quality assurance purposes to prevent inaccurate data output, input, and programming errors.... For example, the two previously referenced citations regarding device failures/complaints and incomplete software validations....

"Failure of the quality assurance program to assure that all quality assurance checks are appropriate and adequate for their purpose and are performed correctly.... For example, signed software release notices for *** version *** release *** and *** system version *** release 1 were not maintained.

"Failure of the device master record (DMR) to include device specifications.... For example, DMRs for *** software versions ***, *** and *** lacked detailed descriptions of modules, functions, data structure, acceptable data input, functional boundaries, files used by the system, reports generated, software/hardware interfaces, image processor hardware, software specifications, and expected outputs."

Medical Device Manufacturer, New Jersey, 1/18/95 -- "1. Your firm does not document that a reprocessed (i.e. repaired) device meets the original specifications prior to re-release. For example, the re-testing of the unit with the *** program was not documented for any repaired devices....

"2. Your firm has failed to validate the *** computer program, used to test the finished *** devices prior to release for packaging."

Medical Device Manufacturer, Japan, 1/20/95 -- "Failure to assure that specification changes shall be subject to controls as stringent as those applied to the original device, and that such changes shall be approved and documented by a designated individual, including an approval date and the date the change becomes effective.... For example, your Engineering Change Orders (ECOs) do not indicate the effective date of the changes, and a revision was performed to the control software for the *** systems.... There was no plan or procedure in place which addressed testing the revised software to assure consistent operation at a new pulse setting.

Medical Device Manufacturer, Ireland, 1/24/95 -- "Failure to validate software programs by adequate and documented testing, when computers are used as part of an automated production or quality system.... For example: The Computer validation document for the spinning machine does not have complete specifications.... Failure to maintain a device history record to demonstrate that the device is manufactured in accordance with the device master record.... Maintenance procedure for the *** does not include verification of acceptable programming nor is there documentation of the programmed settings for the dionizer controller."

Blood Bank, California, 1/25/95 -- "Failure to establish controls over the computer system and maintain documentation to assure that the computer system performs in the manner for which it was designed and to inform and train employees about changes in its use ... in that: (a). The computer system in use has not been validated. A protocol for validation of the computer system used in your blood banking operations must be developed and validation of the system accomplished. Guidance from the Center for Biologics Evaluation and Research (CBER) is available in the document entitled 'Draft Guideline for the validation of Blood Establishment Computer Systems".... (b). Changes in the use of the computer system have not been documented and personnel were not trained regarding changes impacting their responsibilities. The lack of training for the addition of the *** function pertaining to test results for antibody to human immunodeficiency virus, type 1 (anti-HIV-1) contributed to an error whereby a unit of Red Blood Cells repeatedly reactive and later confirmed positive for anti-HIV-1 was released and transfused. (c). There is a lack of security for the database in that the vendor has unrestricted access through a modem.

"Written standard operating procedures (SOP's) are not always followed or adequately maintained ... in that: ... (d) Instructions for entering test data into the computer are not provided in the written SOP's utilized by the Immunopathology lab."

Medical Device Manufacturer, Denmark, 1/30/95 -- "(2). Failure to establish and implement specification control measures to assure that the design basis for the device, components, and packaging is correctly translated into approved specifications.... For example: ... (e). The computerized temperature monitoring system utilized to monitor the temperatures in all significant refrigerated production storage areas has not been validated....

"4. Failure to establish that production equipment is suitable for its intended purpose and capable of producing valid results.... For example: (a). The computer software used for product quarantine and final release functions has not been validated."

Medical Device Manufacturer, Arizona, 1/31/95 -- "3. Failure to assure that established procedures are followed. For example, (a). There is no record to demonstrate that the *** required by procedure 1001 was performed for the ***

software project..... 6. Failure to establish a component acceptance specification/procedure for blank computer tapes."

Drug Manufacturer, Switzerland, 2/1/95 -- "Laboratory computer systems used to control and/or monitor *** and *** analysis have not been validated. It was indicated that validation had begun on the *** system and that the other systems would follow. Please provide more detailed information as to what computer validation has been completed and what work remains to be done before meeting your projected completion date...."

Medical Device Manufacturer, Sweden, 2/9/95 -- "1. Failure of the quality assurance program to assure the review of production records.... For example, although testing was done, the PC board test data sheets ... were not completed....

"7. Failure to have specification changes approved and documented by a designated individual(s), including an approval date and the date the change becomes effective.... For example, the change control procedure is deficient because the changes made to devices or PC boards are not always traceable, and the change control procedure is not approved (signed or dated) by a designated person.

"8. Failure to subject any change in the manufacturing process of a device to a formal approval process.... For example, changes in software or manufacturing processes/procedures are not required to be formally approved."

Medical Device Manufacturer, Italy, 2/13/95 -- "Failure to have specification changes approved and documented by a designated individual(s), including an approval date and the date the change became effective.... For example, the following specification procedure changes were not formally approved, documented, and controlled prior to being released to production/distribution: (a). Specification changes for the *** software upgrade...." After review of manufacturer's response: "Your response is not adequate. Your response includes a ... Software Upgrade procedure. However, the document you provided does not indicate if the procedure has been approved. In addition, you did not indicate what steps you are taking to assure that all specification procedure changes are formally approved, documented, and controlled prior to being released to production/distribution."

Also: "The GMPs do not require a formal procedure for software validation. The current GMPs in regards to software only require that a software version be validated when the version is changed or modified."

Medical Device Manufacturer, U.K., 2/28/95 -- "Failure of the device master record to include adequate device specifications.... For example, you lack complete documentation for some of your software used in some of your devices." After review of manufacturer's response: "Your response appears to be adequate. You submitted for our review a copy of your overview documentation of the software."

Also: "Failure to routinely calibrate measurement equipment.... For example: (a). You have not documented the validation of your in-process computerized PCB testing activities (equipment). (b). Your programmers have not been calibrated in approximately four years." After review of manufacturer's response: Your response to (a.) is not complete. You agree to validate the manufacturing process again. Until this is done and reviewed by us, we are unable to sate that the correction is adequate. Your response to (b.) appears to be adequate. You claim to have sent the first of three EPROM Programmers to a calibration agent and submitted for our review an internal calibration procedure for your digital programmer. In a later response, you provided for our review a copy of the certificates of calibration for these three EPROM Programmers."

Medical Device Manufacturer, Oregon, 3/1/95 -- "Failure to have manufacturing specification changes subject to controls as stringent as those applied to the original design specifications of the device.... For example, no documented validation exists for specification changes ... of the maximum heparin bolus allowed to be delivered." After review of manufacturer's response: "Your response indicates that validation of the heparin bolus feature was inadvertently omitted, but it does not indicate that validation of this feature, allowing a heparin bolus delivery from *** mL to *** mL, in software v 4.6, will be conducted and documented."

Drug Manufacturer, Michigan, 3/3/95 -- "Laboratory records and controls do not assure that components, in-process materials, labeling, and drug products conform to appropriate standards of identity, strength, quality, and purity.... Contract Analytical Laboratory[:} ... The analytical data generated by computer does not document the proportionality between a standard and a sample assay. Analytical work is not reviewed by a second, qualified individual. Errors were observed in the transfer of raw data to the computer-generated worksheets."

Blood Bank, Indiana, 3/6/95 -- "Failure to adequately validate the *** update ... in that not all system parameters were validated, such as sample tracking and test result calculations, prior to placing the updated system on-line"; "Failure to maintain written records of the *** computer system validation protocol and data"; "Failure to exercise appropriate computer controls ... in that: (a). Validation data generated for the *** Software update was not reviewed or approved for accuracy, completion, and acceptance. (b). Current employee passwords are listed in the *** System validation manual which is accessible to all blood bank and laboratory staff."

Medical Device Manufacturer, Denmark, 5/15/95 -- "The software used to control the sterilization processes has not been evaluated. The validation protocol sent in your response for the software appears to be adequate, however, it cannot be fully evaluated until the validation has been completed."; also: "Failure of the device master record to include authorization of any changes by signature of a designated individual(s) as required by 21 CFR 820.181. For example, the device master records are not dated, or lack the written signatures of the responsible and authorized person(s) approving the device master record. The response is not adequate. While a firm may maintain the device master record as an electronic record, there needs to be a written record containing the authorization for each change."

Blood Bank Computer Software Manufacturer, Nebraska, 5/2/95 -- "the blood bank software ... is adulterated within the meaning of Section 501(h) of the Act in that the methods used in, or the facilities or controls used for manufacturing, packaging, storage, or installation are not in conformance with the Good Manufacturing Practice (CGMP) Regulations for Medical Devices.....

"Significant deviations to 21 CFR 820 were observed and include, but are not limited to the following:

"1) failure to maintain a complete Device Master Record in that it lacks a reference to the location of required documents....

"2) failure to have in place various written procedures covering manufacturing specifications and processes such as structural testing of software, program configuration, regression testing, finished device inspections, and signed and dated functional specifications among others....

"3) failure to have in place quality assurance and audit procedures designed to assure that a formally established and documented quality assurance program is performed....

"4) failure to validate the *** Compiler to assure it is suitable for its intended use...."

Medical Device Manufacturer, North Carolina, 4/21/95 -- "You have failed to establish adequate controls to maintain labeling integrity and to prevent labeling mixups. No formalized written procedure has been established addressing the printing of labels by computer."

Medical Device Manufacturer, Massachusetts, 3/16/95 -- "Failure to assure that specification changes are subject to controls as stringent as those applied to the original design.... For example, software version *** is currently utilized the *** pacemaker analyzer. Documented testing of this software is not available to assure its proper performance."

Drug Manufacturer, Florida, 4/11/95 -- "Your response is inadequate because ... written procedure (SOP ***, Finished Product Specification) specifies that microbiological analysis of its product, ***, be conducted using USP XXII-61. This method does not provide for another or alternative test method to be used as stated in your response. USP XXII-61 allows for automated methods that are validated to be equal or better to the USP. The Standard Method for Water and Waste Water that you reference is not considered an automated method nor is it considered an equivalent or better method than USP XXII-61."

Medical Device Manufacturer, Wisconsin, 4/10/95 -- "Failure to develop software specifications;" "Failure to validate the software."

Blood Bank, New York, 3/17/95 -- "Failure to maintain complete and accurate records providing the entire history of all work performed in that printout-tapes containing unit results and generated by *** are not always maintained. This automated instrument is used for processing ... enzyme immunoassays...."

Medical Device Manufacturer, Germany, 3/17/95 -- "Failure to maintain an adequate written record of each investigation made.... For example: Information recorded on computer printouts does not correlate to data in the complain records"; "Failure of the device master record to include device specifications.... For example, a review of software documents noted there is no software functional description for the power-on reset feature which is available in the *** software with a special password."

Also: "Failure to implement appropriate controls to assure the security of data with regard to computerized operations. (a). The level of security and access to functions of the computer operating systems ... have not been adequately defined in all areas. For example, laboratory technicians were assigned security levels 3, 4 or 5 for the *** Data Management System. Supervisory personnel were unable to provide information to our investigators regarding the basis for assignment of those security levels to employees or the specific system functions that could be accessed by the defined security levels. (b). When data has been changed or re-read by peripheral testing devices, security controls are insufficient to substantiate these activities through the use of audit trails or other appropriate means."

Also: "Failure to maintain complete and adequate written standard operating procedures (SOPs) that reflect current operations. For example: ... (f). There is not procedure describing technicians editing of data entries for test plates and sample tubes when bar codes are misread by automated instruments."

After review of the blood bank's response to the investigational findings: "Evaluation of computerized system testing should be performed continuously. Appropriate steps should be in place to verify system hardware ... are operating as specified and that no unauthorized changes have been made. There also should be an audit trail which documents changes made to the data, including changes made a peripheral automated testing devices."

Blood Bank Computer Software Manufacturer, Texas, 3/20/95 -- "(1). Failure to maintain adequate device master records for Blood Bank Systems software versions *** which include, or refer to the location of, device specifications, production process specifications, quality assurance procedures and specifications, and packaging and labeling specifications.... The user manuals or reference guides distributed with software do not meet this requirement."

Medical Device Manufacturer, Illinois, 3/30/95 -- "Failure to review and evaluate all complaints involving ***, in that returned goods processed through the 'computer billing system' have not been evaluated since October 1994."

"(2). Failure to establish, implement, and control written manufacturing specifications and processing procedures to assure that the device conforms to its original design or any approved changes in that design ... in that: (a). There is no documented validation of software version ***. (b). Program changes and customized routines issued to users were completed and distributed without documented validation or regression testing of the changes. (c). Coding standards have not been developed to assure uniformity with original device specifications. As such, there is not assurance that newly developed software and software modifications will meet the original software design or any approved changes in that design. (d). There are no written procedures for documentation and approval of changes in specifications for the programs of the Blood Bank Systems software. (e). There is no documentation that program changes and customized routines were approved by a designated individual, including the approval date and the date the change became effective.

"(3). Failure to maintain an adequate record of complaint investigations including any control number used, the nature of the complaint, and the reply to the complainant ... in that: (a). The *** lacks documentation of all information necessary to conduct a detailed investigation of a failure. Reports do not specify the specific routine, table settings, or variables of the user system at the time the failure occurred. (b). The *** lacks consistent documentation of the conclusion of the investigation, follow-up, and response to the complainant.

"(4). Failure to establish a quality assurance program which identifies, recommends, or provides solutions for quality assurance problems and verifies the implementation of such solutions.

"(5). Failure to perform planned and periodic audits of a quality assurance program, in accordance with written audit procedures, by appropriately trained individuals not having direct responsibility for the matters being audited....

"(6). Failure of all personnel to demonstrate that they have the necessary training to perform their assigned responsibilities adequately.... For example, the following individuals have not had the appropriate medical device GMP training: (a). The Manager of QA and Documentation, responsible for developing, implementing, and maintaining the quality assurance system. (b). The *** Administrator, responsible for providing training and administering the database, which is used to maintain complaint and failure investigation records nd record software changes"

After review of the manufacturer's response to the investigational findings: "Your response states that you will recreate specifications for software version *** if customers are still using that version after December 31, 1995. Your response further states that you do not intend to formally validate software version ***. Please be advised that it is your responsibility to assure that software in distribution is validated and defined by specifications." Also: "Your response states that you plan to utilized volume and stress tests performed by customers *** software version ***. If you plan to rely on *** sites to conduct volume and stress testing, you must assure that testing is performed according to protocol and that the results are documented." Also: "Your response sates that you have implemented a quality assurance program for software version *** validation. Please be aware that a quality assurance program includes functions in addition to validation activities...."

Medical Device Manufacturer, Arizona, 3/15/95 -- "Failure to establish and implement specification control measures to assure that the design basis for the device is correctly translated into approved specifications.... For example, your firm has no documented evidence which provides a high degree of assurance that the computer software for your heart monitoring device will consistently produce a product meeting its pre-determined specifications and quality attributes (traditionally termed as process validation). Additionally, your firm has no documented evidence that your duplication process for the software installation for the *** has been validated."

Medical Device Manufacturer, Texas, 8/17/94 -- "Failure to provide adequate directions for use of your *** Automated Blood Grouping Instrument ... in that information contained in the users' manual for the instrument addresses specimen requirements and preparation of patient blood samples when in fact the agency has, in response to your firm's request, withdrawn the 510(k) notification relating to the use of this device for the testing of patient samples."

Blood Bank Computer Software Manufacturer, Texas, 3/29/95 -- "(1). Failure to maintain adequate device master records for *** software which include, or refer to the location of, device specifications, production process specifications, quality assurance procedures and specifications, and packaging and labeling specifications....

"(2). Failure to establish, implement, and control written manufacturing specifications and processing procedures to assure that the device conforms to its original design or any approved changes in that design ... in that: (a). Program changes issued to users were completed and distributed without documented validation or regression testing of the changes. (b). There is no documentation that coding standards were utilized when software changes were made. (c). There is no documentation that program changes were approved by a designated individual, including the approval date and the date the change became effective. (d). Review of Development Action Requests (DARs) indicate that numerous software defects have been identified by users.

"(3). Failure to establish a quality assurance program consisting of procedures adequate to assure that the following functions are performed: (1) review of production records; (2) approval or rejection of all components, manufacturing materials, in-process materials, packaging materials, labeling, and finished devices; (3) identifying, recommending, or providing solutions for quality assurance problems and verifying the implementation of such solutions; and (4) assuring that all quality assurance checks are appropriate and adequate for their purpose and are performed correctly....

"(4). Failure to perform planned and periodic audits of quality assurance program, in accordance with written procedures, by appropriately trained individuals not having direct responsibility for the matters being audited....

"(5). Failure to document that personnel have the necessary training to perform their assigned responsibilities adequately."

Medical Device Manufacturer, California, 3/27/95 -- "(1) ...There is no documentation describing any testing of the *** component(s) for the *** volume ventilators produced with newly programmed automated machine shop equipment to ensure that the device conforms to its specifications. The programmed equipment produced device components which failed to meet their specifications resulting in finished device field failures. There are not written procedures ensuring that the computer software source code for each model of critical ventilator is executed at least once during prospective validation. Furthermore, employees engaged in software development advised our investigators that comments and remarks in the source code files may not represent the actual source code specification function....

"(3). Failure to control written procedures for finished device inspection of devices to assure that all device specifications are met.... For example, some alarm and alert indicators of the *** infant/pediatric ventilators are not tested prior to acceptance in that there are no written procedures for the evaluation of CPU/Software error alerts or unknown mode setting alarm(s)....

"(6). Failure to ensure that device master records for devices contains or refers to the location of all device specifications and production procedures.... For example, the device master record for the *** ventilator does not contain or refer to the location of all the electrical specifications for printed circuit board(s) and assemblies, equipment specifications for the *** EPROM erasing device used to erase programmed devices for reprogramming, and the module level software specifications for the *** critical care ventilator.

"(7). Failure to document critical operations conducted in the production of ventilators.... For example, the programming of EPROM and other programmable devices is not considered a critical operation whereas if it was improperly performed it could be reasonably expected to cause failure of a critical device or affect its safety or effectiveness. Additionally, the specific programming devices used for programming of these critical components is not documented.

"(8). Failure to assure all personnel have the necessary training to perform their assigned responsibilities and have a thorough understanding of their jobs.... For example, personnel engaged in the programming of EPROM's used in the production of CPU assemblies were unaware of the defects and errors which could arise from the improper usage of EPROM programming software. When a senior programmer inadvertently used an incorrect software program while programming production EPROM's, our investigator(s) advised the programmer of his error.

"(9) Failure to establish written procedures for the maintenance and calibration of all automated equipment used in the production and/or processing of components to ensure that the equipment is capable of producing valid results.... For example, there are no written procedures for the maintenance or calibration of the EPROM erasing device. There is no documentation describing any inspection or checking of your firm's printed circuit board tester during validation to ensure it is capable of producing valid results....

"(10). Failure of your designated complaint handling unit to properly identify and evaluate possible failure of ventilators to meet their performance specifications as complaints.... For example, ... failures of two V.I.P. ventilators which were reported as inoperable because of an error code were not evaluated as complaints....

"(11). Failure of the quality assurance program to identify, recommend, or provide solutions for quality assurance problems and verify the implementation of these solutions.... For example, there are no written procedures for erasing computer EPROM components...."

Medical Device Manufacturer, Kansas, 3/27/95 -- "Failure to have specification changes approved and documented by a designated individual(s),including the approval date and the date the change becomes effective, and to document and adequately verify device software revisions through testing. For example, there was no documentation of authorized approval of the control switch wiring changes made in May 1993 and the cosmetic custom software changes made in October 1994, and no documentation of testing to assure that software changes did not adversely affect device performance."

Medical Device Manufacturer, Colorado, 3/22/95 -- "(8). Failure investigations are not always performed on devices or components which fail after a device has been released for distribution. For example: (a) An *** system ... was released on 8/11/94. During installation on 9/28/94 the computer would not work and had to be replaced.... (c) An *** system ... was released on 3/31/94. During installation on 7/7/94 a black box had the wrong connectors, the digitizer software was blank....

"(9). Software is not developed, modified and implemented under an organized and approved set of programming standards. No evidence was found demonstrating that design tools, i.e., flow charts, decision trees, etc., are being used prior to coding. Software is not subjected to structural testing per written procedures prior to release."

Medical Device Manufacturer, Oregon, 4/11/95 -- 510(k) violations: "The *** differs significantly ... in that it was designed specifically to identify tremors associated with Parkenson's disease; has major changes in configuration and materials ... and the software includes unique portions of the ... software and previously cleared*** software in order to perform tests specific to the ***.

"(3). The *** and *** differ significantly from ***. The *** has additional assessment indications by incorporation of ***. The *** does not include ***. Both have hardware and performance changes which differ from the ***. Also, both have software programs which were combined ***."

Medical Device Manufacturer, Florida, 4/7/95 -- "Failure to adequately validate device software by conducting appropriate and adequate tests to assure the software conforms to specifications and performs as intended, e.g., software revisions must be structured, documented and evaluated as revisions are developed, checked to ensure it meets specifications, adequately tested with the respective hardware systems and operated under varied conditions to ensure it operates consistently and correctly for all computerized devices."

Medical Device Manufacturer, Ohio, 4/3/95 -- "Computer software including the version of software used to control the pumps used in the reagent dispensing system is not referenced in the MDR or the DHR." Also: "DMRs for the *** do not contain or reference all production process specifications and quality assurance procedures and specifications. For example, they ... lack specifications for operational modes for the *** computer programming to be used during filling procedures."

Medical Device Manufacturer, California, 5/25/95 -- "The Center for Devices and Radiological Health Products, Office of Device Evaluation has determined that there have been significant changes to the software and hardware of these devices for which no premarket notification has been submitted. These modifications of the software and hardware represent a significant change which could affect the safety or effectiveness of the device and requires a 510(k)." Also: "The software validation procedure for *** is inadequate in [that] it does not identify changes which require validation, the sections of the protocol which must be completed for revalidation and/or the justification for not completing all sections of the protocol during revalidation"; "The software specification changes implemented under *** approved on 12/16/93 are not documented to include the reason for the change or an explanation of the change"; "There were no data available to support that the software specification changes approved under *** are adequate in eliminating the ... lock-up problem"; "Failure to retain all required records pertaining to validation modifications made on the *** for a period of time equivalent to the design and expected life of the device, but in no case less than two years from the date of release for commercial distribution. Records of validation of the modification to the board were no longer available due to a crash of the computer used to store this information and no back up was available."

Medical Device Manufacturer, Puerto Rico, 6/6/95 (Manufactures computerized monitoring systems) --"(1). Failure to validate significant manufacturing processes and quality assurance tests to assure the finished product meets its pre-determined specifications and quality attributes, e.g. the validation of the *** didn't include formal evaluation of the product performance qualification of parameters such as critical alarm function.

"(2). Failure to have a written validation protocol for the *** system.

"(3). Failure to assure that devices meet all product specifications prior to release for distribution, e.g. release of 36 computerized ***systems without testing the time to alarm parameter.

"(4). Failure to investigate the failure of a device or any of its components to meet performance specifications after a device has been released for distribution, and to make a written record of the investigation including conclusions and follow-up. For example,there are no records of completed failure investigations for the **** which were returned because they failed to operate properly.

"(5). Failure to qualify the *** used for printed circuit boards in the firm's devices, except for the ***, to verify confidence in the quality of the process used to manufacture a finished device.

"(6). Failure of the device master record for the *** system, a critical device, to include or refer to the location of full information concerning critical components and critical component suppliers, including the complete specifications of all critical components, the source where they may be obtained, and written copies of any agreements made with suppliers....

"(7). Failure of the quality assurance program to consist of procedures adequate to assure performance of the functions if identifying, recommending, or providing solutions for quality assurance problems and verifying the implementation of such solutions."

After review of the manufacturer's response to the investigational findings: "There is not stated completion date for the prospective validation of the *** system"; "There is no commitment to retrospectively/prospectively validate other products"; "The submitted validation review/summary data does not include quantitative data, only PASS/FAIL data. It is, therefore, not possible for our office to conduct any meaningful evaluation of the data submitted."

Blood Bank, Maryland, 5/24/95 -- "Failure to establish permanent records of visual inspection of blood immediately prior to distribution ..., in that the *** software used by the blood bank does not allow for retrieval of visual check information."

Drug Manufacturer, Analytical Laboratory, Illinois, 4/20/95 -- "There is no data available to assure that all instruments and software systems have been validated to assure proper performance and integrity of data."

Medical Device Manufacturer, California, 6/9/95 -- "Failure to assure that devices conform to their original design specifications, in that changes made to software were not validated for the full disclosure monitoring system. For example, software changes intended to alter a data display, add a screen saver, and perform other functions, were not evaluated to determine their effectiveness and impact on other functions."

Chapter 13

FUTURE DEVELOPMENTS

Predicting the future is always precarious process. Predicting in print, in a major reference work, is doubly so. You can safeguard your position by looking far enough in the future so that no one will remember your accuracy. An astronomer calculating the life expectancy of a star, for example, has little to fear if a error throws off the estimate by a few million years. You can offer a massive array of predictions, hoping that the public will remember only your hits shining through the misses. This, of course, is the approach of most entertainment clairvoyants. Or, following in the path of Nostradamus, you can be sufficiently vague so future readers will read into your predictions whatever they want to see. Finally, I suppose you can predict the future with some security by limiting your visions to areas of great stability. I predict, for example, that the Rocky Mountains will still exist three years from now.

Unfortunately, none of these paths are practical here. As an alternative, forced by necessity rather than preference, I will identify the validation trends and developments expected to emerge over the immediate future, in as much specificity as possible. I provide only one, generic, disclaimer -- things change.

Continuation of Trends

The FDA will continue to press for validation of systems, processes, and products. The extension of that validation requirement is the current expression of the fundamental thrust of the agency since its inception almost ninety years ago. No one will build a reputation for foresight by predicting that the FDA, or any other US federal agency, will reverse its demand for greater control. Current press reports about Republican pressure to limit regulation ignore the context of the pharmaceutical industries. There is no evidence to suggest that the American public will accept a situation in which medical devices, foods, or pharmaceuticals are offered for sale without some assurances of safety, cleanliness, and effectiveness. That assurance may come *pre hoc* through the approval and regulatory process, or *post hoc* through law suits and a more lengthy (and costly)

This chapter was contributed by Dr. Sandy Weinberg, Chairman, Weinberg, Spelton & Sax, Inc., Boothwyn, PA.

legal process. Each constituency -- government, the public, and the courts --- prefers the former situation. As a result, the FDA will be reformed, at least in a cosmetic manner. New regulations may be more thoughtfully deliberated (though most complaints about the FDA center around how lengthy those deliberations are at present), and a shift to greater reliance on evidence provided by the regulatee (or agent) will continue, in validation and elsewhere.

The same trend will be mirrored internationally, particularly in the EU. Centralization of regulation will slow approval processes, ultimately forcing reliance on regulation by review of submitted evidence and conclusion, rather than collection and evaluation of direct evidence. Domestically and internationally, regulators will spend their time reading validation reports rather than examining systems, samples, and processors.

In the system validation area, certain key areas of regulatory focus have approached fad status, as regulatees all too often scramble to gather answers without understanding the logic behind the regulators' questions. Those questions are likely to focus increasingly on three areas: the System Development Lifecycle, Boundary Case Testing, and the Trace Matrix.

The System Development Lifecycle (SDLC)

Regulatory insistence upon the use of a System Development Lifecycle as a method for system development and control is firmly and appropriately rooted in the importance of assuring that a system meets the purposes for which it is intended. Although wide variations in the number and definitions of stages of the SDLC will continue to be acceptable, the requirement that a vendor or user follow a plan that moves from requirement, through development, to continued support will grow in emphasis.

A typical System Development Lifecycle defines a six-stage approach:

Requirements: The requirements documentation serves as a template for future evaluation of the system, defining the purposes to which the system will be applied and the requirements for each of those purposes.

Design: The design documentation answers the question of how the requirements will be met, including an indication of the system hardware and operating requirements necessary to assure proper performance of the software.

Programming: Programming standards and practices documentation serves as a guideline for all coding efforts and as a reference for programmers. Included are guidelines for technical development, internal documentation standards, identification of key and reserve words, and quality assurance guidelines.

Testing: The testing documentation includes test protocols (descriptions) for the appropriate tests to be conducted, the scripts for those tests, and (eventually) the evidence that those scripts have been followed.

Implementation: The implementation documentation includes protocols for the Installation Qualification, Operational Qualification, and Performance Qualification. It also includes the Standard Operating Procedures for system control and use and for the validation audit.

Maintenance: The maintenance documentation includes procedures for change control, versioning, and re-validation.

Regulators will continue to ask for evidence that such an approach is in place; it followed religiously; and is clearly documented. I expect a sharp increase in adverse finding letters citing failure to follow an SDLC; and failure to maintain a chronology of establishing requirements prior to design, development, and implementation.

Boundary Case Testing

For the past nine years, the appropriate degree and type of stress testing for systems have been debated without resolution. Finally, this state of flux is beginning to focus. All systems can be "broken," but the ability to cause a system to crash tells nothing about its performance or reliability under normal use conditions. A system ought to be trustworthy under real stress conditions -- under boundary cases.

Consider the extreme data conditions in which a system is used. Data points may range in value from zero to one thousand, from zero to five decimal places, or from three to nine characters in length. Each of these descriptors defines a range or (graphically) an area. If we were to test that system with cases that all fell in the midranges of those conditions -- values in the hundreds, with two or three decimal places; or with five characters -- we would be confident of normal operation. But to determine the performance under stress, we need to select data cases that define the boundaries of those ranges. Cases should include a zero value with no decimals, a character string with a length of nine, and so forth. These boundary cases, when integrated with other data points, will provide a test frame indicating a system's true performance and actual reliability.

This emergent definition of stress cases will clarify a murky area of validation testing and provide strong guidelines for testers. Expect to see a great deal written and said about boundary case testing. It will be the dominant test topic of the next three years.

Trace Matrix

A Trace (or Traceability) Matrix is a tool that graphically tracks a requirement through the design elements that meet that requirement to the test script that confirms the functionality of that requirement. For years reviewers, both inside the regulatory agencies and within independent validation organizations, have constructed Trace Matrices as a method of evaluating test scripts and the SDLC. The device allows the reviewer to move from requirement through design to script -- or back -- to assure that testing is complete, that the design incorporates all requirements, that all requirements are tested, and that all tests conform to design specifications.

The inclusion of a Trace Matrix will emerge over the test two years as a *de facto* requirement of a validation study. Increasingly, regulators are asking to see a Trace Matrix, and the responsibility has fallen on the validator rather than the auditor. Over the next year, the first adverse findings to cite failure to construct a Trace Matrix will be written; already, medical device reviewers have been requesting and requiring their inclusion in 510(k) documentation.

These three trends -- regulatory insistence upon the use and documentation of a System Development Lifecycle, the emergence of a clear definition of stress testing using boundary cases, and the requirement of a Trace Matrix as a tool for system analysis -- will continue to emerge as the international regulatory "hot buttons" for validation over the next three years.

Cybernetic Validation Monitoring

For the past ten years, corporations in the pharmaceutical, medical device, and biologics industries worldwide have responded with widely varying levels of enthusiasm to regulatory pressures to prove control over the computer systems upon which they rely. This evidence of control, termed "validation," has been gathered at great expense through the manual review of code, testing of performance, and the development of extensive documentation.

Where diligence has been applied to validation, a significant level of improvement has resulted. Software used in labeling and decision making is much improved. Recall and tracking systems are now proven prior to need. Standard Operating Procedures have helped to bring the rigor of the laboratory to the control of the computer systems that control those labs. Clinical systems have tighter data control. Manufacturing systems can be subject to the same validation requirements as were the manual devices they replaced.

Some companies are just beginning to institute system validation protocols, and they have yet to see any control improvements. Other companies, now committed to validation, have a greater confidence -- perhaps proportional to the investment required. A small minority of companies are now fully compliant with validation guidelines and are seeing the tangible benefits of superior systems under superior organizational and operational control.

But despite this improvement, the regulatory agencies are not satisfied -- nor should they be. For while progress has been made from the "unacceptable" to the "passable" level through carefully controlled validation and system quality assurance standards in at least the most progressive companies, computer performance and control still remains the weakest regulated link in most organizations. Reliance upon complex systems continues to expand even as sophistication in other aspects of research, laboratory operations, manufacturing, distribution, and clinical testing continues to grow. Current validation practices however, cannot advance the state of the art beyond the current "minimally acceptable" level. In an "A+" organization, the best validation still only rates a "C".

That "C" is a significant improvement over pre-validation controls. In many areas of health care it may be tempting to accept that "C" level, particularly in light of growing economic pressures. But accepting a validation "C" has two important strategic flaws.

First, today's "C" level is tomorrow's "F" for failure, as computers and computer applications grow in sophistication. Computers are increasingly being used not just to store and analyze data, but to make and implement decisions based upon that analysis. Blood systems no longer merely store test data; they actively label the blood product as safe or unsafe based upon a consideration of those test results. Newborn incubators do not merely report sensed oxygen levels; they adjust the concentration based upon analyses of infant respiration, blood chemistry, and other factors. Clinical systems no longer merely record study results; they generate the test protocols, assign subjects to the double blind groups, and develop delivery strategies. Electronic signatures, Computer Assisted New Drug Applications, automated adverse reaction reporting systems, and other technical and regulatory advances speak to an elevating curve of reliance on systems. In all areas, the level of performance and the sophistication of system uses has increased.

A second flaw is equally unsettling. Modern computer systems are all but "living" in their continuous growth of code and function. The previous generation of systems expanded as developers wrote code in response to user demands. Adding testing and validation requirements to that development process was difficult, but progress to date in system control has demonstrated that it is possible. But the next generation of systems, now in increasing use, literally write their own code in response to the most casual user request. These (still primitive) "artificially intelligent" systems seem to preclude any practical attempts at human code review or meaningful independent testing.

At the same time that the reliance upon computer systems is growing in complexity and criticality, the ability of systems to be utilized remote from the scope of existent controls has increased.

Today's "classical" validation methods have a flawed limitation that prohibits ever keeping abreast of these advances. Even as best practiced, a validation study can only certify that a given system was performing as intended at a given snapshot in time. As that system evolves over time, the validation evidence of control -- and that control itself -- is weakened in proportion to the temporal distance.

The reality of this flaw does not negate the importance of a classical validation study, any more than the need for a college education suggests the inappropriateness of a high school education. The documentation, developmental testing, and human controls demanded by and demonstrated in a validation study are still of immense value and importance, and they will not be replaced by more advanced validation technologies. But the time and the tools have come to move beyond this original validation strategy and address the fundamental question of validation future.

Once a controlled (and validated) system is in place, how is it possible to demonstrate that control on a continuous basis?

Repeated validation studies are not practical both from a financial viewpoint and from a time-requirement frame. Even if validation studies (or mini reviews) were repeated on a cost-prohibitive three-month basis (probably the shortest possible interval, given the time required for each validation study), what assurances would exist for the application runs, product labels, analyses, sortings, and other myriad of functions performed in that ninety-day period?

We presume a consistency in human beings. If annual audits show a reasonable adherence to performance requirements, it is reasonable to assume that that level is indicative of the intervening periods. But on what basis can we assume the same for a computer system? The strength of a system is that it follows its directions exactly as written every time those directions are exercised. The inherent weakness, however, is that each following of those directions is unique, without pattern memory or learned habit. If the direction is flawed, perhaps in one out of twenty sets of directions, or in one particular and unusual combination of directions, circumstances, and electronic environments, those flawed instructions will be followed just as perfectly as the appropriate pathways.

Can the creation and review of audit trails correct this problem? An audit trail can help with two important problems. Fraud, particularly of the sort characterized by data changes, may be discovered through the examination of an audit file -- assuming the perpetrator is clever enough to change the data, but not so clever as to modify the trail. And a variation from a result expected by the system may be detected through a review of the audit trail. While these kinds of problems are of sufficient importance to justify an inclusion and examination of the audit file, they will not find the flaw unanticipated by the designer of the system. The audit trail will, in effect, confirm that the system is working, but not find the most significant disease within it. The schizophrenic patient does not notice problems within him/herself.

This audit trail limitation also precludes the use of conventional self diagnostics in computer operating systems, BIOS, and application software. This conventional diagnostic can only confirm that the system is working consistent to its own rules; it can not tell whether those rules are inappropriate, or if the system is functioning erroneously in an area outside of the rule-making arena.

Yet if even impossibly frequent periodic reviews are insufficient, and conventional internal audit trails and self diagnostics are blind to fundamental problems, how can this all-important next level of validation confidence be achieved? The FDA strategy calls for expanding its view beyond the users of these complex systems, to encompass the systems themselves. Dr. David Kessler, FDA Commissioner, has suggested that computer systems are themselves medical devices, subject to 510(k) registration. In line with that decision, the Center for Biologics Evaluation and Research has demanded system validation and registration information for every new biologic product. Regulating the software, and the software vendors, however, will still not solve the continuity of analysis problem. Vendors may provide assurances that their systems are compliant as they leave the factory door, but users must rely on some other mechanism to monitor functionality when in applicational use. If user review is progressive but insufficient to achieve the level of controls and assurance that increasingly sophistical system utilization entails, and if vendor responsibility is fundamental but not sufficient to maintain an on going assurance of compliance, some new strategy is required.

That strategy, building upon the foundation of the still important classical validation study, is the inclusion within all regulated systems of a continuous data review: a cybernetic validation module.

The classic model of a cybernetic system is a thermostat. The electronic or mechanical device is set to a particular temperature, used as an valuative criteria in opening or closing a valve. In cold weather, the thermostat opens the valve to increase furnace activity until the set temperature is reached. At that homeostatic level, the valve is closed and equilibrium is maintained. The cybernetic system makes constant checks (or makes periodic checks scheduled in sufficient proximity as to be functionally constant) on the ambient temperature, adjusts itself in response to the need for heat, and hence establishes a consistent equilibrium.

A cybernetic system, then, has three relevant characteristics. First, it is dynamic. A continuous flow of data input, evaluation, and resulting output allows on-going monitoring. Second, the cybernetic system is recursive; it acts upon itself. Frequency of monitoring may vary with encountered result, type of monitoring is adjusted to appropriate decision level, and decision criteria may be flexibly determined based upon situational variables. Finally, the cybernetic system is homeostatically sensitive. That is, the system is established in preference toward stability and equilibrium, and acts with warning or adjustment when sensing a challenge to that homeostasis.

The jump from a thermostat to a validation module crosses an immense chasm, with a downside much too serious to rely merely on a leap of faith. The theory of cybernetic systems is based upon establishment of stability. But data monitoring is predicated upon the search for system-induced error in a data stream that is definitionally unstable, varying in response to appropriate external stimuli. Can a cybernetic system be used to monitor a data stream that is varying in response to external realities? In our thermostat model, temperatures vary only randomly or in direct response to intentional adjustment. But if a cybernetic model is to be useful in monitoring the computer systems used in gathering and analyzing laboratory, clinical, or manufacturing data, it must be able to distinguish between data variations caused by drug interactions, clinical changes, and other dependent (and independent) variables, from erroneous variations -- system errors -- resulting from software flaws, algorithm errors, database corruptions, or other causes. That is, the effective data monitoring system must distinguish between variations caused by measured differences and those caused by system error.

The move from thermostatic control to warning monitor is not complex: one can readily replace the adjustment of temperature with the sending of an alarm indicating a falling temperature. In the same way, a cybernetic system can be envisioned that senses even minute variations in the data stream, and provides a warning about those variations. But how can random (noise) variations caused by system error be separated from meaningful, measured changes in data points or trends?

It is "noise" that is the subject of all validations. In information theory, all electronic signals can be divided into two categories: the intended, purposeful, and useful is termed "information," while signals that are random, erroneous, or non-useful are labeled "noise." In a precise sense, validation is the process of assuring that noise is not misinterpreted as information. To be useful, a cybernetic validation system must be able to draw a clear "noise"/"information" distinction, and use that distinction to provide evidence that a particular system is (or is not) misinterpreting the two.

Accomplishing that meticulous distinction and preventing any significant misunderstanding requires a marriage of the dynamic, recursive, homeostatic cybernetic theory with the statistical concept of cross-validation. Cross-validation is a category of statistical methodologies designed to provide operational control of a data stream by identifying variations from complex patterns that violate the functional assumptions of that data. The cross-validation violations provide an early warning system that can sensitively identify potential hardware, software, algorithmic, or other problems.

Note that the cybernetic cross-validation model will provide conservative Type II warnings. That is, a minority of those generated warning statements will prove to be false signals, resulting from unexpected, but non-erroneous, data variations. Consider, for example, a crude monitoring system added to a crude blood-typing device. Such a simplistic cybernetic monitor might be programmed with expected percentages of Type O, A, B, and AB blood in a so-called "normal" human population, providing a warning if those distributions vary significantly within a donor set. Such a warning might indicate a problem with reagent, device, or process, but the variation might also be a result of an unusual ethnic or blood type population or a coincidental anomaly of distribution. In more sophisticated cross-validation systems, these kinds of "false positive" warnings will be reduced significantly, but ultimately all a cross-validation warning truly indicates is a significant variation from expected norms; those variations may be caused by factors extraneous to the computer system.

On the other hand, the likelihood of not detecting a variation caused by system problems can be controlled to a level of non-significance. This control is accomplished through the use of a series of non-parametric analyses, avoiding the potentially flawed parametric assumptions that would bury real data stream problems. The clinical, laboratory, or manufacturing data stream under review is likely to be parametric, but the continuous search for problems indicating variations is not a parametric process. Non-parametric cross-validation technology can, however, provide a cybernetic methodology for the continuous monitoring of regulated data systems.

Using these cross-validation cybernetic techniques, it is possible to construct an algorithmic function that can be utilized to continuously auto-validate any system. The algorithm can be integrated as a module in a data system or controller system, or it can be used in a free-standing external monitoring system plugged into a device. Once connected to the system, the algorithm can "sense" the most commonly encountered noise errors in the data stream -- changes in signal strength, changes in measurement accuracy, changes in population medians, changes in sample population, changes in auto-correlationality, and changes in variability. In the process, the algorithm continuously "learns," incorporating encountered non-critical variances into its interpretation of acceptable variability. If encountered data points fall outside the range of acceptability, as defined by regulatory guidelines, the system can be programmed to sound or flash a warning signal, and to print a diagnostic message indicating the likely source of the discrepancy. When no serious discrepancies are found, the system can be programmed to print a periodic log of validation evidence demonstrating continuous control of the data stream.

When this evidence log is teamed with the result of classical validation analysis, including documentation of installation and performance testing, development of standard operating procedures, and appropriate user training, an extremely strong validation evidentiary package is created. That evidentiary package is sufficiently powerful to withstand the most rigorous demands of regulatory concern, even with an increase in the complexity of systems and of the applications to which those systems are applied.

The cybernetic cross-validation algorithm developed for the control of complex computer systems utilizes minute changes in the data stream as a continuous probe, testing for sources of error in the primary applicational software, files, and hardware. This testing focuses on four fundamental continuous and simultaneous checks that serve as flag factors of error: 1) problems of auto-correlationality, 2) problems of hetero-scedasticity, 3) problems of multi-collinearity, and 4) problems of symmetry. In weighted composite, these four checks provide evidence of the validation of the system by determining whether any errors have been encountered in the software, functional hardware (including transmission and network), or files. Individually, they provide diagnostic clues identifying any of twelve (12) potential sources of error within the system.

These four problem checks are derived from the theoretical process of cross-validation, which utilizes a series of non-parametric statistical tools to determine whether fluctuations in the data stream fall either within the norm of random variations, or within the norm of intentional (independent or dependent) variations. In composite, these four checks provide a continuous monitoring of virtually every issue of control not included in a classical validation study (documentation, development, installation testing, etc.).

By applying the methodologies of cross-validation to the data stream of a regulated data processing, sensing, or controlling system, a classical validation study of that system is significantly augmented with the addition of continuous variant testing. Cross-validation will not detect problems of insufficient documentation, inappropriate decision rule, or inadequate disaster recovery planning, as a well-focused classical study will. But unlike the classical study, the cross-validation process will note all but immediately variations in the accuracy of measurement, in the degree and direction of variability, in the precision, and in the pattern that are indicative of potentially serious system errors.

A close monitoring of the cross-validation process will provide system managers with either a periodic "pass" log or a diagnostic code recommending specific categorized investigation, as is appropriate. The "pass" log can serve as powerful and credible testimony to assure regulators of system control, while diagnostic recommendations can serve to allow rapid identification and possible correction of any detected anomaly.

Since the cross-validation data set is "rolling," analyses are repeated on a continuous basis (after the entry of a base set of twenty data points) for on-going monitoring. Hence, the cross-validation process is statistical in nature, it is not based upon a sampling technique. All data points in each data stream are analyzed for a statistically significant effect upon the data pattern.

This continuous, rolling concept of data analysis defies the standard statistical assumptions associated with traditional parametric analysis, necessitating the use of the non-parametric methods described above. But by taking the rolling approach described above, the chance that an important data anomaly will be inadvertently missed due to the chance of sampling is eliminated. The cross-validation methodology described herein is truly continuous, validating in effect every data entry, calculation, decision point, and processing step.

The cybernetic validation methodology based upon cross-validation is contained in an integrated software module within a software system. The module is written in the language of the software system, using the original variable names and structures to avoid complicated twists of transition and reference. The module, however, is constructed as a self-contained software unit, carefully designed and classically validated.

The module "intervenes" at all and each of five action points defined specifically for the system under review. These action points are: 1) transmission to storage; 2) retrieval from storage; 3) transmission to output; 4)receipt from input; and 5) post manipulation. A system that involves all five of these actions, cross-validating according to the four formulae, handling one thousand data points per second (arbitrarily selected), would be conducting a total of 1,200,000 validation analysis per minute.

The module itself is divided into three sections. Part A receives and organizes the data stream or streams, creating the continuous "rolling" data set. Note that the actual system is in effect duplicating the data points, sending identical copies to the module and to the original system function, so no delay is caused by module interaction.

Part B applies the four cross validation formulae to the rolling data set or sets, and uses the weighing formula to diagnose any encountered anomalies.

Part C sends to a designated file, printer, and alternate reporting devices a continuous result of the cross-validation. This result can be printed periodically (hourly, daily, or whatever interval is appropriate), or it can continually output an "accept" symbol or analog (for example, continuously lighting a green light) and can print diagnostic messages (presumably changing the light to "red") as an anomaly is detected.

The output of the module can be designed either as a log of normal activity with noted discrepancies, or as a warning system printing a message only as errors are detected.

Finally, the module can be designed as a free-standing device that intervenes between system and system output (presumably by plugging into the RS232 interface) for short term monitoring of the system. This approach is considered most appropriate as a short term diagnostic tool, presumably used by regulators or internal quality assurance monitoring staff.

The module thus has three basic applications. Ideally, it can be built into automated systems by the in-house or outside developers or vendors of those systems. For example, a LIMS vendor or developer of a blood analyzer may build a cybernetic cross-validation module into the software entity. Alternately, a user organization may build the module into or on top of its installed systems, providing a continuous test of those systems. Finally, a free-standing module unit could be used by regulators or internal reviewers as a short term check on a specific system or systems. In each of these applications, the module will provide the cross-validation supplement to traditional validation checks of documentation, support, development, and operations.

Summary

In the future, complete validation of a complex computer system requires a marriage between the techniques of classical validation and the methodological monitoring of the cybernetic validation module. The classical review collects and analyzes evidence of appropriate documentation and Standard Operating Procedures, of development approaches and testing, and of problem response and planning. The Trace Matrix, Boarder Case Testing, and a System Development Lifecycle approach will be the major regulatory emphases in classic validation over the next few years. The cybernetic validation module provides a constant non-parametric testing of the data stream or streams, comparing all new data with emergent recursive patterns in search of anomalies that may indicate sources of system error. Together, the two approaches will provide on-going evidence of system control.

And that control represents the future of validation.

Appendix A

GLOSSARY

A to D (See Analog-to-Digital Converter)

ABNORMAL TERMINATION or ABEND Any unexpected interruption of processing. A key factor in the validation process is the ability to determine the cause of such events. The system should not simply freeze and require a restart. If a problem develops somewhere in the system, there should be a flag indicator and a way for the user to bypass or correct the problem, including tracing the fault. If the system simply freezes, then data in the process loop may be corrupted. Input not fully defined should not cause an abend. An abend is considered to be an Adverse Event (AE). Proper response to an abend includes diagnosis of the cause (with appropriate response) as well as correction of the problem and retesting prior to on-line production use.

ACCELERATED VALIDATION PROGRAM A program of Hewlett-Packard Co. in cooperation with Weinberg, Spelton & Sax, Inc., in which the vendor of HP LIMS makes available to purchasers a validation package and a Weinberg, Spelton & Sax, Inc. certificate. These tools provide an advanced starting point for users in developing the systems validation of their integrated laboratory. Users are then responsible for on-site validation.

ACCEPTANCE TEST Formal test conducted to determine whether a system satisfies its acceptance criteria and to enable the customer to determine whether to accept the system. (ANSI/IEEE Std. 729-1983)

ACCEPTANCE VALIDATION Testing, analysis, and review of a system incorporated into the formal initial approval process for live use of that system to prove control of that system. Acceptance Validation is the "present tense," compared to Developmental Validation (applying to future systems), and Retrospective Validation (of existing or past systems). A Validation Standard Operating Procedure (SOP) should require an Acceptance Validation of all new systems prior to or as part of the user acceptance process. Validation is a process of testing the testers and of making sure that the criterion set by the end users is appropriate. Given the current state of regulation, there is no easily defensible rationale for accepting a new or replacement system without including an Acceptance Validation in that process.

ACCESS, AUTHORIZED The parameters of system resource utilization given to end users. In many systems, various levels of access are used to restrict certain users.

These restrictions, and procedures for documenting the access granted, should be described in detail in an appropriate SOP.

ACCESS CONTROL Setting parameters for end user utilization of a system. Automated teller machines which require both an artifact (card) and a personal code are examples of clear and effective access control devices. For some systems, access is restricted physically by housing a limited system in a locked or controlled room. The key determining factor in access control is accountability; the unique identities of all users should be clearly determinable.

ACCESS PROTECTION The devices and/or procedures designed to control all levels of utilization of system resources. The most common access protection is a differential level password control system, often appropriately coupled with limitations on physical access and restrictions on remote (modem) access.

ACCESS, UNAUTHORIZED/UNINTENTIONAL Inappropriate use of a system or a portion of system resources. Depending upon the critical or hazardous nature of the system, that unauthorized access (hacking) may be of life-threatening significance. In a validated system, the users at each level should not be able to review their own work. For example, a system manager should not be able to overwrite or alter the numerical output of a process without appropriate authorization. Limited access to system locations, restrictions on modems, and general environmental security precautions are appropriate. Physical access to a system should be restricted to authorized individuals who are trained in proper operation. Security precautions should be designed that are appropriate to the vulnerability of the system to prevent unintentional system access. Unauthorized persons should be physically restricted from system access.

ACCURACY In order to be validated, the system as configured must provide appropriate and unerring processing of entered data and activated programs when operated within established standard procedures. Operating instructions should be prepared to reflect current and appropriate action sequences to assure safe and effective use of the system and its components.

ADC (See ANALOG-TO-DIGITAL CONVERTER)

ADDRESS A character or group of characters that identifies a register, a particular part of storage, or some other data source or destination.

ADVERSE DRUG REACTION Any undesirable effect, or interaction with human physiology, that is caused by a drug. This includes abuse, dependency, and interaction with drugs and/or foods.

ADVERSE EVENT (AE) 1) In the validation context, any unexpected interruption of processing anywhere in the system may be considered to be an AE. Systems are particularly vulnerable to environmental factors including power surges, temperature, humidity, dust, and radiation levels which fail to meet hardware manufacturers' requirements. To control AEs, a log should be maintained which indicates the date and time of the occurrence of each Adverse Event, the diagnosis, and the actions taken. Evidence that such a log is regularly reviewed and analyzed to identify trends or hidden problems is important to the validation process. 2) In the context of FDA guidelines, an adverse experience/adverse event is any undesirable experience occurring to a subject, whether or not it is considered to be related to the drug(s) being tested.

AFTERIMAGE Magnetic media may retain a ghost image, or afterimage, even after they have been erased. There are two solutions to this potential source of contaminated data in a high-hazard environment: either constantly use new media or constantly fill all fields with some null code. For example, consider a computer-controlled medical device which releases a chemical when three conditions are met and/or the chemical passes three tests. In this environment, start with x-filled fields rather than blanks or empty ones to prevent the computer from reading previous results.

AI (See ARTIFICIAL INTELLIGENCE)

ALGORITHM 1) A finite set of well-defined rules for the solution of a problem in a finite number of steps; for example, a complete specification of a sequence of arithmetic operations for evaluating sin x to a given precision. 2) An algorithm may be an equation, a recipe, an instruction manual, a part of a computer program, or an entire computer program. An algorithm can be written in a natural language like English, though relatively imprecisely. The source code review includes an analysis of the algorithms utilized in a program to assure conformity to the user's intended approach or method. For the purpose of validation, formulae and decision algorithms employed by the system must be accurate and appropriate. Users cannot assume that the test or decision criteria are correct; those formulae must be inspected and verified.

ALU (See ARITHMETIC LOGIC UNIT)

AMERICAN STANDARD CODE FOR INFORMATION INTERCHANGE (ASCII) A byte-oriented coding system based upon the use of seven-bit codes and used as a standard coding format for data communications.

ANALOG 1) Representation by means of continuously variable physical quantities; for example: a dial speedometer or tachometer. 2) Continuous signal having a voltage which corresponds to the monitored value.

ANALOG-TO-DIGITAL CONVERTER (ADC or A to D) Digital computers perform operations in binary (or equivalent) code. Information that is to be input into a digital system must be converted to binary form to be processed by the digital circuits. Outputs of a digital system are in some type of a binary code and often must be converted to a different form.

Many devices are used on either side of a digital computer in order to communicate with people and/or other computer systems. Input or output devices like keyboards, temperature probes, flow-meters, or pressure sensors provide the means for a digital computer to monitor and control a physical process. On the input side, measurements of process parameters that are analog in nature are usually transduced (changed to a proportional electronic voltage or current) and sent to an analog-to-digital converter, ADC, which then changes the analog quantity into a corresponding digital code. Some process parameters are already digital in nature, but must be changed to the appropriate form for a particular computer system.

Analog-to-digital converters (ADCs) and digital-to-analog converters (DACs) function as devices for communicating between a completely digital system or device, such as a computer, and people. This function is becoming increasingly more important as microcomputers move into areas of process control where computer control was previously not feasible. The conversion of a signal from A to D (or back) creates an area of high-error vulnerability. Careful testing of the converter and the conversion is recommended in all high-hazard environments. (See also Digital-to-Analog Converter.)

ANALYSIS PHASE 1) The stage in the validation process during which existing operations are observed and evaluated. Particular attention is paid to the identification of any differences between Standard Operating Procedures (SOPs), what actually happens, and what the system is supposed to do. 2) In the System Development Life Cycle, the process of determining the expected or intended performance parameters of the system.

APPLICATION (or application program) Software that does relatively complex tasks and that lets the user create and modify data. Some common types of applications are word processors, databases, graphics programs, spreadsheets, and statistical analysis packages. In high-hazard environments, critical applications should be validated.

APPLICATION INDUSTRY Regulated environment or environments in which the system under review will be utilized.

APPLICATION PLATFORM Refers to hardware necessary to support a system. An Application Platform may incorporate firmware. Platforms are generally designated by the name of a specific mainframe or minicomputer manufacturer.

APPLICATION SOFTWARE Programs specifically produced for the functional use of a computer system; for example: software for inventory, data analysis, temperature monitoring, general ledger, etc.

ARCHITECTURE Program architecture design involves identifying the software components, decoupling and decomposing them into processing modules and conceptual data structures, and specifying the interconnection among components. An architectural design specification may include data flow diagrams for the software; conceptual layout of data structures and data bases; name and functional description of each module; interface specifications for constraints; exception conditions; and names, dimensional units, and other attributes of data objects.

ARCHIVE Central repository for all information and records concerning a validated system. The archive should also include testing documentation, change control documentation, and validation records for all phases of SDLC. The archive is dynamic, since it contains current and previous copies of system design specifications and SOPs. The archive is both a record and a template. Regulatory requirements for the retention of an archive vary somewhat: in some circumstances, it may be necessary to retain data seven, ten, or even more years. Archives may be retained in any of a number of media, including paper, microfiche, and magnetic media.

ARCHIVE COPY A historical record of system components maintained for proof of regulatory compliance. Such a record may include (in addition to current version master copy) previous versions of software data-files, SOPs, etc., but does not need to include actual hardware. The validation concern is that the regulated site must be able to access information in a timely fashion. To facilitate such access in light of changes in hardware and software, a reliable retrieval method should be used; for example: historical systems may be saved on microfilm or dumped into ASCII files.

ARCHIVIST The original Good Laboratory Practice (GLP) standards and adapted Good Automated Laboratory Practice (GALP) standards require the archiving of a variety of records for periods of time ranging up to ten years. The archivist is responsible for the safe storage and retrieval of these records. [In the GALP Standards section

5.0, the code letter "A" is used to identify all specific and direct responsibilities of the Archivist.]

ARITHMETIC LOGIC UNIT (ALU) A computational subsystem of the central processing unit (CPU) that performs the mathematical and logical operation of a digital system.

ARTIFICIAL INTELLIGENCE (AI) Refers to programs which emulate human thought processes. Such programs have the potential to make decisions and take initiatives. AI is built into control programs and decision matrices. Relatively primitive AI programs are currently used in a few regulated areas, including molecular modeling in both pharmaceutical and biotech R&D efforts. (See Fifth-Generation Languages.) AI components of systems in high-hazard environments should be validated.

ASCII (See AMERICAN STANDARD CODE FOR INFORMATION INTERCHANGE)

ASSEMBLE To translate a program expressed in an assembly language into a machine language, and perhaps to link subroutines. Assembling is usually accomplished by substituting machine language operation codes for assembly language operation codes and by substituting absolute addresses, immediate addresses, relocatable addresses, or virtual addresses for symbolic addresses. (ANSI/IEEE)

ASSEMBLY LANGUAGE A computer-oriented language whose instructions are written in a one-to-one correspondence with computer instructions and that may provide facilities such as the use of macro instructions.

ASSISTANCE REQUESTS The developer, vendor, or supporter of the system has developed a mechanism for responding to user requests for additional training or assistance.

ASYNCHRONOUS COMMUNICATION The term used to describe the irregular, serial transmission of data in bursts or segments. During asynchronous transmission, at the point that the burst is sent, there is vulnerability due to the possibility that the transmission could be terminated or lost before it is confirmed by the receiver. While it is true that the error detection routine used determines the data integrity, and hence the reliability, of the transmission, it is also true that synchronous transmissions, using a bisync protocol, are theoretically more reliable. Since checking can be both continuous and simultaneous, they are preferable in high-hazard environments. (See Bisync.)

ATTEMPTED ACCESS Many systems will disconnect after three improper access attempts; some will log the improper attempts for later review. These features may be particularly valuable in systems of high criticality. A validated system should have a means of logging and documenting any unusual and/or unsuccessful resource utilization for regular review by the systems administrator.

AUDIT A review of the system validation plan and study results, conducted by a quality assurance group or an outside, independent agency or a regulator. The audit confirms conformity to validation standards and the appropriateness of those standards.

AUDITABILITY The successful inclusion within a system of automated or enforced manual mechanisms to allow tracking of the utilization and performance of that system. Common mechanisms include error trapping routines, console logs, audit trails, activity files, and similar records of functionality. A system may require manual auditing, or may create its own electronic audit trail. (See Audit Trail.)

Validation is a tool for identifying the existence of automatic or manual audit capabilities or the need to devise audit trails.

AUDIT TEST Independent check on the output and/or processing of a system, usually based on comparing controlled data input, with expected results, to the actual output. The audit test is a subset of functionality testing. (See Function Test.)

AUDIT TRAIL Sequential tracking of changes to data records or files that allows an oversight function to determine what the changes were, when they were made, and who made them. An audit trail is a key factor used to determine data integrity. The trail may also indicate unauthorized access or unauthorized modification of data or file contents.

AUDIT TRAIL REVIEW The automatic log of the audit trail is reviewed regularly by management personnel and annually by validators or regulators.

AUDIT TRAIL TEST The audit trail generated by the system has been tested in a "black box" (expected outcomes) test and conforms to the following guidelines:

a) Testing is conducted in a special test library or other environment independent of the actual system's real database; in other words, "off-line."

b) A minimum of 20 test changes have been included in the testing methodology.

c) A perfect conformity of the audit trail to the 20 test changes has been observed.

d) The test is repeated annually, whenever a new version or update is installed, whenever a new hardware component is installed, and immediately upon discovery of any audit trail discrepancy.

AUTHORIZATION A system control feature that requires specific approval before processing and/or increased access can take place. The authorization feature may create an audit trail, or indicate the electronic code of the authorizing agent.

AUTOMATIC AUDIT TRAIL The system shall have a mechanism in place for generating an audit trail of system changes or modifications.

AUTOMATIC CODE GENERATOR The ACG may be used to create source code, particularly in inventory and material requirements planning (MRP) systems. It is not necessary to validate the ACG if the actual source code produced by the process is subject to thorough review. (See Fourth- and Fifth-Generation Languages).

AUXILIARY STORAGE A device for holding data or programs, other than main memory/RAM (e.g., disks and tapes). All auxiliary storage devices have a specific shelf life, after which time data stored on them should not be considered reliable (see archive). The manufacturer-specified parameters for storage life assume location in an appropriately controlled environment. In any case, stored data should be transferred to fresh media well within those parameters.

AVAILABILITY (OF DATA) For regulatory purposes, clinical raw data must be readily available to regulators for review of system(s). Data, including the tests used to test the system, must be available to regulators. Availability of Data should be defined within a site's SOPs. Although no consistent time parameters have been applied to the "availability issue," in most regulatory situations the time frame of 48 hours has emerged as the consensus minimum limit on "ready access" or "availability."

BACKGROUND The electronic and radiation energy environment. In certain parts of the world, naturally occurring radon and other radioactive gases may be present in sufficient concentrations to distort sensitive system-data storage-media. In addition, artificial radiation from nearby high-voltage equipment and power lines and certain other apparatus may have the potential to cause data distortion. In such cases, the vulnerable parts of a system may have to be hardened to the radiation in order for the system to be validated.

BACK-OUT The procedures used to remove software input changes, program versions, and/or input data (files) if serious problems are encountered during program execution and/or system processing. The validation concern is that data integrity be maintained. The system user can demonstrate that the process is still under control if he/she can provide documentation of a Back-Out process that got the system back to a point before the processing problems began.

BACK-UP 1) Alternative systems, procedures, methods, and data files that can be used to restore service in the event of system malfunction or the loss of data resources. Frequently, large operations with many sites rely on back-up sites in the event that the main system goes down. These back-up sites need to be tested and validated as the system changes. There should be back-ups for data, software, critical operating supplies, and hardware spares. Both the location of, and the media chosen for, storage must be appropriate and compatible. The process of backing up should be habitually or automatically cycled every 15 to 30 minutes to avoid the loss of data in the event of power modulation. Only data entered prior to the last back-up can be considered to be pure after an adverse event. 2) Full: a copy of the current version of all files on disk. 3) Incremental: a copy of the current version of all files on disk that have been changed since the previous backup.

BAR CODE Series of solid bars of different widths and intermittent spaces used to encode data for reading by light-emitting scanning devices. As long as bar codes are uniquely and positively affixed to the sample or data point, the use of bar coding can eliminate or mitigate concerns of confusion of identity. Bar codes are also utilized to protect confidentiality (particularly in blood processing systems) and to provide double entry for sensitive data.

BASIC Beginner's All-Purpose Symbolic Instruction Code: A high-level programming language.

BASIC INPUT OUTPUT SYSTEM (BIOS) Firmware that is not modifiable by systems people. Different brands of otherwise compatible personal computers ("clones") may utilize differing BIOS, causing rare and generally harmless compatibility problems. These problems are generally related to transfer speed and to "overflow" conditions of large data sets. In particularly sensitive and hazardous environments, BIOS differences may prove significant and problematic.

BATCH 1) A group of data records processed together as a unit. 2) A specified quantity of a drug or other material that is intended to have uniform characteristics and quality within specified parameters and is produced according to a single manufacturing order during a particular manufacturing cycle.

BATCH INTERFACE A link for the transfer of data between systems in large file segments. In batch transfer, internal control codes may be used to assure that the batch sent and the batch received are identical. (See checksum.)

BATCH JOB 1) A group of programs that are run at controlled intervals. These may be controlled by an operating system and be run based on the availability of system resources. Also refers to programs run at regular intervals, traditionally overnight. Example: Updating data files. 2) Execution of programs serially with no interactive processing. Batch Job processing allows for easier auditing, because each batch can be checked before processing as well as after. In the event of an interrupted batch job, great care must be exercised upon restart to avoid loss or unintended duplication of data.

BAUD A unit for expressing the speed of transmitting data over distances, usually in terms of bits per second. One baud is one bit per second.

BENCHMARK Standard acceptable measure of performance of hardware and software. Independent reviews published in reputable journals and magazines may provide reliable "benchmark" information about a system or software.

BENCHMARKING A process whereby a company compares its strategies, operations, processes, and procedures to those of companies generally regarded as best, in the same class, and uses the results of those comparisons to improve its own performance.

BETA TESTING PHASE In this phase, in the Software Development Life Cycle (SDLC), the Beta Test Plan is executed by the user of the system. The testing is performed on an actual operational system, but not using the system as a live system (online). The results of the testing and any problems encountered with the system or the test plan are reported and documented.

A review of the output from Beta Testing is part of this phase. Documentation of the results of the Beta Testing must be analyzed, and every system or test plan inadequacy must be addressed. Also, suggested system enhancements need to be reviewed and considered as possible additions to the system for inclusion in a later revision.

BETA TEST PLAN DEVELOPMENT PHASE This phase in the Software Develop ment Life Cycle (SDLC) defines instructions for the user to follow to verify the system is functioning as expected. The Beta Test Plan is written from approved functional specifications and the system test plan.

A review of the Beta Test Plan, to ensure that the plan defines all the necessary steps required to test all system functionalities as perceived by the user, is part of this phase. If applicable, the plan must address system configurations. The plan must be written specifically so that a user, not a programmer or tester, can execute the test plan.

BINARY 1) Refers to the base-two number system. Permissible digits are 0 and 1. 2) A characteristic or property involving a choice or condition in which there are two possibilities.

BIOS (See Basic Input Output System)

BISYNC (Binary Synchronous Communication) Bisync protocol is a product of IBM Corporation. It is character-oriented, meaning that each character has a specified boundary. Since synchronous transmission is a continuous process, whereas asynchronous data transmission is discontinuous, the use of a Bisync protocol is preferable in high-hazard environments. (See Asynchronous Communication.)

BLACK BOX TEST A test of a system which compares input data with known and predictable outcome, with actual output produced by the system. In accounting, for example, manually checking the balances of an electronic spreadsheet, without considering the methods which the spreadsheet program used to arrive at those balances, is an example of black box testing. In the crudest sense, a system that has passed the black box test can be said to be functional. The black box test, however, does not carry any implied or explicit assumption that the system will continue to work reliably under varying real-world conditions.

BLOCK DIAGRAM A diagram that shows the operation, interrelationships, and interdependencies of components in a system. Boxes, or blocks (hence the name), represent the components; connecting lines between the blocks represent interfaces. There are two types of block diagrams: a functional block diagram, which shows a system's subsystems and lower-level products and their interrelationships, interfaces, and interdependencies with other systems; and a reliability block diagram, which is similar to the functional block diagram except that it is modified to emphasize those aspects which influence reliability.

BMD A statistical analysis package of programs (Biomedical Data) originally used in the medical field. BMD is similar to SAS in purpose and function, though the package has been criticized as overly "user friendly." The fact that it is difficult for a user to tell when statistical assumptions and accepted methodological conventions have been violated suggests that, in high-hazard environments, its use should be validated.

BOOTSTRAP A short computer program that is permanently resident or easily loaded into a computer, whose execution brings another, larger program, such as an operating system or its loader into memory. (ANSI/IEEE)

BRANCH An instruction which causes program execution to jump to a new point in the program sequence rather than execute the next instruction. In the event of a program modification, tracking functionality back to the last branch will allow determination of whether or not a change has affected other parts of the program.

BRANCH ADDRESS Refers to a specific location in a decision tree within a computer program. If a problem is discovered in part of the program, tracking up to the branch address will allow isolation of that problem and the program parts that may have been affected.

BRANCH COVERAGE Area affected by branch. If a problem is discovered in a part of the program, tracking down through the branch coverage will allow identification of aspects of functionality that may have been affected.

BUFFER An area in memory or auxiliary storage used to temporarily hold data for further processing. Buffers used in high-hazard environments should be secure.

BUG 1) An accidental condition that causes a functional unit to fail to perform its required function. 2) A manifestation of an error in software (ANSI/IEEE). In the validation process, the proper response to a discovered bug is a three-step sequence of: a) identification, b) correction, and c) retest.

BUS Electrical pathway by which information flows to different devices.

BYTE A binary character string operated upon as a unit and usually shorter than a computer word.

CAD (See Computer-Assisted Design)

CALIBRATION The purpose of calibration is to make sure that a device is measuring reality accurately. If a device is calibrated, then its measurements agree with some accepted standard or benchmark. The issue in the context of the validation process is that there must be documentation of appropriate calibration of devices that are part of the system being validated. During the validation process, measurements may be taken from the devices themselves.

CANDA Computer-Assisted New Drug Application.

CAPLA Computer-Assisted Product Licensing Application.

CCRO Clinical Contract Research Organization.

CE MARK The European Community (EC) mark of conformity, indicating that a product is considered to meet the essential requirements of EC Council Directives. Once the mark is given, a product can be marketed throughout the European Union.

CEN European Committee for Standardization. An organization based in Brussels that has established industrial standards for the European Community.

CENELEC European Committee for ElectroTechnical Standardization. An organization based in Brussels that has established electrical standards for the European Community.

CERTIFICATION In the context of system validation, certification means documentation affirming that the software development process was followed, that quality assurance procedures were adhered to, and that testing demonstrates that the requirements were met and the software and system requirements were fulfilled.

CHANGE CONTROL Change control exists when all programming changes and modifications show evidence of consistent oversight, including documentation of the following: effective date of change, the responsible programmer, and retesting certification. Evidence of change control may include an audit trail of patches. The ability to assign responsibility for modification of data or software or any element of a system is also essential.

CHANGE CONTROL COORDINATOR A person who analyzes each change request to ensure that no conflict exists with other requests, coordinates with DP personnel to develop a scheduled date for completion of each change task, monitors all change reports to the change committee, maintains the change files, and assures that the change libraries are maintained. This employee helps support the activities of the project administrator, project manager or phase/project leader, and quality assurance personnel.

CHANGE CONTROL LOG Usage of this form will document all changes made to software. Refer to SOP of software change control.

CHANGE IMPACT STATEMENT Use this document to compile information relative to the impact of a proposed software change. Refer to SOP of software change control.

CHANGE REQUEST Formalized method of notification to system support organization for any modification of system functionality. In a validated system the process will be defined within change control SOPs.

CHECKLIST 1) A series of steps which are to be followed in sequence and crossed off as they are completed. 2) A special tool developed by Weinberg Seplton & Sax, Inc. to provide comprehensive documentation. This report is based upon information provided by the professional and other personnel of the site being validated. It includes observations and field and laboratory tests conducted by Weinberg, Spelton & Sax, Inc., in cooperation with the site managers and the vendor of the system used at the site (if applicable). The standards upon which this report is based have been developed in cooperation with the Institute of Biomedical Science and Engineering of Drexel University; they conform to validation standards of the Institute of Pharmaceutical Sciences of the Center for Professional Advancement and industry standards of system validation as developed by the American Institute of Industrial Engineers and the International Institute of Industrial Engineers. All information, observations, tests, and conclusions are effective as of the date of the final report signature.

CHECKSUM A method for detecting errors in a series of bytes. The checksum is a one-byte number that can be computed fairly simply and appended to the series much as a parity bit is appended to a single byte. Errors are detected by computing a new checksum and comparing it to the checksum that came with the data. Checksums can detect some errors but there are many error combinations that leave the checksum unaltered. It is safest to assume that the checksum is useful for detecting single byte errors and not useful for error correction.

CIOMS (See Council for International Organizations of Medical Science)

CIS Clinical Information System.

CLARKE'S LAW Any sufficiently advanced technology is indistinguishable from magic.

"CLEAN COMPILE" A "clean compile" is a computer-generated output free of error messages or syntax errors. All final printouts should be "clean compile."

CLINICAL TRIAL Clinical trials in the context of FDA guidelines mean systematic studies in humans in order to discover or verify the effects of and/or adverse reactions to drugs (pharmacodynamics) and studies of the absorption, distribution, metabolism and/or excretion of drugs (pharmacokinetics).

CODING AND MODULE TEST PLAN DEVELOPMENT PHASE In this phase in the Software Development Life Cycle (SDLC), the source code is developed and the program/module test plan is defined. The test plan contains detailed instructions for testing the internal logic of each program/module. The test plan must ensure that each statement in the program/ module will be executed at least once, and that eachdecision in the program/module will achieve each possible outcome at least once (although not necessarily in every combination).

This phase includes a review of the source code and program/module test plan to verify their completeness and consistency with the applicable standard operating procedure for coding and testing software.

CODING PHASE The step in system design when computer-language instructions are written to meet the specifications required.

COMMUNICATION INTERFACE The methodology, software, and equipment chosen to allow users to communicate with a system and vice versa.

COMPILE To translate a higher-order language program into its relocatable or abso lute machine code equivalent.

COMPILER A computer program that produces machine language from a source program that is usually written in a higher-level language by a computer user. The compiler is capable of replacing single source-program statements with a series of machine language instructions or with a subroutine.

COMPREHENSIVE SYSTEM PERFORMANCE TEST As an initial validation, Weinberg, Spelton & Sax, Inc. accelerates and enhances the ongoing system performance test by following a statistically valid sample of cases (up to 200) from entry into a processing center through labeling and shipping, checking for system accuracy, reliability, and consistency. This procedure has been used effectively in blood processing environments.

COMPUTER-ASSISTED DESIGN (CAD) When CAD programs are used to design or test designs of medical devices, the Computer-Assisted Design software should be tested and validated.

COMPUTER-CONTROLLED A computer-controlled device is one whose operation is influenced or directly controlled by a program-controlled mechanism that retrieves, decodes, and executes instruction to perform its designed tasks, or whose operation is controlled by the program itself.

CONDITION TEST Data processing term signifying checking for a yes/no, armed/disarmed condition.

CONFIDENTIALITY An agreement for the protection of proprietary and/or person al information.

CONFIGURATION CONTROL The process of evaluating, coordinating, approv ing, disapproving, and implementing changes to identified configuration items. (See configuration item list.)

CONFIGURATION ITEM LIST A list that identifies all the collections of hardware and software elements that are treated as units (configuration items) for the purpose of configuration control of a specific product or system.

CONGRUOUS ENVIRONMENTS The standards of validation apply to the specific user environment, independent of the system supplier, manufacturer, or vendor. These standards apply to each individual application site within a given user organization, regardless of whether or not multiple setting applications are operating under a uniform FDA license or corporate structure. Multiple systems operating in the same laboratory or congruous environment may be validated simultaneously, but a given inspection or investigation is expected to apply only to those systems specifically examined on the occasion of that inspection.

CONNECTIVITY Refers to the ability of various components in a system to communicate with each other. It may also refer to the ability of systems to talk with each other. The validation concern is the maintenance of accuracy, reliability, and integrity through each point of connection.

CONSERVATIVE DEFAULT A regulatory concept which says one should always err on the safe side. Always minimize the chance of a more serious problem. The FDA would rather not release a medication that will have side effects, even if it might cure someone. Consider a substance poisonous until you prove that it is not, rather than safe until someone gets sick from the substance. The concept of conservative default may be summarized by the phrase: when in doubt, delay.

CONTENT ERROR Refers to wrong data; for example a wrong phone number. The common kinds of content errors include: truncation, concatenation (incorrect combination), transposition (order change), and substitution. The way to control most content errors is to list the data fields separately and cross reference them.

CONTINUED ACCURACY AND RELIABILITY Refers to an essential characteristic of a validated system. If a system has continued accuracy and reliability, then the system as configured will continue to provide appropriate and unerring processing of entered data and activated programs when operated within established standard procedures and will monitor and provide warning of deviation from appropriate performance.

CONTRACT RESEARCH ORGANIZATION (CRO) Regulators within certain countries may allow specific tests and/or regulatory requirements to be carried out and/or verified by an organization other than the manufacturer who is seeking regulatory approval. A sponsor may transfer responsibility for any or all of its obligations to a contract research organization. Any such transfer must be in writing and must describe each of the obligations assumed by the CRO.

CONTROL Refers to a characteristic of systems, processes, and organizations that is a minimal requirement of regulatory agencies. Control requires the ability to ensure reliability, to calibrate appropriately, and to recognize abnormal events within the contexts of regulated systems, processes and/or organizations. The utilization of a system without understanding its function and/or without concern for its appropriate use reflects a lack of control. Users should retain conceptual checks upon results and should utilize systems as effective tools.

CONTROL OF ENVIRONMENT The physical environment of a validated system must conform to hardware, firmware, and equipment manufacturers' requirements concerning such things as temperature, power surges, dust, radiation, and humidity.

COSTART A system for reporting to the FDA on adverse reactions as they are documented by manufacturers and distributors.

COUNCIL FOR INTERNATIONAL ORGANIZATIONS OF MEDICAL SCIENCE (CIOMS) Provides a forum for regulators and manufacturers to meet and discuss the development of standard procedures. CIOMS is under the auspices of the World Health Organization (WHO) and UNESCO.

CPR Component Production Rules.

CPU Refers to the Central Processing Unit of a computer where the logic circuitry is located. The CPU controls the entire computer; it sends and receives data through input-output channels, retrieves data from memory, and conducts all program processes.

CRC (See Cyclic Redundancy Check)

CRO 1) Chief Regulatory Officer. 2) See Contract Research Organization.

CRT Cathode Ray Tube; an output device.

CRITICAL SOFTWARE Software whose failure could have an impact on safety, or could cause large financial or social loss.

CUMULATIVE DOCUMENTATION TRAIL Refers to the mass of evidence (written procedures, test data, electronic records) that support the certification of a system as validated.

CURRENT ACCURACY AND RELIABILITY A system has the necessary charac teristic of current accuracy and reliability if, at the time of review, the system as configured provides appropriate and unerring processing of entered data and activated programs when operated within established standard procedures.

CURRENT GOOD MANUFACTURING PRACTICES (CGMP) These are the most recent modifications of Good Manufacturing Practices or a plan for total quality control in the manufacture of pharmaceuticals. (See Good Manufacturing Practice.)

CYCLIC REDUNDANCY CHECK (CRC) A method for detecting errors in a series of bytes. Two extra bytes are computed and appended to the series. The computation is complex, and there are fewer undetectable combinations of errors than for checksum or parity. The Cyclic Redundancy Check technique does not burden the transmitter with the continuous sending of parity bits, and thus offers more efficient data transmission, combined with a powerful error-detection technique.

DAC (See DIGITAL-TO-ANALOG CONVERTER)

DATA ACCESS Refers to the ability to retrieve, view, analyze and/or modify data. In a validated system, access must be controlled (i.e., monitored, reviewed, and approved) to assure accuracy, integrity and the ability to audit. (See Data Integrity.)

DATA ACCESS DIAGRAM Refers to a graphic representation of the organization of and access paths between data files.

DATA ANALYSIS Periodic testing review.

DATABASE 1) A set of data, part or the whole of another set of data, and consist ing of at least one file that is sufficient for a given purpose or for a given data processing system. 2) A collection of data fundamental to a system. (ANSI.)

DATABASE DESIGN PHASE This phase in the Software Development Life Cycle (SDLC) defines the structure of the database and the fields. The database design includes a graphical representation of its structure in conjunction with documentation describing each field within the database. This design must be consistent with the global system design defined in the system architecture phase and the system features defined in the functional specifications phase.

A review of the database design and fields to ensure that they meet the database requirements defined above is included in this phase.

DATA CHANGE LOG Whenever data is changed, an entry must be made into a log indicating the data change, reason for the change, and the various approvals for the change. Refer to SOP of system data changes.

DATA DICTIONARY A definition list of all fields, elements, and individual com ponent elements. It must include a description of what they mean, their purpose, the range of appropriate values, and each element's location.

DATA FIDELITY The data is accurate; that is, there is true correlation between the data and reality. Data fidelity should be checked independently of data integrity.

DATA-FLOW DIAGRAM Refers to a graphic tool which represents a system with emphasis on data movement, processing functions (transformations), and the files that are used to support processing.

DATA INTEGRITY The ability to view and/or utilize data should be controlled, i.e. monitored, reviewed and approved. The system should receive, store, transmit, operate upon, process, recall, sort, and otherwise manipulate user-provided or generated data without system-imposed inappropriate modification, distortion, addition, or deletion.

DDR Donor Deferral Registry.

DEBUG The process of locating, analyzing, and correcting suspected faults (bugs). (ANSI/IEEE.) In validated systems, debugging activities should be fully documented.

DESIGN A system design should be elegant -- the carefully executed result of a planned approach, rather than a patchwork reflecting trial-and-error-without-method approach. Design control implies: organization; a selection of components to reflect function, efficiency, and reliability; and a construction to minimize emergent problems.

DESIGN ANALYSIS The evaluation of a design to determine correctness with respect to stated requirements, conformance to design standards, and other criteria.

DESIGN ANALYSIS REPORT A written report of the proceedings and recommen dations of a design analysis.

DESIGN GOALS PHASE This phase in the Software Development Life Cycle (SDLC) defines the overall design objectives of the product. This phase primarily addresses the needs of the user and the system. The design goals document generated should include: the description of the system environment, overall system functions, identification of any other system(s) that will interface, input/output data requirements, any known constraints of the system, and a description of the target user.

DESIGN PHASE The period of time in the Software Development Life Cycle (SDLC) during which designs for architecture, software components, interfaces, and data are created, documented, and verified to satisfy requirements (See ANSI/IEEE Std 729-1983)

DESIGN REQUIREMENTS SPECIFICATION A document that prescribes com pletely and precisely the design of a system, a system component, or software. Typical contents may include, but are not limited to: algorithms, control logic, data structures, input/output formats, interface descriptions, physical requirements, structural requirements, and functional requirements. {Applicable algorithms may or may not be part of software design requirements. Whether they are included depends on the particular data structures involved. Numeric structures may require specific equations that closely specify algorithms, but lists and strings may occupy different data structures depending on the performance required. Search and match within X seconds, is a requirement; element-by-element search of a matrix table would specify a general algorithm and search a matrix table, starting with element 0,0 followed by 0,1 followed by ...0,n followed by...m,n would be a very tight algorithm specification for a search.}

DESIGN REVIEW Refers to a formal evaluation of a pre-coding plan, measuring the proposed steps against the system design specification and/or the software requirement specification.

DESIGN REVIEW REPORT A written report of the proceedings and recommenda tions of a design review.

DETAILED DESIGN SPECIFICATIONS PHASE This phase in the Software Development Life Cycle (SDLC) defines the overall design, behavior, and data flow of a program or module. The Detailed Design Specifications include the name, functional description, and variables used within the program/module/data file. For each applicable program/module, the interactive relationship(s) with other programs/modules, data input/output, database usage, and file access are defined. For each applicable data file, the format and program/module access is defined.

This phase also includes a review of the Detailed Design Specifications to verify their completeness and consistency with the applicable standard operating procedure for software development documentation.

DEVELOPMENT PHASE This phase in the Software Development Life Cycle (SDLC) develops a system test plan which defines all the necessary steps required to test the system's functionality as defined in the functional specifications. The system test plan must include setting up the system's environment, must provide valid and invalid input data, and must specify the appropriate output. The system test plan must be specific, so that a tester, not a programmer, can execute the test plan.

DEVELOPMENTAL TOOLS Usually refers to 4GL based programming technology; e.g., screen generators.

DEVELOPMENTAL VALIDATION Refers to a system review which is integrated into the process of customizing, configuring, adapting, or inventing a unique system or system component. In the case of new systems being designed or under construction, appropriate features and functions can be "built in" so as to appropriately satisfy defined acceptance criteria.

DEVICE HISTORY RECORD This is the production record which documents that the DMR specs were properly followed in the manufacturing process. It is the key document when a recall is ordered for tracing problems.

DEVICE MASTER RECORD This is the total design history, including the current and prior versions, required for approval of a computerized medical device.

DHR (See Device History Record.)

DIAGNOSTIC 1) Hardware-related self-check equipment or routines. 2) A message generated by a computer program pertaining to the detection and the isolation of errors, faults or failures.

DIGITAL In a digital system or device, quantities are represented by discrete whole numbers, nearly always base 2 (binary), as opposed to continuously variable numbers. To oversimplify, digital computers generate "stepped" data curves, while analog computers generate smooth data curves.

DIGITAL-TO-ANALOG CONVERTER (DAC) Process-related output devices translate the computer's outputs to appropriate actuating signals needed to control the process. These actuating signals might be simply opening and closing of switch contacts, or pulses to a stepping motor. Many times, however, the required actuating signal has to be analog in nature, such as a voltage for controlling the speed of a dc motor. In these cases, a digital-to- analog converter (DAC) is needed to convert the digital system output to the required analog form. (See also Analog-to-Digital Converter.)

DIRECT CONNECT MODEM Communication device allowing an immediate computer-to-computer communication. These raise control concerns with sensitive systems. Ideally, modem access should be through a callback system.

DIRECT DATA ENTRY Refers to the on-line submission of forms, clinical data, etc. into the system by field personnel without the use of a paper form submitted to a data entry function. While economical in time, direct data entry raises control concerns by removing one of the traditional documentation steps from the system.

DISASTER An incident that causes loss of data processing services. Possible disaster causes include fire, explosion, lightning, power disruption/surge, flood, sabotage, and wind damage.

DISASTER RECOVERY PLAN The goal of a Disaster Recovery Plan is to identify probable disasters such as flood, fire, civil unrest, etc., and to detail corrective action. Since the purpose of corrective action is to eliminate confusion and recover, key people and their back-ups must know what the plan is and where to find it.

DISK A circular rotating electromagnetic storage device. Disks come in different sizes and can be hard or flexible. Because of likely degeneration of the substrate, data and applications stored on disk should be replicated every five years.

DISK DRIVE A device used to read from or write to a disk or diskette. Since disk drives are mechanical and therefore relatively vulnerable, disaster recovery SOPs should account for crashes.

DISKETTE A flexible or semi-flexible plastic disk storage medium, also called a floppy disk. Data and applications stored on diskettes should be replicated every three years. (See Disk.)

DISK OPERATING SYSTEM (DOS) The system that controls the input and output to and from a disk.

DMR (See Device Master Record.)

DOCUMENTATION UPDATE Any and all written materials that explain the use, function, and/or evolution of a system. These include, but are not limited to: design specifications, authorizations, procedures, training materials, work aids, and various historical logs, disks, or systems.

DOWNLOAD Passing information (usually data or documents) from a host system to another computer, usually from a mainframe or mini to a workstation or PC.

DRIVER PROGRAM Triggers or enables other programs in a system. Specialized software designed to allow the CPU to deal with peripherals, such as printers.

DUAL MATRIX PROCESS OF VALIDATION The process of validation requires the identification of potential problem areas and the design of two methods of checking each area. For example, to check audit-trail accuracy, the validation process would include both a functionality test and a source code review. The Dual Matrix Methodology was developed to assure maximal practical quality assurance to control, for example, the release of AIDS-contaminated (or other impure) blood products for human transfusion. The methodology is based upon theoretical principles incorporated in the standards developed for the DEP auditing field (Halper, et al., 1985), now considered to be "commonly acceptable accounting practices," and upon hardware and software testing approaches evolving from the system design and development industries. Four key principles form the foundation of the approach: 1) Potential Vulnerability Analysis; 2) Test Design Redundancy; 3) White, Grey, and Black-Box Test Matrix Mix; and 4) Independence of Validation Procedure.

The Dual Matrix Methodology begins with an environment-specific, configuration-specific, version-specific, and implementation-specific analysis of the areas of potential vulnerability for the system to be tested. Following established auditing practices (Baird, 1974), researchers began with a comprehensive system analysis of the test site, including the development of a data-flow analysis, hardware inventory analysis, and structured needs analysis (Semprevivo, 1982). This analysis includes an examination of the computer system hardware (all components), software (operating and application), personnel (training, background, and standard operating procedures), and user requirements (inquiries, reports, and outputs). Inputs to the analysis include direct inventory of components and software; observed processes; interviews with users, management, DP management, DP personnel, and supervisory personnel; examination of operating logs; examination of output reports; and a review of supporting documents, including error logs, repair records, regulatory reports, and internal analyses. The analysis of the data collected focuses upon potential vulnerability of the data entered, stored, and manipulated by the system; of the programmed decision matrix acting upon that data; and upon the system-generated disposition decision constituting the significant output of the system. The areas of potential vulnerability are further analyzed through a review of the actual problems encountered during the previous 24 months of operation and are expanded as new vulnerabilities emerge as secondary results of the tests conducted (below).

For each of the identified areas of potential vulnerability, a total of two distinct tests are designed and implemented, representing two of the three general testing categories. These tests are situationally modified adaptations of the standardized tests designed to assure system accuracy, including source-code analysis tests (Stark, 1978), throughput analysis (Willingham and Carmichael, 1975), stress analysis (Goren, 1989), error-log analysis (Figarole, 1985), and continuous testing (Garwood, 1984).

The redundancy concept described and implemented provides more than just the quality-assurance levels discussed above and required as a result of the sensitivity of data in this environment (Bronstein, 1986). The use of redundant tests also adds a new level of managerial control, providing a divergent approach that, in effect, can recheck the work of in-house teams and provide a control on the efforts of outside validators. Particularly in light of the personal and organizational liability in a compliance-regulated industry, this level of oversight provides a significant "added value."

As generally used in the industry, "White-Box" tests refer to specific line-by-line, or signal-by-signal, traces through a system; "Black-Box" tests examine the input and output of a system, or expected-versus-actual output, without examining the actual system itself; and the compromise "Grey Box" tests blend the two, building internal "flags" or "checkpoints" into a system that provide intermediate Black-Box points. Together, the three-pronged approach constitutes the industry norm level of testing recommended for system acceptance (Swanson, 1988).

In the environment described herein, the FDA is concerned that management tends to rely primarily upon the "black box" result. All too often, managerial focus begins and ends with the question of whether products labeled as safe do indeed meet established and appropriate safety standards. But the consistency of that confidence can only be based upon the internal expert procedures and standards used by the system to make that labeling determination. A mixture of White, Black, and Grey-Box tests is appropriate to assure that consistency of confidence.

In accordance with the redundancy value described above, the Dual Matrix methodology's effectiveness is maximized when the duality provides for the use of two tests representing two of the three testing categories, to check for the designed corrections and protections against the identified vulnerabilities. For each identified area of potential weakness, two tests are implemented: either a White and Black-Box test, a White and Grey, or a Grey and Black-Box test.

Standard managerial data-processing controls call for an acceptance test of all new systems, and a constant "debugging" check on all developed and functioning software. These tests are generally built into the responsibilities of the internal personnel responsible for the implementation of systems and the development and maintenance of software.

The testing procedures described herein, of necessity, are designed, implemented, and interpreted by personnel outside of the normal process, in much the same way that independent auditors annually certify a company's financial records. This independence protects against the carelessness of assumptive acceptance; the shortcutting encouraged by performance pressures; the perceptual difficulties of self-proofing; as well as the purposeful problems of sloth, fraud or misplaced loyalty (Gilhooley, 1985).

The Dual Matrix Methodology calls for the use of a validation team of individuals not permanently employed by the organization to be validated; reporting directly to executive or senior managerial-level personnel; prepared to testify under oath concerning procedures and findings; and operating on a pre-approved, unconditional, fixed-fee, grant or contract. These safeguards assure complete independence of validation staff and findings.

EBCDIC (Extended Binary Coded Decimal Interchange Code) Eight-bit code used to represent data in most computers.

ECHO Display of characters on a terminal output device as each character is keyed.

ECHO CHECK Refers to a means of verifying the accuracy of a data transfer op eration by transmitting received data back to the source and comparing it with the original data.

EFTA European Free Trade Association. As part of its program, the EFTA runs the Pharmaceutical Inspection Convention (PIC), an agreement among member countries for mutual recognition of national Good Manufacturing Practice inspections.

ELECTRONIC DATA INTERCHANGE (EDI) Direct, paperless transmission of requirements from system to system. For example, an MRP system places a JIT order with a vendor system, which responds with an electronic bill of lading and invoice, receiving an electronic funds transfer authorized by the customer's accounts payable system. These paperless systems raise control issues that need to be addressed with more rigorous auditing procedures.

ELECTRONIC IDENTIFICATION Indicates the identity of the individual who created the document. When electronic identification is used, a system thinks it knows who is utilizing its resources.

ELECTRONIC SIGNATURE Refers to a method of imprinting an I.D. code to a specific document within the system, which provides evidence that an individual has reviewed and approved a document that has not been subsequently modified.

EMBEDDED A term that usually refers to software in a read-only memory chip (ROM); that is, computer memory that cannot be changed by the computer. Embedded data usually means data that appears as part of a program.

EMERGENT CONTROLS Acceptable documentation and training for a validated system includes the establishment of procedures to deal with emerging situations, including updates and version changes, hardware recalls, user-discovered problems, and assistance requests.

ENCRYPTION A process by which information once entered into a system is then unrecoverable to all system users without the machine-language encryption key. For all practical purposes, only the computer can recognize and translate encrypted data. Encryption can be used to protect confidentiality; for example, the identification of potential blood donors whose samples test positive for HIV or some other sexually-transmitted disease. Encryption provides the highest level of security.

END USER A person who has the direct use of an automated system to perform their work function (client is a synonym). In a validated system, documentation of end-user training and adherence to SOPs are critical.

ENVIRONMENTAL CONDITIONS The range of temperature, power surges, dust, humidity, and radiation which can have an impact on the reliability of a system. The validation concern is the maintenance of control of such conditions.

ERASABLE PROGRAMMABLE READ-ONLY MEMORY (EPROM) A firm ware chip that can be programmed a single time, then is burned in as firmware. The switch pattern in circuit can be erased and reprogrammed using an EPROM programmer.

ERRONEOUS DATA TEST The "Grey-Box," flagged operations test designed to confirm the system's ability to appropriately reject bad data. The test uses a special test file to avoid corruption of the live data base in the event that inaccurate input is erroneously accepted. Erroneous data testing involves the input of data violating protocols of field size, composition, range, complexity, and coherence, as well as testing of major logical flows and decisions within a representative sub-program. A comparison of the actual results to expected results, discrepancy reports, and a summation of critical and non-critical issues identified during the test are all provided.

ERROR An error is a reported malfunction generated automatically by hardware or system software.

ERROR DETECTION Error detection schemes are described in computer science literature. Among them are parity, checksum, cyclic redundancy, and longitudinal redundancy. A validated system usually employs several of these schemes throughout the system.

In a validated system, the source code supporter shows evidence of appropriate system maintenance procedures, including error detection and correction, such as: internal syntax error detection tools; internal logic error detection tools; error reporting system; non-critical error correction procedure through revisions and updates; critical error correction procedure through immediate user notification; and on-line, hands on, or telephone user assistance procedure, with generated error log.

ERROR LOGS Official documentation of problems (malfunctions, deficiencies, failures) encountered in the operation of a system.

ERROR RECOVERY ROUTINES System software programs for handling prob lems and resuming work flow. In validated systems, the concern is that the error recovery software includes the ability to trace the problem and that the problem does not threaten data integrity.

ETSI European Telecommunication Standards Institute.

EVIDENCE OF VALIDATION The collection of documentation, test results, observations, expert opinion, or other supporting materials demonstrating that an independent review has determined conformity to regulatory guidelines and/or a convincing demonstration of a system's current accuracy and reliability, expected continued accuracy and reliability, auditability, managerial control, and maintenance of data integrity.

EXECUTABLE CODE Successfully compiled instructions that the CPU can follow to perform a task ("execute a program").

EXPERT SYSTEM An Expert System generates code every time it operates; therefore, it is impossible to review source code prior to use. Application source code does not exist prior to use. The solution in the validation context is to validate the expert system rather than the end code of the expert system and to make sure that the end code is regularly audited within SOPs. Example 1: a system that sorts and delivers ingredients. The manager requires different sorts at different times. Therefore, the sorting system must be validated rather than individual user sorts. There must be an audit trail that is regularly reviewed by management. The validator must have evidence of that. Example 2: Statisticians using SAS are constantly writing code which generates different statistical tests every time they use the system. Therefore, the validator must determine that SAS does what it is supposed to do, and that the statistician is following appropriate protocols. Most of the real use of Expert Systems is in the development of systems. Applications have fixed code. Examples of Expert Systems include SAS, SPSS, and BMD. (See also Artificial Intelligence.)

FAILSOFT Refers to a programming concept or practice that allows a system to continue to operate even when there has been a failure, error, or problem. From a validation point of view, the existence of failsoft path within a system is extremely negative. In any high-hazard system, one always wants some notification of a failure, problem, or error. For example, cardiac pacemakers have double batteries so that if one fails, the pacemaker continues to operate. However, if the user has no indication that one battery is gone, then there is no backup (a failsoft condition). When the second one dies, so does the user.

FAILURE MODE ANALYSIS Refers to a procedure that determines which mal function symptoms appear immediately before and/or after a failure of a critical parameter in a system. After all the possible causes are listed for each symptom, the product is designed in such a fashion that the problems are eliminated.

FAILURE MODE EFFECTS ANALYSIS A procedure in which each potential failure mode in every sub-item of an item is analyzed to determine its effect on other sub-items and on the required function of the item.

FAILURE MODE EFFECTS AND CRITICALITY ANALYSIS (FMECA) A procedure that is performed after a failure mode effects analysis to classify each potential failure according to its severity and probability of occurrence. The determination of how a software-failure analysis was conducted. Were failures derived by tracing "backward" from safety system faults to software (and components) where the fault could originate (top down), or by tracing fault propagation from a postulated failure origin to all possible outcomes (including system faults), classifying the criticality of each (bottom up)?

FAST TRACK VALIDATION This is a program developed by Pharmacia Bioprocess in cooperation with Weinberg, Spelton & Sax, Inc., and other consultants. Under a Fast Track system, clients are provided with assistance in any and all phases of validating biotech processes, including controlling and analyzing computer systems.

FDA-483 An indication of an adverse finding by the Food and Drug Administration. A 483 is often used in conjunction with a regulatory letter which contains a detailed description of the finding and the action proposed.

FEATHERSTONE'S STEPS IN SYSTEM DEVELOPMENT

1) wild enthusiasm
2) disillusionment
3) total confusion
4) search for the guilty
5) punishment of the innocent

6) promotion of non-participants

FIBER DISTRIBUTED-DATA INTERFACE (FDDI) Refers to a standard for 100 Mbps token ring LAN, based on a fiber-optic medium. The validation concern is that data be transferred without being corrupted.

5GL (See Fifth-Generation Language)

FIFTH-GENERATION LANGUAGES (5GLs) A fifth-generation language allows the end user to interact with a computer system in conversational English. It may incorporate artificial intelligence and may seem, to the user, to be forgiving of mistakes. A system using a 5GL expects anomalies and corrects according to set parameters by using interactive querying. (See 4GLs and Generation of Languages.) Because of the difficulty in reviewing source code in a situation where users, in effect, create new code every time they utilize the system, many regulated environments have opted for a viable alternative: they concentrate upon the language mechanism for creating the appropriate code from user commands. The review of that "internal language code", coupled with careful documentation of the user commands issued and appropriate review of the accuracy of those commands, has generally met regulatory acceptability guidelines.

FILE ERROR A file error is a combined set of fields that make up a record. A file is reading the wrong entry for a particular record. For example, reading a telephone book and dialing the number above or below the correct name. The correct data is there; you simply assigned it wrongly.

FINAL COMPILATION All final printout reports or data-sets designed for archive, decision point, regulatory report, or review meet criteria including "clean compile," vendor and version identification, and completeness.

FINAL INSPECTION CERTIFICATE After a system has been completely validated, a certificate is issued which says so.

FINGERPRINT A time/date stamp which proves that a document has not been changed since the time of the stamp. In some jurisdictions, a Fingerprint is acceptable evidence that a document has not been modified. It may be accepted as an electronic signature.

FIRMWARE 1) Machine instructions burned into a chip. The instructions are then part of a machine and is not user modifiable (e.g., BIOS chips). 2) Computer programs and data loaded in a class of memory that cannot be dynamically modified by the computer during processing. (ANSI/IEEE) 3) Hardware that contains a computer program and data that cannot be changed in its user environment. (ANSI/IEEE)

FIRST LAW OF HARDWARE Plug it in.

FIRST VALIDATION The initial attempt to validate a system. This usually requires greater resources than subsequent attempts. It is often much more efficient to use external expertise for this first attempt.

510(k) A section of the Food, Drug and Cosmetic Act requiring medical device manu facturers to submit a premarket notification to the FDA to allow the FDA to determine whether the device they intend to market is equivalent to a device already legally marketed. The 510(k) submission is a thorough description of the medical device, including computer hardware and software components. The 510(k) indicates whether the device is new, unique, bypasses human intervention, is potentially hazardous, and/or meets certain other technical measures that classify the (amount of) supporting materials that may be required. It is recommended that evidence of validation be included in the 510(k) for all computer- controlled or computer-produced medical devices and for all hazardous medical devices. Since some system and application software is now classified as medical devices, many vendors are appropriately filing 510(k) submissions for their software. Obtaining a copy of this 510(k) filing (either through the Freedom of Information Act or from the vendor) may help in the process of selecting a system and may provide important data to be included in the validation package for appropriate systems.

FLAG An indicator that signals the occurrence of an error, state, or other specified condition. (ANSI/IEEE)

FLOWCHART A graphical representation of the definition, analysis, or solution of a problem, in which symbols are used to represent operations, data flow, and equipment. (ISO)

FMECA (see Failure Mode Effects and Criticality Analysis)

FOOTPRINT Refers to the physical space taken up by a piece of equipment. The term is sometimes used to indicate a trail or signs followed in an audit trail.

FORMWARE Refers to firmware with data entry gaps. BIOS-controlled password system in which firmware requires correct user response before advancing.

4GL (See Fourth-Generation Language)

FOURTH-GENERATION LANGUAGE (4GL) Refers to a code generator which produces an application language code such as COBOL. The application code is then compiled to executable or object code for the computer. The purpose of fourth-and fifth-generation languages is to enable an end user who is not a programmer to utilize the resources of a system. A 4GL like NATURAL, FOCUS, or SAS tends to trade efficient use of system resources for "user friendliness."

FRACTIONATION The process of separating blood or plasma into its component parts.

FUNCTION 1) General application or purpose of the system. 2) A process that is performed on a number or character string. 3) A precoded routine.

FUNCTIONAL DESIGN The specification of the working relations between the parts of a system in terms of their characteristic actions.

FUNCTIONAL ENHANCEMENTS Upgrades or improvements in software to provide new activities or operations. The validation concern revolves around how any new or improved functions interact with other parts of the system. Changes made to a validated system require a thorough testing of the changed module(s), as well as of the modules that interface with it (them).

FUNCTIONALITY Refers to the ability of a system to consistently provide the quality results required in a specific environment.

FUNCTIONAL SPECIFICATION PHASE This phase in the Software Development Life Cycle (SDLC) defines the system features necessary to fulfill the needs of users. Specifications must be consistent with the design goals and must define the system function, the system usage and operations, the input domain, and procedures for generating output. These specifications must also serve to generate the user's manual, continue software definition and development, and be reviewable and understandable by the project and user personnel. A review of the functional specifications to ensure they are consistent with the design goals is a part of this phase.

FUNCTION TEST The primary system "Black-Box" functionality test, which is designed to determine the degree of accuracy with which the system correlates to the reality it is intended to describe. The function test compares known outputs with obtained outputs from a set of selected inputs. Function tests are usually designed to verify system response times, system throughput, memory utilization, system execution, and traffic rates on data channels and communication links. Functional tests execute typical operating conditions of the software and system with typical input values and results. These tests should test boundary conditions, system defaults, ordering, and special values.

FUZZ Refers to the range of acceptability in specifications. The tighter the FUZZ the more rigorous the requirements for the validation process must be. The range of acceptability in specifications would be greater in a system that controlled the manufacture of bandages as compared to one which controlled the manufacture of cardiac pacemakers. In the hazard analysis procedure, establishing the probability of a hazard is a way of determining FUZZ.

GALEF'S DEDUCTION Fallible people design fallible computers.

GALP (see Good Automated Laboratory Practice)

GCP (see Good Clinical Practice)

GENERATION OF LANGUAGES The first-generation language that activates a computer system consists of positive and negative electrical charges. The second-generation language contains the zeroes and ones of Assembly. Third-generation (3GLs) are higher-level, more user-friendly languages like BASIC, COBOL, and FORTRAN. Each third-generation language has a code with its own unique syntax and logic. The computer takes the source code from a programmer in a 3GL and generates object code to the machine. Fourth-Generation Languages (4GLs) are more like the conversational native language of the user. A word in a 4GL represents a string of 3GL code which is then compiled to the object code that the machine uses. Fifth-Generation Languages (5GLs) are interactive with or query the user in a conversational manner. There is usually some level of artificial intelligence (AI) active with 5GLs.

GLP (see Good Laboratory Practice)

GMP (see Good Manufacturing Practice)

GOOD AUTOMATED LABORATORY PRACTICE (GALP) Since laboratories are relying increasingly upon computer systems to collect, analyze, summarize, and report data, a wide range of situations has arisen for which the GLPs provide little or ambiguous guidance. To avoid the confusion and potential problems that such uncertain situations can create, and to deal with the newly arising problems of possible data corruption, loss, or inappropriate modification that computerized systems may inadvertently generate, the Environmental Protection Agency has developed a series of supplemental guidelines: the Good Automated Laboratory Practice (GALP). These supplemental guidelines are intended to provide clarification of expectations of performance and control for laboratories electing to utilize computer systems for data collection, storage, manipulation, or reporting. The GALPs, representing an addition to, but not a replacement of, GLPs, have been distributed to the industry and represent current EPA compliance policy.

The ultimate issue in the extension of the GLP concepts to an automated laboratory through the implementation of GALPs is that of control. Effective management and operations of an automated laboratory cannot be assured unless use and design of that system are consistent with standards intended to assure system control.

The GALP standards are based upon a foundation of six concepts, detailed below. Together, these concepts define the necessary control issue that underlies the GALPs. Recognizing that wide variations in computer system designs, technologies, laboratory purposes, and applications are likely to create situations in which appropriate and successful alternate control strategies may evolve that were not anticipated in the GALP standards, these six principles are provided as guidelines for evaluating proposed equivalencies. They are intended to serve as guideposts in the interpretation of the standards and as evaluation criteria in the consideration of proposed equivalencies.

1. DATA: The system must provide a method of assuring the integrity of all entered data. Communication, transfer, manipulation, and the storage/recall process all offer potentiality for data corruption. The demonstration of control necessitates the collection of evidence to prove that the system provides reasonable protections against data corruption.

2. FORMULAE: The formulae and decision algorithms employed by the system must be accurate and appropriate. Users cannot assume that the test or decision criteria are correct; those formulae must be inspected and verified.

3. AUDIT: An audit trail, allowing the tracking of data entry and modification to the responsible individual, is a critical element in the control process. The trail generally utilizes a password system or equivalent to identify the person or persons entering a data point and generates a protected file logging all unusual events.

4. CHANGE: A consistent and appropriate change-control procedure capable of tracking the system operation and application software is a critical element in the control process. All software changes should follow carefully planned procedures, including a pre-install test protocol and appropriate documentation update.

5. SOP: Control of even the most carefully designed and implemented system will be thwarted if appropriate user procedures are not followed. This principle implies the development of clear directions and Standard Operating Procedures, the training of all users, and the availability of appropriate user support documentation.

6. DISASTER: Consistent control of a system requires the development of alternate plans for system failure, disaster recovery, and unauthorized access. The principle of control must extend to planning for reasonable unusual events and system stresses.

GOOD CLINICAL PRACTICE (GCP) Good Clinical Practice (GCP) was made law in Japan and went into effect on October 1, 1990. All clinical studies for Product License (PL) application now have to comply with this GCP. The four main issues in Japanese GCP are: Institutional Review Board (IRB), contract between sponsor and research center, informed consent, and the maintenance of records.

GOOD LABORATORY PRACTICE (GLP) The Environmental Protection Agency's performance and operational guidelines, summarized as the Good Laboratory

Practice (GLP) have been widely accepted and implemented by agency laboratories, independent laboratories contracting with the EPA, and by industry laboratories submitting findings and results to the Environmental Protection Agency. These standards have weathered the practical tests of field experience and continue to represent the expectations and requirements for laboratory management and performance.

GOOD MANUFACTURING PRACTICE (GMP) The purpose of the Good Manufacturing Practice regulations of the FDA (21 CFR) is to assure that all pharmaceutical products and biologic products and medical devices meet the requirements of the Federal Food, Drug and Cosmetic Act as to safety and efficacy and have the identity and strength to meet the quality and purity characteristics which they purport to have, as required by section 501(a)(2)(b) of the Act. Nonconformity with GMP, therefore, establishes adulteration prohibited by section 301.

The Food and Drug Administration strives to ensure that the regulated industries comply with a total quality control concept through participation in voluntary CGMP compliance seminars and workshops sponsored jointly with the industries or with educational institutions. The concept of a total quality control system is neither limited in scope to the analytical methods of assay, control charts, product inspections made during the manufacturing processes and prior to finished dosage form distribution, nor to the statistical techniques utilized in these discrete operations.

GRACEFUL DEGRADATION Refers to the characteristics built into a computer system to assure that system failures default conservatively; that is, system characteristics that assure that failures do not produce critical problems.

GREY BOX TEST Testing which builds flags or interim-result points into a system and compares expected to obtained results at those predetermined control points. The "grey-box" test is, in effect, a compromise between the more comprehensive but cumbersome "white box" and the efficient but less complete "black box."

GROUP TEST The process of determining if something fits within the Fuzzy Set.

GUIDELINE Regulatory agency requirements that are framed as suggestions without the force of law. Failure to respond effectively to them may trigger negative responses from their source agency.

HARDWARE Physical equipment used in data processing, as opposed to computer programs, procedures, rules, and associated documentation.

HARDWARE CONSTRAINTS Limitations imposed upon operations by equipment itself. The most common limitations are capacity and speed. If a data set is too large for a machine, then a malfunction could occur and some data could be lost.

HARDWARE INTERFACE Refers to the physical equipment point where peripherals link into the hardware platform. In a validation effort, the concern is for standard interfaces that can readily be tested, maintained and/or replaced.

HARDWARE REPLACEMENTS All replacements of interacting hardware components with non-identical components, differing in BIOS, in chip, or in configuration, as in non-PROM ROM, should result in an immediate revalidation study of the system or sub-system isolate.

HASH TOTAL Refers to a method of determining whether a transmission has been completed. Taking a total of ASCII values and confirming them. (See Checksum.)

HAZARD ANALYSIS A determination of the potential degree of human harm represented by failure of a device.

HIGH-LEVEL LANGUAGE A programming language that usually includes langu age features such as nested expressions, user-defined data types, and parameter passing not normally found in lower-order languages; that does not reflect the structure of any given computer or class of computers; and that can be used to write machine independent source programs. (ANSI/IEEE)

HIGH-LEVEL TESTING Managerial or executive testing which is a more cursory review of results and/or general performance test. A regulatory analysis is generally a high-level review which is looking for, among other elements, evidence of low level-testing. Validation of a system, process, or product incorporates high-level testing.

HOROWITZ'S RULE A computer makes as many errors in two seconds as 20 men working 20 years make.

ID CARD Identification card; required for access to any secure facility.

IMPLEMENTATION PHASE The period of bringing a system into actual use; it includes end-user training and has traditionally led to the discovery of the need for patches, requests for functional enhancements, or for both.

INFO-GLUT An excess of information that has the effect of slowing down a deci sion process. Any time input speed exceeds analysis speed.

INITIAL HAZARD ASSESSMENT The initial classification and estimation of an identified system hazard. It includes the hazard description, an estimate of its criticality, and an estimate of probability that it will occur. The initial probability estimate may be qualitative or quantitative. Initial Hazard Assessment applies to hazards identified any time during the life cycle of a product or system. Preliminary hazard analysis is an important part of Initial Hazard Assessment.

INITIAL HAZARD ASSESSMENT REPORT A complete list of potential system hazards identified throughout product design and development, a qualitative estimate of criticality (cost) of each, a qualitative (or quantitative, if data are available) estimate of the probability of each, and recommendations for resolving any items that seem likely to pose problems.

INPUT PARAMETER The range of values that will be accepted by a system when data is entered.

INSPECTION A formal evaluation technique wherein a person or group other than the author examines software requirements, design, or code in detail to detect faults, violations of development standards, and other problems.

INSTALLATION QUALIFICATION (IQ) The process of checking the system against the vendor/developer standards of operating environment, physical connection, safety parameters, and functional parameters prior to the initial utilization of the system in production. An IQ test generally includes functional testing of system outputs, analog devices, and connections.

INSTRUCTION (1) A program statement that causes a computer to perform a particular operation or set of operations. (ANSI/IEEE) (2) In a programming language, a meaningful expression that specifies one operation and identifies its operands, if any.

INSTRUMENT CONTROL Refers to supervision of a process through a data collec tion device that can also initiate corrective action without human intervention being necessary. Processes controlled by instruments, in high-hazard environments, should be validated.

INTEGRATED SOLUTION A system linking several software (and possibly hardware) platforms to solve a business problem in a way that makes them transparent to the end user. Example: A document control system that provides word processing, publishing, version control, electronic distribution, and archiving within the same shell.

INTEGRATION TESTING An orderly progression of testing in which software elements, hardware elements, or both, are combined and testeduntil the entire system has been integrated (ANSI/IEEE Std. 729-1983).

INTERACTIVE PROCESSING An application in which each entry calls forth a response from a system or program, as in a ticket reservation system.

INTERFACE A shared boundary. An interface might be a hardware or software component to link two devices, or it might be a portion of a storage or registers accessed by two or more computer programs. (ANSI)

INTERNATIONAL ORGANIZATION FOR STANDARDIZATION (See ISO)

INTERPRETER (1) Software, hardware, or firmware used to interpret computer programs. (ANSI/IEEE) (2) A computer program used to interpret. (ISO) (3) A program which translates a high-level language into machine code, one instruction at a time. Each instruction in the high-level language is executed before the next instruction is interpreted.

INTERRUPT 1) Refers to a signal which causes the hardware to transfer program control to some specific location in internal storage, thereby breaking the normal flow of the program being executed. After the interrupt has been processed, program control is again returned to the interrupted program. 2) A signal that breaks or halts a program. These can be initiated by the user, the program, or elsewhere in the system. Uncontrolled interrupts are AEs.

INVENTORY CONTROL Procedures for tracking materials required for a commer cial activity, especially in a production environment. These activities can be automated and are subject to validation as they relate to materials supplied for a product that can be subject to recall. They also track where a released product subject to recall is sent.

I/O PORT Input/Output connection port.

ISO (International Organization for Standardization) A worldwide federation of national standards bodies (ISO member bodies). The work of preparing international standards is normally carried out through ISO technical committees. Each member body interested in a subject for which a technical committee has been established has the right to be represented on that committee. International organizations, governmental and non-governmental, in liaison with ISO, also take part in the work. ISO collaborates closely with the International Electrotechnical Commission (IEC) on all matters of electrotechnical standardization.

ISO 9000 SERIES STANDARDS A series of international standards involved with internal quality management and external quality assurance. ISO 9000 series standards are compatible with, but do not replace, the obligation to validate a system. For vendors developing computerized medical devices, including software products and diagnostic systems, ISO standards serve as an excellent definition of appropriate operating protocols likely to lead to a validated end product.

ITERATION 1) Refers to automatic repetition, under program control, of the same series of processing steps until a predetermined stop or branch condition is reached. 2) Refers to the current stage of development within the gestation period of policies, procedures, documentation, or systems, prior to approval of a formal version.

JOB 1) Set of data completely defining a unit of work for a computer. 2) A collection of specified tasks constituting a unit of work for a computer; e.g., a program or related group of programs used as a unit.

JOB DESCRIPTION Identification of the tasks associated with a job and the skills required to perform those tasks satisfactorily.

JOINT EUROPEAN STANDARDS INSTITUTE (JESI) which consists of CEN, CENELEC, and ETSI.

JUST-IN-TIME MANUFACTURING (JIT) Refers to an optimal material require ment planning system for a manufacturing process, in which there is little or no manufacturing material inventory on hand at the manufacturing site and little or no incoming inspection. The lack of incoming inspection raises control concerns in high-hazard sites. At a manufacturing site where a JIT system is in place, the computer system is the major source of component information that would be critical in the event of a recall.

KERBEROS A network authentication system developed at the Michigan Institute of Technology for its Project Athena. It is an emerging industry standard for security in client-server computing environments. KERBEROS provides a means of identifying individuals on a network through the use of secret encryption key technology.

KERNEL A portion of a program remaining resident in the main storage (RAM) of a computer, which enables that program to be called when needed. This resident portion can cause unforeseen conflicts with applications unless there is stringent prior testing to insure compatibility.

KOSEISHO The Koseisho is the government Ministry of Health and Welfare responsible for drug registration in Japan. It is similar to the Department of Health (DOH) in the UK, the Bundesgesundheitsamt (BGA) in Germany, and the Food and Drug Administration (FDA) in the U.S. The minister of the Koseisho is a member of parliament and may be changed when the cabinet is reorganized by the prime minister. There are nine bureaus and a secretariat in the central office. The pharmaceutical affairs bureau is responsible for the various aspects pertaining to drugs. All clinical studies for a product license (PL) application to the Koseisho have to comply with Japanese Good Clinical Practice (GCP). (See Good Clinical Practice.)

LABORATORY INFORMATION MANAGEMENT SYSTEM (LIMS) A computerized system through which data can be retrieved, manipulated, analyzed and reported on; e.g., toxicology, reproductive, and pathology studies.

LABORATORY MANAGEMENT In compliance with statutes and/or regulations, the laboratory management has ultimate responsibility for compliance with all GALP standards. Laboratory management may designate responsible person(s) to meet the individual requirements and unique set of circumstances inherent in specific operating environment(s) and system(s). However, laboratory management is both accountable and responsible for compliance with statutes and/or regulations. (See Responsible Person.) [In the GALP Standards, Section 5.0, the code letter "M" is used to identify all specific and direct responsibilities of Laboratory Management.]

LANDAU'S PARADOX When system programmers declare that a system works, the system has worked and will work again some day.

LANGUAGE Refers to any symbolic communication media used to furnish information to a computer. Examples are PL/1, COBOL, BASIC, FORTRAN, and ASSEMBLY.

LAUNEGAYER'S OBSERVATION Asking dumb questions is easier than correcting dumb mistakes.

LIFE CYCLE Refers to a period of time that starts when a product is conceived (as an idea) and ends when it is no longer available for use. The life cycle has several phases, typically designated in order as: requirements phase, design phase, implementation phase, test phase, installation and checkout phase, operation and maintenance phase, and, sometimes, retirement phase. Readers should note that the preceding phase designations represent only one Life Cycle model; other models may have more or fewer phases, and tasks that characterize one phase may overlap the tasks of one or more subsequent phases.

LIMITED ACCESS The specific locations of systems equipment should be carefully restricted through the use of limited access doors, restricted hallways, restricted buildings, or other appropriate limitations.

Specifically:

All data processing systems rooms should be locked, with access limited to currently authorized personnel.

Rooms housing minicomputers, mainframes, and their data storage peripherals (disk and tape drives) are included.

All laboratories in which computerized systems are actively in use but not constantly monitored by authorized personnel should have locked-door limited access, locked keyboard restrictions, or both.

LIMS (see LABORATORY INFORMATION MANAGEMENT SYSTEM)

LIVE ENVIRONMENT ERROR LOG REVIEW Over the course of one or more weeks, a real-time audit log of all programs is generated. For validation purposes, review that log to trace the source and disposition of all encountered errors and to identify any unusual programs (not normally accessed) that may require special consideration.

LOADER A routine that reads an object program into main storage prior to its execution. (ANSI/IEEE)

LOG A record documenting all system activity. These must be complete, signed, available, reviewed, audited, and, where appropriate, include corrective actions taken.

LOGGING The documentation of detected errors and problems, so that review and audit trails are created and maintained in hard copy.

LONGFELLOW'S OBSERVATION It takes less time to do something right than it takes to explain why you did it wrong.

LONGITUDINAL REDUNDANCY A method for detecting errors in a series of bytes. Error detection is done as in the checksum method, but the computation differs slightly. The method can detect single-bit errors, and when used with parity it can localize single-bit errors for correction. There are, however, many combinations of errors that can defeat this method for both correction and detection.

LOW-LEVEL TESTING More intensive, detailed testing, looking "down" into a system through comprehensive performance or stress testing. Includes verification of instruments, periodic recalibration of sensing and measurement devices, and complete user acceptance testing of new systems. Development and performance of IQ, OQ, and/or PQ is low-level testing.

MACHINE LANGUAGE A representation of instructions and data that is directly executable by a computer. (ANSI/IEEE)

MAINFRAME A large computer.

MAIN MEMORY A non-moving storage device utilizing one of a number of types of electronic circuitry to store information.

MAIN STORAGE SIZE The amount of memory available to the CPU, usually described in number of bytes. Originally done in thousands of bytes (K or kilobytes), the more current number now is in millions (megabytes), rapidly approaching the billion (gigabytes) range.

MAINTAINABILITY A characteristic of a computer system that refers to its ability to remain functional over extended time.

MAINTENANCE The task of updating software (and hardware) to keep it current. Based upon vendor recommendation and/or common operational parameters, clear maintenance standards and schedules must be established. There is a qualitative difference between controlled maintenance, which is essential to a validated system, and having a programmer simply provide a patch immediately.

MAINTENANCE CONTRACT An agreement with any vendor, especially a software provider, to keep code functionally current and to provide enhancements for an agreed-upon time and fee. Inability to maintain a system will preclude its use in any business environment, not just a regulated one.

MAINTENANCE LOG Refers to documentation of all routine, extraordinary, and recommended maintenance procedures. A system-maintenance log must be designed and established in order to demonstrate control.

MAINTENANCE ORDER A written request which provides a record of mainte nance work requested and performed. These orders, properly logged, provide a part of the evidence trail of control.

MAINTENANCE QUALIFICATION A review of the maintenance schedule, log, and repair list. Analysis confirms that the schedule is met according to the log and that the repair list is accurate according to log entries. Attach the maintenance schedule, repair list, and sample of log.

MAINTENANCE SCHEDULE For each main component part in the system, a maintenance schedule has been established. This schedule identifies the frequency of regular and routine maintenance, identifies trigger points (chronometric, performance, or alternate) for additional maintenance, and provides trouble-shooting guidelines identifying special maintenance procedures.

MANAGEMENT AWARENESS AND CONTROL The organization and/or system management maintains appropriate control mechanisms to assure proper operations, use, maintenance, and procedural supervision of the system. Management awareness and control is generally maintained and demonstrated through the development, distribution, and enforcement of clearly designed standard operating procedures (SOPs).

MANAGER OF SOFTWARE ENGINEERING The Manager of Software Engineering is responsible for appointing and supervising the product manager, approving the project plan, and ensuring that all standard operating procedures (SOPs) are being followed. The manager also completes the product approval phase.

MANUFACTURER In the context of regulated environments, the terms "manufacturer" or "firm" refer to a manufacturer seeking market clearance of a product, a distributor seeking market clearance of a product, or an applicant representing a manufacturer or distributor who is seeking market clearance of a product.

MARK TWAIN'S POSTULATE Always do right. This will gratify some people, and astonish the rest.

MATERIAL REQUIREMENTS PLANNING (MRP) Refers to a computerized system which provides for manufacturing, planning, and control for all activities from acquisition of raw materials to delivery of completed products and is designed to support key management interfaces in various functions of the corporation. The validation concerns come from the fact that the MRP would be utilized for potential product recalls.

McFADDEN'S LAW The trouble with experience as a teacher is that the test comes first, and the lesson afterward.

MEDICAL CLASS I, II, III / IV Refers to three FDA-defined levels of medical devices. Class I requires practically no regulatory action. It includes diapers, bandages, etc. Class II includes some testing kits (glucose, pregnancy tests) and some devices that require testing of reliability (condoms) or sterility. Class III / IV are the most highly regulated. These devices directly impact human health, either with or without human intervention. Prior to being combined, Class III included incubators and monitors, while Class IV included pacemakers, infusion pumps, etc. There is some indication that the FDA will be including laboratory and manufacturing devices as Class II or III / IV devices in the near future.

MEDICAL DEVICE - GENERAL REGULATORY ASSESSMENT Medical device GMP regulations are found in 21 CFR Part 820. The fact that process validation is mandated under this regulation is clearly noted.

Section 820.5 requires the medical device manufacturer to "prepare and implement a quality assurance program that is appropriate to the specific device manufactured."

Relevant questions then, focus on (or should reveal) whether a quality assurance program (those activities required to assure confidence in the quality of the process used to manufacture a finished medical device) has been established that is either specific for a particular medical device or general for all medical devices from the company.

A generalized QA program must identify specific issues related to specific medical devices.

Section 820.100 states that process validation itself is an essential element in establishing confidence that a process will consistently produce a product meeting the designated quality characteristics, specifications and goals.

Written procedures are in place that specify methodologies of control that assure that the medical device conforms to its original design.

Written procedures are in place that assure that the original basis for the medical device design falls within current approved specifications (and required regulatory guidelines).

Written procedures are in place that specify methodologies of control that assure that the medical device conforms to any design changes.

Written procedures are in place that specify methodologies of control that assure that any design changes are fully tracked and show evidence of supervisory approval at all stages of development including: design and assessment stage, prototype production and testing stage, finished product and testing stage, finished product validation stage, and the final review stage.

Written procedures are in place that call for the description of the medical device by physical, electrical, performance, and safety characteristics.

MEGABYTE 1024K bytes.

MENU A computer display listing a number of options. The operator selects one of the options. Sometimes used to denote a list of programs.

MERSKEY'S RULE Do a silly test and you get a silly answer.

MICROFICHE Microfilm record, usually 3" by 5," which can be used for permanent storage of text and/or data archives.

MICROPROCESSOR A microprocessor executes encoded instructions to perform arithmetic operations, internal data transfer, and communications with external devices. It is usually a single integrated circuit on a chip. Logic circuitry of a microcomputer; frequently synonymous with a microcomputer.

MILESTONES Most complex projects are divided into measurable sub goals. These subdivisions of a large system or project can usually be utilized independently. Such independent subdivisions are called Milestones.

MILITARY SAMPLING A method of sampling used in quality control in which a predetermined table (usually randomly generated) is used to select a specified number of representative individual cases.

MINIMAX A theory of efficiency that says minimize effort in order to maximize output.

MIRACLE ERROR An error in data and/or processing for which no known cause has been identified. Validation does not allow for continued utilization of a system that has been found to have a Miracle Error.

MNEMONIC A code or device designed to aid someone's memory. A positive characteristic of well-designed source code is the use of meaningful mnemonics as variable names and internal labels.

MOBIL'S MAXIM Bad regulation begets worse regulation.

MODEL A Model is like an algorithm in one sense. An algorithm can Model a physical process (an equation), or it can Model some aspect of human reasoning (a so-called expert-system computer program). In another sense, a Model (data) can be a collection of numbers that represent the properties of an object. A Model (computer program) can operate on a Model (data) to estimate the effects of a real process on a real object.

MODEMS (modulator-demodulator units) Permit outside access to systems via telephone lines. Careful modem security is vital to the restriction of unauthorized access. Specific security procedures may vary according to the environment, and may include any or all of the following:

Removal of modem at times other than those necessary to authorized use.

Use of "bit check" or "bisync" protocols.

Use of a modem printer to track all access attempts.

Use of a "call back" system.

Use of dedicated or restricted phone lines.

Use of originating phone number verification equipment.

Use of other emerging restrictive technology.

MODULARIZATION The process of modularization allows the designer to decompose a system into functional units, to impose hierarchical ordering on function usage, to implement data abstractions, and to develop independently useful subsystems. It can be used to isolate machine dependencies; to improve the performance of a software product; or to ease debugging, testing, integration, tuning, and modification of the system.

MODULARIZATION CRITERIA These include the conventional criterion in which each module and its sub modules correspond to a step in the execution sequence; the information-hiding criterion, in which each module hides a difficult or changeable design decision from other modules; the data abstraction criterion in which each module hides the representation details of a major data structure behind functions that access and modify the data structure; levels of abstraction, in which modules and collections of modules provide a hierarchical set of increasingly complex services; coupling-cohesion, in which a system is structured to maximize the cohesion of elements in each module and to minimize the coupling between modules; and problem modeling, in which the modular structure of the system matches the structure of the problem being solved.

MODULE TESTING PHASE This phase in the Software Development Life Cycle (SDLC) performs the module test plan and generates all supporting documentation.

A review of the output, to ensure the results are as expected, is a part of this phase. Any errors found are reported in the supporting documentation.

MONITORING The ongoing survey of the functioning of system components.

MRP (See MATERIAL REQUIREMENTS PLANNING)

MULTIPLEXER A device which takes information from any of several sources and places it on a single line.

NASSI-SCHNEIDERMANN CHART An alternate method of flowcharting.

NATURAL LANGUAGE 1) Computer code that mimics conversational human speech in its commands. 2) A 4GL developed for use with ADABASE.

NETWORK A system that ties together several remotely located computers via telecommunications.

NETWORK SECURITY Refers to the protection of network resources against unauthorized disclosure, modification, utilization, or destruction.

NODE In a series or network of interacting computers, each computer represents a node or communications link in the network.

OBJECT CODE Compiled, assembly-language code. A fully compiled or assembled program that is ready to be loaded into the computer. (ANSI/IEEE)

OBJECTIVITY AND INDEPENDENCE Traditionally, regulated industries have relied upon an in-house quality assurance or quality control division with a separate line of reporting authority to provide the necessary objectivity and independence. The same industries follow a different set of standards in situations of financial reporting to the SEC, relying upon outside CPA firms despite the high level of in-house expertise, recognizing the added value of the independence and objectivity this approach provides.

The area of system validation more closely parallels the SEC model since, in effect, it is reviewing the appropriate role of quality assurance, as well as data processing and information systems, and because an unusually high level of expertise is required to follow the intricacies of various source-code and hardware components as well as the regulatory issues involved.

Based upon these requirements, an underlying assumption of appropriate system validation standards is the total independence, high level of expertise, and complete objectivity of the signatory validators. Should the validator have a reporting relationship within the organizational structure, or a real or perceived direct or indirect benefit from the outcomes of the validation, the validation process and results could be severely compromised.

OEM Original Equipment Manufacturer (i.e. maker of computer hardware).

OFF-LINE A term describing equipment, devices, or persons not in direct communi cation with the central processing unit of a computer. Function testing of systems should be done Off-Line.

ONGOING SYSTEM PERFORMANCE TEST Refers to system tests conducted by in-house staff on an ongoing basis, providing immediate indication of problems and an annual base of at least 50 cases.

An example of an ongoing system performance test within an environment that distributes blood products might include a daily comparison of whole blood samples to computer-generated switch status, labeling, and attribute matrix labeling protocol.

ON-LINE 1) Direct accessing of a computer system. 2) A device in current communication with the CPU, versus off-line. Function or stress testing of systems in high-hazard environments should not be done on-line. (See Off-Line.)

OPEN SYSTEMS INTERCONNECTION A model developed by the ISO as a means of simplifying information exchange. Designing communication hardware and software in accordance with OSI standards allows users to customize a system by selecting the combination of products which offers the exact features required. The ISO has divided communications into seven sub-processes called layers and has written OSI specifications for each one. OSI specifications are functional only; the tasks to be performed in each layer are defined exactly, but the methods used to accomplish these tasks are left to the designer. Any method is acceptable as long as the same method is used at both ends of the communications link. The International Organization for Standardization (ISO) defines the model, in 1979, as a framework for defining network protocols.

OPERATING PERSONNEL Refers to staff who use the system. They should have GMP, GLP and/or GALP training.

OPERATING QUALIFICATION (OQ) Refers to a check of process, equipment, and control-system prior to use in a live production environment. The OQ is often combined with the PQ (Performance Qualification). In effect, the OQ is the "present" low level test, in direct parallel to the acceptance validation high-level review.
Decision Criteria: Because of the sensitive nature of an OQ test, the decision criterion is generally one of ACCEPTANCE = 100% conformity to established norms. Any deviation from 100% constitutes a failure. If any means of determining acceptance

is met by an equivalent means, or additional comments are needed, a separate document should be attached with an explanation of reasoning or results.

OPERATING SYSTEM (OS) Essentially, the software that enables users to interact with a computer. Operating systems generally handle interactions with peripheral equipment, programming languages, and so forth. Operating systems generally include software that controls the execution of programs. An operating system may provide services such as resource allocation, scheduling, input/output control, and data management. Although operating systems are predominantly software, partial or complete hardware implementations are possible.

OPERATIONAL LIMITS Prescribed ranges within which a system is expected to function reliably.

OPERATOR 1) The individual responsible for the physical use of a mainframe com puter system. 2) A mathematical and algorithmic term for the subject of an operation. For example, if c=axb then a and b are the operators of the multiplicative function.

PARALLEL Term to describe transmission of data eight bits (one byte) at a time.

PARITY An error-detection method for small units of data (usually 8 bits, a byte), wherein an extra bit is appended to the data unit. It is useful in detecting odd numbers of incorrect bits in a single byte. Even numbers of incorrect bits leave parity unchanged. By itself, Parity is not useful for error correction.

PARITY BIT An extra bit within a byte, used to verify the coded information in the byte itself. The extra bit is either a one or zero, so as to make the total number of ones in a byte equal either an odd or even number (odd or even parity).

PASSWORD A unique string of alphanumeric characters used to verify an end-user's identity. To maintain proper control, these must be secret and changed according to a controlled cycle.

PASSWORD CONTROL LOG The system administrator or individual in charge of system security must keep track of which users have access to various secure database levels. This document will help organize such information, as well as keep track of the passwords of the various users. Refer to SOP on system security and passwords.

PASSWORD FILE A password or user file should restrict unauthorized users and log access attempts. The password file should be characterized by:

1. Maintenance procedures to delete passwords of individuals no longer authorized, and to change passwords periodically.
2. An active log generated automatically by every unusual event.
3. Periodic managerial review of the password log.

PASSWORD SYSTEM A password system should be included in all appropriate devices to uniquely identify the system user. The password system should include the following characteristics:

- Secret Passwords (random preferred over user-selected),
- Passwords of not fewer than four digits.
- Passwords changed at least annually.
- Password system with automatic cut off and log generation after three erroneous

sign-on attempts.

Password systems are not necessary in environments in which a single authorized user has physical access to a device, or with devices that do not allow for user input, modification, or analysis.

PATCH A change to the code of a program that is usually done quickly and for a very specific purpose, most often to correct an immediate problem. In a high-hazard environment, Patches should be validated.

PATH 1) A logical chain of instructions to accomplish a single program action with in a module. 2) A link between modules or programs.

PERFORMANCE Refers to intended specific results or objectives of a system.

PERFORMANCE QUALIFICATIONS Refers to a process during which all automatic systems logs, error messages, variable data, process attributes, and product characteristics (QC and QA) will be examined for validity and accuracy. The purpose of performance qualifications is to ensure that the system will operate and produce expected production-quality results under load conditions. This will give a representation of long-term performance, verification of vendor guarantees for availability and failure rates, and a better debugged and more reliable system.

PERFORMANCE TEST (See Function Test).

PERIPHERAL EQUIPMENT All computer system equipment other than the computer processor circuits and memory circuits; for instance: disk drives, tape drives, and terminals.

PHANTOM PHASE Refers to a tongue-in-cheek label for the retrospective failure to perform a system analysis prior to designing a system. The process of defining the system's needs and objectives may be a phantom or non-existent phase in the SDLC.

PHARMACEUTICAL RESEARCH AND MANUFACTURERS OF AMERICA (PhRMA) Formerly the Pharmaceutical Manufacturers Association (PMA). A national association representing more than 135 major manufacturers. It keeps close contact with the industry and technological developments, specializing in the guidelines to follow in the manufacturing of pharmaceuticals. PhRMA puts out more than 70 publications in the field and a complete listing is available free of charge. The publications cover such topics as regulatory issues of the FDA, de-ionized water for the water industry and water treatment, clinical guidelines and manufacturing guidelines, and the proceedings of seminars. Some of the publications are in a report format of four or five pages; others are 200 pages in length.

PHASE DEFINITION FORMAT The software methodology documentation is based on a standard format used to define all of the software development life cycle. The phases within the development cycle include the following (A definition of each phase is listed alphabetically in this glossary):

1) DESIGN GOALS PHASE
2) FUNCTIONAL SPECIFICATION PHASE
3) PROJECT PLAN PHASE
4) SYSTEM ARCHITECTURE PHASE

5) USER'S MANUAL AND DOCUMENTATION PHASE
6) DEVELOPMENT PHASE
7) DATABASE DESIGN PHASE
8) BETA TEST PLAN DEVELOPMENT PHASE
9) DETAILED DESIGN SPECIFICATIONS PHASE
10) CODING AND MODULE TEST PLAN DEVELOPMENT PHASE,
11) TRAINING SUPPORT PERSONNEL PHASE
12) MODULE TESTING PHASE
13) SYSTEM INTEGRATION TESTING PHASE
14) SYSTEM TESTING PHASE
15) BETA TESTING PHASE
16) PRODUCT APPROVAL PHASE
17) RELEASE PHASE

PHYSICAL INSPECTION Observation and review of system installation and physical site for factors relating to safety, continual operation, and disaster recovery capability.

PORTABILITY Ability of an application code to run under different operation systems and/or on different hardware platforms.

PRELIMINARY HAZARD ANALYSIS (PHA) Refers to an initial assessment of system hazards undertaken early in the product life cycle. It includes describing the hazard, estimating its criticality, and estimating the probability of its occurrence. The estimate may be qualitative or quantitative.

PRELIMINARY HAZARD ANALYSIS REVIEW When the preliminary hazard analysis (PHA) has been performed for the medical device and is available for review, it will meet the following criteria which demonstrates at least four areas of safety criticality[1]:

Class 1 = Catastrophic (may cause death; may cause total system or medical device loss)

Class 2 = Critical (may cause severe injury or occupational illness; may cause severe damage to the system or medical device)

Class 3 = Minor (may cause minor injury or occupational illness; may cause minor damage to the system or medical device)

Class 4 = Negligible (a problem that does not result in injury, occupational illness; problems that do not damage system or medical device)

PRE-MARKETING APPLICATION (PMA) A submission to the FDA required for Class III Medical Devices.

PRIVACY 1) Anonymity guaranteed to the personal details of a clinical subject. 2) A right which is potentially compromised by a company under the Freedom of Information Act once a regulatory filing is made.

PROBLEMS A reported malfunction of a computer component observed by end users of an application.

PROCEDURE A set of steps to be followed to accomplish a task each time the task

---

[1]M. Frank Houston, "Designing Safter, More Reliablee Software Systems," U.S. Food and Drug Administration, Center for Devices and Radiological Health, 1989.

is to be done.

PROCESS CONTROL SYSTEM Automated management of an industrial production system. Even though a system may be used in other areas, such as petroleum and chemicals, validation is required when it is adapted for use in the pharmaceutical industry.

PROCESS PERFORMANCE CHARACTERISTICS Those characteristics which identify components that need constant re-calibration, are easily breakable, and are otherwise essential for proper process performance.

PROCESS PROCEDURE The step in a production process that needs to be documented within a standard operating procedure (SOP).

PROCESS QUALIFICATION (PQ) Refers to a check of the ongoing process equipment and control system as it is utilized in the live production environment. A PQ generally includes testing a known sample or known data against the system performance, and it may include stress testing through a non-live library replica or analysis of Error File stress reactions.

PROCESS VALIDATION The accumulation of evidence in support of expert testimony demonstrating the appropriate functionality of process equipment utilized in the manufacture of a biologic or pharmaceutical product. This evidence includes the development and performance of an installation qualification, a process qualification, and an operating qualification. When the process equipment under review is computer-controlled, computer-regulated, or computer-activated, or when said equipment reports to or through a computer or receives data from or through a computer, a system validation is incorporated into the Process Validation. Because process validation is a current requirement of the FDA's "Current Good Manufacturing Practice Regulations for Finished Pharmaceuticals and Medical Devices," it becomes necessary and applicable to the manufacture of pharmaceuticals and medical devices.

The FDA requires that each of the following three principle be followed:

1. Quality, safety and effectiveness must be designed and built into the product itself.
2. Quality cannot be inspected or tested into the finished product (i.e., quality must begin with manufacturing).
3. Each step of the manufacturing process must be controlled to maximize the probability that the finished product meets all quality and design specifications.

The apparent key to meeting these three criteria is Process Validation.

<u>The FDA Definition:</u>

"Process validation is establishing documented evidence which provides a high degree of assurance that a specific process will consistently produce a product meeting its predetermined specifications and quality characteristics."

It is therefore of the utmost importance that the manufacturer (vendor) composes a validation protocol which clearly states how system validation will be conducted. This must include testing parameters, product characteristics, inventories, acceptance testing of production equipment, and a checklist of acceptance criteria.

In order to maintain quality assurance of a particular product, the manufacturer must be able to track the production stages and current configuration of the product itself. A version-control plan is equally important.

PRODUCT AND PROCESS TESTING CONSIDERATIONS Prior to executing the initial OQ (Operational Qualification Acceptance Test) and rigorous PQ (Product and Performance Qualification Test), product characteristics and specifications must be established. All products intending to be manufactured must have the following characteristics defined: physical, chemical, electromagnetic, and product performance.

PRODUCT APPROVAL PHASE This phase in the Software Development Life Cycle (SDLC) determines whether or not the product can be released. The review committee reviews portions of documentation from each phase. The product is approved if all phases leading to this phase have been performed following established standard operating procedures.

PRODUCT DATABASE A product database contains information maintained on a regulated product that tracks released products for adverse events and possible recall.

PRODUCT MANAGER The Product Manager is responsible for the planning and coordinating of all aspects of a specific product. The product manager defines the work responsibilities for the project and allocates the resources required for completion. The product manager also ensures that all the requirements for product release are met.

PROGRAM 1) A computer program. 2) A schedule or plan that specifies actions to be taken. 3) To design, write, and test computer programs.

PROGRAM CHANGE REQUEST FORM This form is used to initiate the program change process. Refer to SOP of software change control.

PROGRAM LOGIC SOURCE CODE REVIEW An examination of the internal logic of all operating programs, based upon a detailed trace of source code and a review of program documentation.

PROGRAM PATH Refers to the order in which a computer executes its program med instructions under given internal and environmental conditions.

PROGRAM PATH ANALYSIS The identification of data and conditions that will cause a computer to execute a given Program Path.

PROGRAM REQUEST PROCESSING CHECKLIST This form will help track the progress of a requested programming change. Refer to SOP of software change control.

PROJECT ADMINISTRATOR Refers to a person who reviews the initial change request form from the requestor. He/she records the request in log (or begins log page). He/she participates in ultimate approval/rejection of the request. This person also enters/checks logbook for future updates/enhancements.

PROJECT LEADER The Project Leader is responsible for the completion of all development and testing aspects of a specific software project. The Project Leader assigns the tasks of the project to the appropriate individuals and coordinates the testing of the product. Also, the Project Leader is responsible for the technical quality of the project. All steps related to the project are reported to the Project Leader.

PROJECT MANAGER OR PHASE/PROJECT LEADER May share several responsibilities. They assess required action, prepare impact statement, participate in approval/rejection decision, obtain approval of change from end user (if applicable). and place the actual change into work.

PROJECT PLAN Refers to a management document that describes the approach to be taken for a project (e.g., developing a system). The plan typically describes the work to be done, resources needed, methods to be used, configuration management and quality assurance procedures to be followed, schedules to be met, project organization, and so forth.

PROJECT PLAN PHASE This phase in the Software Development Life Cycle (SDLC) develops the Project Plan for the system. The plan may include the assignment of roles defined in this methodology, resource requirements, hardware requirements, scheduling of milestones, and a definition of budgetary considerations. Any assumptions made about the project are also listed in this document.

PROM Programmable read-only memory; once programmed, the switch pattern on a PROM cannot be changed. Special equipment, separate from the computer, is usually used to "burn in" the switch pattern.

PROTECTED STORAGE Refers to storage locations which are protected from heat, dust, humidity, and electromagnetic interference. Their intended use is usually for long-term archiving.

PROTOCOL 1) A set of conventions or rules that governs the interactions of processes or applications within a computer system or network. (ANSI/IEEE) 2) A set of rules that governs the operation of functional units to achieve communication. 3) Within FDA guidelines, protocol refers to a document which provides the background, rationale, and objectives of the clinical trial and describes its design and organization and the conditions under which it is to be performed and managed.

PSEUDO CODE Refers to user language which describes the operation of a program by having each native language statement equivalent to one computer statement. Example:

1. Get 3 numbers.
2. Add 3 numbers together.
3. Divide the total by 3.
4. Print out total.

The preceding is a pseudo code program for averaging three numbers. One could write that program in any computer language.

PURGE The permanent removal of data from a file or archive, whether physical or electronic. Unlike "delete," the purge function wipes a magnetic record clean, rather than simply erasing the directory listing.

QUALITY ASSURANCE (QA) 1) In the context of FDA guidelines, Quality Assurance refers to the implementation of appropriate planned and systematic actions necessary to provide adequate confidence that the required quality will be obtained. 2) In the context of EPA guidelines, the inclusion of a data and procedural "double check" through a Quality Assurance Unit or individual is an established and widespread GLP principle, extended here to automated lab systems. The legitimacy and credibility of that checking function, of course, rests with the independence of QA, assured through a separate reporting relationship. While it is possible that QA may have additional responsibilities in the organization, those responsibilities should not compromise that independence: the QA should not be and should not report to or through the RP. Specific QA responsibilities include review of system SOPs, inspection and audit of the system, review of final reports for data integrity, and review of record archives. [In the GALPs Standards section 5.0, the code letter "Q" is used to identify all specific and direct responsibilities of QA.]

QUALITY ASSURANCE PERSONNEL Should be involved in all phases of configuration management control, but the involvement should be restricted to that of consultant in the early stages (prior to revision release), and active participant thereafter. QA should periodically review and update the change control logbook. QA also must act as liaison in those cases of change initially requested by end users.

QUALITY ASSURANCE SYSTEM Refers to a set of principles followed to allow effective control of quality throughout a process.

QUALITY CHARACTERISTIC 1) Refers to those characteristics of a product or service that impact on its ability to satisfy stated or implied needs. 2) A particular, measurable characteristic examined to test product quality. 3) A product or service free of deficiencies.

QUALITY CONTROL 1) The operational techniques and procedures that are used to fulfill requirements for quality. 2) May refer to any technique which evaluates the quality of a product being processed by comparing it to a predetermined standard and taking the proper corrective action if the quality falls below that standard.

QUALITY CONTROL UNIT Refers to a separate, independent organization, including staff (with a reporting responsibility), equipment, and physical location which has the task of ensuring that the required second check is carried out prior to release of product.

QUALITY PROOF Evidence that the entity (system, process, or product) meets preestablished standards. A product, process, or system is not considered to be of adequate quality until it has been proven to be so.

QUICK AND DIRTY A phrase commonly used to describe 1) A one-time ad hoc report, where statistical rigor is sacrificed to the interest of speed. The likelihood that this report will be used to support long-range strategic decisions is inversely proportionate to the likelihood of ever replicating its results. 2) Programmers' favored approach to producing a patch, where documentation and convention are also sacrificed to the god of speed.

RAM Random Access Memory; internal storage device containing volatile information which can be changed; read-write memory. When electrical power is cut off from a RAM integrated circuit (IC), its memory is lost.

RAW DATA 1) Data that cannot be easily derived or recalculated from other information (US-EPA, GALP-7.11, DEC 1990) and may include:

Handwritten lab worksheets/records/notes of original observation.

Exact copies or transcripts of the above.
Photographs, microfilm, microfiche.
Computer printouts.
Magnetic tapes of dictated observations.
Database files and reports stored on floppy diskettes or tape.
Version-specific notes, source code for developed software.

The specific definition is still subject to some regulatory controversy as of this writing. (see 21 CFR 58.3(k) 2) In the context of FDA guidelines, Raw Data means all records of original observations or activities in a clinical trial necessary for the construction and evaluation of the trial. Such material includes laboratory notes, memoranda, calculations, documents, recorded data from automated instruments etc., or exact verified copies in the form of photocopies, microfiches, etc. thereof. The term can also include photographic negatives, microfilm or magnetic media. In the case of the latter, dated and verified written transcriptions may replace the originals.

REAGENT A substance involved in a chemical reaction used to test a blood or plasma product.

RECALIBRATION Refers to the process of making a comparison of system outputs to predetermined output standards and subsequently adjusting that system to conform to the expected, normalized output.

RECALL OF INFORMATION The ability to retrieve data from archives. Recall is a mandatory ability for demonstrating control.

RECORD Collection of related data or words treated as a unit.

RECOVERY Restoration of an automated system (hardware, software, and/or data) to the point immediately before an AE or a system failure.

RECOVERY CAPABILITY Since regulatory guidelines require the retention of records for a period of time, systems generating, storing, or analyzing that data must be capable of maintenance and continued functionality despite the ravages of reasonably anticipated natural and man-made disasters.

REGULATORY QUALITY Refers to the need for an independent, second check that must be configured to meet specialized regulatory requirements which go beyond corporate and/or industry standards.

RELEASE PHASE This phase in the Software Development Life Cycle (SDLC) releases the system to a limited number of users as a live system. The number and identification of users the system is released to is determined by the product manager.

Review the limited release to ensure the goals from the design goal phase are complete, that the system meets all functional specifications, and analyze all problem reports. If there are no existing problem reports that are considered to be related to a hazardous situation, then the product can be released to all other customers. Otherwise, the problems need to be addressed by identifying and fixing the faulty phase and following through the methodology.

RELEASE PROTOCOL Standardized procedures required to be followed before any GMP regulated product is release for market use.

REPAIR PARTS LIST A list of all the new parts added or replaced in the course of maintenance must be available and current. The Repair Parts List may, in effect, be the current inventory list, identifying major system components.

REQUIREMENTS DEFINITION Those conditions or capabilities that a system or component must meet or possess to satisfy a contract, standard, specification, or other formally imposed document. Subsequent development of a system or component is based on the set of all requirements for that system or component.

REQUIREMENTS INSPECTION REPORT A report of the proceedings and recommendations of an inspection review of a (product, software, or system) Requirements Specification.

REQUIREMENTS TRACE MATRIX Refers to a document that records or traces the links or correspondence between requirements identified at each stage of product or system development. A matrix should at least document individual system requirements, elements of the design intended to fulfill a requirement, and test procedures and data needed to verify meeting the requirement.

REQUISITE STANDARD OPERATING PROCEDURES AND LOGS Studies have shown that nearly 70% of all 483s written by FDA inspectors are related to SOP or log-maintenance issues. Good system and medical device documentation and record keeping is a requirement of a regulatory audit. Our research has identified 35 particulars which must appear as individual, or a part of, corporate or departmental standard operating procedures.[2]

RESILIENCE (See SOP Resilience).

RESISTANT TO STRESS Refers to design features within the system that take into account the reality of conditions exceeding input parameters and operational limits.

RESPONSE TIME The interval elapsing before a computer accepts an input. Failure to allow for elongated response times can lead to unintentional double entry of data or use of inappropriate functions during an otherwise normal job processing.

RESPONSIBLE PERSON (RP) Laboratory management is ultimately responsible for the operations conducted by computerized equipment, laboratory and information systems, and associated operating environments. As such, laboratory management may directly supervise and/or assign/delegate its authority to meet the individual requirements and unique set of circumstances inherent in specific operating environment(s) and system(s). By permitting laboratory management to delegate its authority and responsibility, appropriate personnel (i.e., Responsible Persons) will have the opportunity to put into place those policies, plans, and practices that will provide the required evidence of control.

The RP is generally a professional, with some computer background, in a position of authority related to the control and operation of the automated data system. The RP's responsibilities include training of users, implementing appropriate security measures, developing or reviewing SOPs for system use, enforcing change-control procedures, and responding to emergent problems. [In the GALP Standards section, 5.0, the code letter "R" is used to identify all specific and direct responsibilities of the RP.]

RESTART Refers to the process of reestablishing the execution of a routine.

RETROSPECTIVE VALIDATION System review which determines the reliability, accuracy, and completeness of a system in current or historical use.

REUSABILITY The ability to recycle routines, modules and even entire programs in

---

[2]For more details on items to include in these SOPs, see S. Weinberg, System Validation Standards (Kendall Hunt, 1990).

the development of new software systems.

REVALIDATION Recheck of the reliability of a system which is undertaken at regular intervals. It is recommended that these intervals be no longer than every two years. It is also performed after any major system enhancements or failures.

REVIEWER CREDENTIALS The credentials held by a system reviewer are of critical importance, because the judgments upon which conclusions are based should pass expert-witness credibility review for technical and theoretical expertise, appropriate experience, and complete independence.

ROBOTICS Use of automated equipment to replace human labor and/or operators in repetitive industrial functions.

ROM Read only memory; internal storage device in which information is permanent.

ROUTINE TESTING OF SOFTWARE LOG Any time a change is made to the software, at least module testing must be performed. Such information must be documented. Refer to SOP on retesting/revalidation of software.

RP (See Responsible Person)

RS 232 INTERFACE 1) An Electronic Industries Association (EIA) standard for connecting electronic equipment; data is transmitted and received in serial format. 2) Standard communications connection for most personal computers in use today.

SAFETY ANALYSIS AS RELATED TO COMPUTER SOFTWARE OR SYSTEM FIRMWARE When computer software (or EPROM firmware) is used in the control of a medical device system, its inherent safety becomes a regulatory concern. The concern itself emanates from the links between the software and safety-critical functions. Therefore, as soon as the software functionality is identified, a Software Safety Analysis (SSA) should begin.

SAFETY-CRITICAL SOFTWARE Computer programs that control functions that affect system safety. Such programs include those causing the host computer to directly command accessories that affect safety, or that may communicate with (provide data for) other safety-critical software systems.

SAMPLE The Sample Size Issue

In many ways the performance testing of computer systems parallels the QA testing common to the pharmaceutical industry. The goal--identifying and diagnosing either error sources or results--is consistent. The importance of random selection of test access is congruous. Also, obviously, the criticality of documenting procedures, results, and analyses is also identical. In a very important way, though, the testing of systems differs from the testing of pills on an assembly line or laboratory animals injected with a new drug. System testing uses different rules for defining sample size, largely because it uses a different conceptualization of "sample."

Both the "Military Standard" and the "Power Formula" methods of selecting sample size are inappropriate for system testing, potentially resulting in samples either significantly too small, or unnecessarily and redundantly too large. These statistical approaches are designed to assure that sufficient cases are included to provide confidence that "outlyers" -- unusual cases and abnormal reactions -- will, in effect, be canceled out by mirror-image outlyers on the opposite end of the scale. The results, a smoothed normal (or other unbiased sampling distribution) is then suitable for complex and sophisticated statistical analysis.

If we want to judge average heights in a population, then, we use a sizing rule that will provide reasonable confidence that all the basketball players will be "canceled out" by the jockeys, providing a sample unskewed by the unusual extremes. To create a sample of 20, for example, we might randomly select 20 people from the population and measure each individual's height. It would, of course, be inappropriate to select only two people, write down each of their heights 10 times, and pretend that we were sampling 20 different individuals. Yet that procedure is, in effect, the process of testing a computer system with a random population of 20 test cases: the computer may actually be exercising the same two programming pathways 10 times each. The difference is simple: in product testing (or in most social science and scientific research) a sample is used to test the results on a data population, utilizing a data sample. But in testing computer systems, we are instead testing not a sample of the data, but a processing sample; the sample is designed to be representative of the system pathways, not representative of the data set upon which the system operates. Ideally, statistical sampling will, in appropriate sizings, generate a high level of assurance that most pathways are explored. But the possibility that some pathways are neglected, creating the potential that the system is inaccurate or inoperative (not 1% inoperative, but 100% unreliable) is very real. At the same time, testing multiple cases through the same pathway is an unnecessary time and resource expenditure. Program testing must define its sample through careful stratification of the data set, instead of relying merely on the rules of large numbers to carry the day.

What, then, is the guideline for sample size in system performance testing? A sample is constructed by defining the relevant system pathways and constructing one data case to follow each of those paths. The sample size is determined not by the population of the data to be manipulated nor by arbitrary rules limiting "enough is enough." Instead, the sample size is defined by an extremely simple but sophisticated formula: $Y = X$, where Y is the sample size, and X represents the number of paths to be tested.

The sample size should never be a number smaller than X, and gains nothing in reliability, credibility, or value by being any larger. In system performance testing, the test sample size should be equal to, and constructed to reflect, the relevant system pathways to be tested.

SAS (Statistical Analysis Software) Refers to widely used analytic software support ed by the SAS Institute. though never formally validated, it is still used by the FDA for analytic work and is widely used in the industry.

SAVE The process of making a backup copy of text or data as a precaution against the next system failure and/or to maintain an historical record.

SCENARIO ANALYSIS Application users are responsible for the identification and description of likely, probable, or potential disaster scenarios and should have developed a reasoned response plan to assure data integrity in the event of these scenarios. Possible scenarios include, but are not limited to: power interruption, flood, fire, and sabotage.

SCHEMA Specific and preestablished procedure. In effect, a standard for activity or action connected to a particular environment.

SCHROEDER'S CAT A philosophical conundrum that posits a cat being both alive and dead simultaneously. The Hawthorne Effect is a more formal reference to the phenomena. The validation concern addressed is the fact that the process of measuring something may change it. This problem necessitates unobtrusive observation on the part of validators.

SDLC (SOFTWARE DEVELOPMENT LIFE CYCLE) The process of producing a system from design to implementation and maintenance. The FDA has used the 8-Step or waterfall model in its training, but the key is to have an SDLC that works for you, and to follow it.

SECURITY LOG Used by security personnel, this document indicates who may currently be found on the premises. Refer to System Validation Standards Checklist.

SEGAL'S LAW A person with one watch knows the time; a person with two watches is never sure.

SELF-CHECKING CODE Computer code that automatically reviews itself for errors against certain internal rules and may make corrections to compensate. These correction procedures must be understood and confirmed to be sure they generate acceptable results.

SELF EXPLANATORY A documentation principle, especially applied to source code, requiring that a properly trained professional be able to easily and quickly understand the activity explained in the text. This is a key principle in maintaining a compliant system.

SERIAL Term to describe handling of data one bit at a time.

SHARED LOCK Refers to a security principle which requires that two "keys" be used simultaneously to open a "lock." The "keys" and "lock" may be physical, software related, or a combination of the two.

GEORGE BERNARD SHAW'S PRINCIPLE Build a system that even a fool can use, and only a fool will want to use it.

SITUATION-UNIQUE TESTS Special tests of security, of override access potential, of log accuracy, etc., are designed and conducted as needs arise during the context of other testing. These situational tests are developed, conducted, and interpreted by Weinberg Associates, Inc., or by in-house staff, as appropriate.

SIX SIGMA (6Σ) An alternate notation for expression of a quality-level goal of 99.999% accuracy, the general research standard of the pharmaceutical industry. A 6Σ level of quality represents a minimum of 4 defects per million items. In comparison, the current estimate of software quality in the United States is 3Σ, about three to four defects per thousand lines of code. The airline industry utilizes a standard of 6.4Σ, or approximately three errors per 2.4 million items (or incidents).

SKUNK WORKS A project or activity that uses a trial-and-major-error approach. One does a very rough model and keeps correcting until it works. A skunk works is outrageously difficult to validate.

SMART MODEM 1) A communication device which manages most of the connec tion process and provides error checking transparent to the user. 2) A modem with a CPU and programmable memory.

SOFTWARE Programs, procedures, rules, and any associated documentation pertaining to the operation of a computer system.

SOFTWARE COMPLEXITY METRICS A measure of the degree of complication of a system or system component: number of branches, nesting, data structures, lines of code, program size, etc.

SOFTWARE CONFIGURATION CONTROL PROCEDURES A document that prescribes the procedures to be followed in evaluating, coordinating, approving, and disapproving changes to software.

SOFTWARE DEVELOPER Refers to an individual or organization that provides software for commercial purposes. Also referred to as vendors. Until recently, these were unregulated by the FDA. A new interpretation has classified certain pharmaceutical industry software as medical devices, now requiring them to follow validation procedures.

SOFTWARE DEVELOPMENT PLAN A software development plan should include requisite inspections, reviews, audits, and tests for verification and validation of software. (See Project Plan.)

SOFTWARE ENGINEER The Software Engineer is responsible for translating the design from documentation into code. The code generated is documented in such a way that it can be audited for conformance. This individual also writes module test plans and performs module and integration test plans.

SOFTWARE ERROR LOG Usage of this document is found in cases of errors reported and corrected in a particular piece of software at a specific site.

SOFTWARE HAZARD ANALYSIS The identification of safety-critical software, the classification and estimation of potential hazards, and the creation of a program path analysis to identify hazardous combinations of internal and environmental program conditions.

SOFTWARE HAZARD ANALYSIS REPORT A document that records the proceedings and recommendations of a Software Hazard Analysis.

SOFTWARE INTERFACE Refers to code written specifically to allow otherwise incompatible programs or systems to pass or share data. Example: Programming to allow the QA LIMS to pass its release decision to the MRP or order management systems, bypassing manual input of a paper report.

SOFTWARE REQUIREMENTS SPECIFICATION A document that prescribes completely and precisely the conditions and/or capabilities that system or product software must possess or meet to satisfy the system or product requirements specification. All typical inclusions of requirements specifications apply here. If a product or system is comprised of software alone, then the System or Product Requirements Specification replaces the Software Requirements Specification.

SOFTWARE, REUSABLE Refers to programs that are able to be recycled for re lated purposes. This is a key premise for the current interest in CASE --Computer-Aided Software Engineering. (Also see Reusability.)

SOPs (See Standard Operating Procedures)

SOP ACCURACY All identified SOPs conform to operating requirements, recom mendations, and procedures developed and distributed by system suppliers, designers, vendors, or supporters.

SOP ADHERENCE Exists when current SOPs conform to and accurately reflect actual system use and function, as confirmed through unobtrusive observation, managerial report, and user personnel report.

SOP RESILIENCE The characteristics of a standard operating procedure that permit it to apply over a period of time despite minor situational changes, as opposed to the non-resilience of "working procedures," guides, or other less formal and more technically specific procedures.

SOURCE CODE Software language instructions and logic that are produced by programmers. These can be read and reviewed relatively easily by other programmers and auditors. This code must be passed through a compiler to produce executable code or object code.

SOURCE-CODE REVIEW Examination of software programs to determine their availability, internal documentation, display of evidence of change control, accuracy of algorithms and formulae, logical construction, error detection and correction, internal organization, and consistency of terminology.

SOURCE-CODE TEST RECORD This is a checklist that will aid a programmer in evaluating developed source code.

SOURCE PROGRAM A computer program that must be compiled, assembled, or interpreted before being executed by a computer. (ANSI/IEEE)

SPAGHETTI CODE Refers to highly unreadable and hence virtually unmaintainable source code produced by programmers to guarantee job security. (See Weinberg's Law.) There are many systems currently operating which contain varying amounts of Spaghetti Code. Some level of control of such systems, or parts of such systems, may be demonstrated by historical documentation of their reliability. In high-hazard environments, Spaghetti Code should be replaced as soon as possible.

SPSS Statistical Package for the Social Sciences

STANDARD OPERATING PROCEDURES (SOPs) 1) In the context of FDA guidelines, SOPs consist of written guidelines for dealing with any defined situation occurring during the conduct of clinical trials. 2) Formalized documentation describing and prescribing steps to be taken in normal and defined conditions. Standard Operating Practices refer to the actual activities with which procedures ought to correlate. Effectively controlled systems are managed through a series of SOPs that define security and disaster recovery procedures; data archiving, storage, and retrieval; software versioning and change control; user training; safety; and related operations.

STRESS TEST (See Erroneous Data Test).

STRUCTURE TESTS An examination of the internal processing logic of the soft ware. The goal is to achieve an optimum level of test coverage, both branching and path-traversing.

STUBS Temporary modules placed in a computer program until the real module has been written. In high-hazard environments, stubs should be validated.

SUPPORT The existence of support control implies a built-in clarity of training materials, manuals, and diagrams; the accessibility of experts and assistance; and a concern for finding the cause of problems rather than merely patching holes as they occur. Both system support and user support are intended to solve to solve problems as they emerge.

SUPPORT PERSONNEL The Support Personnel are responsible for providing custo mer support and informing the product manager and other appropriate personnel of specific problems with the product in the field.

SUPPORT PERSONNEL TRAINER The Support Personnel Trainer is responsible for developing a training manual and training the support personnel.

SYNTAX Rules of grammar and structure that must be followed to compile a program in a particular programming language.

SYNTAX REVIEW A second check of the adherence to rules by a programmer. This is facilitated by the fact that virtually all programming languages have a syntax check built in that will point out errors which must be corrected. The major concern in system validation is dealing with vendor proprietary languages that may not have checks built in.

SYSTEM 1) Totality of resources (personal, material, procedural) required to meet an objective. 2) Hardware, firmware, and software components combined to accomplish a variable task, such as data analysis, control, manipulation, transmission, storage, and retrieval.

The significant, defining factor of a computer system is the task performed by the system. This factor may lead to the definition of a system as a smaller, self-contained set of hardware, software and components. In more complex cases, it may lead to the definition of system as a group of hardware, software, and component sets functioning together to perform a given task.

SYSTEM ARCHITECT The System Architect is responsible for interpreting the functional specifications and translating them into a system design or architecture. This System Architecture is the global design of the system. The System Architect also ensures consistency among the modules as they are developed. This individual identifies the parts of the system that can utilize development tools, communicates system interdependencies to the system integrator and makes recommendations to the project manager.

SYSTEM ARCHITECTURE PHASE This phase in the Software Development Life Cycle (SDLC) defines the global system design. The global design must be consistent with the functional specifications and the project plan. The System Architect defines which modules need to be designed, how they interact, and how they interface. The System architect also identifies the need for development tools. The detailed design protocol is defined so any subsequent detailed design that follows the protocol is deemed correct. A review of the System Architecture Document to ensure the concepts in the functional specifications and project plan are addressed is a part of this phase.

SYSTEM BACKUP LOG Used to document system backups as a regularly performed activity done to safeguard existing data, as well as to minimize loss of data in the event of a system and/or application failure. Refer to SOP on system backup and restoration.

SYSTEM DEVELOPMENT PLAN (See Project Plan).

SYSTEM ERROR Failure by the automated system to perform according to expectations. This failure constitutes an AE and must be entered in an error log for review and possible maintenance.

SYSTEM HAZARD ANALYSIS The identification of specific internal and environ mental conditions that may cause hazards in a system. The analysis should identify components and subsystems in which faults or failures would cause hazardous conditions or actions, should describe the fault or failure, and should estimate the probability and severity of the hazard.

SYSTEM HAZARD ANALYSIS CROSS CHECK The System Hazard Analysis Cross Check (SHACC) is complied later in the development process, usually during the design phase. It is used to identify which hazards occur during which particular medical device functions. A matrix format is most commonly used, listing those hazards identified previously (in the PHA) vertically, and the medical device functions horizontally. A check mark is placed in the cell that correctly matches a potential hazard with a relevant function.

SYSTEM HAZARD ANALYSIS REPORT A document that records the proceedings and recommendations of a system hazard analysis.

SYSTEM INTEGRATION TESTING PHASE This phase in the Software Develop ment Life Cycle (SDLC) defines the system integration test plan. The test plan is performed to ensure that the modules are functioning together.

A review of the testing to ensure that the interface logic of the units, rather than the internal logic of the units, is sound is part of this phase. The internal logic of the units is tested in the implementation and module test plan development phase. Any errors found are reported with the supporting documents.

SYSTEM INTEGRATOR The System Integrator is responsible for ensuring that all the interdependencies between modules are consistent. The System Integrator writes a test plan to ensure that any data passed between programs is consistent. The test plan is performed on a system that is representative of a real-life system.

SYSTEM MAINTENANCE LOG This document may be used to demonstrate preventive maintenance of a particular piece of equipment at a certain time by qualified individuals. Refer to SOP on general physical safety of the system.

SYSTEM NETWORK ARCHITECTURE (SNA) A software protocol designed by IBM to allow networking of the various generations of IBM computers. Most major computer manufacturers have developed a network architecture protocol to support their own computer equipment, and many of these designs include a means to include computers from other manufacturers so as to increase the versatility of the network.

SYSTEM OPERATOR'S LOG This form is used for logging all activities related to the operation of a system at a particular site. These system-level activities refer to all actions involving the actual operation of the system and include regularly scheduled activities (such as backups and periodic maintenance) and unexpected activities (such as malfunctions).

SYSTEM PROBLEM LOG This document is used to chronicle all problems (and related fixes) to general system operations. Refer to SOP on system log --procedures and review.

SYSTEM REPAIR LOG Used to document all repairs (and related sign-offs) made to hardware related to a particular system. Refer to SOP on general physical safety of the system.

SYSTEM REQUIREMENTS SPECIFICATION (DOCUMENT) A document that prescribes completely and precisely the conditions and/or capabilities necessary for the system or product to fulfill its intended use and/or to satisfy any additional contracts, standards, specifications, or other formal agreements with the firm's customers. System Requirements Specifications typically should include safety requirements, functional requirements, and development standards. Synonymous with product requirements specification.

SYSTEM SECURITY Sensitive equipment may be damaged or erroneous results reported if the system is purposefully or accidentally subject to tampering. System security standards are designed to permit tracing of a product from raw materials to final product, and from final product back to raw materials, in the event of a recall or serious problem; to track all data input or modification back to the individual responsible for that input or modification; and to restrict unauthorized remote or direct access to confidential data and programs.

SYSTEM SPECIFIER The System Specifier is responsible for writing the functional specifications with input from the user advocate. The System Specifier is also responsible for reviewing the system test plan and the user's manual.

SYSTEM TESTER The System Tester is responsible for performing the system test plan and for documenting each step taken during testing. Any errors found are reported with supporting documentation. The System Tester is also responsible for verifying the technical accuracy of the user's manual.

SYSTEM TESTING PHASE In this phase of the Software Development Life Cycle (SDLC), the system tester performs the system test plan. The output documentation is generated to verify the system performs as described in the system test plan. A review of the output documentation to ensure it meets the expected output described in the system test plan is part of this phase.

SYSTEM TEST PLAN Refers to a test plan oriented to the system's environment, providing valid and invalid input data, specifying appropriate output, and writing detailed test instructions. The code is reviewed in a "black box" format, in which the specifics of the module functionality are ignored.

SYSTEM VALIDATION The accumulation of evidence in support of expert testimony demonstrating present and continuing control of a computer system utilized in manufacturing, research, quality assurance, or the function of a medical device. This evidence includes testing of the overall performance of the system, a review of the source code controlling the system applications, an analysis of the support documentation and standard operating procedures, and a review of the system security and disaster recovery policies.

TAKING OUT OF PRODUCTION Removing an automated system from the work flow, or a specific product or recipe/design from the manufacturing process.

TECHNICAL OPERATING PROCEDURES (TOPs) These procedures support SOPs. However, they need not have the same level of resilience as SOPs because they are more likely to change, and the process for their approval is less stringent. TOPs might be, for example, the instructions in an operating manual of some specific piece of equipment, or the reference guide of application software.

TECHNICAL WRITER The Technical Writer is responsible for writing and formatting the user's manual for the system. The Technical Writer obtains information from the functional specifications, test plans, on-line training, and interviews with the software engineers. The Technical Writer continually edits and updates the user's manual to ensure its validity during the product development cycle. This individual is also responsible for reviewing the manual for consistency, readability, and ease of use.

TEST CASE One or more sets of test data and associated procedures developed to exercise a particular function or program path, or to verify compliance with a specific requirement.

TEST DATA Refers to data-sets created to enable a judgement of system performance. It should include normal, unusual, and abnormal values for possible range of entries.

TEST-FILE STRESS ANALYSIS Using a test file in which programs can be examined without affecting the system, each significant program is tested with expected and unexpected stress input. Programs of special vulnerability, and programs not normally accessed in other tests, receive special consideration.

TESTING Designing, conducting, analyzing, and interpreting a series of white, black, and grey box tests to assure the current accuracy and reliability of computer hardware and software system components. These tests generally include a comprehensive source code review, performance test, stress test, and other tests as appropriate.

TESTING COMPONENTS As part of the overall validation process, a series of tests to assure the current accuracy and reliability of computer hardware and software system components. In this study, specific testing includes:

Ongoing System Performance Testing
Comprehensive (50 Case) System Performance Testing
Test File Stress Analysis
Program Logic Source Code Review
Live-Environment Error Log Review

TEST PLAN A document prescribing the approach to be taken for intended testing activities, typically including: items to be tested, tests to be performed, schedules, personnel requirements, evaluation criteria, risks, and contingency plans.

TEST PROCEDURES DOCUMENT A document that contains detailed instructions (Test Procedures) to set up and conduct one or more test cases and to evaluate the results.

TEST REPORT A document that records the conduct and results of test procedures carried out for a system or system component. (See Test Procedures Document).

TEST REQUIREMENTS SPECIFICATION A document that prescribes completely and precisely the test cases, including correct results, necessary to verify that the system, product, or software possesses, meets, or fulfills the appropriate requirements specification. A Test Requirements Specification may stand alone or may be part of a test procedures document.

THROUGHPUT That which is processed. The status of data before it is output. In effect the vulnerability of data to things like truncation or erroneous algorithms during throughput needs to be checked.

TICKIT Refers to a specific set of software quality-assurance guidelines required in the United Kingdom and similar to the ISO 9000 series.

TOOLS Refers to automated aids for the development of systems.

TQM Total Quality Management

TRACKING Inventory, case, subject, condition, and code identifiers should permit dual-directional tracking in the event that recall necessity emerges or source questions arise.

TRAINING The system developer, supporter, or vendor has designed and implemented a system of user training designed to provide basic operations and interpretation background. The training program includes identification of sub-skills or lessons; identification of appropriate user background; and identification of appropriate user authorization, approval, testing or certification examinations.

TRAINING LOG All user training must be documented. The training supervisor should track employee training and follow-up review of the users' skills.

TRAINING SOPs A user organization (in-house) developed system of training and authorizing procedures for user access, which includes the following elements:

1) Identification, outline, or listing of specific training modules, skills, parameters, or tasks.

2) Log of training activities identifying delivery of above listed training elements, specific by user.

3) System of testing, checking, or approving user mastery of the training elements listed above.

4) System of periodic review, recheck, or reinforcement of user skills.

TRAINING SUPPORT PERSONNEL PHASE This phase in the Software Develop ment Life Cycle (SDLC) defines a training program and trains the support personnel on the new system. Written documentation must be provided to support personnel to aid them in understanding and supporting the system. This phase includes a review of the support training documentation to ensure its consistency with the developed product.

UNOBTRUSIVE OBSERVATION Refers to the process of observing a system or personnel without creating subconscious demands; observation without modifying the entity being observed.

UPDATES AND VERSION CHANGES All new versions and updates of software utilized in or in connection with the system under review is subject to careful retesting prior to implementation, meeting appropriate verification standards.

USER The User is responsible for performing the limited release phase and for reporting any problems encountered with supporting documentation. The User is the person who uses the final product on a day-to-day basis. Responsibility for familiarity with and conformity to SOPs is shared by all system users. Similarly, while responsibility for the enforcement of security controls and for the delivery of adequate training may be vested elsewhere, all users are expected to cooperatively comply with and support management policies. [In the GALP Standards section 5.0 the code letter "U" is used to identify all specific and direct responsibilities of the system users.]

USER ADVOCATE The User Advocate, with the user(s) input, is responsible for proposing the system design goals. The User Advocate reviews the functional specifications and the user's manual to ensure compatibility with the system design goals.

USER APPLICATION DOCUMENTATION All appropriate system documents should be carefully logged in an accessible location. These documents should include current and historical user manuals; source code or source code access information; current and historical technical manuals; validation and testing studies; maintenance records; component inventories; and appropriate regulatory, managerial, or quality control reviews.

USER IDENTIFICATION The system should permit ready identification of the unique user responsible for data input, modification, interpretation, or storage. Such identification is dependent upon a reliable password system and upon the maintenance of a user log and access file.

USER PROBLEM LOG This form is used in the logging of user-reported problems related to the operation of a system and associated applications software, located at a specific site. The User Problem Log differs from the system log in that The User Log pertains to problems related to specific applications. The system log entries are more related to the computer system as a whole. Refer to SOP on user-reported problem log.

USER'S MANUAL AND DOCUMENTATION PHASE This phase in the Software Development Life Cycle (SDLC) develops the user documentation for the system. This documentation includes detailed instructions of the system features and operations and may also contain user validation test plans. The documentation is generated from the functional specifications, test plans, on-line training, and interviews with software engineers. The manual must be technically accurate. The format, readability, ease of use, and consistency must be preserved between different revisions. The user documentation must be continually reviewed and edited to reflect changes to the system.

USER VALIDATION A high-level review conducted to determine the conformity of a system, process, or product to internally established standards relating to its use in a specific (user site) environment; and the appropriateness of those standards to that specific site.

VALIDATION A high-level review to determine the conformity of a system, process, or product to internally established standards and the appropriateness of those standards. Validation is a second "quality assurance" examination by an independent party designed to provide a high level of confidence of control of the system or process for managerial and regulatory personnel.

VALIDATION CERTIFICATE Written document certifying that a system has successfully been through the Validation process. This is the first proof that an inspector will expect to see if s/he asks if a system is validated. FDA field investigators have stated that the signature on the certificate may be as far as they go in assessing the validity of the system.

VALIDATION COMPONENTS Systems validation is the comprehensive process of determining the reliability, accuracy, and completeness of a computer system and the ability of that system to track, recall, preserve, and manipulate data free of system-induced variance. The validation process includes an examination of: standard operating procedures (SOPs), system testing procedures; vendor related verification studies; system technical and user documentation, security protocols, and the planning and procedures for recovery in the event of disaster.

VALIDATION SOPs Standard operating procedures outlining a corporate-wide or organization-wide validation policy should be included and available for review.

VENDOR The organization or individual that designs, codes, supports, licenses, and/or distributes automated systems has some responsibilities under the GALP standards. These responsibilities generally impose design, support, and notification requirements. If the vendor is an outside source, providing an awareness of these guidelines may be a necessary role of the laboratory (via its management). In an in-house developed system, these controls may fall under the general purview of the RP. [In the GALP standards section 5.0 the code letter "V" is used to identify all specific and direct responsibilities of the system vendor.]

VENDOR VALIDATION A high-level review conducted to determine the conformity of a system, process, or product to those internally established standards in use by the developer, distributor, or seller and the appropriateness of those standards. The vendor package, supplied to a user, becomes the core of the user validation.

VERIFICATION A review of the vendor supplied documentation and support for a new system, designed to determine the degree of conformity of the vendor's procedures with the standards established in the user's validation policy. The verification may include spot checks on the accuracy of the data, as well as an examination of the procedures used collecting that test data.

VERIFIER Person who re-keys information to ensure its accuracy.

VERSION, UPDATE

PROPOSED POLICY: All functional user and developmental areas (working with the specific software system) shall endorse and adopt a uniform and consistent methodology for defining software versions. This methodology shall be based upon a three-part designation: **X Y Z** with the following denotations:

1. The *X* designation shall represent the primary version. All software in development (prior to initial release) shall carry an *X* designation of "0;" the initial release of that software shall carry the *X* designation of "1." The *X* designation shall increase (to "2," "3," etc.) in the following circumstances:

a. Software is converted to run on a new hardware platform, within or between competing vendor brands;

b. Software is converted to run under a new operating system, including a new release of the existing operating system if that release carries an increased *X* designation itself;

c. Software modifications of any extent or degree directly affecting fundamental safety issues surrounding the function of the computer software system or its final product, including but not limited to modifications prompted by: newly instituted regulatory policies, when those policies interactively determine the suitability of the end product for human use; in-house specified new release criteria or restrictions;

FDA-mandated changes under formal 483 notification; and modifications resulting from discovered major functional errors in previous versions;
d. Software modifications rendering the current (replacement) version incompatible with the new
version;
e. Software modifications redefining or restructuring files, tables, or case definitions.
2. The *Y* designation shall represent the secondary version. The *Y* designation increases according to the following rules within each *X* number; a new primary version resets the *Y* designation at "0". Initial release of new software systems carries the designation "1.0.0," regardless of the *Y* designation under the development version. The *Y* designation increases incrementally under the following conditions:
a. A release of a correction version in which the errors modified did NOT affect the fundamental safety of the final product processing operation;
b. Software is subject to minor modifications to run on a new compatible hardware platform;
c. Software is subject to minor modifications to run under a modified operating system not designated by a new *X* version number;
d. Software is enhanced with a modification including any or all of the following:
-- expansion or expanded definition of file, decision table, or case structure;
-- a new feature or capacity not fundamental to or critical to the basic operation of the system, nor to conformity with FDA (or other regulatory) mandated changes;
-- a new feature intended as a user aid, help tool, or other support attribute;
3. The *Z* designation shall represent the tertiary version. The *Z* designation increases incremental whenever one of the listed modifications is released, reverting to "0" whenever the *Y* designation is changed. Changes affecting only the *Z* designation include:
a. Enhancements in internal ("comment") documentation;
b. Software modifications to screen layout not affecting the input data or variable assignment;
c. The addition of compatible, independently-validated or accepted, non-customized additional software programs. For example, adding a word processor package to a system residing within the software shell, but not interacting with the files, tables, cases, or other applications programs in that system, would result in an increased *Z* designation.
4. In all "borderline," ambiguous, or questionable cases, judgement shall default in a conservative direction, including:
a. Changes that may represent either *Z* or *Y* modifications shall be considered *Y* modifications, resulting in an incremental increase in the *Y* version designation;
b. Changes that may represent either *Y* or *X* modifications shall be considered *X* modifications, resulting in an incremental increase in the *X* version designation.

WALKTHROUGH 1) Physical tour of an installation or system to review its compli ance to regulations, SOPs, etc. 2) Logical tour of software where a coder explains to one or more peers what has been done and why.

WEINBERG'S LAW If builders built buildings the way programmers program programs, the first woodpecker to wander by would destroy civilization.

WHITE BOX TEST Testing through a system, examining the process of flow at every stage of operation. In system validation, a source code review serves as the primary White Box Test.

WORLD HEALTH ORGANIZATION (WHO) Based in Geneva, Switzerland, the WHO assists countries researching their health needs. Many publications are available, including over 50 on pharmacology. A free publications directory is available upon request.

WRITE-ACCESS TO LIBRARIES Ability to manipulate/create/modify programs in a given system section. Developmental personnel should be confined to the test libraries and should be unable to move programs into production or access the programs once they are in production.

X.25 Communication protocol for packet switching.

X.400 Communication protocol for E-mail.

YURNAUKAS' OBSERVATION To err is human; to really foul things up takes a computer.

Appendix B

VALIDATION BIBLIOGRAPHY

Accomando, William P. CANDA Overview: What Should a CANDA Look Like? *Drug Information Journal*, Vol. 27 (April-June 1993).

Ackoff, Russell L. Management Misinformation Systems. *Management Science*, December 1967, pp. 147-156.

Adamson, J.R. An Approach to Validation. *Pharmaceutical Engineering*, Vol. 12 (May 1992).

Advanced Micro Devices. The AM7990 Family Ethernet Node. Sunnyvale, CA: Advanced Micro Devices, 1983.

Agalloco, James. Validation of Existing Computer Systems. *Pharmaceutical Technology*, Vol. 11 (January 1987).

Agalloco, James. The Validation Life Cycle. *Journal of Parenteral Science & Technology*, Vol. 47 (May-June 1993).

Agalloco, James and Vittorio Mascherpa. Validation of a Computerized System for Autoclave Control. *Pharmaceutical Technology*, Vol. 19 (January 1995).

AICPA. Audit Approaches for a Computerized Inventory System. New York: American Institute of Certified Public Accountants, 1980.

AICPA. Audit Considerations in Electronic Fund Transfer Systems. New York: American Institute of Certified Public Accountants, 1979.

AICPA. Auditing Standards and Procedures. Committee on Auditing Procedure Statement No. 33. New York: American Institute of Certified Public Accountants, 1963.

AICPA. Guidelines to Assess Computerized General Ledger and Financial Reporting Systems for Use in CPA Firms. New York: American Institute of Certified Public Accountants, 1979.

AICPA. Management, Control and Audit of Advanced EDP Systems. New York: American Institute of Certified Public Accountants, 1977.

AICPA. Statement on Auditing Standards No. 3, The Effects of EDP on the Auditor's Study and Evaluation of Internal Control. New York: American Institute of Certified Public Accountants, 1974.

AICPA. The Auditor's Study and Evaluation of Internal Control in EDP Systems. New York: American Institute of Certified Public Accountants, 1977.

Aitman, Mark. Validation of High Integrity Systems. *Drug Information Journal*, Vol. 27 (April-June 1993).

Almgren, Raymond. Software Development Strategies for PC-Based Testing. *Evaluation Engineering*, January 1991.

Allocca, John A. and Allen Stuart. Transducers Theory and Applications. Reston, VA: Reston Publishing Company, Inc., 1984.

American Chemical Society Committee. Guidelines for Data Acquisition and Data Quality Evaluation in Environmental Chemistry. *Analytical Chemistry*, Vol. 52, 1980.

Alter, Steven. Decision Support Systems: Current Practice and Continuing Challenges. Reading, MA: Addison-Wesley Publishing Company, 1980.

Alwitt, Josh and John Kinney. The Impact of Document Image Management on Clinical Data Processing. *Drug Information Journal*, Vol. 27, (October-December 1993).

AMF Potter & Brumfield Catalog. Princeton, IN: Potter & Brumfield, 1984.

Analog Devices, Inc. Analog Devices 3B Industrial Control Series Data Sheet. Norwood, MA, 1981.

Anon. Products-Information Management. *Laboratory Practice*, Vol. 38 (May 1989).

Andersen, Arthur. A Guide for Studying and Evaluating Internal Controls. Chicago: Arthur Andersen and Co., 1978.

Andersen, Niels Erik, *et al*. Professional Systems Development. New York: Prentice-Hall International, 1990.

Anderson, C.A. Approach to Data Processing Auditing. *The Interpreter*, Insurance Accounting and Statistical Association, Durham, NC, April 1975, pp. 23-26.

Anderson, John. Object-Oriented Software Development: Method Assessment. George Mason University, Spring 1990.

Anisfeld, Michael H. Validation -- How much Can the World Afford: Are We Getting Value for Money? *Journal of Pharmaceutical Science and Technology*, Vol. 48 (January/February 1994).

Ariv, Gad and Michael J. Ginzberg. DDS Design: A Systematic View of Decision Support. *Communications of the ACM*, Vol. 28 (October 1985), pp. 1045-1052.

Arthus, L.J. Measuring Programmer Productivity and Software Quality. New York: John Wiley & Sons, 1985.

Aron, Joel. The Program Development Process, Part II: The Programming Team. Reading, MA: Addison-Wesley, 1983.

Auerbach Staff. What Every Auditor Should Know About DP. Data Processing Management Service, Auerbach Publishing Co., Portfolio 3-09-03, 1975.

Auslander, David M. and Paul Sagues. Microprocessors for Measurement and Control. Berkeley: Osborne/McGraw-Hill, 1981.

Awad, Elias M. Introduction to Computers in Business. Englewood Cliffs, NJ: Prentice-Hall, Inc., 1977.

Axner, David H. and Fonnie H. Regan. Alphanumeric Display Terminal Survey. *Datamation*, Vol. 24, (June 1978), pp. 183-219.

Baber, Robert L. Error-Free Software: Know-How and Know-Why of Program Correctness. John Wiley and Sons, Ltd., 1991.

Baer, Robert M. The Digital Villain. Reading, MA: Addison-Wesley Publishing Co., Inc., 1972.

Baird, Lindsay L. Identifying Computer Vulnerability. *Data Management*, June 1974, pp. 14-17.

Bar-Hava, Ne. Training EDP Auditors. *Information and Management*, No. 4, 1981, pp. 30-42.

Barnard, Chester I. The Functions of the Executive. Cambridge: Harvard University Press, 1968.

Barr, David B. FDA's Aseptic Processing: Proposed Regulation. *Journal of Parenteral Science & Technology*, Vol. 47 (March-April 1993).

Barrett, M.J. Education for Internal Auditors. *EDP Auditor*, Spring 1981, pp. 11-20.

Barry, Anne R. and Robert Chojnacki. Biotechnology Product Validation: Chromatography Media and Column Qualification. *BioPharm*, Vol. 7 (September 1994).

Basalla, George. The Evolution of Technology. Cambridge: Cambridge University Press, 1988.

Basden, A. and E.M. Clark. Data Integrity in a General Practice Computer System (CLINICS), *International Journal of Bio-Medical Computing*, Vol. 11 (1980), pp. 511-519.

Bauer, Nancy, Bruce Binkowitz, and Wanda Bidlack. The Progression and Integration of the Statistical CANDA at Merck. *Drug Information Journal*, Vol. 27 (January-March, 1993).

Beckman Instruments. Hall Effect Manual. 2nd ed. Fullerton, CA: Helipot Division of Beckman Instruments, Inc., 1964.

Beckman Instruments. Interfacing Liquid Crystal Displays in Digital Systems, Application Note AN-B. Beckman Instruments, Inc., Scottsdale, AZ, 1980.

Beizer, Boris. The Frozen Keyboard. Blue Ridge Summit, PA: PAB Books, 1988.

Beizer, Boris. Software System Testing and Quality Assurance. New York: Van Nostrand Reinhold, 1984.

Bellin, David and Gary Chapman, eds. Computers in Battle. New York: Harcourt, Brace, Javanovich, 1987.

Bennett, John L., ed. Building Decision Support Systems. Reading, MA: Addison Wesley Publishing Company, 1983.

Bentley, Jon. More Programming Pearls: Confessions of a Coder. Reading, MA: Addison-Wesley, Publishing Company, 1988.

Benton, F. Warren. EXECUCOMP: Maximum Management with the New Computers. New York: John Wiley & Sons, 1983.

Bequai, August. How to Prevent Computer Crime. New York: John Wiley & Sons, 1984.

Berkeley, Peter E. Computer Operations Training: A Strategy for Change. New York: Van Nostrand Reinhold Company, 1984.

Berkland, Richard D. Validatable Software Design. *Pharmaceutical Technology*, Vol. 15 (September 1991).

Biggerstaff, Ted. J. and Alan J. Perlis, eds. Software Reusability. Vol. 2, Applications and Experience. New York: ACM Press/Addison-Wesley, 1989.

Birnbarum, N., J. Hains, J. Farkas, and P. Cruz. A LIMS Spectrophotometer Interface. *American Laboratory*, March 1990.

Black, Henry C. Black's Law Dictionary. Revised Fourth Edition. St. Paul: West Publishing Co., 1968.

Blake, Chuck (American Red Cross Vice President for Computer and Information Systems), private interview, March 1989.

Bliem, R.F. Impact of Research and Development on Validation, Good Manufacturing Practice, and Registration of Biopharmaceuticals. *Pharmaceutical Engineering*, Vol. 15 (May-June 1995).

Block, Robert. The Politics of Projects. Englewood Cliffs, NJ: Yourdon Press/Prentice-Hall, 1983.

Boar, Bernard. Application Prototyping. New York: John Wiley & Sons, 1984.

Board of Governors of the Federal Reserve System. Electronic Fund Transfers, Regulation E (12 CFR 205), Effective March 30, 1979 (as amended effective May 10, 1980).

Boehm, B.W. Software Engineering Economics. Englewood Cliffs, NJ: Prentice-Hall, 1982.

Bolt, Richard A. The Human Interface. London: Wadsworth, Inc., 1984.

Bonczek, Robert H., Clyde W. Holsapple, and Andrew B. Whinston. Foundations of Decision Support Systems. New York: Academic Press, Inc., 1981.

Boskin, Michael I. Bashing the Myths About LIMS. *Plastics Technology*, October 1989.

Boskin, Michael I. and Rick A. Ziebell. LIMS: An SPC Tool. *Quality*, May 1990.

Bottenberg, R.A. and R.E. Christal. An Iterative Technique for Clustering Criteria Which Retains Optimum Predictive Efficiency. Lackland Air Force Base, Texas: Personnel Laboratory, Wright Air Development Division, March 1961 (WADD-IN-61-3, ASTIA Document AS-261 615).

Bouldin, Barbara. Agents of Change. Englewood Cliffs, NJ: Prentice-Hall, 1989.

Bozzo, Tom. Biotechnology Product Validation: A Former FDAer's View. *BioPharm*, Vol. 8 (January 1995).

Braff, Terry. A Test Program -- Actual Costs and Benefits. 5th International Conference on Testing Computer Software, Measuring Testing Success, 1988.

Breitenberg, Maureen. Questions and Answers on Quality, the ISO 9000 Standard Series, Quality System Registration, and Related Issues. NISTIR 4721. Gaithersburg, MD: U.S. Department of Commerce, National Institute of Standards and Technology, Standards Code and Information Program, April 1993.

Bretzin, M.J. and M.L. Hess. Quality Assurance of the Computer Validation Process in the Control Laboratory. *Pharmaceutical Engineering*, Vol. 12 (May 1992).

Bronstein, Robert J. The Concept of a Validation Plan. *Drug Information Journal*, Vol. 20 (1986), pp. 37-42.

Brooks, Fred. The Mythical Man-Month. Reading, MA: Addison-Wesley, 1975.

Brooks, F.P. No Silver Bullet: Essence and Accidents of Software Engineering. *Computer*, Vol. 20 (April 1987).

Brown, Elizabeth H. Procedures and Their Documentation for a LIMS in a Regulated Environment. In R.D. McDowall, ed. Laboratory Information Management Systems. Wilmslow, U.K.: Sigma Press, 1987, pp. 346-358.

Bruckheimer, Michael. FDA's Inspection of Clinical Investigators. *Drug Information Journal*, Vol. 27 (January-March 1993).

Buxton, J.M., P. Naur, and B. Randell. Software Engineering, Concepts and Techniques. New York: Petrocelli/Charter, 1976.

Buyse, Marc. Regulatory Versus Public Health Requirements in Clinical Trials. *Drug Information Journal*, Vol. 27 (October-December 1993).

Callahan, John J. Needed: Professional Management in Data Processing. Englewood Cliffs, NJ: Prentice-Hall, 1983.

Campbell, D.T. and J.C. Stanley. Experimental and Quasi-Experimental Designs for Research. Chicago: Rand McNally, 1963.

Campbell, S.K. Flaws and Fallacies in Statistical Thinking. Englewood Cliffs, NJ: Prentice-Hall, 1974.

Campo, Jose A. Design Transfer. *Medical Device & Diagnostic Industry*, Vol. 16 (February 1994).

Canadian Institute of Chartered Accountants Committee. Competence and Professional Development in EDP for the CA. *CA Magazine*, September 1974, pp. 26-70.

Canning, Richard G. COBOL Aid Packages. *EDP Analyzer*, January 1972.

Canning, Richard G. Computer Security: Backup and Recovery Methods. *EDP Analyzer*, January 1972.

Canning, Richard G. That Maintenance "Iceberg." *EDP Analyzer*, October 1972.

Canning, Richard G. Project Management Systems. *EDP Analyzer*, September 1976.

Caputo, C.A. Managing the EDP Audit Function. *COM-SAC, Computer Security, Auditing and Controls*, Vol. 8, January 1981, pp. A-1 to A-8.

Card, David N. with Robert L. Glass. Measuring Software: Design Quality. Englewood Cliffs, NJ: Prentice-Hall, 1990.

Cardarelli, Joseph S., Sami A. Halaby, Chris Gopal, and Art Mlodozeniec. Initiating CIM: Manufacturing vs. Pilot-Plant Operations. *Pharmaceutical Technology*, Vol. 15 May 1991).

Carr, Jim, Michael Eddy, and Albert Rego. How to Implement an Electronic Document Storage and Retrieval System. *Medical Device & Diagnostic Industry*, Vol. 15 (February 1993).

Cash, James I., Jr., F. Warren McFarlan, and James L. McKenney. Corporate Information Systems Management: Text and Cases. Homewood, IL: Richard D. Irwin, 1983.

Casti, John L. Paradigms Lost. New York: William Morrow, 1989.

Chamberlain, Richard. Computer Systems Validation for the Pharmaceutical and Medical Device Industries. 2nd Edition. Buffalo Grove, IL: Interpharm Press, 1994.

Chapman, Kenneth G. Challenges, Transitions,and Paradigm Shifts for Information Technology. *Pharmaceutical Technology*, Vol.19 (October 1995).

Chapman, Kenneth G. A History of Validation in the United States: Part I. *Pharmaceutical Technology*, Vol. 15 (October 1991).

Chapman, Kenneth G. A History of Validation in the United States: Part II, Validation of Computer-Related Systems. *Pharmaceutical Technology*, Vol. 15 (November 1991).

Chapman, Kenneth G. and Paul F. Winter. Electronic Identification and Signatures: A Response. *Pharmaceutical Technology*, Vol. 16 (November 1992).

Chapman, M. Audit and Control in Data Base/IMS Environment. *COM-SAC, Computer Security, Auditing and Controls*, Vol. 6, July 1979, pp. A-9 to A-14.

Charette, Robert N. Software Engineering Risk Analysis and Management. New York: McGraw-Hill, 1989.

Cheney, P.H. Educating the Computer Audit Specialist. *EDP Auditor*, Fall 1980, pp. 9-15.

Chew, Nancy. Validation: When Too Much is Not Enough. *BioPharm*, Vol. 6 (September 1993).

Christ, Oliver P. Software Validation: Requirements for Programmable Electronic Medical Systems, Part I, How to Comply with the Essential Requirements of the EC Directives. *Medical Device Technology*, March 1994.

Christian, Kaare. The UNIX Operating System. New York: John Wiley & Sons, 1983.

Chudzicki, Mark J. LIMS: Family Designs for Growth, Flexibility, and Corporate Productivity. *Research & Development*, November 1989.

CICA. Computer Audit Guidelines. Canadian Institute of Chartered Accountants, Toronto, Ontario, Canada, 1975.

CICA. Computer Control Guidelines. Canadian Institute of Chartered Accountants, Toronto, Ontario, Canada, 1971.

Circuit Court of Tennessee. Integrated Medical Systems, Inc. vs. Methodist Health-System, Inc., case number 16316-2. Circuit Court of Tennessee, Thirteenth Judicial Circuit at Memphis, July 1986.

Clark, A. Samuel. Computer Systems Validation: An Investigator's View. *Pharmaceutical Technology*, Vol. 12 (January 1988).

Clark, C.T. and L.L. Schkade. Statistical Methods for Business Decisions. Cincinnati: South-Western Publishing, 1969.

Clark, Robin and Paul Redstone. Establishing the Functions to be Validated. *Drug Information Journal*, Vol. 27 (April-June 1993).

Clary, J.B. and R.A. Sacane. Self-Testing Computers. IEEE Computer, October 1979, pp. 45-59.

Clinical Laboratory Improvement Act of 1967 (P.L. 90-174, December 5, 1967).

Clinical Laboratory Improvement Amendments of 1988 (P.L. 100-578, October 31, 1988).

Clocksin, W.F. and C.S. Mellish. Programming in PROLOG. New York: Springer-Verlag, 1981.

Coady, P.J. and A.P. DeClaire. Best Practice Engineering for Validation of Process Control Systems. *Pharmaceutical Engineering*, Vol. 15 (July-August 1995).

Cochran, W.G. Sampling Techniques. New York: John Wiley, 1963.

Cochran, W.G. and G.M. Cox. Experimental Designs. 2nd ed. New York: John Wiley, 1957.

College of American Pathologists. Standards for Laboratory Accreditation. Skokie, IL: Commission on Laboratory Accreditation, College of American Pathologists, 1988.

Cole, Jon, Tyler W. Mainquist, and James R. Cichoracki. Computerization and Computer Validation: Managing the Risks. *Applied Clinical Trials*, Vol. 2 (December 1993).

Collins, Mary, ed. CANDA, 1995: An International Regulatory and Strategy Report. Waltham, MA: Parexel International Corp., 1994.

Connell, John L. and Linda Brice Shafer. Structured Rapid Prototyping. Englewood Cliffs, NJ: Prentice-Hall, 1989.

Connolly, Thomas M. EPA Investigations of Laboratory Fraud in the CLP and GLP Programs. Presentation at the Society of Quality Assurance 1993 Annual Meeting, San Francisco, October 6, 1993.

Connor, J.E. and B.H. De Vos. Guide to Accounting Controls -- Establishing, Evaluating and Monitoring Control Systems. Boston: Warren, Gorham & Lamont, Inc., 1980.

Conover, W.J. Practical Nonparametric Statistics. New York: John Wiley, 1971.

Contrac Corporation. Raster Graphics Handbook. Covina, CA: Conrac Corporation, 1980.

Cooke, John E. and Donald H. Drury. Management Planning and Control of Information Systems. Hamilton, Ontario: Society of Management Accountants of Canada, April 1980.

Cummings, Stuart W. Distributed Databases for Clinical Data Processing. *Drug Information Journal*, Vol. 27 (October-December 1993).

Crosby, P.B. Quality is Free. New York: McGraw-Hill, 1979.

Cutaia, Al. Technology Projection Modeling of Future Computer Systems. Englewood Cliffs, NJ: Prentice-Hall, 1990.

Cutter, Michael. Effective GMP Training. *Pharmaceutical Technology Europe*, Vol. 7 (January 1995).

Dahl, O.J., Edsger Dijkstra, and C.A.R. Hoare. Structured Programming. Englewood Cliffs, NJ: Prentice-Hall, 1972.

Daniel, W.W. Applied Nonparametric Statistics. Boston: Houghton Mifflin, 1978.

Daniels, Michael W., Douglas O. Fegenbush, and Edward F. Frees. An Alphanumeric and Graphic LIMS for Routine Physical Chemistry Analyses. *American Laboratory*, July 1992.

Data Acquisition Telecommunications. Local Area Networks 1983 Data Book, Sunnyvale, CA: Advanced Micro Devices, Inc., 1983.

Datapro Research. Datapro Reports on Information Security. Delran NJ: McGraw- Hill, Inc., 1989.

Date, Chris. An Introduction to Database Systems. Vol. 1, 4th ed. Reading, MA: Addison-Wesley, 1986.

Davey, Judith. Electronic Data Transfer for Clinical Trials. *Drug Information Journal*, Vol. 28 (April-June 1994).

David, F.N. Games, Gods, and Gambling. London: Charles Griffin and Company, 1962.

Davies, O.L., ed. The Design and Analysis of Industrial Experiments. New York: Hafner Publishing, 1956.

Davis, G. Ensuring On-Line System Integrity Using Parallel Simulation on a Continuing Basis. *COM-SAC, Computer Security, Auditing and Controls*, Vol. 7 (July 1978), pp. A-9 to A-11.

Davis, Gordon W. Early Development of the Clinical Information System of a Start-Up Pharmaceutical Company. *Applied Clinical Trials*, Vol. 3 (June 1994).

Davis, Keith. Human Behavior at Work. New York: McGraw-Hill, 1972.

Davis, Randal, Bruce Buchanan, and Edward Shortliffe. Production Rules as a Representation for a Knowledge-Based Consultation Program. *Artificial Intelligence*, Vol. 8 (1977).

Davis, Stanley M. Future Perfect. Reading, MA: Addison-Wesley, 1987.

Davis, W.P., W.J. Erhard and J.C. McKenney. Computer System Validation: Value of System Prototyping and Vendor Acceptance Testing. *Pharmaceutical Engineering*, Vol. 15 (July-August 1995).

Dean, R.N. A Comparison of Decision Tables Against Conventional COBOL as a Programming Tool for Commercial Applications. *Software World*, Spring 1971.

Dearden, John. MIS is a Mirage. *Harvard Business Review*, January-February, 1972, pp. 90-99.

De Feu, Marc. ISO 9000 and its Relation to System Validation. *Drug Information Journal*, Vol. 27 (April-June 1993).

DeMarco, Thomas. Controlling Software Projects. New York: Yourdon Press, 1982.

DeMarco, Thomas. Structured Analysis and System Specification. Englewood Cliffs, NJ: Yourdon Press/Prentice-Hall, 1978.

DeMarco, Thomas and Tim Lister. Peopleware. New York: Dorset House, 1987.

DeMarco, Thomas and Tim Lister. Software State-of-the-Art: Selected Papers. New York: Dorset House, 1991.

Dent, Nigel J., ed. Implementing International Good Practices: GAPs, GCPs, GLPs, and GMPs. Buffalo Grove, IL: Interpharm Press, 1993.

DeSain, Carol. Drug, Device and Diagnostic Manufacturing: The Ultimate Resource Handbook. 2nd Edition. Buffalo Grove, IL: Interpharm Press, 1992.

DeSain, Carol and Charmaine Vercimak. Documentation Basics that Support Good Clinical Practices: The Master Plan. *Applied Clinical Trials*, Vol. 2 (June 1993).

DeSain, Carol and Charmaine Vercimak. Documentation Basics that Support Good Clinical Practices: Part II. *Applied Clinical Trials*, Vol. 2 (July 1993).

DeSain, Carol and Charmaine Vercimak. Documentation Basics that Support Good Clinical Practices: Part III, Protocols and SOPs. *Applied Clinical Trials*, Vol. 2 (September 1993).

DeSain, Carol and Charmaine Vercimak. Documentation Basics that Support Good Clinical Practice: Part IV, Clinical Data Collection Documents. *Applied Clinical Trials*, Vol. 2 (October 1993).

DeSain, Carol and Charmaine Vercimak. Documentation Basics that Support Good Clinical Practice: Part V, Clinical Systems of Accountability and Traceability. *Applied Clinical Trials*, Vol. 2 (November 1993).

DeSain, Carol and Charmaine Vercimak. Implementing International Drug, Device, and Diagnostic GMPs. Buffalo Grove, IL: Interpharm Press, 1994.

DeSain, Carol and Charmaine Vercimak Sutton. Process Development that Supports Process Validation. *Pharmaceutical Technology*, Vol. 19 (October 1995).

Dessy, Raymond E. The Electronic Laboratory. Washington, D.C.: American Chemical Society, 1985.

Dewoskin, Robert S. and Stephanie M. Taulbee. International GLPs. Buffalo Grove, IL: Interpharm Press, 1993.

Dice, Barry, Operations Manager, Sovran Financial Corp., Telephone Interview, April 25, 1990 (Hyattsville, MD).

Dietz, David L. and Christie J. Herald. Reconciling a Software Development Methodology with the PMA Validation Life Cycle. *Pharmaceutical Technology*, Vol. 16 (June 1992).

Dijkstra, Edsger. A Discipline of Programming. Englewood Cliffs, NJ: Prentice-Hall, 1976.

Dixon, W.J. and F.J. Massey, Jr. Introduction to Statistical Analysis. 3rd ed. New York: McGraw-Hill, 1969.

Dolotta, T.A., *et al.* Data Processing in 1980-1985: A Study of Potential Limitations to Progress. New York: John Wiley & Sons, Inc., 1976.

Dominy, K.S. and M.A. Fazio. Integrated Valiation Project Approach. *Pharmaceutical Engineering*, Vol. 15 (July-August 1995).

Dorf, Richard C. Robotics and Automated Manufacturing. Reston, VA: Reston Publishing Company, 1983.

Dorfmann, Joan. Data Center Housekeeping. *Infosystems*, July 1984.

Dorricott, K.O. Organizing a Computer Audit Specialist. *CA Magazine*, May 1979, pp. 66-68.

Double, Mary Ellen and Maryann McKendry. Computer Validation Compliance: A Quality Assurance Perspective. Buffalo Grove, IL: Interpharm Press, 1993.

Draper, N. and H. Smith. Applied Regression Analysis. New York: John Wiley, 1966.

Drug Information Association. Computerized Data Systems for Nonclinical Safety Assessment: Current Concepts and Quality Assurance. Maple Glen, PA: Drug Information Association, 1988.

Ducate, M.J., J.P. Eycken, L.E. Rivera and S. Edmonds. Framework for Developing Global Standard Operating Procedures. *Drug Information Journal*, Vol. 28 (October-December 1994).

Dunn, O.J. and V.A. Clark. Applied Statistics: Analysis of Variance and Regression. New York: John Wiley, 1974.

Dunn, Robert. Software Defect Removal. New York: McGraw-Hill, 1984.

Dunn, Robert and Richard Ullman. Quality Assurance for Computer Software. New York: McGraw-Hill, 1982.

EDP Auditors Foundation. Control Objectives -- 1980. Streamwood, IL: EDP Auditors Foundation, 1980.

Edwards, Perry and Bruce Broadwell. Data Processing. Belmont, CA: Wadsworth Publishing Co., Inc., 1979.

Egol, Len. Putting PCS to Work in the Lab. *Chemical Engineering*, August 1990.

Ehrmeyer, S.S. and R.H. Laessig. Monitoring Clinical Trials in Multiple Testing Sites: Can Good Laboratory Practice and Proficiency Testing Assure Uniform Testing? *Drug Information Journal*, Vol. 28 (October-December 1994).

Electronic Industries Association. EIA Standard RS-422, Electrical Characteristics of Balanced Voltage Digital Interface Circuits. Washington, D.C.: EIA, Engineering Department, 1975.

Electronic Industries Association. EIA Standard RS-232-C, Interface Between Data Terminal Equipment and Data Communication Equipment Employing Serial Binary Data Interchange. Washington, D.C.: EIA, Engineering Department, 1969.

Electronic Industries Association. Equipment Employing Serial Binary Data Interchange. Washington, D.C.: EIA, Engineering Department, 1969.

Electronic Fund Transfer Act. 15 USC (1979), sec. 1693 et seq.

Emens, K.L. A Survey of Internal EDP Audit Activity Among EDPAC Member Companies. *EDP Auditor*, Fall 1976, pp. 11-17.

Enger, Norman L. and Howerton, Paul W. Computer Security: A Management Audit Approach. New York: AMACOM, 1981.

English, John T. ISO 9000 and System Validation. Presentation at Pharmaceutical and Medical Device Manufacturing Seminars, San Juan, Puerto Rico, March 1-19, 1994.

Epstein, R.A. The Theory of Gambling and Statistical Logic. New York: Academic Press, 1967.

Fagan, M.E. Design and Code Inspections to Reduce Errors in Program Development. *IBM Systems Journal*, Vol 15, (No. 3, 1976).

Feigenbaum, Edward, Pamela McCorduck, and H. Penny Nii. The Rise of the Expert Company. New York: Times Books, 1988.

Figarole, Paul L. Computer Software Validation Techniques. DIA Conference on Computer Validation, January 21-23, 1985.

Fike, John L. and George E. Friend. Understanding Telephone Electronics. Dallas: Texas Instruments, 1984.

Firebaugh, Morris, *et al.* A Feast of Microcomputers. *Personal Computing*, Vol. 2 (November 1978), pp. 60-81.

Fischer, Renate and Karl Schick. The Role of Quality Assurance in Good Clinical Practice. *Drug Information Journal*, Vol. 27 (July-Sept. 1993).

Fisher, F.E. Fundamental Statistical Concepts. New York: Canfield Press, 1973.

Fisher, Royal P. Information Systems Security. Englewood Cliffs, NJ: Prentice-Hall, Inc., 1984.

Flavin, Matt. Fundamental Concepts of Information Modelling. New York: Yourdon Press, 1981.

Forcinio, Hallie. Reaping the Benefits of Bar Coding: Quick Response Comes to Healthcare. *Pharmaceutical Technology*, Vol. 19 (January 1995).

Forsyth, A. An Approach to Audit an On-Line System. *COM-SAC, Computer Security, Auditing and Controls*, Vol. 7, January 1980, pp. A-1 to A-4.

Frank, E. Integrating Reviews of EDP Systems with Regular Audit Project. *EDP Auditor*, Summer 1974, pp. 10-11.

Freedman, Daniel and Gerald Weinberg. Handbook of Walkthroughs, Inspections and Technical Reviews. Boston: Little, Brown, 1982.

Freedman, David H. Is the Micro-Mainframe Link Connecting with MIS? *Infosystems*, February 1984.

Freedman, David H. Stress and the MIS Manager. *Infosystems*, March 1984.

Freeman, Peter. Software Perspectives: The System Is the Message. Reading, MA: Addison-Wesley, 1987.

Fribush, Howard M., Gary Robertson and Joseph Vagaggini. CLP Quick-Turn- Around Method. *Environmental Testing & Analysis*, March/April 1994.

Friedson, Eliot. The Social Construction of Illness. In Profession of Medicine. New York: Harper and Row, 1970, pp. 203-332.

Friend, George E., *et al.* Understanding Data Communications. Dallas: Texas Instruments, 1984.

Friis, M. William. Putting Information Systems Theory into Practice. *Infosystems*, February 1984.

Gagnon, Steven R. and Jesse Rodriguez. Applying Quality Assurance Concepts to Ensure the Performance of Pharmaceutical High-Purity Water Treatment Systems. *Microcontamination*, January 1994.

Gallegos, Frederick and Doug Bieber. What Every Auditor Should Know About Computer Information Systems. Available as Accession Number 130454 from the General Accounting Office (GAO) and reprinted from pp. 1-11 in EDP Auditing, Auerbach Publishers, 1986.

Gane, Chris and Trish Sarson. Structured Systems Analysis: Tools and Techniques. New York: Improved System Technologies, 1977.

Gardine, Thomas. Automation of Import Procedures. *RAPS News*, October 1993.

Gardner, Elizabeth. Computer Dilemma: Clinical Access vs. Confidentiality. *Modern Healthcare*, November 3, 1989, pp. 32-42.

Gardner, Elizabeth. Secure Passwords and Audit Trails. *Modern Healthcare*, November 3, 1989, p. 33 (sidebar).

Gardner, Elizabeth. System Assigns Passwords, Beeps at Security Breaches. *Modern Healthcare*, November 3, 1989, p. 34 (sidebar).

Gardner, Elizabeth. System Opens to Physicians, Restricts it to Others. *Modern Healthcare*, November 3, 1989, p. 38 (sidebar).

Gardner, Elizabeth. "Borrowed" Passwords Borrow Trouble. *Modern Healthcare*, November 3, 1989, p. 42 (sidebar).

Gardner, Elizabeth. Recording Results of AIDS Tests Can be a Balancing Act. *Modern Healthcare*, November 3, 1989, p. 40 (sidebar).

Garwood, R.M. FDA's Viewpoint on Inspection of Computer Systems. Course Notes from "Computers in Process Control: FDA Course," March 19-23, 1984.

Gaul, Gilbert M. A Threat to the Safety of Red Cross' Blood Supply. *The Philadelphia Inquirer*, March 21, 1989, page 1.

Gauthier, Richard. Using the UNIX System. Reston, VA: Reston Publishing Company, Inc., 1981.

GE-Intersil. Hot Ideas in CMOS Data Book. Cupertino, CA: GE-Intersil, 1984.

George, Joseph M. Lessons From the Field: Real-Life Experiences in Computer System Validation. *Pharmaceutical Technology*, Vol. 18 (November 1994).

Gerrity, T.P., Jr. The Design of Man-Machine Decision Systems: An Application to Portfolio Management. *Sloan Management Review*, Vol. 112 (Winter 1971), pp. 59-75.

Gibbons, J.D. Nonparametric Methods for Quantitative Analysis. New York: Holt, Rinehart and Winston, 1976.

Gild, Tom. Principles of Software Engineering Management. Reading MA: Addison-Wesley, 1988.

Gilhooley, Ian. Improving the Relationship Between Internal and External Auditors. *Journal of Accounting and EDP*, Vol. 1 (Spring 1985), pp. 4-9.

Ginsbury, Karen and Gil Bismuth. Compliance Auditing for Pharmaceutical Manufacturers: A Practical Guide to In-Depth Systems Auditing. Buffalo Grove, IL: Interpharm Press, 1994.

Glass, R.L. Real-Time: The "Lost World" of Software Debugging and Testing. *Communications of ACM*, Vol. 23, (May 1980), pp. 264-271.

Glass, R.L. Software Reliability Guidebook. Englewood Cliffs, NJ: Prentice-Hall, 1979.

Glover, Donald E., Robert G. Hall, Arthur W. Coston, and Richard J. Trilling. Validation of Data Obtained During Exposure of Human Volunteers to Air Pollutants. *Computers and Biomedical Research*, Vol. 15 (Number 3, 1982), pp. 240-249.

Goodman, Daniel B. Standardized and Centralized Electrocardiographic Data for Clinical Trials. *Applied Clinical Trials*, Vol. 2 (June 1993).

Goodwin, Connie. Considerate, Carefully Planned Office Environments Reduce Employee Anxiety. *Data Management*, June 1985.

Goren, Leonard J. Computer System Validation, Part II. *BioPharm*, Vol. 2 (February 1989), pp. 38-42.

Grady, Robert and Deborah Caswell. Software Metrics: Establishing a Company- Wide Program. Englewood Cliffs, NJ: Prentice-Hall, 1987.

Grimes, J. and E.A. Gentile. Maintaining International Integrity of On-Line Data Bases. *EDPACS Newsletter*, February 1977, pp. 1-14.

Guenther, William C. Analysis of Variance. Englewood Cliffs, NJ: Prentice-Hall, 1964.

Guilford, J.P. Fundamental Statistics in Psychology and Education. New York: McGraw-Hill, 1965.

Guynes, Steve. EFTS Impact on Computer Security. *Computers & Security*, Vol. 2 (1983), pp. 73-77.

Haider, Trong and Kjell Skaug. Forming an Understanding of GCP and a Validation Strategy at Nycomed Imaging. *Drug Information Journal*, Vol. 27 (April-June 1993).

Hall, Douglas V. Microprocessors and Digital Systems 2nd ed., New York: McGraw-Hill, Inc., 1983.

Halper, Stanley D., Glen C. Davis, P. Jarlath O'Neil-Dunne, and Pamela R. Pfau. Handbook of EDP Auditing. Chapter 28 of Warren, Gorham, & Lamont, Testing Techniques for Computer-Based Systems, 1985, pp. 28-1 to 28-26.

Halstead, Maurice. Elements of Software Science. New York: Elsevier, 1977.

Hamrell, Michael R. The Use of Computers in Multinational Drug Development. *Regulatory Affairs*, Vol. 5 (Winter 1993).

Harnett, D.L. and J.L. Murphy. Introductory Statistical Analysis. Reading MA: Addison-Wesley, 1975.

Harston, S.J. Implementation of Good Laboratory Practices by European Pharmaceutical Companies. *Drug Information Journal*, Vol. 28 (October-December 1994).

Hartmann, David C. Software Safety in Medical Devices. McLean, VA: The MITRE Corporation, 1993.

Hartwig, Frederick and Brian E. Dearing. Explanatory Data Analysis. Beverly Hills: Sage Publications, 1979.

Harz, Nan S. Telecommuting -- There's No Place Like Home. *Data Management*, June 1985.

Hatley, Derek and Imtiaz, Pirbhai. Strategies for Real-Time System Specification. New York: Dorset House, 1987.

Havard, J., J. Weaver, B. Gray, and P. Granade. An Automated In-Plant Analytical System. *American Laboratory*, October 1990.

Hayes Microcomputer Products, Inc. Smartmodem 1200 Hardware Reference Manual. Norcross, GA: Hayes Microcomputer Products, Inc., 1983.

Hayes, John R. The Complete Problem Solver. Philadelphia: The Franklin Institute Press, 1981.

Hayes-Roth, Frederick, Donald A. Waterman, and Douglas B. Lenat. Building Expert Systems. Reading, MA: Addison-Wesley Publishing Company, 1983.

Hedman, Gregory A. Increases in Laboratory Productivity Through LIMS Migration: Part I. *American Laboratory*, October 1992.

Hedman, Gregory A. Increases in Laboratory Productivity Through LIMS Migration: Part II. *American Laboratory*, January 1993.

Heidorn, G.E. Automatic Programming through Natural Language Dialogue: A Survey. *IBM Journal of Research and Development*, Vol. 20 (July 1976), pp. 302-313.

Heir, R.S. Good Manufacturing Practice: Historical Overview and Actual Status. *Drug Information Journal*, Vol. 28 (October-December 1994).

Henke, Cliff. International Electronics Standards: Compatibility or Interference? *Medical Device & Diagnostic Industry*, Vol. 15 (February 1993).

Herrick, Steven S. and Peter Berthrong. Validation of Computer Systems. *Scientific Computing and Automation*, September 1989.

Hershey, P. and K.H. Blanchard. Management of Organization Behavior. Englewood Cliffs, NJ: Prentice-Hall, 1972.

Hetzel, William. The Complete Guide to Software Testing. 2nd ed., Wellesley, MA: QED Information Sciences, 1988.

Hewlett-Packard. Improving Measurements in Engineering and Manufacturing -- HP-1B. Palo Alto, CA: Hewlett-Packard, 1976.

Hewlett-Packard. Optoelectronics Designer's Catalog. Palo Alto, CA: Hewlett-Pack-ard, 1985.

Highland, Harold Joseph. Protecting Your Computer System. New York: John Wiley & Sons.

Hinton, M.D. ACS-LIMS: Manual of Operation. Mobile: Applied Computer Systems,1994.

Hinton, M.D., M.E. Sawyer, and D. Davis. LIMS and the Quality Assurance Laboratory. *American Laboratory*, October 1992.

Hirsch, Allen F. Good Laboratory Practice Regulations. New York: Marcel Dekker, Inc., 1989.

Hoaglin, David C., Frederick Mosteller, and John W. Tukey. Understanding Robust and Exploratory Data Analysis. New York: John Wiley & Sons, 1983.

Hollingworth, Dennis. Minicomputers: A Review of Current Technology, Systems, and Applications. Rand Report R-1279. Santa Monica, CA: The Rand Corporation, July 1973.

Holscher H. and J. Rader. Microcomputers in Safety Technique. Bayern, Germany: Verlag, 1984.

Honan, Patrick. Telecommuting: Will It Work for You? *Computer Decisions*, June 15, 1984.

Hopper, E.L. Staffing of EDP Auditors on the Internal Audit Staff. *The Interpreter*, August 1975, pp. 24-26.

Hordeski, Michael. Computer Integrated Manufacturing. Tab Professional and Reference Books, 1988.

Horwitz, G. Needed: A Computer Audit Philosophy. *Journal of Accountancy*, April 1976, pp. 69-72.

House, William C., ed. Data Base Management. New York: Petrocelli Books, 1974. Part V, pp. 330-97.

Houston, M. Frank. Designing Safer, More Reliable Software Systems. U.S. Food and Drug Administration, Center for Devices and Radiological Health, 1989.

Houston, M. Frank. What Do the Simple Folk Do? Software Safety in the Cottage Industry. *Supplemental Proceedings, Compass '87*, 2nd Annual Conference on Computer Assurance (June 1987).

Huang, Linda G. *et al*. The Guide to Biomedical Standards. 18th Edition. Brea, CA: Quest Publishing Company, 1992.

Hubbert, J. Data Base Concepts. *EDP Auditor*, Spring 1980.

Huff, D. How to Lie with Statistics. New York: W.W. Norton, 1954.

Hulme, K. and M.E. Aiken. The Normative Approach to Internal Control Evaluation of On-Line/Real-Time Systems. *The Chartered Accountant in Australia*, July 1976, pp. 7-16.

Humphrey, Watts S. Managing the Software Process. Reading, MA: Addison- Wesley, 1989.

Hunter, Ronald P. Automated Process Control Systems Concepts and Hardware. Englewood Cliffs, NJ: Prentice-Hall, 1978.

Huntsberger, D.V. Elements of Statistical Inference. 3rd ed. Boston: Allyn and Bacon, 1971.

Hussain, Donna and K.M. Hussain. Information Resource Management. Homewood, IL: Richard D. Irwin, 1984.

Hutchinson, R.L. Is Good Laboratory Practice Compatible with Total Quality Management? *Drug Information Journal*, Vol. 28 (October-December 1994).

IEEE. Fiber Optic Applications in Electrical Substations. IEEE 83TH0104 PWR, IEEE Service Center, Piscataway, NJ, 1983.

IIA. Establishing the Internal Audit Function. Altamonte Springs, FL: Institute of Internal Auditors, 1974.

IIA. Hatching the EDP Audit Function. Altamonte Springs, FL: Institute of Internal Auditors, 1975.

IIA. How to Acquire and Use Generalized Audit Software. Altamonte Springs, FL: Institute of Internal Auditors, 1979.

IIA. Systems Auditability and Control -- Audit Practices. Altamonte Springs, FL: Institute of Internal Auditors, 1977.

IIA. Systems Auditability and Control -- Control Practices. Altamonte Springs, FL: Institute of Internal Auditors, 1977.

IIA. Systems Auditability and Control -- Executive Report. Altamonte Springs, FL: Institute of Internal Auditors, 1977.

Ince, Darrel, ed. Software Quality and Reliability. London: Chapman and Hall, 1991.

Inhorn, Stanley L., ed. Quality Assurance Practices for Health Laboratories. Washington, D.C.: American Public Health Association, 1978.

Inlander, Charles B., Lowell S. Levin, and Ed Weiner. Medicine on Trial, Prentice-Hall, 1988; "Blood, Sweat, and Fears," pp. 92-111.

Intel Corporation. A History of Microprocessor Development at Intel. Intel Article Reprint/AR-173. Intel Corporation, Santa Clara, CA.

Intel Corporation. An Intelligent Data Base System Using the 8272. Application Note AP-116. Intel Corporation, Santa Clara, CA, 1981.

Intel Corporation. ASM286 Assembly Language Reference Manual. Intel Corporation, Santa Clara, CA, 1983.

Intel Corporation. 8086/8087/8088 Macro Assembly Language Reference Manual for 8086-Based Development Systems. Intel Corporation, Santa Clara, CA, 1980.

Intel Corporation. Error Detecting and Correcting Codes, Application Note AP-46. Santa Clara, CA: Intel Corporation, 1979.

Intel Corporation. Getting Started With the Numeric Data Processor, Application Note AP-113. Intel Corporation, Santa Clara, CA, 1981.

Intel Corporation. iAPX 86/88, 186/188 User's Manual Programmer's Reference, Intel Corporation, Santa Clara, CA. 1983.

Intel Corporation. iAPX 286, Hardware Reference Manual, Intel Corporation, Santa Clara, CA. 1983.

Intel Corporation. iAPX 286, Operating Systems Writer's Guide, **Intel Corporation.** Santa Clara, CA. 1983.

Intel Corporation. iAPX 286, Programmer's Reference Manual, Intel Corporation, Santa Clara, CA. 1983.

Intel Corporation. iAPX 386, High Performance 32-bit Microcomputer Product Preview, Advance Information, Intel Corporation, Santa Clara, CA.

Intel Corporation. iRMX 86, Introduction and Operator's Reference Manual. Intel Corporation, Santa Clara, CA. 1984.

Intel Corporation. Intel SBC 202 Double Density Diskette Controller Hardware Reference Manual. Intel Corporation, Santa Clara, CA, 1977.

Intel Corporation. Introduction to the 80186 Microprocessor Application Note AP-186, Intel Corporation, Santa Clara, CA, 1983.

Intel Corporation. Introduction to the iAPX 286. Intel Corporation, Santa Clara, CA, 1985.

Intel Corporation. Microsystem Components Handbook Microprocessors and Peripherals, vols. 1 and 2. Intel Corporation, Santa Clara, CA, 1985.

Intel Corporation. Multibus Handbook. Intel Corporation, Santa Clara, CA, 1983.

Intel Corporation. Multibus Interfacing. Application Note AP-28A, Intel Corporation, Santa Clara, CA, 1979.

Intel Corporation. Multiprogramming with the iAPX 88 and iAPX 86 Microsystems. Application Note AP-106, Intel Corporation, Santa Clara, CA, 1980.

Intel Corporation. SDK-86 MCS-86 System Design Kit User's Guide. Intel Corpora-tion, Santa Clara, CA 1981.

Intel Corporation. Using the 8273 SDLC/HDLC Protocol Controller, Application Note AP-36. Intel Corporation, Santa Clara, CA 1978.

Intel Corporation. Using the 8292 GPIB Controller, Application Note AP-66, Intel Corporation, Santa Clara, CA, 1980.

Jackson, Barbara Bund. Computer Models in Management. Homewood: Richard D. Irwin, Inc., 1979.

Jackson, Michael. System Development. Englewood Cliffs, NJ: Prentice-Hall, 1983.

Jackson, Michael. Principles of Program Design. New York: Academic Press, 1975.

Jancura, E.G. Developing Concepts of Technical Proficiency in EDP Auditing. *The Ohio CPA*, Spring 1979.

Japanese Ministry of Health and Welfare. Guide to Medical Device Registration in Japan. 5th Edition. Tokyo: Yakuji Nippo, LTD., 1994.

Jenkins, A. Milton. MIS Design Variables and Decision Making Performance. Ann Arbor: UMI Research Press, 1983.

Johnson, Curtis D. Process Control Instrumentation Technology. 2nd ed., New York: John Wiley & Sons, 1982.

Johnson, R.R. Elementary Statistics. N. Scituate, MA: Duxbury Press, 1973.

Johnston, D. and N.H. Anderson. Global Perspective on Technology Transfer for Analytical Sciences. *Drug Information Journal*, Vol. 28 (October-December 1994).

Johtatsu, Y. System of Pharmaceutical Inspection and Official Testing in Japan. *Drug Information Journal*, Vol. 28 (October-December 1994).

Jones, T. Capers. Programming Productivity. New York: McGraw-Hill, 1986.

Justice, Richard M., Jr., Jole O. Rodriguez, and William J. Chiasson. Ten Steps to Ensure a Successful PreNDA Approval Inspection. *Journal of Parenteral Science & Technology*, Vol. 47 (March-April 1993).

Kahan, Jonathan S. Medical Device Reclassification: An Opportunity Ignored. *Medical Device & Diagnostic Industry*, Vol. 15 (January 1993).

Kaisler, Stephen H. The Design of Operating Systems for Small Computer Systems. New York: John Wiley & Sons, 1983.

Kaitin, Kenneth I. and Helen L. Walsh. Are Initiatives to Speed the New Drug Approval Process Working? *Drug Information Journal*, Vol. 26, 1992.

Kawamaura, K. Good Manufacturing Practice for Investigational Pharmaceutical Products in Japan. *Drug Information Journal*, Vol. 28 (October-December 1994).

Kay, Robert L. The Systems Engineering Approach to Quality Software. *Medical Device & Diagnostic Industry*, Vol. 16 (June 1994).

Keane, John Austin. Computers and Quality Assurance in a Regulated Industry. *Medical Device & Diagnostic Industry*, Vol. 3 (October 1981).

Kearns, J. Are We Ready for Continuous Process Auditing. *CA Magazine*, September 1980, pp. 68-71.

Keeler, R. Manage Your Data Better With a LIMS System. *R&D Magazine*, April 1991.

Keen, Peter G.W. "Interactive" Computer Systems for Managers: A Modest Proposal. *Sloan Management Review*, Fall 1976, pp. 1-17.

Keen, Peter G.W. and Michael S. Scott Morton. Decision Support Systems: An Organizational Perspective. Reading, MA: Addison-Wesley, 1978.

Kelly, F.J., D.L. Beggs, K.A. McNeil, T. Eichelberger, and J. Lyon. Multiple Regression Approach. Carbondale, IL: Southern Illinois Press, 1969.

Kendrick, John. Implementation of a Software Quality Assurance Program. Final Project, Fall 1989.

Kennedy, Colleen N., Gail Softer, John McEntire and James Akers. Biotechnology Product Validation: Managing Equipment and Procedural Change Control. *BioPharm*, Vol. 8 (March 1995).

Kenney, Donald P. Minicomputers. New York: AMACOM, 1973.

Kent, Phyllis. Internal Clinical Audits of Computerized Data for NDA Submissions. *Clinical Research Practices & Drug Regulatory Affairs*, Vol. 22 (No. 3 1984).

Kerlinger, F.N. Foundations of Behavioral Research. New York: Holt, Rinehart and Winston. 1965.

Kernighan, B.W. and P.J. Plauger. The Elements of Programming Style. New York: McGraw-Hill, 1978.

Kernighan, B.W. and P.J. Plauger. Software Tools. Reading, MA: Addison-Wesley, 1976.

Kimbrell, John Y. CANDA as a Strategic Initiative for the Pharmaceutical Industry. *Pharmaceutical Technology*, Vol. 16 (October 1992).

King, J.P. A Vendor's Contribution to Computer System Software Validation -- Change Control. *Pharmaceutical Engineering*, Vol. 12 (May 1992).

Kish, L. Survey Sampling. New York: John Wiley, 1965.

Kleijnen, Jack P.C. Computer and Profits: Quantifying Financial Benefits of Information. Reading, MA: Addison-Wesley Publishing Company, 1980.

Koller, Alexander J. and Gerard W. Liesegang. Information Management Architec-ture Within the Analytical Laboratory: Part I. *American Laboratory*, September 1992.

Korteweg, Marijke. Computer Validation in Clinical Research: Regulatory Requirements in the European Community. *Drug Information Journal*, Vol. 27 (April-June 1993).

Korteweg, Marijke. GCP Audit Certificates and Compliance Statements in the European Community. *Applied Clinical Trials*, Vol. 2 (June 1993).

Korteweg, Marijke. General Concepts with Regard to Quality Control/Quality Assurance in Data Management. *Drug Information Journal*, Vol. 27 (October-December 1993).

Kosmala, Rupert M. The Relationship Between Functional Requirements and Software Design. *Pharmaceutical Technology*, Vol. 14 (November 1990).

Kraak, Frans J. and Karel J. de Neef. In-Process Data Quality Control: A Practical Implementation. *Drug Information Journal*, Vol. 27 (October-December 1993).

Kuhn, Thomas. The Structure of Scientific Revolutions. 2nd ed. Chicago: University of Chicago Press, 1970.

Kull, David. Demystifying Ergonomics. Computer Decisions, September 1984.

Kuong, J.F. A Framework for EDP Auditing. *COM-SAC, Computer Security, Auditing and Controls*, Vol. 3 (1976), pp. A-1 to A-8.

Kuong, J.F. Advanced Tools and Techniques for Systems Auditing. Management Advisory Publication, 1978.

Kuong, J.F. Approaches to Justifying EDP Controls and Auditability Provisions. *COM-SAC, Computer Security, Auditing and Controls*, Vol. 7, July 1980, pp. A-1 to A-8.

Kuong, J.F., ed. Audit and Control of Advanced/On-Line Systems. Management Advisory Publications (MAP-7), 1983.

Kuong, J.F. Audit and Control of Computerized Systems. Management Advisory Publications (MAP 6), 1979.

Kuong, J.F. Auditor Involvement in System Development and the Need to Develop Effective, Efficient, Secure, Auditable, and Controllable Systems. Keynote Speech at the First Regional EDP Auditors Conference, Tel-Aviv, Israel, June 3, 1982.

Kuong, J.F. Checklists and Guidelines for Reviewing Computer Security and Installations. Management Advisory Publications (MAP-4), 1976.

Kuong, J.F. Computer Auditing and Security: Manual-Operations and System Audits. Management Advisory Publications (MAP-5), 1976.

Kuong, J.F. Computer Security, Auditing, and Controls: Text and Readings. Management Advisory Publications, 1974.

Kuong, J.F. Controls for Advanced/On-Line/Data-Base Systems, vols. 1 and 2. Management Advisory Publications, 1985.

Kuong, J.F. Managing the EDP Audit Function. Paper Presented at the First Regional EDP Auditors Conference, Tel-Aviv, Israel, May 29-June 3, 1982.

Kuong, J.F. Organizing, Managing, and Controlling the EDP Auditing Function. Seminar Text, Management Advisory Publications, 1980.

Kuong, J.F. Organizing and Staffing for EDP Auditing. *COM-SAC, Computer Security, Auditing and Controls*, Vol. 2 (1975).

Lagan, J.J. The Development of an EDP Audit Function in a Complex Environment. *Employment Research Management*, 680, 1978.

Lainhart, J.W. and B.R. Snyder. A Simultaneous-Parallel Approach to Testing Computerized Systems. *EDPACS Newsletter*, October 1977, pp. 7-12.

Lamprecht, James L. Implementing the ISO 9000 Series. New York: Marcell Dekker, Inc., 1993.

La Niece, Chris and Dai Bedford. Best Practice in the Validation of Business Critical Systems. *Drug Information Journal*, Vol. 27 (April-June 1993).

Langmead, J.M. and R.V. Boos. How Do You Train EDP Auditors? *Management Focus*, September-October 1978, pp. 6-11.

Laposki, Edward J. Planning and Managing the Computer Automation Project. *Pharmaceutical Manufacturing*, August 1985.

Larhammar, H.M. Development of Quality Management at the Medical Products Agency in Sweden. *Drug Information Journal*, Vol. 28 (October-December 1994).

Laurel, Brenda, ed. The Art of Human-Computer Interface Design. Reading, MA: Addison-Wesley, 1990.

Lawrence, Gill. On-Line to Central Database Processing at Searle. *Drug Information Journal*, Vol. 27 (October-December 1993.

Lee, John Y. Documentation Requirements for Preapproval Inspections. *Pharmaceutical Technology*, Vol. 17 (March 1993).

Lesea, Austin and Rodnay Zaks. Microprocessor Interfacing Techniques. 2d ed., Berkeley: Sybex Inc., 1977.

Leveson, Nancy G. Software Safety: Why, What and How. *ACM Computing Surveys*, Vol. 18 (June 1986).

Leveson, Nancy G. Software Safety in Embedded Computer Systems. *Communications of the ACM*, Vol. 34 (February 1991).

Leveson, Nancy G. and J.L. Stolzy. Safety Analysis Using Petri Nets. *IEEE Transactions on Software Engineering*, Vol. 13 (March 1987).

Leveson, Nancy G. and P.R. Harvey. Analyzing Software Safety. *IEEE Transactions on Software Engineering*, Vol. 9 (September 1983).

Levey, Brian and William Leonard. LIMS Operational Qualification. *American Laboratory*, March 1993.

Levy, C.J. Case Study of ISO 9002 Versus cGMPs of Devices and Drugs. *Drug Information Journal*, Vol. 28 (October-December 1994).

Lientz, B.P. and E.B. Swanson. Software Maintenance Management. Reading, MA: Addison-Wesley, 1980.

Lindauer, Richard F. An Official, Electronically Searchable Version of USP-NF. *Pharmaceutical Technology*, Vol. 17 (March 1993).

Linder-Ciccolunghi, S.N. How International Quality Assurance Audits can Contribute to Quality Management. *Drug Information Journal*, Vol. 28 (October-December 1994).

Linger, R.C., H.D. Mills and B.L. Witt. Structured Programming: Theory and Practices. Reading, MA: Addison-Wesley, 1979.

Liscouski, Joseph G. Issues and Directions in Laboratory Automation. *Analytical Chemistry*, Vol. 60 (January 15, 1988).

Litecki, C.R. and J.E. McEnroe. EDP Audit Job Definitions: How Does Your Staff Compare? *The Internal Auditor*, April 1981, pp. 57-61.

Littlewood, Bev and Lorenzo Strigini. The Risks of Software. *Scientific American*, November 1992.

Lumato, Ann Nicols and Nico Lumato. A UNIX Primer. Englewood Cliffs, NJ: Prentice-Hall, 1983.

Lyles, Richard I. Practical Management Problem Solving and Decision Making. New York: Van Nostrand Reinhold, 1982.

Macchiaverna, P.R. Auditing Corporate Data Processing Activities. New York: The Conference Board, Inc., 1980.

Mackintosh, Douglas R. Building Quality Assurance Into Clinical Trials. *Applied Clinical Trials*, Vol. 2 (April 1993).

Mackintosh, Douglas R. Quality Assurance of Data Management and Statistical Analysis. *Applied Clinical Trials*, Vol. 3 (June 1994).

Mader, Chris and Robert Hagin. Information Systems: Technology, Economics and Applications. Chicago: Science Research Associates, Inc., 1974, pp. 291-309.

Maddox, M., C. Higgins, and D. Mathews. Client-Server LIMS Solutions: Part I. *American Laboratory*, March 1993.

Maggon, Krishan K. and Daniel Brandt. Standard Operating Procedures and the Conduct of Clinical Trials. *Applied Clinical Trials*, Vol. 3 (July 1994).

Magnani, Harry H. and Guus A Sturk. Centralized Laboratory Testing in Europe. *Applied Clinical Trials*, Vol. 3 (March 1994).

Mair, W.C., K. Davis, and D. Wood. Computer Control and Audit. Altamonte Springs, FL: Institute of Internal Auditors, 1976.

Mallory, Steven R. Building Quality into Medical Product Software Design. *Biomedical Instrumentation & Technology*, Vol. 27 (March/April 1993).

Manell, Paul. CANDAs: A New Standard for Electronic Regulatory Submissions in Europe. *Applied Clinical Trials*, Vol. 2 (April 1993).

Manell, Per. Computer Validation Applied in the Regulatory Environment. *Drug Information Journal*, Vol. 27 (April-June 1993).

Mannion, James C., Anthony Abruzzini, George Nichoalds, Siegrid Hessenthaler, and Christine Uhlinger. Predictable Challengers in the Preparation of the Clinical Sections of NDAs and PLAs. *Applied Clinical Trials*, Vol. 3 (March 1994).

Marks, R.C. Performance Appraisal of EDP Auditors. Speech Given at the 11th Conference on Computer Auditing, Security, and Control, ATC/IIA, New York, May 4-8, 1981.

Marks, William E. Evaluating the Information Systems Staff. *Information Systems News*, December 24, 1984.

Martin, J. Accuracy, Security and Privacy in Computer Systems. Englewood Cliffs, NJ: Prentice-Hall, 1973.

Martin, J. Principles of Data Base Management. Englewood Cliffs, NJ: Prentice-Hall, Inc., 1976.

Martin, James. Strategic Data Planning Methodologies. Englewood Cliffs, NJ: Prentice-Hall, 1982.

Martin, James. The Wired Society. Englewood Cliffs, NJ: Prentice-Hall, Inc., 1978.

Martin, James and Carma McClure. Software Maintenance: The Problem and its Solutions. Englewood Cliffs, NJ: Prentice-Hall, 1983.

Martin, James and Carma McClure. Structured Techniques: The Basis for CASE. Englewood Cliffs, NJ: Prentice-Hall, 1987.

Mason, Richard O. and E. Burton Swanson. Measurement for Management Decision. Reading, MA: Addison-Wesley, 1981.

Masters, George and Paul Figarole. Validation Principles for Computer Systems -- FDA's Perspective. *Pharmaceutical Technology*, Vol. 10 (November 1986).

Matsuda, Takehiko. Guideline on Control of Computerized Systems in Drug Manufacturing. *Journal of Pharmaceutical Science and Technology*. January/February 1994.

Mattes, D.C. LIMS and Good Laboratory Practice. In R.D. McDowall, ed., Laboratory Information Management Systems. Wilmslow, U.K.: Sigma Press, 1987, pp. 332-345.

Maynard, David W. Validation Master Planning. *Journal of Parenteral Science & Technology*, Vol. 47 (March-April 1993).

McClure, Carma. CASE is Software Automation. Englewood Cliffs, NJ: Prentice- Hall, 1989.

McClure, Carma. Managing Software Development and Maintenance. New York: Van Nostrand Reinhold, 1981.

McCorduck, Pamela. Machines Who Think. San Francisco: W.H. Freeman, 1979.

McDonald, Thomas F. TechnoAcceptance Cycle Soothes Automation Blind Spots. *Data Management*, May 1985.

McDowall, R.D., ed. Laboratory Information Management Systems. Wilmslow, U.K.: Sigma Press, 1987.

McFarlan, F. Warren and James L. McKenney. Corporate Information Systems Management: The Issues Facing Senior Executives. Homewood, IL: Richard D. Irwin, 1983.

McGuire, P.T. EDP Auditing -- Why? How? What? *The Internal Auditor*, June 1977, pp. 28-34.

McKibbin, Wendy Lea. A Solution to PC Management? *Infosystems*, February 1984.

McMenamin, Steve and John Palmer. Essential Systems Analysis. Englewood Cliffs, NJ: Yourdon Press/Prentice-Hall, 1984.

McNamara, John E. Technical Aspects of Data Communication. Digital Equipment Corporation, Maynard, MA, 1977.

Megargle, Robert. Laboratory Information Management Systems. *Analytical Chemistry*, Vol. 61 (May 1989), pp. 612/a-621A.

Mendenhall, W. An Introduction to Linear Models and the Design and Analysis of Experiments. Belmont, CA: Wadsworth, 1967.

Mendenhall, W., L. Ott, and R.L. Scheaffer. Elementary Survey Sampling. Belmont, CA: Wadsworth, 1971.

Mendenhall, W. and James E. Reinmuth. Statistics for Management and Economics, 3rd ed. N. Scituate, MA: Duxbury Press, 1978.

Merrer, Robert J. and Peter G. Berthrong. Academic LIMS: Concept and Practice. *American Laboratory*, Vol. 21 (March 1989), pp. 36-45.

Merrett, T.H. Relational Information Systems. Reston, VA: Reston Publishing Co., 1984.

Metzger, Philip. Managing a Programming Project. 2nd ed. Englewood Cliffs, NJ: Prentice-Hall, 1983.

Meyer, P. Future GMP Inspection System (Inside and Outside the European Community): What Does Subsidiarity Really Mean? *Drug Information Journal*, Vol. 28 (October-December 1994).

Michie, Donald, ed. Introductory Readings in Expert Systems. New York: Gordon and Breach Science Publishers, Inc., 1982.

Mick, John and Jim Brick. Bit-Slice Microprocessor Design. New York: McGraw- Hill, Inc., 1980.

Miller, T.L. EEDP -- A Matter of Definition. *The Internal Auditor*, July-August 1975, pp. 31-38.

Mills, Harlan, Richard Linger, and Alan Hevner. Principles of Information Systems Analysis and Design. New York: Academic Press, 1986.

Minsky, Marvin. The Society of Mind. New York: Simon & Schuster, 1986.

Mitchell, Tony. Computer-Assisted Clinical Research. *Drug Information Journal*, Vol. 27 (April-June 1993).

Moehle, Frederick L. W. A Knowledge Base Model for CANDAs, Part I. *Applied Clinical Trials*, Vol. 2 (June 1993).

Moller, H. Quality Design of Medicinal Products in Light of Current European Community Regulations. *Drug Information Journal*, Vol. 28 (October-December 1994).

Moore, Richard W. Introduction to the Use of Computer Packages for Statistical Analyses. Englewood Cliffs, NJ: Prentice-Hall, 1978.

Morel, Guillemette. Validation: From Theory to Practice, A Quality Assurance Perspective. *Drug Information Journal*, Vol. 27 (April-June 1993).

Morozoff, Paul E. Specifying Software Validation and Verification for a Biomedical Application. *Biomedical Instrumentation & Technology*, Vol. 28 (May/June 1994).

Morris, R., III. The Internal Auditors and Data Processing. *The Internal Auditor*, August 1978.

Morrissey, John. Automated Patient Record in Spotlight. *Modern Healthcare*, Vol. 25 (June 19, 1995).

Motise, Paul. FDA Considerations on Electronic Identification and Signatures. *Pharmaceutical Technology*, Vol. 16 (November 1992).

Motise, Paul. What to Expect When FDA Audits Computer-Controlled Processes. *Pharmaceutical Manufacturing*, July 1984.

Motorola Semiconductor Products, Inc. Optoelectronics Device Data Book, DL118R1. Phoenix: Motorola Semiconductor Products Inc., 1983.

Motorola Semiconductor Products, Inc. Power Mosfet Selector Guide and Cross Reference SG56R4. Phoenix: Motorola Semiconductor Products, Inc., 1983.

Murty, Ram and Sreemantula Satyanarayana. Evaluation of an Automated Washer for Processing Parenteral Closures. *Pharmaceutical Technology*, Vol. 17 (March 1993).

Musa, John, Anthony Iannino, and Kazuhira Okumoto. Software Reliability: Measurement, Prediction, Application. New York: McGraw-Hill, 1987.

Myers, Glenford. The Art of Software Testing. New York: Wiley-Interscience, 1979.

Myers, Glenford. Digital System Design with LSI Bit Slice Logic. New York: John Wiley and Sons, 1980.

Myers, Glenford. Software Reliability. New York: John Wiley & Sons, 1976.

Myers, Glenford. Reliable Software Through Composite Design. New York: Petrocelli/ Charter, 1975.

Myers, Ware. Build Defect-Free Software, Fagan Urges. *IEEE Computer*, August 1990.

Naisbitt, John and Patricia Aburdene. Megatrends 2000. New York: William Morrow, 1990.

Nally, J. and R. Kieffer. The Future of Validation: From QC/QA to TQ. *Pharmaceutical Technology*, Vol. 17 (October 1993).

Nelson, D.C., V. Dauciunas, R. Lysakowski, and M. Duff. The Standard for Chromatography Data Communication from the Analytical Instruments Association. *American Laboratory*, September 1990.

Neter, J. and W. Wasserman. Applied Linear Statistical Models. Homewood, IL: Richard D. Irwin, 1974.

Neuman, Anita. Get a Head Start: Perform Process Validation in Pilot Production. *Medical Device & Diagnostic Industry*, Vol. 16 (January 1994).

Neumann, Kurt. Some Aspects of Total Quality Management in the Computer System Validation Process in Clinical Research. *Drug Information Journal*, Vol. 27 (April-June 1993).

Niehoff, Ken. Using Rovira Diagrams to Specify the User Interface. *Medical Device & Diagnostic Industry*, Vol. 16 (January 1994).

Noether, G.E. Introduction to Statistics: A Nonparametric Approach. 2nd ed. Boston: Houghton Mifflin, 1976.

Nolan, Richard L. Managing the Crises in Data Processing. *Harvard Business Review* (March-April 1979), pp. 115-126.

Nolan, Richard L. Managing the Data Processing Function. 2nd ed. St. Paul, MN: West Publishing, 1982.

Norman, Donald A. The Design of Everyday Things. New York: Doubleday/Currency, 1990.

Norris, P.M. EDP Audit and Control – A Practitioner's Viewpoint. *EDP Auditor*, Winter 1976, pp. 8-14.

Nowak, Horst. A Systematic Approach for Handling Adverse Events. *Drug Information Journal*, Vol. 27 (October-December 1993).

O'Brien, James A. Computers in Business Management, An Introduction. Homewood, IL: Richard D. Irwin, Inc., 1979, pp. 150-74.

O'Donnell, Peter. Computers and Numbers. *Applied Clinical Trials*, Vol. 3 (July 1994).

O'Donnell, Peter. GCP and Computers: Mixing People with Processes. *Applied Clinical Trials*, Vol. 3 (March 1994).

O'Shea, Tim and Marc Eisenstadt, eds. Artificial Intelligence. New York: Harper and Row, Publishers, 1984.

Odiorne, George S. Management Decisions by Objectives. Englewood Cliffs, NJ: Prentice-Hall, Inc., 1969.

Olivier, Daniel P., Michael D. Konrad, and Markus Weber. Software Development: Comparing Software Assessment Standards. *Medical Device and Diagnostic Industry*, Vol. 16 (June 1994).

Orr, Ken. Structured Requirements Definition. Topeka, KS: Ken Orr & Associates, 1981.

Orr, Ken. Structured Systems Development. Englewood Cliffs, NJ: Yourdin Press/ Prentice-Hall, 1977.

Outterson, Dave. Color Visual Inspection in Pharmaceutical Manufacturing. *Pharmaceutical Technology*, Vol. 18 (January 1994).

Pabst, C.A. What's All the Fuss About EDP? *California CPA Quarterly*, June 1975, pp. 10-16.

Page-Jones, Meilir. The Practical Guide to Structured Systems Design. 2nd ed. Englewood Cliffs, NJ: Yourdon Press/Prentice-Hall, 1988.

Parikh, Girish. The Politics of Software Maintenance. Infosystems, August 1984.

Parker, Marilyn and Robert J. Benson. Information Economics. Englewood Cliffs, NJ: Prentice-Hall, 1988.

Parnas, David Lorge and Paul C. Clements. A Rational Design Process: How and Why to Fake It. *IEEE Transactions on Software Engineering*, Vol. SE-12 (February 1986).

Patrick, R.L. Performance Assurance and Data Integrity Practices. Washington, D.C.: National Bureau of Standards, 1978.

Paulson, Daryl S. and Dennis G. Vogel. Preparatory Software Documentation for Validation of Computer-Controlled Manufacturing Operations. *Pharmaceutical Manufacturing*, May 1984.

Pava, Calvin H.P. Managing New Office Technology: An Organizational Strategy. New York: Free Press, 1983.

Peach, Robert W., ed. The ISO 9000 Handbook. 2nd Edition. Fairfax, VA: CEEM Information Services, 1994.

Pearson, E.S. and M.G. Kendall. Studies in the History of Statistics and Probability. Darien, CT Hafner Publishing Company, 1970.

Peine, Ira C. Quality Assurance Compliance: Procedures for Pharmaceutical and Biotechnology Manufacturers. Buffalo Grove, IL: Interpharm Press, 1994.

Pency, Laura, ed. Interpharm Pharmaceutical Manager's Compliance Guide. Buffalo Grove, IL: Interpharm Press, 1994.

Perry, W.E. Adding a Computer Programmer to the Audit Staff. *The Internal Auditor*, July 1974, pp. 1-7.

Perry, W.E. Auditing Computer Systems. Melville, NY: FAIM Technical Products, Inc., 1977.

Perry, W.E. Career Advancement for the EDP Auditor. *EDPACS Newsletter*, February 1975, pp. 1-6.

Perry, W.E. Ensuring Data Base Integrity. New York: John Wiley & Sons.

Perry, W.E. Internal Control. Melville, NY: FAIM Technical Products, Inc., 1980.

Perry, W.E. The Making of a Computer Auditor. *The Internal Auditor*, November-December 1974.

Perry, W.E. Snapshot -- A Technique for Tagging and Tracing Transactions. *EDPACS Newsletter*, March 1974, pp. 1-7.

Perry, W.E. Trends in EDP Auditing. *EDPACS Newsletter*, December 1976, pp. 1-6.

Perry, W.E. Using SMF as an Audit Tool -- Accounting Information. *EDPACS Newsletter*, February 1975.

Perry, W.E. and Donald L. Adams. SMF -- An Untapped Audit Resource. *EDPACS Newsletter*, September 1974.

Perry, W.E. and J.F. Kuong. Developing an Integrated Test Facility for Testing Computerized Systems. Management Advisory Publications (MAP-12), 1979.

Perry, W.E. and J.F. Kuong. EDP Risk Analysis and Controls Justification. Management Advisory Publications, 1981.

Perry, W.E. and J.F. Kuong. Effective Computer Audit Practices Manual (ECAP). Management Advisory Publications.

Perry, W.E. and J.F. Kuong. Generalized Computer Audit Software-Selection and Application. Management Advisory Publications (MAP-14) 1980.

Pharmaceutical Manufacturers Association. PMA Computer Systems Validation Committee Report. *Pharmaceutical Technology*, Vol. 10 (May 1986), pp. 24-26.

Phipps, Gail. Practical Application of Software Testing. Computer Sciences Corporation, presented October 8, 1986.

PhRMA, Computer Systems Validation Committee. Computer System Validation: Auditing Computer Systems for Quality. *BioPharm*, Vol. 7 (September 1994).

Pieterse, Herman, *et al*. Source Documents: Definitions, Verification Procedures, and Archiving. *Applied Clinical Trials*, Vol. 3 (September 1994).

Pirsig, Robert. Zen and the Art of Motorcycle Maintenance. New York: Bantam Books, 1975.

Pinkus, Karen V. Financial Auditing and Fraud Detection: Implications for Scientific Data Audit. *Accountability in Research*, Vol. 1 (1989), pp. 53-70.

Plettenberg, H.D. Quality-Related Terminology in Good Practice Regulations and ISO Standards. *Drug Information Journal*, Vol. 28 (October-December 1994).

PMA, Clinical Data Management Group Task Force. Electronic Exchange of Laboratory Data: Recommendations for Improving Quality and Reducing the Hidden Costs. *Drug Information Journal*, Vol. 27 (July-Sept. 1993).

PMA, Computer System Validation Committee. Computer System Validation -- Staying Current: Installation Qualification. *Pharmaceutical Technology*, Vol. 14 (September 1990).

PMA, Computer System Validation Committee. Computer System Validation -- Staying Current: Introduction. *Pharmaceutical Technology*, Vol. 13 (May 1989).

PMA, Computer System Validation Committee. Computer System Validation -- Staying Current: Security in Computerized Systems. *Pharmaceutical Technology*, Vol. 17 (May, 1993).

PMA, Computer System Validation Committee. Computer System Validation -- Staying Current: Software Development Testing Strategies. *Pharmaceutical Technology*, Vol. 13 (September 1989).

PMA, Computer System Validation Committee. Computer System Validation -- Staying Current: Vendor-User Relationships. *Pharmaceutical Technology*, Vol. 17 (September 1993).

PMA, Computer System Validation Committee. Validation Concepts for Computer Systems Used in the Manufacture of Drug Products. *Pharmaceutical Technology*, Vol. 10 (May 1986).

Polanis, M.F. Choosing an EDP Auditor. *Bank Administration*, January 1973, pp. 52-53.

Polya, G. How to Solve It. Garden City: Doubleday and Company, 1957.

Pratt, Terrence W. Programming Languages: Design and Implementation. Englewood Cliffs, NJ: Prentice-Hall, 1975.

Pressman, Roger S. Making Software Engineering Happen. Englewood Cliffs, NJ: Prentice-Hall, 1988.

Prutch, Shirley F. In Praise of Operations Managers. *Datamation*, June 15, 1984.

Quinn, Thomas. Benefits-Driven Phased Implementations for EBR/MES Projects, Part II. *Pharmaceutical Technology*, Vol. 17 (November 1993).

Rabbitt, John T. and Peter A. Bergh. The ISO 9000 Book: A Global Competitor's Guide to Compliance & Certification. White Plains, NY: Quality Resources, 1993.

Raiffa, Howard. Decision Analysis. Reading, MA: Addison-Wesley Publishing Company, 1970.

Ramsey, James B. and Gerald L. Musgrave. APL-STAT: A Do-It-Yourself Guide to Computational Statistics. Belmont, CA: Lifetime Learning Publications, 1981.

Ravden, Susannah and Graham Johnson. Evaluating Usability of Human-Computer Interfaces. New York: John Wiley & Sons, 1989.

Rawling, Alan K. Computer-Aided Report Auditing. *Drug Information Journal*, January-March 1994.

Ray, Curtis. An Advanced Chromatography Data System. *American Laboratory*, September 1990.

Reber, S. Laboratory Information Management Systems. *American Laboratory*, February 1985.

Reilly, R.F. and J.A. Lee. Developing an EDP Audit Function: A Case Study. *EDPACS Newsletter*, May 1981, pp. 1-10.

Reimann, Bernard C. and Allen D. Warren. User-Oriented Criteria for the Selection of DDS Software. *Communications of the ACM*, Vol. 28 (February 1985), pp. 166-79.

Rhodes, Wayne L., Jr. The "Mythical" Datacom Manager. *Infosystems*, March 1984.

Rockart, John F. Chief Executives Define Their Own Data Needs. *Harvard Business Review* (March-April 1979), pp. 81-93.

Roland, John. The Microelectronic Revolution. *The Futurist*, Vol. 13 (April 1979), pp. 81-90.

Romano, Carol A. Privacy, Confidentiality, and Security of Computerized Systems: The Nursing Responsibility. *Computers in Nursing*, May/June 1987, pp. 99-104.

Roth, M.J. Validation Tools and Technologies for Computer Systems Development. *Pharmaceutical Engineering*, Vol. 15 (July-August 1995).

Rothman, Lawrence J. Sharing the Informational Wealth. *Pharmaceutical Technology*, Vol. 14 (November 1990).

Rubinstein, Moshe F. Patterns of Problem Solving. Englewood Cliffs, NJ: Prentice-Hall, 1975.

Ruff, P. SQL*LIMS -- Providing Information Management Solutions for the Pharmaceutical Industry. *Laboratory Practice*, June 1990.

Rugg, Tom. LANtastic. Berkeley, CA: Osborne McGraw-Hill, 1992.

Ryan, James. LIMS Aid Environmental Labs. *R&D Magazine*, July 1991.

Ryan, James. Data Quality with LIMS. *Quality*, May 1988.

Sackman, Harold. Mass Information Utilities and Social Excellence. Princeton: Auerbach Publishers, 1971.

Salazar, Joanne M., Chris Gopal, and Arthur Mlodozeniec. Computer Migration and Validation: A Vendor's Perspective. *Pharmaceutical Technology*, Vol. 15 (June 1991).

Sammet, Jean E. The Use of English as a Programming Language. *Communications of the ACM*, vol. 9, (March 1966), pp. 228-30.

Sandberg, Jim. Design Output. *Medical Device & Diagnostic Industry*, Vol. 15 (December 1993).

Sandhu, H.S. Hands-On-Introduction to ROBOTICS -- The Manual for the XR-Series Robots. Champaign, IL: Rhino Robots, 1983.

Sandowski, C. and G. Lawler. A Relational Data Base Management System for LIMS. *American Laboratory*, Vol. 21 (March 1989), pp. 70-79.

Savich, R.S. The Care and Feeding of an EDP Auditor. *EDP Auditor*, Summer 1974, pp. 12-13.

Sayre, James E., Jr. Good Clinical Practices (GCP) Quality Audit Manual. 2nd Edition. Buffalo Grove IL: Interpharm Press, 1994.

Sax, Philip E. Compatibility of Environmental Quality Requirements and ISO 9000 Recommendations. Presentation at 1st Annual International Conference on ISO 9000, Lake Buena Vista, FL, November 1-2, 1993.

Sax, Philip E. Product and Process Inspections Using ISO 9001. Proceedings of the HIMA Conference on Medical Device Software Good Manufacturing Practices. HIMA Publication 93-6, September 1993.

Schatt, Stan. Understanding Local Area Networks. Third Edition. Carmel, IN: SAMS.

Scheffe, Henry. The Analysis of Variance. New York: John Wiley, 1959.

Schindler, Max. Computer-Aided Software Design. New York: John Wiley & Sons, 1990.

Schmidt, Ian. Client/Server ODBMSs. *Software Development*, October 1993.

Schneidermann, B. Improving the Human Factors Aspect of Database Interactions. *ACM Transactions on Database Systems*, Vol. 3, (December 1978), pp. 417-39.

Schneidermann, B. Designing the User Interface: Strategies for Effective Human-Computer Interaction. Reading, MA: Addison-Wesley, 1987.

Schneidman, A. A Need for Auditors' Computer Education. *The CPA Journal*, June 1979, pp. 29-35.

Schroeder, Frederick J. Developments in Consumer Electronic Fund Transfers. *Federal Reserve Bulletin*, Vol. 69 (1983), pp. 395-403.

Schuber, Stefan. Product, Not Paper: A Manufacturing Execution System Installation at SmithKline Beecham. *Pharmaceutical Technology*, Vol. 17 (November 1993).

Schuetze, Guenter. Computerized Production of Artwork for Labeling. *Drug Information Journal*, Vol. 27 (April-June 1993).

Schulmeyer, G. Gordon. Zero Defect Software. New York: McGraw-Hill, Inc., 1990.

Schuyler, Michael. Now What? How to Get Your Computer Up & Keep it Running. New York: Neal-Schuman Publishers, 1988.

Scientific American. Microelectronics. San Francisco: W.H. Freeman and Co., 1977.

Scoma, I., Jr. The EDP Auditor. *Data Management*, May 1977, pp. 14-17.

Sculley, John. Odyssey. New York: Harper & Row, 1987.

Seippel, Robert G. Transducers, Sensors, and Detectors. Reston, VA: Reston Publishing Company, Inc., 1983.

Sellers, G.R. Elementary Statistics. Philadelphia: W.B. Saunders Co., 1977.

Semprevivo, Philip C. Systems Analysis, Science Research Associates, 1982.

Sharp, John. European Pharmaceutical Technical and Regulatory Compendium. Buffalo Grove, IL: Interpharm Press, 1994.

Sheingold, Daniel H., ed. Transducer Interfacing Handbook -- A Guide to Analog Signal Conditioning. Norwood, MA: Analog Devices, Inc., 1981.

Sheridan, Robert. How to Manage Regulatory Submissions for Device Modifications. *Medical Device & Diagnostic Industry*, Vol. 15 (October 1993).

Shibota, M. Systems Implemented by Sponsors to Assure Conduct of Clinical Trials in Compliance With Japanese GCP. *Drug Information Journal*, Vol. 28 (October-December 1994).

Shlaer, Sally and Steve Mellor. Object-Oriented Systems Analysis: Modeling the World in Data. Englewood Cliffs, NJ: Yourdon Press/Prentice-Hall, 1988.

Shore, John. The Sachertorte Algorithm. New York: Penguin Books, 1986.

Siegel, S. Nonparametric Statistics for the Behavioral Sciences. New York: McGraw-Hill, 1956.

Silva, Marianne A. A Personal Experience with Validation and Regulation. Presentation at the 1st Annual Validation Forum, Williamsburg, VA, May 3, 1994; published in *Weinberg Validation Review*, Vol. 3 (No. 4, 1994).

Simon, H.A. The New Science of Management. New York: Harper and Row, Publishers, 1960.

Simon, H.A. The Science of the Artificial. Cambridge: Massachusetts Institute of Technology Press, 1981.

Simon, H.A. What Computers Mean for Man and Society. *Science*, Vol. 195, (March 18, 1977), pp. 1186-91.

Singer, Miller, and Witt. Structured Programming, Theory and Practice. Reading, MA: Addison-Wesley Publishing Co., 1979.

Smith, Martin R. Commonsense Computer Security. London: McGraw-Hill Book Company, 1989.

Smith, K. and J. Farrie. Data Management in a New Generation Chromatography System. *American Laboratory*, March 1990.

Snyder, Dean E., ed. FDA-Speak: The Interpharm Glossary of Acronyms and Regulatory Terms. Buffalo Grove, IL: Interpharm Press, 1992.

Spelton, Janie. Validation Requirements for CANDA Systems. Presentation at Drug Information Association Workshop, "Clinical Data Management," Brussels, Belgium, November 8-10, 1993.

Sprague, Ralph H., Jr. and Eric D. Carlson. Building Effective Decision Support Systems. Englewood Cliffs, NJ: Prentice-Hall, Inc., 1982.

Spriet, Alain and Therese Dupin-Spriet. Imperfect Data Analysis. *Drug Information Journal*, Vol. 27 (October-December 1993).

Stallings, William. Local Networks -- An Introduction. New York: Macmillan, 1984.

Stark, Peter A. Computer Programming Handbook. Tab Professional and Reference Books, 1978, pp. 229-232.

Stearns, E.W. The Information Center: The Best Answer? *Infosystems*, June 1984.

Steifel, M. Floppy Disk System. *Mini-Micro Systems*, Vol. 11 (November 1978), pp. 37-51.

Stein, Gary C. A Comparison of Validation Aspects of Statistical Data Analysis Systems. Presentation at Drug Information Association Workshop, "Statistical Issues in the Pharmaceutical Industry," Hilton Head, SC, March 30, 1993.

Stein, Gary C. Document Architecture in a Global Document Management System. Presentation at Drug Information Association, 29th Annual Meeting, Chicago, IL, July 11-15, 1993.

Stein, Gary C. Ensuring Computer System Validation: The Regulatory Imperative. Presentation at IBC Computer Assisted Product License Applications Workshop, Washington, DC, March 24-25, 1994.

Stein, Gary C. The Good Automated Laboratory Practices (GALPs) as a Guide to Laboratory Regulatory Compliance. Presentation at Analytical Laboratory Exposition and Conference, San Francisco, CA, October 5-7, 1993.

Stein, Gary C. Issues in the Regulated Laboratory and the Demand for Validation of Analytical Systems. Presentation at Hewlett-Packard Analytical Forum, Chepstow, UK, October 26-28, 1993.

Stein, Gary C. Validation of Computer Systems: What FDA Needs and Expects. Presentation at Drug Information Association Workshop, "FDA Pre-NDA Approvals in Europe," London, UK, April 28-29, 1993.

Stevens, J. Lawrence. Design Verification. *Medical Device & Diagnostic Industry*, Vol. 16 (January 1994).

Stevenson, Robert. HPLC '93: Environmental Applications of HPLC. *American Environmental Laboratory*, August 1993.

Stokes, Theresa E. Clinical Computing and GCP Impact Study. *Applied Clinical Trials*, Vol. 2 (April 1993).

Stokes, Theresa E. Hardware and Software System Validation Issues Under EC Regulations: Vendor Support for Validation. *Drug Information Journal*, Vol. 27 (April-June 1993).

Stokes, Theresa E., *et al*. Good Computer Validation Practices. Buffalo Grove, IL: Interpharm Press, 1994.

Sturm, Robert M. Technology Update: Introduction to a Global Document Management System. *Applied Clinical Trials*, Vol. 3 (February 1994).

Sudman, S. Applied Sampling. New York: Academic Press, 1976.

Sulcas, P. Planning Timing of Computer Auditing. *The South African Chartered Accountant*, July 1975, p. 232.

Svenson, Debra and Michael McLean. Computer-Integrated Electronic Batch Records and Pharmaceutical Documents: A First Step Toward Paperless Factories. *Pharmaceutical Technology*, Vol. 14 (November 1990).

Swanson, E. Burton. Information System Implementation. Irwin, 1988, "Evaluation: Taking the Measure of Performance." pp. 74-83.

Swanson, E. Burton and Cynthia Mathis Beath. Maintaining Information Systems in Organizations. New York: John Wiley & Sons, 1989.

Sweeney, Fergus. Merging GCP and ISO 9000 Requirements -- Source of Synergy in Quality Management of Clinical Research. *Drug Information Journal*, Vol. 28 (October-December 1994).

Sweeney, Fergus. Practical Application of ISO 9000 and EC Good Clinical Practice in Validating Clinical Data Management Systems. *Drug Information Journal*, Vol. 27 (April-June 1993).

Taylor, J. Monitoring of GMP Compliance in the Development of Medicinal Products in the United Kingdom. *Drug Information Journal*, Vol. 28 (October-December 1994).

Taylor, John K. Validation of Analytical Methods. *Analytical Chemistry*, Vol. 55 (May 1983).

Teagarden, C.J. A Stepwise Approach to Software Validation. *Pharmaceutical Technology*, Vol. 13 (September 1989).

Tektronix Inc. Essentials of Data Communications. Beaverton, OR: Tektronix Inc., 1978.

Tetzlaff, Ronald F. GMP Documentation Requirements for Automated Systems. Part I. *Pharmaceutical Technology*, Vol. 16 (March 1992).

Tetzlaff, Ronald F. GMP Documentation Requirements for Automated Systems. Part II. *Pharmaceutical Technology*, Vol. 16 (April 1992).

Tetzlaff, Ronald F. GMP Documentation Requirements for Automated Systems. Part III: FDA Inspections of Computerized Laboratory Systems. *Pharmaceutical Technology*, Vol. 16 (May 1992).

Tetzlaff, Ronald F. Validation Issues for New Drug Development: Part I, Review of Current FDA Policies. *Pharmaceutical Technology*, Vol. 16 (September 1992).

Tetzlaff, Ronald F. Validation Issues for New Drug Development: Part II, Systematic Assessment Strategies. *Pharmaceutical Technology*, Vol. 16 (October 1992).

Tetzlaff, Ronald F. Validation Issues for New Drug Development: Part III, Systematic Audit Techniques. *Pharmaceutical Technology*, Vol. 17 (January 1993).

Tetzlaff, Ronald F., Richard E. Shepherd, and Armand J. LeBlanc. The Validation Story: Perspectives on the GMP Inspection Approach and Validation Development. *Pharmaceutical Technology*, Vol. 17 (March 1993).

Thiel, Carol Tomme. Managing the Highly Skilled. *Infosystems*, March 1984.

Thierauf, Robert J. Decision Support Systems for Effective Planning and Control. Englewood Cliffs, NJ: Prentice-Hall, Inc., 1982.

Thompson, P. William. Advances in the Automation of Bioprocessing. *Pharmaceutical Technology*, Vol. 14 (November 1990).

Tomlinson, Edward A. Automating a Manufacturing Operation: Selecting the Right Software. *Pharmaceutical Technology*, Vol. 12 (February 1988).

Traisnel, M. and A.T. Gayot. Practice of Validation. *Drug Development and Industrial Pharmacy*, Vol. 21 (January 1995).

Trill, A.J. EU GMP Requirements and the Good Automated Manufacturing Practice (GAMP) Supplier Guide for the Validation of Automated Systems for Pharmaceutical Manufacturing. *Pharmaceutical Engineering*, Vol. 15 (May-June 1995).

Tsokos, C.P. Probability Distributions: An Introduction to Probability Theory with Applications. Bellmont, CA: Duxbury Press, 1972.

Tucker, Jane and Iain Menzies Smith. Electronic Data Transfer for Clinical Trials. *Drug Information Journal*, Vol. 28 (April-June 1994).

Turn, Klein. Computers in the 1980s. New York: Columbia University Press, 1977.

Tussing, R.T. and G.L. Helms. Training Computer Audit Specialists. *Journal of Accountancy*, July 1980, pp. 71-74.

Ungar, Wendy. Good Clinical Practices: Principles and Application of the European Community Note for Guidance. *Drug Information Journal*, Vol. 28 (January-March 1994).

United Kingdom, GLP Compliance Programme, Department of Health. Good Laboratory Practice, Advisory Leaflet Number 1: The Application of GLP Principles to Computer Systems. London: Department of Health, 1989.

UK Pharmaceutical Industry Computer Systems Validation Forum. Pharmaceutical Industry Supplier Guidance: Validation of Automated Systems in Pharmaceutical Manufacture. First Draft (Consultation Document), February 1994.

U.S., Department of Commerce, National Bureau of Standards. Glossary for Computer Systems Security. FIPS Publication 39.

U.S., Department of Commerce, National Bureau of Standards. Guidelines for Automatic Data Processing Risk Analysis. FIPS Publication 65, 1979.

U.S., Department of Defense. Defense System Software Development. DOD-STD-2167A. February 29, 1988.

U.S., Department of Defense. Defense System Software Development Data Item. Description for the Software Design Document. DI-MCCR-80012A. Washington, D.C.: Government Printing Office, February 29, 1988.

U.S., Department of Defense, National Computer Security Center. Glossary of Computer Security. NCSC-TG-004-88, Version 1, 1988.

U.S., Department of Health and Human Services. Mandatory Guidelines for Federal Workplace Drug Testing Programs; Final Guidelines. Federal Register, Vol. 53, No. 69, April 11, 1988, pp. 11969-11989.

U.S., Department of Health and Human Services. Medicare, Medicaid and CLIA Programs; Final Rule with Comment Period. Federal Register, Vol. 55, No. 50, March 14, 1990, pp. 9537-9610.

U.S., Department of Health and Human Services, Food and Drug Administration. Compliance Program Guidance Manual, Chapter 28, "Human Drugs and Biologics: Bioresearch Monitoring," Document 7348.808/TN-FY 87-2, July 29, 1986.

U.S., Department of Health and Human Services, Food and Drug Administration. Compliance Review Team (for American Red Cross), March 1989.

U.S., Department of Health and Human Services, Food and Drug Administration. Report of the FDA Task Force on International Harmonization. December 1992.

U.S., Department of Transportation. Procedures for Transportation Workplace Drug Testing Programs; Final Rule. Federal Register, Vol. 54, No. 230, December 1, 1989, pp. 49854-49884.

U.S., Environmental Protection Agency, Office of Compliance Monitoring, Pesticide Enforcement Branch. Enforcement Response Policy for the Federal Insecticide, Fungicide, and Rodenticide Good Laboratory Practice (GLP). n.d.

U.S., Environmental Protection Agency, Office of Information Resources, Management. Automated Laboratory Standards: Current Automated Laboratory Data Management Practices (Final, June 1990).

U.S., Environmental Protection Agency, Office of Information Resources, Management. Automated Laboratory Standards: Evaluation of the Standards and Procedures Used in Automated Clinical Laboratories (Draft, May 1990).

U.S., Environmental Protection Agency, Office of Information Resources, Management. Automated Laboratory Standards: Evaluation of the Use of Automated Financial System Procedures (Final, June 1990).

U.S., Environmental Protection Agency, Office of Information Resources, Management. Automated Laboratory Standards: Good Laboratory Practices for EPA Programs (Draft, June 1990).

U.S., Environmental Protection Agency, Office of Information Resources, Management. Automated Laboratory Standards: Survey of Current Automated Technology (Final, June 1990).

U.S., Environmental Protection Agency, Office of Information Resources, Management. EPA LIMS Functional Specifications (March 1988).

U.S., Environmental Protection Agency, Office of Information Resources, Management. EPA System Design and Development Guidance, Vols. A, B, and C (1989).

U.S., Environmental Protection Agency, Office of Information Resources, Management. Good Automated Laboratory Practices: EPA's Recommendations for Ensuring Data Integrity in Automated Laboratory Operations, With Implementation Guidance (Draft, December 28, 1990).

U.S., Environmental Protection Agency, Office of Information Resources, Management. Survey of Laboratory Automated Data Management Practices (1989).

U.S., Environmental Protection Agency, Office of Prevention, Pesticides, and Toxic Substances. Questions and Answers: Enforcement Action Against Bio-Tek Industries and Microbac Labs for Violations of Good Laboratory Practice Standards. For Your Information Bulletin, June 26, 1992.

U.S., General Accounting Office. Bibliography of GAO Documents, ADP, IRM, & Telecommunications, 1986. Washington, D.C.: GAO, 1987.

U.S., General Accounting Office. Evaluating the Acquisition and Operation of Information Systems. Washington, D.C.: GAO, 1986.

U.S., General Accounting Office. Improvements Needed in Managing Automated Decision-making by Computers Throughout the Federal Government. The Comptroller General of the United States, Report FGMSDD-76-5, 1976.

U.S., Office of Management and Budget. Guidance for Preparation and Submission of Security Plans for Federal Computer Systems Containing Sensitive Information. OMB Bulletin No. 88-16, July 6, 1988.

Van Bruwaene, Ben, Luc Dekeyser, and Marleen Stevens. Integrated Study Automation: Computerization at the Site of Investigation. *Applied Clinical Trials*, Vol. 2 (October 1993).

Van Manen, Robbert P. and Erik Houthoff. A Central Database with Distributed Off-Line Data Entry and Correction: The Best of Both Worlds? *Drug Information Journal*, Vol. 27 (October-December 1993).

Van Manen, Robbert P. and Karel J. de Neef. Validating a Clinical Data System: Technical, Organizational, and Managerial Aspects. *Drug Information Journal*, Vol. 27 (April-June 1993).

Vasarhelyi, M.A., C.A. Pabst, and I. Daley. Organizational and Career Aspects of the EDP Audit Function. *EDP Auditor*, 1980, pp. 35-43.

Von Tilsit, H. Information -- Main Ingredient. *Pharmaceutical Executive*, Vol. 14 (December 1994).

Votrax. SC-01 Speech Synthesizer Data Sheet. Troy, MI: Votrax, 1980.

Walsh, James E. Handbook of Nonparametric Statistics. 3 vols. New York: Holt, Rinehart and Winston, 1953.

Warburton, David. Establishing an Effective Calibration and preventive Maintenance Program. *Medical Device & Diagnostic Industry*, Vol. 15 (January 1993).

Ward, J.H., Jr. and E. Jennings. Introduction to Linear Models. Englewood Cliffs, NJ: Prentice-Hall, 1973.

Ward, Paul and Steve Mellor. Structured Development for Real-Time Systems, Vols. 1-3. Englewood Cliffs, NJ: Yourdon Press/Prentice-Hall, 1986.

Warnier, Jean-Dominique. The Logical Construction of Programs. New York: Van Nostrand Reinhold, 1976.

Webster, Edward. Figuring the Economics of the New Page Printers. *Datamation*, Vol. 24 (May 1978), pp. 171-77.

Wegner, Peter. Programming Languages-the First 25 Years. *IEEE Transactions on Computers*, Vol. C-25 (December 1976), pp. 1207-25.

Weinberg, Gerald. An Introduction to General Systems Theory. New York: John Wiley & Sons, 1975.

Weinberg, Gerald. Rethinking Systems Analysis and Design. Boston: Little, Brown, 1982.

Weinberg, Gerald. The Psychology of Computer Programming. New York: Van Nostrand Reinhold, 1971.

Weinberg, Sanford B. Case Studies in System Validation: Sixteen Theses. *Drug Information Journal*, Vol. 27 (April-June 1993).

Weinberg, Sanford B. Development of a Central Database for Pharmaceutical and Environmental Results: Technical Considerations. Presented at ACHEMA Conference, Frankfurt, Germany, June 7, 1994; published in *Weinberg Validation Review*, Vol. 3 (No.3, 1994).

Weinberg, Sanford B. FDA Issues in System Validation of the QC Laboratory. Address to the Investigations Branch, Compliance Branch, IBr Steering Committee, U.S. Food and Drug Administration, December 10, 1993; published in *Weinberg Validation Review*, Vol. 3 (No. 1, 1994).

Weinberg, Sanford B. Good Automated Laboratory Practices. *Quality Assurance: Good Practice, Regulation, and Law*, Vol. 2 (March-June 1993).

Weinberg, Sanford B. Impact of GLP and GALP Requirements on Customers and Laboratory Users of Instruments. Presentation at Analytical Instrument Association, Spring Meeting, April 25-27, 1993.

Weinberg, Sanford B. Methodology for the Validation of a Laboratory Information System. Presented at Pittcon '94, Chicago, IL, February 27-March 4, 1994.

Weinberg, Sanford B. 1993 International LIMS Conference. *American Environmental Laboratory*, August 1993.

Weinberg, Sanford B. System Validation: The Regulatory Environment. Presentation at the Neu Ulm Conference on the Evolution to Paperless Clinical Trials, Neu Ulm, Germany, March 10-11, 1993.

Weinberg, Sanford B. System Validation Standards. Dubuque: Kendall/Hunt Publishing Company, 1990.

Weinberg, Sanford B. Toward Harmonization of Raw Data. *American Environmental Laboratory*, April 1993.

Weinberg, Sanford B. Validation of Computerized Biotechnology Systems. Presentation at Drug Information Association Workshop on "Biotechnology: Meeting the Challenges of the 1990s," May 19-21, 1993.

Weinberg, Sanford B. Why the GALPs? *Environmental Lab*, February/March 1993.

Weinberg, Sanford B., Ronald M. Romoff, and Gary C. Stein. Handbook of System Validation. Philadelphia: Weinberg, Spelton & Sax, Inc., 1993.

Weinberg, Sanford B., *et al*. System Validation Checklist. Copyrighted Monograph, 1988.

Weinberg, Sanford B., *et al*. Testing Protocols for the Blood Processing Industry. Copyrighted Monograph, 1989.

Weinberg, Spelton & Sax, Inc. GALP Regulatory Handbook. Boca Raton, FL: Lewis Publishers, 1994.

Weissinger, J. Systems Approach to Quality Auditing. *Drug Information Journal*, Vol. 28 (October-December 1994).

Weizenbaum, Joseph. Computer Power and Human Reason. San Francisco: W.H. Freeman and Company, 1976.

Wells, Heather J., Terry M. Price, and Caroline A. Evans. Design and Construction of an Automated Validation Procedure for a Clinical Data Management System. *Drug Information Journal*, Vol. 27 (April-June 1993).

Wesley, Roy L. and John A. Wanat. A Guide to Internal Loss Prevention. Stoneham, MA: Butterworth Publishers.

Wickelgren, Wayne A. How to Solve Problems. San Francisco: W.H. Freeman and Company, 1974.

Wiederhold, Gio. Database Design. New York: McGraw-Hill, 1983.

Wilkins, B.J. The Internal Auditor's Information Security Handbook. Altamonte Springs, FL: Institute of Internal Auditors, 1979.

Williams, Michael H. Good Computer Validation Practice is Good Business Practice. *Drug Information Journal*, Vol. 27 (April-June 1993).

Willingham, J. and D.R. Carmichael. Auditing Concepts and Methods. New York: McGraw-Hill, 1975.

Winer, B.J. Statistical Principles in Experimental Design. 2nd ed. New York: McGraw-Hill, 1991.

Wingate, G. Computer-System Validation: Historical Perspective. *Pharmaceutical Engineering*, Vol. 15 (July-August 1995).

Wirth, Niklaus. Algorithms + Data = Programs. Englewood Cliffs, NJ: Prentice- Hall, 1976.

Wirth, Niklaus. Systematic Programming. Englewood Cliffs, NJ: Prentice-Hall, 1983.

Wood, Bill J. and Julia W. Ermes. Applying Hazard Analysis to Medical Devices; Part II: Detailed Hazard Analysis. *Medical Device & Diagnostic Industry*, Vol. 15 (March 1993).

Wood, Bill J. and Julia W. Ermes. Applying Hazard Analysis to Medical Devices; Part I: Initial Hazard Analysis. *Medical Device & Diagnostic Industry*, Vol. 15 (January 1993).

Wood, Janie and Denise Silver. Joint Application Design. New York: John Wiley & Sons, 1989.

Wonnacott, T.H. and R.J. Wonnacott. Introductory Statistics for Business and Economics. 2nd edition. New York: John Wiley and Sons, 1977.

Written, Ian H. Communicating with Microcomputers. New York: Academic Press, Inc., 1980.

Wyrick, Michael L and Rafik H. Bishara. Guidelines for Quality Assurance Lead Auditors. *Pharmaceutical Technology*, Vol. 19 (October 1995).

Yourdon, E. Structured Walkthroughs. New York: Yourdon Press, 1982.

Zaffarano, Joan. Calculators: The Programmables and Prepos. *Administrative Management*, August 1973, pp. 40-48.

Zeisel, H. Say It with Figures. New York: Harper & Row, 1957.

Zimmerli, Wesley D., Gale P. DeHaven and Frank R. Funderburk. Computerized Behavioral Testing: Implications for Quality-of-Life and Outcomes Assessment. *Applied Clinical Trials*, Vol. 2 (July 1993).

Zuboff, Shoshana. In the Age of the Smart Machine. New York: Basic Books, 1988.

INDEX

Abbreviated New Drug Application (ANDA), 50, 57, 148-157,
AIMD (*see* Active implantable medical devices)
Acceptance
  criteria, 3, 129, 133, 167-169, 229, 243, 648, 650, 652, 747, 762, 787
  procedures, 133,
  testing (*see* Testing)
Access
  authorized, 250
  unauthorized, 285, 394, 748, 752, 772, 781
Active implantable medical devices (AIMD), 174
Adverse reactions, 119, 121, 221, 256-258, 260, 269, 274, 317, 319, 372, 465, 470, 512, 526, 592, 742, 757, 759
Alarms, 468, 477, 491, 500, 546, 593, 733, 735, 744
Algorithms (*see also* Formulas), 3, 13, 24, 167, 209, 210, 229, 460, 462, 482, 584, 592, 744, 745, 749, 761, 762, 772, 781, 797, 802
American Association of Blood Banks, 12, 89, 99, 211, 216, 254, 331, 335, 336, 494, 512, 521, 626, 738, 749, 751, 757
Archives, 162, 229, 367, 562, 750, 752, 769, 781, 789-791
Audit trails, 4, 229, 253, 269, 272, 283, 285, 382, 387, 731, 742, 743, 751, 752, 756, 767, 770, 772, 778
Australia, 5, 118, 253, 268
Automated data collection, 160, 161, 163
Automated instruments, 165, 273, 731, 791
Automation, 2, 6, 9, 16, 27, 129, 132, 133, 136, 160-166, 169, 171, 172, 207-209, 220, 228-230, 235-238, 242, 249, 253, 273, 306, 347, 349, 350, 357, 358, 367, 390, 440, 452, 458, 464, 467-470, 476-478, 486, 501, 503, 504, 508, 509, 514, 527, 529, 534, 546, 547, 554, 582, 635, 645, 655, 657, 660, 723, 726, 727, 730-734, 742, 747, 750, 751, 766, 771, 772, 776, 787, 790-793, 799, 800, 802, 804

Back-up, 219, 285, 356, 388, 394, 395, 438, 468, 724, 753, 768, 794, 798
Biologics, 1, 5, 7, 10, 20, 24, 25, 27, 28, 111, 208, 209, 213-215, 217, 218, 220-222, 224, 225, 234, 239, 254, 335, 336, 362, 375, 458, 463, 488, 493, 503, 510, 512, 516, 521, 524, 538, 556, 568, 602, 626, 727, 741, 743
Blood banks (*see* Blood establishments)
Blood establishments, 209, 212-216, 239, 458, 459, 463, 465, 476, 483, 488, 489, 491, 493, 494, 512-516, 537, 539, 544, 545, 549, 551, 722-724, 727, 729-732, 736

Calibration, 1, 3, 133, 196, 200, 204, 219, 220, 240, 245, 265, 467, 488-490, 492, 493, 495-497, 508, 590, 625, 656, 729, 734, 756, 787
CANDA (see Computer-Assisted New Drug Application)
Center for Biologics Evaluation and Research (CBER), 25, 28, 208, 212-215, 218, 219, 224, 336, 337, 463-465, 468,

469, 471-473, 475, 476, 478-488, 493, 501, 503, 504, 506, 508, 510-512, 521, 522, 524, 525, 527-530, 535, 536, 538-546, 548, 549, 551, 553, 554, 556-562, 568, 569, 727, 743
Center for Devices and Radiological Health (CDRH), 28, 126, 208, 213, 223-225, 337, 338, 459, 493, 564, 568, 581, 583, 591, 598, 599, 611, 612, 616, 617, 619, 620, 622-624, 627, 631, 632, 636, 638, 648, 663-674, 683, 686-688, 690, 691, 693, 714, 720, 735, 786
Center for Drug Evaluation and Research (CDER), 28, 124, 139, 145, 148, 151, 155, 161, 218, 233, 336, 337, 348, 349, 353, 355, 356, 359-364, 366, 390, 395, 397, 398, 415, 435, 437, 568
Center for Veterinary Medicine, 29, 225, 335, 568
Certification, 8, 9, 12, 13, 17, 20, 21, 23-25, 27, 48-50, 53, 55, 57, 58, 60-63, 83, 89, 97-99, 102, 103, 105, 107, 110, 118, 121, 142-155, 157, 236, 237, 474, 481, 558, 559, 561, 562, 578, 617, 618, 673, 683, 684, 712, 717, 718, 721, 756, 760, 802
CGMP (*see* Good Manufacturing Practice)
Change control, 167, 169, 249, 250, 252, 392, 393, 491, 549, 608, 633, 634, 639-641, 649, 721, 728, 739, 750, 756, 788, 790, 797
Complaint Files, 137, 221, 602, 604, 608-610, 614, 622, 623, 625, 627, 635, 637, 641, 645, 646, 649, 685-687, 690, 691
Computer-Assisted New Drug Application (CANDA), 347,348, 350-354, 357, 358, 361, 362, 368, 377-379, 382-384, 388-399, 403, 407, 411, 417, 425, 430, 434, 435, 438-457, 756
Computer-Assisted Product License Application (CAPLA), 557, 756
Control, evidence of, 1-6, 8-12, 23, 24, 38, 51-53, 63, 68, 77, 78, 82, 87, 102-104, 120-122, 125, 128, 142, 146, 152, 156, 157, 161, 167, 170, 208, 215, 217, 221-224, 226, 229, 233, 234, 238-240, 245, 259, 278, 304, 306, 324, 393, 435-538, 452, 464, 469, 475, 484, 490, 507, 548, 574, 603, 608, 610, 613, 618-621, 633, 640, 732, 734, 738, 739, 741, 742, 744, 745, 747, 748, 756, 760, 766-770, 772, 774, 779, 780, 787, 790, 792, 797, 800
Corrective actions, 10, 88, 129, 131, 136, 137, 166-168, 198, 199, 213, 246, 248, 253, 469, 481, 491, 511, 537, 609, 616, 621, 652, 669, 673, 680, 684, 686, 695, 697, 699, 701, 702, 704, 705, 707, 710, 712, 763, 775, 778, 790
Cosmetics, 28, 115, 123, 233, 329, 330, 375, 584

Data entry (*see also* data input), 229, 253, 254, 356, 415, 417, 423, 491, 731, 746, 763, 770, 772
Data input, (*see also* data entry), 252, 459, 657,726, 743, 752, 762, 800, 803
Data integrity, 4, 160, 163, 165, 166, 229, 530, 560, 722, 751-753, 760, 761, 767, 790, 794
Design:
changes, 130, 243, 577, 585, 586, 780
controls, 125, 129, 225, 577
Device design, 126, 130, 575-579, 639, 780
Device History Record (DHR), 127, 131, 133-136,614, 635, 644, 645, 647, 649, 656-658, 687, 724, 727, 735, 762
Device Master Record (DMR), 127, 133, 136, 577, 614, 635, 639, 640, 643, 645, 649, 650, 656, 694, 696, 698, 700, 703, 706, 709, 711, 726-730, 733, 736, 762, 763
Disaster recovery, 18, 208, 251-253, 395, 746, 763, 772, 786, 794, 797, 800, 804
Documentation (*see also* records), 3, 6, 8, 10, 11, 23, 24, 70, 71,130, 133, 134, 136, 137, 146, 152, 160, 165, 167-169, 171, 185, 194, 197, 199-202, 204, 205, 207, 209, 210, 215, 219, 227, 238, 239, 243, 244, 249-252, 254, 270, 288, 289, 298, 308, 313, 326, 328, 388, 389, 393, 394, 396, 428, 453, 460, 462, 466, 468-472, 478, 488-490, 493, 506, 507, 528, 530, 537, 552-554, 561, 579-581, 592, 619-621, 643, 649, 656, 668, 672, 674, 675, 683, 693, 721, 722, 724, 725, 727, 728, 731-734, 739, 740, 741, 742, 745-747, 750, 753, 754, 756, 760, 762, 763, 766, 767, 769, 772, 773, 776, 778, 779, 782, 785, 786, 788, 790, 792, 795-797, 800, 802-805

Drug Master File (DMF), 426, 431
Drug products, 1, 7, 10, 24, 110, 114, 123-125, 174, 207, 209, 217, 222, 223, 231, 238, 241, 252-254 264, 336, 364-366, 398, 407, 411, 440, 441, 444, 446-451, 453, 729, 738, 760, 785, 787

Establishment License Application (ELA), 503, 554, 557-560
Electronic records, 760
Electronic signatures, 220, 741
Environmental conditions, 132, 169, 196, 548, 585, 654, 766, 788
Environmental laboratories, 1, 5, 20, 25, 228-230
Environmental Protection Agency (EPA), 1, 5-7, 9, 16, 17, 20, 27, 159-165, 167-172, 228-232, 235-237, 771-773, 789, 790
European Union (EU), 5, 6, 174, 178, 179, 183, 185, 187, 234, 239, 241, 254, 264, 268, 336-338, 738, 756, 766, 776

Firmware, 126, 585, 750, 753, 759, 766, 769, 770, 775, 793, 798
510(k) *see* premarket notification)
Food and Drug Administration (FDA), 1, 2, 4-7, 9-11, 17-20, 23, 25, 27, 28, 30-34, 47, 109-111, 114, 122, 123, 125, 126, 135-142, 144-161, 208, 209, 211-216, 218-227, 231-239, 249, 254, 255, 329-331, 334-349, 351, 352, 357-363, 366, 368, 377-383, 386, 387, 389-398, 405, 407, 415-417, 425-436, 438-442, 444, 447, 448, 451, 458, 459, 463-465, 467, 468, 471-473, 476, 481, 483, 485, 493, 501-505, 507-509, 511, 513, 514, 518, 522, 529, 536-538, 541-545, 549, 551, 553, 556, 558, 561-575, 577, 578, 583, 587, 590, 595, 598, 600, 601, 604-613, 615-617, 619-628, 633, 636, 638, 641, 644, 654, 659-661, 664, 669, 674, 680, 686-691, 694-710, 712-714, 720-722, 738, 743, 748, 757-759, 765, 768, 769, 773, 777, 780, 785-787, 789, 791, 792, 794-797, 803, 805
Food, Drug and Cosmetic Act, 25, 28, 30, 32, 34, 35, 37, 116, 123, 125, 126, 147, 212, 214, 215, 225, 329, 488, 505, 512, 563, 575, 577, 584, 591, 600, 625, 694, 696, 698, 700, 703, 706, 709, 711, 769, 773
Foreign inspections, 612, 674
Formulas (*see also* Algorithms), 3, 24, 33, 40, 42, 43, 168, 226, 229, 234, 265, 309, 746, 749, 772, 793, 794, 797
483 (*see* Inspectional observations)
France, 5, 118, 253, 723
Functional requirements, 167, 462, 761, 800

GALP (*see* Good Automated Laboratory Practice
GCP (*see* Good Clinical Practice)
Germany, 5, 777
Good Automated Laboratory Practices (GALPs), 16, 27, 160-166, 171, 172, 229, 230, 235-238, 240, 750, 771, 772, 777, 783, 790, 792, 802, 804
Good Clinical Practices (GCPs), 249, 254, 255, 262, 268-270, 272-276, 278, 280, 284, 286, 287, 290, 292-294, 300, 327, 771, 772, 777
GLP (*see* Good Laboratory Practice)
GMP (*see* Good Manufacturing Practice)
Good Laboratory Practice 13, 15, 27, 28, 159, 161-165, 169-171, 207, 226, 228, 230, 236, 249, 250, 252, 254, 750, 771-773, 783, 790
Good Manufacturing Practice, 1, 7, 13, 27, 28, 39, 89, 103, 117, 121-125, 129, 207, 209, 211-221, 223-227, 236, 237, 239, 249, 251, 254, 275, 288, 289, 338, 347, 466, 483, 511, 517, 521, 549, 564, 574, 576, 577, 579, 581, 586, 588, 589, 598, 600, 601-612, 614-623, 625-627, 629, 631-638, 642-645, 654, 659-686, 688, 690, 691, 694-701, 703, 704, 706, 707, 709-712, 714-721, 723, 725, 728, 729, 732, 760, 766, 771, 773, 780, 783, 787, 791

Hardware (*see also* Firmware), 1, 4, 6, 12, 13, 17, 18, 161, 163, 168, 170, 220, 232, 245, 249, 251, 252, 254, 350, 352, 353, 355, 356, 360-362, 366, 369, 370, 390-392, 395, 415, 430, 434, 445, 459, 491, 549, 723, 724, 726, 731, 735, 739, 744, 745, 748, 750, 752-754, 758, 759, 762, 764, 766, 767, 769, 770, 773, 775, 779, 783, 784, 786, 789, 791, 798, 800, 801, 804, 805
Hazard analysis, 223, 459, 461, 771, 774,

786, 796, 799

Investigational New Drug (IND), 138-141, 145, 236, 263, 353, 362, 425, 426, 557
Inspectional Observations (Form FDA-483), 4, 501, 504, 505, 507, 511, 518, 522, 536, 538, 542, 551, 598, 605, 606, 609-613, 615, 616, 620-624, 633, 636, 637, 638, 641, 659, 660, 664-667, 674-678, 687, 688, 690, 691, 694-701, 703, 704, 706, 707, 709, 710, 712, 714, 72, 768, 805
Inspections, (*see also* Foreign inspections), 8, 10, 11, 19, 25, 32, 34, 37, 38, 65, 103, 104, 106, 107, 109, 111, 126, 131-133, 135, 136, 166, 168, 185, 197-201, 204-207, 213-215, 218, 219, 221, 223-225, 232, 234, 242, 245, 246, 248-251, 254, 271, 289, 293, 320, 338, 343, 347, 458, 463-466, 471, 472, 480, 482, 483, 501-508, 510-512, 515, 521, 522, 524, 525, 535-538, 542-545, 547, 548, 553-555, 563-565, 568, 598-612, 614-621, 623, 626-628, 631-642, 645, 648, 649, 652, 654, 655, 657, 659-667, 670, 673-675, 683-685, 687, 688, 690, 691, 694-701, 703, 704, 706, 707, 709-712, 714, 721, 723, 725, 730, 733, 734-, 736, 758, 766, 769, 773, 774, 776, 786, 790, 792, 796
Installation qualification, 12, 220, 221, 739, 775, 787
Institutional Review Board (IRB), 271, 275-284, 287-290, 292-294, 301, 325, 772
Intended use, 14, 25, 27, 30-33, 39, 40, 42-44, 47, 53, 61, 62, 65-73, 81, 83-85, 87, 88, 90, 93, 95-97, 99, 101, 116, 125-130, 132, 133, 135, 137, 138, 160, 163, 167, 171, 175-177, 180-182, 184, 189, 190, 191-193, 195-198, 201, 212, 214, 218-220, 223, 226, 227, 229, 230, 264, 268, 269, 281, 298, 299, 303, 309, 311, 312, 314, 315, 320, 331, 332, 339, 340, 343, 350, 359, 360, 372, 373, 377, 393, 399, 407, 427, 436, 441, 461, 463, 467, 476, 487, 490, 491, 494, 501, 505, 512, 514, 535, 537, 545, 547-549, 553, 556, 557, 565, 571-579, 581-583, 585, 586, 588-594, 601, 607, 626, 640, 643, 645, 651, 654, 662, 695, 697, 699, 701, 704, 707, 710, 712, 727, 730, 735, 736, 739, 742, 744, 749, 750, 753, 771, 772, 785, 789, 792, 798, 800, 801, 805
Internal checking, 3-5, 13, 15-18, 24, 44, 128, 178, 196, 208, 210, 223, 224, 235, 236, 238, 246, 248, 254, 308, 333, 345, 393, 429, 478, 494, 529, 558, 564-566, 568, 570, 576, 577, 606, 633, 652, 660, 695, 697, 729, 739, 743, 747, 753, 757, 764, 765, 767, 769, 775, 776, 781, 788, 790, 793, 795-797, 799, 805
ISO 9000, 9,, 23, 125, 171, 236, 237, 240-243, 247, 248, 335, 336, 338, 588, 770, 775, 776, 783, 802

Japan, 5, 19, 23, 118, 249-251, 254, 264, 268, 336-338, 727, 772, 777

Labeling, 8, 30-32, 34-36, 38, 40-43, 48, 51, 52, 54, 56, 59, 62, 63, 64, 66, 68, 73, 76-79, 81, 88, 119, 124, 125, 126, 127, 133, 135-137, 143, 150, 156, 159, 160, 210, 212, 216, 222, 223, 225, 226, 288, 306, 315, 329, 400, 405, 408, 427, 453, 461, 481-483, 485, 486, 494-496, 504, 507, 511, 513, 523, 525, 530, 536, 544, 552, 574, 575, 578, 580, 581-585, 587, 589, 591, 592, 614, 626, 635, 645, 650, 652, 653, 656-658, 689, 692, 720, 729-733, 741, 758, 765, 783
Laboratory Information Management System (LIMS), 3, 15, 25, 162-172, 209, 229, 236, 239, 747, 777, 778, 796
LANS (*see* Local area networks)
Levels of concern (*see also* Hazard analysis), 6, 7, 9, 11, 18, 20, 24, 25, 27, 179, 190-192, 203, 209, 223, 257, 267, 293, 296, 319, 320, 330, 339, 377, 446, 451, 453, 459, 490, 511, 564, 567, 586, 622, 730, 741-743, 745, 764-766, 771, 782, 783, 794, 795, 797, 798, 801, 803,
Local area networks (LANS), 455, 371, 557

Maintenance, 3, 25, 26, 46, 54, 62, 73, 83, 91, 111, 128, 132, 133, 136, 164, 166, 168-171, 179, 196, 206, 212, 214, 219, 244, 249-251, 253, 269, 295, 394, 395, 460, 462, 468, 469, 488, 489, 491, 493, 495, 498, 500, 508, 525, 530, 546, 560, 591, 654, 655, 692, 693, 722, 726, 727, 734, 739, 758, 765-767, 772, 777, 779,

784, 791, 792, 795, 799, 803
Medical Device Amendments, 93, 213, 337, 563, 575
Medical Device Reporting (MDR), 137, 212, 214, 549, 599-602, 604-606, 610-612, 614, 622-625, 627, 630, 632, 633, 634, 636-638, 659, 660, 685-701, 703, 704, 706, 707, 709-712, 714-719, 735
Medical Devices:
 Class I, 67-72, 81, 82, 86, 88, 94, 95, 125, 129, 224, 459, 564, 577, 598, 601, 604, 605, 607-609, 616, 627, 630, 661, 663, 670, 674, 780,
 Class II, 65, 67-72, 74, 76, 81, 88, 94, 95, 129, 224, 337, 577, 603, 604, 780
 Class III, 39, 40, 68-72, 74-76, 78, 81, 88, 94, 95, 129, 189, 564, 577, 603, 604, 780, 786
 Fitness-for-use, 132, 135, 651, 654, 655, 658
 predicate device, 72

New Drug Application (NDA), 9, 19, 31, 48, 50, 53-57, 60, 62, 93-95, 110, 117, 124, 138, 142, 144-146, 148-157, 236, 263, 315, 347-350, 353, 357, 359, 360, 362, 366-368, 370, 373, 377, 378, 381, 382, 384, 386, 387, 389-393, 395, 397-400, 405-408, 414, 425, 426, 427, 428, 433-442, 444, 446, 447, 557, 569, 742, 756, 793
Not substantially equivalent (NSE) (*see also* Substantial Equivalence (SE), premarket notification), 580, 582, 591

Performance:
 qualification, 12, 221, 735, 739, 783, 788
 specifications, 128, 137, 138, 490, 574, 580, 581, 584, 585, 588, 589, 590, 592, 594, 646, 649, 694, 696, 698, 700, 703, 706, 709, 711, 726, 734, 735
 testing (*see* Testing)
Pharmaceutical products (*see* Drug products)
Premarket approval, 39, 40, 68, 71, 75, 76, 92-94, 213-215, 564, 577, 580, 582, 591
Premarket notification, 25, 35, 43, 52, 67, 71, 72, 208, 212-215, 217, 222-225, 239, 459, 461, 549, 564, 572-591, 598, 601, 603, 605, 611, 612, 616, 627, 630, 634, 636, 661-663, 666, 667, 670, 671-674, 676-684, 701, 704, 720, 722, 732, 734, 735, 740, 743, 769, 770
Product License Application (PLA), 221, 503, 534, 538, 543, 550, 554, 556-560,
Protocols, 11, 39, 40, 78-80, 92, 97, 132, 139-142, 222, 230, 261, 268-270, 272, 273, 275, 277-280, 282-287, 289, 290, 291-294, 296-300, 302, 304, 308-314, 325, 326, 372, 386, 401, 402, 409, 410, 436, 490, 549, 722, 724, 727, 729, 732, 735, 751, 754, 772, 783, 787, 789, 791, 798, 799, 806
Public Health Service Act (PHS), 329, 483, 553

Quality:
 assurance, 3, 4, 6, 7, 11, 14-20, 24, 33, 125, 127, 136, 159, 161, 166-172, 184, 197, 198, 200, 203, 204, 208, 213, 215, 218-220, 229, 239, 242, 249, 250, 273, 284, 285, 292, 296, 308, 338, 463, 488, 492, 493, 502, 513, 537, 554, 599, 600, 605, 607, 611, 614, 617, 629, 633, 641, 642, 648, 650, 652, 653, 655-657, 672, 683, 685, 694, 695-701, 703, 704, 706, 707, 709-712, 722, 723, 725, 726, 728, 730-736, 739, 741, 747, 756, 764, 776, 780, 782, 783, 787, 789, 790, 800, 803
 control, 8, 14, 18, 20, 26, 33, 196, 198, 225, 238, 242, 249, 251, 253, 273, 284, 285, 292, 296, 308, 338, 467, 489, 492-496, 498, 506, 510, 591, 626, 722, 760, 773, 781, 782, 790, 803
 system, 125-128, 132, 134, 136, 137, 197-201, 204, 205, 242, 243, 248, 338, 727

Raw data, 166-172, 229, 413, 429, 445, 561, 729, 752, 790, 791
Recalls, 4, 26, 85, 186, 198, 221-223, 289, 522, 523, 553, 581, 608, 610, 616, 618, 619, 633, 672, 683, 721, 722, 741, 761, 762, 766, 772, 776, 780, 788, 791, 800, 802, 804
Recordkeeping (*see also* Archives), 285, 296, 614, 645
Records (*see also* Documentation), 4, 8, 35, 37, 45, 46, 52-54, 60, 62, 63, 77, 78, 81, 85, 86, 91, 100, 106, 107, 118, 125-127, 130-137, 158, 161, 166-170, 185, 199, 204, 206, 219, 221, 229, 243, 245-248, 254, 261, 270, 271, 273, 275, 276, 278,

280, 282, 283, 289, 291, 305, 352, 392, 416, 417, 418-423, 432, 437, 466, 467, 469, 471, 474, 477, 479, 482-485, 487, 503, 505-508, 510, 527, 529, 530, 535, 537, 538, 541, 544, 545, 549-552, 577, 579, 581, 606-608, 622, 623, 627, 633, 635, 637-639, 641, 642, 645, 649, 651-653, 655-660, 685-691, 694, 696, 698, 700, 703, 706, 709, 711, 722-726, 728-733, 735, 736, 750-753, 760, 764, 765, 772, 788, 790-792, 796, 799, 801, 803
Responsible person, 16, 229, 250, 505, 777, 792, 793
Risk assessment (*see also* Hazard analysis), 259, 312

Safe Medical Devices Act, 215, 218, 337, 563, 577, 600,
Safety, requirements, 207, 800
Safety and effectiveness, 31, 32, 53-56, 58-60, 62, 67-73, 79, 81, 85-87, 89-93, 95-97, 111, 119, 132, 142, 225, 226, 304, 343, 345, 346, 437, 461, 575-579, 581-590, 659, 723, 733, 735, 787
Security, 18, 20, 107, 109, 160-166, 168-172, 189, 208, 210, 220, 228, 229, 232, 249, 252-253, 285, 331, 351, 352, 354, 355, 358, 390-395, 491, 520, 530, 549, 560, 562, 721, 723, 727, 730, 731, 738, 748, 766, 776, 781, 782, 784, 792, 795, 797, 800, 802, 804
  physical, 69, 110, 126, 159, 164, 170, 191, 192, 274, 304, 307, 321, 324, 394, 469, 471, 479, 503, 506, 510, 516, 522, 527, 539, 540, 593, 594, 610, 621, 748, 749, 759, 761, 770, 773, 775, 781, 784, 785, 786, 788-790, 795, 799, 800, 805
Software, 3, 4, 6, 10, 12, 13, 16, 17, 23-25, 126, 127, 130, 132, 133, 136, 161, 163, 165, 167-171, 177, 181, 208-210, 212-217, 219, 220, 222-225, 228, 229, 232, 236, 238-240, 242, 245, 250-252, 254, 350, 352, 355, 356, 358, 360, 361, 366, 367, 369-371, 373, 388, 390-394, 396, 415, 426, 428, 430, 431-434, 438, 439, 443, 449-451, 454-462, 491, 494, 500, 511, 513, 549, 557, 559-561, 585, 593, 659, 722-736, 739, 741, 743-747, 750, 753-762, 764-768, 770-772, 775, 776, 779, 782-786, 788, 789, 791-805
  application, 25, 163, 228, 229, 549, 743, 750, 770, 772, 801
  development life cycle (SDLC), 739, 740, 750, 754, 757, 760-762, 770, 782, 785, 788, 789, 791, 795, 798-800, 802, 803
  statistical, 433, 439
SOPs (*see* Standard Operating Procedures)
Source Code (*see also* Algorithms), 3, 4, 6, 8, 18, 24, 136, 733, 749, 752, 757, 764, 767, 769, 771, 781, 788, 791, 797, 800, 801, 803, 806
  review of, 6, 749, 764, 788, 801, 806
Standard Operating Procedures (SOPs), 3, 6, 8-11, 14, 27, 39, 43, 60, 66, 68, 70, 72-74, 76-78, 82, 88, 92, 95, 101, 102, 103-107, 109, 114, 131, 133, 166-169, 171, 172, 179, 203, 208, 210, 218, 219, 224, 228, 229, 236, 238, 242, 250, 251, 253, 255, 259, 260, 264-266, 268-270, 274, 284, 285, 292, 307-309, 319, 326, 330-338, 340-343, 345, 346, 351, 371, 372, 395, 398, 437, 441, 445, 463, 464, 466, 467-469, 472, 474, 476, 480, 482, 484-486, 489, 494, 501, 506-508, 523, 526, 530, 534, 536-541, 546-548, 550-554, 560, 561, 576, 586, 590, 607, 608, 626, 718, 721, 722, 724, 727, 729-731, 739, 741, 745-752, 754, 756, 757, 759, 760, 762-767, 769, 772, 773, 776, 779, 784, 785, 787, 788, 790, 792-805
Substantial equivalence (SE), 47, 48, 64, 69-72, 75, 93, 461, 574, 580, 582, 584-586, 590, 591, 636, 663, 692
System:
  development methodology, 750
  requirements, 125, 128, 415, 423, 756, 792, 800
  specifications, 250, 251, 460, 462

Telecommunications, 166, 356, 360, 387, 390, 394
Testing, 3, 5, 6, 8, 10-20, 23-25, 53, 62, 73, 76, 78, 91-93, 96, 100, 101, 103, 104, 106, 107, 122, 130, 159, 162, 167-171, 181, 184, 189, 193, 195-199, 202, 203, 205, 207, 208, 210-213, 215, 219, 222-225, 229, 232, 235, 242, 245, 248, 250, 253, 254, 264, 336, 344, 346, 355, 378, 393, 395, 415, 441, 460, 462, 465, 466, 473, 474, 476, 480-486, 488-494, 497, 502, 505-507, 509-511, 513-516,

521, 522, 525, 528, 530, 534-538, 543, 546, 548-550, 552, 565, 577, 579, 582, 584, 586, 588-590, 606, 608, 613, 628, 630, 633, 640, 641, 643, 648, 651, 653, 660, 721-728, 730-735, 739-742, 745-747, 749, 750, 752, 754-757, 760, 764, 765, 767, 770, 773-775, 777, 778, 780, 782, 783, 786-788, 793-795, 799-804, 806
acceptance, 168, 169, 747, 765, 778, 787, 788
erroneous data, 766, 767, 797
functional, 24, 725, 775
performance, 745, 793, 794, 801
Training, 10, 12, 14, 18, 31, 32, 52, 53, 62, 68, 89, 90, 95, 99, 103, 107, 119, 129, 165, 166, 194, 195, 206, 218, 219, 221, 232, 233, 239, 246-249, 252, 253, 275, 278, 286, 293, 308, 326, 350, 353, 356, 361, 369, 373, 382, 393, 396, 468, 489, 491, 508, 522, 529, 540, 541, 546, 554, 566, 625, 626, 639, 644, 653, 720, 722, 725, 727, 732-734, 745, 751, 763, 764, 766, 772, 774, 783, 786, 792, 795, 797, 798, 801-803

United Kingdom, 5, 28, 30, 87, 115, 116, 119, 122, 220, 239, 251, 252, 466, 543, 777, 802
User fees, 234

Validation:
cleaning, 127, 132, 196, 221, 493, 525, 635, 645, 654, 655,
computer system, 1-3, 5, 10-13, 15, 17, 19, 23-26, 28, 111, 124, 129, 132, 160-165, 169-172, 207, 208, 210-212, 214-223, 227-230, 232, 235, 238, 239, 249-254, 298, 328, 347-351, 353, 356, 357, 359-362, 366-373, 375, 377, 378, 380, 388-391, 393-396, 398, 399, 407, 415, 425, 429, 433, 434, 438, 441-444, 447, 448, 451-453, 458, 459, 461, 463, 469, 484, 493, 494, 500, 511, 513, 521, 522, 530, 549, 556-561, 628, 721-735, 741-745, 747, 749-751, 755-759, 762-764, 766-779, 781-789, 791-801, 83, 804, 806
methods, 432
process, 2, 12, 132, 208, 217, 218, 347, 490, 494, 577, 614, 625, 640, 732, 780, 787
prospective, 733, 736
retrospective, 17, 250, 278, 549, 640, 747, 785, 792
system, 5, 6, 8, 10-14, 18-20, 23, 27, 28, 174, 209, 216, 219, 235, 236, 238, 242, 249, 253, 347, 396, 722, 729, 739, 741, 743, 747, 751, 756, 757, 782, 783, 787, 792, 795, 798, 800, 806
Validation study, 2-4, 6-9, 11, 12, 16, 209, 740, 742, 743, 746
Vendors, 6, 12, 14, 16, 24, 25, 165, 212, 213, 215, 216, 219, 224, 236, 250, 252, 351, 352, 360, 361, 393, 446, 722, 743, 747 770, 776, 796
Verification, 34, 52, 77, 124, 128-131, 133, 165, 180, 184, 185, 187, 190, 202, 203, 205, 206, 243-245, 273, 275, 282, 308, 460, 462, 488-490, 492, 493, 499, 554, 579, 605, 612, 716, 727, 778, 781, 785, 796, 802, 804
Versioning, 10, 127, 166, 167, 171, 353, 438, 443, 449, 450, 454-457, 491, 549, 575, 625, 659, 722-724, 726, 728, 730-732, 735, 739, 750, 752, 753, 764, 766, 769, 775, 776, 788, 791, 797, 802, 804, 805
Veterinary products, 28, 29, 189, 209, 225, 227, 329, 330, 335, 336, 376, 568

Walkthroughs, 13, 805
Warning letters, 10, 11, 227, 458, 511, 512, 517, 521-523, 527, 565, 570, 598, 599, 605, 611, 614, 616-620, 622-624, 627, 660, 664-669, 671, 672, 674-682, 684, 694, 696-701, 703, 704, 706, 707, 709-712, 715, 716, 718, 719, 721, 722